Z° Physics
Cargèse 1990

NATO ASI Series

Advanced Science Institutes Series

A series presenting the results of activities sponsored by the NATO Science Committee, which aims at the dissemination of advanced scientific and technological knowledge, with a view to strengthening links between scientific communities.

The series is published by an international board of publishers in conjunction with the NATO Scientific Affairs Division

A	**Life Sciences**	Plenum Publishing Corporation
B	**Physics**	New York and London
C	**Mathematical and Physical Sciences**	Kluwer Academic Publishers
D	**Behavioral and Social Sciences**	Dordrecht, Boston, and London
E	**Applied Sciences**	
F	**Computer and Systems Sciences**	Springer-Verlag
G	**Ecological Sciences**	Berlin, Heidelberg, New York, London,
H	**Cell Biology**	Paris, Tokyo, Hong Kong, and Barcelona
I	**Global Environmental Change**	

Recent Volumes in this Series

Volume 255—Vacuum Structure in Intense Fields
    edited by H. M. Fried and Berndt Müller

Volume 256—Information Dynamics
    edited by Harald Atmanspacher and Herbert Scheingraber

Volume 257—Excitations in Two-Dimensional and Three-Dimensional Quantum Fluids
    edited by A. F. G. Wyatt and H. J. Lauter

Volume 258—Large-Scale Molecular Systems: Quantum and Stochastic Aspects—
    Beyond the Simple Molecular Picture
    edited by Werner Gans, Alexander Blumen, and Anton Amann

Volume 259—Science and Technology of Nanostructured Magnetic Materials
    edited by George C. Hadjipanayis and Gary Prinz

Volume 260—Self-Organization, Emerging Properties, and Learning
    edited by Agnessa Babloyantz

Volume 261—Z° Physics: *Cargèse 1990*
    edited by Maurice Lévy, Jean-Louis Basdevant, Maurice Jacob,
    David Speiser, Jacques Weyers, and Raymond Gastmans

Series B: Physics

Z° Physics

Cargèse 1990

Edited by

Maurice Lévy and Jean-Louis Basdevant

Université Pierre et Marie Curie
Paris, France

Maurice Jacob

CERN
Geneva, Switzerland

David Speiser and Jacques Weyers

Université Catholique de Louvain
Louvain-la-Neuve, Belgium

and

Raymond Gastmans

Katholieke Universiteit Leuven
Leuven, Belgium

Plenum Press
New York and London
Published in cooperation with NATO Scientific Affairs Division

Proceedings of a NATO Advanced Study Institute on
Z° Physics,
held August 13–25, 1990,
in Cargèse, France

Library of Congress Cataloging-in-Publication Data

NATO Advanced Study Institute on Z° Physics (1990 : Cargèse, France)
    Z° physics : Cargèse, 1990 / edited by Maurice Lévy ... [et al.].
        p.   cm. -- (NATO ASI series. Series B, Physics ; v. 261)
    "Proceedings of a NATO Advanced Study Institute on Z° Physics,
held August 13-25, 1990, in Cargèse, France"--T.p. verso.
    "Published in cooperation with NATO Scientific Affairs Division."
    Includes bibliographical references and index.
    ISBN 0-306-43934-4
    1. Z bosons--Congresses.   I. Lévy, Maurice, 1922-   . II. North
Atlantic Treaty Organization. Scientific Affairs Division.
III. Title. IV. Series.
QC793.5.B62N38   1990
539.7'21--dc20   91-22692
                                                            CIP

ISBN 0-306-43934-4

PREVIOUS CARGÈSE SYMPOSIA PUBLISHED IN THE NATO ASI SERIES B: PHYSICS

Volume 223   PARTICLE PHYSICS: *Cargèse 1989*
edited by Maurice Lévy, Jean-Louis Basdevant, Maurice Jacob,
David Speiser, Jacques Weyers, and Raymond Gastmans

Volume 173   PARTICLE PHYSICS: *Cargèse 1987*
edited by Maurice Lévy, Jean-Louis Basdevant, Maurice Jacob,
David Speiser, Jacques Weyers, and Raymond Gastmans

Volume 156   GRAVITATION IN ASTROPHYSICS: *Cargèse 1986*
edited by B. Carter and J. B. Hartle

Volume 150   PARTICLE PHYSICS: *Cargèse 1985*
edited by Maurice Lévy, Jean-Louis Basdevant, Maurice Jacob,
David Speiser, Jacques Weyers, and Raymond Gastmans

Volume 130   HEAVY ION COLLISIONS: *Cargèse 1984*
edited by P. Bonche, Maurice Lévy, Phillippe Quentin, and
Dominique Vautherin

Volume 126   PERSPECTIVES IN PARTICLES AND FIELDS: *Cargèse 1983*
edited by Maurice Lévy, Jean-Louis Basdevant, David Speiser,
Jacques Weyers, Maurice Jacob, and Raymond Gastmans

Volume 85   FUNDAMENTAL INTERACTIONS: *Cargèse 1981*
edited by Maurice Lévy, Jean-Louis Basdevant, David Speiser,
Jacques Weyers, Maurice Jacob, and Raymond Gastmans

Volume 72   PHASE TRANSITIONS: *Cargèse 1980*
edited by Maurice Lévy, Jean-Claude Le Guillou, and
Jean Zinn-Justin

Volume 61   QUARKS AND LEPTONS: *Cargèse 1979*
edited by Maurice Lévy, Jean-Louis Basdevant, David Speiser,
Jacques Weyers, Raymond Gastmans, and Maurice Jacob

Volume 44   RECENT DEVELOPMENTS IN GRAVITATION: *Cargèse 1978*
edited by Maurice Lévy, and S. Deser

Volume 39   HADRON STRUCTURE AND LEPTON–HADRON INTERACTIONS: *Cargèse 1977*
edited by Maurice Lévy, Jean-Louis Basdevant, David Speiser,
Jacques Weyers, Raymond Gastmans, and Jean Zinn-Justin

Volume 26   NEW DEVELOPMENTS IN QUANTUM FIELD THEORY AND STATISTICAL
MECHANICS: *Cargèse 1976*
edited by Maurice Lévy and Pronob Mitter

Volume 13   WEAK AND ELECTROMAGNETIC INTERACTIONS AT HIGH ENERGIES:
Cargèse 1975 (Parts A and B)
edited by Maurice Lévy, Jean-Louis Basdevant, David Speiser,
and Raymond Gastmans

PREFACE

The 1990 Cargèse Summer Institute on Z^0-Physics was organized by the Université Pierre et Marie Curie, Paris (M. Lévy and J.-L. Basdevant), CERN (M. Jacob), the Université Catholique de Louvain (D. Speiser and J. Weyers), and the Katholieke Universiteit te Leuven (R. Gastmans), which, since 1975, have joined their efforts and worked in common. It was the ninth Summer Institute on High Energy Physics organized jointly at Cargèse by these three universities.

Because of the start-up of LEP in the summer of 1989, we broke with our tradition of having our Summer Institutes in the odd years. Indeed, it seemed to us that the many new data from LEP had to be presented in detail as soon as possible in order to prepare the young researchers in particle physics better for the experimental results with which they will be confronted in the coming years.

The main theme of the school was therefore Z^0-physics, with particular emphasis on the way the experiments at LEP are analyzed. We had one lecturer from each LEP experiment: they agreed among each other to present different topics in e^+e^- physics. Nevertheless, they made sure that all the major topics were discussed and that the results could be critically compared.

Professor L. Maiani summed up the comparison between the standard electroweak model and experiment, while Professors T. Sjöstrand and V. Khoze discussed the new developments in QCD and jet physics in connection with LEP. Finally, Professor F. Berends described in detail the finer points of making radiative corrections, which are essential for refined experimental analyses.

Of course, more information on electroweak interactions is obtained from other accelerators. Thus, Professor R. Cashmore reviewed the $p\bar{p}$ collider physics, Professor M. Swartz reviewed the SLC results, and Professor R. Eichler presented the case for a B-factory.

Finally, Professor D. Schramm presented the latest developments in astrophysics based in part on the new findings in particle physics.

We owe many thanks to all those who have made this Summer Institute possible!

Special thanks are due to the Scientific Committee of NATO and its President for a generous grant. We are also very grateful for the financial contribution given by the Institut National de Physique Nucléaire et de Physique des Particules (IN^2P^3).

We also want to thank Ms. M.-F. Hanseler for her efficient organizational help, Mr. and Ms. Ariano for their kind assistance in all material matters of the school, and, last but not least, the people from Cargèse for their hospitality.

Mostly, however, we would like to thank all the lecturers and participants: their commitment to the school was the real basis for its success.

M. Lévy	D. Speiser
J.-L. Basdevant	J. Weyers
M. Jacob	R. Gastmans

CONTENTS

LEP Results: Measurement of the Z^0 Line Shape in Hadrons and
Leptons and of the Lepton Forward-Backward Asymmetries 1
G. Sauvage

Study of Z^0 Couplings to Quarks at LEP 27
D.R. Ward

QCD Studies at LEP 47
D.R. Ward

Searches at LEP 69
D. Treille

Polarization at LEP 115
A. Blondel

Physics at the SLC 141
M.L. Swartz

Physics from TRISTAN 201
Young-Kee Kim

W & Z Physics at $p\bar{p}$ Colliders 229
R.J. Cashmore

Theory of the Electroweak Interactions 237
L. Maiani

Electroweak Radiative Corrections for Z Physics 307
F.A. Berends

QCD and Jets at LEP 367
T. Sjöstrand

Colour-Coherence Physics at the Z^0 419
V.A. Khoze

The Physics Program and Accelerator Properties of a B-Meson Factory 449
    R.A. Eichler

Unstable Particles 483
    A. Martin

The Particle-Cosmology Connection: Neutrino Counting, Dark Matter
        and Large Scale Structure 495
    D.N. Schramm

Index 527

LEP RESULTS: MEASUREMENT OF THE Z^0 LINE SHAPE IN HADRONS AND LEPTONS AND OF THE LEPTON FORWARD-BACKWARD ASYMMETRIES

Gilles Sauvage (L3)

Laboratoire d'Annecy-Le-Vieux de Physique des Particules (LAPP)

ABSTRACT

The measurement of the Z^0 line shape in hadrons and leptons is described. The determination of the forward backward asymmetry in each of the 3 lepton channels (e,μ,τ) is given. The results quoted here, based on hadron samples of about 100000 events and on lepton channels of about 3000 events for each LEP experiment, are the results presented at the Singapore Conference (2-8 August 1990). The different fits performed to extract physical quantities, Z^0 mass, peak cross section, partial widths, weak couplings, are described and compared to the standard model predictions.

INTRODUCTION

The first Z^0's produced by LEP have been observed by the four LEP experiments in August 1989. Less than one year after, each experiment has been able to collect hadron samples $(e^+e^- \rightarrow Z^0 \rightarrow hadrons)$ of about 100000 events and lepton samples $(e^+e^- \rightarrow Z^0 \rightarrow e^+e^-, \mu^+\mu^-, \tau^+\tau^-)$ of about 2000-3000 events. This amount of experimental data allow already to do very accurate determinations of Z^0 parameters (mass, partial widths, weak couplings...) and to perform precise tests of the standard electroweak theory. The results of the 4 experiments are presented together, as well as the combined results.

In the first chapter a quick overview of the standard electroweak theory allows to introduce the notations used for Z^0 weak couplings and to give the formulae for the cross sections and asymmetries.

The second chapter is dedicated to the description of LEP performances. The emphasis is put on the achieved luminosity and on the beam energy measurement. The description of the measurement of the luminosity by the experiments is given elsewhere in this book (A.Blondel).

The event selection for each studied channel, as well as the acceptances and contaminations are described in the third chapter.

The presence of t channel in $e^+e^- \rightarrow (\gamma, Z^0) \rightarrow e^+e^-(\gamma)$ has to be taken into account. Different scenarios, t channel subtraction or global fit, are described in the 4th chapter.

The line shape fits are presented in the 5th chapter. Global fits of line shapes and asymmetries are described. The experimental results of each experiment are compared between themselves and also to the prediction of the standard model.

Z° Physics - Cargèse 1990, Edited by M. Lévy *et al.*
Plenum Press, New York, 1991

I. STANDARD MODEL OF THE ELECTROWEAK THEORY

A complete description of the standard model of the electroweak theory can be found for example in (1). We recall here the basic features of the model, to introduce the notations used throughout this lecture and to define the quantities which can be determined through the measurement of the Z^0 line shape and of the forward backward asymmetries.

In $SU(3)_{color} \otimes SU(2)_{\vec{T}} \otimes U(1)_Y$ the leptons and quarks are grouped in left handed doublets and right handed singlets of the weak isospin $\vec{T}$. The following table summarizes the values of the quantic numbers, weak isospin $(\vec{T}, I_3)$, hypercharge Y and of the electric charge $Q = I_3 + Y/2$.

$$\text{leptons} \quad Y = -1, \vec{T} = 1/2 \quad \left\{ \begin{array}{ll} I_3 = 1/2 & Q = 0 \\ I_3 = -1/2 & Q = -1 \end{array} \right\} \quad \begin{bmatrix} \nu_e \\ e \end{bmatrix}_L \begin{bmatrix} \nu_\mu \\ \mu \end{bmatrix}_L \begin{bmatrix} \nu_\tau \\ \tau \end{bmatrix}_L$$

$$\text{quarks} \quad Y = +1/3, \vec{T} = 1/2 \quad \left\{ \begin{array}{ll} I_3 = 1/2 & Q = 2/3 \\ I_3 = -1/2 & Q = -1/3 \end{array} \right\} \quad \begin{bmatrix} u \\ d \end{bmatrix}_L \begin{bmatrix} c \\ s \end{bmatrix}_L \begin{bmatrix} t \\ b \end{bmatrix}_L$$

The gauge couplings between the 3 gauge bosons γ, $W^{\pm}$, Z^0 and the fermions are :

$$-e\, Q_f \gamma^\mu \qquad \frac{e}{2\sqrt{2}\,\sin\theta_W}\, \gamma^\mu(1-\gamma^5) \qquad \frac{e}{2\sin 2\theta_W}\, \gamma^\mu(v_f - \gamma^5 a_f)$$

where the vectorial and axial couplings, v_f and a_f, of the Z^0 are equal to :

$$v_f = 2 I_{3f} - 4\, Q_f \sin^2\theta_W, \qquad\qquad a_f = 2\, I_{3f}$$

This formula leads to the following couplings for the different fermions :

	ν	e	u	d
v_f	1	$4\sin^2\theta_W - 1$	$1 - 8/3 \sin^2\theta_W$	$-1 + 4/3 \sin^2\theta_W$
a_f	1	-1	1	-1

The gauge couplings depend on 2 parameters, e, charge of the electron and $\sin^2\theta_W$, the so called Weinberg angle. Very often, instead of v and a, one uses g_v and g_a which are given by :

$$g_v = v/2 \quad \text{and} \quad g_a = a/2$$

The partial decay width, Γ_f, of Z^0 decaying into a fermion-antifermion pair, $f\bar{f}$, is then given by :

$$\Gamma_f = C\,N\,\frac{G_F\,M_Z^3}{24\pi\,\sqrt{2}}\left[a_f^2 + v_f^2\right](1 + \delta_f)$$

In this formula, M_Z is the Z^o mass and $G_F/\sqrt{2}$ the Fermi coupling constant can be expressed as : $G_F = \pi\alpha/M_W^2 \sin^2\theta_W$. C is a space phase factor depending on the mass or on the velocity, β_f, of the final fermion f, and is equal to :

$$C = \left[1/2\,\beta_f\left(3-\beta_f^2\right)v_f^2 + \beta_f^2\,a_f^2\right]\,/\,\left(v_f^2 + a_f^2\right)$$

This factor is only slightly different from the unity for the τ lepton and the quarks c and b. [$1 - C_\tau = 2.2\ 10^{-3}$, $1 - C_b = 1.2\ 10^{-2}$ for $m_b = 5$ GeV]. The term $1+\delta_f$ represents the electroweak radiative to the partial decay and is also small ($\delta \sim 10^{-3}$). The factor N is different for leptons and quarks :

$$N = 1 \quad \text{for leptons}, \quad N = 3(1+\alpha_s/\pi + ...) \quad \text{for quarks.}$$

For quarks, one recognizes the factor 3 for the color and the QCD corrections governed by the α_s coupling constant. This last correction is of the order of 4 %. Up to the corrections listed below the measurement of the partial decay widths allow to determine the combination $a_f^2 + v_f^2$ of the Z^o gauge couplings.

The total cross section $\sigma_f(s)$, around the Z^o peak, of $e^+e^- \rightarrow f\bar{f}$ can be written as (2):

$$\sigma_f(s) = \frac{12\,\pi}{M_Z^2}\,\frac{s\,\Gamma_e\,\Gamma_f}{(s-M_Z^2)^2 + s^2\Gamma_Z^2/M_Z^2}\,(1 + \delta(s)) \qquad (1)$$

This formula corresponds, for the lowest order, to the production of a fermion antifermion pair through the annihilation of e^+e^- in a photon or a Z^o (s channel).
The term $1+\delta(s)$ represents the electroweak radiative corrections. In this process the radiative corrections in the initial state, emission of a photon by the incoming electron (positron) are important. At the Z^o peak, $s = M^2{}_Z$, the cross section is reduced by about 30 %. The peak cross section can be expressed as :

$$\sigma_{peak} = \sigma_f^o\,(1+\delta(m_Z^2)) \quad \text{with} \quad \sigma_f^o = 12\pi\,\frac{\Gamma_e\Gamma_f}{M_Z^2\,\Gamma_Z^2} \qquad (2)$$

The measurement of $\sigma_f(s)$ for different values of s around the Z^o peak, the so called line shape measurement, allows to determine the Z^o mass, M_Z, the total width, Γ_Z, and the partial width Γ_f through $\sigma^o{}_f)$. The measurement of Γ_e by the process $e^+e^- \rightarrow e^+e^-$ is important as Γ_e enters all $\sigma^o{}_f$ formulae. Due to the presence of the t channel for this reaction, the extraction of Γ_e is more difficult as one will see later.

The differential cross section, $e^+e^- \rightarrow f\bar{f}$, for unpolarized electrons and positrons corresponding to the s channel with an exchange of a γ or a Z^o, can be written as (2) :

$$\frac{d\,\sigma_f}{d\Omega} = \frac{\alpha^2}{4\pi}\left\{1+2\,g_V^e g_V^f\,\mathrm{Re}(\chi) + (g_V^{e\,2}+g_A^{e\,2})(g_V^{f\,2}+g_A^{f\,2})|\chi|^2\right\}(1+\cos^2\theta)$$

$$+\left\{2\,g_A^e\,g_A^f\,\mathrm{Re}(\chi) + 4\,g_V^e\,g_A^e\,g_V^f\,g_A^f\,|\chi|^2\right\}2\cos\theta) \tag{3}$$

$$= A\left(1+\cos^2\theta + \frac{8}{3}A_{FB}^f\,\cos\theta\right)$$

In this formula χ is the Z^0 propagator :

$$\chi = \frac{G_\mu}{2\sqrt{2}}\,\frac{M_Z^2}{\pi\alpha}\,\frac{s}{(s-M_Z^2 + i\,M_Z\Gamma_Z)}$$

The presence of a $\cos\theta$ term induces a forward backward asymmetry. A_{FB}^f defined as :

$$A_{FB}^f = \frac{\displaystyle\int_0^1 \frac{d\sigma}{d\Omega}\,d\cos\theta - \int_{-1}^0 \frac{d\sigma}{d\Omega}\,d\cos\theta}{\displaystyle\int_{-1}^{+1} \frac{d\sigma}{d\Omega}\,d\cos\theta}$$

At the Z^0 peak, the Z^0 exchange dominates and $\mathrm{Re}(\chi) = O$, the asymmetry is then equal to :

$$A_{FB}^f = 3\,\frac{g_V^e g_A^e g_V^f g_A^f}{(g_V^{e\,2}+g_A^{e\,2})(g_V^{f\,2}+g_A^{f\,2})}$$

In case of lepton asymmetry which is reported here and assuming lepton universality, A_{FB} is finally equal to :

$$A_{FB}^f = 3\,\frac{g_V^2\,g_A^2}{(g_V^2 + g_A^2)^2} = 3\,\frac{v^2 a^2}{(a^2+v^2)^2} \tag{4}$$

This formula is symmetric in g_A^2, g_V^2). The measurement of the asymmetry at the Z^0 peak and of the lepton decay width gives rise to an ambiguity between two solutions $(g_A^{2\,m}, g_V^{2\,m})$ or $(g_V^{2\,m}, g_A^{2\,m})$. The variation of the asymmetry with the energy has a slope mainly governed by g_A^2 as can be seen in the formula (1). The only unknown which remains, is then the sign of g_A and g_V as one measures only the squares of this quantities. One uses the sign determined in experiments at lower energies (see later).

To summarize, the measurement of the different line shapes, $Z^0 \rightarrow$ hadrons, $Z^0 \rightarrow$ leptons (e,μ,τ) and of the corresponding lepton forward backward asymmetries, allows to determine the Z^0 mass M_Z, the total width Γ_Z, the partial decay widths Γ_f and the vectorial and axial coupling constants g^ℓ_A, g^ℓ_V. All these quantities can be expresses in terms of fundamental parameters of the electroweak theory, like $\sin^2\theta_W$. One will see in the next chapters, the improvement already brought by the first year of LEP functioning to the accuracy with which these quantities are known.

II. LEP PERFORMANCES

A brief description of the LEP machine is given here (a complete description can be found, e.g. in (3). The values obtained in 1989 and 1990 for the peak and integrated luminosities are given and discussed. The measurement of the beam energy at LEP, a very important parameter for the line shape determination, is presented.

<u>LEP description</u>

LEP, the large electron-positron storage ring built at CERN, has a circumference of about 27 km. It is presently the largest storage ring in the world. The ring is composed of 8 arcs interleaved with 8 straight sections.

The RF cavities necessary to accelerate the beams and to keep them at the working energy are installed in the straight sections. For LEP phase I these cavities are copper cavities (operated at room temperature). The maximal beam energy is 55 GeV, well above the Z^0 peak: the scan of the whole Z^0 peak is thus possible. In a second phase, superconducting RF cavities will be added to the copper cavities, and eventually will replace them. For LEP phase II the maximal beam energy will be around 100 GeV, id est above the W^+W^- production threshold. At the middle of 4 straight sections, 4 experiments are installed, ALEPH, DELPHI, OPAL, L3. To get the smallest beam cross section at the crossing point, normal and superconducting quadrupoles are placed very close to the experimental setups. The best focusing, or the smallest β functions can be obtained.

Presently 4 bunches are circulating in LEP, which correspond to 8 crossing points, at the middle of each straight section. The e^+e^- beams are electrostatically separated in the straight sections without experiments. To increase the luminosity of LEP, a scheme with more bunches, 8 to 32 is discussed (see D.Treille in this book).

The LEP beams are bent in the arcs by dipolar magnets (more than 3000 in total). These dipoles are interleaved with quadrupoles and sextupoles to get the necessary beam focussing. In order to diminish the energy loss by synchrotron radiation, which is proportional to the fourth power of the energy and inversely proportional to the curvvature radius (E^4/R), a very high value of the radius has been chosen ($R \simeq 3096$ m). This implies very low values of the magnetic field, 0.0215 T at the injection energy of 20 GeV and about 0.048 T at the Z^0 peak. The maximal magnetic field of these specially designed (4) magnets corresponds to a maximal beam energy of 125 GeV.

<u>Luminosity</u>

The main parameter of a collider is, of course, the luminosity $\mathscr{L}$, as this parameter determines the counting rate of any given process $e^+e^- \rightarrow X$, with cross section σ, by the simple formula :

$$dn/dt = \sigma\mathscr{L}$$

For example the counting rate of $e^+e^- \rightarrow Z^0 \rightarrow$ hadrons at the Z^0 peak is expected to be, for the design luminosity $\mathscr{L}_0 = 1.6 \ 10^{31} \ cm^{-2}s^{-1}$

$$dn/dt = 30 \text{ nb} \times 1.6 \ 10^{31} \text{ cm}^{-2}\text{s}^{-1} = .48 \text{ s}^{-1}$$

The simplest way to express $\mathcal{L}$ in function of machine parameters is :

$$\mathcal{L} = k \ n^+ n^-/S$$

where k is the number of crossings per second, $n^{+(-)}$ the number of positrons (electrons) in a bunch and S the sections of the beams. This formula can be rewritten in a more elaborated way as :

$$\mathcal{L} = \frac{T_R}{4\pi \ e^2 \ n_b} \ \frac{I^+ I^-}{\sigma_X \ \sigma_Y} \tag{5}$$

where T_R is the revolution time, n_b the number of bunches, $I^{+(-)}$ the positron (electron) beam intensity, $\sigma_{X(Y)}$ the rms of the distribution of the horizontal (vertical) density of the beam at the interaction point. In this formula, one sees that $\mathcal{L}$ increases as the square of the beam intensity, which is true only when the beam-beam forces does not limit the stored intensity. In that case the luminosity is proportional only to the beam intensity :

$$\mathcal{L} \propto \gamma \xi \ I/2r_e \ e \ \beta_y^*$$

($\mathcal{L}$ is proportional to the beam energy through the relativistic γ factor and ξ is the limiting value of the beam-beam tune shift).

In (5) the beam cross section are related to the beam emittances $\varepsilon_{X(Y)}$ by :

$$\sigma_{X,Y} = \sqrt{\varepsilon_{X,Y} \ \beta_{X,Y}^* (0)}$$

β^* is related to the focusing.

$\beta^*(0)$ is the value of the beta function at the interaction point. This function describes the focusing properties of LEP in each point of the ring. In order to get the maximal focusing, and therefore the smallest value of β^*, superconducting quadrupoles have been installed very close to the interaction point. As the vertical emittance ε_Y is small (the momentum dispersion increases essentially the horizontal emittance) only β_Y^* (0) is chosen small.

In the following table, the values of the main parameters entering the luminosity formulae are given. In particular the values obtained in 1989 and in the first half of 1990 can be compared to the design values [4].

	1989		1990		
	peak	average	peak	average	design
Total current at 20 GeV (mA)	2.85	2.20	4.05	2.95	6
Total current at 45 GeV (mA)	2.64	1.66	3.50	2.40	6
β_Y^* (10) (cm)	7	7	4.3	7	7
$\mathscr{L}_{peak}$ (10^{30}cm^{-2}s^{-1})	4.25	1.59	7.70	3.76	16
Integrated luminosity (pb^{-1})		1.74		4.10	

The achieved peak luminosity is about half the design value, and this only one year after the start of LEP. Some more progresses have been made at the end of the 1990 beam period and a peak luminosity of 1.1×10^{31} cm^{-2}s^{-1} has bean achieved and a total integrated luminosity of 12.4 pb^{-1} has been obtained. The figure 1 shows the integrated luminosity as function of time in 1989 and 1990. One can notice the improvements in the peak luminosity (slope of the curve) and in the integrated luminosity. Among the recent achieved progresses, the unequality of luminosities between LEP experiments has been understood (most of the time, e.g. $\mathscr{L}_{L3} \simeq .75 \, \mathscr{L}_{ALEPH}$). The β^* function, around its minimum, is described by :

$$\beta^*(z) = \beta^*(0)(1 + \frac{z^2}{\beta^{*2}(0)}) \quad (z = 0 \text{ interaction point})$$

This formula shows that β^* has to be greater than the longitudinal extension of the bunches ($\sigma_z \sim 15$ mm), to have for the whole bunch the minimal value $\beta^*(0)$. The unequality of the luminosities between experiments were due to a shift of the minimum of β^* with respect to the interaction point. For example the shift in L3 was about 16 mm. The corrections of these shifts have suppressed these unequalities and have also increased the luminosities of each experiments.

The prospect for 1991 is therefore optimistic. Due to a longer beam period, a better reliability of the machine and also some improvements in the peak luminosity, one expects a factor 4 in the integrated luminosity. Each experiment should collect more than 500000 Z$^0 \rightarrow$ hadrons.

<u>Beam momentum measurement</u>

One can measure the momentum of the colliding beams very accurately, therefore the total initial energy in fact much more accurately than any LEP experiment can measure the total energy of the decay products of the Z^0. Up to now two different methods have been used, one consisting in the measurement of the total integral field $\int Bdl$, and the second one in the measurement of the difference of the revolution frequencies of positrons and protons of same momentum circulating in the machine.

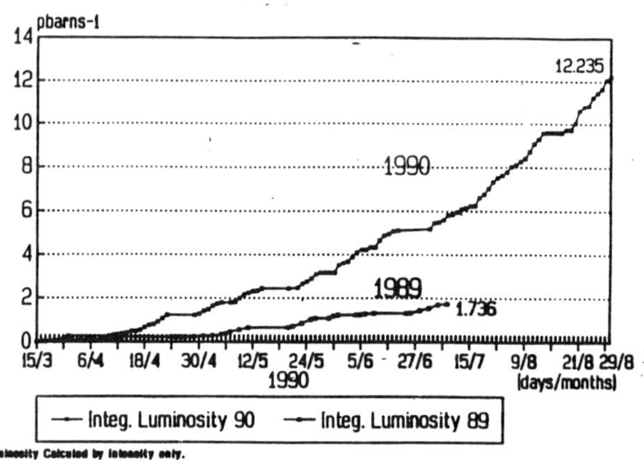

Figure 1 .   Integrated luminosity at LEP in 1989 and 1990.

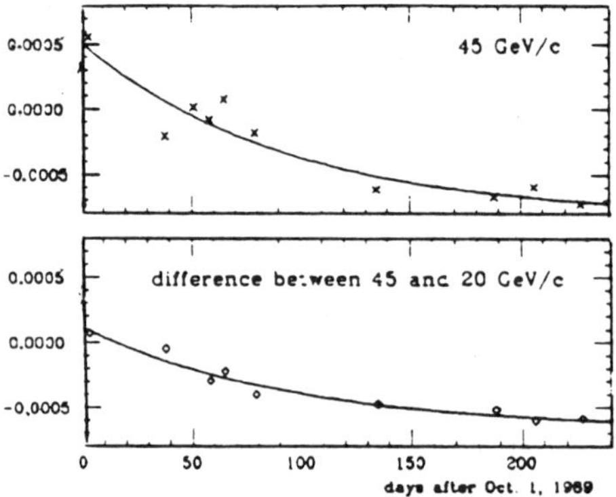

Figure 2 .   Field decrease with time of the LEP dipole as measured by the flux loop calibration.

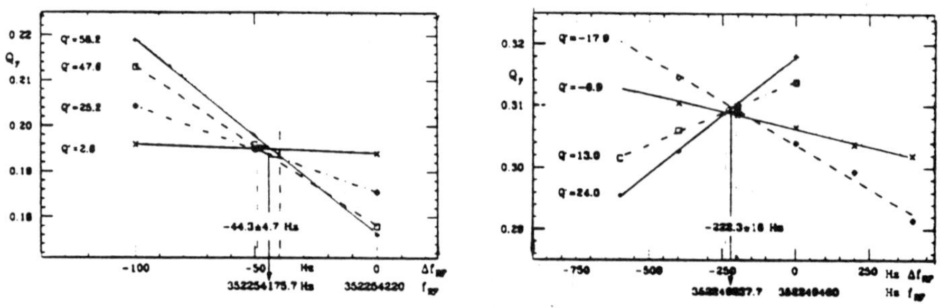

Figure 3 .   Finding central orbit

with positrons.

Figure 4 .   Finding central orbit

with protons.

- Field display and flux loop method (5)

The momentum of a circulating particle is given by :

$$p = .3 \int Bdl / 2\pi$$

where $\int Bdl$ is the total integral field seen by a particle during one revolution. For a particle on the nominal orbit, id est on the orbit going through the axis of the LEP quadrupoles and sextupoles, this integral field is only given by magnet dipoles. The LEP magnets, made of iron laminations embedded in mortar are known to age (figure 2 shows a field diminution of $\sim 10^{-3}$ in about 200 days) and need to be frequently measured. One reference magnet, identical to the LEP dipoles but without mortar, is powered in series with all the LEP dipoles. The measurement of the magnetic field of this reference magnet ("Field display") (6) is performed by the flip coil method. A movable coil of known area inserted in the magnet gap is rotated by $\pm 180°$ and the variation of flux precisely measured. This method allows to measure static fields but cannot of course be used for the LEP dipoles. To calibrate this reference magnet with respect to the LEP dipoles, or to measure the ageing of those dipoles, a flux loop has been mounted on the lower pole piece of each LEP dipole. The flux variation can then be measured where ramping the LEP. Polarity reversals are done to take into account the presence of remanent fields (earth magnetic field... etc) which are not negligible for these low field magnet ($\sim .05$ T at 50 GeV). The field display and flux loop methods determine the momentum with an accuracy of $\pm 5 \ 10^{-4}$.

- Proton measurement (7)

The revolution time for a particle is given by :

$$T_R = L_{LEP} / \beta c$$

where L_{LEP} is the LEP circumference and b_c is the particle velocity. The revolution time is related to the RF frequency used to compensate beam energy losses by synchrotron radiation (or to accelerate the beam) by :

$$T_R = h/f$$

where h is the chosen harmonic number (integral number) and f the frequency of the RF. In the case of electrons where β is very close to 1 ($\beta \simeq 1 - 3 \ 10^{-10}$ for $p = 20$ GeV/c) the measurement of f gives directly the LEP circumference.

If the same measurement is performed with less relativistic particles, like protons, one gets the following relations :

$$L_{LEP} = \beta_e c \frac{h_e}{f_e} = \beta_p c \frac{h_p}{f_p}$$

$$\text{or } \beta_p = \beta_e \frac{h_e}{f_e} \times \frac{f_p}{h_p}$$

The measurement of f_e and f_p allows to compute β_p and therefore the momentum p. One has of course to check that the electron and proton orbits are exactly the same for the two measurements.

The accuracy on the momentum is given by :

$$\frac{dp}{p} \simeq \gamma^2 \, d\beta = \gamma^2 \left(\frac{df_p}{f_p} - \frac{df_e}{f_e} \right)$$

$$\text{or } \frac{\Delta p}{p} = \gamma^2 \left(\frac{\Delta f_e^2}{f_e^2} + \frac{\Delta f_p^2}{f_p^2} \right)^{1/2}$$

In this method, the measurement of two RF frequencies gives the absolute momentum of the LEP beam. The term γ^2 in the formula shows that the accuracy increases when the less relativistic particle is heavier. In practice the measurement is done at the LEP injection energy, id est at 20 GeV. To go to the working energy, around 45 GeV, some relative corrections to the momentum have to be applied. These corrections are determined by using the preceding method, flip coil + flux loop.

The measurement of the electron and proton orbits are of a crucial importance here. The most precise measurement of the orbit is done for the central orbit, the orbit going through the axis of the quadrupoles and sextupoles. For such an orbit the betatron tune is independent of the sextupole excitation as the magnetic field vanishes along the axis. Therefore the betatron tune is measured as a function of RF frequency for different sextupole excitation. The variation of RF frequency is equivalent to a variation of the orbit. The figures 3,4 show the results of the measurements, the different curves cross at a given point corresponding to the central orbit. The error on the frequence f_e, f_p is derived from the figures 3,4. The following table shows, as an example, the results of the momentum determination made in May 1990 (7).

Field display value	20.009 GeV/c
e central frequency, f_e	352 254 172.9 $\pm$ 9 Hz, $h_e = 31324$
p central frequency, f_p	352 249 242.6 $\pm$ 20 Hz, $h_p = 31358$
Momentum (central orbit)	20.0038 GeV/c $\pm$ 0.3 10^{-4}
Correction to field display : at 20 GeV/c	$(-2.6 \pm 0.3)\ 10^{-4}$

The accuracy at 20 GeV/c is rather impressive, $3\ 10^{-5}$ or 0.6 MeV. For the scaling to 45 GeV/c one uses the differences between the flux loop calibrations at these two energies. This scaling introduces a much bigger error on the energy determination. Some other small corrections, due to the fact that the working orbit is not the nominal one are summarized in the following table (corrections and errors are given in units of 10^{-4})

Correction at 20 GeV/c	-2.6 $\pm$ 0.3
Effect of orbit correctors	0.0 $\pm$ 0.5
Scaling to 45 GeV	-0.7 $\pm$ 2.4
Correction for working orbit	-3.5 $\pm$ 0.7
Total correction	-6.8 $\pm$ 2.6

The error on Z mass, for 1990 data, is therefore 23,6 MeV. (One uses generally a rounded systematic error of 30 MeV in 1989 and 20 MeV in 1990).

The acceleration of protons to 45 GeV/c requires some beam gymnastics, like a change of harmonic number or (and) a rather important variation of the RF frequency f_p during ramping. Due to the γ^2 term the accucary $(0.3\ 10^{-4})$ will be degraded by more than a factor 4, giving finally only an improvement of a factor 2 on the present achieved accuracy. Therefore it is envisaged to accelerate, instead, deuterons up to 45 GeV/c, to reach an accuracy comparable to what is obtained with 20 GeV/c protons.

The transverse polarization of the beams due to synchrotron radiation has been observed at the end of the 1990 beam period. The method used to measure very precisely the Y and Y' masses at others circular e^+e^- colliders (8, 9) can therefore be applied at CERN. In this method one depolarizes the previously polarized beam by a weak time dependent magnetic field. The measurement of the frequency of this variable magnetic field determines the $e^+(e^-)$ energy to an accuracy of $5\ 10^{-5}$ (8). Such a method applied at CERN should allow to get an error on the Z mass in the range 4 to 5 MeV.

III. EVENT SELECTION

<u>General considerations</u>

The cross section σ_f of a process $e^+e^- \to Z^0 \to f\bar{f}$ is experimentally given by :

$$\sigma_f = \sigma_{\mathscr{L}}\ n_c^f / n_c^{\mathscr{L}}$$

where n_c^f is the number of $f\bar{f}$ observed events corrected for background contamination and for acceptance, $n_c^{\mathscr{L}}$ is the corresponding corrected number of detected Bhabha events inside the luminosity detector of a LEP experiment. $\sigma_{\mathscr{L}}$ is the known Bhabha cross section $e^+e^- \to e^+e^-$ inside the fiducial cuts of the luminosity detector. At small angle the Bhabha cross section is mainly a QED process and the influence of the Z^0 exchange very small. The statistical error on $n_c^{\mathscr{L}}$ and the systematic errors on $n_c^{\mathscr{L}}$ and $\sigma_{\mathscr{L}}$ are of course common to the 4 different line shapes (q,e,μ,τ). They are discussed by A.Blondel elsewhere in this book.

The corrected number n_c^f can be written as :

$$n_c^f = n_{obs}^b \cdot \varepsilon_f \cdot A_f$$

where n_{obs}^f is the observed number of events, ε_f the correction factor for the remaining background contamination and A_f the acceptance including the effect of the cuts used to isolate the $f\bar{f}$ channel. These two factors are determined by MC simulations. The comparison of MC and data distributions of different kinematic distributions is used to estimate the systematic error on ε_f, A_f and therefore on n_c^f.

The determination of the forward backward asymmetry does not need the knowledge of the integrated luminosity. The observed forward backward asymmetry is simply given by :

$$A_{fb}^{obs} = \frac{n_f - n_b}{n_f + n_b}$$

where $n_{f(b)}$ is the number of observed particles (leptons in the case studied here) of a given sign in the forward (backward) region. To go to the real lepton asymmetry A_{fb}^{exp}, MC simulations are also used to correct A_{fb}^{obs} for the remaining background and for in complete acceptance. In the same way than above, comparisons between MC and data are used to

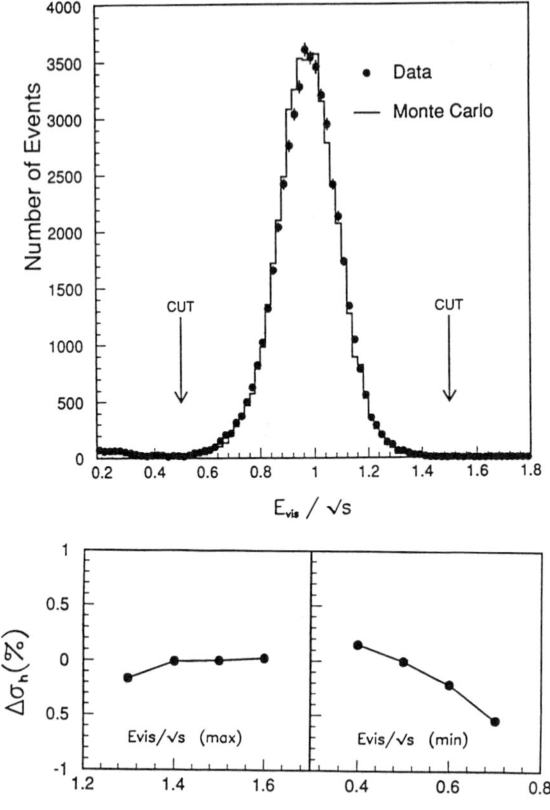

Figure 5 .   Visible energy distribution ($E_{vis}/\sqrt{s}$) of $Z^o \to q\bar{q}$ candidates (L3).

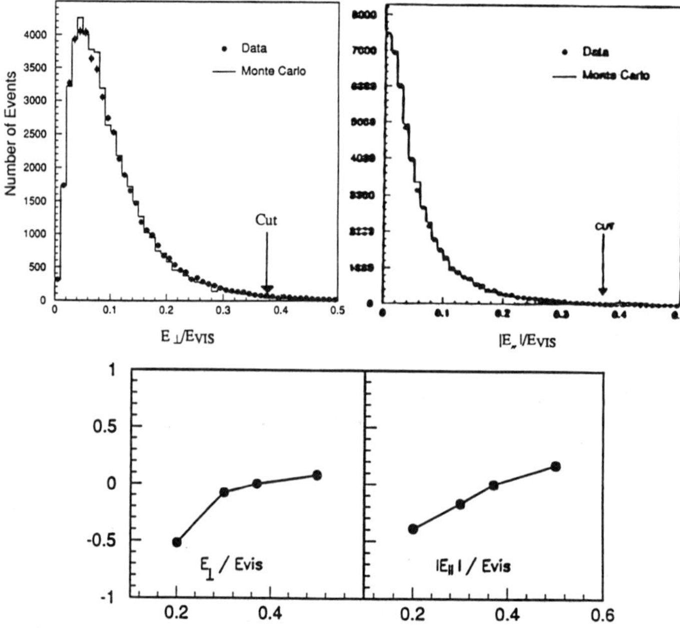

Figure 6 .   Ratio of the transverse energy to the visible energy $E_\perp/E_{vis}$ and ratio of the parallel energy to the visible energy $E_{//}/E_{vis}$ for $Z^o \to q\bar{q}$ candidates (L3).

acceptance. In the same way than above, comparisons between MC and data are used to estimate the systematic error. The forward backward asymmetry is also determined by fitting the whole experimental angular distribution.

For a given Z^0 decay channel, the different sources of background are :
- the other Z^0 decay channels,
- events coming from 2γ processes $e^+e^- \to e^+e^- + X$
- beam gas interactions : the $e^+(e^-)$ beam can interact with the residual gas in the beam pipe in the vicinity of the interaction point,
- beam wall interactions : particles lost by the beams can hit the walls of the beam pipe and then interact,
- cosmic rays in coincidence with the passage of the beam at the interaction point.

Different cuts are necessary to isolate the decay channel under study and to eliminate the different sources of background. In particular a cut on the energy seen in the detector will eliminate the main part of the non Z^0 background. We give here the main cuts used for the different decay channels.

$Z^0 \to q\bar{q}(\gamma)$

This is the main decay channel (branching ratio $\sim$ 70 %) and is characterized by the presence of two or more jets in the final state and almost no missing energy. For this mode, all the 4 experiments have an acceptance close to 4π steradians, which diminishes the systematic error on this parameter. The systematic errors for this mode will come mainly from the background contamination.

Figures 5 to 7 show the effects of the cuts used by L3 to select the $Z^0 \to q\bar{q}(\gamma)$ channel. On these figures the Monte Carlo events have been obtained by using the LUND generator (10). The distribution of the visible energy (fig.5) the energy seen in the calorimeters, shows a clear peak corresponding to the $Z^0 \to q\bar{q}$, e^+e^-, $\tau^+\tau^-$. The other sources of background are already strongly suppressed. As there is almost no unseen particles in a $q\bar{q}$ event a cut on the total transverse and longitudinal momentum is applied. The computation of these quantities is done using the energy instead of the momentum, assuming that all particles are relativistic. Figure 6 shows the distribution of the normalized transverse energy $E_\perp/\sqrt{s}$ and of the normalized longitudinal energy $E_{_\parallel}/\sqrt{s}$. The variation of the cross section when one changes the cuts, is also shown. This variation is used to estimate the systematic error due, to the selection procedure. To remove the $Z^0 \to \tau^+\tau^-$ and $Z^0 \to e^+e^-$ events, a cut on the number of clusters in the calorimeter is applied (Fig.7). The number of clusters is related to the number of particles produced in the e^+e^- collision.

The cut on the number of clusters can be replaced by a cut on the number of tracks as done in the other LEP experiments. In the case of ALEPH, two independent selection procedures have been applied, one mainly based on the calorimeter information and the other one on the track information given by the TPC. This allows ALEPH to precisely estimate their systematic error. Table 1 gives the number of hadronic events $Z^0 \to q\bar{q}$ presented at the Singapore Conference together with the acceptance and the integrated luminosity.

Leptonic samples (e, μ, τ, ℓ)

The leptonic branching ration of the Z^0 is small, about 3.4 % for each lepton. The main Z^0 decay channel, $Z^0 \to q\bar{q}$, will constitute a major contamination source. This is specially true for the τ channel, when both τ's decay hadronically. Nevertheless the topology of the events, characterized by a small number of charged particles (2 for e and μ channels) helps to isolate the leptonic channels. The cosmic ray background is important in the case of the muon channel : time of flight information, when available, is often used to reject it.

The separation of the 3 different lepton channels is complicated by the different τ decays. The electron (muon) τ decay channel makes difficult the isolation of the electron (muon) Z^0 decay channel. This gives rise to systematic errors for the 3 lepton channels. To overcome this difficulty, ALEPH and DELPHI have also defined a selection procedure able to select at the same time the 3 lepton flavors.

To reduce the influence of the t channel for the electron channel, $e^+e^- \rightarrow Z^0(\gamma) \rightarrow e^+e^-$, generally one considers only a reduced acceptance, as shown in Table 1. The same reduced acceptance is also taken for the lepton sample. As we will see in the nest chapter the t channel contribution is important at forward angles.

ee channel

The event has a simple topology with the presence of almost two back to back electrons. Therefore the selection criteria require an important fraction of the available energy seen in the electromagnetic calorimeter, the presence of two charged tracks in the event with a small acollinearity angle. These two main cuts are adjusted to diminish the contamination of the $q\bar{q}$ and $\tau\tau$ channels. In the case of L3, the requirement of the presence two tracks is replaced by a requirement of the presence of two high energy clusters, which oblige to subtract the contribution of $e^+e^- \rightarrow \gamma\gamma$ events (this correction amounts at ~ 1.5 % at the Z^0 peak) (11).

$\mu\mu$ channel

The main background for this simple topology – 2 muon tracks with a small acollinearity – is either a cosmic muon or a $\tau\tau$ event where both τ's decay in a muon. To reject the cosmic ray background the muon tracks are required to have a small distance of closest approach to the interaction point and to have a good timing with respect to beam crossing. To reject $\tau\tau$ events a cut is applied on the momentum of the muon tracks. To sign the presence of a muon, the experiments use the information of dedicated subdetectors, like the muon spectrometer of L3 e.g., or the characteristic energy deposition of a muon in the hadron calorimeter (e.g. ALEPH or both information (DELPHI, OPAL).

$\tau\tau$ channel

The decay of the τ particle is characterized by a small number of particles. Therefore the $\tau\tau$ events are separated from the dominant $Z^0 \rightarrow q\bar{q}(\gamma)$ decay by requiring a moderate number of tracks seen in the central chambers of the detectors or (and) a moderate number of clusters is done by applying the "opposite" cuts used to select these channels. For example the energy seen in the electromagnetic calorimeter is required to be less than a given threshold (ee contamination) or the momentum of the muon to be not too high ($\mu\mu$ contamination). To improve the ee rejection, ALEPH applies also a cut on the missing mass computed using the momenta of the charged particles. To eliminate the background coming from 2γ process, the presence of a minimal energy or of a track with a minimal momentum is required. A cut on the acollinearity between the two jets, helps to reject this background.

Leptonic sample : The remaining contamination in the above selections from the other leptonic channels is in the range 0.5 - 3% giving rise to systematic error up to 1%. To get rid of this systematic error ALEPH and DELPHI have also defined a selection procedure which is flavor independent. In the preceeding selections one keeps essentially the cuts used to reject the $Z^0 \rightarrow q\bar{q}(\gamma)$ events, the 2γ processes, the beam-gas events and the cosmic rays. One has also to estimate the fraction of ee events due to the t-channel. This subtraction is explained in the next chapter. The remaining sample can then be fitted with the formula of chapter 1.

Table 1. Number of selected events (18)

	ALEPH	DELPHI	L3	OPAL
$\int \mathcal{L} dt (pb^{-1})$	2.5	3.3	2.8	5.2
Acceptance (% of 4π)	.98	.93	.97	.98
$N_{q\bar{q}}$	56000 (84000)	68000	61000	112000
cos θ range	+0.7, −0.9	± 0.64	± 0.74	± 0.7
N_{ee}	2400	1400	2600	3200
cos θ range	± 0.95	± 0.73	± 0.7	± 0.95
$N_{\mu\mu}$	2000	1600	1800	4600
cos θ range	± 0.9	± 0.73	± 0.7	± 0.9
$N_{\tau\tau}$	2100	1000	1100	3400
cos θ range	+ 0.7, − 0.9	± 0.64		
$N_{\ell\ell}$	63000	3200		

Remark :

- The integrated luminosity for the selected leptonic events was generally smaller than for the $q\bar{q}$ events.

- In the case of ALEPH the real number of $q\bar{q}$ events analysed for the Singapore Conference was 84000.

Table 1 summarizes the number of events selected by the 4 experiments. The angular range for each leptonic sample is also indicated.

IV.   e⁺e⁻ CHANNEL

At lowest order, the elastic process $e^+e^- \to e^+e^-$ is described by two diagrams, an s and an t channel diagram :

At small angles the second diagram with the exchange of a γ in the t channel dominates and the luminosity in all experiments is measured through this nearly pure QED process. At large angles, however, the s channel with a Z^0 exchange dominates and is identical to the s channel diagram describing any fermion pair production $e^+e^- \to f\bar{f}$. There still remains a contribution from the t channel and from the interference between the s and t channels as shown in Fig.8. To reduce these contributions, the 4 experiments have studied the $Z^0 \to e^+e^-$ channel in a reduced angular range ($|\cos \theta| < \sim .7$ for Delphi, L3, OPAL) or in an asymmetric

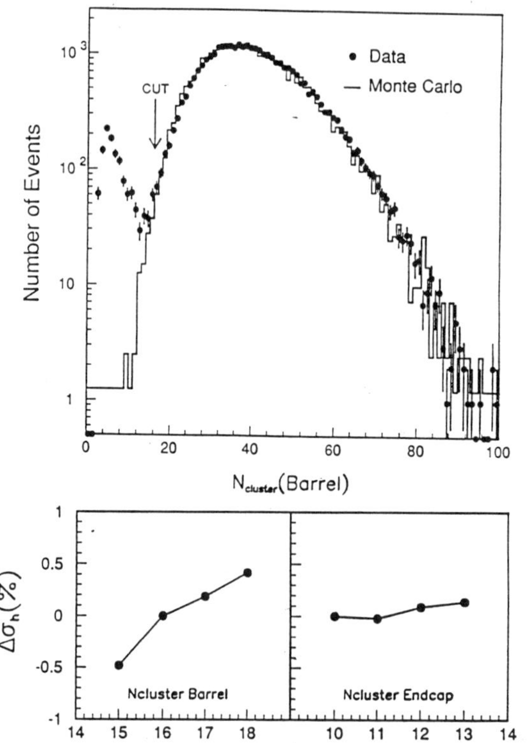

Figure 7 .   Cluster multiplicity of $Z^o \to q\bar{q}$ candidates (L3).

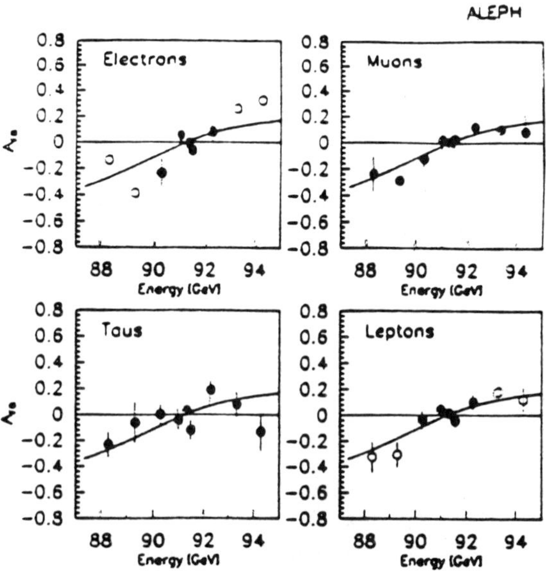

Figure 8 .   s and t contributions to $e^+e^- \to e^+e^-$ cross section.

angular range for ALEPH $(-.9 < \cos \theta < .7)$. The interference term vanishes at Z^0 peak and the contribution of the t channel in the quoted angular range $(|\cos \theta| < .7)$ amounts to about 16 %.

Two different procedures have, up to now, been applied to extract the electronic width Γ_{ee} from this process : subtraction of the t channel and its interference to the experimental cross section in order to fit with the same formula the line shapes of any fermion pair, or a separate fit of the experimental cross section with a complete formula describing the e^+e^- channel. Some technical complications arise from the radiative corrections to the e^+e^- channel. Up to now it exists only a MC generator BABAMC (12) including radiative corrections at first order $O(\alpha)$. The more general formulae, including first and second order $O(\alpha)$, $O(\alpha^2)$ and soft photon exponentiation may imply also new kinematic cuts not easily done experimentally. This is the case, e.g., of the Greco analytic expression (13) and of the Caffo-Remiddi program (14) where a cut on the energy on the radiated γ in the final state has to be done. The preceding observations imply that the determination of the systematic error, e.g. on the acceptance, has to be done very carefully. Γ_{ee} can then be extracted in three ways :

• from the peak cross section (see formula 2). The t channel contribution is taken out using either the Greco-Caffo-Remiddi formula (14) or the ALIBABA formula (15).

• from the whole line shape itself and by subtracting the t channel contribution and the contribution from the interference between the s and t channels. Both contributions are evaluated with the help of (14) or (15). The remaining "s" line shape can then be fitted by a common program like ZFITTER (16). Of course in this case an iterative procedure is used as the interference term depends on the value of the Z^0 parameters like $M_Z, \Gamma_Z \ldots$. The advantage of this method is to allow a simultaneous fit of all ff line shapes $(f = e,\mu,\tau,q)$.

• from the whole line shape without subtraction. The electronic width Γ_{ee} is extracted by a fit by the complete formula like the ones given by (14) and (15).

The same procedures can be applied to the angular distribution $e^+e^- \to e^+e^-$. But in order to extract the forward-backward asymmetry A_{FB} defined earlier (formulae 3 or 4) it is more suitable to subtract the contributions from the t channel and from the interference between the s and t channel. The consequence of the subtraction is, of course, that the asymmetry for the electron is determined with an accuracy less good than the asymmetry for the muon or the τ (see Fig.9 where the asymmetries for the 3 leptons e,μ,τ are displayed).

V.   FITS AND RESULTS

To fit the experimental line shapes, two different methods have been used. The first one is a step by step fit of the different line shapes : one begins by fitting the $Z^0 \to q\bar{q}$ data, as the hadronic line shape is the most precisely measured, to extract the Z^0 mass, M_Z, the total width, Γ_Z, and the peak cross section, σ_0^h. The different leptonic data sets are then fitted, using as inputs parameters the previously obtained values of M_Z and Γ_Z and thus determining each leptonic width Γ_ℓ $(\ell = e,\mu,\tau)$. The second method consists in a global fit of all line shapes, to determine simultaneously $M_Z, \Gamma_Z, \sigma_0^h$ and Γ_ℓ. The analytic formula, like formula 2, used to describe the line shapes are given in (17).

The first step of the first method is a 3 parameter fit $(M_Z, \Gamma_Z, \sigma_0^h)$ if one ignores the relations given by the standard model. Using them it becomes a 2 parameter fit $(M_Z$ and invisible width $\Gamma_{inv})$ as can be seen with the relations :

$$\Gamma_Z = \Gamma_h + \Gamma_e + \Gamma_{inv} \quad (\Gamma_h, \Gamma_e \text{ given by the standard model})$$
$$\sigma_0^h = 12 \, \pi \Gamma_e \Gamma_h / M_Z^2 \Gamma_Z^2$$

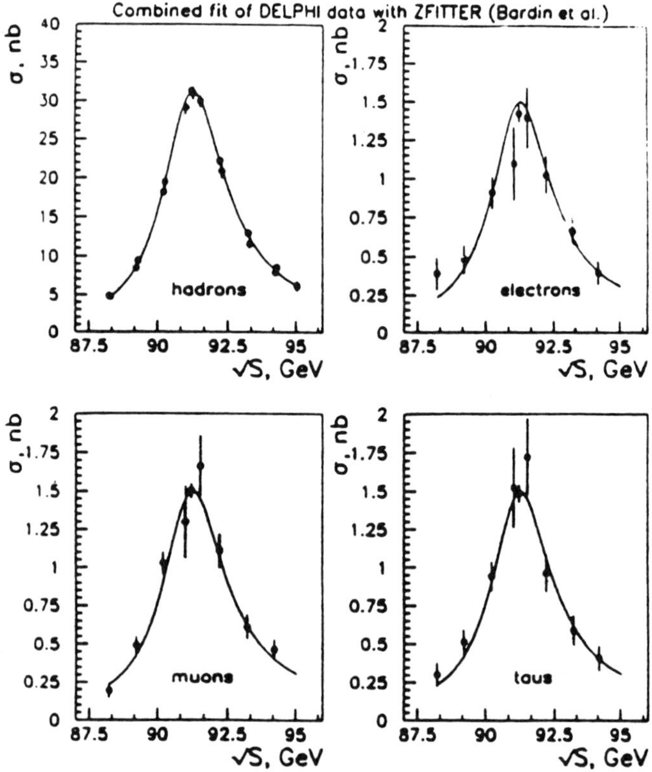

Figure 9 .   Lepton asymmetries (ALEPH).

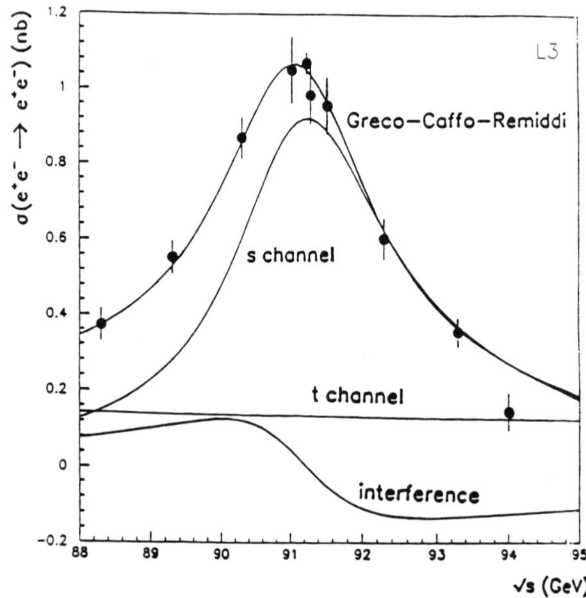

Figure 10 .   Z⁰ line shapes (DELPHI).

In the case of the leptonic data, one usually first checks the universality of the coupling $Z^0\ell\ell$, by not assuming $\Gamma_e = \Gamma_\mu = \Gamma_\tau$. The leptonic peak cross sections determine Γ^2_e, $\Gamma_e\Gamma_\mu$, $\Gamma_e\Gamma_\tau$ allowing to extract Γ_e, Γ_μ, Γ_τ. The universality being experimentally well verified as we will see, the fit is redone imposing $\Gamma_e = \Gamma_\mu = \Gamma_\tau$. Therefore, for the global fit, the parameters determined are either 6 (M_Z, Γ_Z, σ_0^h and Γ_ℓ, $\ell = e,\mu,\tau$) or 4 (M_Z, Γ_Z, σ_0^h, Γ_ℓ) without universality or imposing it.

Table 2 gives the combined results of the 4 experiments, presented at the Singapore Conference (18). To combine the results, the common systematic error is first removed of each experimental error, then a weighted average is computed, using as a weight the remaining experimental error. The common systematic error is added (quadratically) at the end to the error of the combined value. The most evident examples of common systematic errors are the LEP scale energy or the theoretical error on the Bhabha cross section used to compute the luminosity. Table 2 gives also the predicted values of the Standard Model of the measured parameters, taking as input the Z mass, $M_Z = 91.177$ GeV and values for the top mass, m_T, the Higgs mass, m_H and the strong coupling constant, α_s, in a reasonable range :

$$50 < m_T < 250 \text{ GeV}, \quad 20 < m_H < 1000 \text{ GeV}, \quad .09 < \alpha_s < .15.$$

Figure 11 displays the values of each experiment for different parameters, M_Z, Γ_Z, σ_0^h, Γ_ℓ, $\ell = e,\mu,\tau$... . The agreement between the experiment is good as indicated by the normalized chi square (expected value 1.0). Figure 10 shows the line shapes from one experiment, as an illustration. One notices also the good agreement with the predicted values of the standard model.

When averaging the results of the 4 experiments, one notices that, already after one year of LEP operation, the systematic errors begin to limit the accuracy with which the Z^0 parameters are determined. This is specially true for the Z mass, M_Z, where the main error comes from the absolute LEP energy scale. It is hoped, thanks to the ransverse polarization of the beam, or, thanks to a deuteron measurement, to reduce substantially this error. Nevertheless the accuracy already achieved is impressive : $3 \cdot 10^{-4}$. The other systematic errors (Bhabha cross section, t channel subtraction, ...) will be studied in more details with the increased statistics of the coming years and should also be reduced.

The assumption of the lepton universality, $\Gamma_e = \Gamma_\mu = \Gamma_\tau$, is well verified as can be seen in Table 2. This allows, then, to perform a global fit imposing this universality.

For the comparison with the standard model predictions, it is convenient to consider some ratios where the dependence on the top mass vanishes or is strongly reduced (19). This is the case of the peak cross section, σ_0^h and of the ratio of the hadronic partial width to the leptonic partial width, h_Z :

$$\sigma_0^h = \frac{12\,\pi}{M_Z^2}\,\frac{\Gamma_e\Gamma_h}{\Gamma_Z^2} \qquad\qquad R_Z = \Gamma_h/\Gamma_e$$

Table 2 shows for these two quantities a good agreement :

$$\sigma_0^h \text{ (exp)} = 41.78 \pm 0.52 \text{ nb}, \qquad \sigma_0^h \text{ (SM)} = 41.30 \pm .10 \text{ nb}$$
$$R_Z \text{ (exp)} = 21.08 \pm 0.20, \qquad R_Z \text{ (SM)} = 20.86 \pm .20$$

The error on the predicted values is rather conservative, with the allowed range of variation for m_T, m_H and α_s. Even with a more restricted range ($100 < m_T < 250$ GeV, $42 < m_H < 1000$ GeV, $0.11 < \alpha_s < 0.13$) (20) the agreement is still good :

$$\sigma_0^h \text{ (SM)} = 41.43 \pm .07 \text{ nb}, \qquad R_Z \text{ (SM)} = 20.79 \pm 0.08$$

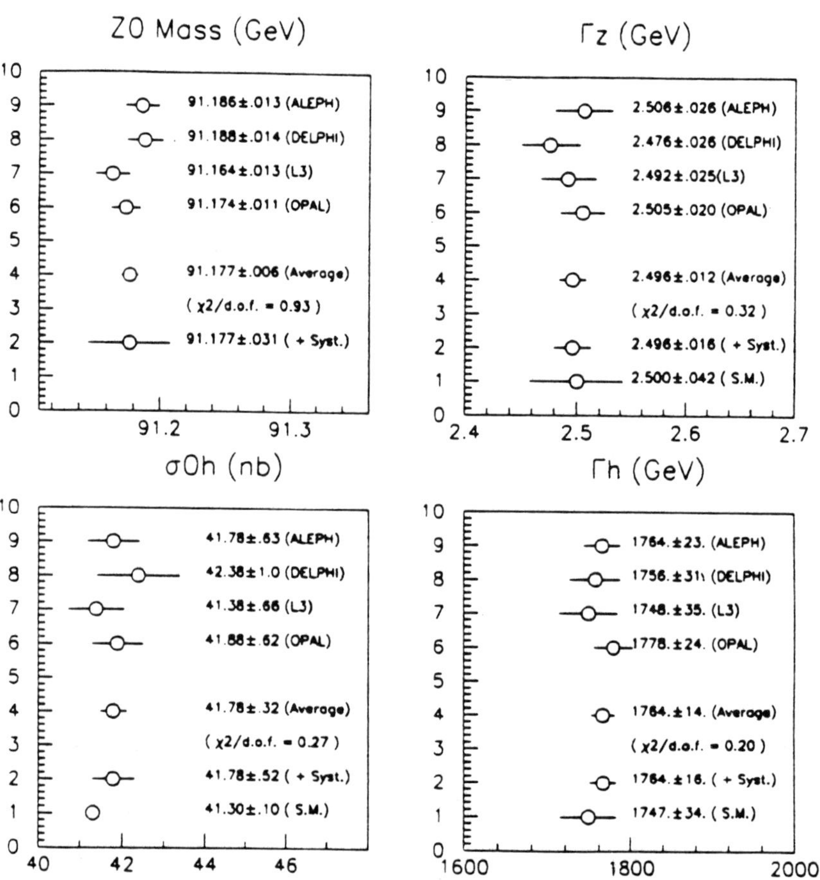

Figure 11a. Results of the 4 LEP experiments.

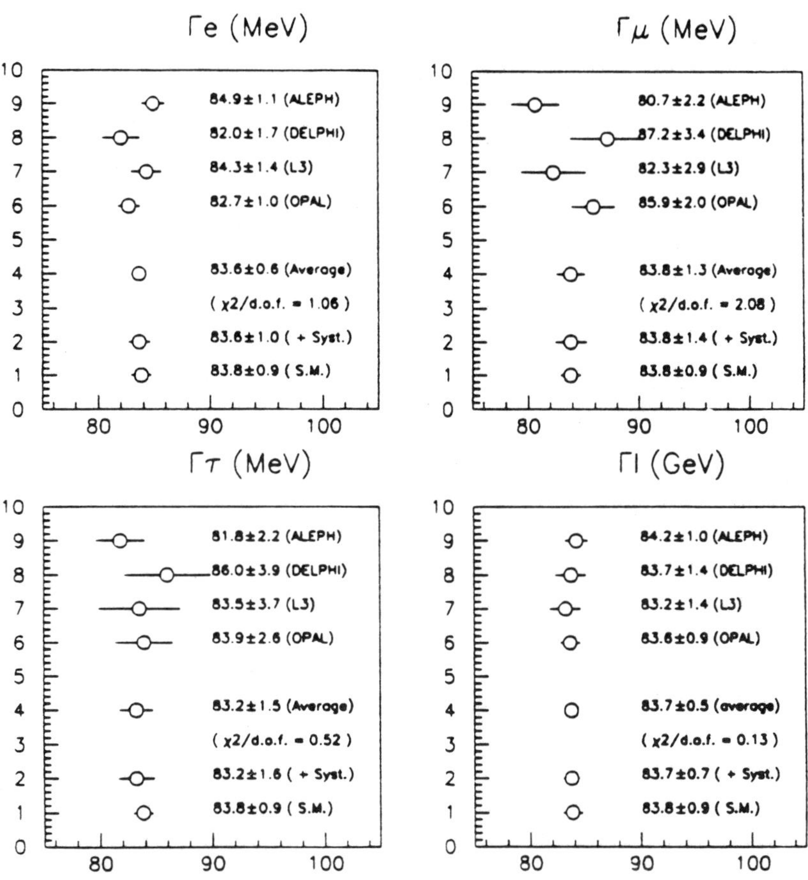

Figure 11b. Results of the 4 LEP experiments.

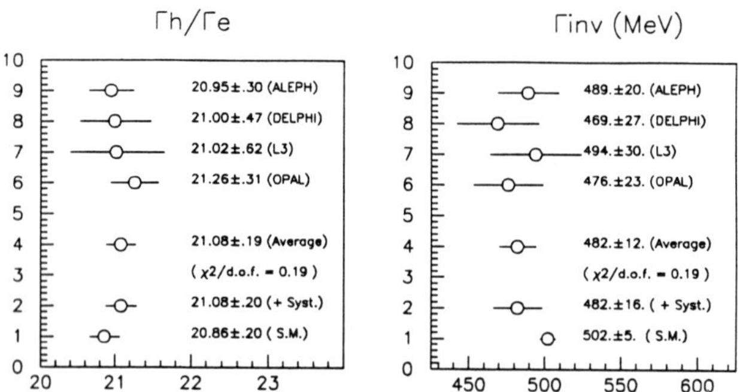

Figure 11c . Results of the 4 LEP experiments.

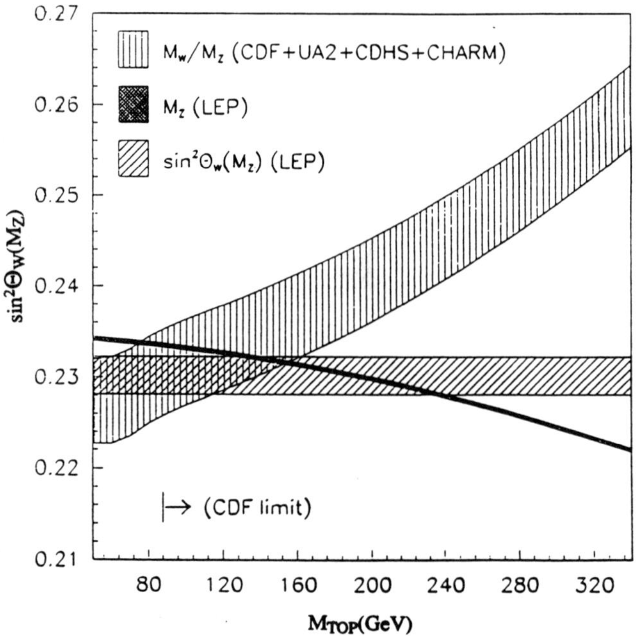

Figure 12 . m_T, $sin^2\bar\theta_W$ plot for $m_H = 200$ GeV.

The number of neutrinos can be derived directly from the invisible width Γ_{inv}. But is is again possible to eliminate the top mass dependence (19) of the standard model by expressing the number of neutrinos in the following way :

$$N_\nu = \frac{\Gamma_{inv}}{\Gamma_\nu{}^{SM}} = \frac{\Gamma_Z - \Gamma_h - 3\Gamma_e}{\Gamma_\nu{}^{SM}} = \frac{\Gamma_e{}^{SM}}{\Gamma_\nu{}^{SM}} \left((\frac{12\,\pi\,R_Z}{\sigma_o{}^h\,M_Z{}^2})^{1/2} - R_Z - 3 \right)$$

The ratio $(\Gamma_e/\Gamma_\nu)^{SM}$ is also top mass independent. From the Table 2 one extracts :

$$N_\nu = 2.89 \pm .10$$

This result completely rules out the existence of a fourth light neutrino (with a mass $M_\nu < M_Z$).

The forward backward asymmetry is expressed easily in terms of the effective neutral weak couplings $\bar{g}_A$, $\bar{g}_v$ (see Chapter 1). Therefore to extract them, the leptonic width Γ_e has also to be expressed with these two quantities (see Chapter 1). The best way to use the experimental asymmetries is to perform a global fit with the line shapes with five parameters, M_Z, Γ_Z, Γ_h, $\bar{g}_A$, $\bar{g}_v$. The changes in the values of M_Z, Γ_Z, Γ_h are rather small as this parameters are mainly determined by the line shapes. As already seen, only the square of $\bar{g}_A$ and $\bar{g}_v$ are determined. Therefore the sign is taken from experiments at low energies (21). One gets :

$$\bar{g}_A = -\,0.501 \pm .002 \qquad\qquad \bar{g}_v = -\,.045 \pm .006$$

One can express the leptonic width Γ_e and the asymmetry A_{FB} in terms of the effective Weinberg angle $\sin^2 \bar{\theta}_W$ and the effective $\bar{\rho}$ parameter (one has $\bar{g}_A = -\,.5\,(\bar{\rho})^{1/2}$ and $\bar{g}_v = \bar{g}_A\,(1 - 4\sin^2 \bar{\theta}_W)$) and redo the complete fit. The effective Weinberg angle is then :

$$\sin^2 \bar{\theta}_W = .2302 \pm .0021$$

Finally, one can try to extract limits on the top mass from the LEP measurements. This is shown on figure (12) which represents $\sin^2 \bar{\theta}_W$ in function of the top mass : three bands are represented, one corresponding to the above value of $\sin^2 \bar{\theta}_W$, one corresponding to the constraint given by the value of the Z^o mass, M_Z and one corresponding to the ratio of the W mass to the Z mass which is measured either directly in $p\bar{p}$ experiments (22) or indirectly in neutrino experiments (23). Allowing a Higgs mass $20 < m_H < 1000\,GeV$ one then finds :

$$m_t = 137 \pm 40\,GeV.$$

This high value for the top mass, shows that the top is outside of the range of LEP phase 2.

CONCLUSION

After one year of operation, LEP has allowed already a precise determination of the Z^o parameters. The predictions of the Standard Model are well verified. It is hoped that with increased statistics one can either find a failure of the Standard Model or at least to set better limits on unknown parameters like the top mass or the Higgs mass.

**Table 2. Combined experimental results and
standard model prediction**

Parameter	Combined value	S.M.value	χ^2/d.o.f**
M_Z	91.177 ± .006 GeV ± .031 (± 30 MeV)*		0.93
Γ_Z	2.496 ± .012 GeV ± .016 (± 10 MeV)	2.500 ± .042 GeV	0.32
σ_0^h	41.78 ± 0.32 nb ± 0.52 (± 1 %)	41.30 ± .10 nb	0.27
Γ_h	1764 ± 14 MeV ± 16 (± 0.5 %)	1747 ± 34 MeV	0.20
Γ_e	83.6 ± .6 MeV ± 1.0 (±1 %)	83.8 ± .9 MeV	1.06
Γ_μ	84.1 ± 1.2 MeV ± 1.4 (0.5 %)	83.8 ± .9 MeV	2.08
Γ_τ	83.2 ± 1.5 Mev ± 1.6 (0.5 %)	83.8 ± .9 MeV	0.52
Γ_e(univ)	83.7 ± .5 MeV ± .7 (0,6 %)	83.8 ± .9 MeV	0.13
Γ_h/Γ_e	21.08 ± 0.19 ± .20 (.3 %)	20.86 ± .20	0.19
Γ_{inv}	482 ± 12 MeV ± 16 (± 10 MeV)	502 ± 5. MeV	0.19

*The systematic error is quoted in parenthesis and comes from the LEP energy scale (M_Z), radiative correction and point to point energy error (Γ_h,Γ_{inv}), Bhabha cross section for the luminosity (σ_0^h, Γ_h), from the t channel subtraction ($\Gamma_e,\Gamma_\mu,\Gamma_\tau,\Gamma_\ell,\Gamma_h$).

**It is the chi square value divided by the number of degree of freedom of the 4 experimental values used to compute the combined value.

ACKNOWLEDGMENT

The results of the LEP experiments have been kindly given to me in advance by A.Blondel (ALEPH), D.Treille (DELPHI) and D.Ward (OPAL). I thank them also for many discussions on these results. I thank also D.Plane for illuminating discussions on the LEP energy measurement.

REFERENCES

(1)   Z.Physics at LEP1, CERN 89-03, Vol.1.
(2)   R.N.Cahn, Phys. Rev. D36 (1987) 2666.
(3)   LEP design report, Vol.2, CERN-LEP/84-01
(4)   J.Billan et al., Magnetic performance of the LEP bending magnets
       paper presented at the 1989 Particle Accelerator Conference.
(5)   R.Bailey et al., LEP energy calibration, CERN SL/90-95
       paper presented at the 2nd European Particle Accelerator Conference
(6)   J.Billan et al., Field display system, LEP-MA/89-41
       XIV Int. Conf. on High Energy Accel., Tsukuba, Japan (August 1989).
(7)   A.Hoffmann, T.Risselada, CERN LEP Note 383, 1982.
(8)   A.S.Artamonov et al., Phys. Lett. 118B (1982), p.225.
(9)   D.P.Barber et al., Phys. Lett. 135B (1984), p.498.
(10)   LUND parton shower program, JETSET 7.2
       T.Sjöstrand and M.Bengtsson, Comput. Phys. Commun., 43 (1987) 367.
       T.Sjöstrand in Z Physics at LEP1, CERN 89-08 Vol.3, p.143.
(11)   F.A.Berends and R.Kleiss, Nucl. Phys. B186 (1981) 22.
(12)   M.Böhm, A.Denner and W.Hollik, Nucl. Phys.B 304 (1988) 687.
       F.A.Berends, R.Kleiss and W.Hollik, Nucl. Phys. B304 (1988) 712.
(13)   M.Greco, Phys.Lett. B177 (1986) 97 ; Riv. Nuovo Cimento 11 (1988) 1.
(14)   M.Caffo, E.Remiddi and F.Semeria, Z.Phys. at LEP1, CERN report 89-08
       Vol.1 (1989) 171.
       F.Aversa et al., Phys. Lett. B247 (1990) 93
(15)   W.Beenacker, F.A.Berends and S.C.Van der Marck, ALIBABA, Leiden
       preprint (1990) (See also F.A.Berends Contribution in this school).
(16)   D.Bardin et al., Z.Phys. at LEP1, CERN report 89-08 Vol.3 ;
       Berlin-Zeuthen preprint PHE 89-19 (1989).
(17)   R.N.Cahn, Phys. Rev. D36 (1987) 2666.
       A.Borelli et al., Nucl. Phys. B333 (1990) 357.
       G.Burgers in Z Physics at LEP1, CERN 88-06, p.121.
(18)   25th International Conference on High Energy Physics,
       Singapore (August 2-8, 1990). Communications from the 4 LEP experiments
       and summary talk of F.Dydak.
(19)   C.Verzegnassi : Theory faces experiment at High Energy Colliders,
       LAPP-TH-31(1990).
(20)   W.Hollik : Contribution to the Workshop "High precision tests in the Standard
       Model : where do we stand and where do you want to go ?".
       Annecy, January 17-18, 1991.
(21)   CHARM Collab., J.Dorenbosch et al., Z.Phys. C41 (1989) 567.
       K.Abe et al., Phys. Rev. Lett. 62 (1989) 1709.
       CHARM II Collab., D.Geiregat et al., Phys. Lett. B232 (1989) 539.
       F.Avignone et al., Phys. Rev. D16 (1977) 2383.
       U.Amaldi et al., Phys. Rev. D36 (1987) 1385.
(22)   CDF Collab., Phys. Rev. Lett. 65 (1990) 2243.
       UA2 Collab., Phys. Rev. B241 (1990) 150.
(23)   G.L.Fogli and D.Haidt, Z.Phys. C40 (1988) 379.

Study of Z^0 couplings to quarks at LEP

D.R. Ward

Cavendish Laboratory, Madingley Road, Cambridge, U.K.

1 Introduction

A large proportion, $\sim 70\%$, of Z^0 decays lead to hadronic final states. The LEP experiments have therefore by now accumulated large numbers of multihadronic events, typically 10^5 per experiment. My two lectures will be concerned with two aspects of these data:

a) The study of electroweak couplings of the quarks, especially their couplings to the Z^0.

b) The study of QCD.

Many of the questions to be addressed have been discussed in ref [1], where a good outline of the underlying theory is given. A few points will be briefly reviewed here. In the Standard Model the (left-handed) quark states form three weak isospin doublets:

$$\begin{pmatrix} u \\ d' \end{pmatrix} \quad \begin{pmatrix} c \\ s' \end{pmatrix} \quad \begin{pmatrix} t \\ b' \end{pmatrix}$$

where the weak eigenstates d', s', b' are related to the strong eigenstates d, s, b through the Kobayashi-Maskawa mixing matrix. The charged weak bosons $W^\pm$ induce transitions within these doublets. Such processes are typically studied through the decays of hadrons containing heavy quarks. Such studies can be performed at LEP and in many other experiments.

At present the particular contribution of LEP is, however, to study the couplings of quarks to the Z^0. Because of electroweak mixing these couplings depend on the electric charge of the quark. The current for the Z^0 coupling to a fermion-antifermion pair is of the form:

$$\frac{ie\gamma_\mu}{4 sin\theta_{\mathrm{W}} cos\theta_{\mathrm{W}}}[v - a\gamma_5]$$

Z° Physics - Cargèse 1990, Edited by M. Lévy *et al.*
Plenum Press, New York, 1991

where the vector and axial-vector couplings, v and a, are given by

$$v = 2I_3 - 4Q \sin^2\theta_W \qquad a = 2I_3$$

Q and I_3 being the charge and third component of weak isospin for the (left-handed) fermion [1]. Our objective is therefore to measure these couplings. Note that the couplings of the b-quark are of particular interest because of its connection with the t-quark. The couplings are already sufficiently well known from experiments at PETRA to tell us that the b-quark has $I_3 \simeq -\frac{1}{2}$ and that the t-quark must therefore exist.

The basic measurements we must try to make are the same as for the leptonic decays of the Z^0 – the partial width for Z^0 decay into each species of quark, and the forward-backward asymmetry. To lowest order the partial width for Z^0 decay to $q\bar{q}$ is given by:

$$\Gamma(Z^0 \rightarrow q\bar{q}) = \frac{G_\mu M_Z^3}{8\sqrt{2}\pi}(v_q^2 + a_q^2)$$

Many higher order radiative corrections need to be taken into account. Many of these are directly analogous to those for the leptonic final states ([2], [3]), e.g. initial state photon radiation or corrections to the Z^0 propagator. Corrections which are new in the hadronic case include:

- Mass effects are not negligible for the b-quark. The effect is roughly to multiply the a_q^2 term by a factor β^3, where $\beta^3 = 0.984$ for the case of $Z^0 \rightarrow b\bar{b}$.

- QCD corrections have been computed to $O(\alpha_s^3)$ [4]. The partial width is multiplied by a factor R_{QCD}:

$$R_{QCD} = 1 + \alpha_s/\pi + 1.41(\alpha_s/\pi)^2 + 64.84(\alpha_s/\pi)^3$$

  which, for $\alpha_s(M_Z) \sim 0.12$, yields a correction of about 4%. However, there are now known to be problems with the $O(\alpha_s^3)$ calculation.

- Electroweak vertex corrections, especially for the $b\bar{b}$ final state. Loops involving the t-quark and $W^\pm$ bosons give rise to small corrections which depend on the t-quark mass, m_t. The effect is approximately to add $\frac{2\sqrt{2}G_\mu m_t^2}{(4\pi)^2}$ to the couplings v_b and a_b, which corresponds to less than a 1% effect on $\Gamma_{b\bar{b}}$.

The forward-backward asymmetry A_{FB} is given (to leading order, on the Z^0 peak) by:

$$A_{FB} \equiv \frac{N_F - N_B}{N_F + N_B} = \frac{3}{4} \frac{2v_e a_e}{v_e^2 + a_e^2} \frac{2v_q a_q}{v_q^2 + a_q^2}$$

where N_F and N_B are the numbers of events in which the **quark** lies in the region $cos\theta > 0$ and $cos\theta < 0$ respectively. On the Z^0 peak these asymmetries are greater for quarks than for leptonic final states, because of the larger vector couplings of the quarks, amounting to $\sim 6\%$ for $c\bar{c}$ and $\sim 8\%$ for $b\bar{b}$.

Figure 1 (taken from [1]) shows the Standard Model predictions for the partial widths and A_{FB} for $d\bar{d}$, $c\bar{c}$ and $b\bar{b}$ production, as a function of the masses of the Higgs boson and the t-quark. Note the difference between $d\bar{d}$ and $b\bar{b}$, which results from the vertex corrections in the $b\bar{b}$ case – it turns out that $\Gamma_{b\bar{b}}$ is almost independent of the mass of the t-quark.

---

[1] Other conventions are often encountered for these couplings, for example differing by a factor of 2. They are also frequently referred to as g_v and g_a.

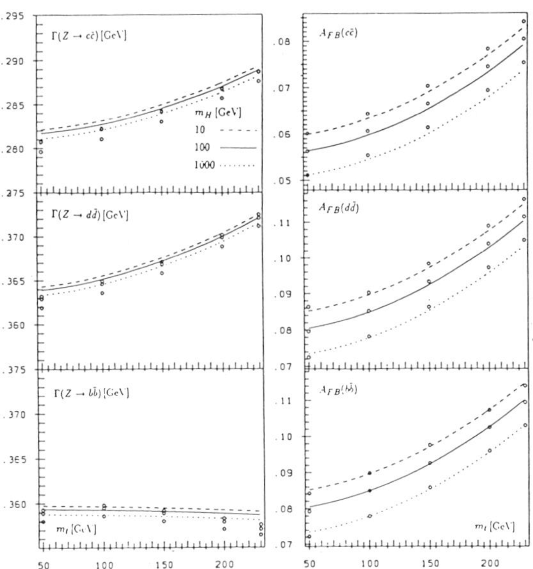

Figure 1. Standard Model predictions for the partial widths and $A_{\rm FB}$ for $d\bar{d}$, $c\bar{c}$ and $b\bar{b}$ production, as a function of the masses of the Higgs boson and the t-quark.

2 Techniques for quark tagging

We have seen in an accompanying lecture [2] that the selection of a clean sample of multihadronic events with high efficiency is a straightforward matter for the LEP experiments. However, it is much less easy to separate these into individual quark flavours. There are many techniques in use (most of them pioneered at PETRA and PEP), none of which provides a completely clean sample, despite low efficiency. All of the techniques are therefore subject to significant systematic errors, but these errors tend to be different for each method. The different techniques are therefore complementary, and it is important to pursue several of them. It should also be noted that some of the techniques are appropriate for determining the partial widths, or the asymmetries, but not necessarily both.

In this section we discuss the techniques in rather general terms, and in the next section proceed to see how these have been applied at LEP so far. Broadly there are two approaches:

a) Tag the *quark* by identifying the *hadron* which contains it. This method is most appropriate for c- and b-quarks, because these quarks are unlikely to be formed in the QCD showering or subsequent hadronization process. It may perhaps be possible to extend this approach to s-quarks. The first four methods below fall into this general category.

b) Identify the charge of the quark, and assume universality between the couplings of d, s, b and of u, c. This is the main way to gain information on the light u- and d-quarks, and thus complements the first technique. The last two methods discussed are of this type.

In addition one exploits kinematic differences between the quarks, such as the greater mass of the b-quark, and knowledge/prejudices about their fragmentation properties.

2.1 Lepton Tagging

Hadrons containing b-quarks are believed to decay predominantly by the *spectator mechanism*:

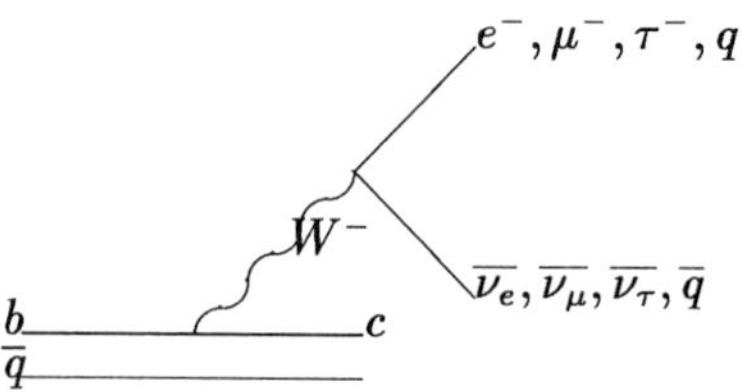

whereby the b-quark decays to cW^-, the W^- then forming $e^-\overline{\nu_e}$, $\mu^-\overline{\nu_\mu}$, $\tau^-\overline{\nu_\tau}$ or $q\overline{q}$. The branching ratio for $b \to \mu$ or $b \to e$ is about $10 \pm 1\%$, and is expected to be roughly the same for $B^\pm, B^0, \Lambda_b$ etc. Likewise hadrons containing c-quarks also have significant semi-leptonic branching ratios for decay, but the spectator mechanism is not so dominant, and the branching ratios are therefore different for different hadrons.

The idea is therefore to look for **prompt** leptons emanating from the main interaction point in hadronic events. Other sources of leptons are decays of $K^\pm$ and $\pi^\pm$ and Dalitz decays. A number of kinematic tricks can be used to enhance the b- and c-quark contributions:

- The fragmentation function for $b \to B$ is known to be very hard, i.e. the momentum of the B meson tends to be a large fraction of that of the b-quark. Thus a lepton resulting from B decay will generally have a rather high momentum, compared to one from π decay. The same is true, though to a lesser extent, of $c \to D$ fragmentation.

- The large mass of the b-quark means that a lepton from its decay will tend to have a large transverse momentum (p_T) relative to the quark direction, and therefore relative to the axis of the observed jet of hadrons.

Note that the charge of the lepton determines the charge of the quark, e.g.

though e^+ can also originate from b-quark decay via the cascade process:

2.2 D^* Tagging

One may try to reconstruct an exclusive decay of a hadron containing b- or c-quarks. The most notable example where this is practicable is the decay chain:

$$D^{*+}(2010) \to D^0 \pi^+ \quad ; \quad D^0 \to K^+ \pi^- (\pi^+ \pi^-)$$

which occurs with probability $\sim 5\%$. The kinetic energy release in the D^* rest frame is only 6 MeV, so that the π^+ emitted in the D^* decay is very distinctive. One consequence is that the mass difference between the D^* and the D can be measured with much better resolution (and less background) than the invariant mass of either D^* or D meson separately.

Of course, D^* mesons are also produced from c-quarks formed in the decay of b-quarks, so the mere observation of a D^* does not necessarily signify the formation of a primary $c\bar{c}$ pair. However the D^*'s formed via the $b \to c \to D^*$ cascade have lower momentum than those from direct $c\bar{c}$ formation, so the observation of a D^* with more than half the beam energy, say, is a good signature for a $c\bar{c}$ event.

The charge of the quark can be taken from the charge of the D^*; a D^{*+} originating from a c-quark and D^{*-} from a $\bar{c}$.

2.3 Vertex Tagging

Weakly decaying hadrons containing b- or c-quarks have lifetimes $\sim 10^{-12}$ sec, and therefore their decay paths at LEP are typically a few mm. In contrast, strange particles such as K_S^0 or Λ^0 travel on average a few cm. Therefore the identification of a secondary vertex close to the interaction point can be used to tag b- or c-quarks.

A number of techniques based on this principle are possible. One can either reconstruct complete vertices, or one can devise algorithms based on impact parameters[2], e.g. the mean of the three largest impact parameters has been proposed [1]. The vertex tagging method is likely to be most effective for b-quarks, because of the higher multiplicities in their decays, and the higher p_T values which result from the large b-quark mass, and lead to greater impact parameters, and better resolution on flight paths.

There are no results from LEP to date on vertex tagging, but the method is expected to be an important one in the future, especially when microvertex detectors are fully operational and understood. Many Monte Carlo studies have been performed, and tagging efficiencies of up to 25% are anticipated. Clearly very good understanding of detector alignment and calibration is needed before useful results can be obtained. A significant uncertainty in determining the tagging efficiency is knowledge of the lifetimes of B mesons, known to $\sim 10\%$ at present.

2.4 Event Shapes

As discussed earlier, the mass of the b-quark is much greater than for the other quarks. In consequence, the jets induced by b-quarks will tend to be fatter, having particles of higher p_T. This difference is much greater at lower energies, but there is still some effect which can be used at LEP. The main uncertainties in this technique are our lack of knowledge about QCD and fragmentation effects.

---

[2] The impact parameter of a track is its distance of closest approach to the interaction point.

2.5 Charge Asymmetry

In the Standard Model, the quark couplings are supposed to depend on their charge. One can expect the quark or antiquark to end up in one of the highest momentum hadrons in the corresponding jet. One can therefore attempt to measure the charge of the quark by summing the charges of the hadrons, taking some appropriate weighting to favour the leading hadrons. One then studies the forward-backward charge asymmetry. The principal uncertainty in this technique is our imperfect understanding of the fragmentation process in QCD whereby quarks and gluons turn into hadrons.

2.6 Prompt Photon Production

The principle of this method is that the probability of final state radiation of a photon depends on the $(charge)^2$ of the produced quark. Therefore the probability of observing direct photons in the final state (i.e. not coming from decays of hadrons such as π^0) is proportional to the average $(charge)^2$ of the quarks, which therefore depends on the relative couplings of the Z^0 to charge $\frac{1}{3}$ and charge $\frac{2}{3}$ quarks.

3 Results from LEP

It should be noted that many of these results have appeared very recently, and that most are as yet unpublished, and thus should be regarded as preliminary.

3.1 The Hadronic Width

The partial width for Z^0 decay to hadrons, Γ_{had}, or the ratio $R_{had} = \Gamma_{\mathrm{had}}/\Gamma_{\mu\mu}$, constitutes a global test of the electroweak couplings of the quarks. As seen in [2] [5] the data are in perfect agreement with the Standard Model. The values (averaged over the four LEP experiments) are:

$$\Gamma_{\mathrm{had}} = 1764 \pm 16 MeV \quad ; \quad R_{had} = 21.08 \pm 0.20$$

to be compared with the Standard Model expectations:

$$\Gamma_{\mathrm{had}} = 1747 \pm 34 MeV \quad ; \quad R_{had} = 20.86 \pm 0.20$$

In the case of R_{had} where the theoretical error is smallest the agreement is good to about 1.5%.

3.2 Prompt Muons

There are results here from L3 [6] [7], ALEPH [8] and OPAL [9]. The basic signature of a muon is a track from the main vertex penetrating many (at least 6) interaction lengths of material. In the case of OPAL and ALEPH the procedure is to extrapolate tracks from the central tracking detector, and associate them with "tracks" of hits in the external muon chambers or the outer layers of the hadron calorimeter. In L3 the muon chambers are located in a magnetic field, so the track and its momentum can be reconstructed there, and then projected inwards to check it comes from the interaction point.

The main problem is to reduce, and understand, the backgrounds to the signal of interest. The possible sources of "muons" are:

Prompt: These form the signal which we are interested in

- $b \rightarrow \mu$

- $b \rightarrow c \rightarrow \mu$ (cascade formation)

- $c \rightarrow \mu$

"Fakes": These are the backgrounds which we need to reduce

- $\pi^{\pm}$ and $K^{\pm}$ decays.

- Hadron "punch-through" – a particle formed in a hadronic shower (very often a muon) penetrates to the outer part of the detector.

- Hadron "sneak-through" – a hadron simply fails to interact.

The L3 detector should be less susceptible to fakes because of the smaller central detector (which reduces the decay path for π/K decay) and because of the improved determination of the angles and momentum of the "muon" (which reduces the likelihood of a punch-through muon being accepted).

Both OPAL and ALEPH obtain muon identification efficiencies in the region of 80-85 %, with fake rates of typically 0.7-1.4 % (depending on momentum, p, and p_T). The fake rates can be estimated either from Monte Carlo simulations or by using a source of guaranteed pions in data, those coming from $K_S^0 \rightarrow \pi^+\pi^-$ decays. Figure 2 shows the typical composition of the "muon" sample, as a function of p and p_T.

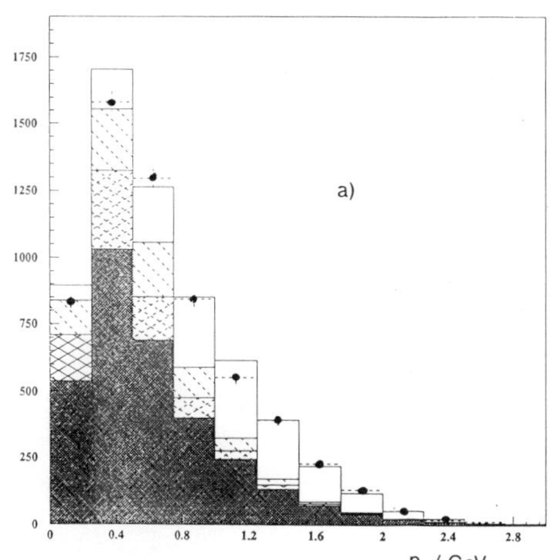

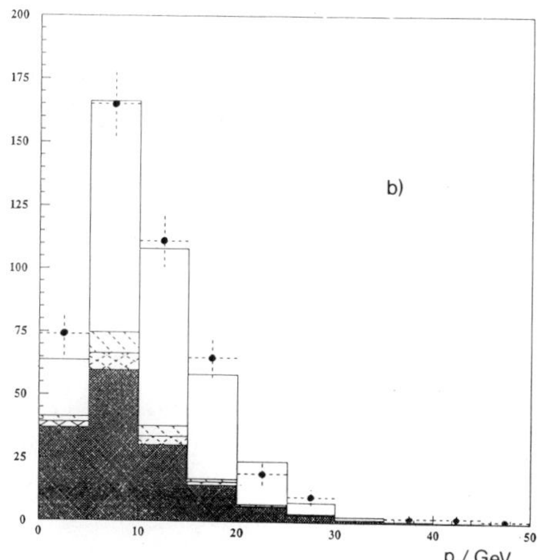

Figure 2. Composition of the "muon" sample in OPAL, taken from a Monte Carlo simulation, as a function of p and p_T.

The proportion of $b \rightarrow \mu$ decays can clearly be enhanced by cuts on p and p_T [3]. Typical cuts used by the experiments are:

ALEPH	$p > 3GeV$	$p_T > 2GeV$
L3	$4 < p < 25GeV$	$1.6 < p_T < 3.5GeV$
OPAL	$p > 4.5GeV$	$p_T > 1GeV$

---

[3] Note there are differences between the experiments in the definition of p_T. It is usually defined with respect to the axis of the jet closest to the muon, but the muon itself may or may not be included in defining the jet axis.

With such cuts, roughly 71% (ALEPH), 55% (OPAL) or 85% (L3) of the remaining muons are estimated to originate from the direct $b \to \mu$ decay. In defining such cuts there is a compromise between efficiency and purity – OPAL for example choose a lower cut on p_T than ALEPH, which gives a higher efficiency and thus smaller statistical errors, but lower purity, and thus greater systematic errors.

To determine the number of $b\bar{b}$ events one then subtracts the residual background and divides by the efficiency for muons from $b\bar{b}$ events to pass one's cuts. This efficiency is sensitive to the fragmentation function of the b-quark. Typically this is determined from data by assuming the "Peterson" form of the fragmentation function [10]:

$$ f(z) = \frac{1}{(1 - \frac{1}{z} - \frac{\epsilon}{1-z})^2} $$

in which the parameter ϵ (expected to be of order $(m_q/m_Q)^2$ where m_q and m_Q are the masses of light and heavy quarks respectively) is treated as a free parameter to be determined from data by fitting the momentum distribution for the leptons. The results, before correcting for the branching ratio for the $b \to \mu$ decay, are:

ALEPH	$br(b \to \mu)\Gamma_{b\bar{b}}/\Gamma_{\mathrm{had}} = 0.0238 \pm 0.0028(stat) \pm 0.0010(syst)$
L3	$br(b \to \mu)\Gamma_{b\bar{b}}/\Gamma_{\mathrm{had}} = 0.0248 \pm 0.0008(stat) \pm 0.0012(syst)$
OPAL	$br(b \to \mu)\Gamma_{b\bar{b}}/\Gamma_{\mathrm{had}} = 0.0206 \pm 0.0010(stat) \pm 0.0018(syst)$

The main sources of systematic error are in the estimation of the muon tagging efficiency and the fake rate. To convert these values into $\Gamma_{b\bar{b}}/\Gamma_{\mathrm{had}}$ for comparison with the Standard Model, we need a value for the semileptonic branching ratio $br(b \to \mu)$. OPAL and ALEPH use the values 0.10 ± 0.01 and 0.102 ± 0.010 respectively, based on recent CLEO and ARGUS measurements (see ref [8] for details), whilst L3 use 0.118 ± 0.011 based on averaging the data in the 1988 Particle Data Book [11]. The values finally quoted are:

ALEPH	$\Gamma_{b\bar{b}}/\Gamma_{\mathrm{had}} = 0.233 \pm 0.028(stat) \pm 0.010(syst) \pm 0.023(br)$
L3	$\Gamma_{b\bar{b}}/\Gamma_{\mathrm{had}} = 0.210 \pm 0.007(stat) \pm 0.010(syst) \pm 0.019(br)$
OPAL	$\Gamma_{b\bar{b}}/\Gamma_{\mathrm{had}} = 0.206 \pm 0.010(stat) \pm 0.018(syst) \pm 0.020(br)$

The experiments are mutually compatible, and agree with the Standard Model expectation of 0.217. Note that the dominant systematic error is the uncertainty in the branching ratio $br(b \to \mu)$, so it is important to try to measure this at LEP, for the correct mixture of b-hadrons.

3.3 Prompt Electrons

Here there are results from ALEPH [8] and recently from L3 [7]. Taking ALEPH as an example, the basic techniques one uses to distinguish electrons from hadrons are:

- The energy measured in the electromagnetic calorimeter (ECAL), E should match the momentum p measured in the central tracking detector. ALEPH cut on a variable :

$$ R_T = \frac{E/p - \langle E/p \rangle}{\sigma(E/p)} $$

  where the expected mean value $\langle E/p \rangle$ and its r.m.s. $\sigma(E/p)$ are taken from test beam data.

- The longitudinal shower profile (based on three samplings in depth in the ECAL, in ALEPH) should match expectations for an electron (again based on test beam data).

- The ionization, dE/dx measured in the central tracking chamber should match the expectation for an electron.

The possible sources of "electrons" are:

Prompt: These form the signal which we are interested in

- $b \to e$

- $b \to c \to e$ (cascade formation)

- $c \to e$

Unwanted: These are the backgrounds which we need to reduce

- $\pi^0 \to \gamma e^+ e^-$ "Dalitz" decays.

- $\gamma \to e^+ e^-$ conversions in material of the detector.

- Hadronic "fakes" – a hadron happens to pass the electron cuts.

ALEPH estimate electron identification efficiencies of $\sim 80\%$ using ECAL alone, and $\sim 40\text{-}75\%$ (depending on p and p_T) using both ECAL and dE/dx. The hadron misidentification probability is typically $\sim 0.4\text{-}0.7\%$ and $<0.25\%$ respectively. Contributions from Dalitz decays and conversions can be reduced by cuts on the $e^+ e^-$ invariant mass, and on the impact parameter of the tracks. Otherwise, the analysis proceeds much as for the muon case.

ALEPH's result is:

$$\boxed{\text{ALEPH} \quad br(b \to e)\Gamma_{b\bar{b}}/\Gamma_{\text{had}} = 0.0217 \pm 0.0019(stat) \pm 0.0010(syst)}$$

in excellent agreement with their independent determination from inclusive muons:

$$\boxed{\text{ALEPH} \quad br(b \to \mu)\Gamma_{b\bar{b}}/\Gamma_{\text{had}} = 0.0238 \pm 0.0028(stat) \pm 0.0012(syst)}$$

The ALEPH electron sample (when both ECAL and dE/dx are used) is clean enough for them to fit the full p_T spectrum and determine $\Gamma_{b\bar{b}}$ and $\Gamma_{c\bar{c}}$ separately. The p and p_T distributions (figure 3) are fitted with four free parameters, the $c\bar{c}$ and $b\bar{b}$ fractions and the fragmentation parameters ϵ_c and ϵ_b. The results are:

$$\boxed{\begin{array}{l} \text{ALEPH} \quad br(b \to e)\Gamma_{b\bar{b}}/\Gamma_{\text{had}} = 0.0219 \pm 0.0017(stat) \pm 0.0010(syst) \\ \text{ALEPH} \quad br(c \to e)\Gamma_{c\bar{c}}/\Gamma_{\text{had}} = 0.0133 \pm 0.0040(stat)^{+0.0038}_{-0.0031}(syst) \end{array}}$$

The average momentum fractions of the b- and c-hadrons determined from the fit are $0.67^{+0.04}_{-0.03}$ and $0.52^{+0.16}_{-0.15}$ respectively. The result for $\Gamma_{b\bar{b}}$ is almost identical to that determined from the high-p_T electrons. We see that $\Gamma_{c\bar{c}}$ has a very large error from this method, because the electron spectrum from c-quark decay is very close to that of the remaining background. Nonetheless, inserting the value $br(c \to e) = 0.090 \pm 0.013$ yields $\Gamma_{c\bar{c}}/\Gamma_{\text{had}} = 0.148 \pm 0.044(stat)^{+0.045}_{-0.038}(syst)$, in good agreement with the Standard Model expectation, 0.171.

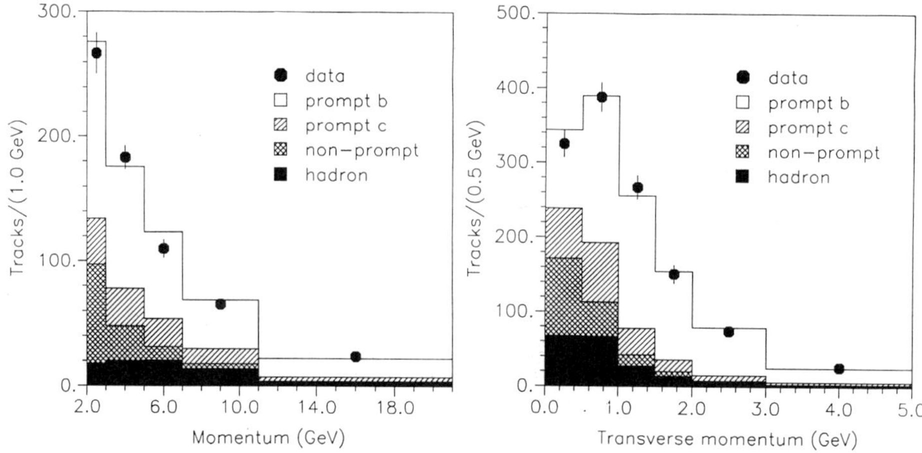

Figure 3. Composition of the electron sample in ALEPH, as a function of p and p_{T}, based on a fit to determine $\Gamma_{b\bar{b}}$ and $\Gamma_{c\bar{c}}$.

3.4 A_{FB} Using Leptons

The forward-backward asymmetry , A_{FB}, is defined by:

$$A_{\mathrm{FB}} \equiv \frac{N_{\mathrm{F}} - N_{\mathrm{B}}}{N_{\mathrm{F}} + N_{\mathrm{B}}}$$

where N_{F} and N_{B} are the numbers of events in which the **quark** lies in the region $cos\theta > 0$ and $cos\theta < 0$ respectively. One uses the Thrust [12] axis to define the direction of the original $q\bar{q}$ pair. By using high-p_{T} leptons one has a sample principally composed of b-quark decays, in which case the charge of the lepton may be used to determine whether one has a quark or an antiquark ($b \to \ell^-$; $\bar{b} \to \ell^+$). Care must be taken however in subtracting or correcting for residual background, because the processes $c \to \ell^+$ and $b \to c \to \ell^+$ will give opposite asymmetries to $b \to \ell^-$. The present statistics are barely sufficient to see a non-zero asymmetry (see Figure 4 for typical data). The results quoted[4] are [13] [7] [9]:

ALEPH	$A_{\mathrm{FB}}^{b\bar{b}} = 13.4 \pm 5.6 \pm 4.2 \ \%$
L3	$A_{\mathrm{FB}}^{b\bar{b}} = 8.4 \pm 3.3 \ \%$
OPAL	$A_{\mathrm{FB}}^{b\bar{b}} = 1 \pm 8 \ \%$

to be compared with the Standard Model expectation of $\sim 8\%$.

However, the interpretation of these data is complicated by the existence of $B^0\overline{B^0}$ mixing. This means that some b-quarks will turn into $\bar{b}$-antiquarks before decaying, which will then lead to leptons of the wrong sign. If the probability of $b \to \bar{b}$ is given by χ ($\chi = 0.5$ corresponding to complete mixing) then the observed asymmetry will be related to the real $b\bar{b}$ asymmetry by

$$A_{\mathrm{FB}}^{obs} = A_{\mathrm{FB}}^{b\bar{b}}(1 - 2\chi)$$

---

[4]with no correction for $B^0\overline{B^0}$ mixing

The value of χ for B_d^0 mesons has been determined by CLEO and ARGUS to be 0.17 ± 0.05 [14] , and theoretically one expects the mixing for B_s^0 mesons to be greater: $\chi > 0.45$. UA1 have measured $\overline{\chi} = 0.12 \pm 0.05$ [14] for some mixture of b-hadrons in high energy jets. So, roughly one expects the observed $A_{\mathrm{FB}}^{b\overline{b}}$ to be reduced to around 6%. But really one needs to measure χ at LEP, for the correct mixture of $B_d^0, B_s^0, B^{\pm}, \Lambda_b$ etc. The first results are just appearing.

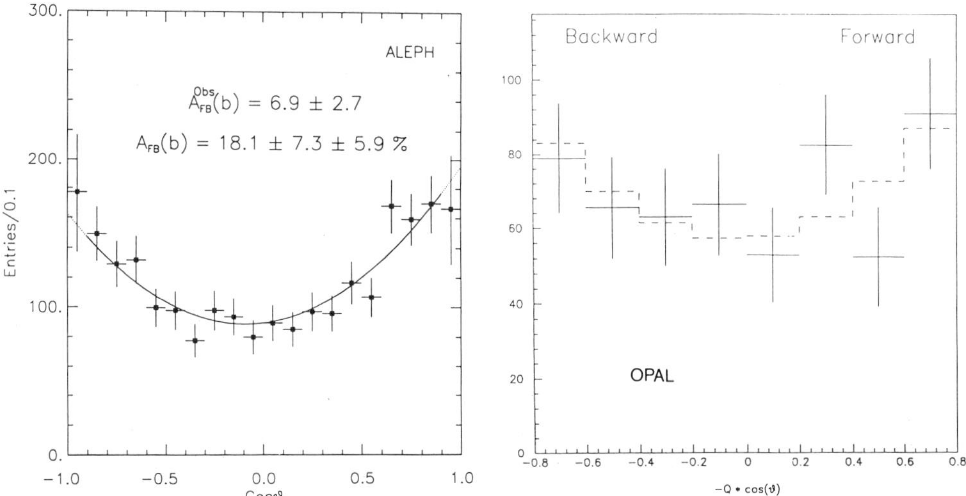

Figure 4. Angular distributions for inclusive leptons from ALEPH and OPAL. The ALEPH data represent all leptons passing the p and p_{T} cuts, whilst the OPAL data have had background subtraction performed.

3.5 $B^0\overline{B^0}$ Mixing

Measurements have recently been reported by ALEPH [13] and L3 [7]. The basic method is to look for events where both the b-quark and $\overline{b}$-antiquark decay to leptons. If all leptons can be assumed to come from b-decays then in the absence of mixing all dilepton events should have opposite charges. In the presence of mixing where the probability of $b \to \overline{b}$ is given by χ the probability of observing two leptons of the same sign will be $2\chi(1 - \chi)$. In practice, even with cuts on p and p_{T} the lepton sample is not totally comprised of $b\overline{b}$ events, and so corrections have to be applied. Furthermore, the cascade decay $b \to c \to \ell^+$ gives muons of the opposite sign from $b \to \ell^-$ and can thus fake mixing.

ALEPH, using events with either electrons or muons and with $p > 5GeV$ and $p_{\mathrm{T}} > 1\mathrm{GeV}$, find 135 $\ell^+\ell^-$ events and 67 $\ell^{\pm}\ell^{\pm}$ events. They estimate that 66% of the sample originates from $b\overline{b}$ events. Based on this they obtain $\chi = 0.129^{+0.045+0.016}_{-0.039-0.020}$. L3, using 130 dimuon events, quote $\chi = 0.11^{+0.08}_{-0.06}$.

Based on these measurements, ALEPH and L3 quote corrected values for $A_{\mathrm{FB}}^{b\overline{b}}$:

ALEPH	$A_{\mathrm{FB}}^{b\overline{b}} = 18.1 \pm 7.3 \pm 5.9$ %
L3	$A_{\mathrm{FB}}^{b\overline{b}} = 10.9 \pm 4.4$ %

3.6 D^* Tagging

DELPHI have presented results [16] based on the observation of the pion from the decay chain $D^{*+} \rightarrow D^0\pi^+$. In this decay the π^+ is almost at rest in the D^* rest frame. Typically the D^* will have $z = p/p_{beam} \sim 0.5$, and thus momentum $\sim$23GeV. The momentum of the decay pion will thus be roughly $p_\pi \sim \gamma_{D^*} \cdot m_\pi \sim 1.8$GeV (from a Lorentz transform). The decay pion will also have a low p_T, around 65 MeV, compared to typically 300 MeV for pions from other sources.

The procedure is therefore to look at $\pi^\pm$ in the momentum range $1.5 < p_\pi < 2.5$ GeV [5] , and plot the p_T^2 distribution. In a Monte Carlo study $c\bar{c}$ events show a clear excess at low values of p_T^2, which is not seen for other quark flavours (Figure 5). The data show a distinct excess for $p_T^2 < 0.01$GeV^2 when the pion has $1.5 < p_\pi < 2.5$ GeV, but no such signal for higher momenta (Figure 6). There is thus clearly a $c\bar{c}$ signal in the data; to determine $\Gamma_{c\bar{c}}$ one fits the p_T^2 distribution to a mixture of signal and background. An estimate of the error is obtained by trying different functional forms for the spectrum. The result is a signal of 381 ± 76 events. After correcting for the efficiency of the selection and fitting procedures, and for the probability of the decay chain $c\bar{c} \rightarrow D^{*\pm} \rightarrow D^0\pi^+/\overline{D}^0\pi^-$ (0.31 ± 0.05 – the largest source of systematic error in this analysis) the result is:

$$\boxed{\text{DELPHI} \quad \Gamma_{c\bar{c}}/\Gamma_{\text{had}} = 0.162 \pm 0.032(stat) \pm 0.046(syst)}$$

in good agreement with the Standard Model expectation of 0.171. A similar analysis by ALEPH just reported [13] gives:

$$\boxed{\text{ALEPH} \quad br(c \rightarrow \pi_{soft}D^0) \cdot \Gamma_{c\bar{c}}/\Gamma_{\text{had}} = 0.0290 \pm 0.0035(stat) \pm 0.0023(syst)}$$

which is very compatible with DELPHI's result, if we use their value for $br(c \rightarrow \pi_{soft}D^0) = 0.155 \pm -.025$.

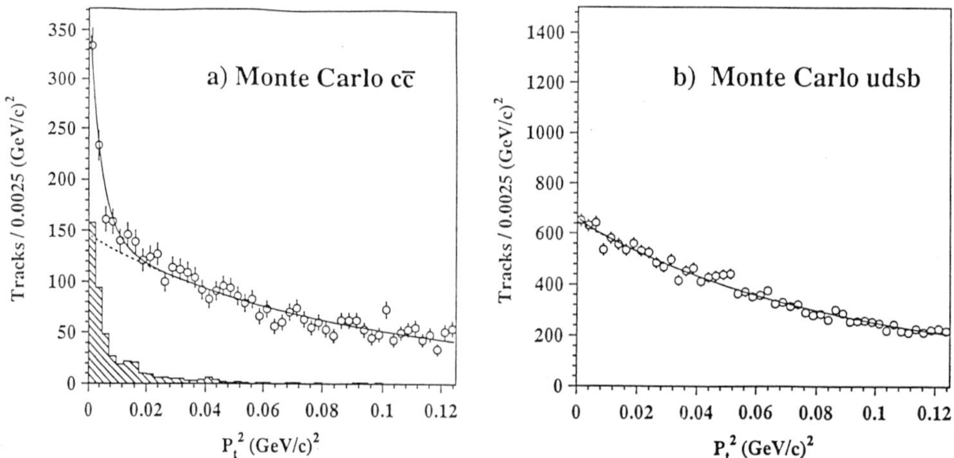

Figure 5. Distributions of p_T^2 for $\pi^\pm$ in $c\bar{c}$ and in other hadronic events (Monte Carlo – DELPHI).

---

[5]The lower cut at 1.5 GeV has the purpose of reducing $b \rightarrow c$ contribution; $c\bar{c}$ events dominate the momentum spectrum of D^*'s for $p_{D^*} > 20$GeV.

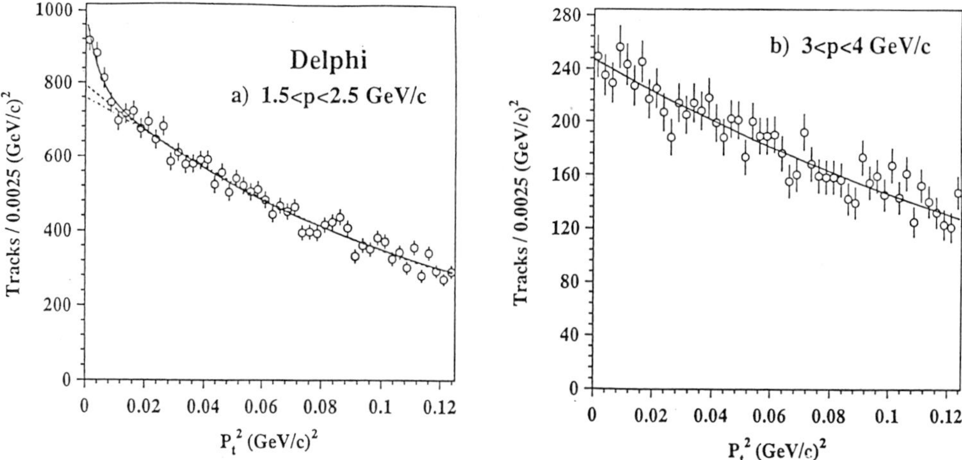

Figure 6. Distributions of p_T^2 for $\pi^\pm$ in data (DELPHI). The momentum range $1.5 < p_\pi < 2.5 GeV$ should contain the D^* signal, while the region $3 < p_\pi < 4 GeV$ lies too high for any significant D^* contribution.

3.7 Sphericity Product

DELPHI have applied [17] a technique for distinguishing $b\bar{b}$ events based on event shapes, exploiting the fact that the larger mass of the b-quark should give higher p_T to its fragmentation products. In outline their method is as follows:

- Select 2-jet events. This is done in order to reduce the contribution of gluon radiaton to the fattening of jets.

- Divide into two hemispheres using the plane orthogonal to the Sphericity [12] axis.

- In each hemisphere boost the observed particles by a velocity β. This is supposed to roughly correspond to a boost into the rest frame of a b-hadron, if one were produced. Calculate the Sphericities, S_1, S_2 in each hemisphere in these boosted frames.

The product $S_1 \cdot S_2$ is used as a tagging variable. The value of β is chosen so as to optimize the separation of $b\bar{b}$ events from others; the value chosen was $\beta = 0.96$.

Figure 7(a) shows Monte Carlo simulations of the $S_1 \cdot S_2$ distribution for $b\bar{b}$ events and for other flavours. The $b\bar{b}$ sample clearly favours higher values of $S_1 \cdot S_2$, though there is nowhere a clear separation. Nonetheless, it might be useful to cut on $S_1 \cdot S_2$ to enrich the proportion of $b\bar{b}$ events before pursuing one of the other tagging techniques. Alternatively, DELPHI have derived the fraction of $b\bar{b}$ events in their data by fitting the observed $S_1 \cdot S_2$ distribution in their data to a mixture of the $b\bar{b}$ and non-$b\bar{b}$ Monte Carlo distributions (Figure 7(b)). The result is:

$$\boxed{\text{DELPHI} \quad \Gamma_{b\bar{b}}/\Gamma_{\text{had}} = 0.211 \pm 0.020(stat) \pm 0.031(syst)}$$

in good agreement with the Standard Model value of 0.217. The systematic error comes from uncertainties in the heavy quark fragmentation functions, and in the modelling of QCD (parton shower or matrix element approaches [12] [18]).

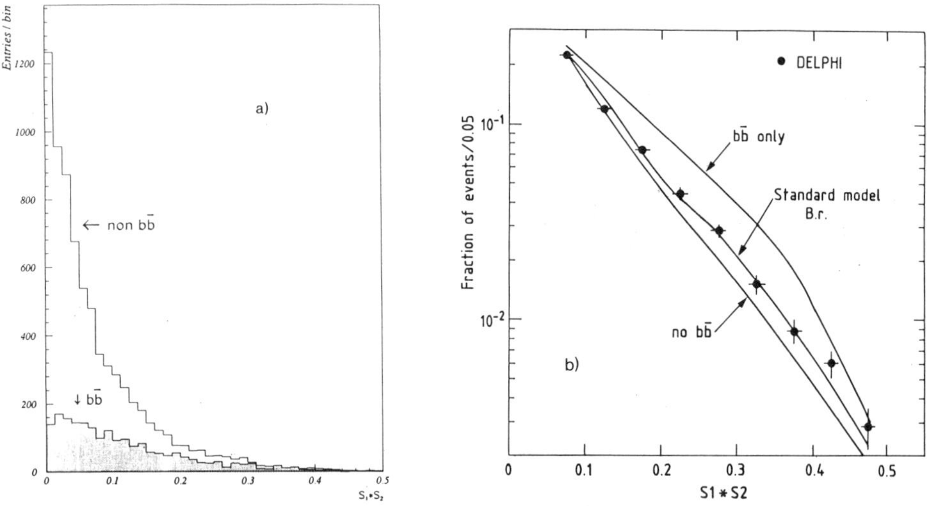

Figure 7. (a) Distributions of $S_1 \cdot S_2$ for $b\bar{b}$ and non-$b\bar{b}$ events (Monte Carlo – DELPHI). (b) Distribution of $S_1 \cdot S_2$ for data compared with expectations if no $b\bar{b}$ produced, only $b\bar{b}$ produced, and for the Standard Model fraction of $b\bar{b}$.

3.8   Charge Asymmetry

ALEPH have shown an analysis [19] based on the forward-backward asymmetry of charge in hadronic events. The event is divided into two hemispheres by the plane orthogonal to the Thrust axis, $\hat{\mathbf{n}}_\mathbf{T}$ (the Thrust axis being taken to lie in the forward c.m. hemisphere; $cos\theta_T > 0$). The charges in each event hemisphere, Q_F, Q_B, are computed from a weighted mean of track charges:

$$Q = \frac{\sum_i q_i (\mathbf{p_i}.\hat{\mathbf{n}}_\mathbf{T})^\kappa}{\sum_i (\mathbf{p_i}.\hat{\mathbf{n}}_\mathbf{T})^\kappa}$$

where for Q_F and Q_B the sum runs over tracks having $\mathbf{p_i}.\hat{\mathbf{n}}_\mathbf{T} > 0$ and $\mathbf{p_i}.\hat{\mathbf{n}}_\mathbf{T} < 0$ respectively. The value of κ is chosen so as to maximize the charge separation between hemispheres, and thus to maximize one's sensitivity to the weak couplings. The value $\kappa = 1$ was chosen. Figure 8 shows the results of a Monte Carlo study of the distributions of $Q_{FB} \equiv Q_F - Q_B$ for $u\bar{u}$ events, separated according to whether the u or the $\bar{u}$ lay in the forward c.m. hemisphere $cos\theta_T > 0$. A fair degree of separation is seen, so the variable Q_{FB} is clearly sensitive, on average, to whether the quark or the antiquark was produced in the forward c.m. hemisphere.

Applying this method to data, using $\sim 93K$ hadronic events, ALEPH find that the average value $< Q_{FB} >$ deviates from zero, increasing with $cos\theta_T$ as expected (figure 9). The overall mean is $< Q_{FB} >= -0.0107 \pm 0.0021(stat) \pm 0.0010(syst)$ in the region $cos\theta_T < 0.9$. $< Q_{FB} >$ can be related to the partial widths and A_{FB} values for each quark species through the relation:

$$< Q_{FB} >= \frac{4}{3} \cdot \frac{cos\theta_{max}}{1 + \frac{1}{3}cos^2\theta_{max}} \sum_{i=u,d,s,c,b} q_i A^i_{FB} \frac{\Gamma_i}{\Gamma_{had}} \quad .$$

The quantity q_i is the mean value of Q_{FB} for quark i as determined from a Monte Carlo (i.e. the mean of the histogram in Figure 8 for the case of u-quarks). The factor preceding the sum serves merely to correct to the full range of $cos\theta_T$. The asymmetry for quark i is given by:

$$A_{\mathrm{FB}}^i = \frac{3}{4}\mathcal{A}_e\mathcal{A}_i \quad ; \quad \mathcal{A} = \frac{2va}{v^2 + a^2} \quad .$$

In the Standard Model model the widths and asymmetries are all determined in terms of $\sin^2\theta_W$, and so ALEPH have used their measurement of $< Q_{FB} >$ to determine

$$\sin^2\theta_W = 0.225 \pm 0.005(stat) \pm 0.002(syst, expt) \pm 0.004(syst, theo)$$

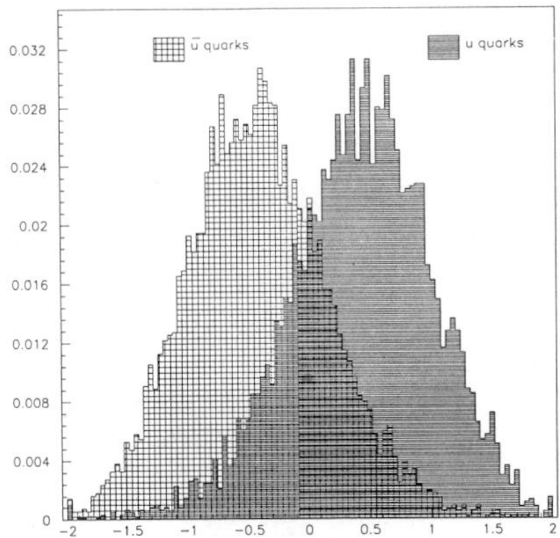

Figure 8. Distributions of Q_{FB} for $u\bar{u}$ events, separated according to whether the quark or the antiquark lies in the region $|cos\theta| > 0$ (Monte Carlo – ALEPH).

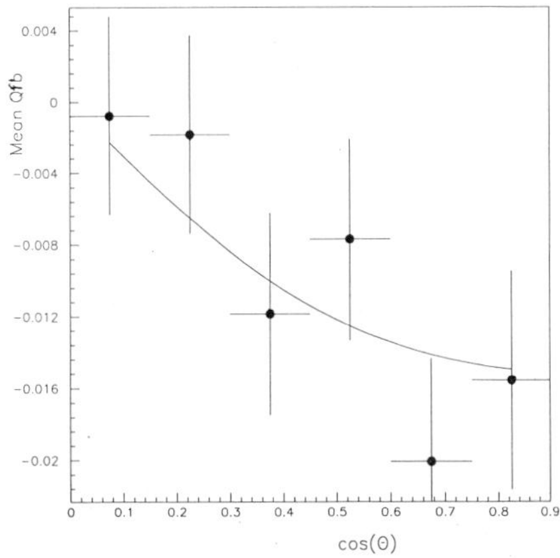

Figure 9. Mean value of Q_{FB} for ALEPH data as a function of $|cos\theta|$.

As an alternative way of analyzing these data, ALEPH have used results from νq and $\bar{\nu}q$ scattering which determine the left-handed and right-handed couplings of u and d. From these one obtains $\mathcal{A}_u = 0.640 \pm 0.080$ and $\mathcal{A}_d = 0.956 \pm 0.051$, and assuming universality of couplings between d, s, b and between u, c one deduces $\mathcal{A}_e = 0.172\pm0.059$, or equivalently $v_e/a_e = 0.090 \pm 0.031$. The significance of this result is that it tells us that the vector and axial-vector couplings of the electron have the same **sign** (to 3 the standard deviation level) – information which is not available from the leptonic final states alone.

3.9 Direct Photons

This technique has been pursued by OPAL [20] (data updated here). The basis of the method is to identify final state photons which can be ascribed to final state bremsstrahlung:

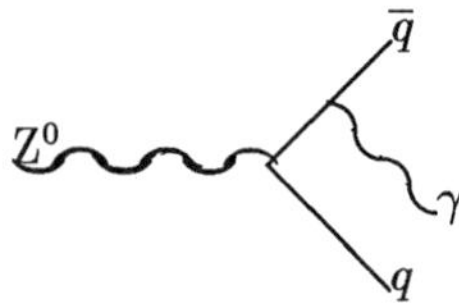

The rate of such events is determined by the relative proportions of charge $\frac{1}{3}$ and $\frac{2}{3}$ quarks.

The experimental procedure was, in outline:

- Require an electromagnetic cluster of energy > 10 GeV.

- Demand no tracks within a cone of half-angle $20°$ around the cluster.

- Require the photon candidate to have $p_T > 5$GeV relative to the Thrust axis of the event. This reduces the contribution of π^0s in jets.

- Make cuts on the shower shape, particularly the transverse shape. A real photon will tend to give narrower showers than the likely sources of background, neutral hadrons or pairs of photons from π^0 decay.

After these cuts there remain 78 candidates in about 77000 multihadronic events. The photon detection efficiency is estimated (from a reference sample of photons in $\gamma\gamma$ or $\ell^+\ell^-\gamma$ final states) to be 84%. The residual background from non-photons is calculated to be 8 ± 5 events, and the contribution from initial state bremsstrahlung is estimated to be 5.1 ± 1.4 events. The signal is therefore 64.9 ± 10.2 events, to be compared with the Standard Model expectation for final state bremsstrahlung of 61.0 ± 2.6. The distributions of energy and p_T (Figure 10) for the photon candidates are in good agreement with expectations.

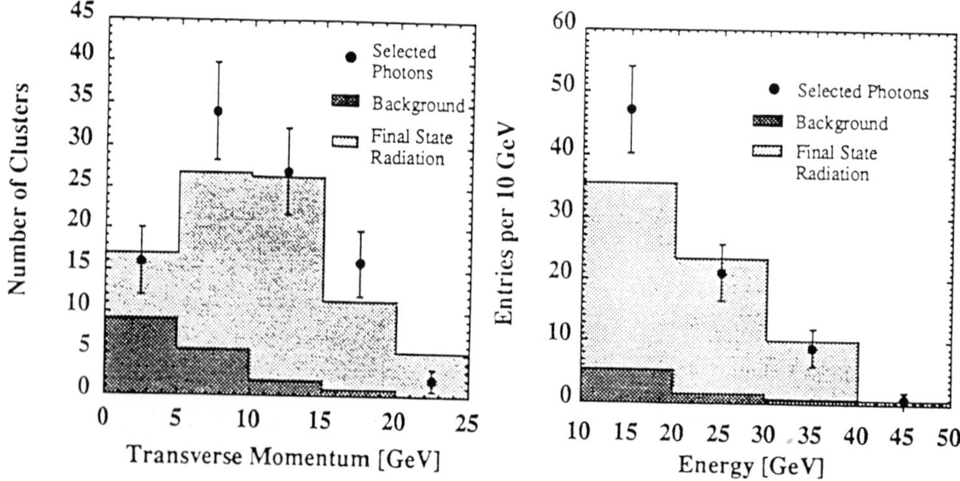

Figure 10. Distributions of energy and p_T for prompt photon candidates (OPAL) compared with expectations for signal and background.

If we write $c_i = v_i^2 + a_i^2$, and assume universality so that $c_u = c_c \equiv c_{\frac{2}{3}}$ and $c_d = c_s = c_b \equiv c_{\frac{1}{3}}$ then we can write the hadronic width as:

$$\Gamma_{\text{had}} \propto \sum_i c_i = \left(3c_{\frac{1}{3}} + 2c_{\frac{2}{3}}\right)$$

and the number of $q\bar{q}\gamma$ events as:

$$N_{q\bar{q}\gamma} \propto \sum_i e_i^2 c_i = \frac{1}{9}\left(3c_{\frac{1}{3}} + 8c_{\frac{2}{3}}\right)$$

(e_i being the charge of quark i). Measuring these two quantities therefore enables one to solve for $c_{\frac{1}{3}}$ and $c_{\frac{2}{3}}$, or equivalently for the partial widths:

$$\Gamma_i/\Gamma_{\text{had}} = \frac{c_i}{3c_{\frac{1}{3}} + 2c_{\frac{2}{3}}}$$

yielding:

$$\Gamma_{\frac{1}{3}}/\Gamma_{\text{had}} = 0.209 \pm 0.038$$

$$\Gamma_{\frac{2}{3}}/\Gamma_{\text{had}} = 0.187 \pm 0.056$$

in agreement with the Standard Model values of 0.217 and 0.171 respectively.

4 Summary

We now have a number of measurements of $\Gamma_{b\bar{b}}$ (or $\Gamma_{\frac{1}{3}}$) and $\Gamma_{c\bar{c}}$ (or $\Gamma_{\frac{2}{3}}$), using a variety of techniques. All agree (to much better than one standard deviation!) with the Standard Model (figure 11). Already the systematic errors exceed the statistical errors in most cases. Typical errors quoted at present, and their main sources are:

- $\Gamma_{b\bar{b}}$ (or $\Gamma_{\frac{1}{3}}$)

    - lepton tagging : $\sim 10\%$ – branching ratio for $b \to \mu$.
    - event shapes : $\sim 15\%$ – QCD / fragmentation.
    - prompt γ : $\sim 10\%$ – QCD / fragmentation.

- $\Gamma_{c\bar{c}}$ (or $\Gamma_{\frac{2}{3}}$)

    - lepton tagging : $\sim 30\%$ – background estimation.
    - D^* tagging : $\sim 25\%$ – branching ratios.
    - prompt γ : $\sim 15\%$ – QCD / fragmentation.

Many of these are capable of improvement, given work and more data. For example, the LEP experiments can measure $br(b \to \mu)$ by double tagging – one tags one b-quark by some means, and determines the probability of a muon (or electron) tag on the other side of the event. The expectation is that the partial widths $\Gamma_{b\bar{b}}$ and $\Gamma_{c\bar{c}}$ can be determined to roughly the 5-10% level eventually [1] with $10^6 Z^0$s.

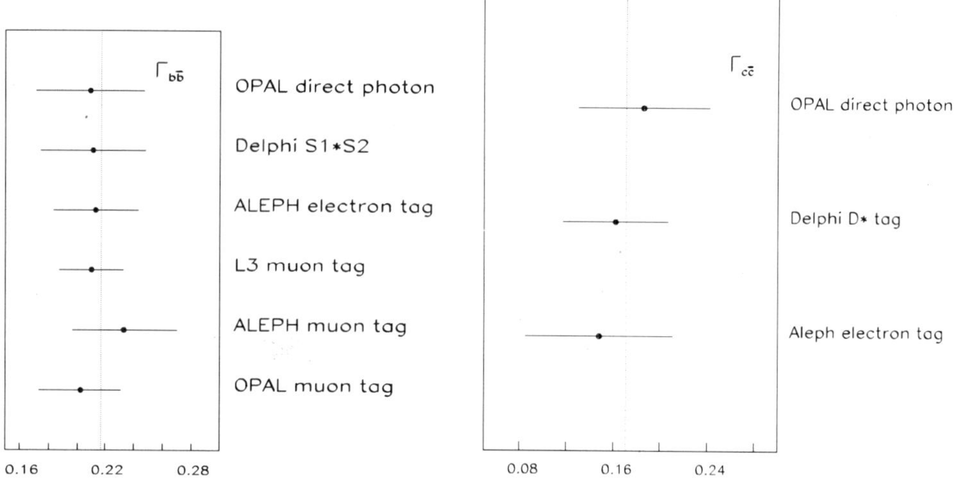

Figure 11. Values of $\Gamma_{b\bar{b}}$ (or $\Gamma_{\frac{1}{3}}$) and $\Gamma_{c\bar{c}}$ (or $\Gamma_{\frac{2}{3}}$) measured at LEP. The dashed lines show the Standard Model expectations.

The first results on A_{FB} are just appearing for the $b\bar{b}$ case. In the longer term, A_{FB} should be less limited by systematic errors than the partial widths, because the error is not determined by one's knowledge of the tagging efficiency, but rather by understanding and control of background. The expectation is that with $10^6 Z^0$s an accuracy in the region of 0.005 (0.010) could be achievable for $A_{\mathrm{FB}}^{b\bar{b}}(A_{\mathrm{FB}}^{c\bar{c}})$. The interpretation of $A_{\mathrm{FB}}^{b\bar{b}}$ requires measurements of $B^0\overline{B^0}$ mixing. First results are just appearing, and this should not be a limiting factor.

There are many other topics to pursue in heavy flavour physics; some of these are discussed in the context of future plans for LEP in an accompanying lecture [21]. A few points can just be mentioned here:

- **QCD studies** – differences are expected between heavy and light quark jets. Also the tagging of a b-hadron is a good way of tagging a quark jet and distinguishing it from a gluon jet.

- **Lifetimes** – measurements are likely to be dominated by systematic errors ($\sim$ 10%?). Of particular interest would be to study B^+, B_s^0, B_d^0 etc. separately. To do this needs the identification of exclusive decay modes, such as $B_d^0 \to D^{*-}\pi^+$, or the use of semileptonic decays like $B_d^0 \to \ell^+ D^- X$. At least $10^6 Z^0$s are likely to be required.

- $b \to u$ decays.

- **Rare decays** e.g. flavour changing neutral currents, such as $b \to s\gamma$.

- $B\overline{B}$ **mixing** – to study B_s^0, B_d^0 separately needs exclusive decays, and thus $\sim$ $10^7 Z^0$s.

- **CP violation** is highly important, but probably out of LEP's reach, needing at least $10^8 Z^0$s.

5 Acknowledgements

It is a pleasure to thank the organisers of the Cargèse Summer School for arranging such an enjoyable meeting. I would also like to thank the members of the LEP collaborations who helped provide information for this talk, and particularly my fellow speakers, A.Blondel, G.Sauvage and D.Treille.

References

[1] J.H.Kühn, P.M.Zerwas *et al.*, CERN 89-08, vol 1, 267.

[2] G.Sauvage, these proceedings.

[3] F.Berends, these proceedings.

[4] S.G.Gorishny, A.L.Kateev and S.A.Larin, Phys. Lett. **212B** (1988) 238.

[5] F.Dydak, 25th International Conference in High Energy Physics, Singapore (1990).

[6] L3 Collaboration: B.Adeva *et al.* Phys. Lett. **241B** (1990) 416.

[7] L3 Collaboration: S.C.C.Ting, 25th International Conference in High Energy Physics, Singapore (1990).

[8] ALEPH Collaboration: D.Decamp *et al.* Phys. Lett. **244B** (1990) 551.

[9] OPAL Collaboration: H.Jawahery, 25th International Conference in High Energy Physics, Singapore (1990).

[10] C.Peterson *et al.* Phys. Rev. **D27** (1983) 105.

[11] Particle Data Group, Phys. Lett. **204B** (1988) 1.

[12] D.R.Ward, "QCD studies at LEP", these proceedings.

[13] ALEPH Collaboration: R.Johnson, 25th International Conference in High Energy Physics, Singapore (1990).

[14] ARGUS Collaboration: H.Albrecht *et al.* Phys. Lett. **192B** (1987) 245;
CLEO Collaboration: M.Artuso *et al.* Phys. Rev. Lett. **62** (1989) 2233.

[15] UA1 Collaboration: C.Albajar *et al.* Phys. Lett. **186B** (1987) 247.

[16] DELPHI Collaboration: *Measurement of the partial width of the Z^0 boson into charm quark pairs*, International Conference in High Energy Physics, Singapore (1990).

[17] DELPHI Collaboration: P.Abreu *et al.* CERN PPE/90-118, submitted to Phys. Lett.

[18] T. Sjöstrand, these proceedings.

[19] ALEPH Collaboration: J.R.Hansen, 25th International Conference in High Energy Physics, Singapore (1990).

[20] OPAL Collaboration: M.Z.Akrawy *et al.* Phys. Lett. **246B** (1990) 285.

[21] D.Treille, these proceedings.

QCD Studies at LEP

D.R. Ward

Cavendish Laboratory, Madingley Road, Cambridge, U.K.

1 Introduction

1.1 How to Study QCD at LEP?

Many of the questions to be addressed in the study of QCD at LEP have been discussed in ref [1], and also in two of the accompanying theoretical lectures [2] [3].

In principle the predictions of QCD are completely specified in terms of the coupling constant α_s. Because of loop diagrams the *effective* coupling constant runs with the energy scale μ^2

$$\mu^2 \frac{d\alpha_s(\mu^2)}{d\mu^2} = -b_0 \alpha_s^2 - b_1 \alpha_s^3 + O(\alpha^4)$$

where

$$b_0 = (33 - 2N_f)/12\pi \quad ; \quad b_1 = (153 - 19N_f)/24\pi^2$$

and thus (to second order[1])

$$\alpha_s(\mu^2) = \frac{1}{b_0 \ln(\mu^2/\Lambda^2)} \left[1 - \frac{b_1 \ln(\ln(\mu^2/\Lambda^2))}{b_0^2 \ln(\mu^2/\Lambda^2)} \right] \cdots$$

Calculations of physical cross-sections depend on the "renormalization scheme". Most commonly used is the "$\overline{MS}$" scheme, so we refer to the value of the QCD scale parameter Λ so obtained as $\Lambda_{\overline{MS}}$.

However, it is clear that QCD studies at LEP can not at present approach the same level of precision as electroweak measurements. This reflects the sheer complexity of QCD calculations, and more particularly the largeness of α_s which means that perturbation theory does not converge rapidly. Typically QCD calculations are available to $O(\alpha_s^2)$ and are reckoned to be reliable to $\sim 10 - 15\%$. Furthermore, there are always effects due to the formation of hadrons from quarks and gluons, which cannot be treated perturbatively in QCD. In correcting for such "hadronization" (or "fragmentation") effects further uncertainties of $\sim 5 - 10\%$ [1] may be introduced.

---

[1] n.b. most of the QCD calculations used at LEP are available to $O(\alpha_s^2)$, so this is the appropriate form for α_s.

The broad strategy in testing QCD is therefore to look at many different observables, which have different (theoretical) systematic uncertainties, and determine α_s (or equivalently $\Lambda_{\overline{MS}}$) from each, in rather the same way that one tests the electroweak model by measuring $\sin^2\theta_W$ in different processes. Unfortunately there is no clear "gold-plated" test of QCD.

Figure 1 (taken from ref [1]) illustrates some of the problems. It shows QCD predictions for three observables, $< 1 - T >$ where T is the Thrust, $< O >$ where O is the Oblateness and the Asymmetry in the Energy-Energy Correlation (AEEC), for lowest order (LO) and next-to-lowest (NL) order QCD, and for three possible choices of renormalization scale μ.

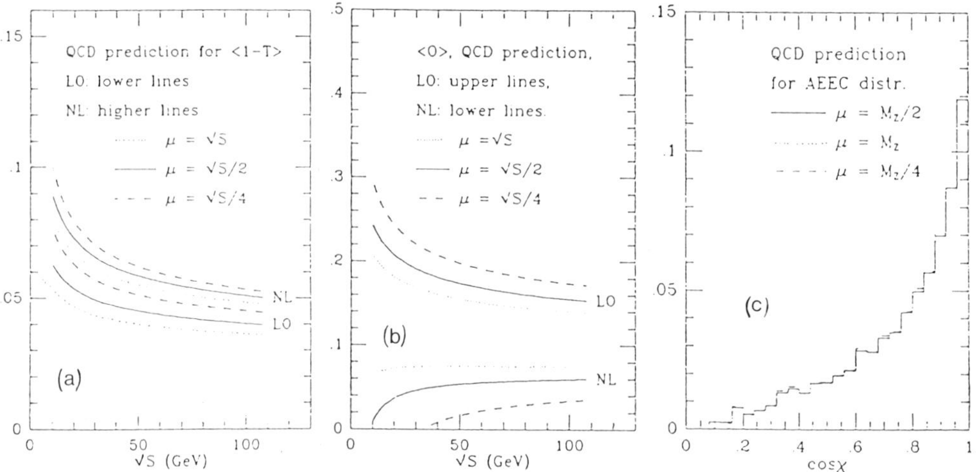

Figure 1. (a) QCD calculations for $< 1 - T >$ for three choices of scale. (b) QCD calculations for $< O >$ for three choices of scale. (c) QCD calculations for AEEC for three choices of scale.

We note that the NL corrections to $< 1 - T >$ reinforce the LO, whilst for $< O >$ NL partially cancels LO. We also note the large scale dependence of $< 1 - T >$ and $< O >$, which is substantially absent from the AEEC. Therefore different observables show different degrees of sensitivity to the various theoretical uncertainties.

Among the particular advantages in studying QCD at LEP are:

- We shall have very large statistics from $Z^0 \rightarrow q\bar{q}$ decays – up to 10^5 events per experiment already.

- The initial $q\bar{q}$ system is very simple, so that QCD studies can be separated from the uncertainties over structure functions which affect similar studies in hadron-hadron collisions(see e.g. [5]).

- We have high energy or high Q^2, which means that α_s is reduced and we are less susceptible to non-perturbative hadronization effects and phase-space.

- QCD processes constitute a major background in many other studies at LEP, particularly searches for new particles. It is therefore important to have a good understanding of the hadronic events.

1.2 Theoretical Tools

The various QCD calculations and Monte Carlo programs available for use at LEP are extensively reviewed in refs [1] and [4], and also in two accompanying lectures [2] [3]. Only a brief outline can be given here. There are two main approaches – starting from exact QCD matrix elements (ME), or using the leading log approximation (LLA).

1.2.1 <u>QCD Matrix Elements (ME)</u>

The QCD matrix elements are fully known to $O(\alpha_s^2)$, corresponding to up to 4-parton final states. The calculations most commonly used at LEP are those of Ellis, Ross and Terrano (ERT) [6]. Also used are the "GKS" matrix elements [7] which are, however, known to neglect some terms. There are a number of implementations interfacing these matrix elements to hadronization models, usually the LUND string fragmentation scheme [8].

The only complete $O(\alpha_s^3)$ calculation is of the total hadronic width [9], yielding:

$$R_{QCD} = 1 + \frac{\alpha_s}{\pi} + 1.41(\frac{\alpha_s}{\pi})^2 + 64.84(\frac{\alpha_s}{\pi})^3 \quad .$$

Thus at LEP where $\alpha_s \sim 0.115$ the third order correction appears to exceed the second order term. There are now known to be some problems with the third order calculation, but until a new calculation is completed there must remain some doubt about the adequacy of the $O(\alpha_s^2)$ matrix elements, since it suggests that the higher order terms may be far from negligible. Note that the measurement of the hadronic width is not an easy way to determine α_s, since a 10% accuracy in α_s requires measuring the hadronic width (or its ratio to the leptonic width) to better than 0.5%. Based on their present measurement of $\Gamma_{had}/\Gamma_{lepton}$ ALEPH have quoted [10]:

$$\alpha_s = 0.13 \pm 0.05(stat) \pm 0.03(syst) \quad .$$

1.2.2 <u>The Parton Shower (PS) Approach</u>

This technique is based on the leading log approximation, and is therefore not guaranteed to give correct results even to leading order. However, it can give an approximation to higher orders, and is well suited to Monte Carlo implementation. One can also include coherence effects in a straightforward way, and these turn out to be important. Several implementations exist, which are broadly similar, though they have many differences in detail. Those used at LEP are:

JETSET [11] is the well known Lund model, which incorporates string fragmentation. This model includes a facility to map the first gluon emission onto the $O(\alpha_s)$ matrix element. JETSET also includes several ME simulation options with string or independent fragmentation as well as the PS model.

HERWIG [12] is the Marchesini-Webber model. It includes probably the most sophisticated treatment of QCD, including spin effects, but uses a much simpler hadronization procedure than JETSET, based on isotropic decay of colour singlet clusters.

ARIADNE [13] again comes from the Lund group, and is based on an alternative formulation of the theory in terms of radiation from colour dipoles.

The PS models are basically controlled by two parameters in the QCD shower stage, a scale Λ_{LLA} and a parameter to cut off the branching at low Q^2. They also need various parameters governing the hadronization phase, which may be adjusted to fit data.

1.2.3 Scale Optimization

The PS calculations indicate that at LEP typically ~ 9 partons with $Q^2 > 1\text{GeV}^2$ are produced. One can therefore expect deficiencies in the ME approach in certain kinematic regions. These deficiencies may be reduced by the use of an "optimized scale". In principle the ME calculations depend on a single parameter, e.g. $\Lambda_{\overline{MS}}$. However, when calculations are only performed to finite order in perturbation theory an additional uncertainty is the scale μ^2 at which the coupling constant is evaluated. Conventionally one would take $\mu^2 = E_{cm}^2$, but more generally one could take $\mu^2 = f \cdot E_{cm}^2$. A calculation to all orders would be independent of f, but to finite order (above first order) the predictions depend on both f and $\Lambda_{\overline{MS}}$. One can therefore optimize f so as to fit the data best, and the hope is that by doing so one is minimizing the (uncomputed) higher order contributions. This idea has somewhat revitalized the ME approach. Clearly it is no real substitute for a proper higher order calculation, but it gives us some means of assessing the possible magnitude of the uncomputed terms. A small value of $f \sim 0.001 - 0.005$ seems to work best (for jet rates); the effect is to increase the effective α_s and thereby increase the average number of partons.

One can also improve the agreement between the ME calculations and data by retuning the hadronization parameters so as partially to compensate for the missing collinear and soft partons. Of course, this illustrates that it can be difficult to disentangle the calculable perturbative QCD effects from the non-perturbative hadronization phenomena.

1.3 Some Experimental Issues

The LEP experiments appear to have followed similar procedures in order to identify and analyse multihadronic events, and where data can be compared between experiments they seem to agree. The selection of multihadrons is rather easy on the Z^0 peak [14] because of the large cross-section and high multiplicity. Typical cuts are:

- Demand high multiplicity (to remove dileptons).

- Require significant visible energy, with some degree of forward-backward energy balance (to remove two-photon and beam-gas events).

- Cut on some region of $\cos\theta$ to ensure good quality data.

Backgrounds are typically at the 0.2-0.3% level, coming mainly from $\tau^+\tau^-$ production.

The analyses have been based on charged tracks (ALEPH, DELPHI) or on clusters of electromagnetic energy (L3) or both together or separately (OPAL). Almost all the results to date are based on the simplest possible correction procedure. In essence one makes two Monte Carlo runs:

1. QCD including initial state QED radiation and full detector simulation.

2. The same QCD model without initial state radiation or detector simulation. Particles with proper lifetimes $> 10^{-9} sec$ are treated as stable.

The raw observed distribution of some observable is then multiplied, bin by bin, by the ratio of the distributions in Monte Carlos 2 and 1. In this way one corrects for detector smearing and inefficiencies, trigger losses and radiative corrections all together, and obtains results corresponding to a hadronic system at fixed energy E_{cm}. In Monte Carlo 2 one can compute the observable using either the hadrons or the partons, and therefore correct the data down to either the "hadron level" (correcting for detector effects only) or to the "parton level" (correcting for hadronization also). This procedure is strictly only correct if the Monte Carlo precisely reflects the true physics, otherwise the corrected data may be biased towards the model used for the corrections. To minimize this effect it is important to choose bins in the observable that are wider than the experimental resolution.

2 Event Shapes

There are results on event shapes from ALEPH [15] (updated in [16]), DELPHI [17] and OPAL [18]. There are many variables available. Those studied at LEP are:

- Thrust **T**, given by the expression

$$T = max \left(\frac{\sum_i |\vec{p}_i \cdot \hat{n}|}{\sum_i |\vec{p}_i|} \right) \quad . \tag{1}$$

  $\hat{n}_T$ is the axis $\hat{n}$ for which equation 1 is satisfied and is taken to be the event axis. The Thrust major value **M** equals expression (1) for directions $\hat{n} = \hat{n}_M$ in the plane perpendicular to $\hat{n}_T$. In an analogous manner, the Thrust minor value **m** is given by expression (1) evaluated for the direction $\hat{n} = \hat{n}_m$ perpendicular to both $\hat{n}_T$ and $\hat{n}_M$. Oblateness **O** is given by $O = M - m$. A two-jet event will tend to have $T \simeq 1, M \simeq 0, m \simeq 0$, a three-jet event will have $\frac{2}{3} < T < 1$, $0 < M < \frac{1}{3}$, $m \simeq 0$ (the three jets have to be coplanar) and a multijet event $\frac{1}{2} < T < 1$, $0 < M < \frac{1}{2}$, $0 < m < \frac{1}{2}$. A completely isotropic event would have $T = M = m = \frac{1}{2}$.

- The momentum tensor $S^{\alpha\beta}$ is defined by

$$S^{\alpha\beta} = \frac{\sum_i p_i^\alpha \cdot p_i^\beta}{\sum_i p_i^2} \quad ; \quad \alpha, \beta = 1, 2, 3 \quad . \tag{2}$$

  The eigenvalues of $S^{\alpha\beta}$ are denoted Q_1, Q_2 and Q_3, for which $Q_1 < Q_2 < Q_3$ and $Q_1 + Q_2 + Q_3 = 1$. Sphericity **S** and Aplanarity **A** are given by

$$S = \frac{3}{2} \cdot (Q_1 + Q_2) \quad \text{and} \quad A = \frac{3}{2} \cdot Q_1 \quad . \tag{3}$$

- The spherocity tensor $\theta^{\alpha\beta}$ is the linear analogue of the sphericity tensor:

$$\theta^{\alpha\beta} = \frac{\sum_i (p_i^\alpha \cdot p_i^\beta)/|\vec{p}_i|}{\sum_i |\vec{p}_i|} \quad ; \quad \alpha, \beta = 1, 2, 3 \quad . \tag{4}$$

  The eigenvalues λ_1, λ_2 and λ_3 of the spherocity tensor are used to construct the **D** variable:

$$D = 27 \cdot (\lambda_1 \cdot \lambda_2 \cdot \lambda_3) \quad . \tag{5}$$

- The 2nd Fox-Wolfram moment H_2 normalized to the 0th moment H_0, given by

$$(H_2/H_0) = \frac{1}{2} \cdot \frac{\sum_{i,j} |\vec{p_i}||\vec{p_j}| \cdot (3\cos^2\theta_{ij} - 1)}{\sum_{i,j} |\vec{p_i}||\vec{p_j}|} \quad . \tag{6}$$

The experimental values of these observables have been compared with Monte Carlo calculations, for both the ME and PS approaches to QCD. In the ME case both optimised and non-optimised scales and hadronization parameters have been considered. The experiments have not all used the same parameters in their Monte Carlos:

ALEPH in their original paper [15] used JETSET PS with parameters optimised by MARK II at $E_{cm} = 29 GeV$ [19], HERWIG with the authors' default parameters and JETSET ME unoptimised. In their updated report [16] improved parameters were used for HERWIG and JETSET ME, though details are not published yet.

DELPHI used JETSET PS with the MARK II parameter set, HERWIG with default parameters, JETSET with GKS matrix elements and JETSET with ERT matrix elements and optimised scale with and without retuned fragmentation parameters.

OPAL used JETSET PS, HERWIG, ARIADNE and JETSET with ERT matrix elements and optimized scale. The parameters for all these models were optimised to fit the OPAL data. The optimisation was performed on two distributions only, H_2/H_0 and M.

The parameters used are summarized below:

	Default	OPAL [18]	TASSO [20]	MARK II [19]
Λ_{LLA}	0.40	0.29	0.26	0.40
Q_0	1.0	1.0	1.0	1.0
a	0.5	0.18	0.18	0.45
b	0.9	0.34	0.34	0.90
σ_q	0.35	0.37	0.39	0.33

Parameter sets used in JETSET. Q_0 defines the cutoff virtuality in the parton shower, a and b govern the longitudinal string fragmentation and σ_q determines the transverse fragmentation. The parameters optimised by TASSO and MARK II are included for comparison.

	Default	OPAL [18]	TASSO [20]	MARK II [19]
Λ_{LLA}	0.20	0.11	0.25	0.20
m_g	0.65	0.65	0.61	0.75
$CLMAX$	5.0	3.0	2.3	3.0

Parameter sets used in HERWIG. m_g is an effective gluon mass, used to truncate the parton shower and $CLMAX$ is the maximum cluster mass which is allowed to decay isotropically. The parameters optimised by TASSO and MARK II are included for comparison – they were based on the EARWIG precursor of HERWIG.

Figure 2 shows the thrust distributions measured by OPAL, DELPHI and ALEPH, compared with QCD Monte Carlos. Likewise figure 3 shows the thrust minor value. Figures 4 and 5 show the distributions of thrust and aplanarity for data at PEP [19], PETRA [20] and LEP(OPAL), compared with the JETSET PS and HERWIG models (with OPAL parameters).

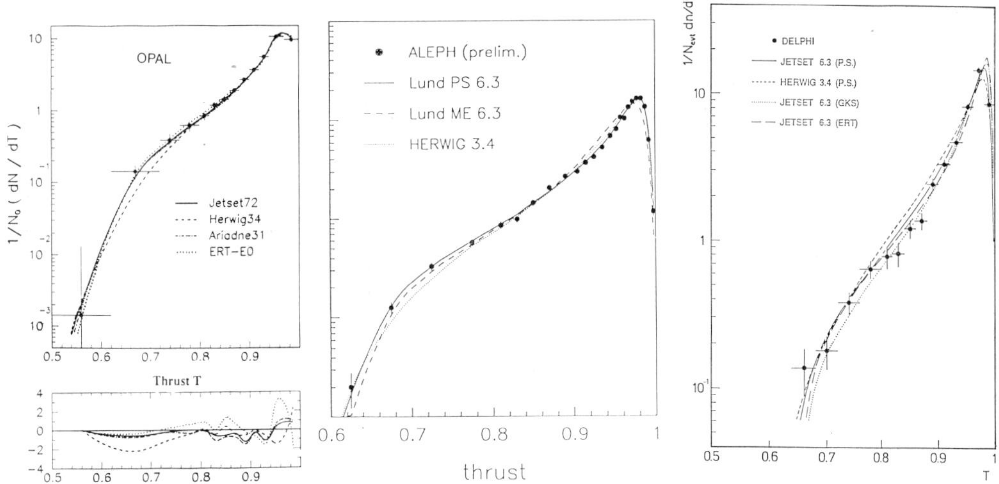

Figure 2. Distributions of Thrust measured by OPAL, ALEPH and DELPHI

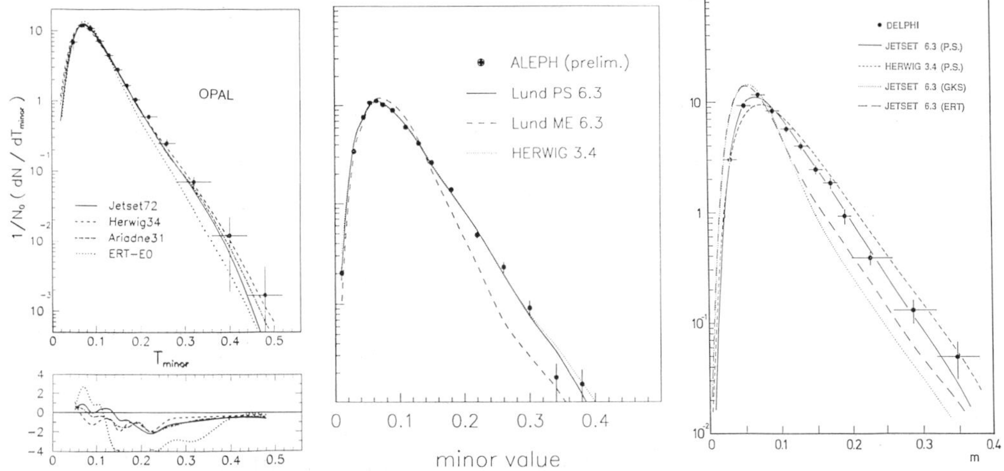

Figure 3. Distributions of the thrust minor value measured by OPAL, ALEPH and DELPHI

The general conclusions one can draw from these and other shape variables are as follows:

- JETSET PS fits all the data extremely well using the parameters derived at lower energies by MARK II or TASSO. The OPAL optimized parameters are very close to the TASSO parameter set.

- ARIADNE (only studied by OPAL) also fits the data extremely well.

- HERWIG fits the data well, but only if the parameters are adjusted. The OPAL optimised parameters differ significantly from the authors' defaults (future versions of HERWIG will use the OPAL parameters as default).

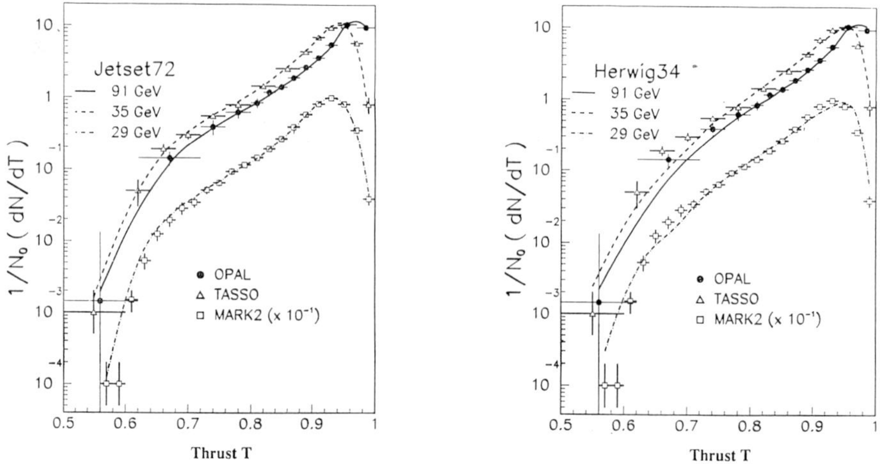

Figure 4. Distributions of Thrust measured at PEP, PETRA and LEP

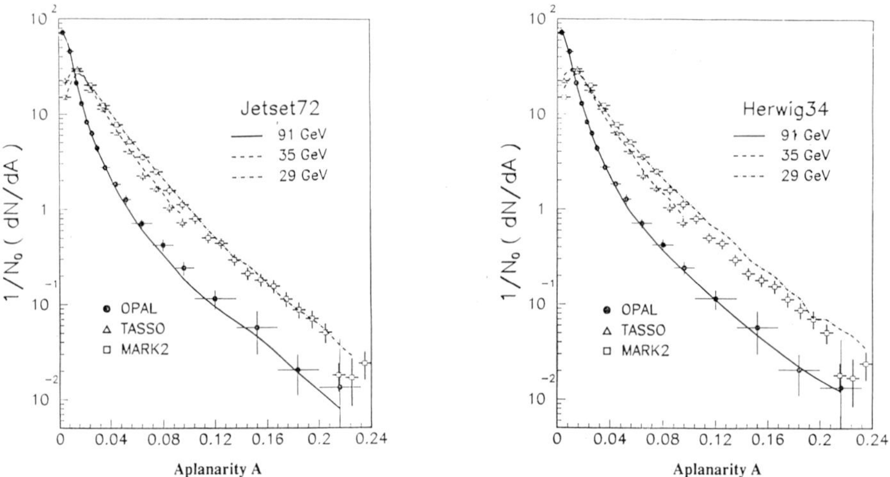

Figure 5. Distributions of Aplanarity measured at PEP, PETRA and LEP

- The HERWIG prediction for thrust around 0.7 is low, after optimizing parameters. This region corresponds to hard single gluon emission, and possibly the trick of mapping the first branching onto the $O(\alpha_s)$ matrix element helps the Lund models here. However, OPAL observe that the discrepancy between data and HERWIG is reduced if HERWIG is used to compute the corrections instead of JETSET, so the apparent discrepancy may not be significant.

- The ME model gives a poor account of the data if the scale and fragmentation parameters are not both optimized. After optimization, many distributions, such as Thrust, are well fitted, but distributions like the minor value, which measure momentum out of the event plane and are therefore sensitive to multiple gluon emission, can still not be adequately described by the ME approach (at least with the OPAL optimized parameter set; DELPHI seem to see rather better agreement, but details are not yet published).

- The energy dependence of the Thrust and Aplanarity distributions can be described by the PS models, with the same parameters at all energies. JETSET is more successful in this respect than HERWIG.

3 Determination of α_s from Jet Rates

There are published results on this topic from OPAL [21] (updated here), DELPHI [22] and L3 [23], and preliminary results from ALEPH [24] . The basic idea is to count jets using the "JADE" algorithm [25]. In brief, one computes the effective masses m_{ij} of all pairs of particles. If the smallest of these satisfies $m_{ij}^2 < y \cdot E_{vis}^2$ then particles i and j are replaced by a pseudoparticle of four-momentum $(p_i + p_j)$ and the process is repeated until all pair masses satisfy $m_{ij}^2 > y \cdot E_{vis}^2$. The particles or groups of particles at this stage are deemed to be the jets. In this way one can determine the probability, R_n, of producing an n-jet event, as a function of the jet resolution parameter y, which governs the minimum jet-jet mass. The introduction of this jet resolution parameter ensures infra-red stability of the jet rates, and therefore their computability in QCD perturbation theory. The data are well described by the PS Monte Carlos, so one can reliably correct the data for detector effects. The effect of hadronization appears to be small for the standard JADE algorithm [21].

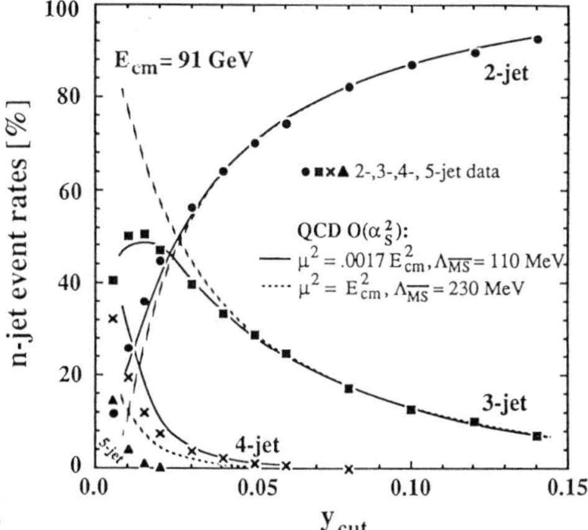

Figure 6. Distributions of jet rates measured by OPAL.

To $O(\alpha_s^2)$ the jet rates are given by:

$$\begin{aligned}
R_2 &= 1 &+&\ C_{21}\alpha_s &+&\ C_{22}\alpha_s^2 \\
R_3 &= &&\ C_{31}\alpha_s &+&\ C_{32}\alpha_s^2 \\
R_4 &= &&&&\ C_{42}\alpha_s^2
\end{aligned}$$

The coefficients C have been computed in QCD by Kramer and Lampe [26] and by Kunszt and Nason [1]. One can therefore determine α_s by fitting these QCD calculations to the data. However, there is some uncertainty about the choice of scale. If one takes $\mu^2 = f \cdot E_{cm}^2$ then the next-to-leading coefficients C_{22} and C_{32} depend on f. DELPHI have also considered the choice $\mu^2 = f \cdot y \cdot E_{cm}^2$. Empirically one finds that the data for $y > 0.05$ can be well fitted for any choice of scale, but the region $0.01 < y < 0.05$ can only be adequately described if an optimised scale is used, with f in the region

0.001-0.01. Typical data from OPAL together with the QCD predictions are shown in figure 6. [2]

The experimental errors on the values of α_s so determined are quite small, typically $\Delta\alpha_s < 0.005$. The main source of error is theoretical, in which there are three important effects:

- The choice of *renormalization scale* factor f. If one uses an optimised scale then the value of $\alpha_s(M_Z)$ comes out lower[3] by about 0.010-0.015 than if one chooses $f = 1$. This can be regarded as a measure of the uncertainty resulting from the neglect of higher order terms.

- The choice of *recombination scheme*. The QCD calculations are for massless partons, but after combining the four-momenta of pairs of hadrons or partons their effective mass is non-zero unless some adjustment is made. There is thus a measure of ambiguity in how this recombination is done. Possible schemes are:

	m_{ij}^2	p_{ij}
E	$(p_i + p_j)^2$	$p_i + p_j$
E0	$2E_iE_j(1 - cos\theta_{ij})$	$p_i + p_j$
P	$(p_i + p_j)^2$	$\mathbf{p}_{ij} = \mathbf{p}_i + \mathbf{p}_j$; $E_{ij} = |\mathbf{p}_{ij}|$
P0	$(p_i + p_j)^2$	As P scheme, but recompute E_{vis}

Aspects of this problem have been studied by OPAL, L3 and DELPHI, OPAL in particular having considered all these possible schemes. The E0 scheme is the standard JADE algorithm, and has the smallest hadronization corrections, while the E scheme (the only one that is Lorentz invariant) is the most sensitive to hadronization and detector effects. One can apply each scheme in turn to both the data and to the QCD calculations, and derive a value of α_s. The spread of the results so obtained is regarded as an additional theoretical systematic error. The different schemes have different degrees of sensitivity to the choice of the optimized scale parameter f – it appears that the E scheme is the most sensitive, and the p scheme the least. The values of $\Lambda_{\overline{MS}}$ obtained from the four schemes agree much better with optimized scale than with $f = 1$, giving some support to the idea that fitting for f does indeed minimize the higher order corrections.

- *Hadronization* corrections. Measurements at the hadron level have to be compared with QCD calculations for partons. These corrections are smallest for the E0 scheme, and largest for the E scheme. The uncertainties in the corrections are studied by using a variety of different QCD models, and the resultant systematic error is found to be $\Delta\alpha_s \sim 0.005$ in the best case (the E0 scheme) or ~ 0.011 in the worst case (the p scheme).

The only straightforward comparison we can make between the experiments is for the case with non-optimised scale, $f = 1$. The values are:

---

[2]The QCD fits are not actually made to these distributions, because they are integral distributions with highly correlated errors, but rather to the corresponding differential distributions, e.g. the distribution of the value of y at which an event changes from 2- to 3-jet.

[3]Of course the value of α_s actually going into the perturbation expansion is larger, typically ~ 0.18, because the scale is smaller. What the experiments have done is to consider Λ as the parameter which is really being determined. They then take the value of Λ from the fits with optimized scale, and re-express this as $\alpha_s(M_Z)$ evaluated at scale M_Z.

ALEPH	$\alpha_s(M_Z) = 0.127 \pm 0.004 \pm 0.003$
DELPHI	$\alpha_s(M_Z) = 0.114 \pm 0.003 \pm 0.004$
L3	$\alpha_s(M_Z) = 0.127 \pm 0.005 \pm 0.005$
OPAL	$\alpha_s(M_Z) = 0.123 \pm 0.003 \pm 0.005$

where the first error is statistical and the second systematic (mainly coming from the hadronization corrections). The values agree well. Some typical values with optimised scale are:

DELPHI	$\alpha_s(M_Z) = 0.104$	$f = 0.001$	(fixed)
L3	$\alpha_s(M_Z) = 0.114$	$f = 0.05$	(fixed)
OPAL	$\alpha_s(M_Z) = 0.106$	$f = 0.0047$	(fitted)

In view of the theoretical ambiguities it is best to quote a value and errors which encompass all the possible scale and scheme uncertainties. The best values quoted by the experiments (by slightly different methods) are:

ALEPH	$\alpha_s(M_Z) = 0.127^{+0.006}_{-0.017}$
DELPHI	$\alpha_s(M_Z) = 0.114 \pm 0.003 \pm 0.012$
L3	$\alpha_s(M_Z) = 0.115 \pm 0.005^{+0.012}_{-0.010}$
OPAL	$\alpha_s(M_Z) = 0.116 \pm 0.003 \pm 0.016$

Note that the OPAL and DELPHI errors include both scale and scheme uncertainties, while L3 just consider the scale uncertainty. Note that the ALEPH value has a highly asymmetric error, so that the range of values straddled is actually quite compatible with the other experiments.

4 Determination of α_s from Energy Correlations

So far one other technique for determining α_s has been reported from LEP – using the Energy-Energy Correlation (EEC) and its asymmetry (AEEC), by DELPHI [27], OPAL [28] and ALEPH [29]. The QCD calculations for AEEC are expected to have smaller uncertainties than most variables [1]. The EEC is defined as follows:

$$EEC(\chi) = \frac{2}{\Delta\chi \cdot N} \int_{\chi - \frac{\Delta\chi}{2}}^{\chi + \frac{\Delta\chi}{2}} \sum_{events} \sum_{i,j}^{N} \frac{E_i E_j}{E_{vis}^2} \delta(\chi' - \chi_{ij}) d\chi'$$

where χ_{ij} is the angle between particles i and j, E_i and E_j their energies, $E_{vis} = \sum_i E_i$, $\Delta\chi$ the width of the histogram bin and N the number of events.. Two-jet events yield a distribution sharply peaked near $\chi = 0°$ and $180°$ whereas events with hard gluon radiation fill the central region, in a non-symmetric fashion, with more entries in the region $90° < \chi < 180°$ than $0° < \chi < 90°$. The EEC asymmetry (AEEC)

$$AEEC(\chi) = EEC(\pi - \chi) - EEC(\chi)$$

removes the two-jet component and so is particularly sensitive to α_s. Also many systematic experimental errors are likely to cancel. The value of α_s may be determined by comparing the data with either $O(\alpha_s^2)$ QCD calculations, or with the ME Monte Carlo programs.

The OPAL analysis uses measurements of both EEC and AEEC, which are compared with four different theoretical calculations [1] [30], and with the ERT and GKS ME implementations[4] in JETSET 7.2. The integral of the EEC for $43.2° < \chi < 136.8°$ and the integral of the AEEC for $28.8° < \chi < 90°$ were compared with QCD in order to determine α_s. The results, for the choice of scale $\mu^2 = f \cdot E^2_{cm}$ with $f = 1$, are

OPAL EEC	$\alpha_s(M_Z) = 0.131 \pm 0.006(exp) \pm 0.006(theo)$	$f = 1$
OPAL AEEC	$\alpha_s(M_Z) = 0.117^{+0.007}_{-0.009}(exp)^{+0.006}_{-0.002}(theo)$	$f = 1$

In the case of the EEC, there are significant discrepancies between the different theoretical calculations, which leads to a systematic error. These uncertainties are less in the case of AEEC, but the experimental errors are a little greater. If the scale optimization parameter f is allowed to vary, the χ^2 of the fit and the value of $\Lambda_{\overline{MS}}$ are rather independent of f for the AEEC, but in the case of EEC a much better χ^2 may be obtained, yielding:

OPAL EEC	$\alpha_s(M_Z) = 0.117^{+0.006}_{-0.008}(exp)$ $\quad f = 0.027$

Again, there is some indication here that the idea of using the optimized scale to reduce the effect of higher order terms may be working – the AEEC, for which higher order terms are expected to be small, shows little dependence on f, and the values of α_s from EEC and AEEC (and jet rates) using the optimized values of f agree much better than those with $f = 1$.

Figure 7 shows the DELPHI data for EEC and AEEC, compared with the JETSET ME prediction, using optimised scale $f = 0.002$ (fixed; not fitted to data). DELPHI determine α_s from a fit to the AEEC distribution for $28.8° < \chi < 90°$, obtaining:

DELPHI AEEC	$\alpha_s(M_Z) = 0.106 \pm 0.006 \pm 0.007$

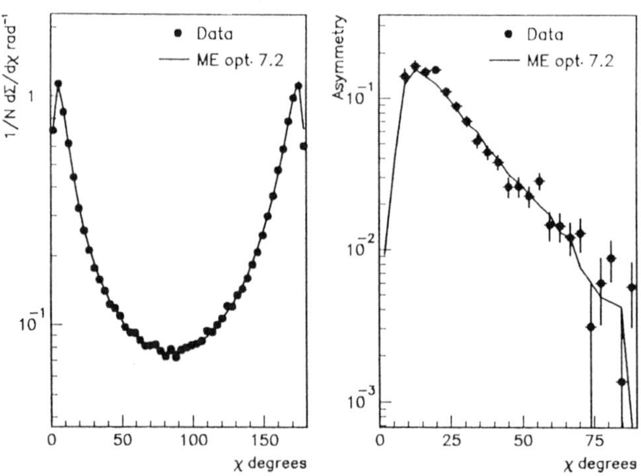

Figure 7. Distributions of the EEC and AEEC measured by DELPHI, compared with QCD expectations.

---

[4]There is a technical problem in using the Monte Carlo implementations of the Matrix Elements – in order to get finite, positive rates for 2-, 3- and 4-parton final states one needs to use a value of the jet resolution parameter $y > 0.01$. The heights of the EEC and AEEC distributions depend on the value of y. The OPAL procedure was to correct the JETSET ME results down to $y = 0.0001$ using a factor derived from the analytic calculations of Falck and Kramer [30]. This correction is $\sim +15\%$ for EEC and $\sim -6\%$ for AEEC.

ALEPH have used a slightly different definition of the EEC. They first form the observed hadrons into "clusters", using a jet finding algorithm with jet resolution parameter y. The EEC is then computed from the parameters of these clusters, and studied as a function of y. The result quoted is:

$$\boxed{\text{ALEPH EEC} \quad \alpha_s(M_Z) = 0.117 \pm 0.002^{+0.007}_{-0.010}}$$

5 Single Particle Distributions and Coherence

A number of single charged particle distributions have been studied by ALEPH [15] [16], DELPHI [17] and OPAL [31]. The variables considered include:

- The momentum components transverse to the sphericity axis, in and out of the event plane: p_T^{in} and p_T^{out}.

- The scaled momentum: $x_p = p/p_{beam}$.

- The rapidity, y, relative to the Thrust axis.

Figure 8 shows some typical data on rapidity and x_p distributions. The general conclusions are:

- The PS models fit the data well. HERWIG, however, gives too few particles at $x > 0.6$ – this is a consequence of its model of hadronization in terms of cluster decay into two particles, which makes it unlikely that a single hadron will acquire a very high momentum.

- The ME models, without optimised scale and without retuning hadronization parameters, fit very poorly, giving too few particles in the central rapidity region, but this can be largely overcome by retuning.

A topic of particular theoretical interest is the shape of the x_p distribution at low x_p. In this region one expects destructive interference which suppresses the emission

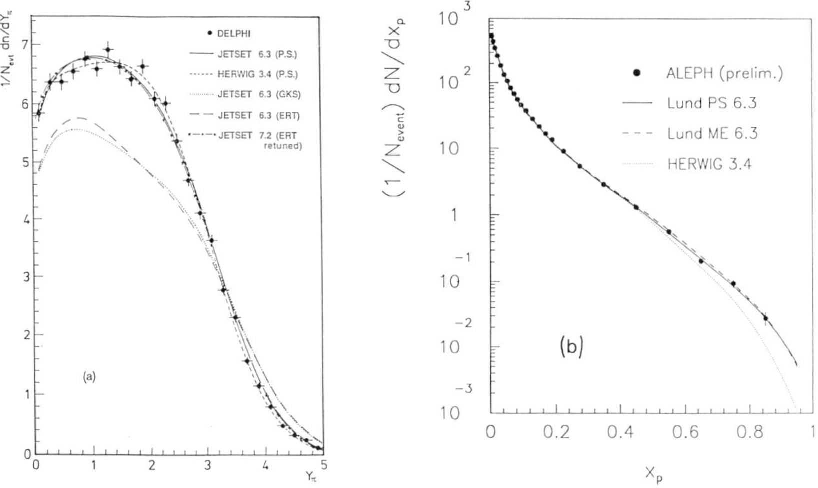

Figure 8. (a) Rapidity distribution (DELPHI) compared with QCD models (b) x_p distribution (ALEPH) compared with QCD models.

of soft gluons. To leading order the leading log approximation (LLA) including coherence predicts[5] a roughly Gaussian distribution of $\ln(1/x_p)$. The position of the peak, $1/x_0$, should evolve with collision energy like $(E_{cm}/2\Lambda)^{0.5}$, where Λ is some effective scale [1] [3]. In contrast, if coherence effects were absent the exponent would be 1.

OPAL has recently investigated this question [31]. Figure 9 shows the $\ln(1/x_p)$ distribution, which exhibits a peak at $\ln(1/x_p) \sim 3.6$. Note that at LEP energy this corresponds to momentum ~ 1.2GeV, so we can expect the peak position to be not too much dominated by phase space effects. Figure 9(a) compares the data with QCD predictions: a Gaussian corresponding to the LLA calculation, and two next-to-leading order calculations; a "modified LLA" (MLLA) [32] calculation, and a recent calculation of the higher moments of the distribution leading to a modified Gaussian form [33]. The next-to-leading order calculations agree well with the data in the peak region, roughly in $2 < \ln(1/x_p) < 5$.

Figure 9(b) compares the data with QCD Monte Carlos:

- JETSET and HERWIG PS models, including coherence, fit the data well.

- JETSET without coherence gives too high a multiplicity, as expected. If one retunes the string fragmentation parameters to get the multiplicity right then the model fits the data well (curve labelled *conv PS* in the figure).

- However, the Lund string fragmentation is known to embody some of the effects of coherence, therefore a better test is to use JETSET without coherence and with independent fragmentation (curve labelled *conv PS + IF*). This completely fails to fit the data.

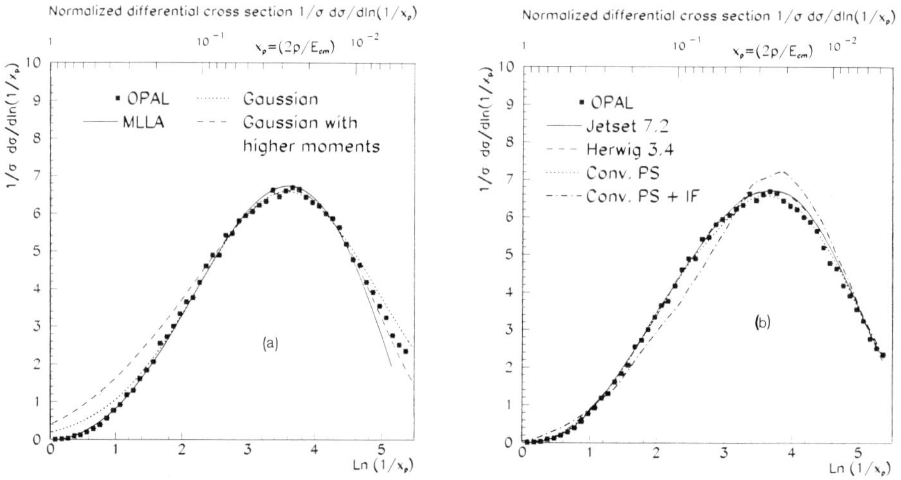

Figure 9. (a) $\ln(1/x_p)$ distribution (OPAL) compared with analytic QCD predictions. (b) $\ln(1/x_p)$ distribution (OPAL) compared with QCD Monte Carlos.

The other characteristic signature of coherence is the energy evolution of the peak position. Figure 10 shows how the peak position varies with c.m. energy. A roughly linear rise is seen, which is well parametrised by:

$$\ln(1/x_0) = (0.637 \pm 0.016) \cdot \ln(E_{cm}) + (0.735 \pm 0.067)$$

---

[5]The QCD calculations are of course for quarks and gluons. The hypothesis of "Local Parton Hadron Duality" [3] suggests that the distributions of quarks and hadrons should be the same, up to a normalization constant.

The table below shows the slope derived from various QCD models. This shows us that the effect of hadronization is to increase the slope, by around 0.03 typically. The analytic MLLA gives an effective slope of 0.627, in good agreement with data (compared to 0.5 for the LLA alone). The coherent parton shower models are all compatible with the data, while the incoherent version of JETSET with string fragmentation is only marginally compatible and the incoherent model with independent fragmentation is quite incompatible with data, but close to the simple QCD expectation of 1 in the incoherent case.

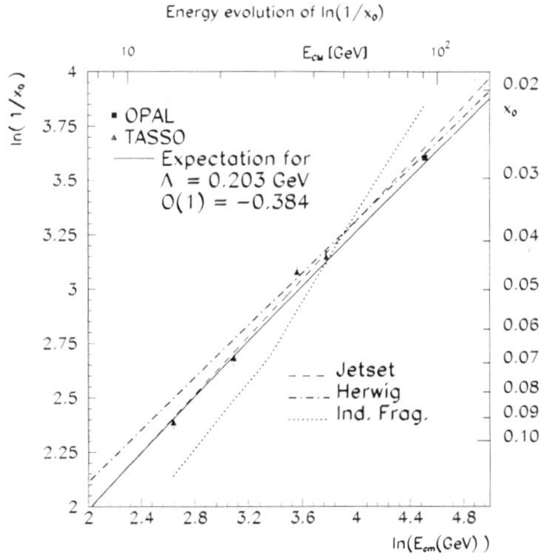

Figure 10: Dependence of the peak of the $\ln(1/x_p)$ distribution on c.m. energy (data from OPAL and TASSO)

Slopes of the $\ln(1/x_0)$ vs $\ln(E_{cm})$ distribution obtained from various Simulation schemes

Model	Coherence	Fragmentation	slope (partons)	slope (hadrons)
Data				0.637 ± 0.016
MLLA	yes		0.627	
HERWIG	yes	cluster	0.600 ± 0.012	0.629 ± 0.036
ARIADNE	yes	string		0.686 ± 0.020
JETSET	yes	string	0.597 ± 0.036	0.663 ± 0.019
	no	string	0.666 ± 0.003	0.692 ± 0.012
	no	independent		0.952 ± 0.017

A recent result which is also connected with coherence comes from DELPHI [34]. They study the "string effect". In this analysis one takes 3-jet events, and orders the jets in energy ($E_1 > E_2 > E_3$); jet 3 tends to be the gluon. The observed tracks are then projected onto the event plane defined by the jets, and the energy flow studied as a function of θ (θ being the polar angle in the plane, measured from jet 1 towards jet 2). The experimental distribution is shown in fig 11. The "string effect" is the depletion of particles around 80° compared to the region around 300°. This effect was first predicted in the LUND string model, as strings are not being stretched in the region between the two quarks [2], but it can also be accounted for in terms of constructive and destructive interference between soft gluons [3]. DELPHI have compared their data with three versions of the JETSET Monte Carlo program. The coherent PS model gives an excellent fit to the data. The ME model with *string* fragmentation does not fit the data quite so well, though it still shows the main effect, the depletion around 80°. The ME model with *independent* fragmentation exhibits little depletion at 80° and completely fails to reproduce the data.

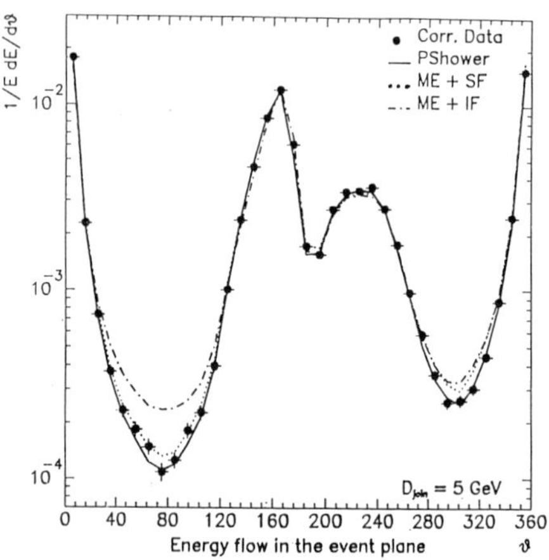

Figure 11. Energy flow in the event plane (DELPHI).

6 Angular Correlations in 4-Jet Events

Recent studies by OPAL [35], L3 [36] and DELPHI [37] have examined 4-jet events to see whether there is evidence for the triple-gluon coupling, which is a characteristic feature of QCD. This is a study for which the high statistics at LEP are needed. The Feynman graphs which can lead to a four parton final state are shown in Figure 12.

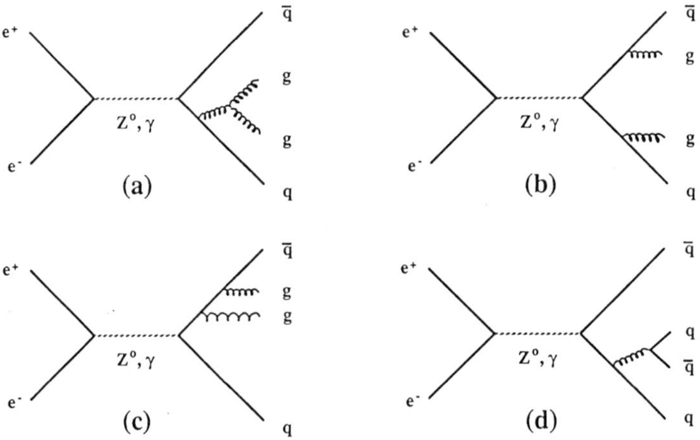

Figure 12. Feynman graphs leading to 4-parton final states in QCD

The method is to take events with four jets (as identified by the JADE algorithm, say), and order the jets in energy so that jet 1 is the most energetic. Several angles may then be computed:

- The Bengtsson-Zerwas angle [38], χ_{BZ}, defined as the angle between the planes defined by $\mathbf{p}_1$ and $\mathbf{p}_2$, and by $\mathbf{p}_3$ and $\mathbf{p}_4$.

- The Nachtmann-Reiter angle [39], θ^*_{NR}, defined as the angle between the vectors $\mathbf{p}_1 - \mathbf{p}_2$ and $\mathbf{p}_3 - \mathbf{p}_4$.

- The Körner-Schierholz-Willrodt angle [40], Φ_{KSW}, is defined for events in which there are two jets in each of the hemispheres (defined by the plane orthogonal to the thrust axis). Φ_{KSW} is the angle between the normals to the planes defined by the pairs of jets in each hemisphere.

- The angle between the two lowest energy jets, 3 and 4, is defined to be α_{34}.

To see whether these angles are sensitive to the non-Abelian nature of QCD, OPAL and L3 have compared the data with QCD and with a "QED-like" Abelian model, in which the triple-gluon coupling is absent. Specifically, three QCD models were tried, HERWIG PS (OPAL only), JETSET PS and JETSET ME $O(\alpha_s^2)$. Two "QED" models were also constructed, one a straightforward adaptation of the JETSET ME $O(\alpha_s^2)$ with modified group constants, and the second based on JETSET PS [6]. Since fragmentation in the "QED" model is not well defined the experimental procedure was to correct the data down to the parton level using the JETSET PS model. Of course the "QED" model is not a serious contender for a theory of the strong interactions, one introduces it merely in order to judge whether the data are sensitive to any alternative to QCD.

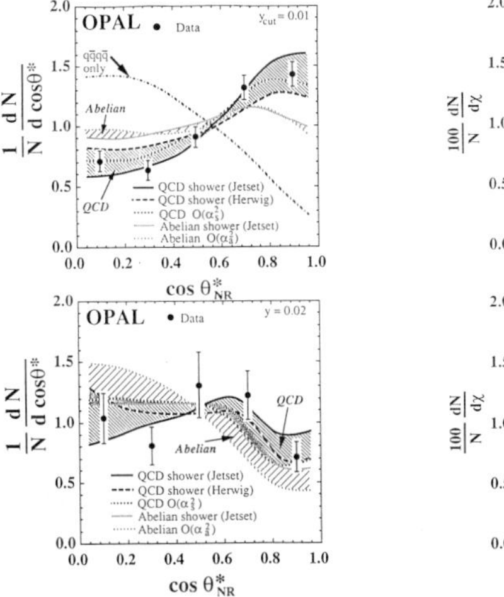
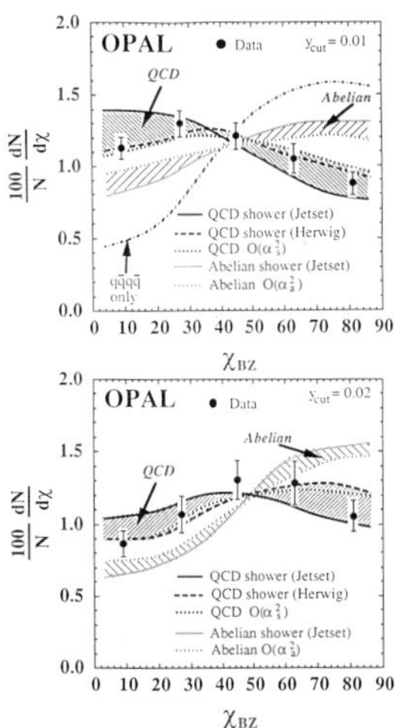

Figure 13. Distributions of the angles χ_{BZ} and θ^*_{NR} for two values of y. The shading shows the areas spanned by the different QCD and "QED-like" models.

The OPAL results are shown in figure 13. The L3 data show essentially the same results. L3 have also studied Φ_{KSW} (not shown by OPAL), but this angle seems less sensitive than the others. The different models do not agree perfectly, so the areas spanned by the various QCD and "QED" models are shaded. The data clearly favour the QCD band. The Bengtsson-Zerwas angle seems the more sensitive. On further investigation the effect appears to originate in the different proportions of $q\bar{q}q\bar{q}$ events

---

[6]The prescription here is by no means unique. What was chosen was to modify the parton branching probabilities appropriately for an Abelian theory, to fix the coupling constant to 0.25, to disable angular ordering and azimuthal interference, but to retain gluon spin and polarization effects.

in the QCD and "QED" theories (4.7% and 31.4% respectively). The distributions just
for $q\bar{q}q\bar{q}$ events are shown in figure 13. This difference arises because of the existence of
diagram 12(a) in the non-Abelian case, but these distributions do not really distinguish
between diagram 12(a) and diagrams 12(b) and (c).

The OPAL data have therefore been used to place limits on the production of $q\bar{q}q\bar{q}$
events (or events with similar spin structure, such as $q\bar{q}\tilde{g}\tilde{g}$ which could have otherwise
escaped detection in a supersymmetric model where the gluino is the lightest supersymmetric particle [41]). The result is that the proportion of $q\bar{q}q\bar{q}$ events in the 4-jet
sample is $< 9.1\%$ at 95% c.l..

DELPHI have taken a different approach. They note that the relative weights of
the $q\bar{q}q\bar{q}$, $q\bar{q}gg$ (double bremsstrahlung) and $q\bar{q}gg$ (triple gluon vertex) diagrams depend
on the group constants C_F, N_C and T_R of the symmetry group of the theory. They
therefore make a fit to their data (using the JETSET 7.2 ERT ME program) treating
the group constants as free parameters, and thus varying the relative weights of different
diagrams. They fit the distributions of two angles, θ^*_{NR} which we have seen is sensitive
to the differences between $q\bar{q}q\bar{q}$ and $q\bar{q}gg$, and α_{34} which gives some discrimination
between the double bremsstrahlung and the triple gluon diagrams. The table below
gives the values of the likelihood function $\mathcal{L}$ for various choices of the group constants:

	N_C/C_F	T_R/C_F	$\log \mathcal{L}$
Fit N_C/C_F and T_R/C_F	2.05 ± 0.40	0.1 ± 1.7	834
Fix $N_C/C_F = 0; \Rightarrow no\ g \to gg$	0	2.8 ± 2.3	815
QCD	2.25	1.875	834
Abelian "QED"	0	15	803

We see that the fitted values of the group constants are compatible with those of QCD,
and that the likelihood function for QCD is as good as that obtained in a free fit.
However, if N_C is set to zero, thus disabling the triple gluon coupling, a much worse fit
is obtained. The particular assumptions of the "QED" Abelian model used by OPAL
and L3 give an even worse fit.

7 Multiplicity Distributions

ALEPH [16] and DELPHI [42] have presented preliminary results on the corrected
charged multiplicity distribution. In this case the simple bin-by-bin correction procedure
is not viable, and a matrix method (DELPHI) or a method based on "reduced entropy"
(ALEPH) was used. The value of $< n_{ch} >$ can be obtained in this way, which we
may compare with other estimates, based not on the corrected multiplicity distribution
but on integrating a corrected rapidity distribution, or even a simple multiplicative
correction obtained from a Monte Carlo. The values are:

ALEPH [16]	$< n_{ch} >= 20.36 \pm 0.06 \pm 0.79$
DELPHI [42]	$< n_{ch} >= 20.83 \pm 0.14 \pm 0.96$
DELPHI [17]	$< n_{ch} >= 20.6 \pm 1.0$
OPAL [18]	$< n_{ch} >= 21.28 \pm 0.04 \pm 0.84$

These determinations agree within errors. For comparison the values predicted by
various QCD models (the ranges reflecting the different parameters used by different
experiments) are

$$
\begin{array}{ll}
\text{JETSET PS} & < n_{ch} > = 21.1 - 21.4 \\
\text{HERWIG PS} & < n_{ch} > = 20.1 - 21.2 \\
\text{ARIADNE PS} & < n_{ch} > = 20.9 \\
\text{JETSET ME not optimized} & < n_{ch} > = 18.0 - 18.2 \\
\text{JETSET ME optimized} & < n_{ch} > = 19.1 - 21.2
\end{array}
$$

There are also analytic calculations of the energy dependence of $< n_{ch} >$ in QCD, based on the LLA approach [1]. To next-to-leading order the prediction is:

$$
\ln < n_{ch} > = \frac{a}{\sqrt{\alpha_s}} + b \ln \alpha_s + O(1)
$$

where

$$
a = \frac{\sqrt{96\pi}}{\beta} \quad ; \quad b = \frac{1}{4} + \frac{10 N_F}{27\beta} \quad ; \quad \beta = 11 - \frac{2}{3} N_F \quad .
$$

The energy dependence all lies in the running of α_s. A fit to pre-LEP data yielded $\Lambda = 80\text{MeV}$, and predicted $< n_{ch} > \sim 20.4$ in excellent agreement with the measurements.

ALEPH and DELPHI find that the multiplicity distribution can be well fitted by a Negative Binomial form with parameter $k = 19.8 \pm 1.8$ (ALEPH) or $k = 24.3 \pm 1.6$ (DELPHI). The data have been compared with lower energy results in terms of the KNO variable $z = n_{ch} / < n_{ch} >$. The data are compatible with KNO scaling.

8 Intermittency

A recent study by DELPHI has addressed the question of "intermittency" [43]. This phenomenon is the apparent occurrence of concentrations of particles ("spikes") in the rapidity distributions of individual events. To quantify the effect one takes some (central) region of rapidity of length Y, divides it into M bins of width $\delta y = Y/M$, and examines the factorial moments of the multiplicity n within a bin:

$$
F_q = \frac{1}{M} \cdot \frac{< n(n - 1) \cdots (n - q + 1) >}{< n >^q}
$$

as a function of the bin width, or M. [7] Figure 14 shows the DELPHI data on $F_2 \cdots F_5$ compared with QCD Monte Carlos. The JETSET PS model describes the data pretty well, while the JETSET ME model seriously overestimates the moments. However, after retuning JETSET ME to fit single particle and event shape data the fit to the factorial moments is improved, and in fact is now too low. This is in some contrast to a similar study at TASSO where no Monte Carlo was found to fit the data. However, TASSO also had models lying both above and below their data; it appears likely that these moments are particularly sensitive to the fragmentation parameters, and therefore it is not yet clear how much they tell us about perturbative QCD.

---

[7]Note that the qth factorial moment filters out multiplicities $n < q$, and therefore these moments can be particularly sensitive to large fluctuations. In a cascade-type model one expects a power law dependence $F_q \sim (\delta y)^{-f_q}$ whilst in random Poissonian emission one should get $F_q \sim constant$.

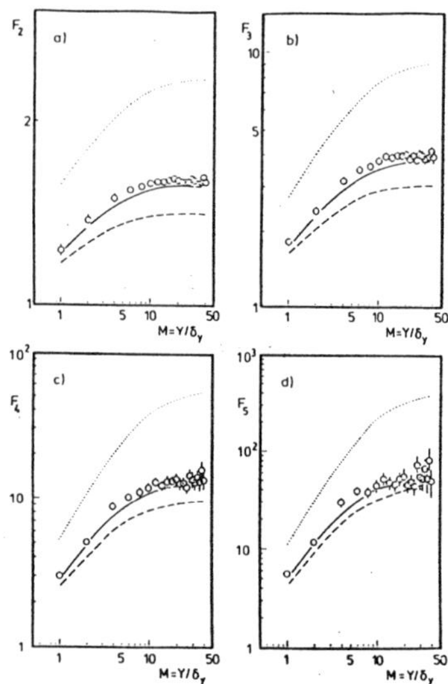

Figure 14. Factorial moments (DELPHI data) compared with JETSET PS (solid) JETSET ME (dotted) and JETSET ME retuned (dashed curve).

9 Summary

The work of studying QCD at LEP has just begun. So far the theory is working extremely well. The data are best fitted by the Parton Shower Monte Carlos, possibly too well fitted in view of the approximations used in the theory. Presumably the fitting of hadronization parameters to data has masked any deficiencies in the theory. Of these models, JETSET PS appears to give the best description of the data, but HERWIG and ARIADNE are also highly successful.

The Matrix Element approach is needed in the determination of α_s or $\Lambda_{\overline{MS}}$. The use of an optimised scale considerably improves the description of data, but there are still $\sim 10-15\%$ theoretical uncertainties in the value of α_s. It can be hoped that the use of a wider range of observables in the measurement of α_s will help us to understand and thus reduce these uncertainties. The determinations of $\alpha_s(M_Z)$ are in the range 0.10 to 0.13. There is some indication that the optimization of the renormalization scale brings the measurements closer together in the region 0.11-0.12.

There is evidence for coherence effects in the $1/x_p$ distribution, angular correlations in 4-jet events support the non-Abelian nature of QCD, and first results on multiplicity distributions and intermittency are becoming available. There are plenty of results for us to digest already, and there is much more that LEP can do in the QCD field in the coming years.

10 Acknowledgements

It is a pleasure to thank the organisers of the Cargèse Summer School for arranging such an enjoyable meeting. I would also like to thank the members of the LEP collaborations who helped provide information for this talk, and particularly my fellow speakers, A.Blondel, G.Sauvage and D.Treille.

References

[1] Z. Kunszt, P. Nason *et al.,* CERN 89-08, vol 1, 373.

[2] T. Sjöstrand, these proceedings.

[3] V. Khose, these proceedings.

[4] T. Sjöstrand *et al.,* CERN 89-08, vol 3, 143.

[5] R.J.Cashmore, these proceedings.

[6] R.K.Ellis, D.A.Ross and A.E.Terrano, Nucl. Phys. **B178** (1981) 421.

[7] F.Gutbrod, G.Kramer and G.Schierholz, Zeit. Phys. **C21** (1984) 235.

[8] B. Andersson *et al.,* Phys. Rep. **97** (1983) 31.

[9] S.G.Gorishny, A.L.Kateev and S.A.Larin, Phys. Lett. **212B** (1988) 238.

[10] ALEPH Collaboration, D.Decamp *et al.,* CERN-PPE 90-104; submitted to Zeit. Phys. C.

[11] T. Sjöstrand, Comp.Phys.Comm. **39** (1986) 347;
T. Sjöstrand and M. Bengtsson, Comp.Phys.Comm. **43** (1987) 367.

[12] G. Marchesini and B.R. Webber, Nucl.Phys. **B310** (1988) 461.

[13] U. Pettersson, LU TP 88-5 (1988);
L. Lönnblad and U. Pettersson, LU TP 88-15 (1988);
L. Lönnblad, LU TP 89-10 (1988).

[14] G.Sauvage, these proceedings.

[15] ALEPH Collaboration, D.Decamp *et al.,* Phys. Lett. **234B** (1990) 209.

[16] ALEPH Collaboration, M.Schmelling, Rencontres de Moriond, March 1990.

[17] DELPHI Collaboration, P.Aarnio *et al.,* Phys. Lett. **240B** (1990) 271.

[18] OPAL Collaboration, M.Z. Akrawy *et al.,* Zeit. Phys. **C47** (1990) 505.

[19] MARK II Collaboration, A.Petersen *et al.,* Phys. Rev. **D37** (1988) 1.

[20] TASSO Collaboration, W.Braunschweig *et al.,* Zeit. Phys. **C41** (1988) 31.

[21] OPAL Collaboration, M.Z. Akrawy *et al.,* Phys. Lett. **235B** (1990) 389.

[22] DELPHI Collaboration, P.Abreu *et al.,* Phys. Lett. **247B** (1990) 167.

[23] L3 Collaboration, B.Adeva *et al.,* L3 Preprint #011.

[24] ALEPH Collaboration, R. Settles, A.Blondel, Private communication.

[25] JADE Collaboration, W. Bartel *et al.* Zeit .Phys. **C33** (1986) 23.

[26] G. Kramer and B. Lampe, Zeit. Phys. **C39** (1988) 101.

[27] J.Drees, Neutrino '90 conference (CERN June 1990);
DELPHI Collaboration, paper in preparation.

[28] OPAL Collaboration, M.Z. Akrawy *et al.*, CERN-PPE 90-121; submitted to Phys. Lett.

[29] ALEPH Collaboration, A.Blondel, Private Communication

[30] A. Ali and F. Barreiro, Phys. Lett. **118B** (1982) 155; Nucl.Phys. **B236** (1984) 269.
D.G. Richards, W.J. Stirling and S.D. Ellis, Phys.Lett. **119B** (1982) 193; Nucl.Phys. **229B** (1983) 317.
N.K. Falck and G. Kramer, Z.Phys. **C42** (1989) 459.

[31] OPAL Collaboration, M.Z. Akrawy *et al.*, CERN-EP/90-94; submitted to Phys. Lett.

[32] Yu.L.Dokshitzer, S.Troyan, Leningrad preprint LNPI No.922 (1984);
Yu.L.Dokshitzer, V.A.Khoze and S.Troyan, in Perturbative Quantum Chromodynamics, ed. A.H.Mueller, World Scientific, Singapore (1989) 241;
Yu.L.Dokshitzer, V.A.Khoze, A.H.Mueller and S.I.Troyan, Rev.Mod.Phys. **60** (1988) 373.

[33] C.P.Fong and B.R.Webber, Phys. Lett. **229B** (1989) 289.

[34] DELPHI collaboration, *String effect and jet properties in 3-jet events at the Z^0 pole* International Conference in High Energy Physics, Singapore (1990).

[35] OPAL Collaboration, M.Z. Akrawy *et al.*, CERN-EP/90-97; submitted to Zeit. Phys.

[36] L3 Collaboration, B.Adeva *et al.*, L3 Preprint #012.

[37] DELPHI collaboration, *Experimental evidence for the Triple-gluon Vertex* International Conference in High Energy Physics, Singapore (1990).

[38] M. Bengtsson and P. Zerwas, Phys. Lett. **208B** (1988) 306.

[39] O. Nachtmann and A. Reiter, Zeit. Phys. **C16** (1982) 45.

[40] J.G.Körner, G.Schierholz and J.Willrodt, Nucl. Phys. **B185** (1981) 365.

[41] G. Farrar, Preprint RU-90-37.

[42] DELPHI collaboration, P.Abreu *et al.*, CERN-PPE/90-117, submitted to Zeit. Phys. C.

[43] DELPHI Collaboration, P.Abreu *et al.*, Phys. Lett. **247B** (1990) 137.

Searches at LEP*

D. Treille

Introduction

After one year of physics at LEP and $\sim 700k$ Z^0 decays registered by its four experiments, no new particle nor new phenomenon has been unveiled. At the per cent level the Z^0 is standard. Several potentially interesting channels have been explored to the 10^{-4} level or less without success.

To conclude that there is nothing to discover at LEP would however be premature and rather short minded. Interesting modes can still appear with much smaller branching ratios: among many other possibilities a classical Higgs of ~ 55 GeV or a manifestation of Z compositeness through its decay into $3\,\gamma$ are not expected to show up before one reaches the 10^{-6} level or so.

Moreover if a theory like the Minimal Supersymmetric Standard Model is a reality one has a good chance to discover something at LEP I + LEP 200 since such a theory absolutely needs a scalar particle, at least, below or not far above the Z mass.

1 Indirect Searches

As it is well known and abundantly illustrated in this school, another kind of systematic exploration is made at LEP through a set of accurate measurements.

If new particles occur at energies which are not accessible directly at LEP, one has still the possibility to observe them indirectly through their manifestation as virtual intermediate states appearing in loop diagrams. Indeed the structure of electroweak interactions is such that, at the difference of what happens in pure electromagnetic ones, heavy particles do not decouple from the lower energy world.

If the top is not accessible directly at LEP, which is now very likely, this particle will then be the most conspicuous example of such a state not produced but influencing considerably several observations. The effect of virtual top exchange on the boson masses or the width of the Z into $b\bar{b}$ (Fig. 1) are well known. Besides the top, which will hopefully be discovered in hadronic machines during LEP lifetime, many other possible virtual effects can show up as deviations of observables from their standard value. A complete strategy to study such deviations and identify their physical origin is now existing: it rests on quite sophisticated linear combinations of several observables (asymmetries, partial widths, masses of IVB) and its power depends crucially on the accuracies with which these quantities will be measured. Figure 2 [1] illustrates this strategy on which we come back in the Appendix 3 summarizing the future options. So, whatever be the issue of direct searches, one is certain with LEP to perform meanwhile, if the luminosity–and in some cases the available energy–are sufficient, a set of very interesting measurements.

2 Direct Searches: Generalities

Let us now come back to this program of direct searches, which are obviously a must for every new machine, as long as possibilities are still open experimentally.

---

*Lectures given at Cargèse Summer School, Cargèse, France, August 90

Z° Physics - Cargèse 1990, Edited by M. Lévy *et al.*
Plenum Press, New York, 1991

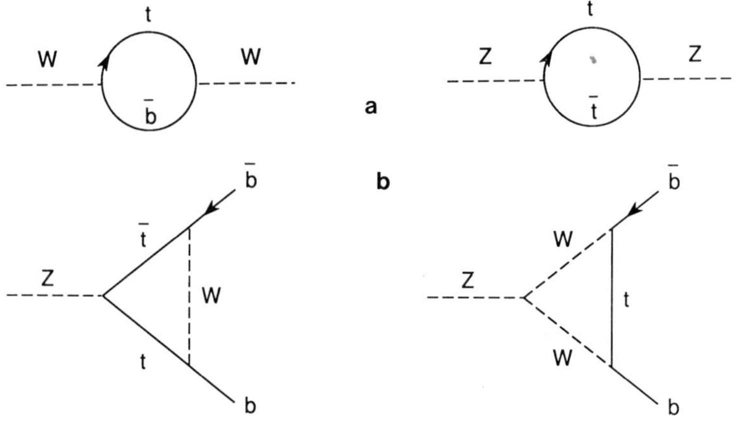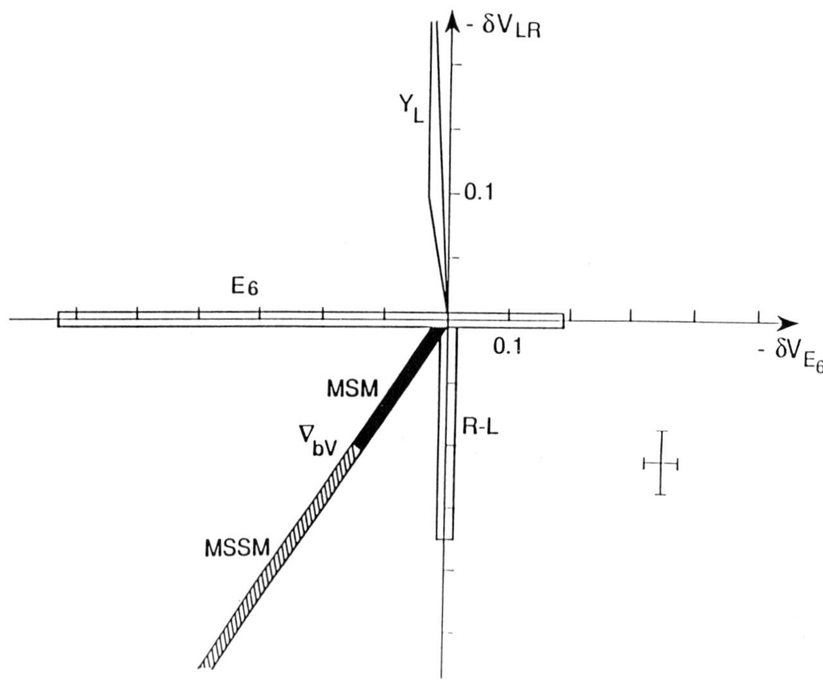

Figure 1. a) Effect of virtual top in IVB propagators, b) the same in Zb$\bar{\text{b}}$ vertex.

Figure 2. The strategy to identify new physics. The δV_i are the deviations, from their Standard Model value, of quantities built up from appropriate linear combinations of LEP observables: $\Gamma_{b\bar{b}}$, $A_{b\bar{b}}$, M_Z, M_W, etc. The cross shows the experimental error expected from High Luminosity LEP.

The occurence of non standard final states or effects can be looked for by an inspection of global quantities describing the Z resonance, like the total width, the invisible width and the total hadronic cross section.

This last quantity can be expressed as:

$$\sigma_{\rm h} \propto \frac{\Gamma_e \Gamma_{\rm h}}{\Gamma_{\rm tot}^2}.$$

Its sensitivity to the presence of a new channel will depend on the type of final state involved. If the channel is of the hadronic type, it will affect both the denominator and numerator and

$$\frac{\Delta \sigma_{\rm h}}{\sigma_{\rm h}} \sim \frac{\Delta \Gamma}{\Gamma}$$

If on the contrary it is an "invisible" one the sensitivity will be twice as large

$$\frac{\Delta \Gamma}{\Gamma} = \frac{1}{2} \frac{\Delta \sigma_{\rm h}}{\sigma_{\rm h}}$$

Presently the experimental error $\Delta \sigma_{\rm h} / \sigma_{\rm h}$ is 1.3% and the theoretical one: 0.5%.

Similarly one can feel the presence of an invisible channel through its contribution to $\Gamma_{\rm inv}$. With the present uncertainty the room left for novelties in $\Gamma_{\rm inv}$ is $\lesssim$ 30 MeV at the 95% CL (per experiment).

The impact of such global inspections is limited but nevertheless quite useful in specific cases.

The bulk of searches is however obtained from the study of specific final states with characteristic topological features.

It would be tedious to list all the cuts used to isolate such final states and a synthetic description is quite preferable. In fact most of the searches belong to one (or more) of three generic categories:

A) Acoplanar two prongs leptonic states: such final states occur when "escaping" particles (ν, $\tilde{\nu}$, lightest supersymmetric particle, LSP) are emitted in leptonic decays. This can concern heavy leptons, supersymmetric states, light Higgses, etc.

The Standard Model background for such a configuration in a perfect detector is negligible. If the detector has "holes" or "cracks" for photon detection the radiative leptonic modes of the Z are a potential background. Initial state radiation on the Z is limited by the resonant situation and gives mostly acolinearity. Final state radiation on the contrary is dangerous: for a heavy radiator, like a τ, the radiation is no more peaked along the lepton trajectory and its angular distribution relative to the emittor can be expressed as:

$$\frac{\mathrm{d}n}{\mathrm{d}\cos\theta} \alpha \frac{\sin^2\theta}{\left(\sin^2\theta + \frac{\cos^2\theta}{\gamma^2}\right)^2}$$

where $\gamma \equiv E/m$ is the Lorentz factor of the emittor. However with such a simple final state and in the case of single γ emission one can protect oneself by using constraints: the missing momentum should not point to a notoriously weak region of the detector.

B) If in the final state under study leptonic or semileptonic decays are supposed to take place, the signature expected is the presence of isolated particles, sticking out of the hadronic jets. The isolation variable usually considered is

$$\rho_i = \min_{\substack{jet\ j \\ j\,=\,1,n}} \sqrt{2E_i(1 - \cos\theta_{\rm ij})}$$

where i denotes the candidate particle, E_i its energy and θ_{ij} its angle with the axis of jet j. The inclusion of particle i in or its exclusion from the nearest jet is a matter of debate.

The standard model background is here at the level of $\sim 10^{-3}$. The isolated tracks of background events are not always leptons and one could improve the situation by requiring a leptonic identification. However many hypothetical states could decay into a τ and such a leptonic requirement would suppress the abundant $\tau \rightarrow {\rm h}^{\pm} + ...$ monotrack source: it is therefore generally not applied.

C) A purely hadronic decay of a pair of particles into four jets is generally the most abundant expected final state. The standard model background is high since 10% of hadronic Z final states will appear as four jets.

One has to use the equality of the masses of the two hypothetical particles (if appropriate). Jet-jet masses must be reconstructed; constraints to the beam energy are exploited by a rescaling of these effective masses, and peaks are looked for in two dimensional plots. The typical jet-jet mass resolution is a few GeV/c^2. In the absence of reference physical signal, one has however to trust the result of Monte Carlo simulations for such an estimate.

In the remaining of the description of searches, I will for each one simply refer to the type of topological configuration it involves (as A, B, C).

3 What are we looking for?

The most obvious candidates are the standard missing objects: the tau neutrino and the top quark.

For the former LEP has not much to say; it has simply confirmed that there seems to be three and only three light neutrino species. It is likely that a dedicated and clever fixed target experiment will be necessary to bring a direct evidence for the ν_τ. For the later we will review the situation in Ch. 4 and see what LEP can bring.

Then comes the standard scalar sector. The standard Higgs search will be described in Section 5 and future prospects discussed in Appendix 3.

A straightforward extension of the SM would be to add further families (with the constraint of the number of light neutrinos). Several kinds of heavy neutral leptons can be envisaged, per se, or to provide solutions to various problems (generation of neutrino mass,...). This will be explored in Section 6.

Finally, as often repeated, we know that the SM suffers from serious defects which have to be cured by new ideas.

The most promising is supersymmetry about which LEP has a lot to tell us (Ch7). Another option, less clearly formulated theoretically, is compositeness: possible hints from LEP in this respect will be described in Section 9.

4 The Top

The top must exist. The studies of the b quark, at various e^+e^- machines including LEP, have demonstrated that it is the lower member of a doublet, waiting for a partner. The present limit on $B \to \ell^+\ell^- X$ cannot be accomodated for in the absence of a GIM mechanism involving the top. Other arguments (cancellation of anomalies, ...) could be used as well.

That the top has not been seen up to now can be the result of its high mass or of a non standard decay pattern.

For a standard top the present mass limit is 89 GeV (95% CL) from CDF. It is expected that a mass limit of 120 GeV can be reached by the end of 1992 if an integrated luminosity of 10 pb^{-1} (10 times more than up to now) is registered. The chance to discover it at the Tevatron are high. LEP–even LEP 200–cannot compete.

If the top decay is non standard the situation is different. A non standard decay means $t \nrightarrow bW^\star$ and can represent many possibilities. One is $t \to bH^\pm$ through a charged Higgs: for $m_t < m_W + m_b$ the standard decay can be negligible, while for $m_t > m_W + m_b$ it is probably not small (2-body phase space). We can note (see Section 7) that for $m_H < m_W$, such a scalar cannot be the Minimal SUSY charged Higgs which is bound to be heavier than the W.

From the hadronic colliders the mass limit for a non standard top, obtained by measuring the width of the W into $l\nu$, is 45 GeV.

LEP has already reached a similar limit. One way is through the exclusion of a $H^\pm$ up to M, which excludes a top up to $M + m_B$. Another method consists in excluding directly a top decaying with such a topology. The search for $H^\pm$ at LEP [2] has used topologies A, B and C, depending on the unknown branching ratio $H \to \tau\nu^\star$. Figure 3 shows the results of DELPHI and OPAL, very similar to the ones of ALEPH and L3. Both show the interplay of the three topologies: charged Higgs are excluded up to 43 GeV when $\tau\nu$ dominates. One can plot this limit on Fig. 4 [3] which represents the most recent exclusion contour of the collider in the $M_t - M_H$ plane. The direct exclusion of a top decaying into $bH^\pm$ can be illustrated by the work of OPAL [4], which is a systematic search for heavier quarks. Figure 5

---

*The cross section for H^+H^- pair production is given by formula A4 of Appendix 1, a particular case of the general formula A1.

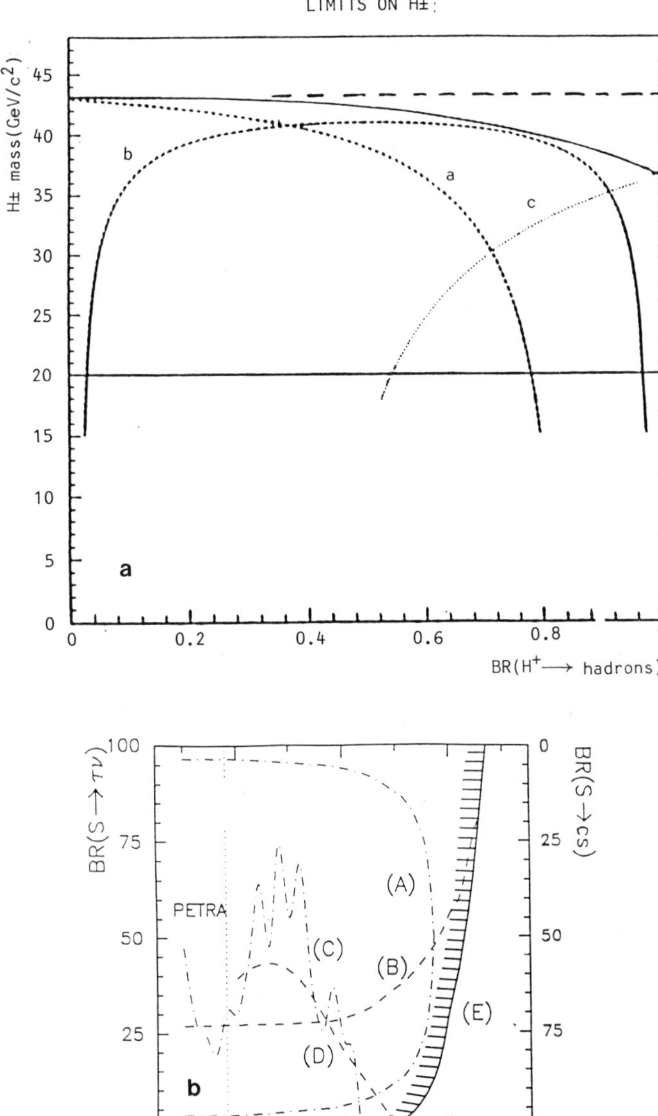

Figure 3. Charged Higgs mass limit: a) from DELPHI, b) from OPAL.

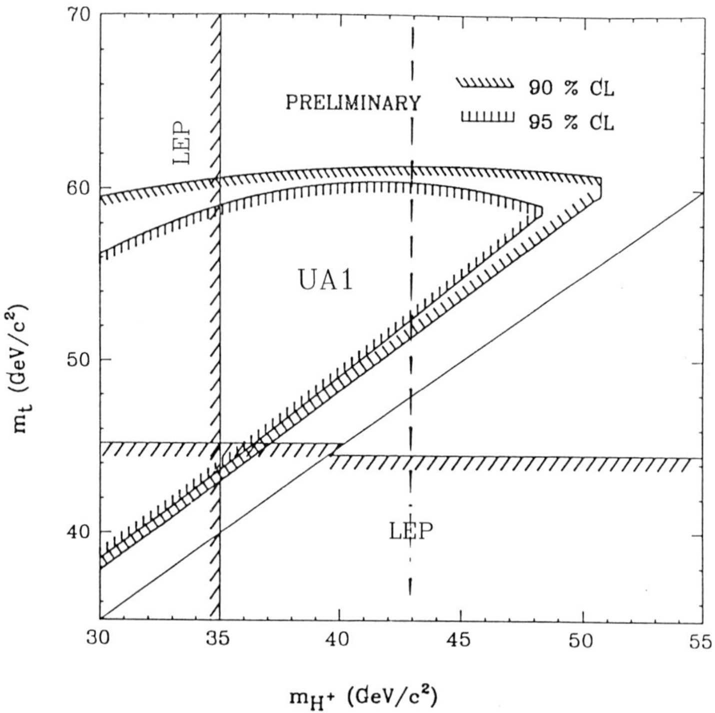

Figure 4. Limits on $M_{H^\pm} - M_{top}$ from UA1.

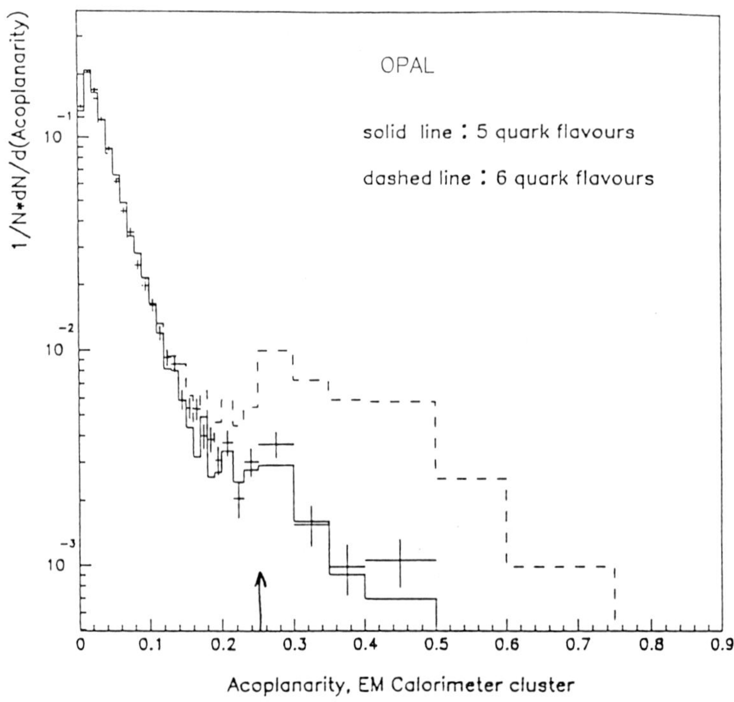

Figure 5. Acoplanarity of hadronic events with and without the 6th flavour.

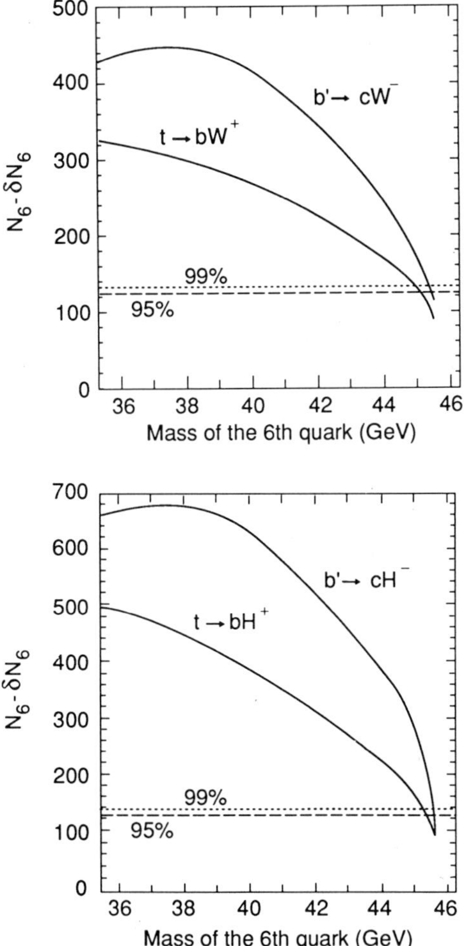

Figure 6. OPAL limit on t and b'.

shows how the acoplanarity of the hadronic Z decays should be modified by the existence of a heavy quark (top or b′, the lower member of the doublet of a fourth family). Acoplanarity is defined here as

$$A \equiv 4\mathrm{Min}(\Sigma\,|p_{i\perp}|/\Sigma\,|p_i|)^2$$

The Monte Carlo curve of Fig. 5 is obtained using ZHADRO for the amount of heavy quark pairs present in Z decay (most important is the QCD correction near threshold) and JETSET + HERWIG for the fragmentation. The experimental results clearly exclude such a production and lead to the mass limits shown in Fig. 6. The exclusion of t → bH has been obtained assuming $m_H = 23$ GeV and 100% of H → c$\bar{\text{s}}$ but the authors have shown that the result is quite independent of these assumptions. No top, even non standard, can exist below 45 GeV.

Hopefully LEP 200 will allow to push that limit to close to 90 GeV [5].

Meanwhile what LEP will bring is a more and more accurate determination of the top mass through its virtual contribution to loop diagrams. We can note from Fig. 6 that a b′ quark is excluded as well even if its decay is dominantly a flavour changing neutral current (FCNC) process.

5　The Standard Higgs

We admit here that the Higgs mechanism is the process which generate masses in the Standard Model; we also assume that the degree of freedom left over after spontaneous breaking of the symmetry and mass generation for the IVB corresponds indeed to an elementary neutral scalar boson.

The production and decay of this boson are quite familiar. The Bjorken mechanism (Fig. 7) leads to a double bump cross section depending on which Z is real (Fig. 8). Up to ~ 50 GeV the Higgs production is larger when one sits on the Z peak. For masses ≥ 50 GeV it is more profitable (at equal luminosity) to sit above the Z, at an energy roughly given by

$$\sqrt{s} \simeq M_H + M_Z + (10 - 15 GeV)$$

Radiative corrections dominated by initial state radiation (see formula of App. 1) are shown in (Fig. 9) and their shape can be intuitively understood quite readily.

The Higgs boson will decay into the most massive fermion–antifermion pair kinematically available. For clarity of the final state the Z (or Z*) produced in association is required to decay into a lepton pair ($\ell^+\ell^-$ or $\nu\bar{\nu}$).

Light Higgs

The Higgs mass is an unknown quantity which can range from zero to ~ 1 TeV or so. We will come back later to the upper mass accessible at LEP.

Below $2m_\mu$ the Higgs boson can only decay, depending on its mass, into $\gamma\gamma$ and ee. Its lifetime and therefore its flight path increases when its mass decreases.

There is a gradual change, with increasing mass, from an invisible objet, escaping detection and only identified as missing momentum, to a long lived one, decaying into an e^+e^- pair in the detector. In the former case a search for topology A gives the exclusion curve labeled A in Fig. 10a. In the later case the search for e^+e^- pairs born "from nothing" in the tracking region of the detectors and either alone (Z → $\nu\bar{\nu}$) or associated to a charged lepton pair gives the exclusion curve B. The combination of both methods definitely excludes the existence of a light Higgs boson, as shown in Fig. 10a-d [6].

Heavier Higgs

As the mass increases other channels ($\mu\mu, \pi\pi, \tau\tau, c\bar{c}$) open up. Such Higgs decays have been excluded as well. Above 10 GeV the b$\bar{\text{b}}$ decay mode dominates. The final state one is looking for is either

$$\mathrm{jet} - \mathrm{jet} - \ell^+\ell^-\ (\ell = \mathrm{e},\mu)$$

or jet-jet acoplanar with missing $\nu\bar{\nu}$, the later case having a cross section $2 \times 3 = 6$ times bigger than each lepton species in the former case.

The exclusion curves of DELPHI, OPAL and L3 [7] are shown in (Fig. 11). Clearly the most efficient channel is: jet jet $\nu\bar{\nu}$, as expected. The horizontal line at ~ 3 is the 95% CL curve corresponding to a Poisson law and the case where no candidate is observed and no background expected.

In Figs. 12 and 13 ALEPH results [8] are described. The curve of visible energy in the ALEPH detector (Fig. 12a), after some harmless cuts against 2γ background, is totally free of any tail on the low side (Fig. 12b), the region where a Higgs of 40 GeV (associated to Z → $\nu\bar{\nu}$) would show up (Fig.

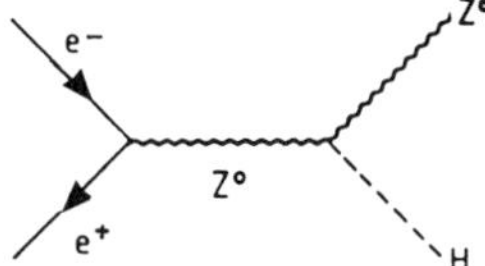

Figure 7. The Higgs production mechanism.

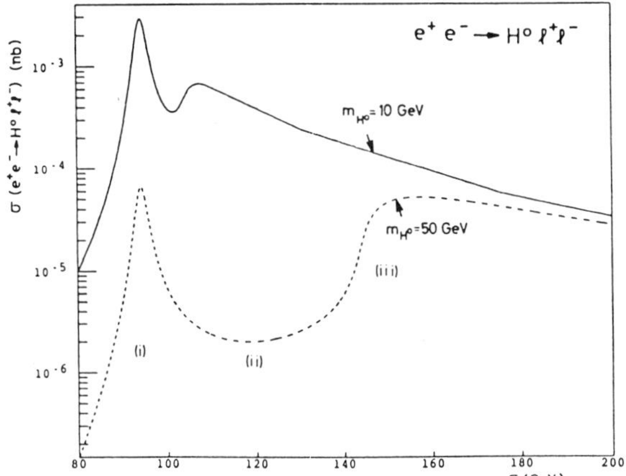

Figure 8. The corresponding cross-section versus $\sqrt{s}$.

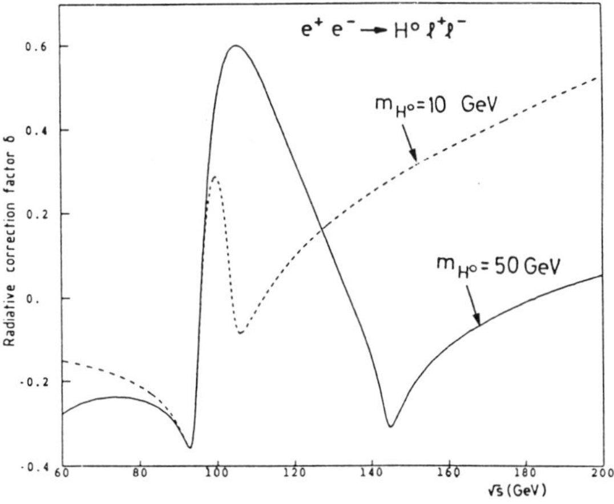

Figure 9. Radiative correction to the Higgs cross-section.

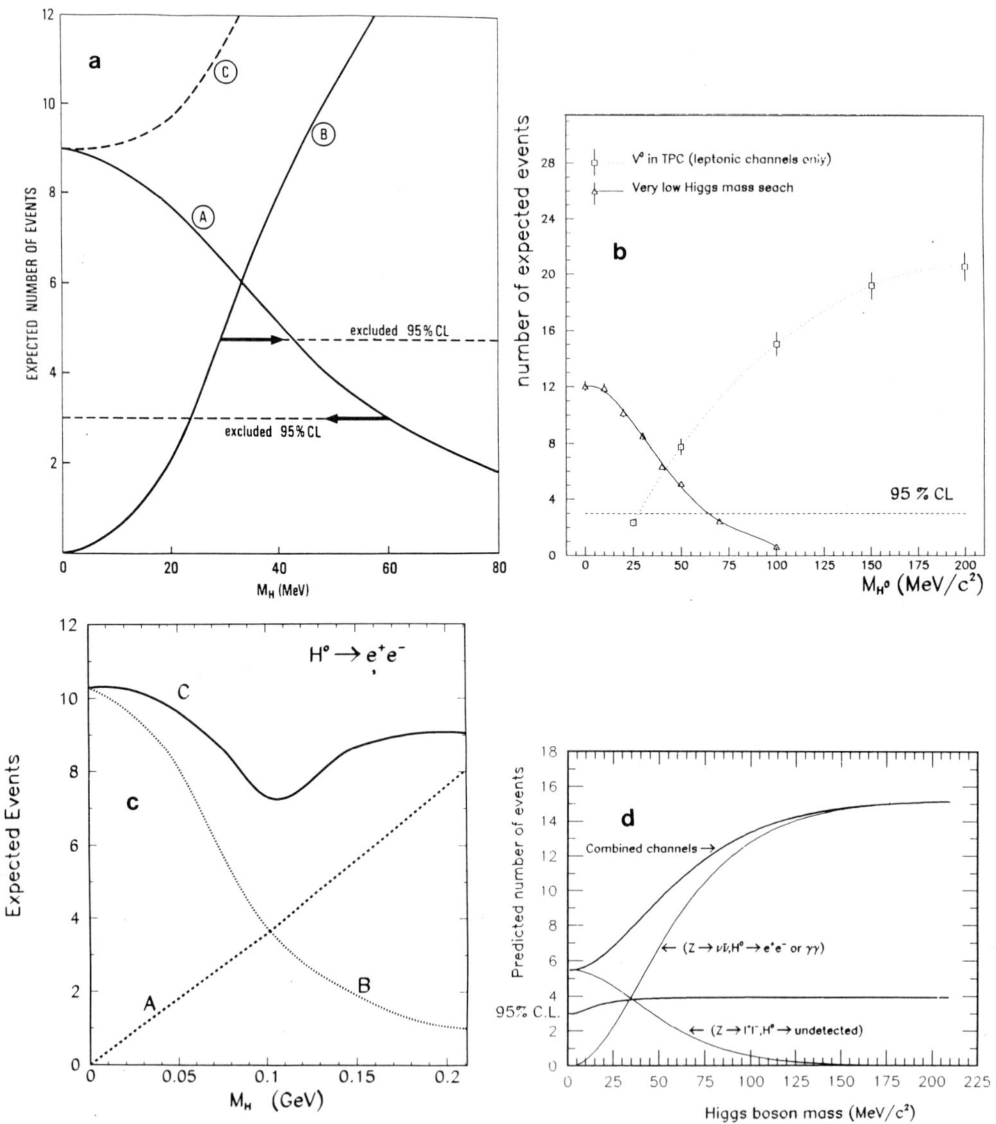

Figure 10. Limit on the light classical Higgs boson: a) ALEPH, b) DELPHI, c) L3, d) OPAL.

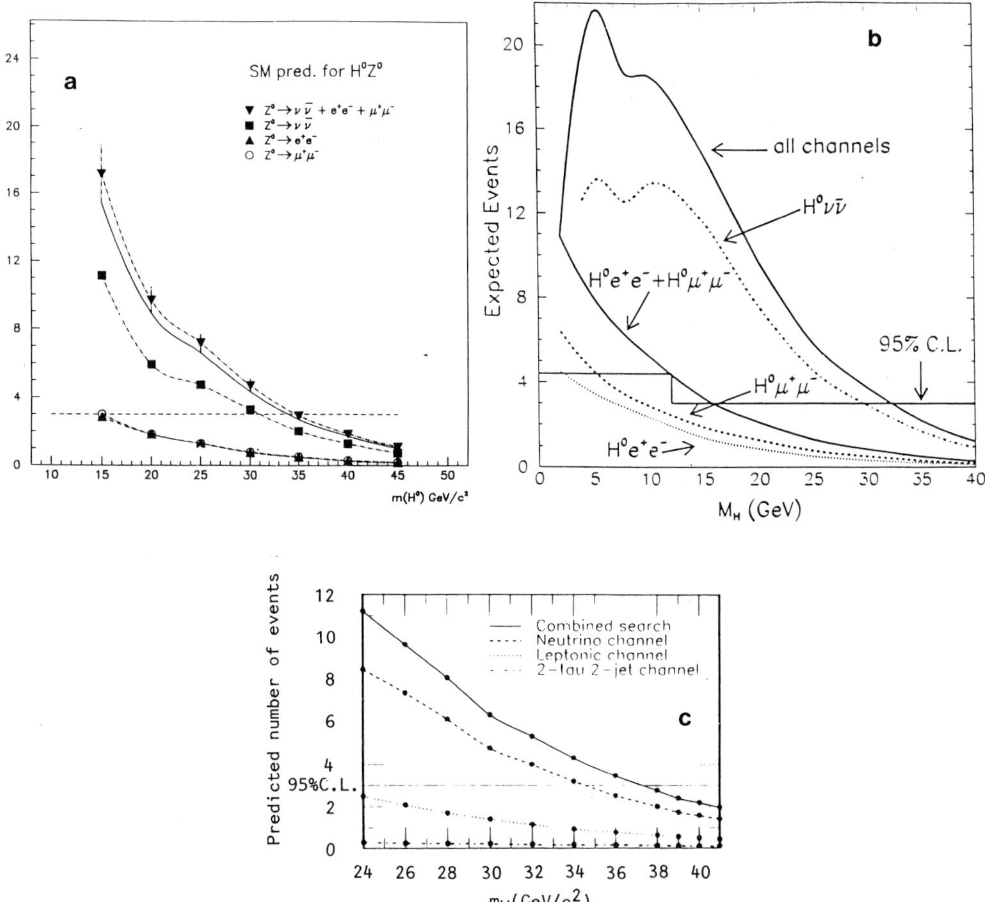

Figure 11. Heavy Higgs limit from: a) DELPHI, b) L3, c) OPAL.

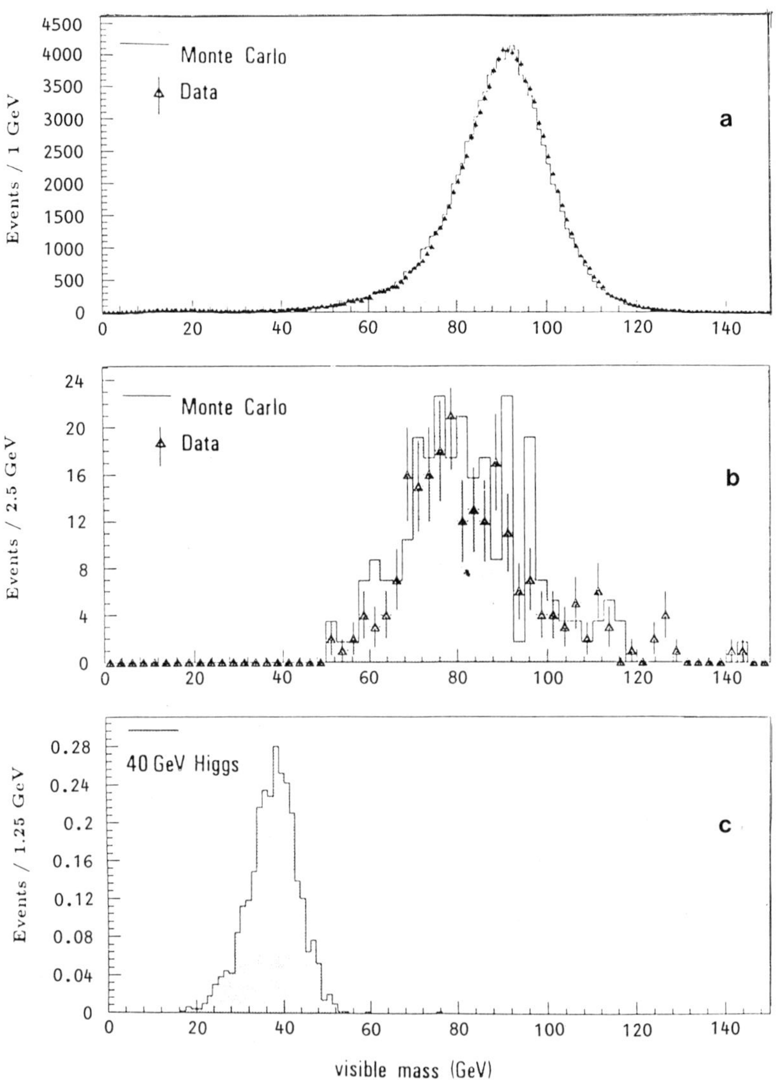

Figure 12. ALEPH calorimetry properties.

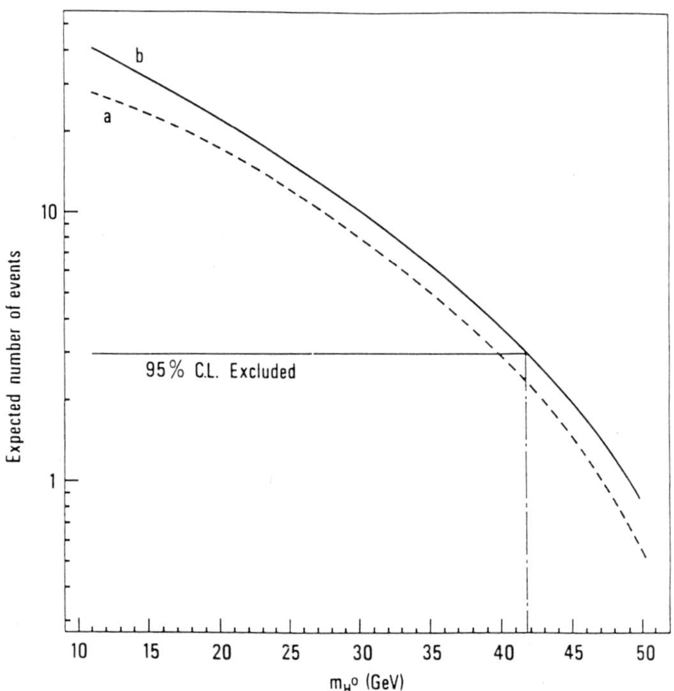

Figure 13. ALEPH mass limit on heavy standard Higgs.

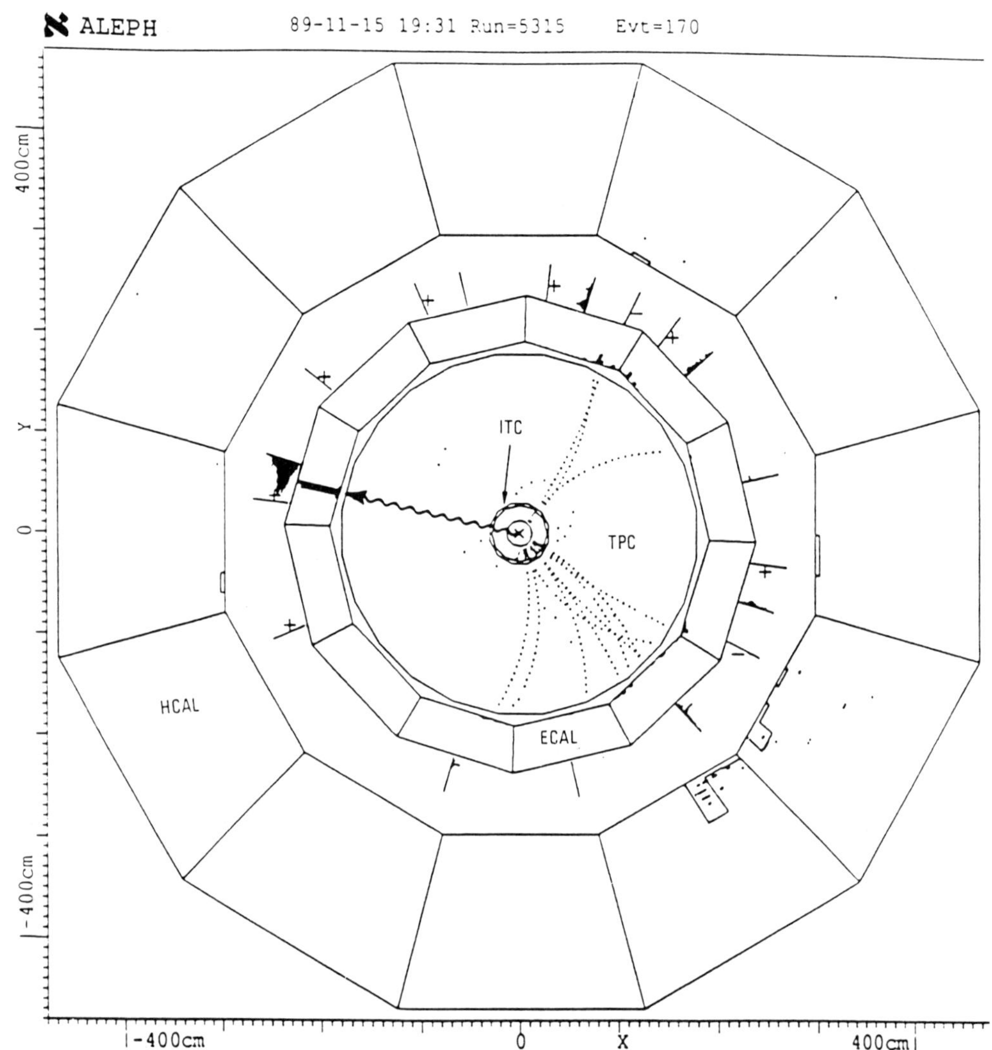

Figure 14. A q$\bar{\text{q}}$ radiative event.

12c). This is a good demonstration of the hermeticity of the ALEPH detector, which allows to set as a mass limit on the Higgs scalar (Fig. 13)

$$m_H > 42 \, \text{GeV}.$$

Beyond this kind of value one should pay attention to two varieties of background. One is due to $q\bar{q}$ radiative final states (Fig. 14) with the photon escaping detection because of non hermeticity of the electromagnetic coverage: this is a detector dependent background, and ALEPH for instance according to Fig. 12 does not seem to be in immediate danger. Here it is unlikely that kinematical constraints could save the situation if the photon is lost. We will see that such a final state, which is only a potential danger for the H $Z^\star$ final state, is a disaster for $Z \rightarrow H\gamma$ and totally forbids any hope to observe the standard value of this mode which would be the dominant one on the Z for high Higgs masses.

Another background for $HZ^\star$ ($Z^\star \rightarrow \ell^+\ell^-$) is direct 4 fermion production in Z decay [9]. Due to the actual process (Fig. 15) the kinematical features are fortunately quite peculiar: the mass of fermion pairs (for instance $q\bar{q}$) peaks either at low values or near the maximum available (Fig. 16). If one superimposes the signal expected for a Higgs (the effective mass resolution is assumed here to be ~ 4 GeV for $q\bar{q}$), it would be drowned in the background at ~ 50 GeV: but a cut on the dilepton mass, in case of $HZ^\star$ ($Z^\star \rightarrow \ell^+\ell^-$), can save the situation by reducing the background without doing much harm to the signal.

So, provided the detector is hermetic, one should be able with enough luminosity to push the Higgs search on the Z to higher masses. A factor ~ 10 in integrated luminosity is needed to go ~ 10 GeV higher in mass.

Above ~ 55 to 60 GeV, given the fact that $Z \rightarrow H\gamma$ is inaccessible, it will be more reasonable to move to LEP 200, provided the luminosity at high $\sqrt{s}$ is sufficient. As discussed in Appendix 3 one should then with enough available energy and integrated luminosity reach ~ 90 GeV.

6 Heavy Neutral Leptons (HNL)

Many reasons can push us to consider the possible existence of HNL [10].

In spite of the limits already set, one cannot exclude the existence of a further generation: one has then to provide a right handed field so that the neutrino can acquire a mass to respect the bound.

Right handed HNL would provide a natural mechanism to give a small mass to the normal neutrinos, by a see-saw mechanism (Fig. 17) [11]. From a mass matrix

$$\begin{pmatrix} 0 & m \\ m & M \end{pmatrix}$$

where m stands for a charged lepton mass ($m_e, m_\mu, ...$) and M is a high mass associated to the HNL one can by diagonalization get the eigenvalues m^2/M (identified to the small neutrino mass) and M.

The idea of Grand Unification by itself calls for HNL since it provides large multiplets with room for such objects.

The phenomenology of HNL depends of their nature. Sequential heavy leptons, members of an isospin doublet, are expected to be pair produced with normal strength. FCNC decay are absent. The decay is $L^o \rightarrow \ell W^\star$ with an amplitude $|U_{\ell L}|$ depending on the mixing matrix element with lower generations.

Extremely small $|U_{\ell L}|$ lead to stable invisible L^0 (that one may call ν_H proper, heavy stable neutrino); the only way of identifying their existence is through their influence on Γ_Z or Γ invisible. For some intermediate domain of $|U_{\ell L}|$ the lifetime is such that the L^0 can be found by its secondary vertex. For larger $|U_{\ell L}|$ normal $L^0 \rightarrow \ell W^\star$ decay occur leading to topologies A or B described previously.

Figure 18 from ALEPH [42] illustrates the result of the three methods, excluding most of the possibilities for sequential HNL up to ~ 43 GeV. Similar results come from other experiments. For instance L3 [13] gives (Fig. 19) limits on heavy neutrinos from Γ_Z and Γ_{inv}.

Right handed HNL are produced in association with a neutrino

$$ee \rightarrow \nu N$$

FCNC decay $N \rightarrow \nu Z^\star$ is competing with the charged current decay $N \rightarrow \ell W^\star$ (ratio NC/CC $\sim 1/2$ and matrix element $|U_{\ell N}| \sim m/M$). OPAL results [14] are shown in (Fig. 20) and we will come back briefly to this topic in Appendix 3 since rare decay modes of the Z can be related to such HNL.

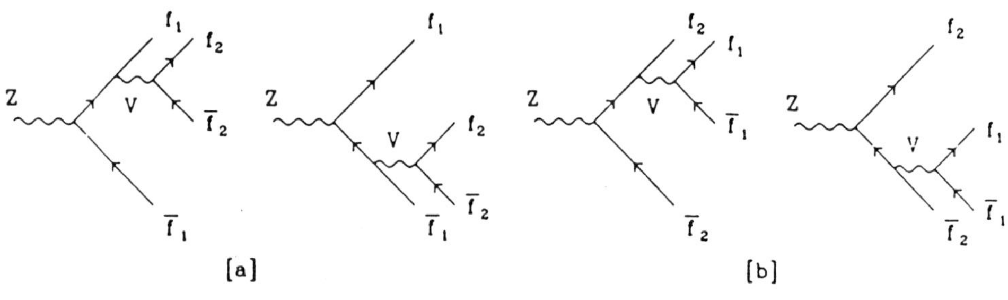
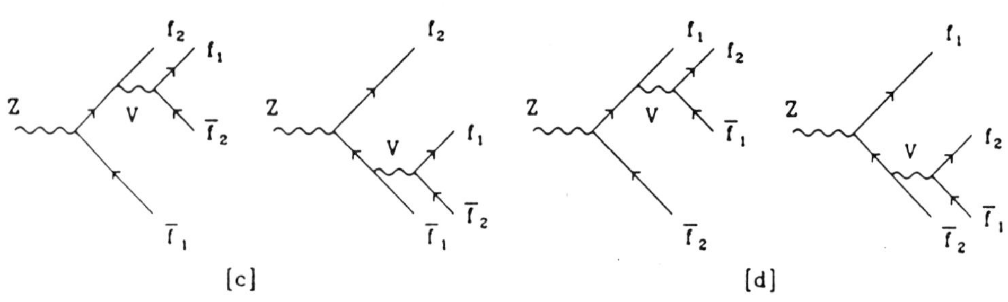

Figure 15. The four-fermion production mechanism.

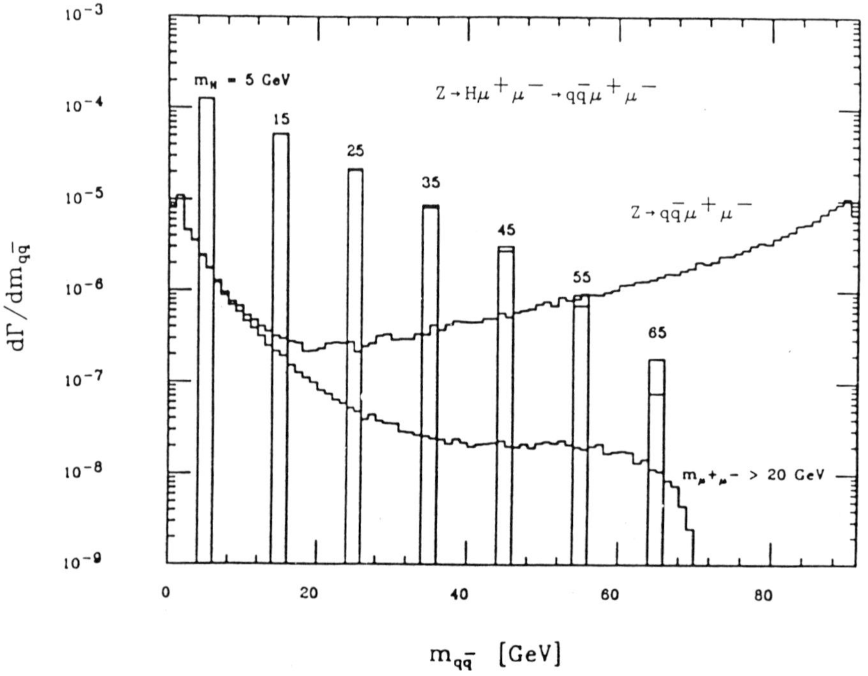

Figure 16. The visibility of heavy Higgs on the 4 fermion background.

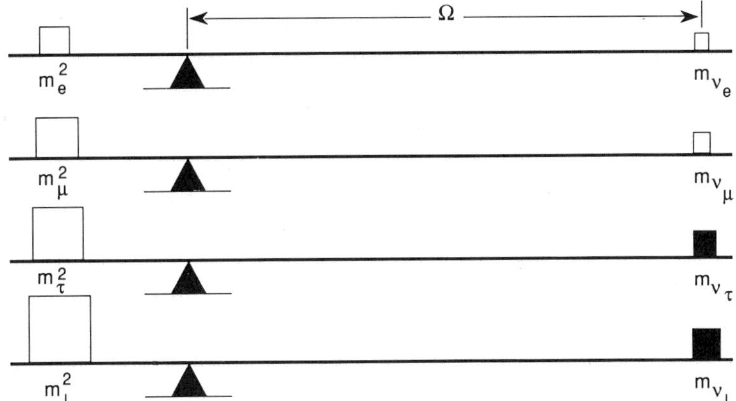

Figure 17. The see-saw mechanism: a simple point of view [11]. The see-saw mechanism proposes that the mass of each neutrino (m_ν) is related to the mass of the associated charged lepton (m) by the formula $m_\nu = m^2/\Omega$; Ω is an unknown mass scale, visualized here as the lever arm of the see-saw. Since, for example, the electron-neutrino mass is known to be less than 16 eV and the electron mass is known to be 0.5 MeV, the see-saw equation requires Ω to be at least 16 GeV. The tau mass is 1.8 GeV and cosmological limits on the tau-neutrino make it less than 65 eV. Running the see-saw equation with these values gives the stricter lower limit on Ω of 5×10^7 GeV. If Ω is related to a large fourth-lepton mass, the see-saw mechanism shows how this large mass could generate the very small neutrino masses.

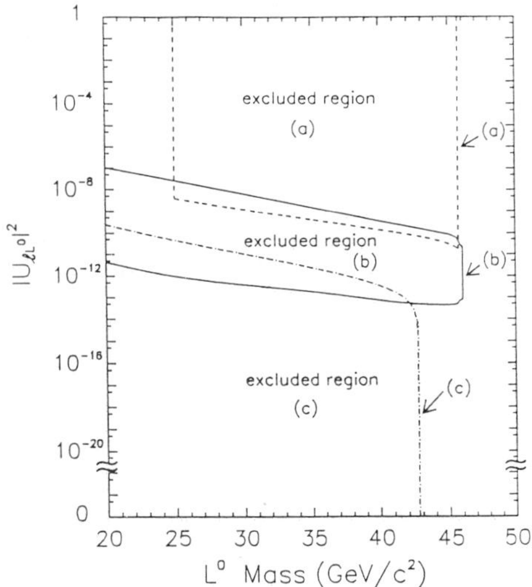

Figure 18. ALEPH neutral lepton exclusion.

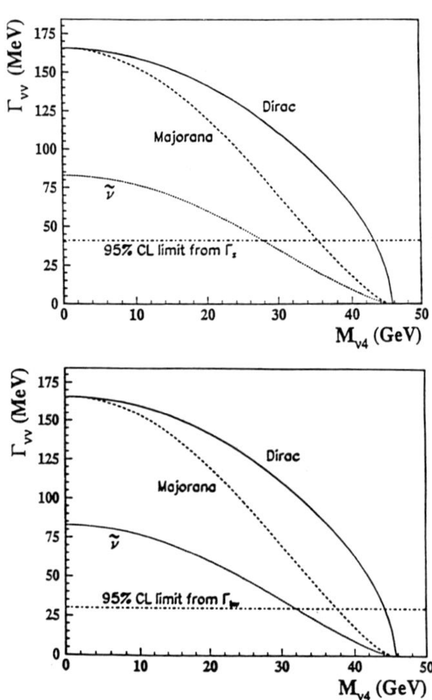

Figure 19. L3 exclusion of Dirac and Majorana L^0 from Γ_Z and $\Gamma_{\text{invisible}}$.

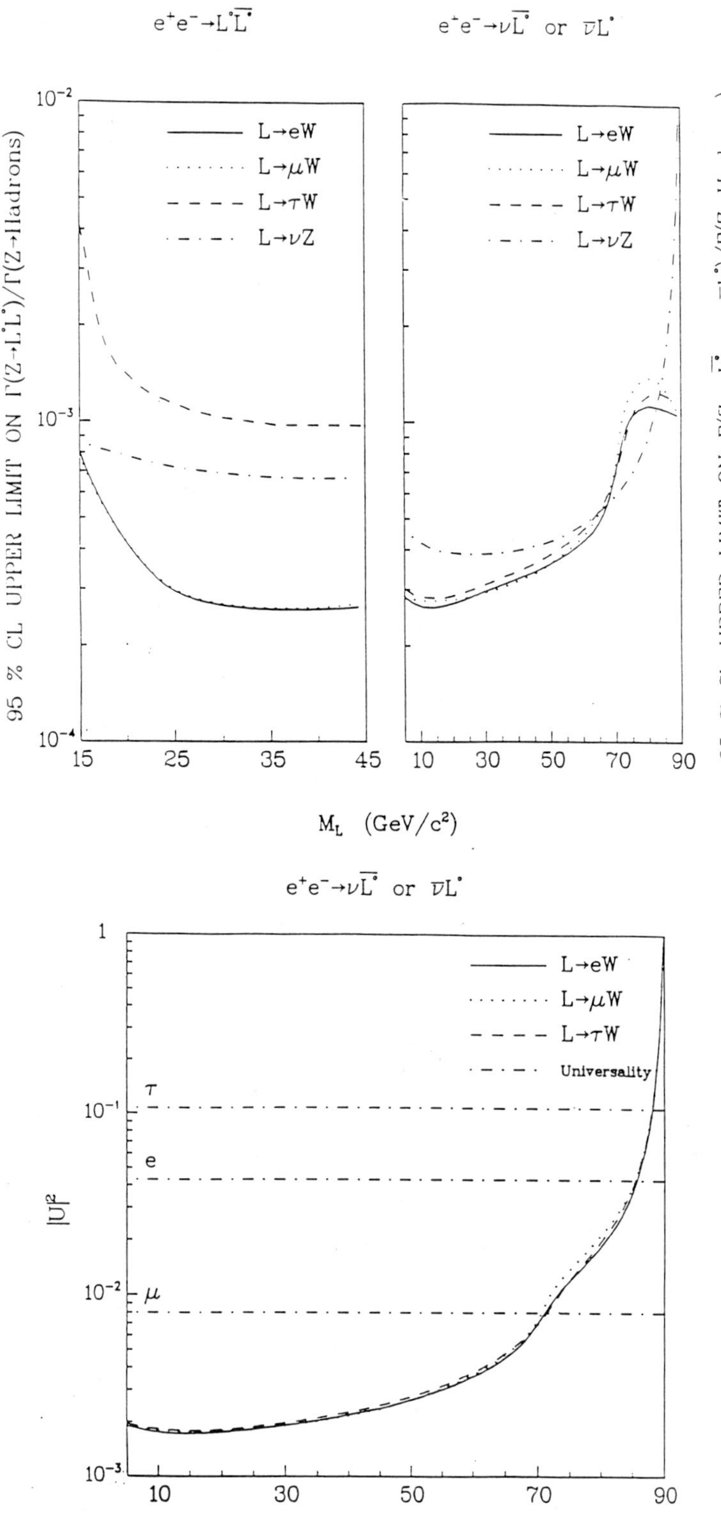

Figure 20. Exclusion curves by OPAL of heavy neutral leptons.

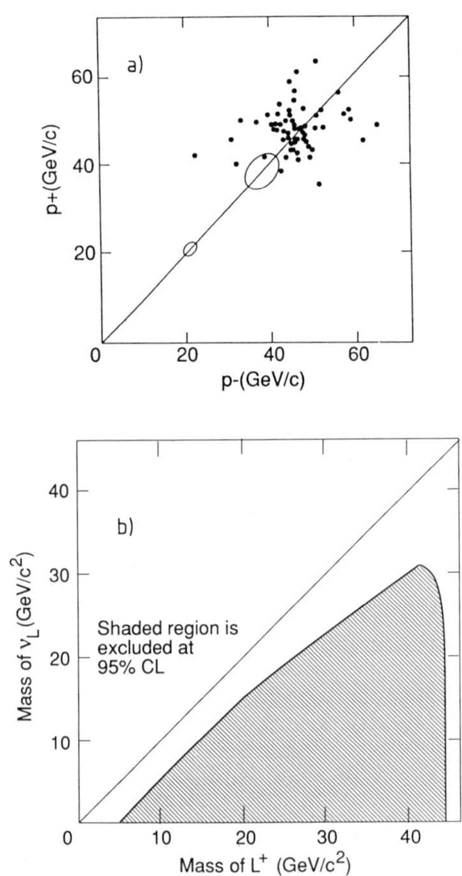

Figure 21. Exclusion of heavy charged leptons: a) stable, from DELPHI, b) from OPAL.

Mirror leptons (i.e. RH doublets and LH singlets) can be considered as well. Their decay modes either to the normal standard world or to "sterile" mirror world channels depend on the details of the theory.

Charged leptons, stable or decaying, have been excluded similarly [15] (Fig. 21).

7 Supersymmetry

I think we need here some clarification about the meaning of the theory which is tested for. We will concentrate on a version of SUSY which stems from the following assumptions:

a) supergravity [16] provides the needed clue for supersymmetry breaking.

b) three simple assumptions define the scenery of SUSY at low energies:

    – Minimal particle content supermultiplets of quarks, leptons and pair of Higgs doublets.

    – Universality i.e. breaking of flavour and/or CP only through properly s-symmetrized Yukawa couplings at some grand mass M_x, universal mass term for all scalars and universal gaugino mass term.

    – R-parity unbroken. The number of Superpartners is conserved modulo 2 and the LSP is stable.

The set of these assumptions describe what is called the Minimal Supersymmetry Standard Model [17]. The important fact is that MSSM is a falsifiable theory: the constraints imposed on its components are such that experimentation can turn it wright or wrong, as expected from a physical theory. In other words, it cannot be saved for ever and be kept as an argument for an escalade in the available $\sqrt{s}$ of future machines.

Appendix 2 helps realizing the constraints which exist within the MSSM. The Supersymmetric Lagrangian valid at some high scale M_x (M_{Pl}?) has universal terms for scalars and gauginos, with mass terms m and M respectively. When energy is downscaled to present ones, the Lagrangian is renormalized and parameters like the top mass enter the game. In the gaugino sector the three terms correspond to gluinos, neutralinos and charginos. For neutralinos the mass matrix is 3×3, with parameters M_1, M_2, β, μ_R. For charginos, it is a 2×2 matrix, with parameters M_2, β, μ_R. The relationships between renormalized mass terms are

$$M_i = M \frac{\alpha_i (M_W)}{\alpha_i (M_x)}$$

with

$$M_{\tilde{\gamma}} = M_1 \cos^2\theta_W + M_2 \sin^2\theta_W \simeq 0.6 M_2$$

being the mass of the LSP (for small values of M_2).

For the Higgs sector the Lagrangian is shown in Appendix 2. When the symmetry breaking mechanism has acted, one gets the masses of the surviving Higgs particles as functions of g, g', m, μ, M. Because of the presence of g, g' in the Lagrangian, Higgs masses and IVB masses are connected. In particular one arrives to the important inequality for the lightest Higgs

$$m_{h^o} \leq |\cos 2\beta| \, m_Z,$$

valid at tree level.

Another concept, naturalness [17], can be described as follows: standard quantities like the Z mass depend on the parameters of SUSY. It is well known that, for SUSY to fulfil its role, the masses of superpartners cannot be too heavy. To keep at their observed values quantities like m_Z –or m_H– there must be fine tuning among the SUSY parameters. A natural theory is one which does not require too much fine tuning. A way to express it is to impose that the relative variation of an observable like M_Z does not depend too much on the relative variation of SUSY parameters a_i. Let us fix for instance:

$$\frac{dM_Z}{M_Z} \bigg/ \frac{da_i}{a_i} < 10$$

Such a condition has the merit to set upper bounds to the masses of SUSY objects: because of the renormalization of couplings when one goes down from high scales to our world (Appendix 2) such bounds depend for instance on the top mass. Nevertheless SUSY partners as charginos and

---

*we call: h^0 the lighter scalar, H^0 the heavier scalar, A^0 the pseudoscalar

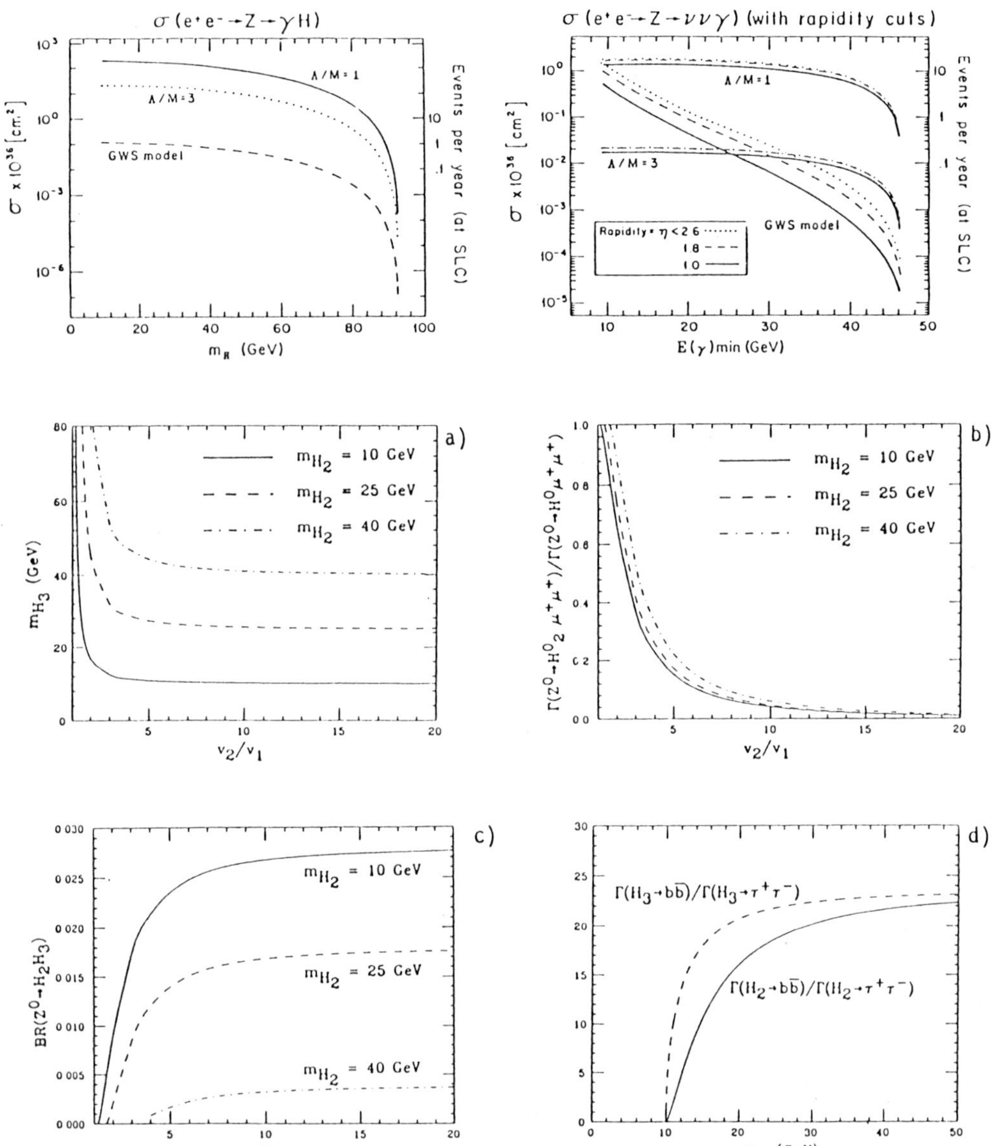

Figure 22. The phenomenology of SUSY Higgses.

neutralinos, within the present limits on top, are bound to be fairly light objects ($\lesssim$ 200 GeV) if the above naturalness condition is required. This is at least an encouragement to look for such objects within the accessible energy domain at LEP.

MSUSY Higgs

As we saw much more restrictive are the bounds in the MSUSY Higgs sector. The phenomenology of MSUSY Higgses on the Z is described in Fig. 22 [18]*. For low values of $\tan \beta \equiv v_2/v_1$, the ratio of expectation values of the two doublets, the mechanism ee $\to$ Zh is operative with appreciable strength (Fig. 22b) and searches for h^0 follow the pattern of standard Higgs ones. When $\tan \beta$ increases (and $\tan \beta \gg 1$ is theoretically favoured because of the high top mass), this mechanism gradually disappears, but the associated production

$$ee \to hA$$

of the lightest higgs with its then nearly mass degenerate pseudoscalar partner takes over (Fig. 22c), providing an extremely useful method of search through topologies A, B and C.

Decay patterns of h and A are governed as well by $\tan \beta$. For $\tan \beta > 1$ the branching ratio into $\tau\nu$, leading to A and B, is appreciable ($\geq$ 5%). For $\tan \beta < 1$ however one has to look to h $\to$ c$\bar{\text{c}}$ modes. Either 4 jets (topology C) or special c tagging through D* can then be used.

The LEP results [19] are shown in Figs. 23 and 24. The first ones give details, in the $\tan \beta$ versus m_h plane, on the strategies used, every experiment bringing one or more specific method. The second one summarizes the global result up to now. The achievement is spectacular and mostly due to the mild ($\beta^3 \equiv (P/E)^3$) dependence on the associated cross section with m_h. However the way to go to explore fully the domain allowed by MSSM ($m_h \leq m_z$) is still long as shown in Fig. 25. LEP I with increased L will close the gap up to $M_Z/2$, but LEP200 with much luminosity will be needed to complete the job (Fig. 25). We come back to this crucial issue in Appendix 3.

8 Neutralinos and Charginos

These are the generic names of the particles resulting from mixing among (photino + zino + higgsinos) and (winos + higgsinos), respectively. The phenomenology is much dependent on the MSUSY parameters, but does not vary too rapidly with $\tan \beta$.

Exclusion of such objects can, as usual, rely on global methods like the inspection of Γ_Z or on specific channels. For these the only assumption is that the lightest neutral object, $\tilde{\chi}$, is stable and escapes detection. Signatures are therefore

$$ee \to \tilde{\chi}'\tilde{\chi}$$

where the heavier $\tilde{\chi}'$ cascades into the lightest one

$$\tilde{\chi}' \to \quad \tilde{\chi} f\bar{f} \quad \text{(Zen one sided events)}$$
$$\tilde{\chi} \gamma \quad \text{(single } \gamma \text{ on the } Z^0 \text{ peak)}$$

with probabilities which again depend on the parameters.

Similarly ee $\to \tilde{\chi}^+\tilde{\chi}^-$ with $\tilde{\chi}^{\pm} \to \tilde{\chi}W^{*\pm}$ and obvious signatures.

The excluded regions are shown in the M, μ_R parameter plane (Fig. 26) [20]. Here M represents for small values the mass of the lightest $\tilde{\chi}$ (essentially a photino when μ is large and negative) while $|\mu|$ for small values again represents the Higgsino mass.

The most recent ALEPH exclusion region [21] for $\tan \beta = 2$ is shown in Fig. 27 where various exclusion methods are no more distinguished (to make the figure more reader friendly). One sees that the horizontal limit on M for negative μ is equivalent to a limit on $m_{\tilde{\chi}}$ (at $\sim$ 20 GeV) or a limit on $\tilde{g}$ at $m_{\tilde{g}} \sim 100$ GeV (Fig. 28). This last limits, valid for $\tan \beta \geq 2$, are quite important since $\tilde{\chi}$ could be suspected to be a good candidate for non baryonic dark matter [22]. However LEP data do not presently allow to set a bound on $m_{\tilde{\chi}}$ when $\tan \beta < 2$: the tree level limit $\tan \beta \geq 1.6$ quoted in the past from the non observation of h^0 (see above) is unfortunately not valid when the effect of loops are included (see Appendix 3). Calculations of this effect are under way.

Chargino exclusion [23] is illustrated by Fig. 29.

Superpartners of quarks and leptons

Even naturalness arguments do not promise that these objects should be light.

The exclusion of sleptons through topology A is a relatively simple matter [24] (Fig. 30).

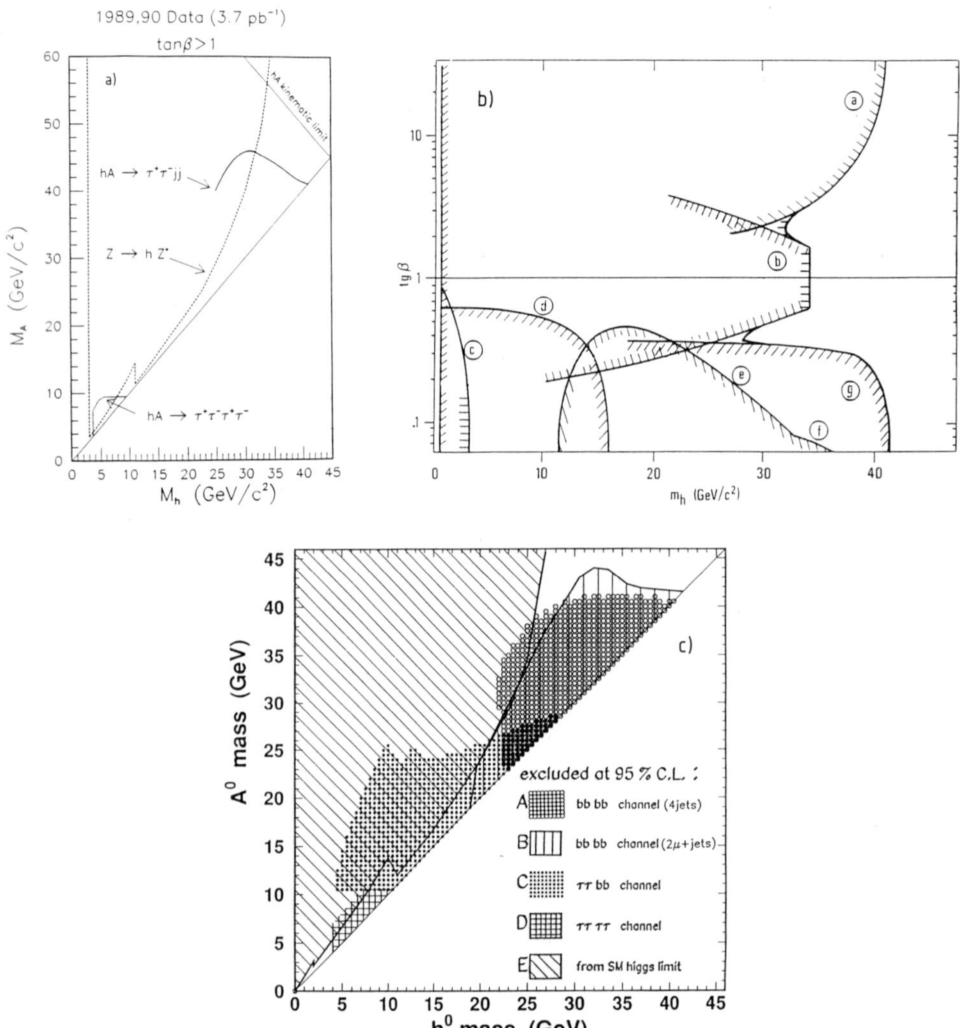

Figure 23. Limits on SUSY Higgses; a) OPAL, b) DELPHI, c) L3.

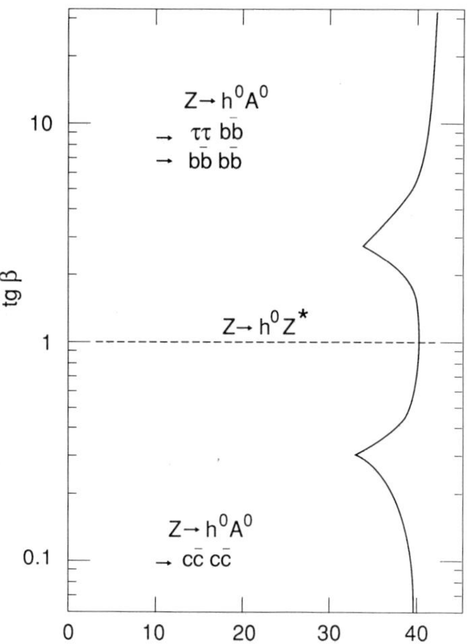

Figure 24. Global limit of LEP on SUSY Higgs (from M.J. Oreglia, talk at the CERN 90 Neutrino Conference, June 90).

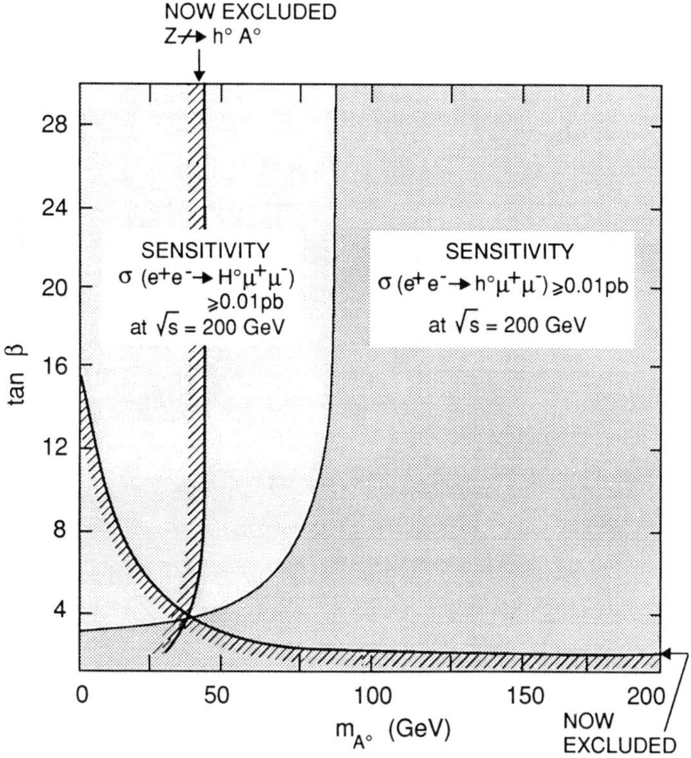

Figure 25. The MSUSY plane, seen in tg β versus M_A (courtesy of F. Zwirner and F. Pauss); LEP 200 with $\sqrt{s} \geq 190$ GeV, $L = 500$ pb^{-1} can exclude the whole plane.

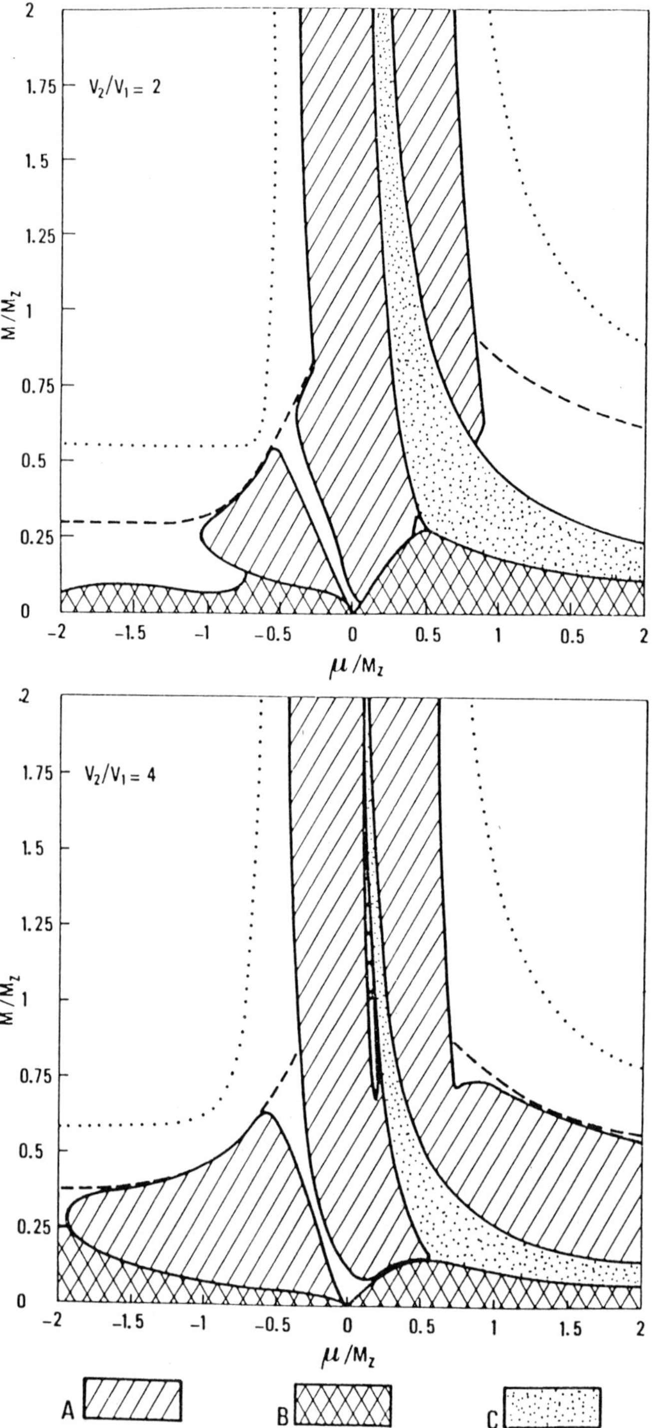

Figure 26. Neutralino limits from ALEPH.

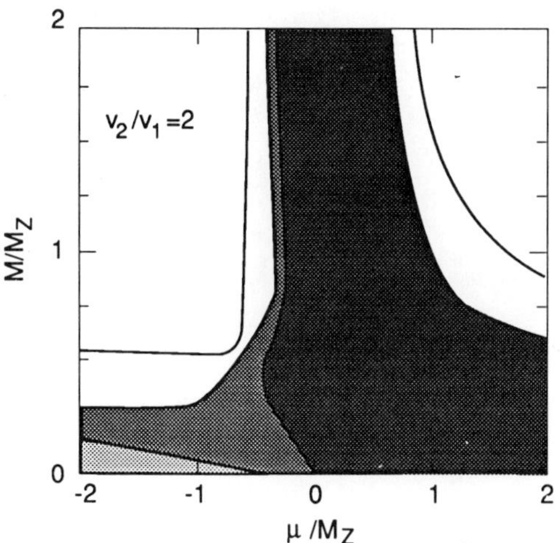

Figure 27. Last limit from ALEPH on neutralinos (Singapore).

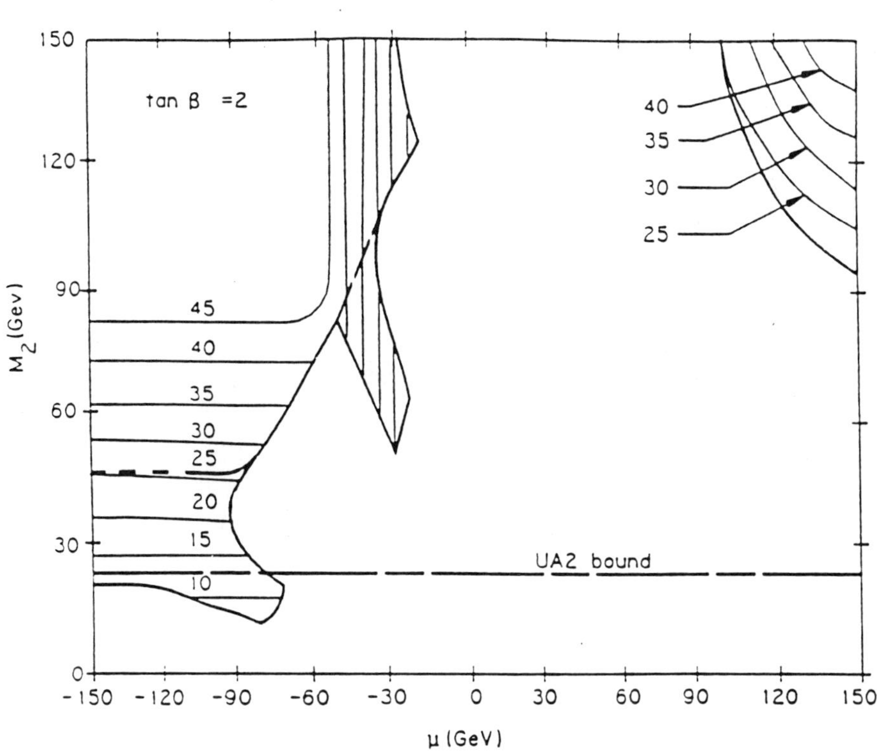

Figure 28. Iso-LSP mass curves in the M_2-μ plane.

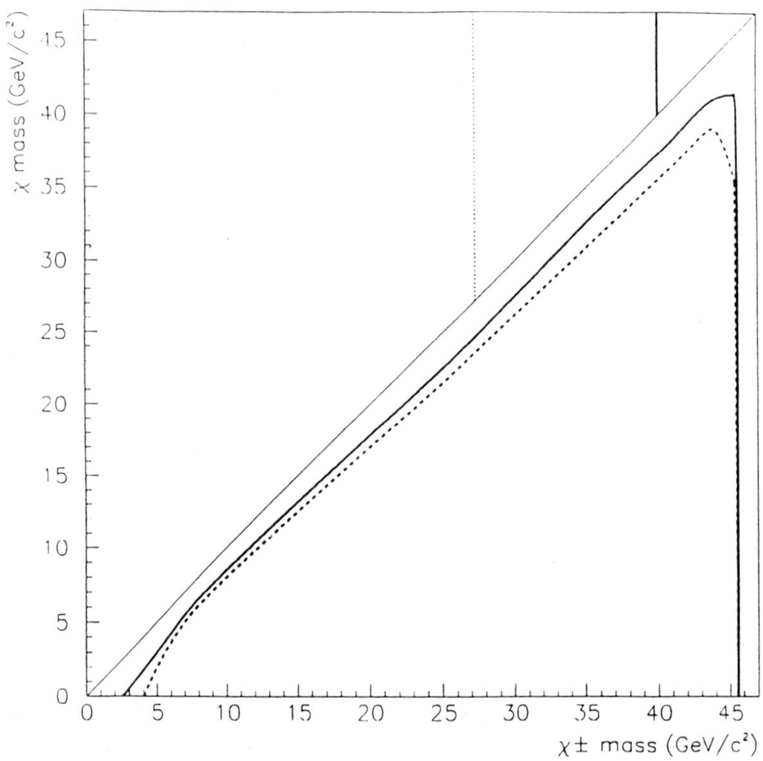

Figure 29. Limits on charginos (DELPHI).

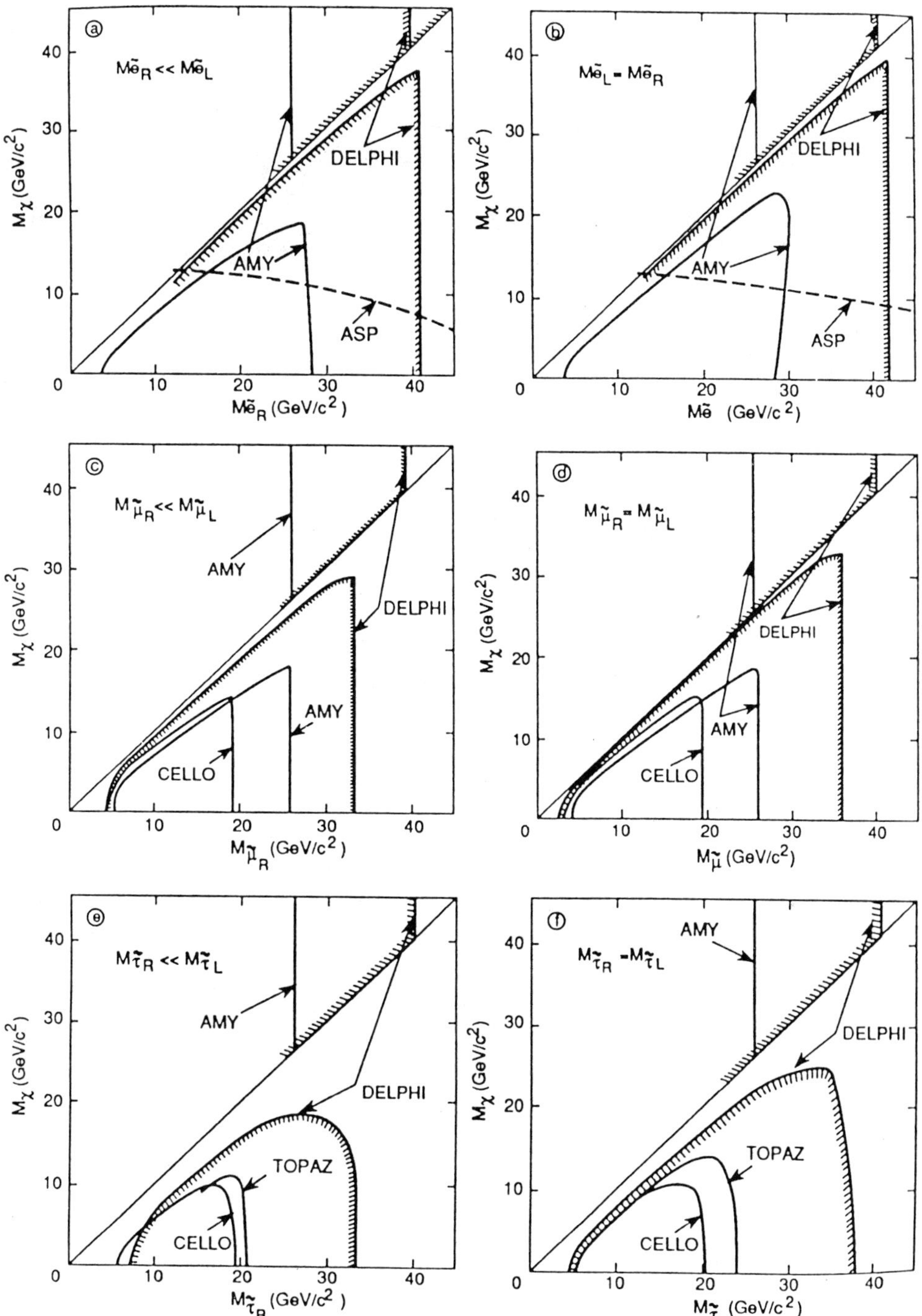

Figure 30. Limits on sleptons (DELPHI).

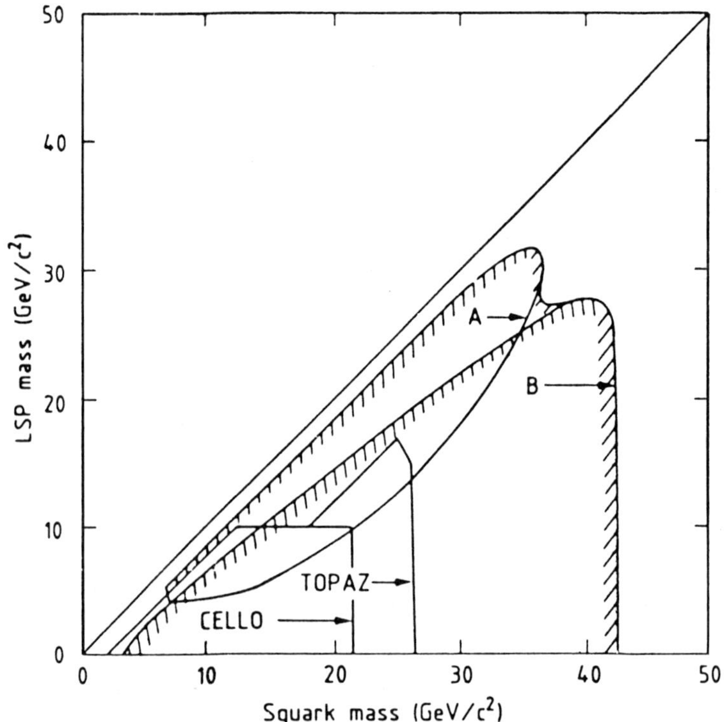

Figure 31. Limit on squarks (DELPIII).

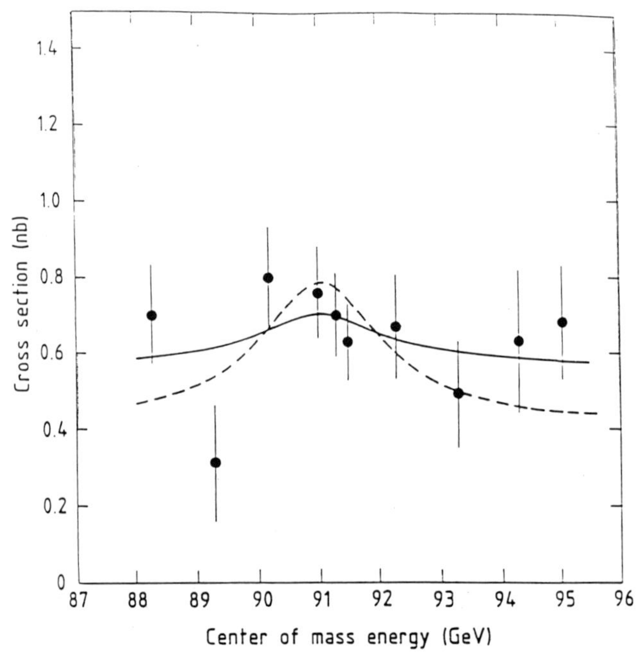

Figure 32. Absence of a resonant signal for low visible-energy events (DELPHI).

For squarks one problem is to go as close as possible to the diagonal of Fig. 31, in case the LSP is nearly as heavy as the squark. This corresponds to events with low visible energy. $\gamma\gamma$ and beam gas interactions are then notorious backgrounds but non resonating ones. DELPHI [25] (Fig. 32) has shown that no resonant component is present among the low visible energy events, thus excluding the region quoted in Fig. 31.

Sneutrinos have been excluded as well.

9   Compositeness

We come here to more speculative–and less theoretically grounded–issues to cope with the deficiencies of the SM. The fact that no theory ever managed to build nearly zero mass objects (light quarks and leptons) from heavy constituents (preons, rishons, ...) is not a reason to drop the idea of a further shell of constituents, explaining the strange features of the SM.

At a very high scale Λ we suppose that basic constituents manifest themselves: they can be created or exchanged, as partons are at present energies. For $E \ll \Lambda$ the only manifestation left is some kind of effective Lagrangian, giving rise to contact terms or excitation patterns. The ambiguity of the definition of Λ which requires that the strength of the interaction is defined as well has been emphasized in the past [26].

On the Z at LEP the manifestation of contact terms is minimal since their real contribution cannot interfere with a purely imaginary amplitude.

One should rather look for excited objects or specific decay modes of the Z manifesting its composite nature.

Excited leptons, foreseen by compositeness, can be pair produced with no ambiguity in the coupling. LEP easily sets a limit of $M_Z/2$ on their mass. They can also be singly produced

$$ee \rightarrow \ell\ell^\star.$$

However the coupling is then an unknown magnetic one which has to be parametrized as follows

$$L_{eff} = \sum_{V=j,Z} e \frac{\lambda_v}{m_{e^\star}} \overline{\psi}_{e^\star} \sigma^{\mu\nu} \left(c_v - d_v \gamma_5\right) \psi_e \partial_\mu V_\nu$$

c_v and d_v are determined by constraints from g-2 and electric dipole of the leptons. The unknown quantity λ can be related to the compositeness scale Λ:

$$\frac{1}{\Lambda_v^2} \simeq \alpha \left(\frac{\lambda_v}{m_{e^\star}}\right)^2$$

and stays as an open parameter. Therefore LEP results [27] can only be expressed in the plane: $\lambda - m_\ell^\star$ as shown in Fig. 33. L3 has up to now published the stronger limits.

Another manifestation of an excited e, e*, would be its t-channel exchange, in the process

$$ee \rightarrow \gamma\gamma$$

This $ee \rightarrow \gamma\gamma$ channel has been explored by all four experiments. Without any interference with an exotic Z component, its cross section can be written as

$$\frac{d\sigma}{d\Omega} = \frac{\alpha^2}{s} \frac{1 + cos^2\theta}{\varepsilon - cos^2\theta} \text{ with } \varepsilon \simeq 1 + \frac{2m_e^2}{s}$$

or:

$$\frac{d\sigma}{d\Omega} / \frac{d\sigma}{d\Omega_{QED}} = 1 \pm \frac{s^2}{2\Lambda_\pm^4} sin^2\theta H \left(cos^2\theta\right)$$

when the explicit contribution of a e* is shown.

$$H \left(cos^2\theta\right) \rightarrow 1 \text{ as } m_{e^\star} \gg \sqrt{s}.$$

$\Lambda^\pm$, the QED cut off parameter, is related to the $m_e^\star$ and the compositeness scale Λ by

$$\frac{1}{\Lambda_\pm^4} \approx \frac{1}{\Lambda^2} \frac{1}{\alpha m_{e^\star}^2}.$$

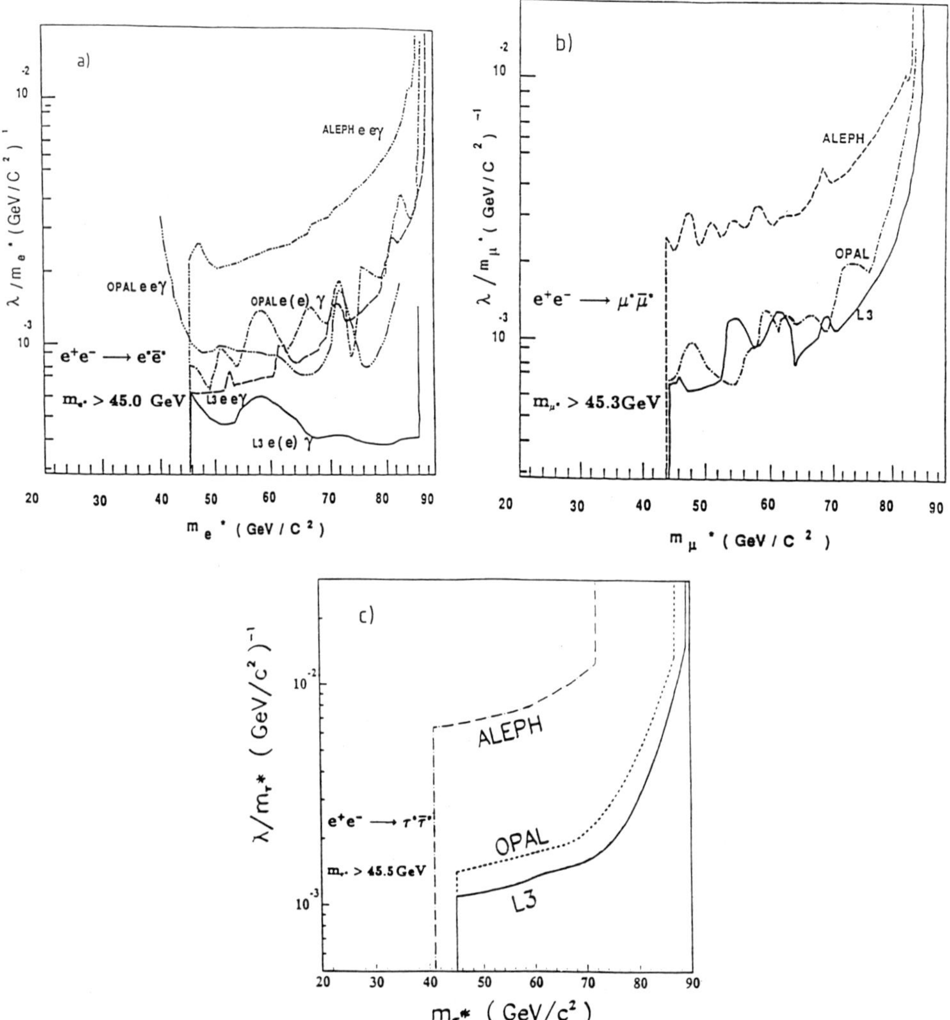

Figure 33. Limits on excited leptons, a) e*, b) μ*, c) τ*.

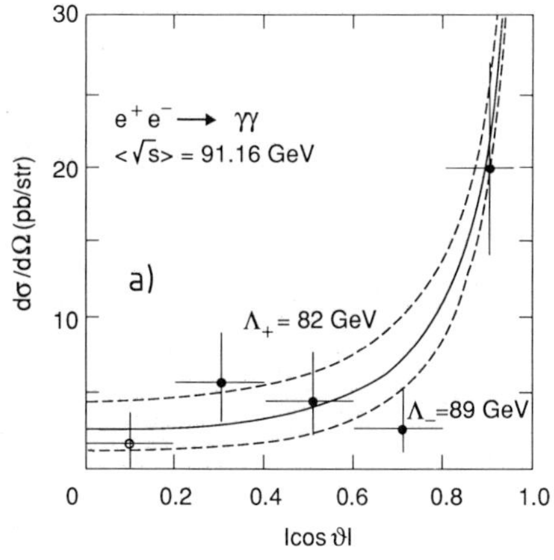

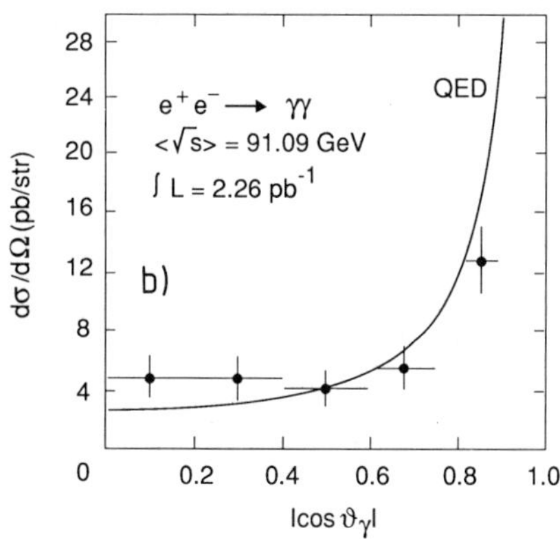

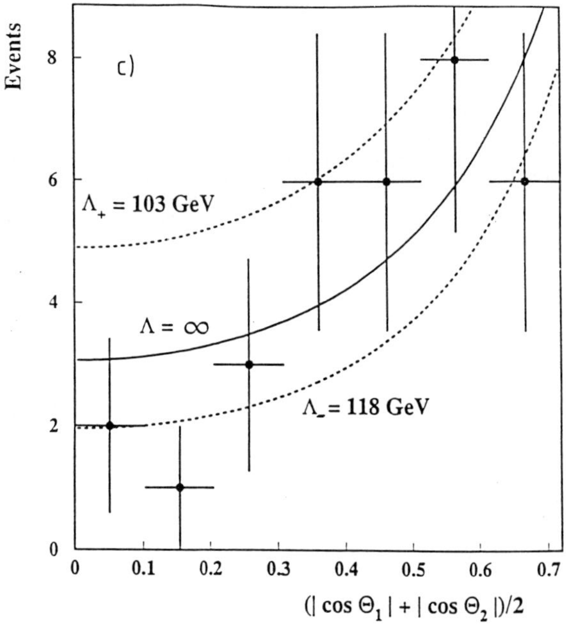

Figure 34. ee→ $\gamma\gamma$ cross-sections from; a) OPAL, b) DELPHI, c) L3.

BIG-BANG HELIUM PRODUCTION AND NEUTRINO FAMILIES

1. Assume that the universe consists only of neutrons (n) and protons (p), with a vastly larger background of electrons (e^-), positrons (e^+), neutrinos and antineutrinos ($\nu_e, \nu_\mu, \nu_\tau, \bar{\nu}_e, \bar{\nu}_\mu, \bar{\nu}_\tau$) and photons ($\gamma$), all indicated below by dots. At times much less than one second after the big bang and temperatures much higher than 10^{10} degrees Kelvin, the n and p appear in almost equal numbers:

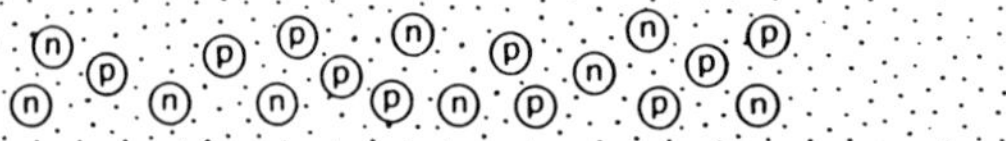

2. Neutrons and protons are constantly transmuted into one another by the so-called weak nuclear reactions:

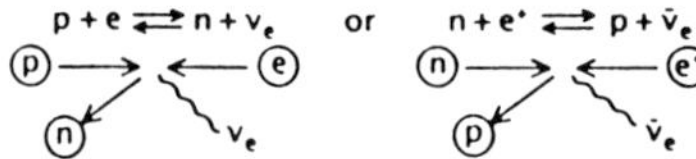

3. Because neutrons are slightly more massive than protons, they are energetically more difficult to produce, and so the n-p transmutations in step (2) result in slightly more protons. As the universe expands and cools, less and less energy is available to produce neutrons, and so the weak reactions result in ever more protons. At about one second after the big bang and a temperature of about 10^{10} degrees K. protons outnumber neutrons by about five to one:

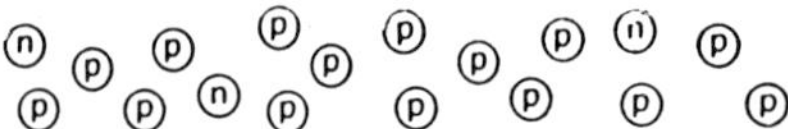

4. At this time the expansion rate of the universe overtakes the ever slowing weak-reaction rates, so that collisions between particles essentially cease:

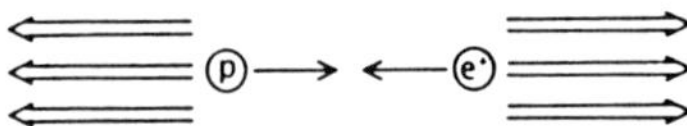

No more neutrons are converted into protons; the 1:5 ratio is "frozen out."

5. Neutrons are radioactive and decay into protons. The lifetime of the neutron is about 15 minutes, so that after three minutes or so about one-third of the neutrons have decayed into protons, leaving one n for every eight p:

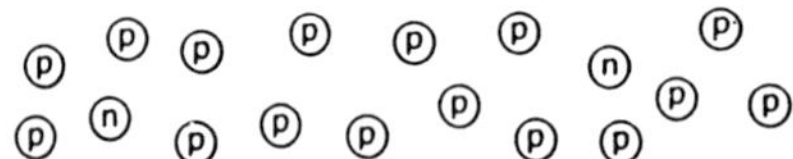

6. At three minutes after the big bang the temperature has dropped to about 10^9 degrees K., which is low enough so that the nucleus of the isotope deuterium (n,p) can stay bound. Deuterium is then rapidly processed into helium ($2n,2p$). Since helium requires equal numbers of p and n, helium formation ceases when all the available neutrons are used up:

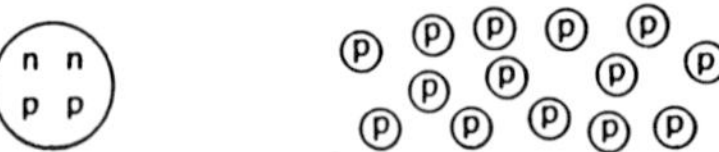

Since neutrons and protons are of almost equal mass, about 4/16, or 25 percent, of the mass of the universe ends up in helium, with 75 percent left over in protons (hydrogen nuclei).

7. The more families of neutrinos there are, the faster is the expansion rate of the universe. Step (4) therefore occurs earlier and at a higher temperature when more neutrons are present; steps (5) and (6), then, proceed in the presence of more neutrons, resulting in the formation of more helium. Astronomical observations, however, limit helium to less than 25 percent of the mass of the universe. This in turn indicates that there are no more than four neutrino families.

Figure 35. A simple way to understand the effect of N_ν on primordial element abundances (Ref.[11]).

The present e* mass limit, coming from the cross sections of Fig. 34 [28], is still rather poor (less than $\sqrt{s}/2$). A scaling law deduced from the above formula reads as follows:

$$\frac{\alpha}{m_{e^*}^4} s^{3/2} \sqrt{\int L dt} = \text{constant}$$

where m_{e^*} is the accessible mass for $\lambda = 1$. This shows that "patience does not pay" and that it is preferable to go to higher energies than insist on the Z.

A striking manifestation of the compositeness of the Z could be the observation of the decay $Z \to 3\gamma$. The expected branching ratio is $\sim 2 \; 10^{-4} Q^6$ where Q is the mean electric charge of the preons. The present limit of OPAL [29] is $\sim 5.2 \; 10^{-5}$. With more statistics one can improve it easily.

10   Conclusion

Many other searches have been performed, generally by all four experiments. For the sake of completeness one can consult the talk of F. Dydak at Singapore which gives all results available.

In fact the first result to come out from LEP, besides M_Z, was the number of light neutrinos. The present value is $N_\nu = 2.89 \pm 0.10$.

This accurate measurement is a good example of an indirect information on our universe. Further light neutrinos are excluded (and heavy ones for masses below ~ 40 GeV as well through the width measurements); sneutrinos are excluded. This information has strong consequences on our view of the post big bang evolution: Figure 35 [30] recalls how it tightly constrains quantities like the He abundance in the Universe. The need for non baryonic dark matter is reinforced. But at the same time, as we saw, searches at LEP could in the future exclude up to ~ 20 GeV one of the best candidates for such a dark matter, the LSP, making the situation still more exciting.

My own conviction is that searches with the guidance of the SM, SUSY, astrophysical ideas and in parallel with accurate measurements, should be a major concern at LEP until the end of its hopefully thorough exploitation.

Acknowledgements

I am grateful to many colleagues for fruitful discussions, in particular F. Richard and R. Barbieri. Previous reviews (G. Wormser, F. Richard, ...) were quite useful as well. I warmly thank the service of M. Jouhet for efficient typing and drawing.

After completion of this review the first numerical results on the effect of loops on SUSY Higgs masses came out [43]. This effect has indeed to be taken into account: it depends dramatically on the top mass.

References

[1] A. Djouadi et al., Preliminary report to the LEPC from the Working Group on High Luminosities at LEP (CERN, Geneva, May 1990), p.121.

[2] H$^\pm$ searches at LEP:
ALEPH, CERN-EP/90-34, Phys. Lett. **B241** (1990) 623,
DELPHI, CERN-EP/90-33 and Phys. Lett. **B241** (1990) 449,
L3, preprint # 18 (Sept. 90) to be published in Phys. Lett. **B**.
OPAL, CERN-EP/90-38 (March 90), and Phys. Lett. **B242** (1990) 299.

[3] M. Felcini, talk given at the DESY Theory Workshop, Hamburg (Oct. 90).

[4] OPAL, CERN-EP/89-154 (Nov. 89) and update for Singapore.

[5] Aachen Workshop on LEP 200, vol.2, p. 251.

[6] Light Higgs at LEP:
ALEPH CERN-EP/90-70 (May 90) Phys. Lett. **B245** (1990) 289
DELPHI CERN-EP/90-44
L3 Preprint # 19 (Sept. 90) to be published in Phys. Lett. **B**,
OPAL CERN-PPE/90-116 (Aug. 90) and Phys. Lett. **B251** (1990) 211.

[7] Higgs at LEP:
DELPHI CERN-EP/90-60 (May 90) updated for Singapore
L3 Preprint # 10 (June 90), Phys. Lett. **B248** (1990) 203,
OPAL EP/90-100 (July 90), PPE/90-150 (Oct. 90).

[8] ALEPH Higgs search: CERN-PPE/90-101 (July 90) Phys. Lett. **B246** (1990) 306.

[9] E.Glover et al., CERN TH 5584/89.

[10] F. Gilman, Comments Nucl. Part. Phys. 16 (1986) 231.

[11] D. Cline, Scient. American (Aug. 85).

[12] ALEPH, Phys. Lett. **B236** (March 90) 4.

[13] S. Ting, presented at Singapore conference, and L3 Preprint # 16 (Aug. 90), Phys. Lett. **B251** (1990) 321.

[14] OPAL results on Heavy Neutral Leptons, CERN-EP/90-72 (May 90) and Phys. Lett. **B247** (1990) 448.

[15] Charged leptons at LEP:
ALEPH Phys. Lett. **B236** (1990) 511.
DELPHI CERN-EP/90-80, 1990.
L3, Preprint # 16 (Aug. 90), Phys. Lett. **B251** (1990) 321.
OPAL preprint CERN-EP/90-09, PPE/90-132 (Sept. 90) and Phys. Lett. **B240** (1990) 250.

[16] H. Nilles, Phys. Rep. **C10** (1984) 1.
R. Barbieri, Riv. Nuovo Cimento **11** (1988).

[17] R. Barbieri, in Physics at LEP I, CERN 89-08, editors G. Altarelli,
R. Kleiss, C. Verzegnassi vol. 2.

[18] G.F. Giudice, Phys. Lett. **B208** (1988) 315.

[19] LEP results on SUSY Higgs:
ALEPH CERN-EP/90-70 (May 90) Phys. Lett. **B237** (1990) 291,
L3 preprint # 15 (Aug. 90), Phys. Lett. **B251** (1990) 311,
OPAL CERN-EP/90-100 (July 90),
DELPHI CERN-EP/90-60 (May 90) updated.

[20] Neutralinos at LEP:
ALEPH CERN-EP/90-63 (May 90) Phys. Lett. **B244** (1990) 541,
DELPHI CERN-EP/90-80, 1990.
L3 Preprint # 15 (Aug. 90), Phys. Lett. **B251** (1990) 311,
OPAL PPE 90-95 (July 90) and Phys. Lett. **B248** (1990) 211.

[21] A. Roussarie, talk at Singapore for the ALEPH collaboration.

[22] LEP and the Universe: CERN TH 5709/90 (April 90).

[23] Charginos at LEP:
ALEPH CERN-EP/89-158 (Dec. 89) Phys. Lett. **B236** (1990) 86,
L3 preprint # 002, Phys. Lett. **B233** (1989) 530,
DELPHI CERN-EP/90-80 (June 90).
OPAL CERN-EP/89-176 (Dec. 89), Phys. Lett. **B240** (1990) 261.

[24] Sleptons at LEP see ref. [23].

[25] DELPHI, CERN EP/90-79 (June 90).

[26] Compositeness at LEP 2: preprint CERN-EP/87-50 (March 87).

[27] Excited leptons at LEP:
ALEPH Phys. Lett. **B236** (1990) 501, and PPE 90-107 (Aug. 90) Phys. Lett. **B250** (1990) 172,
L3 Preprint, # 007 (June 90) Phys. Lett. **B247** (1990) 177, L3 # 0014 (Aug. 90) Phys. Lett. **B250** (1990) 205, L3 # 0021 (Oct. 90) to be published in Phys. Lett. **B**,
OPAL CERN EP/90-49 (April 90) and Phys. Lett. **B244** (1990) 135.

[28] ALEPH Phys. Lett. **B241** (May 90) 635,
L3 preprint # 013 (Aug. 90) Phys. Lett. **B250** (1990) 199,
OPAL CERN EP/90-29 (Feb. 90) and Phys. Lett. **B241** (1990) 133.

[29] OPAL CERN-EP/90-29, updated.

[30] See Ref. [11].

[31] Physics at LEP 200, Aachen, CERN 87-08.

[32] Polarization at LEP, CERN 88-06.

[33] High luminosity at LEP, Preliminary report to the LEPC.

[34] D. Treille, preprint CERN-EP/90-30, March 1990.

[35] P. Mättig, preprint CERN-EP/90-71.

[36] M. Placidi, Int. Symposium on High Energy Spin Physics, Bonn, September 1990.

[37] R. Barbieri and F. Zwirner, private communication.

[38] C. Arnaud et al., preprint CERN/80-AF/90-06, presented at the EPAC Conference, Nice (1990).

[39] J. Jowett, More bunches in LEP, CERN LEP TH/89-17.

[40] J. Drees et al., CERN 88-06, Vol. 1, p. 317.

[41] E. Lieb et al., DELPHI 90-44, Phys. 71 (1990).
See also Ref. [5].

[42] P. Roudeau, LAL 89-21 (1989).
C. Defoix, DELPHI 90-40, Phys. 67 (1990).
H.G. Moser, MPI preprint Exp. 209.

[43] Y. Okada et al., TU-360, Oct. 1990 (+Erratum).
J. Ellis et al., CERN TH 5946, Nov. 1990,
R. Barbieri et al., IFUP-TH 46/90, Dec. 1990.

A Appendix 1

At LEP 1 Workshop (CERN 86-02, p. 297) a very general formula was given, which applies to the production of any spin 1/2 fermion–antifermion pair, or of a pair of charge-conjugate spin 0 bosons produced via γ or Z^0 exchange in the s-channel.

The cross-section reads as

$$\sigma_0^f \left(Q, T_{3L} T_{3R}, \beta, s\right) =$$

$$\frac{4\pi\alpha^2}{3s}\beta \times \left[Q^2 \left\{ \frac{3-\beta^2}{2} \right\} - 2QC_V C_V' \frac{s\left(s-m_Z^2\right)}{\left(s-m_Z^2\right)^2 + m_Z^2 \Gamma_Z^2} \left\{ \frac{3-\beta^2}{2} \right\} \right.$$

$$\left. + \left(C_V^2 + C_A^2\right) \frac{s^2}{\left(s-m_Z^2\right)^2 + m_Z^2 \Gamma_Z^2} \left\{ C_V'^2 \left\{ \frac{3-\beta^2}{2} \right\} + C_A'^2 \left\{ \beta^2 \right\} \right\} \right] \tag{1}$$

where

$$C_V = \frac{1 - 4\sin^2\theta_W}{4\sin\theta_W \cos\theta_W}, \quad C_A = \frac{1}{4\sin\theta_W \cos\theta_W} \tag{2}$$

are the vector and axial-vector couplings of the electron, and

$$C_V' = \frac{-2\left(T_{3L} + T_{3R}\right) + 4Q\sin^2\theta_W}{4\sin\theta_W \cos\theta_W}, \quad C_A' = \frac{2\left(T_{3L} + T_{3R}\right)}{4\sin\theta_W \cos\theta_W} \tag{3}$$

are those of the produced fermion f, whose left- and right-handed components have third components of weak isospin T_{3L} and T_{3R}, respectively. The centre-of-mass energy $E_{cm} = \sqrt{s}$ and $\beta = \left(1 - 4m_f^2/s\right)^{1/2}$ is the centre-of-mass velocity of the produced fermion. The term $\propto C_A'^2$ has a different form from the others because the axial current can only generate fermions in a P-wave, not in an S-wave.

For s-channel pair production of spin-zero particles such as sleptons, squarks, Higgs bosons, we get the special case

$$\sigma_0^s \left(Q, T_3, \beta, s\right) = \frac{\beta^3}{4} \sigma_0^f \left(Q, T_3, T_3, 1, s\right). \tag{4}$$

The main corrections are radiative corrections. The initial-state electromagnetic one yields

$$\frac{\partial\sigma}{\partial x_\gamma} = \frac{\alpha}{\pi} \left(\ell n \frac{s}{m_e^2} - 1\right) \frac{1 + \left(1 - x_\gamma\right)^2}{x_\gamma} \sigma_0 \left[s\left(1 - x_\gamma\right)\right], \tag{5}$$

where $x_\gamma \equiv 2E_\gamma/E_{cm} = E_\gamma/E_{beam}$ and σ_0 is the total leading-order cross-section (A1) or (A4).

The final-state QCD correction leads to

$$\sigma_{QCD} = N\sigma_0 \left(Q, T_{3L}, T_{3R}, \beta, s\right) F_{QCD}, \tag{6}$$

$N = 1$ for lepton, 3 for colour triplets etc... $F_{QCD} = 1 + C\alpha_s\left(\beta\right)$ at leading order. Details can be found in the Yellow Book.

A Appendix 2

At very high scale (Planck scale?) M_x one assumes a Lagrangian

$$L = L_{\text{SUPERSYM}} + L_{\text{BREAKING}} \tag{7}$$

$$L_{\text{SUPERSYM}} = L_{\text{SUPER}}(\text{SU}(3) \times \text{SU}(2) \times \text{U}(1); f)$$

where

$$f = f_Y + \mu\, H_1 H_2. \tag{8}$$

f_Y is the Yukawa part; μ governs the mass coupling between Higgses.

The symmetry breaking Lagrangian can be written as:

$$m^2 \sum_i |\phi_i|^2 + M \sum_\alpha \overline{\lambda}_\alpha \lambda_\alpha + (A m f_Y + B m \mu H_1 H_2 + hc) \tag{9}$$

where m is a universal mass term for scalars and M a universal mass term for gauginos.

When one goes down to our energies, there are effects of renormalization (such $\mu \to \mu_R$) and the gaugino–higgsino sector becomes;

$$L = 1/2 M_3 \overline{\lambda}_a \lambda_a + 1/2 \overline{\chi} M^{(0)} \chi + \left(\psi M^{(c)} \overline{\psi} + \text{h.c.} \right) \tag{10}$$

The first term describes 8 gluinos. The second term describes the neutralino sector: $M^{(o)}$ is a 4×4 matrix whose parameters are M_1, M_2, β, μ_R. The third term describes the chargino sector: $M^{(c)}$ is a 2×2 matrix whose parameters are M_2, β, μ_R.

We have defined

$$M_i = M \alpha_i (M_w) / \alpha_i (M_x), \tag{11}$$

$$\text{tg}\,\beta = v_2/v_1, \quad \text{and} \quad \alpha_i = g_i^2/4\pi. \tag{12}$$

Indices 1, 2, 3 are for $\text{U}(1)_Y$, $\text{SU}(2)_L$, $\text{SU}(3)_C$, respectively. M_2 is the $\text{SU}(2)_L$ gaugino mass, M_1 is the $\text{U}(1)_Y$ gaugino mass, $m_{\tilde{g}}$ is the $\text{SU}(3)_C$ gaugino mass.

All three masses are related, from Grand Unification, by

$$M_1 = \frac{5}{3} \frac{\alpha_1}{\alpha_2} M_2 = \frac{5}{3} \text{tg}^2 \theta_w M_2$$

$$M_2 = \frac{\alpha_2}{\alpha_3} m_{\tilde{g}} = 0.3 m_{\tilde{g}}$$

Charginos are a mixing of Winos ($\tilde{W}^\pm$) and Higgsinos ($\tilde{H}^\pm$). The matrix $M^{(c)}$ reads as:

$$\left[\begin{matrix} M_2 & \sqrt{2} M_w \sin\beta \\ \sqrt{2} M_w \cos\beta & \mu \end{matrix} \right] \begin{matrix} \tilde{W} \\ \tilde{H} \end{matrix}$$

Neutralinos are a mixing of photino $\tilde{\gamma}$, zino $\tilde{Z}^0$ and Higgsinos $\tilde{H}^0$, $\tilde{h}^0$.

The matrix $M^{(o)}$ reads as

$$\left| \begin{matrix} M_1 & 0 & -M_Z S c & -M_Z S s \\ . & M_1 & M_Z C c & M_Z C s \\ . & . & 0 & -\mu \\ . & . & . & 0 \end{matrix} \right| \begin{matrix} \tilde{B} \\ \tilde{W}_3 \\ \tilde{H}_0 \\ \tilde{H}^0 \end{matrix}$$

where $S = \sin\theta_w$, $C = \cos\theta_w$, $s = \sin\beta$, $c = \cos\beta$.

A Appendix 3: Possibilities for the Future: A Summary

I would like to summarize here the main possibilities for the future of LEP. Details can be found in the Proceedings of various workshops [31]–[33] and other publications [34,35].

One may go along three directions:

i) increasing the energy, an already approved option, whose objectives (max $\sqrt{s}$, $\int L\,dt$) need to be more precisely defined;

ii) increasing the luminosity, an option technically related to the previous one;

iii) increasing the accuracy of the SM tests, by getting longitudinal polarization in LEP [32]. This promising option, received a strong encouragement from the recent observation [36] of transverse polarization.

Accurate measurements are the 'guaranteed' part of the LEP programme. Figure 36 summarizes the prospects: option (i) will give access to $m_{\rm W}$ with an accuracy at least equal to and probably better than what the hadron colliders will have achieved.

Options (ii) and (iii) will have major impacts on the third corner of the triangle (Fig. 36): the measurement of electroweak couplings.

The global potential physics output of the three options is shown in Table 1.

Table 1

		STAT.	SYST.
Polarization	$-A_{\rm LR}$ $-A_{\rm CH}^{\rm POL}$	~ 1 MZ	OK from experimental point of view, problems are on the machine side
High L	−Accurate measur. $A_{\rm CH}^{f\bar{f}}$, $\Gamma^{f\bar{f}}$,... −rare modes of Z, searches −B physics	several 10^7Z	heavy work ahead
LEP 200	−WW physics($M_{\rm w}$,...) −accurate measur. −searches	meager, luminosity as well as $\sqrt{s}$ are vital	

A.1 The Energy Increase

The most important fact to be kept in mind is the smallness of the cross-sections beyond the Z peak. At $\sqrt{s} = 190$ GeV the point-like cross-section is ~ 2 pb and the WW pair cross-section is ~ 17 pb. The main background is ee $\rightarrow$ Zγ (~ 70 pb); one should not forget either the quasi-symmetric radiative return to the Z, ee $\rightarrow$ Z($\gamma\gamma$) (~ 10 pb), which can be dangerous for some searches.

A standard Higgs of 90 GeV is produced with a cross-section of ~ 0.45 pb at $\sqrt{s} = 190$ GeV and the threshold is at $\sqrt{s} \sim 180$ GeV. One can remember that to get the full production rate for a Higgs of $m_{\rm H}$ one needs

$$\sqrt{s} \cong m_{\rm H} + m_{\rm Z} + (10\,{\rm to}\,15\,{\rm GeV}).$$

Whilst a mass $m_{\rm H} \sim m_{\rm Z}$ has nothing magic for the Standard Higgs, for the neutral Higgses of Minimal Supersymmetry, the vicinity of $m_{\rm Z}$ is an 'accumulation point', where at least one of them *has to be*[**]. The anticipated total production rate of these scalars always exceeds the production rate of the would-be Standard Higgs of $m_{\rm H} = m_{\rm Z}$, provided that there is no kinematical suppression [37] (i.e. for $\sqrt{s} > 190$ GeV, see Fig. 37).

An energy of $\sqrt{s} > 190$ GeV[**] and an integrated luminosity $\int L\,dt \geq 500$ pb^{-1} would allow us to achieve the following:

i) To bring important information on the WW pair physics: $m_{\rm W}$, W properties and tests of the three-boson couplings. This has been studied in detail in Ref.[31] and the improvement of the test with increasing $\sqrt{s}$ is explicited by Fig. 38.

[**]Radiative corrections could in fact push the Higgs masses somewhat higher; this is under study. One would then need a corresponding increase in $\sqrt{s}$. This shift depends critically on the top mass.

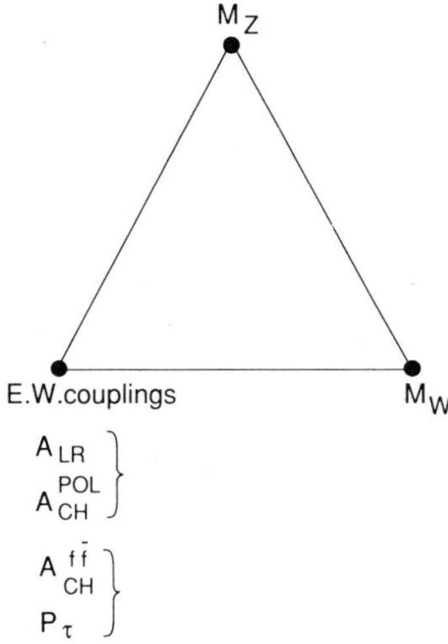

Figure 36. Symbolic representation of the impact of accurate measurements at LEP 200.

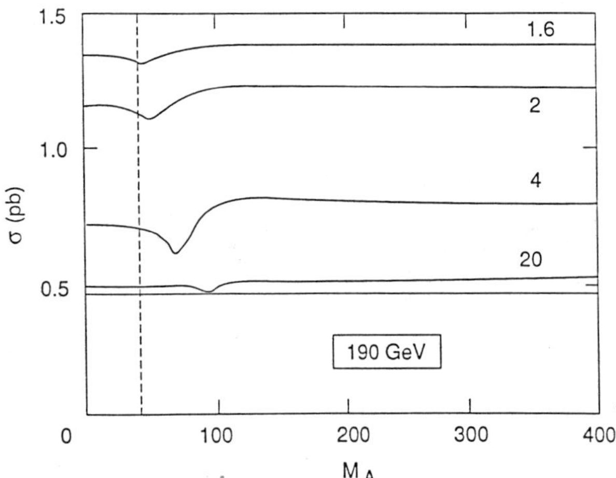

Figure 37. The possibility of excluding all MSUSY parameters plane at LEP 200 ($\sqrt{s} \geq 190$ GeV, $L \geq 500$ pb^{-1}). The curves show for various tg β the sum of the cross-sections for h^0 and H^0 at $\sqrt{s} = 190$ GeV. It is always higher than the cross-section of the standard Higgs ($m_{\rm H} = m_{\rm Z}$) at that energy. A 5σ signal is obtainable with 500 pb^{-1}. (Courtesy of R. Barbieri.)

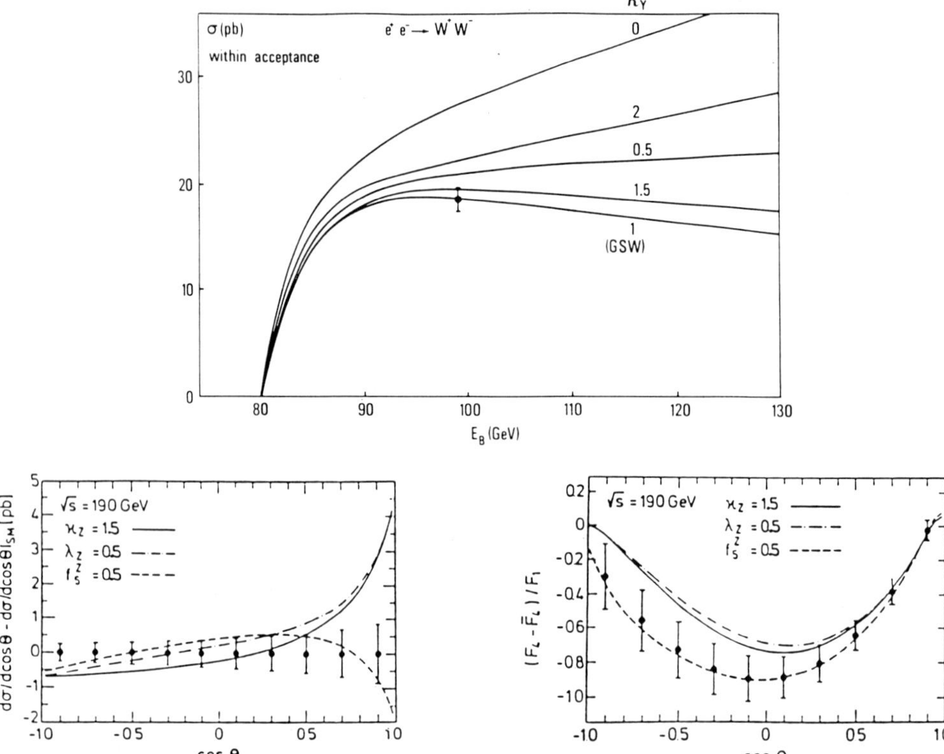

Figure 38. Tests of anomalous W couplings at LEP 200.

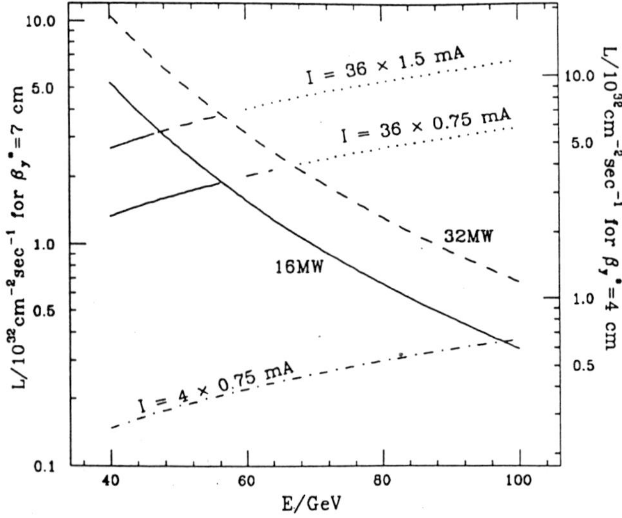

Figure 39. Luminosity versus $\sqrt{s}$ for given beam power values.

ii) To explore the MSUSY neutral Higgs sector. MSUSY (the next theory after the SM in order of growing complexity) and most extensions from it can be tested there in a severe way, since, as we said, a light scalar should be present not far away from M_Z. A good b tagging is a key factor for such a programme. Similar tests at future hadron colliders, the detection of $[H(\sim 90\ \mathrm{GeV}) \to \gamma\gamma]$ are, to say the least, not easy ones.

Other very interesting measurements would be performed, e.g. the fermion asymmetries $A_{CH}^{f\bar{f}}$ at full energy which are major ingredients of the strategies of indirect searches. Various direct searches would push to $\sim \sqrt{s}/2$ the mass limit of pair-produced objects (charginos, sleptons).

Technically, the key point is the performance of a large set of superconducting cavities. Preliminary results [38] on a bunch of them are encouraging. Table 2 gives the energy achievable as a function of the number of cavities for various assumptions on their accelerating fields. Only under the most favourable conditions (7 MV/m, all cavities working altogether) can one approach (but not reach) the desirable $\sqrt{s}$ with the presently approved 192 cavities.

Table 2. Energy that can be achieved as a function of the number of SC cavities (warm cavities being removed); current that can be accelerated as a function of available power.

No. of	MV/m		
cavities	5	6	7
192	~ 84.1		~ 91.4
256	~ 89.6	~ 93.8	~ 98
Beam	Beam energy		
power	84	90	95
16 MW	~ 6 mA	~ 4 mA	~ 3 mA
32 MW	~ 12 mA*	~ 8 mA	~ 6 mA**

* i.e. 8 bunches of 1.5 mA or more bunches
** i.e. 8 bunches of 0.75 mA

The luminosity is obviously of outmost importance. Figure 39 shows, under two assumptions on β^* (i.e. achievable beam focusing), the luminosity expected as a function of the available beam power. A safe way to gain a factor 2 in luminosity is to get 8 bunches in LEP instead of 4. With 8 bunches and the nominal current per bunch (0.75 mA) at $\sqrt{s} = 190$ GeV, 32 MW are needed. The luminosity would be $(5\text{--}10) \times 10^{31}$ cm^{-2} s^{-1} and the time needed to accumulate 500 pb^{-1} would then be ~ 2 to 3 years.

A.2 The Luminosity Increase

Figure 39 shows that if a beam power P is available the luminosity achievable at a given beam energy E varies like

$$L \sim \frac{P}{E^3}.$$

This increase of L can be achieved by a multiplication of the number of bunches n_b in each beam. The pretzel scheme [39], with an induced separation of the beams (Fig. 40), offers such a possibility. Various conditions to be fulfilled lead to

$$n_b = 2, 4, 8, ..., 18, ..., 36, 40$$

as viable options. Taking $n_b = 8$ is an easy choice, which could be readily accepted by the experiments and requires the introduction in the machine of eight new separators. Since $n_b = 36$ is quite demanding from the experiments (although not impossible), $n_b = 18$ appears as the safe upper limit, likely to provide an increase of a factor of 4 in luminosity.

From the machine point of view, however, more studies and machine developments are needed before one can arrive to a real assessment and an optimized solution for a pretzel scheme.

Let us note that from the physics point of view the introduction of horizontal separators, because it leaves the door open to more bunches and because it is *a priori* harmless for polarization, is preferable to the mere mid-arc vertical separation, suggested for the eight-bunch scheme.

Such an increase in luminosity would make it possible to obtain an exposure of $N \simeq 25 \times 10^6$ Z in 2–3 years. Three main topics would greatly benefit from this high statistics (the first two of them being missed for ever if LEP misses them):

a) accurate measurements of SM parameters,

b) rare decays of the Z,

c) fermion–antifermion physics, especially $b\bar{b}$.

a) SM parameters

In this domain we already said that a measurement of A_{LR} should give the most accurate determination of $\sin^2\theta_w$ (Fig.40). However, the availability of longitudinal polarization at a sufficient level ($P \sim 40$–50%) is not yet guaranteed and one may have to look for an alternative way by improving the measurements of $A_{\mathrm{CH}}^{\ell^+\ell^-}$, P_τ, $A_{\mathrm{CH}}^{q\bar{q}}$ and combining them. Furthermore high statistics gives access to better determinations of other quantities such as $\Gamma^{f\bar{f}}$, i.e. information not contained in a polarization programme of 10^6 Z.*

In particular, high luminosity allows us to use, for quark tagging, besides the single-arm method [40] double-tag procedures [41], which are more demanding in statistics but less sensitive to systematic errors (although they still require a good measurement of the contamination). A strategy of flavour tagging with all steps based only on measurements (and not on Monte Carlos) can be devised. It is admitted that the combination of various measurements of unpolarized quantities, can bring with 25 $\times 10^6$ Z an accuracy on $\sin^2\theta_w$ comparable to the one from the standard polarization programme. An accuracy of $\sim 2\%$ can also be reached on $\Gamma_{b\bar{b}}$.

b) Rare decays of the Z

The most outstanding classical decay involves the Higgs boson. While $Z \to H\gamma$ is inaccessible at its SM value because of $ee \to q\bar{q}\gamma$ background, one can possibly push the $Z \to Z^*H$ up to ~ 60 GeV: however, at some stage it is more effective to move to LEP 200 and search for the Higgs scalar in:

$$ee \to Z^* \to ZH\,.$$

Other possible rare decays, if they are observed at rates above the SM expectations, would reveal new physics. Their observation would be of the utmost importance, but we have no guarantee at all that they will show up. Among them, several have been demonstrated to be experimentally accessible down to very low branching ratios such as $Z \to 3\gamma$ (nearby compositeness), $Z \to e\tau, \mu\tau$ (FCNC), etc. [33].

c) Fermion–antifermion physics

Tau physics certainly deserves a special study: besides P_τ and $A_{\mathrm{CH}}^{\tau\tau}$ basic measurements, a general study of τ Lorentz parameters (Michel parameter, chirality parameters, etc.) can be envisaged through the complete measurement of the simple final state

$$ee \to \tau \qquad \tau$$
$$\qquad\ \ \llcorner\!\!\to A+\ldots \quad \llcorner\!\!\to B+\ldots$$

$A, B = \pi, \ell, \rho$, etc.

The competition of eventual τ (and b) factories is obvious and for some aspects (τ properties, $m_{\nu_\tau},\ldots$) completely overwhelming.

Beauty physics is by far the most promising sector. Table 3 summarizes the outstanding features of the Z as a beauty factory.

Table 3. Properties of B production at LEP

Cross-section: $\sigma_{b\bar{b}} = 6.5$ nb.
Percentage, relative to the hadronic Z modes:
$\sigma_{b\bar{b}}/\sigma_{\mathrm{had}} = 0.22$.
Percentage, relative to the visible Z modes:
$\sigma_{b\bar{b}}/\sigma_{\mathrm{vis}} = 0.19$.
Population of various species, from 100 million Z:

$15.5 \times 10^6\ B^0$	$15.5 \times 10^6 B^+$
$4.5 \times 10^6\ B_s^0$	$1.7 \times 10^6\ \Lambda_b$
	$0.35 \times 10^6\ \Xi_b$

Mean number of charged particles per B: ~ 5.
Mean number of charged particles at the primary vertex: ~ 10.
Mean flight path of B: ~ 2.2 mm.

---

*If one day high luminosity *and* polarization are available (a polarized pretzel) so much the better!

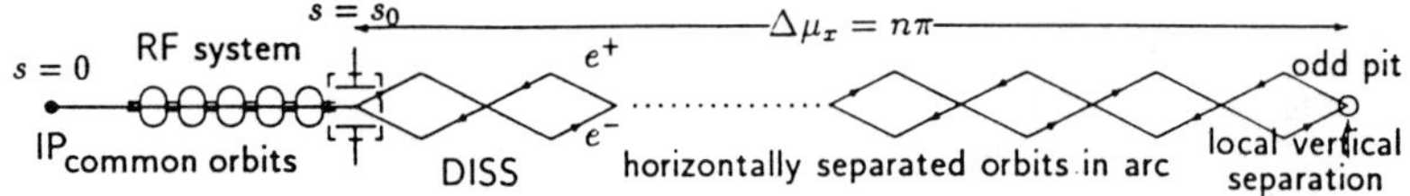

Figure 40. The pretzel scheme.

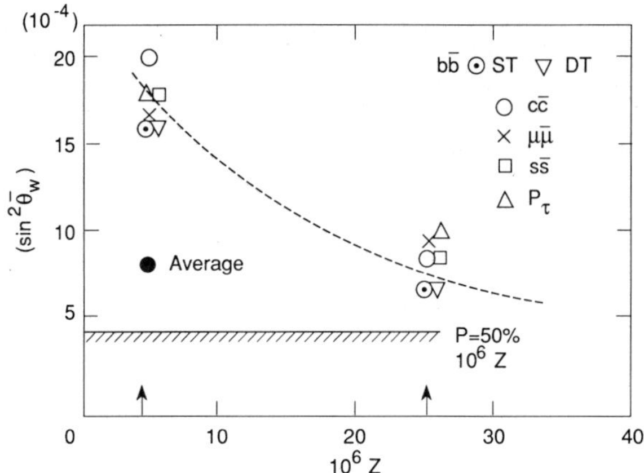

Figure 41. Accuracies on $\sin^2\theta_w$ from polarized and unpolarized measurements.

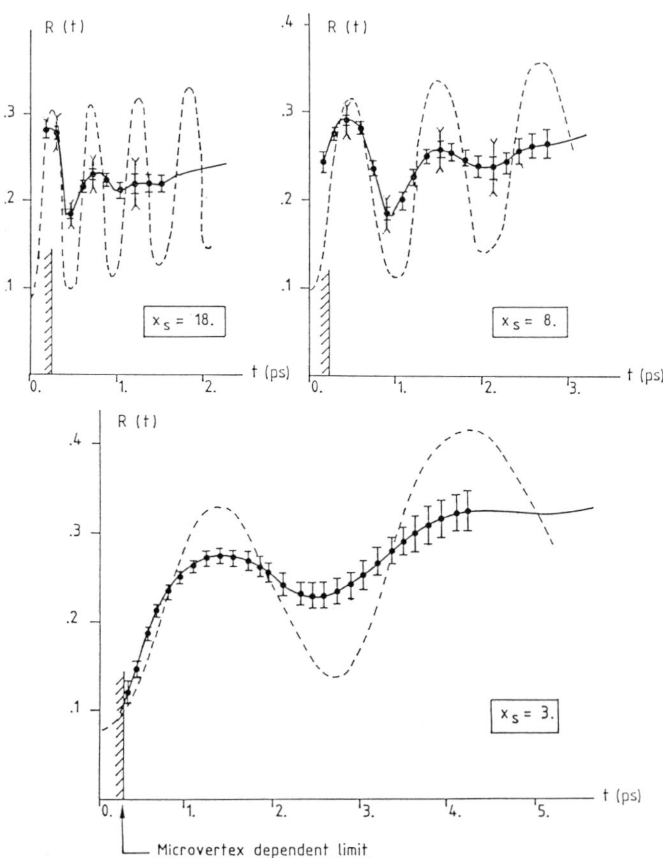

Figure 42. Oscillation pattern for B_{0s} mixing. The error bars are for 100×10^6 Z; a few larger ones are indicated for 25×10^6 Z.

'One-arm' measurements (lifetimes, rare modes of the B, spectroscopy, ...) are quite accessible, but prone to competition (B factories, etc.). In particular, it is becoming clear that the Tevatron collider program under way can probably do a lot in this respect, especially through the $B \to \psi$ ... modes.

However, for 'double-arm' measurements where one B has to be reconstructed and the other one tagged (B oscillations, etc.) I think LEP has great advantages. At the collider the probability to have access to the other b is tiny; at LEP one knows exactly where to look and what to look for in a very clean environment.

One of the main concerns of LEP experiments should be (and indeed is) b tagging: high efficiency and purity should be very rewarding, both for searches ($H \to b\bar{b}$) and for B physics itself. Good microvertices and a clever exploitation of the special features of b fragmentation should make this programme possible with $\sim$ 50% efficiency for $b\bar{b}$ tagging.

Figures 42 and 43 show what one could achieve for instance in the field of B_s mixing [42], a specially rewarding one since it gives access to the phase of the KM matrix, supposed to be the key to the CP problem (Fig. 44). In such a domain LEP with high luminosity should be better than any other competitor.

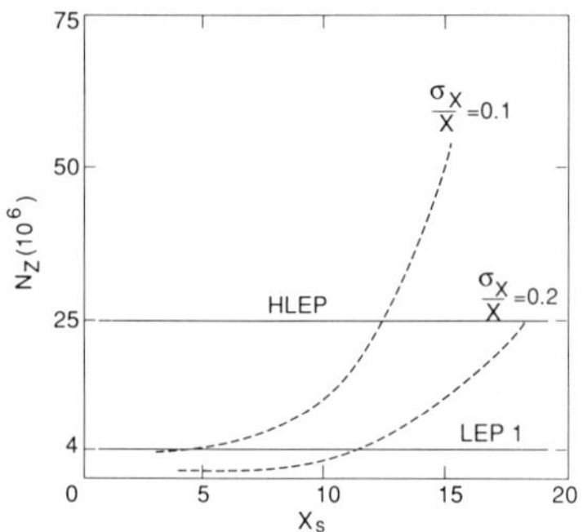

Figure 43. Number of Z needed to measure B_{0s} mixing

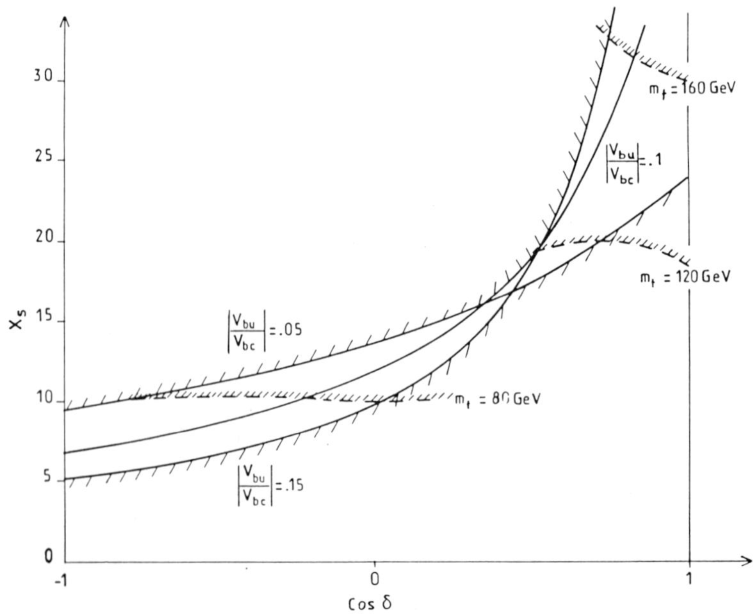

Figure 44. Access to KM phase through the measurement of X_s

POLARIZATION at LEP

Alain Blondel

L. P. N. H. E., Ecole Polytechnique, 91128 Palaiseau Cedex, France

Abstract

Evidence for beam polarization has been recently observed in the Large
Electron Positron Collider (LEP) at CERN. The future of polarized beam
experiments in LEP is reviewed. The main physics motivations, and
foreseeable difficulties, are summarized.

1   Introduction

The Large Electron Positron Collider (LEP) [1] at CERN has had an extraordinarily
successful start. Since middle of August 1989, about 700000 decays of the Z boson
have been observed in the four experiments ALEPH, DELPHI, L3 and OPAL. The
most striking result is that nothing has been found yet that contradicts the Standard
Model of Electroweak interactions. The excitation curve of the resonance has been
measured [2],[3],[4],[5]; an example is shown in figure 1a. The main outcome is a
measurement of the number of light neutrino species, $N_\nu = 2.90 \pm 0.10$ (average
over the four experiments), which rules out the possibility of further families of the
same type as those already known. The measurement of the Z mass, m_Z, fixes
a fundamental parameter of the Electroweak Standard Model. Many searches for
new particles [6] have been performed, unfortunately without success, ruling out in
particular the existence of the Standard Model Higgs boson for masses between 0
$< m_H <$ 44 GeV. Finally, the effective weak mixing angle, $sin^2\vartheta_w(m_Z)$, [7], [8], [9],
has been measured from the electronic partial width of the Z, $\Gamma_{ee} = 83.7 \pm 0.7$ MeV, as

Z° Physics - Cargèse 1990, Edited by M. Lévy *et al.*
Plenum Press, New York, 1991

well as from forward-backward asymmetries, to be $sin^2\vartheta_w(m_Z) = 0.230 \pm 0.002$ [10]. The measurements of m_Z, $sin^2\vartheta_w(m_Z)$ and of the ratio of W and Z masses can be combined to set limits on the mass of the top quark, as shown in figure 1b.

In view of this great start, some optimism is in order. One of the obvious tasks for the future is to improve the precision of our measurements, to reach the level where significant tests of the Standard Model can be performed. Transverse and longitudinal polarization are tools of choice in this endeavor.

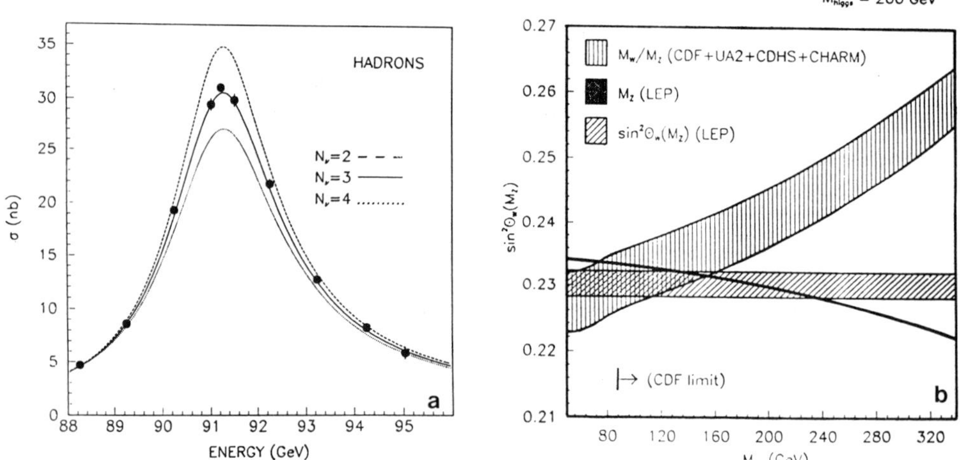

Figure 1. *First results from LEP.*
a)The Z line shape, as measured by the ALEPH collaboration.
b)Electroweak measurements setting the top quark mass to 137±40 GeV.

2 Physics with Polarized Beams

2.1 Transverse Polarization

Transverse Polarization builds up in a storage ring by the Sokolov-Ternov effect[11]: synchrotron radiation emission has a small spin-flip probability, with a large asymmetry in favor of orienting the particles' magnetic moment along the guiding magnetic field. In a perfect machine a large asymptotic transverse polarization (92.4 %) builds up slowly. In LEP at 45.5 GeV per beam, the polarization time is 330 minutes. In a real machine, depolarizing resonances occur, reducing in the same ratio the asymptotic degree of polarization P_∞ and its effective raise time τ_p^{eff}. More specifically, a depolarizing time τ_d competes with the polarization time τ_p so that:

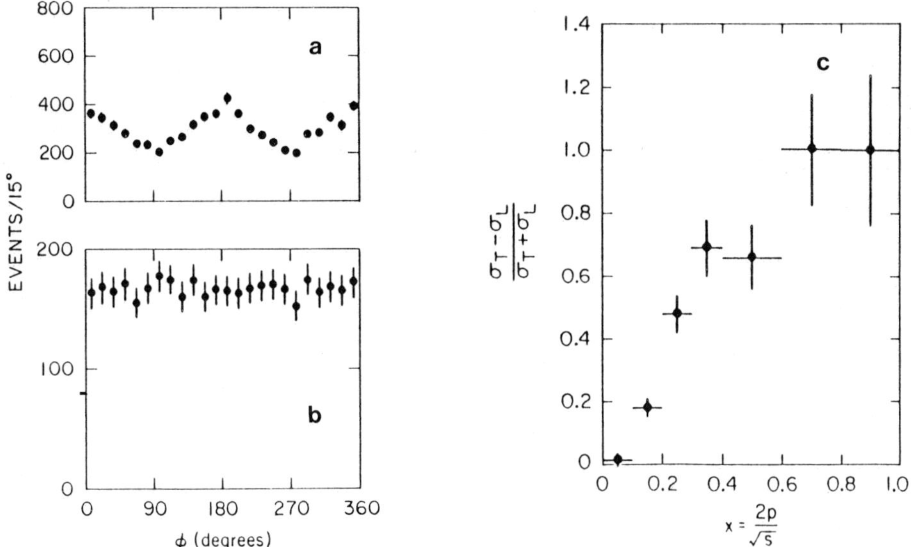

Figure 2. Transverse polarization asymmetry at SPEAR.
a) Observed hadron yield for all particles with x > 0.3 at √s=7.4 GeV;
b) Same at the spin-depolarization resonance at √s=6.2 GeV;
c) Transverse polarization asymmetry as a function of particle momentum, showing leading particle effect.

$$P_\infty = 0.924 \times \frac{1}{1 + \frac{\tau_p}{\tau_d}}, \tag{1}$$

$$\tau_p^{\text{eff}} = \tau_p \times \frac{1}{1 + \frac{\tau_p}{\tau_d}}. \tag{2}$$

Transverse Polarization has been observed in every electron storage ring where it was searched for.

A few beautiful physics experiments were done. At SPEAR, observation of the transverse polarization asymmetry [12] –an azymuthal modulation of particle production, dependent on the spin of the particle– was a clear confirmation of the conjecture that jets of hadrons originate from the production of spin $\frac{1}{2}$ partons, fig.2. This experiment was performed with over 70% beam polarization in collision mode.

Extremely precise calibration, to one part in 10^5, of the beam energy can be done by exciting an artificial depolarizing resonance, using the electrons spins in much the same way as the nuclear spins in a Nuclear Magnetic Resonance probe. The spin precession frequency is related to the energy of the particles by the anomalous magnetic moment of the electron $(g_e\text{-}2)/2$ which is known to one part in 10^{11}. By applying to the beam a small RF field at the proper frequency one can create a resonance with spin precession frequency and destroy polarization completely.

Accurate measurements of the the masses of the J/ψ, ψ', Υ, Υ' resonances, have been performed using this technique, at VEPP4 in Novosibirsk (U.S.S.R.) [13], [14], DORIS in Hambourg (Germany) [15], and CESR in Cornell (U.S.A) [16]. Fig. 3 shows the observation of depolarization in VEPP4 and DORIS. For all these experiments, the measurement of energy was performed while the accelerator was in normal colliding conditions, with detectors taking data, polarization degree of up to 80% and luminosities approaching half the peak luminosity.

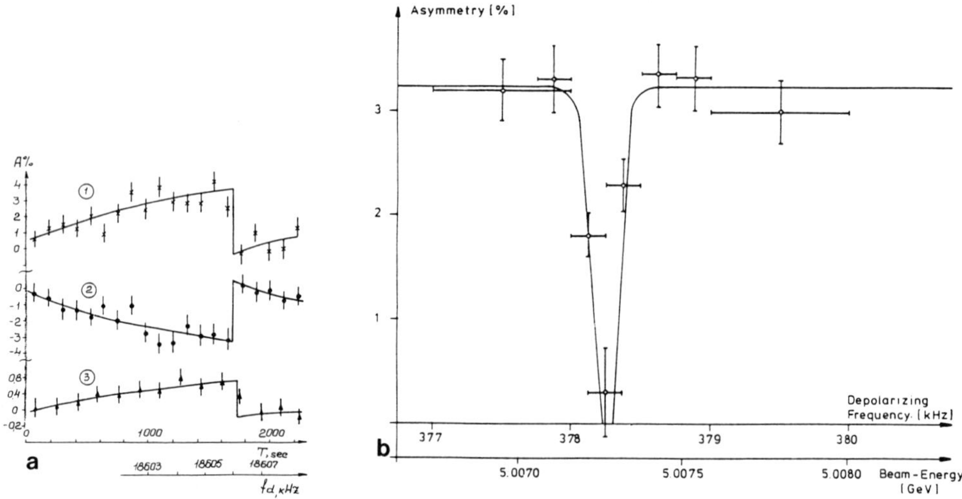

Figure 3. *Examples of depolarization curves: the beam polarization is measured as an up-down asymmetry of backscattered light on the polarized beam. A small, tunable, RF magnetic field is applied on the beam, and polarization measured as a function of its frequency.*
a) at VEPP4 in Novosibirsk [14]; 1 and 2: backscattering of synchrotron radiation light from the e^+ beam on the e^- beam and vice versa; 3: backscattering of polarized laser light on the e^- beam; $\tau_p = 50$ minutes, $P_\infty = 80\%$.
b) at DORIS [15] with backscattered laser light; $\tau_p = 4$ minutes, $P_\infty = 80\%$, luminosity up to $1.5 \; 10^{31}$ cm^2/s.

The Z boson, so copiously produced at LEP, is the only known elementary resonance. The Z mass, a fundamental parameter, is currently measured to be 91.177 ± 0.021 GeV [10]; the error is determined by present uncertainties in the beam energy [17]. In a first step [20], with polarization available in special runs only, this uncertainty could decrease to about $\pm$ 5 MeV. A precision of ± 1 MeV could be obtained if polarization were available at the same time as collisions. Similar arguments hold for the Z width, providing a sensitive test of the theory.

2.2 Longitudinal Polarization Effects

In the energy range of LEP, weak interactions dominate. The Standard Electroweak Model $SU(2)_L \times U(1)$ being left-right asymmetric, helicity effects are expected to play an important role. This is in contrast with lower energy e^+e^- machines where the QED-dominated physics leaves limited interest for longitudinally polarized beams.

The early LEP studies were quite aware of this fact [18]. The exact benefits were not realized, however, and the difficulties in obtaining polarized beams seemed overwhelming. A review of the subject was been published in 1988 [19].

Longitudinal Polarization effects manifest themselves in many reactions. Most of all, longitudinally polarized beams allow measurements of the weak couplings of fermions to the Z in a clean and precise way. Helicity effects in $e^+e^- \to Z \to f\overline{f}$ are sketched in figure 4.

The fact that the Neutral Current couplings are different for left-handed and right-handed fermions leads to polarization and forward-backward asymmetries which can be expressed in terms of the coupling asymmetries:

$$\mathcal{A}_f \equiv \frac{g_{Lf}^2 - g_{Rf}^2}{g_{Lf}^2 + g_{Rf}^2} = \frac{2v_f a_f}{v_f^2 + a_f^2} \tag{3}$$

The couplings are related to the weak mixing angle $sin^2 \vartheta_w(m_Z)$ by the well known relations:

$$g_{Lf} = I_{3f} - Q_f . sin^2 \vartheta_w \tag{4}$$

$$g_{Rf} = -Q_f . sin^2 \vartheta_w \tag{5}$$

$$a_f = 2(g_{Lf} - g_{Rf}) \tag{6}$$

$$v_f = 2(g_{Lf} + g_{Rf}) \tag{7}$$

The values of Neutral Current couplings and their sensitivity to $sin^2 \vartheta_w(m_Z)$ are given in table 1.

Table 1. *Numerical values of quantum numbers, Neutral Current couplings, chiral coupling asymmetry $\mathcal{A}_f$ and sensitivity of $\mathcal{A}_f$ for the four types of fermions. The value of $sin^2 \vartheta_w(m_Z)$ is 0.23.*

Fermion type	I_{3f}	Q_f	a_f	v_f	$\mathcal{A}_f$	$\frac{\partial \mathcal{A}_f}{\partial sin^2 \vartheta_w(m_Z)}$
ν	1/2	0	1	1	1	0
e^-	-1/2	-1	-1	-0.08	0.16	-7.9
u	1/2	2/3	1	0.39	0.69	-3.5
d	-1/2	-1/3	-1	-0.69	0.94	-0.6

Initial state helicity cross-section	Final state helicity
e^- Z e^+ $\propto g_{Le}^2$	Forward: $\bar{f}$ f $\propto g_{Lf}^2$ Backward: f $\bar{f}$ $\propto g_{Rf}^2$
e^- Z e^+ $\propto g_{Re}^2$	Forward: $\bar{f}$ f $\propto g_{Rf}^2$ Backward: f $\bar{f}$ $\propto g_{Lf}^2$
e^- Z e^+ 0	
e^- Z e^+ 0	

Figure 4. *Helicity effects in $e^+e^- \to Z$ production. The arrows $\Leftarrow$ and $\Rightarrow$ indicate the helicity of the particles.*

The observables obtainable with longitudinal polarization are

- The Left-Right asymmetry of Z production [21,22,23,24,25,26,27]:

$$A_{LR} = \frac{\sigma_L - \sigma_R}{\sigma_L + \sigma_R} \simeq \mathcal{A}_e, \tag{8}$$

- The forward-backward polarized asymmetries [28]:

$$A_{FB}^{pol(f)} = \frac{1}{\mathcal{P}} \frac{(N_{\mathcal{P},F} - N_{-\mathcal{P},F}) - (N_{\mathcal{P},B} - N_{-\mathcal{P},B})}{(N_{\mathcal{P},F} + N_{-\mathcal{P},F}) + (N_{\mathcal{P},B} + N_{-\mathcal{P},B})} \simeq \frac{3}{4}\mathcal{A}_f. \tag{9}$$

Without polarization one has at one's disposal the forward-backward asymmetry:

$$A^{(f)}_{FB} \simeq \frac{3}{4} \mathcal{A}_e \mathcal{A}_f. \qquad (10)$$

In either case, for the tau lepton, the polarization of the final state fermion is measurable:

$$\langle \mathcal{P}_\tau \rangle \simeq \mathcal{A}_\tau. \qquad (11)$$

If longitudinal polarization is available, measurements of A_{LR} and $A^{pol(f)}_{FB}$ for the various types of fermions allows a complete determination of the fermion couplings.

In particular, A_{LR} is a real jewel: it is simply obtained by comparing the total cross-section for producing a Z from a left-handed (σ_L) or right-handed (σ_R) e^+e^- system. Intuition –and figure 4– tells us that A_{LR} at the Z pole will not depend on the way the Z decays, but only on the electron Neutral Current couplings [22,24]. Detailed calculations show that this is affected only by minor and well understood corrections [24,29]. Given the large cross-section at the top of the resonance, (34nb for visible decays, 10^6 Z events for a $30pb^{-1}$ exposure typical of a 100 days run under realistic assumptions), a remarkable statistical precision can be envisaged: $\Delta A_{LR} = 0.002$ (statistical) for 10^6 events and 50% polarization. This corresponds to $\Delta sin^2\vartheta_w(m_Z) = 0.0003!!$

The precisions on $sin^2\vartheta_w(m_Z)$ obtainable from measurement of asymmetries without polarized beams are affected by two difficulties: first, the smallness of $\mathcal{A}_e \simeq 8(1/4 - sin^2\vartheta_w)$ for $sin^2\vartheta_w$ in the vicinity of 0.25. For the present value of the Z mass, the effective weak mixing angle, which is relevant for weak couplings, is predicted [30] to be $sin^2\vartheta_w(m_Z) = 0.232 \pm 0.002$. The left-right asymmetry $\mathcal{A}_e$ is then 0.14 ± 0.02, but the muon forward-backward asymmetry at the pole is then 1.5%! In addition these quantities, especially for the leptonic final state, are rapidly varying functions of the center-of-mass energy around the Z pole; this makes them very sensitive to the exact knowledge of the beam energy and to initial state radiation effects. In polarization asymmetries, where parity violating effects are isolated, no such problem occurs, as illustrated in figure 5b.

The $b\bar{b}$ asymmetry seems to be the best alternative at measuring $sin^2\vartheta_w$ assuming the $SU(2)_L \times U(1)$ structure, with a precision of $\Delta sin^2\vartheta_w = 0.0008$ (0.0006 with high luminosity LEP). Of course $A^{(b)}_{FB}$ will be even more interesting if A_{LR} is already available, as it then measures the b quark couplings..

The statistical and systematic errors that one can foresee for the measurement of A_{LR} [31] and $A^{pol(f)}_{FB}$ [32], are summarized in table 2 and figure 5a. The comparison shows that measurements with longitudinally polarized beams are more sensitive, less prone to systematic errors and allow a full and precise determination of the fermion couplings without further assumptions. Even when combining all measurements available, a procedure which precludes any further test, measurements without polarization are at best half as good as the measurement of A_{LR}, which requires much less statistics.

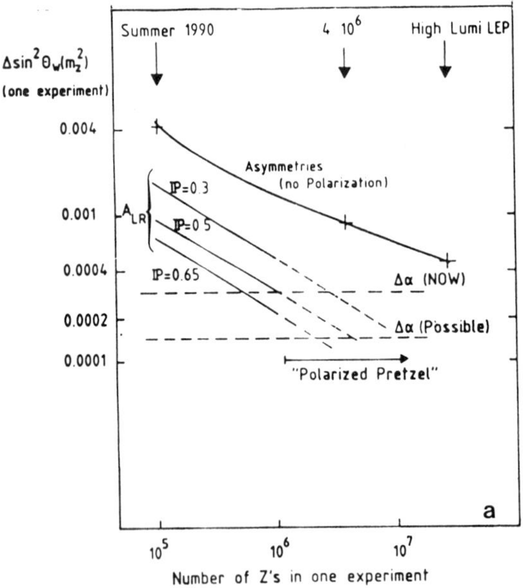

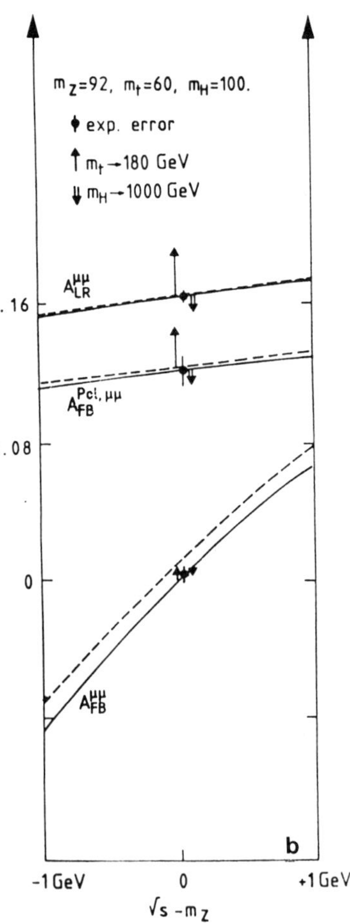

Figure 5. a) *Expected precision on* $sin^2\vartheta_w(m_Z)$
*as a function of the number of Z recorded
in one LEP experiment.*
*b) Behavior of various asymmetries
around the Z pole:* $A_{FB}^{(\mu)}, A_{LR}, A_{FB}^{pol(\mu)}$
Full line: geometrical cuts only
Dashed line: acolinearity cut of 2^0.
Full arrow: effect of a heavy top quark,
$m_t = 180 \ GeV$.
White arrow: effect of a heavy Higgs,
$m_H = 1 \ TeV$.
*Error bars correspond to LEP,
with and without polarized beams.*

The interpretation of the measurements of $sin^2\vartheta_w(m_Z)$ is limited by the knowledge of the effective QED coupling constant at the Z mass, $\alpha(m_Z)$. more specifically,

$$\alpha(m_Z) = \frac{\alpha}{1 - \Delta\alpha} \simeq 128.7^{-1} \tag{12}$$

This important quantity is calculated from the cross-section of $e^+e^- \to hadrons$ from low energies up to the Z mass. The present error corresponds to $\Delta \ sin^2\vartheta_w(m_Z) = \pm 0.0003$, and is limited by the accuracy of the low energy measurements, in particular above the charm threshold [34]. It certainly cannot be excluded that, if need be, more precise measurements of these cross-sections would be performed. This uncertainty would then be reduced by at least a factor of two.

Table 2. *Comparison of errors on the weak fermion couplings combination $\mathcal{A}_f$ obtained from a $200pb^{-}1$ exposure at the Z peak without polarized beams, and from a $30pb^{-}1$ exposure with 50% polarized beams. Some assumptions are necessary to extract information from unpolarized beams experiments, and are labeled as follows: A – e-μ-τ universality; B – τ lepton pure V-A couplings; C – universality of $SU(2)_L \times U(1)$ formulae for fermion couplings.*

quantity	$200pb^{-1}$ no Polarization			$30pb^{-1}$, 50% Polarization		
	reaction used	assumptions	error	reaction used	assumptions	error
$\mathcal{A}_e$	$A_{FB}^{(e)}$	–	0.021	A_{LR}	–	0.003
$\mathcal{A}_\mu$	$A_{FB}^{(\mu)}$	–	0.025	$A_{FB}^{pol(\mu)}$	–	0.009
$\mathcal{A}_\tau$	$A_{FB}^{(\tau)}$	–	0.029	$A_{FB}^{pol(\tau)}$	–	0.015
	P_τ	B	0.012			
$\mathcal{A}_\ell$	$A_{FB}^{(\ell)}$	A	0.013			
$\mathcal{A}_\ell$	$A_{FB}^{(\ell)}, P_\tau$	A+B	0.010			
$sin^2\vartheta_w$	all leptonic	A+B	0.00120	A_{LR}	–	0.00035
$\mathcal{A}_b$	$A_{FB}^{(b)}$	–	0.125	$A_{FB}^{pol(b)}$	–	0.027
		A+B	0.062			
$\mathcal{A}_c$	$A_{FB}^{(c)}$	–	0.100	$A_{FB}^{pol(c)}$	–	0.033
		A+B	0.055			
$sin^2\vartheta_w$	all channels	A+B+C	0.00080	A_{LR}	–	0.00035
	Syst. limit:		0.00080			0.00012
$sin^2\vartheta_w$	from m_Z		0.00033			0.00013
	Th. error $\Delta\alpha$		0.00025			0.00025

2.3 Physics Impact of Longitudinal Polarization Experiments

The predictions of the Standard Model can be tested with great precision by comparing various observables that can be measured at LEP. Radiative effects, sensitive to m_t and m_H, can be gauged as indicated on figure 6.

In particular, once the top quark mass is known, A_{LR} appears as a powerful 'Higgsometer': the range of Higgs masses corresponding to the Strongly Interacting Standard Model, $m_H \geq 1\,\mathrm{TeV}$, is separated from the light Higgs region by 8 standard deviations.

It is of particular interest that one can, to some extent, distinguish the effects of the top quark mass from those of the Higgs boson mass by using A_{LR} and comparing it either to the leptonic partial width of the Z or to the W mass. The comparison of Γ_{l+l^-} with m_W is, on the other hand completely ambiguous in that respect. The Higgs and top effects separate best in the comparison of $R_z = \Gamma_{had}/\Gamma_z$ and $R_b = \Gamma_{b\bar{b}}/\Gamma_{had}$, obtainable without polarized beams, but the expected resolution does not allow much optimism for fixing the Higgs mass.

These comparisons are sensitive to many types of physical effects beyond the Standard Model [23,35]: in contrast with QED where heavy particles decouple from low energy phenomena, $SU(2)_L \times U(1)$ is such that decoupling does not always occur, and particles much heavier than the working energy can have an effect. In addition to

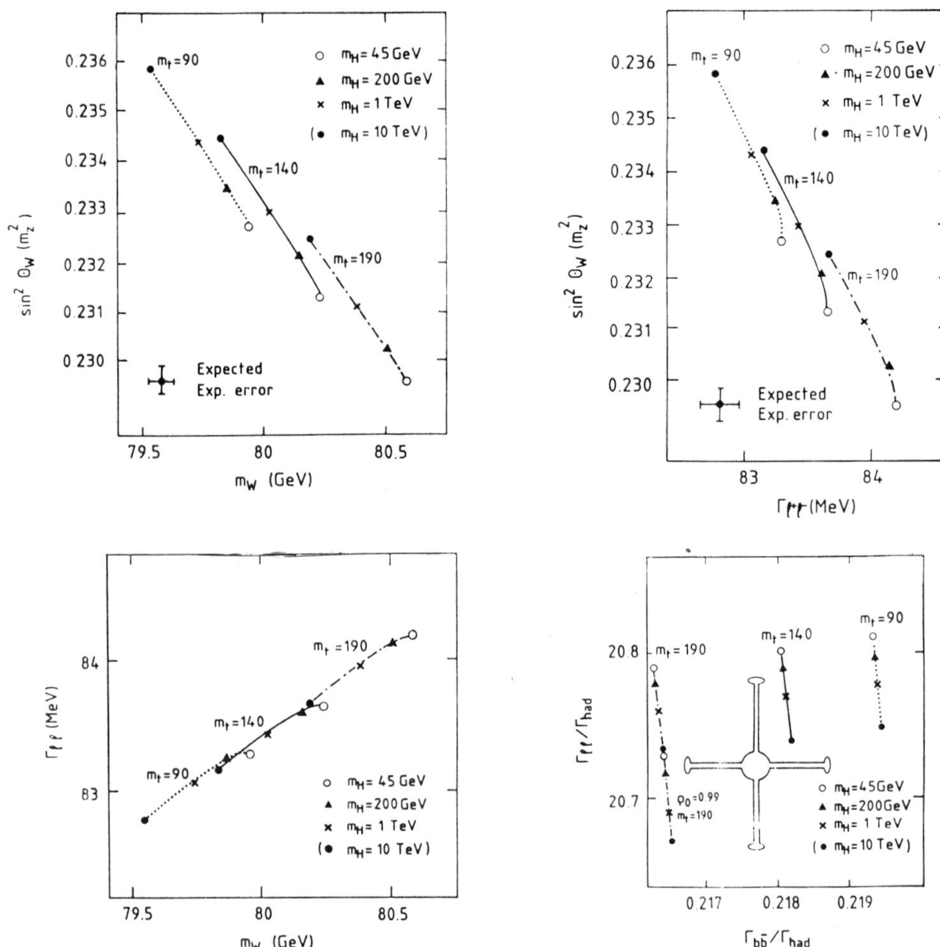

Figure 6. *Dependence of the electroweak mixing angle, $\sin^2\vartheta_w(m_Z)$, as measured from the polarization asymmetries, of the W mass m_W, of the leptonic partial width of the Z, Γ_{l+l^-}, and of the ratios of partial widths $R_z = \Gamma_{had}/\Gamma_z$ and $R_b = \Gamma_{b\bar{b}}/\Gamma_{had}$ upon the values of m_t and m_h. For Higgs masses above 1 TeV the calculations are extremely unreliable and should be taken as indication of a trend towards the 'strongly interacting Higgs' regime. (numerical values from [33])*

new particles, A_{LR} is also sensitive to new structures: Superstring inspired E6, left-right symmetric models, composite models [36], supersymmetry, more complicated Higgs structure, etc. This sensitivity could very well make A_{LR} the first experimental indication that there exist something beyond the Standard Model.

It could be argued that such evidence would be highly ambiguous. Much effort has been devoted to understanding how to disentangle the origin of a possible effect. It turns out that the bulk of Standard Model effects – heavy top, new families, higgs triplets – can be reabsorbed in a redefinition of $\sin^2\vartheta_w$, $\sin^2\vartheta_w(m_Z)$, which is precisely what A_{LR} measures. It is straightforward [37], [38] to then form the linear combination of any quantity with A_{LR} that will be insensitive to Standard Model effects. Any deviation from the Standard Model prediction will be direct evidence for a gauge structure beyond $SU(2)_L \times U(1)$. It thus appears that the additional precision brought about by polarization in the measurements of fermion couplings will play a crucial role to reveal and analyze new physics.

2.4 Experimental Procedure for Longitudinal Polarization Experiments

Two sources of systematic errors are specific to polarized beam experiments: the measurement of the polarization of the beams and the normalization of the data samples taken with opposite e^+e^- helicities. It is possible in LEP to eliminate the need to know these quantities in absolute terms. The running mode envisaged for LEP is inspired by the original suggestion by Placidi and Rossmanith [39]: the natural transverse polarizations of electrons and positrons are opposite, and so will their longitudinal polarization in any practical spin rotator where the beams are transported in the same magnetic channel. The net helicity of the e^+e^- system will then be zero, but can be made either positive or negative by depolarizing one beam or the other. In fact, any one of the bunches in the machine can be depolarized at will to a negligible level [40] offering the following succession of e^+e^- helicity states [41]:

The comparison of the four respective total cross-sections:

$$\sigma_1 = \sigma_u(1 + P^+ A_{LR}) \tag{13}$$

$$\sigma_2 = \sigma_u(1 - P^- A_{LR}) \tag{14}$$

$$\sigma_3 = \sigma_u \tag{15}$$

$$\sigma_4 = \sigma_u(1 - P^+ P^- + (P^+ - P^-)A_{LR}) \tag{16}$$

allows a measurement of A_{LR} but also of P^+ and P^- from the data.

The role of the polarimeter is to monitor the evolution of the polarization with time and the possible differences between one bunch and an other. A suitable polarimeter can be build in more than one way [42]:

- The transverse e^- polarimeter, already build and operated by the LEP polarization collaboration [43], has performed successfully the transverse polarization measurements in 1990 (see next section). It measures the vertical displacement of a backscattered polarized laser beam under reversal of the laser polarization. It can certainly be operational for longitudinal polarization experiments, the magnitude of the spin vector being conserved in the spin rotators. It is straightforward to extend the present design to an e^+ polarimeter.

- A longitudinal polarimeter design is proposed in [19] measuring the change in energy spectrum of a backscattered polarized laser beam under reversal of the laser polarization [42]. This involves a very similar set-up as the transverse polarimeter and should benefit from the experience gained in its operation.

Systematic errors in relative polarization measurements have been estimated for both devices and found to be very small: $\delta P/P \simeq 0.2\%$. The absolute precision is less good, a few percent, but has the virtue of coming directly from the data in the four-bunch scheme.

The four-bunch scheme is very favorable from another point of view: the data in the various helicity configurations are taken simultaneously. Therefore most of the experimental effects will cancel out, including in the notoriously difficult luminosity measurement, up to possible small differences from one bunch type to an other. The normalization of the various samples will be done using the polarization-independent small angle Bhabha rate. It has been shown [45] that this provides adequate normalization, provided the luminosity monitors have a counting rate well in excess of that of Z events at the peak and are insensitive to the small differences between bunches that can occur – mostly from differences in the beam-beam interaction. These differences can be monitored quite well by the LEP instrumentation [46], or in the detectors themselves, leading to sufficiently small systematic errors in the standard luminosity monitors, $\Delta L/L \leq 10^{-3}$.

The recent achievement of 1% absolute precision on the luminosity measurement in the LEP experiments [2], [5], gives confidence that this can be achieved. The breakdown of the systematic errors obtained by the LEP experiments is shown in table 3; the errors relevant to bunch-to-bunch normalization are already smaller than 10^{-3}, without any particular control of the beam parameters.

One can conclude that there is a very strong physics case for polarization. The question is: will there be any, and at what cost?

3   Polarized Beams in LEP

The improvement in precision measurements brought about by the availability of longitudinally polarized beams is qualitatively equivalent to "having a new accelerator" [47]. Obtaining longitudinally polarized beams in LEP seems qualitatively equivalent to "building a new accelerator" [48].

The problems to be solved to obtain a useful degree of polarization in LEP are many:

- The natural polarization time is too long: 310 minutes at $E_{beam} = 46 GeV$ without wiggler magnets. Even if the asymptotic polarization is high, the effective polarization in a physics run is considerably reduced, even though the long beam life time (600 minutes) in LEP is in that respect very good news. However the time necessary to tune the machine to improve the polarization degree could be prohibitive.

- Depolarizing effects increase rapidly with the spin tune, $\nu = \gamma.(g_e - 2)/2 = 103.55$ at the Z pole. In particular, the defects observed in LEP –spurious fields induced by a layer of nickel on the beam pipe, excessive vertical dispersion– were widely expected to simply kill the polarization.

- Depolarizing effects are further increased by the beam's energy spread which becomes uncomfortably large in comparison with the spacing between depolarizing resonances.

Table 3. *Comparison of errors on the luminosity measurement in the four LEP detectors. The DELPHI numbers are expected to improve to the same level as those of the other experiments. Systematics that are relevant for bunch-to-bunch systematics are indicated by an arrow,* $\Rightarrow$, *and summed at the bottom of the table; all other errors cancel in the relative measurement.*

	Systematic error $\Delta\mathcal{L}/\mathcal{L}(\%)$			
source of error	ALEPH	DELPHI*	L3	OPAL
Inner Geometry	0.3	0.5	0.6	1.0
Beam Parameters $\Rightarrow$	0.1	0.5*	0.1	negl.
Trigger efficiency	negl.	0.5	0.1	0.3
Background subtraction $\Rightarrow$	negl.	0.5*	negl.	negl.
Energy scale	0.2	1.0	incl. in 'cuts'	0.4
M.C. Modeling	0.9	1.0	incl. in 'cuts'	incl. in 'cuts'
Effect of cuts			0.7	0.5(energy) 0.2(radius)
Purely exp. error	0.96	1.8	0.93	1.2
M.C. Statistics	0.4	0.2	0.6	0.4
Theory	0.7	1.0	0.7	1.0
Total	1.3	2.1	1.3	1.6
Bunch-to-Bunch $\Rightarrow$	0.1	0.8	0.1	negl.

- The Sokolov-Ternov process [11] creates transverse, and not longitudinal, polarization. Modifications have to be made to implement spin rotators around the interaction points.

- It is not clear whether it is possible to obtain simultaneously high luminosity and high polarization.

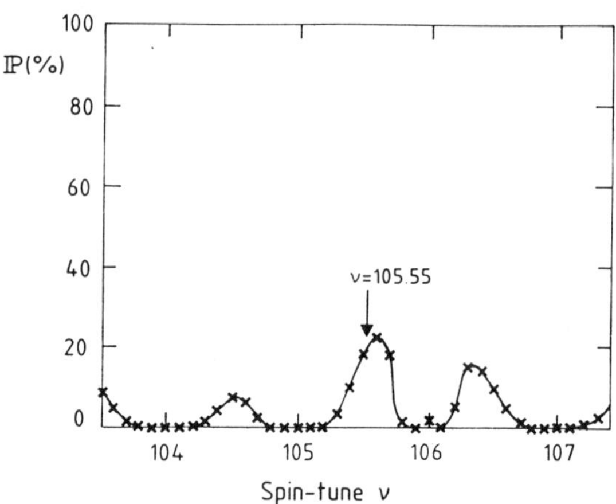

Figure 7. *Expected polarization degree as a function of beam energy in the LEP optics used in the first polarization measurements.*

3.1   First Evidence of Transverse Polarization

The aim of the 1990 polarization experiments was to establish that transverse polarization can indeed be created and measured in LEP. The following sections describe first the preparation work done one the optics of the machine to increase the chances of creating polarization. Then the polarimeter is described, and finally the experimental results.

3.1.1   Optical Conditions

The first step was to eliminate all known sources of depolarization. The experimental solenoids, the vertical separators, low β quadrupoles were turned off. Only e^- were injected in the machine to remove beam-beam depolarization. Each of these items can in principle be corrected for, but... one thing at a time.

The second step [49] was to find a combination of the betatron tunes Q_x and Q_y with the spin tune $\nu = E_{beam}/0.44065$ where the effect of known depolarizing resonances is minimal. The main resonances occur at the following beam energies:

- Integer resonances $\nu = n$;

- betatron resonances $\nu = n' \pm Q_x$ or $\nu = n'' \pm Q_y$, the strongest ones being $n' = 8k$ or $n'' = 8k$, the so-called systematic resonances. The number 8 is related to the eight-fold symmetry of LEP.

The usual LEP operating point was $Q_y = 71.38$, $Q_y = 77.28$. Thus first order resonances occur at $\nu = N, \nu = N + 0.28, \nu = N + 0.38, \nu = N + 1 - 0.38, \nu = N + 1 - 0.2, \nu = N + 1$, etc.. In order to leave more space away from any of these values near $\nu = N + 0.5$, the tunes were slightly modified to $Q_y = 71.12$, $Q_y = 77.20$. These tunes also turned out to be very good for physics. Avoiding the systematic resonances led to chose $\nu = 105.55$, i.e. $E_{beam} = 46.5$ GeV.

This being done, one is left with the unknown depolarizing effects, related to machine imperfections, especially those creating vertical oscillations, out of the machine plane. The average vertical excursion was reduced from the usual $< y_{orbit} > \simeq 1.5$ mm down to 0.5 mm, with an improved correction procedure (this constitutes an important contribution to LEP operation in general). The vertical dispersion $\mathcal{D}_y$ was reduced from around 20 cm down to 13 cm. Depolarizing effects scale proportionally to $< y_{orbit}^2$ and $\mathcal{D}_y{}^2$.

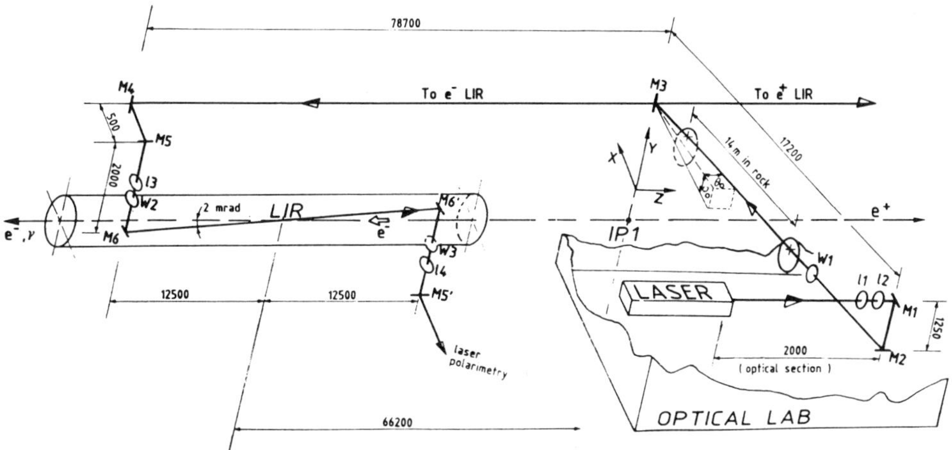

Figure 8. *The LEP polarimeter setup.*

Under those conditions, first order calculations predict a degree of polarization of about 20%, figure 7. Higher order calculations predict an increase of depolarization be a factor of about 2 [48], but this statement is very strongly energy dependent, as additional resonances appear. Even first order predictions are very uncertain, the imperfections being unknown by nature. One thus expected a polarization of the order of 10%, with a distinct possibility of seeing nothing at all.

3.1.2   The Transverse Polarimeter

The measurement of transverse polarization is performed by Compton scattering of polarized photons on the e^- beam, which has a polarization dependent angular distribution.

The polarimeter is described in [44]. The intense beam of low energy photons is provided by a frequency doubled Nd-Yag laser, transported over 114m, and shot at the e^- beam, figure 8.

The backscattered photons are of high energy, 0-30 GeV, and emitted with a very small angle, 10^{-5} radians, with respect to the incident electrons. The backscattered photon beam is separated from the electrons by the first bending magnet of the LEP arc, and detected 300 m downstream in a silicon strip calorimeter. The showers of about 200 photons pile-up at each laser shot, at a rate of 30 shots per second, creating a spot of 4 mm vertical size. Full e^- polarization and circular light result in a shift of the spot by 250 microns.

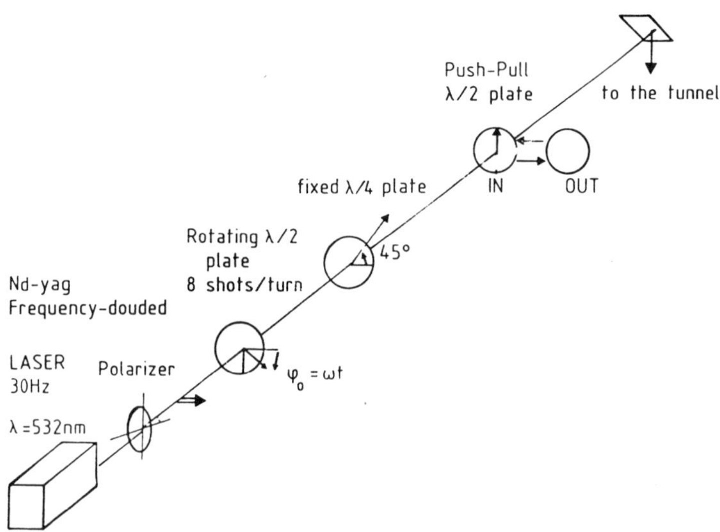

Figure 9. *Generation of alternating light polarization for the LEP polarimeter. A pure linear polarization state is produced in a polarizer, then rotated by the rotating $\lambda/2$ retardation plate. Depending on the orientation of the light axis with respect to the following $\lambda/4$ plate, light is circular or linear. The whole sequence is reversed when inserting the additional $\lambda/2$ plate.*

Given the long lever-arms, small variations in the position or angle of the e^- or photon beams can easily generate much larger effects. In order to fight these systematic shifts, the polarization of the laser was reversed at every shot using a rotating retardation plate as shown in figure 9.

The difference in the center-of-gravity $\Delta < Y >$ or in the height Δh of the spot between two consecutive shots were recorded. Depending on the time offset Δt of the rotating plate with the laser trigger, a sequence of alternating right to left circular

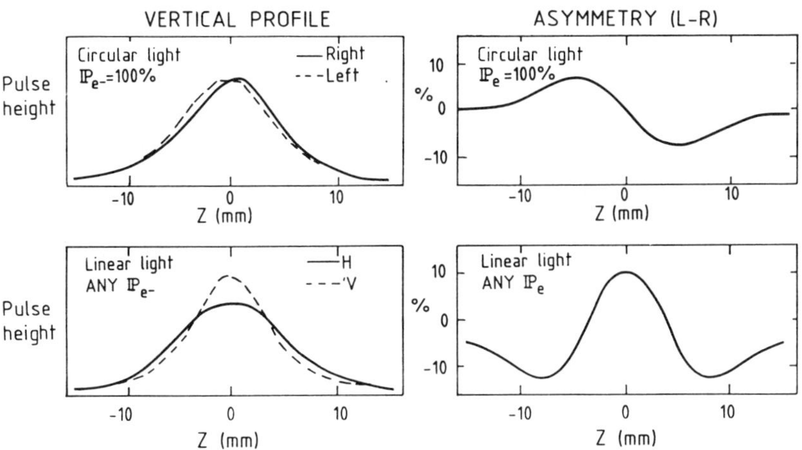

Figure 10. *Effect on the vertical profile of the backscattered photon spot of a reversal of the laser light. If the light is circularly polarized, and the electron beam transversally polarized, the average position of the spot moves, and an up-down asymmetry is generated. If the light is linearly polarized, for any electron polarization, the width or height of the spot changes, and a symmetric asymmetry is generated.*

polarization, or vertical to horizontal linear polarization for cross-check purposes, was generated. The expected changes in the backscattered photon spot are:

- Left to right circular: $\Delta< Y > = 500 \ \mu m \times P_\perp(e^-) \times P_\gamma$
  $\Delta h = 0$

- Horizontal to Vertical linear: $\Delta< Y > = 0$
  $\Delta h/h \simeq 20\% \times P_\gamma$, independent on $P_\perp(e^-)$.

The effects are sketched on figure 10, together with the asymmetry at each point of the vertical distribution. A significant $\Delta< Y >$ when reversing circular polarization constitutes a signal for transverse beam polarization. In order to eliminate the possibility that the helicity-reversing device itself creates optical changes resulting in a spurious $\Delta< Y >$, an additional retardation plate was inserted at times in the light path, providing a second reversal of the polarization. Spurious effects would remain unchanged, while genuine polarization effects would change sign.

The laser-electron interaction point was adjusted carefully, in order to minimize the sensitivity to alignment errors. This was done by local straight section bumps, which, in theory, do not excite depolarizing effects.

These measurement procedures were developed during a few machine development sessions in spring and summer 1990. Parasitic operation during physics runs, were no

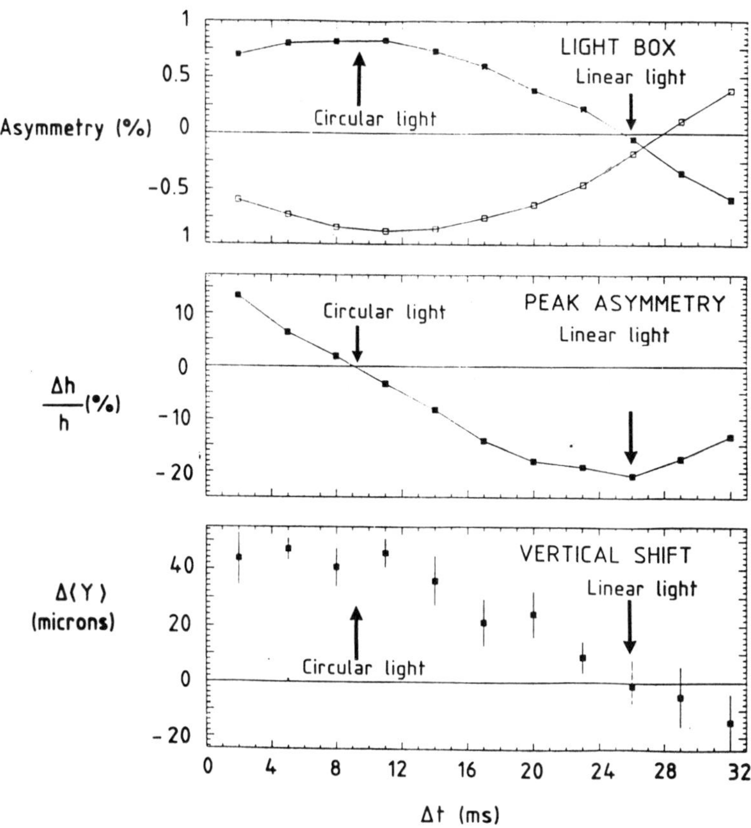

Figure 11. *First evidence for transverse polarization: The time offset of the rotating plate with the laser trigger is varied.*
Top: the asymmetry between two consecutive laser shots in the light polarization measurement: maximum asymmetry takes place for circular light. Full squares: additional $\lambda/2$ out; open squares: additional $\lambda/2$ in.
Middle: the difference in peak height of the backscattered photon spot (additional $\lambda/2$ out).
Bottom: the difference in the center-of-gravity of the backscattered photon spot showing a maximum for circular light and a zero for linear light as expected for a e^- beam polarization of $(10.6 \pm 1.6)\%$.

polarization was really expected, was also very useful to verify the systematic errors of the device.

3.1.3 Results

A few attempts earlier in the year failed to reach the point were beam polarization could actually measured. The last 24 hours of the 1990 LEP campaign were given away by the four LEP collaborations for a last attempt at finding polarization. The LEP machine was carefully set-up with only e^- beams as explained above.

First, the variation of $\Delta < Y >$ and Δh as a function of the laser trigger timing Δt was measured. The laser light polarization was analyzed at the same time, after the collision point, by a set of retardation plates with a maximum asymmetry between the two light polarizations for circular light, and zero asymmetry for linear light. The measurements are shown in figure 11. The behavior of Δh matches expectations perfectly: for circular light, $\Delta t = 9.5$ ms, Δh is 0, while for linear light, $\Delta t = 26.0$ ms, Δh is maximal.

The behavior of $\Delta < Y >$ shows a cosine-like variation, with a maximum amplitude for circular light, $\Delta t = 9.5$ ms, and a zero for linear light, $\Delta t = 26.0$ ms. This curve was the first evidence for transverse polarization of the e^- beam. The observed asymmetries are shown in figure 12, and correspond to this hypothesis. Given the estimated circular polarization of $(85 \pm 10)\%$, the observed shift corresponds to a transverse polarization of $(10.6 \pm 1.5)\%$. The time offset was fixed to $\Delta t = 9.5$ ms, circular light, in the following.

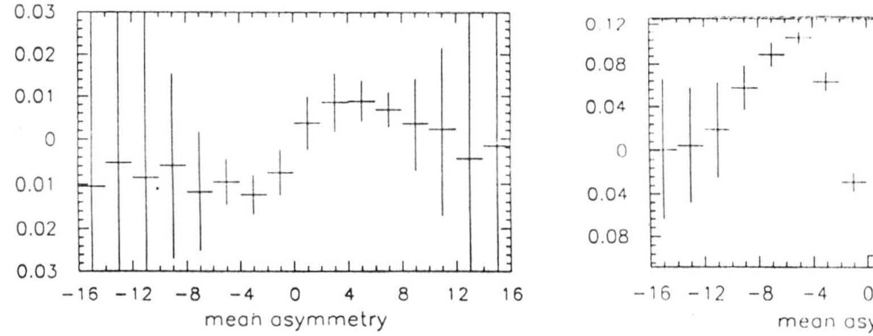

Figure 12. *Observed asymmetry in the vertical profile: each point corresponds to the asymmetry between on laser shot and the next in one strip if the silicon strip detector.*
Left: linear polarization; right: circular polarization.

The following measurements are shown in figure 13. The additional reversal of light polarization was performed a few times, always leading to a reversal of the observed effect, such as in points 4 to 7. This test, together with the Δt curve, can be used to infer that a spurious, systematic, shift in $\Delta < Y >$ can not exceed 5 microns. The following tests aimed at destroying the effect by applying to the electron beam known depolarizing effects with orbit correctors. The difficulty in this procedure lies in the fact that optical changes to the beam modify the closed orbit and the emittances and thus, the conditions of the laser e^- interaction. There is also an uncertainty of a few minutes as to when exactly the corrections are applied

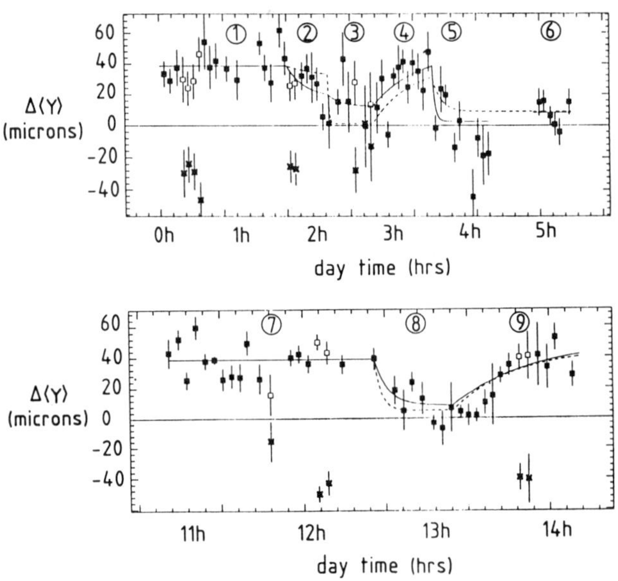

Figure 13. *Following the polarization of the beam over 14 hours on 31 August 1990. Full squares: measurements with additional $\lambda/2$ out; Crosses: additional $\lambda/2$ in. open squares: additional $\lambda/2$ in, point represented with opposite sign. Periods 1 to 9 correspond to various beam conditions as described in the text.*

to the beam. The depolarizing effects at one's disposal were: i) exciting an integer resonance, here $\nu = 106$, with various strengths; ii) turning on the correctors foreseen for spin compensation of the solenoids (the solenoids themselves being turned off, this is equivalent to exciting the solenoids).

The various tests are summarized in table 4. In each of the operations, a new depolarization time is expected, leading to a new asymptotic degree of polarization P_∞. Fits were performed to P_∞ for each period. From the knowledge of the perturbations applied, one can also predict approximately an expected degree of polarization. The fits are generally quite good and in agreement with expectations. Three periods have rather poor χ^2; this could be traced to instabilities in machine operation, RF trips and orbit drifts, which could affect both the response of the polarimeter and the behavior of the beam polarization itself.

From periods 4 and 9, the rise time could be measured, to be $\tau_p^{\text{eff}} = (38 \pm 8.4)$ minutes, in good agreement with the value of (29.5 ± 5.2) minutes corresponding via eq. 2 to the asymptotic degree of polarization $P_\infty = (9.1 \pm 1.6)\%$ seen in periods 1 and 6.

The phenomenon behaves altogether rather well. For the skeptics, the fit to 9 different levels gives a χ^2 per degree of freedom of 165/88 d.o.f. (fit performed with only statistical errors), the fit to expectations gives 170/96 d.o.f., an a fit to a constant gives 386/96 d.o.f..

These observations strongly suggest that polarization has indeed been observed.

3.2 Improving the Degree of Polarization

It was shown in [50] that dedicated wigglers can reduce the polarization time to 36 minutes and would have many other beneficial consequences: machine imperfections can be corrected empirically in a reasonable time; the effective polarization for a given asymptotic degree of polarization is much better (see fig. 14); the depolarizing effects are also relatively reduced –including beam-beam induced depolarization– allowing a higher degree of asymptotic polarization to be reached. Wigglers, however, unavoidably increase the beam energy spread, thereby enhancing higher order depolarizing effects. The wigglers are now build and installed in LEP for operation early in 1991.

Table 4. *Summary of fits to the polarization measurements. The first error is statistical, the second is systematic, and include: the possibility of a 5 microns systematic shift in $\Delta < Y >$ (all periods); uncertainties in the transition time between consecutive conditions (periods 2,3,4,9); and scale factor for the goodness-of-fit (period 5,7,9). The expectations based on the calculations by Koutchouk [49] are also shown for comparison.*

Period	χ^2/D.F. of the fit	P_∞(fit) %	P_∞(exp.)	Conditions and comments
1	0.99	$9.1 \pm 0.6 \pm 1.2$	input	stable beam, asymptotic polarization
2	1.07	$5.1 \pm 0.6 \pm 1.6$	7.4	excitation of $\nu=106 \times 1$
3	0.74	$2.5 \pm 1.6 \pm 1.3$	0.3	excitation of $\nu=106 \times 5$
4	1.26	$11.5 \pm 3.1 \pm 2.1$	9.1	natural rise
5	4.4	$0.6 \pm 0.6 \pm 1.7$	2.1	RF trips, solenoid bumps ON
6	1.3	$1.9 \pm 0.7 \pm 1.2$	2.1	stable beam, solenoid bumps ON
7	2.36	$9.1 \pm 0.3 \pm 1.3$	9.1	asymptotic polarization
8	0.62	$2.0 \pm 0.6 \pm 1.2$	1.2	excitation of $\nu=106 \times 2.5$
9	2.4	$11.7 \pm 2.4 \pm 2.7$	9.1	natural rise

Empirical corrections of depolarizing effects have been studied and developed using various simulation programs [51], [52], [54]. They involve typically 8 correctors, 4 for the vertical orbit harmonics and 4 for the vertical dispersion. The vertical orbit correction is sufficient to reach 85% polarization with dedicated wigglers in the linear theory, but the vertical dispersion must also be corrected to reach a decent degree of polarization when higher order effects are taken into account. A difficulty there is to find uncorrelated correctors. This has only partially been achieved so far, but

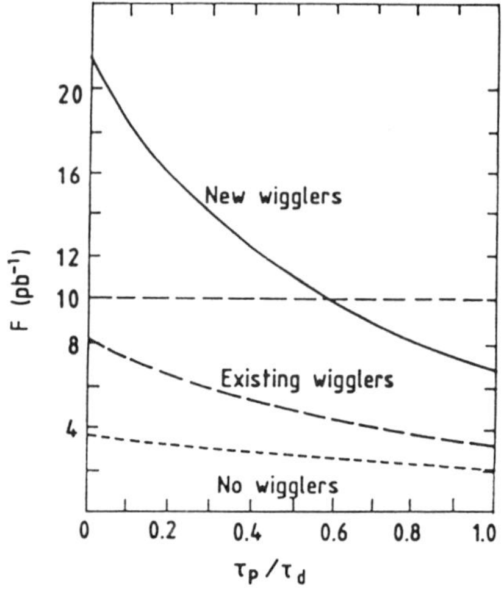

Figure 14. *Polarization figure of merit* $F = \int_0^{T_{fill}} P^2(t)L(t)\,dt.$ *as a function of the relative strength of depolarizing effects* τ_p/τ_d *for different wiggler configurations. A polarization of 65% corresponds to* $\tau_p/\tau_d = 0.4$.

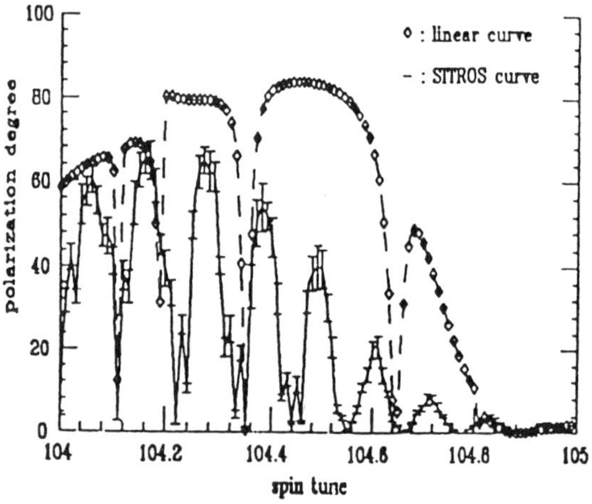

Figure 15. *Polarization as a function of spin tune* $a\gamma = 1.1347E_{cm}$ *in SITROS simulation after harmonic correction of the vertical orbit and partial harmonic correction of the vertical dispersion (nh=104/sine). Standard LEP with dedicated wigglers.*

a polarization of 65% (fig. 15) has been already obtained. The parasitic coupling between vertical and horizontal motion observed in the machine was not taken into account for these calculations, however, and solutions will have to be found to correct for it. A detailed cross-check with analytical calculations of higher order depolarizing effects [55] is being carried out to improve our confidence in these results.

3.3   Spin-rotators

The very simple straight-section spin-rotator, shown in figure 16, originally proposed by Richter and Schwitters [56], can be made exactly spin-transparent and produces almost no depolarization, at least in first order theory [57]. Solutions can be found [58] to reduce the synchrotron radiation to an acceptable level. Keil and Söderström [59] have recently shown that this rotator can be shortened to accommodate radio-frequency cavities in the straight section around all four experiments. This rotator constitutes a cheap and, hopefully, satisfactory solution. It has various advantages, since the spin matching equations can be adapted to all energies by simply varying the amplitude of the vertical bump. It has to be seen that the beam geometry can be restored to the normal, flat, geometry in a reasonable amount of time.

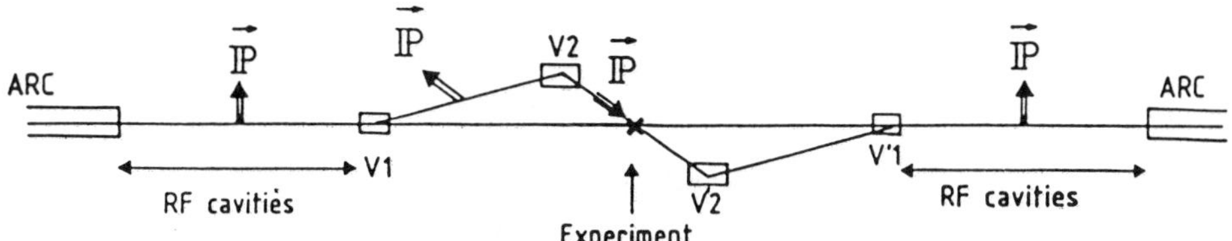

Figure 16. *Schematic layout of the short Richter-Schwitters spin-rotator compatible with radio-frequency cavities.*

3.4 Open Questions, Schedule, Cost

Finally a question that always comes is the following: "What luminosity will be lost by having polarization?". We only have partial answers; the limit to luminosity in polarization running will be dictated by the beam-beam depolarization; there are spin-matching conditions for the beam-beam interaction, but they have not been investigated so far; wigglers will reduce the relative importance of the beam-beam depolarization and help push the limit farther; increasing the number of bunches in the machine does not *a priory* increase the depolarizing effects.

This question and many others will find their answer in the polarization experiments that will be performed in 1991 and 1992. The steps to be taken next are:

- Decision to build the spin rotators, after transverse polarization has been observed reproducibly.

- Construction, installation and commissioning of the spin-rotators. Depending on the moment when the decisions are made, this could take place at the earliest in 1992. In this scenario, longitudinal polarization experiments could take place in 1993.

The costs involved seem reasonable, as far as we can judge: a set of 12 dedicated wigglers costs 2.5 MSF, Richter-Schwitters rotators 2.5 MSF per experiment, each set of polarimeters for positrons and electrons 0.5 MSF [42].

4 Conclusion

Transverse and Longitudinal polarization considerably sharpen the precision tests of the Electroweak theory. Although the extra components are not very expensive, obtaining it in LEP is challenging. The programme is now well under way to implement this very logical improvement to a machine aimed at studying weak interactions.

References

[1] LEP Design Report, CERN-LEP/84-01 (1984).

[2] ALEPH Collaboration, D. Decamp et al., Phys. Lett. B231 (1989) p. 519;
ibid, Phys. Lett. B235 (1989) p. 399;
ibid, Z. Phys. C48 (1989) p. 365.
see also ref. [10] for a review.

[3] DELPHI Collaboration, P. Aarnio et al., Phys. Lett. B231 (1989) p. 539;
ibid, Phys. Lett. B241 (1990) p. 435
see also ref. [10] for a review.

[4] OPAL Collaboration, M. Z. Akrawy et al., Phys. Lett. B231 (1989) p. 530;
ibid, Phys. Lett. B240 (1990) p. 497;
see also ref. [10] for a review.

[5] L3 Collaboration, B. Adeva et al., Phys. Lett. B231 (1989) p. 509;
ibid, Phys. Lett. B249 (1990) p. 341;
see also ref. [10] for a review.

[6] D. Treille in 'Cargese 1990, Z Physics', R. Gastmans ed. (1990).

[7] W. J. Marciano and A. Sirlin, Phys. Rev. Lett. 46 (1981) p. 163;
A. Sirlin, Phys. Lett B232 (1989) p. 123.

[8] D.C. Kennedy and B.W. Lynn, SLAC-Pub 4039 (1986, rev. 1988), Nucl. Phys.
B322 (1989).

[9] M. Consoli and W. Hollik, in 'Physics at LEP1', G. Altarelli, R. Kleiss and
C. Verzegnassi (ed.), CERN 89-08 (1989) p. 7.

[10] E. Fernandez, in 'neutrino 90', Nucl. Phys. B (proc. suppl.) 666(1990) p. 1.

[11] A. A. Sokolov and I. M. Ternov, Sov. Phys. Doklady, 8 (1964) p. 1203.

[12] R. F. Schwitters et al., Phys. Rev. Lett. 35 (1975) p. 1320.
G. Hanson et al., Phys. Rev. Lett. 35 (1975) p. 1611.

[13] A. A. Zholentz et al., Phys. Lett. 96B (1980) p. 214.

[14] A. S. Artamonov et al., Phys. Lett. 118B (1982) p. 225.

[15] D. P. Barber et al., Phys. Lett. 135B (1984) p. 498.

[16] W. W. MacKay et al. Phys. Rev. D29 (1984) p. 2483.

[17] A. Hofmann et al., 2nd European Particle Acc. Conf., Nice 1990.

[18] P. Darriulat, in 'proceedings of the LEP summer study', CERN 79-01 (1979),
p. 219.

[19] 'Polarization at LEP', G. Alexander et al. (editors), CERN 88-06 (1988),
vol. I, II.

[20] C. Bovet, B.W. Montague, M. Placidi and R. Rossmanith, in 'Physics at LEP', CERN 86-02 (1986), p. 58.

[21] C.Y. Prescott, proc. 1980 Int. Symp. on High-Energy Physics with Polarized Beams and Polarized Targets, Lausanne, 1980, eds. C. Joseph and J. Soffer (Birkhäuser Verlag, Basel, 1981), p. 34.
M. Böhm and W. Hollik, Nucl. Phys. B204 (1982), p. 45.

[22] B.W. Lynn and R.G. Stuart, Nucl. Phys. B253 (1985), p. 84.
B.W. Lynn, in [19], vol.I, p. 24.

[23] B.W. Lynn, M.E. Peskin and R.G. Stuart, SLAC-PUB 3725 (1985) and in 'Physics at LEP' CERN 86-02, Geneva,(1986), p. 90.

[24] B.W. Lynn and C. Verzegnassi, Phys. Rev. D35 (1987), p. 3326.

[25] D. Blockus et al., 'Proposal for polarization at SLC', SLAC-Prop-1, Stanford, April 1986.

[26] G. Alexander et al., Working Group Report CERN/LEPC/87-6 LEPC/M81 (1987).

[27] J. Badier et al., ALEPH Note 87-17 (1987).

[28] A. Blondel, B.W. Lynn, F.M. Renard and C. Verzegnassi, Nucl. Phys. B304 (1988), p. 438.

[29] B.A. Kniehl, J.H. Kühn and R.G. Stuart, in [19], vol.I, p. 158.

[30] A. Blondel, CERN-EP/89-84 (1989);
A. Sirlin, Phys. Lett. B232 (1989) p. 123.

[31] D. Treille, in [19], vol.I, p. 265.

[32] J. Drees, K. Mönig, H. Staeck and S. Überschär, in [19], vol.I, p. 317; P.J. Dornan, in [19], vol.I, p. 344.

[33] D.C. Kennedy, J.M. Im, B.W. Lynn and R.G. Stuart, Nucl. Phys. 321 (1989) p. 83.
Computer programm EXPOSTAR 1.0, courtesy of D.C. Kennedy.

[34] H. Burkhardt, F. Jegerlehner, G. Penso and C. Verzegnassi, Z Phys. C43 (1989) p. 497. and in [19], Vol. I, p. 145.

[35] Reviewed by W. Hollik in [19], vol.I, p. 83.

[36] F. Boudjema and F.M. Renard, in [19], vol.I, p. 223.

[37] M. Cvetic and B.W. Lynn, Phys. Rev. D35 (1987), p. 51.

[38] C. Verzegnassi, in [19], vol.I, p. 204; A. Djouadi, in [19], vol.I, p. 215.

[39] M. Placidi and R. Rossmanith, LEP Note 545 (1985)

[40] J. Buon and J.M. Jowett, LEP Note 584 (1987)

[41] A. Blondel, Phys. Lett. 202B (1988), p. 145.

[42] G. Alexander et al., in [19], vol.II, p. 3.

[43] J. Badier, A .Blondel (L.P.N.H.E., Palaiseau), M. Crozon (college de France, Paris), B. Dehning (Max Plank Institute Munich), L. Knudsen, J.-P. Koutchouk, M. Placidi, R. Schmidt (CERN-LEP); M. Placidi, proc. 9 Int. Conf. on High Energy Spin Physics, Bonn, Germany, (1990).

[44] M. Placidi and R. Rossmanith, CERN LEP-BI/86-25 (1986, rev.1988); M. Placidi, R. Rosmanith, NIM A274 (1989) 79-94.

[45] G. Coignet and following articles, in [19], vol.II, p. 83.

[46] G. Von Holtey, in [19], vol.II, p. 89.

[47] G. Altarelli, public comment at the Polarization Workshop.

[48] J. Buon, public comment at the Polarization Workshop.

[49] J. -P. Koutchouk, proc. 9 Int. Conf. on High Energy Spin Physics, Bonn, Germany, (1990).

[50] A. Blondel and J.M. Jowett, in [19], vol.II, p. 216.

[51] A. Chao, Nucl. Instr. Meth. 180 (1981), p. 29.

[52] A. Ackermann, J. Kewisch, T. Limberg, to be published.

[53] J. Kewisch, DESY 83-032 (1983).

[54] J.-P. Koutchouk and T. Limberg, in [19], vol.II, p. 204.

[55] J. Buon, in [19], vol.II, p. 243. S.R. Mane, in [19], vol.II, p. 233.

[56] R. Schwitters and B. Richter, PEP Note 87 (1974).

[57] A. Blondel and E. Keil, in [19], vol.II, p. 250.

[58] D. Treille and G. Von Holtey, in [19], vol.II, p. 258.

[59] E. Söderström, CERN/LEP-TH/89-58, (1989); H. Grote, E. Keil and E. Söderström, European Particle Accelerator Conference, Nice, (1990), CERN-SL/90-22 (1990).

Physics at the SLC

M.L. Swartz

Stanford Linear Accelerator Center

Stanford, California, USA

1. Introduction

The SLAC Linear Collider (SLC) was constructed in the years 1983-1987 for two principal reasons:

1. To develop the accelerator physics and technology that are necessary for the construction of future linear electron-positron colliders.

2. To produce electron-positron collisions at the Z^0 pole and to study the physics of the weak neutral current.

To date, the SLC program has been quite successful at achieving the first goal. The machine has produced and collided high energy electron and positron beams of three-micron transverse size. The problems of operating an open geometry detector in an environment that is more akin to those found in fixed-target experiments than in storage rings have largely been solved. As a physics producing venture, the SLC has been less successful than was originally hoped but more successful than is commonly believed. Some of the results that have been produced by the Mark II experiment with a very modest data sample are competitive with those that have been produced with much larger samples by the four LEP collaborations. At the current time, SLAC is engaged in an ambitious program to upgrade the SLC luminosity and to exploit one of its unique features, a spin polarized electron beam.

These lectures are therefore organized into three sections:

1. A brief description of the SLC;

2. A review of the physics results that have been achieved with the Mark II detector;

3. A description of the SLC's future: the realization and use of a polarized electron beam.

2. The SLAC Linear Collider

2.1. The Importance of Linear Colliders

Circular electron-positron (e^+e^-) storage rings were developed in the 1960's and early 1970's as an inexpensive technique for the production of high energy collisions (the center-of-mass energies were typically a few GeV). They proved to be such extremely important tools in the development of particle physics that two more generations of higher energy machines have been constructed.

Unfortunately, the size and cost of e^+e^- storage rings increase rapidly with the beam energy. This can be illustrated by considering a simple cost model for electron storage rings. The cost $\mathcal{C}$ can be expressed as follows,

$$\mathcal{C} = \alpha R + \beta \frac{E^4}{R}, \tag{2.1}$$

where: R is the radius of the storage ring; E is the beam energy; and α, β are constants. The first term represents the costs that are proportional to the size of the ring (such as the tunnel, magnets, vacuum chamber, etc). The second term represents the cost (size) of the RF acceleration system that is necessary to replace the energy that is lost to synchrotron radiation. The optimal size of the ring for a given beam energy can be found from the zero of the first derivative $d\mathcal{C}/dR$. The optimal radius and cost, R_{opt} and $\mathcal{C}_{opt}$, scale with the square of the beam energy,

$$R_{opt} = \sqrt{\frac{\beta}{\alpha}}E^2, \quad \mathcal{C}_{opt} = 2\sqrt{\beta\alpha}E^2.$$

The unpleasant reality of this scaling law can be illustrated by considering the scaling of LEP200 to 1 TeV. The size and cost of a *LEP1000* are listed in Table I. The cost is comparable to the entire CERN budget since its inception in 1954. The size is roughly eight times larger than that of the SSC. (If the beams were injected into LEP1000 at CERN, the 180° interaction region could be located in Zürich!)

Table I

Machine	CM Energy	Length of Tunnel	Cost
LEP200	180 GeV	27 km	FS 1×10^9
LEP1000	1 TeV	833 km	FS 31×10^9
TLC	1 TeV	14 km	FS 3×10^9

Table I also lists a guess at the size and cost of a 1 TeV linear collider (TLC). The size and cost of the linear machine are smaller than the circular one by factors of 60 and 10, respectively. The size is well within the range of existing machines. The cost, while large, is smaller than those of the next generation of hadron colliders. *It appears that linear colliders are the only practical technique for the building of very high energy e^+e^- machines.*

2.2. LINEAR COLLIDERS

Linear colliders differ from circular machines in that the beams are accelerated to collision energy (in one or two linear accelerators), collided, and discarded after only one use.

The luminosity $\mathcal{L}$ of any collider is given by the following expression,

$$\mathcal{L} = \frac{N_+ N_- f}{4\pi\sigma_x\sigma_y},\tag{2.2}$$

where: N_+ and N_- are the number of positrons and electrons in each bunch; f is the frequency of collision of the bunches; and $4\pi\sigma_x\sigma_y$ is the overlap area of the Gaussian bunches of size σ_x by σ_y.

Circular machines have rather high frequencies of bunch collisions (from 4×10^4 at LEP to $\sim10^6$ at small storage rings). Unfortunately, non-superconducting linear accelerators are limited to collision frequencies in the range 10^2 to 10^3. In order to produce comparable luminosity, a linear collider must compensate the low collision frequency with increased bunch population or with reduced beam size. The former leads to a number of technical difficulties and would require a tremendous amount of RF power. The latter approach is the choice of all linear collider designers. Since the beams are to be discarded after a single use, they can be subjected to extreme perturbations from the transport system or from the other beam. It is therefore quite natural to use very small beams in a linear collider. The SLC has produced beams of 3 μm transverse size which is substantially smaller than the LEP design value of 12 μm by 300 μm (vertical by horizontal size). Future linear colliders are expected to utilize beams that are smaller than those of the SLC by nearly two orders of magnitude.

2.3. TERMINOLOGY

It is clear that the designer of a linear collider must concern himself (herself) with those aspects of the machine design that affect the beam size. The size of a charged particle beam within a magnetic transport system is determined by the focusing strength of the transport system and by the phase space that is occupied by the beam particles.

The concept of the phase space of a beam is an important one. Let us assume that we have a beam that consists of a large number of particles moving principally in the z-direction. We can then use Liouville's theorem to write the following expression,

$$\frac{1}{N}\sum_{\text{beam}} P_y y = \gamma\beta m\frac{1}{N}\sum_{\text{beam}} y' y = \pi\gamma\beta m\varepsilon_y = \text{constant},\tag{2.3}$$

where: N is the number of particles in the beam; P_y is the momentum of a beam particle along one of the directions that is transverse to the beam direction; y is the displacement of the particle with respect to the beam axis; γ and β are the usual Lorentz quantities; m is the particle mass; and y' is the derivative dy/dz. The sums in equation (2.3) are taken over all particles in the beam. It is customary to define the *emittance* of the beam (in the

y direction) ε_y as follows,

$$\varepsilon_y \equiv \frac{1}{\pi N} \sum_{\text{beam}} y'y. \qquad (2.4)$$

Liouville's theorem and the definition of the emittance lead to the following properties of the horizontal and vertical emittances:

1. The horizontal and vertical emittances, ε_x and ε_y, are invariant in *conservative* fields. Note that magnetic transport systems are composed of conservative fields.

2. The products $\gamma\varepsilon_x$ and $\gamma\varepsilon_y$ are invariant under acceleration (we assume that $\beta \simeq 1$).

3. The emittances ε_x and ε_y are intimately related to the transverse beam sizes. It can be shown that the variances of the (Gaussian) horizontal and vertical beam particle distributions, σ_x and σ_y, are related to the horizontal and vertical emittances by the following expressions,

$$\sigma_x^2 = \varepsilon_x \beta_x(z)$$
$$\sigma_y^2 = \varepsilon_y \beta_y(z), \qquad (2.5)$$

   where $\beta_x(z)$ and $\beta_y(z)$ are functions which describe the focussing strength of the transport system (a complete description of the β functions formalism can be found in Reference 1).

4. The variances of the angular distributions of the beam particles, $\sigma_{x'}$ and $\sigma_{y'}$, are also related to the emittances and the β functions,

$$\sigma_{x'}^2 = \varepsilon_x / \beta_x(z)$$
$$\sigma_{y'}^2 = \varepsilon_y / \beta_y(z). \qquad (2.6)$$

2.4. Linear Accelerators

The main element(s) of a linear collider is a linear accelerator. The first linear accelerators were Cockroft-Walton and Van der Graaf accelerators. They consist of a linear drift space across which a large voltage difference V is maintained. This generates a strong axial electric field which is used to accelerate charged particles to kinetic energies that are equal to the voltage V. Unfortunately, it is not possible to maintain an arbitrarily large voltage across the accelerating structure. As the voltage is increased, one inevitably exceeds the dielectric strength of the insulators being used and discharges to ground occur. The maximum voltage that is typically obtained is 10-20 MV.

This limitation was overcome by the development of the traveling wave accelerator in the late 1940's.[2] The traveling wave accelerator is based upon the observation that the TM_{01} mode of an electromagnetic field in cylindrical waveguide has a longitudinal electric field. The electric field is oriented along the axis of the cylinder which is the direction of propagation of the electromagnetic field (this is quite different from the case of a freely propagating em field). Charged particles can therefore be accelerated by a moving pulse of RF power and there is no need to produce a huge voltage along the entire length of the accelerator. Unfortunately, the phase and group velocities of the TM_{01} mode conspire to complicate the design of the accelerator. The phase velocity of the cylindrical waveguide is larger than the speed of light. A bunch of charged particles would see a longitudinal electric field of constantly changing sign and no acceleration would take place. The solution to this problem is to *load* the cylinder with annular disks. The phase velocity can be adjusted to be equal to the speed of light.

The group velocity of a radio frequency pulse of electromagnetic energy in a cylindrical accelerating structure is much less than the speed of light (in the SLC it is 1%-2% of c). A bunch of electrons cannot travel along with an electromagnetic pulse of energy. The solution to this problem is to feed-in RF power at short intervals along the length of the accelerator. In the SLC, this is done at intervals of 3 m.

2.5. A Description of the SLC

A layout of the SLAC Linear Collider is shown in Figure 1. A cycle of the machine begins when one bunch of positrons and two bunches of electrons are extracted from the damping rings and are accelerated down the linac structure. As the after the positron bunch and the first electron bunch pass the 2/3 point of the linac (the 3 km linac is composed of 30 sectors), a pulsed kicker magnet diverts the second electron bunch onto the positron production target to make more positrons. The positrons are returned to the beginning of sector 1 by a long return line. The electron gun at the front end of the machine fires to produce two electron pulses which are coaccelerated with the positrons to 1.15 GeV in the first linac sector. The beams are then injected into their respective damping rings.

The positron bunch and the first electron bunch continue to be accelerated to $\sim$46.5 GeV at the end of the linac. The bunches pass through a large dipole magnet which sends them into the north (electron) and south (positron) machine arcs. The beams each lose approximately 1.0 GeV as they traverse the 1.5 km arcs. They then enter the final focus regions which cause them to be demagnified to sizes of a few microns at the interaction point. After the collision, the beams are ejected into beam dumps by kicker magnets.

The following sections describe each of the machine susbsytems in more detail.

<u>The Linac</u>

The actual linac has been modified substantially for use in the SLC. The klystron power supplies which provide the s-band RF power (2860 MHz) have been upgraded from 20 MW devices to 67 MW devices. The energy upgrade along with the implementation of a pulse compression technique have increased the maximum energy of the linac from 20 GeV to more than 50 GeV.

The increased energy has required that the focussing strength of the of the quadrupole lattice be increased. Improved quadrupole magnets are placed at 12 m intervals along the machine.

As we have already mentioned, the production of very small beams is a critical design feature of any linear collider. This requires that the emittance of the beam be kept a small as is possible. Unfortunately, there are several effects that can increase the emittance of a beam as it is accelerated in a linac. Collectively, they are known as wakefield effects. They fall into two categories:

1. Transverse wakefield effects are caused by the interaction of a bunch with its own image fields or with the image fields of other bunches. Within a single bunch, the image fields of the head of the bunch can affect the transverse positions of the

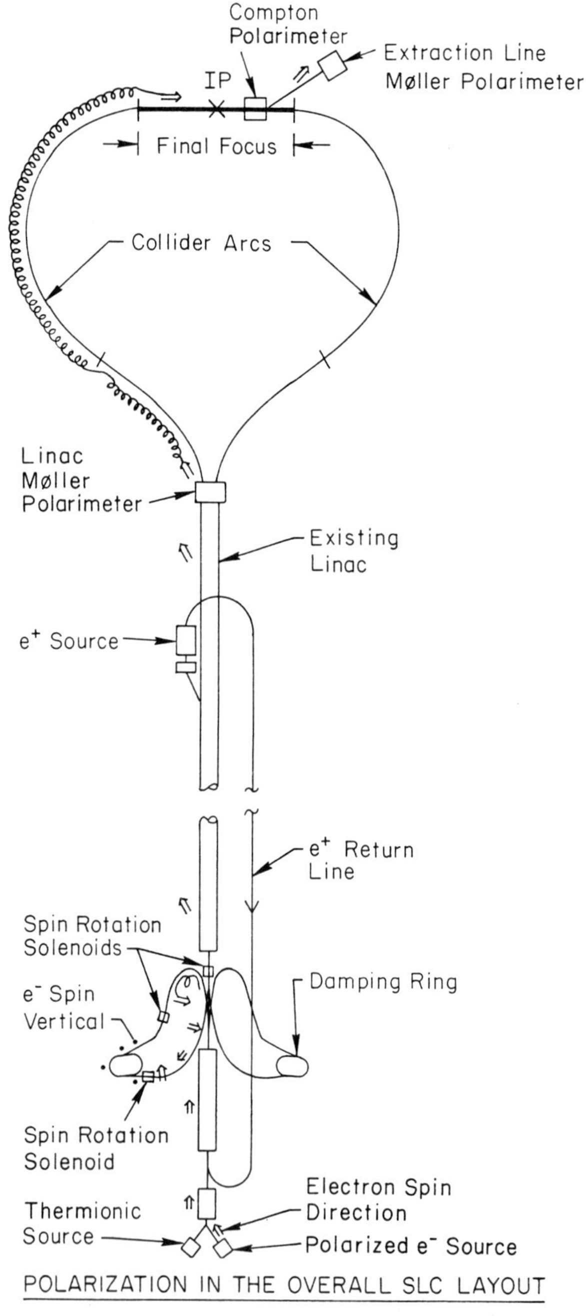

Figure 1

A layout of the SLAC Linear Collider. The orientation of an electron spin vector is shown as the electron is transported from the electron gun to the interaction point.

particles in the tail of the bunch. The ensuing rotation of the bunch causes an
effective increase in the transverse emittance of the beam. The solutions to this
problem are to make the bunches as short as possible and to steer the beam as
close to the axis of the accelerator structure as is possible (the effect vanishes on
the accelerator axis).

2. Longitudinal wakefields effects are caused by the intrabunch electrostatic fields.
   Fields from the bunch head tend to decelerate the particles in the bunch tail. This
   causes an increase in the energy spread of the beam as it is accelerated in the linac.
   This problem is minimized by making the beam bunches as long as is possible.

In order to control wakefield effects, a number of changes to the linac and its mode of
operation have been implemented:

1. The linac is operated with a short (optimized) bunch length of 2 mm.

2. Beam position monitors and corrector magnets have been installed at intervals of
   12 m along the linac. The system can control the trajectory of a single beam bunch
   to ±100 um of the accelerator axis.

3. The linac RF phases are adjusted to introduce and to remove an energy spread
   as the beam is transported down the linac. This causes the beam to decompose
   into filaments of different momenta which follow different orbits. The transverse
   wakefield effects are reduced by this technique (which is called BNS damping).

The Electron Source

The electron source consists of several components. An electron gun produces two
2-ns pulses of up to 2×10^{11} electrons from a hot cathode. The 175 kV electron pulses are
separated in time by 61 ns.

A system of three RF bunchers is then used to reduce the bunch length from $\sim$20 cm
at 175 kV to 2 mm at 40 MeV. The bunchers make use of the non-relativistic velocity of
the electrons that are emitted from the gun. A long wavelength axial electric field is used
to accelerate the tail of the bunch and to decelerate the head of the bunch. This velocity
dispersion decreases the bunch length until the increasing energy causes the velocity to
saturate at c along the entire bunch.

Finally, an accelerator section is used to increase the beam energy to 200 MeV at
the entrance to the first linac sector. The electron bunches are then coaccelerated with
positrons returning from the positron production target to 1.15 GeV for injection into the
damping rings.

Damping Rings

The electron and positron bunches that are produced by the respective sources have
invariant emittances that are too large for the high luminosity operation of a linear collider.
It is therefore necessary to make use of a phenomenon that doesn't conserve energy to
reduce the beam emittances.

The emission of synchrotron radiation is particularly useful for this purpose. Let us
consider a bunch of electrons circulating in a storage ring. Let the z-axis define the instantaneous direction of a particle that is traveling along the central orbit of the bunch.
Most of the remaining particles have some momentum components that are perpendicular
to the central orbit (let x and y be the horizontal and vertical directions in the perpendicular plane). As the average beam particle passes through the machine arcs, it radiates

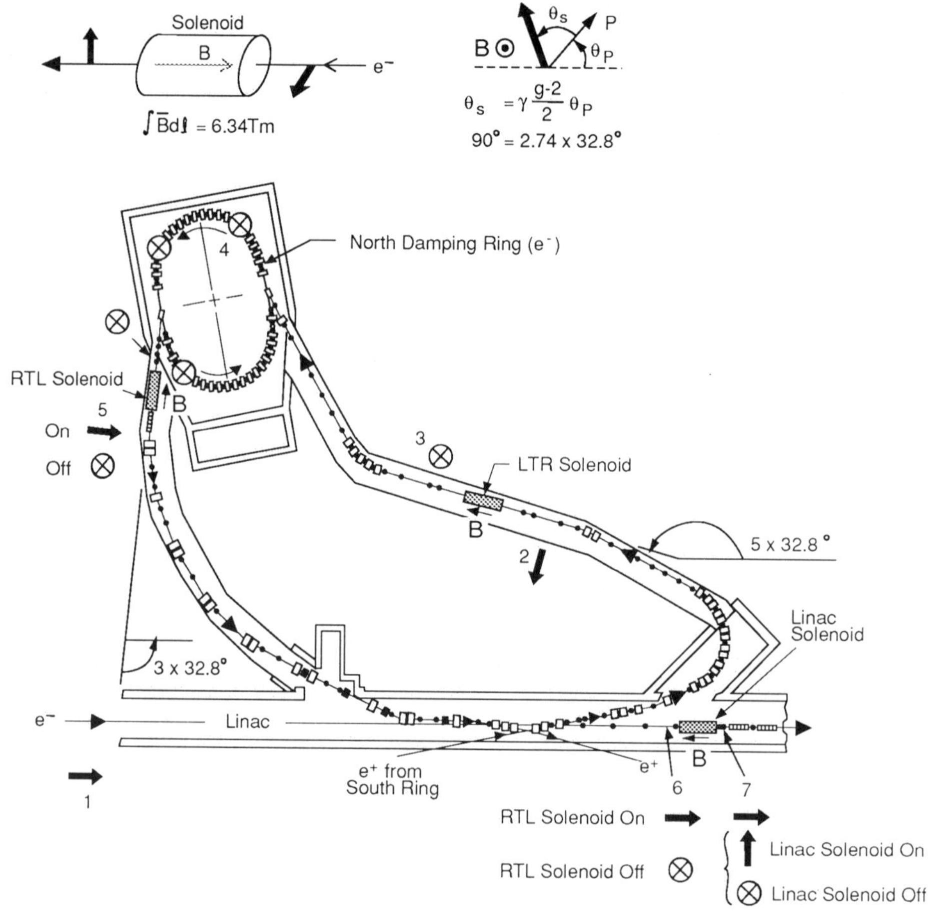

Figure 2

The spin rotation system as incorporated into the north damping ring complex. The orientation of the polarization vector at several points is shown by the double arrow.

photons that are collinear with its instantaneous direction of motion. The momenta along the three directions, p_x, p_y, and p_z, are reduced by the emission process. This lost energy is replaced in an RF accelerating cavity. *Note, however, that the energy is replaced along the z direction only.*

In the vertical direction (which is orthogonal to the bending plane of the arcs), the transverse momentum of the beam is reduced without affecting the spatial distribution of the particles. *The vertical emittance, $\gamma\varepsilon_y$, therefore becomes smaller.*

In the horizontal direction, the transverse momentum components are also reduced. Unfortunately, the particle trajectory moves horizontally when the particle energy changes (the radius of the orbit becomes smaller when energy is lost). Thus, we have two competing effects: one that reduces the transverse momentum components, and one that increases the spatial distribution of the beam. The machine lattice can be designed to enhance the damping effect and to reduce the horizontal emittance, $\gamma\varepsilon_x$, at a cost in longitudinal energy spread.

The SLC has two small storage rings that are designed to reduce the vertical and horizontal emittances of the electron and positron bunches. The north (electron) damping ring is shown in Figure 2. The ring is designed to reduce the emittance of a particle bunch with a characteristic $1/e$ time of 3 ms. Since the positron bunches are produced with larger emittances, they must be stored for two machine cycles (16 ms) to be sufficiently damped.

The Positron Source

As was described at the beginning of this section, a positron bunch and two electron bunches are extracted from the damping rings and accelerated in the linac. As they pass the 19th sector of the machine, the trailing electron bunch is diverted onto the positron production target by a pulsed kicker magnet and a short beam transport system. The electrons produce electromagnetic showers in the 6 radiation length target. Positrons between the energies of 2 MeV and 20 MeV are captured by a solenoidal focusing system and are accelerated to 200 MeV. The system is designed to capture 2 positrons for each incident electron. The positrons are then returned to the front end of sector 1 for acceleration and storage in the south damping ring. Including losses enroute, the system is designed to store 1 positron in the damping ring for each electron striking the positron production target.

The Arcs

After the positron and leading electron bunches are accelerated to full energy (which is approximately 1 GeV larger than the interaction point energy), they are transported to the final focus systems in 1.5 km, S-shaped arcs. In our discussion of radiation damping, we noted that particles traversing a magnetic field lose energy along three coordinates. Since we have no RF system to replace the lost energy, the emittance of a beam bunch that is transported through a large arc must increase. The horizontal emittance is also increased by the horizontal displacement caused by the energy change (this is called dispersion).

The increase in the horizontal emittance can be minimized by keeping the beam strongly focused as it is transported through the arcs. In order to do this, the SLC arcs consist of three-pole, combined-function magnets. The fields produced by these magnets have strong dipole, quadrupole, and sextapole components. Each arc is constructed of 460 such magnets arranged to alternately focus and defocus in each plane. The arcs are designed to be achromatic to second order and to be capable of transporting a beam with an energy spread of 0.5%.

Final Focus

The last 150 m of each (electron and positron) beamline is called the final focus. Each final focus system is a transport system that consists of 8 bending magnets, 26 quadrupole lenses, 8 sextapole magnets, and a number of correction and monitoring devices. These systems are designed to demagnify the 250 μm$\times$30 μm beams that leave the arcs to spots of 2 μm$\times$2 μm at the interaction point. The final focus systems are designed to cancel all geometrical and chromatic aberrations to second order.

Beam Monitoring

The beams that are stored in storage rings have stable orbits and energies. After stored electron and positron beams are brought into collision, they will generally remain in collision for some macroscopic time. Unfortunately, this convenient behavior is not necessarily true in a linear collider.

The problem of energy stability in the linac is confronted with feedback systems and by making the arcs and final foci fairly achromatic (they can transport momenta over a range $\Delta p/p = 0.5\%$). Residual energy drifts are measured on each pulse by spectrometers that are placed in the extraction lines. These will be described in chapter 4.

Linear colliders must rely heavily on sensors and feedback systems to control the orbits of the beams. The SLC makes use of several techniques to bring the beams into collision:

1. A system of beam position monitors is used to measure the positions and directions of the beams at the interaction point. These devices measure the beam centroid position by comparing the beam induced signals in pickup loops that are placed on either side of the vacuum chamber. They are capable of steering the beams to within 20 μmof each other.

2. The phenomenon of beam-beam deflection provides the single most important technique for establishing collisions.[3] The fields of each beam deflect the other in a manner that depends upon their transverse sizes, the distance of closest approach, and number of particles in the bunches. The deflection angle θ of an infinitely narrow beam by a target beam of finite size is given by the following expression,

$$\theta = \frac{-2r_e N_e}{\gamma} \cdot \frac{1 - e^{-\Delta^2/2\sigma^2}}{\Delta}, \tag{2.7}$$

   where: r_e is the classical radius of the electron; γ is gamma factor of the deflected beam; N_e is the number of particles in the target bunch; Δ is the (signed) miss distance of the beams; and σ is the size of the target beam. Note that the deflection angle has maxima at the miss distances, $\Delta \simeq \pm 1.6\sigma$. As the beams are moved closer together, the deflection angle becomes smaller. It passes through zero when the beams collide and changes sign as the original beam positions are interchanged. Using the system of beam position monitors to measure the deflection angles (which are of the order of 100 μrad), it is straightforward to target the beams to within a small fraction of a beam size.

3. The strong magnetic fields that are associated with the beam-beam deflection process (up to 100 T) also produce a large quantity of synchrotron radiation (10^6 to

10^{10} photons of energy larger than 20 MeV). This radiation, called *beamsstrahlung*, is separated from the electron beam by a large bending magnet in the final focus. Since there is a large background of lower energy photons (typically 2 MeV) from the focusing elements of the beam transport system, the beamsstrahlung photons are converted into e^+e^- pairs with a radiator and detected in a cerenkov counter.

4. The transverse profiles of the electron and positron beams are measured by the devices called *wire scanners*. These devices work by passing the beams through a very fine carbon filament (the smallest has a 4 μmdiameter) and by detecting scattered radiation. The wire scanner used in the interaction point produces bremsstrahlung photons that are detected by the beamsstrahlung monitor.

2.6. SLC PERFORMANCE

The performance of the SLC has not yet approached the level that was intended when the machine was designed. The technical capabilities of the machine were recently assessed by a committee of experts.[4] They conclude that it is possible to improve the performance of the SLC to produce 10^4 to 10^5 Z events per year. The correctness of this assessment will be established in the next several years. In chapter 6, we shall see that if this level of performance is achieved, the implementation of the polarized electron beam should provide an interesting and unique test of the Standard Model.

3. Physics with the Mark II Detector

The Mark II detector was originally constructed to study e^+e^- collisions at the PEP storage ring. It was upgraded for use at the SLC by the replacement of the tracking system. The detector has operated at the SLC in 1989 and 1990. Beginning in 1991 it will be replaced by the new SLD detector.

3.1. THE MARK II DETECTOR

A schematic diagram of the Mark II detector is shown in Figure 3. The detector consists of a system to reconstruct the tracks of charged particles, a calorimeter to measure the energies of charged and neutral particles, and a system to identify penetrating charged particles (which are presumed to be muons).

The cylindrically symmetric tracking system consists of three distinct devices. A three-layer silicon strip detector (SSD) occupies the region from a radius of 2.5 cm from the beam axis to a radius of 5.0 cm from the beam axis. Each of the three measurements is made in the azimuthal direction with a precision of approximately 10 μm. The SSD is followed by a twelve-layer high pressure drift chamber microvertex detector (DCVD). The DCVD occupies the region between the radii 5.0 cm and 17 cm. It is capable of measuring tracks in the azimuthal direction with a precision of approximately 40 μm per measurement. The DCVD is followed by a large cylindrical drift chamber that spans the region between the radii 19 cm and 147 cm. The sense wires of the chamber are organized into 6-wire vector cells. There are 12 layers of vector cells which provide 72 measurements of each track. Alternate layers are oriented parallel to the beam axis or are tilted by $\pm 3°$ with respect to the axis (the axial coordinate is provided by this small-angle stereo arrangement). The typical precision of each measurement is 145 μm.

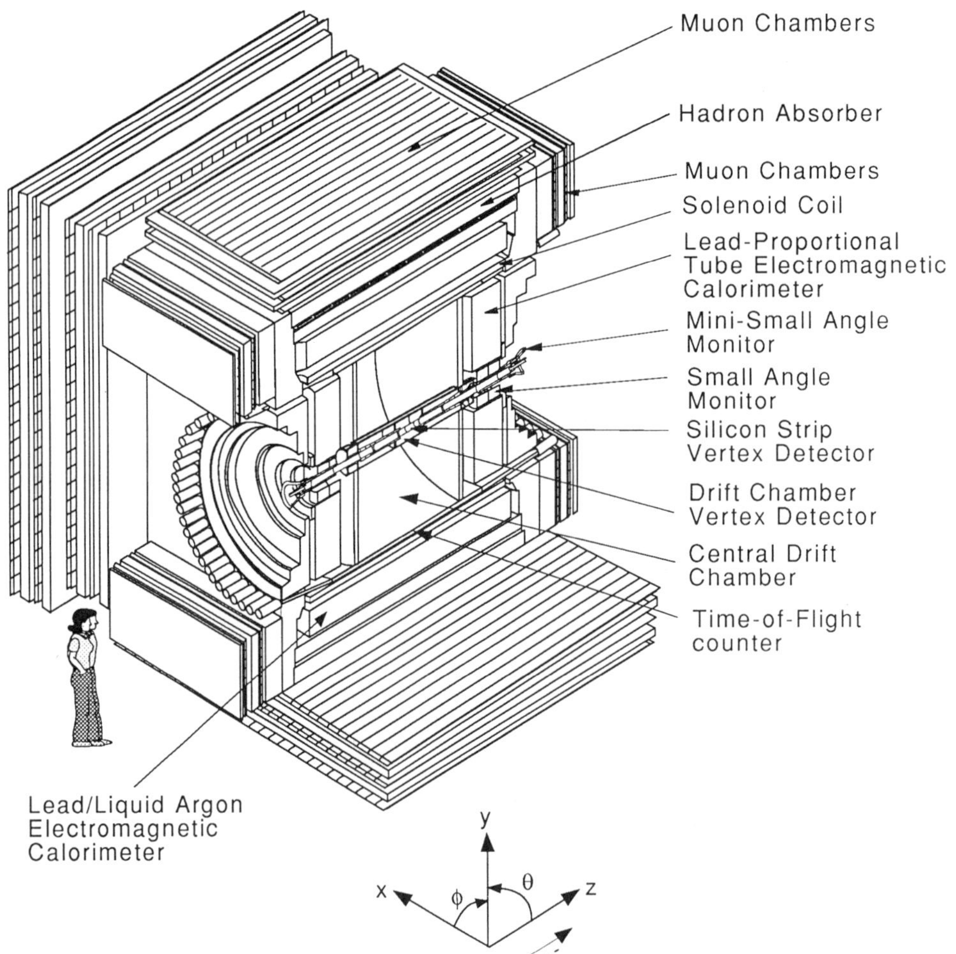

Figure 3. The Mark II detector.

The tracking system is capable of reconstructing tracks in the region of polar angle $|\cos\theta| < 0.8$ with an efficiency of 99%. The momentum resolution of the combined system is given by the following expression, $\sigma_p/p \sim 0.002p$, where p is the momentum in GeV. The large drift chamber has the capability to measure the ionization energy loss in the gas with a precision of roughly 7%.

The tracking system is surrounded by barrel and endcap calorimeters. The barrel calorimeter is a lead-liquid argon device that covers the region of polar angle $|\cos\theta| < 0.7$. Since it is only 14 radiation lengths thick, the energy resolution is given by the following expression,

$$\frac{\sigma_E}{E} = \begin{cases} 0.12 \cdot \text{GeV}^{1/2}/\sqrt{E} & E < 15 \text{ GeV} \\ 0.15 \cdot \text{GeV}^{1/2}/\sqrt{E} & 15 \text{ GeV} < E < 50 \text{ GeV}. \end{cases}$$

The resolution of the barrel calorimeter in azimuthal and polar angle is approximately 3.5 mRad. The endcap calorimeter is a lead and proportional wire chamber device that covers region of polar angle $0.7 < |\cos\theta| < 0.98$. It has a thickness of 20 radiation lengths and is sampled each 0.5 radiation length of thickness. The energy resolution is given by the following expression,

$$\frac{\sigma_E}{E} = 0.20 \cdot \text{GeV}^{1/2}/\sqrt{E}.$$

The muon system is a sandwich structure composed of iron of the flux return of the solenoid and an triangular proportional tubes. The system covers 60% of the solid angle and has a total thickness of 1.05 m. The probability that a hadron be misidentified as a muon is less than 1%.

The luminosity is determined from the number of small angle Bhabha scattering events that are detected in two different detectors. The minisam consists of a pair of cylindrical tungsten-scintillator calorimeters that detect both legs of Bhabha events in the region of polar angle between 15 and 25 milliradians. Each calorimeter is segmented into four quadrants. The small angle monitor (SAM) is comprised of a pair of lead-proportional wire chamber calorimeters that detect both legs of events in the region of polar angle between 50 and 160 milliradians.

4. The Lineshape of the Z^0 Boson

The measurement of the Z^0 lineshape provides a great deal of information about the Standard Model. In order to fully appreciate the relevance of the mass measurement, we must discuss the Standard Model briefly.

4.1. PARAMETERS OF THE STANDARD MODEL

The minimal Standard Model contains some 21 empirical parameters. They are listed in Table II with their approximate values.

Table II

Parameter	Description	Approximate Value
g_s	SU(3) coupling constant	1.3 @ 34 GeV
g	SU(2) coupling constant	0.63
g'	U(1) coupling constant	0.35
$\langle \phi \rangle$	VEV of the Higgs field	174 GeV
M_H	Higgs boson mass	?
m_{ν_e}	electron neutrino mass	< 12 eV
m_{ν_μ}	muon neutrino mass	< 0.25 MeV
m_{ν_τ}	tau neutrino mass	< 35 MeV
m_e	electron mass	0.511 MeV
m_μ	muon mass	106 MeV
m_τ	tau mass	1.78 GeV
m_u	up-quark mass	5.6 MeV
m_d	down-quark mass	9.9 MeV
m_s	strange-quark mass	199 MeV
m_c	charm-quark mass	1.35 GeV
m_b	bottom-quark mass	5 GeV
m_t	top-quark mass	?
$\sin\theta_{12}$	K-M Matrix parameter	0.217-0.223
$\sin\theta_{23}$	K-M Matrix parameter	0.030-0.062
$\sin\theta_{13}$	K-M Matrix parameter	0.003-0.010
$\sin\delta$	K-M Matrix parameter	?

The dynamics of electroweak physics are determined (at tree level) by three of the parameters: the SU(2) coupling constant (g), the U(1) coupling constant (g'), and the vacuum expectation value of the Higgs field ($\langle\phi\rangle$). The complete specification of the electroweak sector of the Standard Model requires that all three parameters be *precisely* known. The values of these quantities are extracted from the measurement of three related quantities: the electromagnetic fine structure constant (α), the Fermi coupling constant (G_F), and the mass of the Z^0 boson (M_Z). The current values of these quantities are listed in Table III.

The value of α is extracted from a very precise measurement of the anomalous magnetic moment of the electron.[5] The value of G_F is derived from the measured value of the muon lifetime.[6] The first precise measurements of the Z^0 mass have been made quite recently at the Tevatron collider, the SLC, and at LEP. The quoted value is determined mostly by the remarkable measurements of the LEP collaborations.[6] Although M_Z is determined with far less accuracy than are α and G_F, it is expected to remain the most well-determined Standard parameter for the foreseeable future. It is clear that the measurement of a fourth physical quantity should overconstrain the determination of the electroweak parameters. We should therefore be able to *test* the electroweak sector of the Standard Model.

The current values of the physical parameters that determine the determine the electroweak sector of the Standard Model.

Quantity	EW Parameters	Current Value	Precision (PPM)
α	$\frac{1}{4\pi}\frac{g^2 g'^2}{g^2+g'^2}$	$[137.0359895(61)]^{-1}$	0.045
G_F	$\frac{1}{\langle\phi\rangle^2\sqrt{8}}$	$1.16637(2)\times 10^{-5}$ GeV^{-2}	17
M_Z	$\sqrt{\frac{g^2+g'^2}{2}}\,\langle\phi\rangle$	91.16(3) GeV	320

Unfortunately, the expression given in Table III that relates M_Z to g, g', and $\langle\phi\rangle$ is valid only at tree-level. Since M_Z is measured at a substantially larger energy scale than are α and G_F, we must include virtual electroweak corrections in order to extract accurate values for the electroweak parameters. In principle, this requires a knowledge of all of the parameters listed in Table II. In practice, a dispersion relation is used to determine the dominant correction (due to low mass fermion loops) from the low energy e^+e^- total cross section. The largest remaining corrections depend upon the top quark mass (strongly) and the Higgs boson mass (weakly). A reasonably precise test of the Standard Model therefore requires at least two more experimental measurements (ideally a measurement of m_t would be one of them).

4.2. MASS AND WIDTH OF THE Z^0

We have already discussed the importance of a high precision measurement of the mass of the Z^0. The width of the Z^0 has a tree-level dependence upon the parameters of the Standard Model and the particle content of the theory. The total width is the sum of the partial widths for the decay into each fermion-antifermion final state,

$$\Gamma_Z = \sum_f \Gamma_{f\bar{f}} = \frac{G_F M_Z^3}{24\pi\sqrt{2}} \sum_f C_f(v_f^2 + a_f^2), \qquad (4.1)$$

where: $\Gamma_{f\bar{f}}$ is the partial width for the decay $Z^0 \rightarrow f\bar{f}$; v_f and a_f are the vector and axial vector coupling constants,

$$v_f = \tau_3^f - 4Q_f\frac{g'^2}{g^2+g'^2}\ ; \qquad (4.2)$$
$$a_f = -\tau_3^f$$

τ_3^f is twice the third component of the fermion weak isospin; Q_f is the fermion charge; and the constant C_f is defined as

$$C_f = \begin{cases} 1 + \frac{3\alpha}{4\pi}Q_f^2 & \text{for leptons} \\ 3\cdot\left[1 + \frac{3\alpha}{4\pi}Q_f^2 + \frac{\alpha_s}{\pi}\right] & \text{for quarks.} \end{cases}$$

Note that the expression of each partial width in terms of M_Z has the advantage that the m_{top} and m_{Higgs} dependences are minimized.

Table IV

Final State	$\Gamma_{f\bar{f}}$
$\nu\bar{\nu}$	166 MeV
$\ell^+\ell^-$	83 MeV
$u\bar{u}$	297 MeV
$d\bar{d}$	383 MeV
2.75 Generations	2.481 GeV

The partial widths for a generation of quarks and leptons are listed in Table IV. The last line shows the expected total width for three lepton flavors and five quark flavors. A small phase space suppression factor is included for the $b\bar{b}$ final state.

The actual measurement of M_Z and Γ_Z is made by measuring the cross section for the process $e^+e^- \rightarrow Z^0 \rightarrow f\bar{f}$ for a number of center-of-mass energies about the Z^0 pole. The theoretical Z lineshape is then fit to the measured cross section points to extract the desired parameters. The theoretical lineshape has been the subject of much analysis.[7] It can be shown that the tree-level lineshape for the process $e^+e^- \rightarrow Z^0 \rightarrow f\bar{f}$ is well-approximated by a relativistic Breit-Wigner form,

$$\sigma_f^0(s) = \frac{12\pi}{M_Z^2} \cdot \frac{s\Gamma_{ee}\Gamma_{f\bar{f}}}{(s - M_Z^2)^2 + \Gamma_Z^2 s^2/M_Z^2}. \tag{4.3}$$

Equation (4.3) does not apply to the process $e^+e^- \rightarrow e^+e^-$ which occurs via both s-channel and t-channel subprocesses.

The electron and positron radiate real photons rather copiously in a hard collision. The lineshape is strongly affected by the initial state radiation. This effect can be treated in a Drell-Yan-like formalism by introducing an electron structure function. The electron structure function $D(x, s)$ is defined as the probability that an electron (positron) radiates a fraction $1 - x$ of its initial energy during the collision (of cm energy $\sqrt{s}$). The radiatively corrected cross section can then be written as,

$$\sigma_f(s) = \int dx_1 dx_2 D(x_1, s) D(x_2, s) \sigma_f^0(\hat{s} = x_1 x_2 s), \tag{4.4}$$

where x_1 and x_2 the electron and positron energy fractions. The leading term of the electron structure function has the form,

$$D(x, s) \simeq \frac{\beta}{2}(1 - x)^{\frac{\beta}{2}-1}, \tag{4.5}$$

where the dimensionless constant β is the effective number of radiation lengths for the process,

$$\beta \equiv \frac{2\alpha}{\pi}\left[\ln\left(\frac{s}{m_e^2}\right) - 1\right] \simeq 0.11.$$

The effect of the convolution described in equation (4.4) is to reduce the peak cross section by $\sim 25\%$ and to shift the peak of the cross section by roughly 120 MeV from the pole position.

It is convenient to write the radiatively corrected cross section in a form that is close to the underlying Breit-Wigner form,

$$\sigma_f(s) = \frac{12\pi}{M_Z^2} \cdot \frac{s\Gamma_{ee}\Gamma_{f\bar{f}}}{(s - M_Z^2)^2 + \Gamma_Z^2 s^2/M_Z^2} \cdot [1 + \delta_{RC}(s)], \tag{4.6}$$

where the effects of the radiative corrections are contained in $\delta_{RC}(s)$. Using equation (4.1), we can expression all of the quantities that appear in equation (4.6) in terms of a single parameter, M_Z. Note that this choice of parameters minimizes the sensitivity of the lineshape to higher-order terms in m_{top} and m_{higgs}.

Equation (4.6) is the basis for the measurement of a number of Z resonance parameters. The Mark II analysis was performed with several sets of constraints:

1. All resonance parameters are constrained to their Standard Model values. In this case, the only free parameter is M_Z. The measurement was performed by summing all of the final states except the electron-positron final states.

2. The visible partial widths are constrained to their Standard Model values and the invisible width is allowed to vary as a free parameter. The total width Γ_Z is decomposed into visible and invisible portions,

$$\begin{aligned}
\Gamma_Z &= \sum \Gamma_{q\bar{q}} + 3\Gamma_{\ell^+\ell^-} + 3\Gamma_{\nu\bar{\nu}} \\
&= \Gamma_{vis} \qquad\qquad + \Gamma_{inv},
\end{aligned} \tag{4.7}$$

where the visible width Γ_{vis} contains all hadronic final states and all charged lepton pairs, and Γ_{inv} contains the neutrino decays and any additional unobserved particles. All of the final states except the electron pairs are used to perform the measurement. The data are therefore fit to a function of two parameters (M_Z and Γ_{inv}),

$$\sigma_f(s) = \frac{12\pi}{M_Z^2} \cdot \frac{s\Gamma_{ee}\Gamma_{f\bar{f}}}{(s - M_Z^2)^2 + (\Gamma_{vis} + \Gamma_{inv})^2 s^2/M_Z^2} \cdot [1 + \delta_{RC}(s)]. \tag{4.8}$$

3. The resonance parameters of the total hadronic cross section are not constrained to their Standard Model values. The hadronic cross section is described by the model-independent form,

$$\sigma_{had}(s) = \frac{s\Gamma_Z^2 \sigma_{had}^0}{(s - M_Z^2)^2 + \Gamma_Z^2 s^2/M_Z^2} \cdot [1 + \delta_{RC}(s)], \tag{4.9}$$

where the free parameters are: M_Z, Γ_Z, and the tree-level hadronic peak cross section σ_{had}^0. The Standard Model prediction for the tree-level peak cross section is,

$$\sigma_{had}^0 = \frac{12\pi}{M_Z^2} \cdot \frac{\Gamma_{ee}\Gamma_{had}}{\Gamma_Z^2} \simeq 41.5 \text{ nb}^{-1}. \tag{4.10}$$

<u>Scanning Theory</u>

A hadron collider gives the experimenter a free energy scan. The hadron structure functions are quite broad in that reasonable quark-quark luminosity is produced over a large range of energies. The electron structure functions have an integrable singularity at $x = 1$. Most of the e^+e^- luminosity is produced near the nominal value of $\sqrt{s}$. The experimenter can therefore choose the most efficient energy scan to optimize the measurement he/she wishes to measure. Note that an optimal scanning strategy requires some *a priori* knowledge of the parameters that one desires to measure. In the earliest runs of the SLC, the Z^0 mass was not well known and it was necessary to search for an enhancement in the event rate. Once M_Z became somewhat constrained, it was possible to choose very efficient operating points. The presence of the Standard Model as a predictor of widths and couplings made this task much easier.

Let us consider a hypothetical scan of N energy-luminosity points:

$$E_b = E_1, E_2, ..., E_N$$

$$\int \mathcal{L} dt = L_1, L_2, ..., L_N.$$

We assume that a cross section σ_i is measured at each point,

$$\sigma_{measured} = \sigma_1, \sigma_2, ..., \sigma_N.$$

The M parameters a_j $(j = 1, M)$ of our theoretical lineshape $\sigma(E)$ can be extracted from a χ^2 fit to the measured points. The quantity χ^2 is defined as,

$$\chi^2 \equiv \sum_{i=1}^{N} \frac{[\sigma_i - \sigma(E_i)]^2}{(\delta\sigma_i)^2}, \tag{4.11}$$

where $\delta\sigma_i$ is the error on the ith measurement.

The best estimate of the parameters $(\bar{a}_j)$ is the one that minimizes χ^2. The parameter errors are found from a Taylor expansion of χ^2 about the minimum value,

$$
\begin{aligned}
\chi^2 &= \chi^2(\bar{a}) + \frac{1}{2} \sum_{j,k=1}^{M} \frac{\partial^2 \chi^2}{\partial a_j \partial a_k}(a_j - \bar{a}_j)(a_k - \bar{a}_k) \\
&= \chi^2(\bar{a}) + \sum_{j,k=1}^{M} (\mathbf{C}^{-1})_{jk}(a_j - \bar{a}_j)(a_k - \bar{a}_k)
\end{aligned}
\tag{4.12}
$$

where the matrix $\mathbf{C}^{-1}$ is the inverse of the parameter covariance matrix. The error hyperellipsoid is determined by changing χ^2 by one unit about the minimum value. It is straightforward to show that the parameter errors are given by the diagonal elements of the covariance matrix $\mathbf{C}$,

$$(\delta a_j)^2 = \mathbf{C}_{jj}. \tag{4.13}$$

Averaging equation (4.12) over many experiments, the inverse matrix can be expressed in the following form,

$$(\mathbf{C}^{-1})_{jk} = \sum_{i=1}^{N} \frac{1}{(\delta\sigma_i)^2} \cdot \left[\frac{\partial\sigma}{\partial a_j}(E_i)\right] \cdot \left[\frac{\partial\sigma}{\partial a_k}(E_i)\right]. \tag{4.14}$$

Although equation (4.14) is quite general, it is useful to express the cross section errors in terms of the luminosity and the theoretical cross section. Ignoring the statistical errors on the luminosity measurements,[*] we can express the cross section errors as $(\delta\sigma_i)^2 = \sigma(E_i)/L_i$. Equation (4.14) can then be written as,

$$(\mathbf{C}^{-1})_{jk} = \sum_{i=1}^{N} \frac{L_i}{\sigma(E_i)} \cdot \frac{\partial\sigma}{\partial a_j}(E_i) \cdot \frac{\partial\sigma}{\partial a_k}(E_i) = \sum_{i=1}^{N} L_i \cdot S(E_i, a_j) \cdot S(E_i, a_k), \tag{4.15}$$

where we define the so-called *sensitivity function* $S(E, a_j)$ as

$$S(E, a_j) \equiv \frac{1}{\sqrt{\sigma(E)}} \cdot \frac{\partial\sigma}{\partial a_j}(E). \tag{4.16}$$

If the lineshape is a function of a single parameter or if the off-diagonal elements of the inverse matrix $\mathbf{C}^{-1}$ are small, the parameter errors have a particularly simple form,

$$(\delta a_j)^{-2} \simeq \sum_{i=1}^{N} L_i \cdot \left[S(E_i, a_j)\right]^2. \tag{4.17}$$

Equation (4.17) implies that the error δa_j is minimized when the integrated luminosity is concentrated in regions of scan energy where $|S(E, a_j)|$ is large. Note that $|S(E, a_j)|$ is large where the derivative $|\partial\sigma/\partial a_j|$ is large and where the cross section is small.

The correlations between the parameters are described by the off-diagonal elements of the matrices $\mathbf{C}^{-1}$ and $\mathbf{C}$ (the error ellipsoid is unrotated if they vanish). *The presence of non-zero correlation always increases a parameter error beyond the value given in equation (4.17).*[†] It is clearly important to minimize the off-diagonal elements by our choice of the scan point luminosities.

Equations (4.15) and (4.13) predict the complete parameter error matrix in terms of the theoretical lineshape and the scan point luminosities. *Note that it is assumed that χ^2 is well-defined $(N > M)$ and that a sufficient number of events is collected at each point that the errors are Gaussian.*

Since any cross section measurement has an associated normalization uncertainty, it is important to consider the sensitivity of the final result to systematic shifts in the measured cross sections. Expanding the theoretical cross section in parameter space about the best

---

[*] This assumption is quite valid for the measurement of non-resonant cross sections.

[†] The presence of non-zero correlation allows the error associated one parameter to *leak* into the error associated with another parameter.

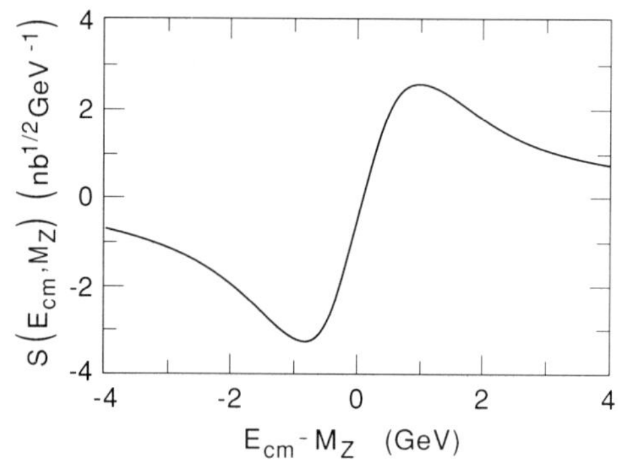

Figure 4

The sensitivity function for M_Z as a function of center-of-mass energy about the Z pole, $E - M_Z$.

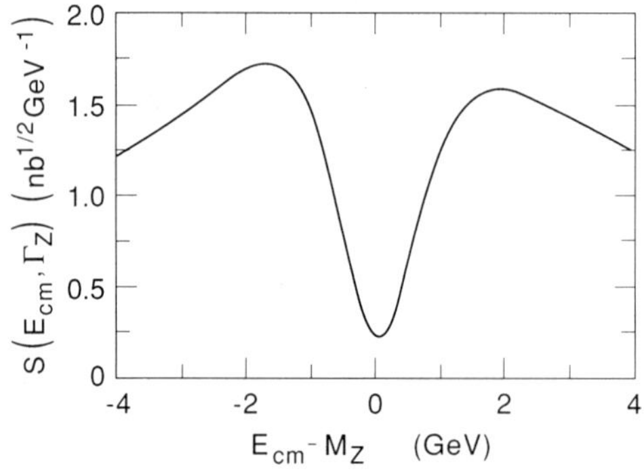

Figure 5

The sensitivity function for Γ_Z as a function of center-of-mass energy about the Z pole, $E - M_Z$.

estimates $\bar{a}_j$, it is straightforward to derive the average shift in a parameter Δa_j caused by shifts in the measured cross sections $\Delta\sigma_i$,

$$\langle\Delta a_j\rangle = \sum_{k=1}^{M} \mathbf{C}_{jk} \cdot \sum_{i=1}^{N} L_i \cdot \frac{\Delta\sigma_i}{\sigma_i} \cdot \frac{\partial\sigma}{\partial a_k}(E_i). \qquad (4.18)$$

It is clear that we would like to choose the energies and luminosities to minimize the parameter errors and the correlations between the parameters. We can be guided in this task by examining the energy dependence of the functions $S(E, a_j)$.

As an example of the usefulness of the sensitivity functions, let us consider the measurement of the model-independent parameters of the hadronic cross section. For simplicity, we assume that values of M_Z, Γ_Z, and $\sigma^0_{had}(M_Z^2)$ are 91 GeV, 2.5 GeV, and 40 nb, respectively. The sensitivity functions for M_Z, Γ_Z, and $\sigma^0_{had}(M_Z^2)$ are plotted in Figures 4-6 as functions of $E - M_Z$. The maximum sensitivity to M_Z occurs at the scan energies -0.8 GeV and +1.0 GeV about the pole. Note that there is little sensitivity to Γ_Z at these points. The maximum sensitivity to Γ_Z occurs at points that are approximately ± 2 GeV about the pole. If we choose our energy-luminosity points symmetrically about the pole, the sum of the products $S(E_i, M_Z) \cdot S(E_i, \Gamma_Z)$ will tend to cancel since $S(E, M_Z)$ is odd about the pole and $S(E, \Gamma_Z)$ is even about the pole. The maximum sensitivity to σ^0_{had} occurs at the pole. The same odd-even effect that cancels the M_Z-Γ_Z correlation will cancel the M_Z-σ^0_{had} correlation. The Γ_Z-σ^0_{had} correlation cannot be cancelled by a choice of scan energies. However, it is not intrinsically large since $S(E, \Gamma_Z)$ is small in the energy region where $S(E, \sigma^0_{had})$ is large.

In general, a scan strategy that is based upon equations (4.15) and (4.13) is a problem in linear programming. The scan planner must decide how important various parameters are and what constraints must be satisfied. Nevertheless, fairly simple considerations lead to the conclusion that a minimal Z-pole scan should include points at 0, ± 1, and ± 2 GeV about the pole.

4.3. THE EXPERIMENTAL ANALYSIS

In order to appreciate the selection criteria that must be applied to the data, we must first discuss the signatures and the relative rates of various processes that occur in an electron-positron collider.

The Electron-Positron Environment

Unlike the situation with hadron colliders, the most copious processes in a high energy e^+e^- collider are also the most interesting ones. The signatures and relative sizes of the various processes are indicated in Table V. The most serious background to Z^0 production is due to the various two-photon processes. The two-photon background is rather trivial to remove from the data sample (a total energy cut is sufficient to suppress it by several orders of magnitude).

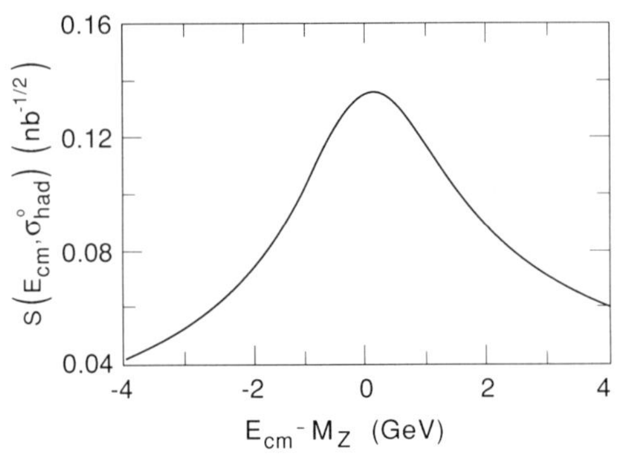

Figure 6

The sensitivity function for $\sigma^0_{had}(M_Z^2)$ as a function of center-of-mass energy about the Z pole, $E - M_Z$.

Table V

Event Type	Signature	$\sigma(\sqrt{s} = M_Z)$
$e^+e^- \rightarrow Z^0 \rightarrow$ hadrons	2-3 jets $\gtrsim 20$ charged tracks	$\sim$30 nb
$e^+e^- \rightarrow e^+e^-$ (small angle)	45 GeV clusters in small angle *tagger*	$\sim$50-200 nb (dep on acceptance)
$e^+e^- \rightarrow e^+e^-\ell^+\ell^-$ $e^+e^- \rightarrow e^+e^-h^+h^-$	Transversely balanced low energy track pairs	$\sim$7-8 nb (dep on acceptance)
$e^+e^- \rightarrow Z^0 \rightarrow \mu^+\mu^-$	back-to-back high energy tracks	$\sim$1.5 nb
$e^+e^- \rightarrow Z^0 \rightarrow \tau^+\tau^-$	acolinear track pairs 1-3 combinations	$\sim$1.5 nb

<u>Event Selection</u>

The Mark II Z^0 mass and width measurement[8] was performed with a data sample that corresponds to a total integrated luminosity of 19.3 nb^{-1} that was collected at 10 center-of-mass energies. The hadronic final states were selected with the following criteria:

1. Each event is required to contain three or more charged tracks. Each charged track must originate from within a cylindrical volume of 1 cm radius and 6 cm length that is centered upon the nominal interaction point. The momenta of all reconstructed tracks must be larger than 110 MeV/c and the reconstructed polar angle must fall within the region $|\cos\theta| < 0.92$.

2. The visible energy (track momenta and/or calorimeter energy) that is observed in each of the forward and backward hemispheres must be larger than 5% of the center-of-mass energy. The criterion suppresses beam-gas and two-photon events.

3. Any events which are also identified as $\tau^+\tau^-$ pairs are removed from the sample.

A total of 450 events passed the selection criteria. The detection efficiency for hadronic events is 95.3±0.6%. The residual background contamination is at level of a few parts in 10^3 (mostly from $\tau^+\tau^-$ events).

Leptonic events were selected with the following criteria:

1. Each event is required to contain between two and six charged tracks.

2. The polar angle of the thrust axis must be contained within the region $|\cos\theta_{thrust}| < 0.65$.

3. The energy measured in the calorimeters must be less than 80% of the center-of-mass energy. (This criterion eliminates e^+e^- pairs.)

4. Each event must satisfy either of the following:

   (a) If the momenta of each of two tracks are larger than 50% of the beam momentum, the event is categorized as a muon pair.

   (b) If the event fails the previous criterion and has a total visible energy that is larger than 10% of the center-of-mass energy, the event is categorized as a $\tau^+\tau^-$ pair.

A total of 30 events passed the lepton selection criteria. The efficiencies of the selection criteria are 99±1% and 96±1% for muon and tau pairs produced within the fiducial region $|\cos\theta| < 0.65$, respectively.

<u>Luminosity Measurement</u>

The luminosity at each scan point is inferred from the measured rate of small angle (15-160 milliradian) Bhabha scattering. In the small angle region, this process is dominated by t-channel exchange of photons and is independent of the parameters of the Z^0 system. The tree-level differential cross section has the form,

$$\frac{d\sigma_{lum}}{d\theta} \simeq \frac{4\pi\alpha^2}{s} \cdot \frac{1}{\theta^3}, \tag{4.19}$$

where the scattering angle θ is assumed to be small. An accurate determination of the luminosity requires that the radiative corrections be included in equation (4.19). Nevertheless, equation (4.19) does illustrate one of the difficulties in the measurement of the luminosity. The measured cross section σ_{lum}^{meas} is a sensitive function of the angular acceptance of the detector edges,

$$\sigma_{lum}^{meas} \simeq \frac{2\pi\alpha^2}{s}\left(\frac{1}{\theta_1^2} - \frac{1}{\theta_2^2}\right), \tag{4.20}$$

where θ_1 and θ_2 are the angles of the inner and outer detector edges.

It is clear from equation (4.20) that the very small angle luminosity monitor, the minisam, has a much larger rate than does the larger angle SAM (the counting rate of the former is six times larger than the latter). However, the larger angle device has good spatial (angular) resolution whereas the minisam has no segmentation in polar angle. The systematic error that is associated with the detector acceptance is substantially smaller for the SAM than for the minisam. The strategy that was used to determine the luminosity of each scan point was therefore to use the minisam to determine the relative luminosities of different scan points and to use the SAM to determine the overall normalization.

Each SAM event was required to satisfy the following selection criteria:

1. The measured energies of the scattered electron and positron were required to be larger than 40% of the beam energy.

2. The measured scattering angles had to satisfy one of the following criteria:

    (a) Both scattering angles were larger than 65 milliradians. Events in this category were assigned a weight of 1.0.

    (b) One of the scattering angles was larger than 60 milliradians and the second was larger than 65 milliradians. Events in this category were assigned a weight of 0.5.

The weighting procedure reduced the sensitivity of the result to detector misalignments and to radiative corrections. A total of 485 events satisfied the selection criteria.

The theoretical cross section for accepted SAM events is given by the following expression,

$$\sigma_S(E_{cm}) = 25.2 \text{ nb} \cdot \left[\frac{91.1 \text{ GeV}}{E_{cm}}\right]^2,$$

where E_{cm} is the center-of-mass energy. The systematic error due to uncertainties on the detector resolution and position is 2%. The systematic error due to uncertainties on the radiative corrections are taken to be 2%. The combined systematic error is therefore 3%.

The total integrated luminosity for the 10 scan points is therefore evaluated to be,

$$\int \mathcal{L} dt = 19.3 \pm 0.9 \ \text{ nb}^{-1},$$

where the 5% error includes both statistical and systematic effects.

Each minisam event was required to satisfy the following selection criteria:

1. The measured energies of the scattered electron and positron were required to be larger than 25 GeV in diagonally opposite quadrants.

2. The timing of the minisam signals was required to be consistent with that expected for a scattering process.

A total of 4299 minisam events were recorded during the energy scan. The theoretical cross section for accepted minisam events is given by the following expression,

$$\sigma_M(E_{cm}) = 230 \text{ nb} \cdot \left[\frac{91.1 \text{ GeV}}{E_{cm}}\right]^2 .$$

The luminosity of each point, L_i, is given by the following expression,

$$L_i = \frac{(N_S^i + N_M^i)/(\sigma_S(E_i) + \epsilon_M^i \sigma_M(E_i))}{\sum_i (N_S^i + N_M^i)/(\sigma_S(E_i) + \epsilon_M^i \sigma_M(E_i))} \cdot \int \mathcal{L} dt, \qquad (4.21)$$

where: N_M^i and N_S^i are the number of SAM and minisam events recorded at energy E_i, respectively; and where ϵ_M^i is the minisam efficiency for scan point i (the minisam was sensitive to radiation background during some runs).

Experimental Results

In order to improve the statistical accuracy of the mass and width measurements, the $q\bar{q}$ final states are combined with the leptonic final states ($\mu^+\mu^-$ and $\tau^+\tau^-$ within the region $|\cos\theta| < 0.65$) to calculate the cross section at each scan point. The average energy, integrated luminosity, and measured cross section are listed in Table VI.

The measured cross sections and the results of the one, two, and three parameter fits are shown in Figure 7. The parameter estimates for the three fit hypotheses are listed in Table VII.

The measured value of M_Z agrees well with the results of the LEP measurements. The number of neutrino species is consistent with the expected number of 3. Including the systematic errors, the upper limit on the number of light neutrinos is 3.9 with 95% confidence.

Systematic Errors

The errors listed in Table VII include systematic uncertainties on the cross section normalization and on the energy scale of the SLC. The various resonance parameters vary in their sensitivity to the energy scale and normalization uncertainties.

The determination of M_Z depends completely on the accelerator energy scale. The energies of the SLC beams are measured by a pair of energy spectrometers after they have collided. A schematic diagram of the north (electron) energy spectrometer is shown in Figure 8. The beam is focused by a quadrupole doublet to a point at the detector plane. The beam passes through a small horizontal bend dipole magnet, a large vertical bend dipole magnet, and a second small horizontal bend magnet. The passage of the beam through the horizontal bend magnets produces flat distributions of synchrotron radiation which are detected by a phosphor screen detector. The separation of the flat distributions is proportional to the beam energy. The spectrometers have sufficient resolution to determine the center-of-mass energy (and the value of M_Z) to ±40 MeV.

The model-independent determinations of M_Z are completely insensitive to the normalization uncertainty. The model constrained determinations of M_Z have a slight sensitivity to the normalization uncertainty. These uncertainties are typically a few MeV or less (even with the model constraints, most of the M_Z information is derived from the resonance shape).

The peak cross section and the invisible width are strongly affected by normalization uncertainty. This can be seen from an inspection of equation (4.8). The invisible width enters the cross section as a component of the total width. The influence of the total width

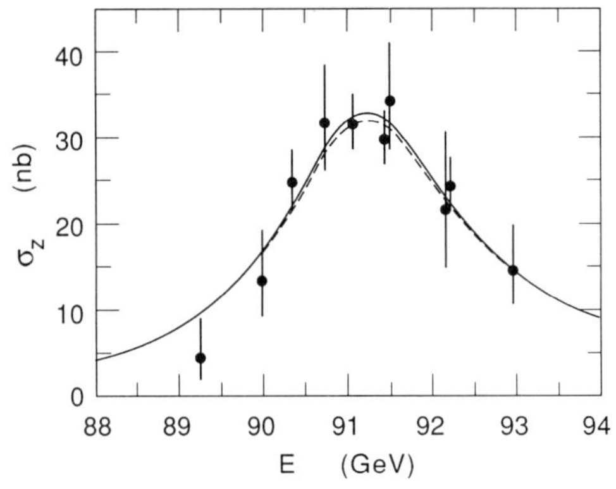

Figure 7

The Z^0 lineshape as measured by the Mark II Collaboration.[8] The dashed curve is the result of a single parameter fit (for M_Z). The results of two and three parameter fits are indistinguishable and are shown as the solid curve.

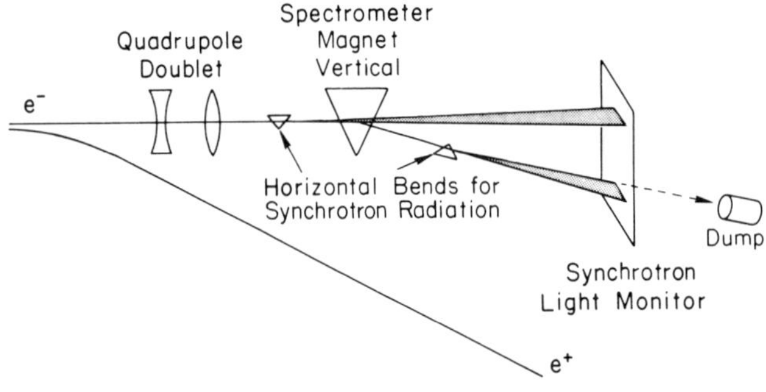

Figure 8

The north (electron) energy spectrometer of the SLC. The beam is focused by a quadrupole doublet to a point at the detector plane. The beam passes through a small horizontal bend dipole magnet, a large vertical bend dipole magnet, and a second small horizontal bend magnet. The passage of the beam through the horizontal bend magnets produces flat distributions of synchrotron radiation which are detected by a phosphor screen detector. The separation of the flat distributions is proportional to the beam energy.

Table VI

Average energy, integrated luminosity, number of events, MiniSAM efficiency and σ_Z for each energy scan point. The luminosity for each scan point is given by Lum $= (N_S + N_M)/\sigma_L$, where $\sigma_L = \sigma_S + \epsilon_M \sigma_M$. The given error is the statistical error on N_S and N_M only; there are additional statistical errors on σ_L due to the scaling errors on σ_S and σ_M (see text). The total luminosity is calculated from the 485 "precise" SAM Bhabha events and has an overall 2.8% systematic error.

Scan Point	$<E>$ (GeV)	N_S	N_M	ϵ_M	Lum. (nb^{-1})	Z Decays Had.	Lep.	Tot.	σ_Z (nb)
3	89.24	24	166	0.99	0.68±0.05	3	0	3	$4.5^{+4.5}_{-2.5}$
5	89.98	36	174	0.99	0.76±0.05	8	2	10	$13.5^{+6.0}_{-4.3}$
10	90.35	116	617	1.00	2.61±0.10	60	2	62	$24.8^{+3.8}_{-3.3}$
2	90.74	54	266	0.96	1.21±0.07	33	3	36	$31.7^{+6.8}_{-5.5}$
7	91.06	170	923	0.99	4.08±0.12	114	6	120	$31.6^{+3.4}_{-3.1}$
8	91.43	164	879	0.91	4.12±0.13	108	6	114	$29.8^{+3.3}_{-2.9}$
4	91.50	53	275	0.99	1.23±0.07	33	6	39	$34.3^{+7.0}_{-5.7}$
1	92.16	31	105	0.97	0.54±0.05	11	0	11	$21.5^{+9.2}_{-6.6}$
9	92.22	128	680	0.98	3.05±0.11	67	4	71	$24.3^{+3.4}_{-3.0}$
6	92.96	39	214	0.98	1.00±0.07	13	1	14	$14.6^{+5.4}_{-4.0}$
Totals		815	4299		19.3±0.9	450	30	480	

Table VII

Z resonance parameters. The three fits are described in the text.

Fit	m_Z GeV/c^2	N_ν	Γ GeV	σ_0 nb
1	91.14±0.12	–	–	–
2	91.14±0.12	2.8±0.6	–	
3	91.14±0.12	–	$2.42^{+0.45}_{-0.35}$	45±4

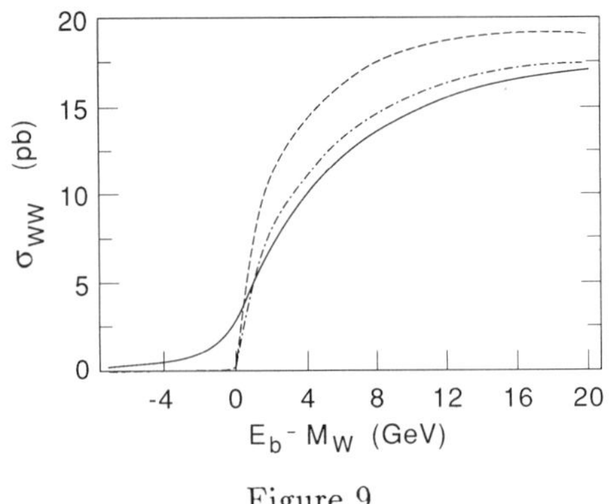

Figure 9

The cross section for the process $e^+e^- \to W^+W^-$ as a function of $E_b - M_W$. The mass and width of the W are assumed to be 80 GeV and 2.1 GeV, respectively. Note that three curves are plotted: the dashed curve is the basic tree-level cross section; the dashed-dotted curve is the cross section including the effect of initial state radiation; and the solid curve is the cross section including initial state radiation and the effect of a finite W width.

is maximized when the center-of-mass energy is $s = M_Z^2$. The effect of the normalization uncertainty $\delta\sigma$ upon the invisible width is approximately,

$$\delta\Gamma_{inv} \simeq 1.5 \text{ GeV} \cdot \left(\frac{\delta\sigma}{\sigma}\right) = 75 \text{ MeV}$$

$$\delta N_\nu \simeq 9 \cdot \left(\frac{\delta\sigma}{\sigma}\right) = 0.45.$$

The measurement of Γ_Z depends almost entirely upon the measurement of the resonance shape. It is therefore insensitive to the absolute energy and normalization errors. It is sensitive to point-to-point errors in the energy and luminosity. These are much smaller than the absolute errors.

The effects of the theoretical uncertainties on the Z lineshape upon the value of the extracted parameters are small as compared with the energy scale and normalization uncertainties.

4.4. Mass and Width of the W

The measurement of the W boson mass and width will become possible in the second phase of LEP operation. The installation of superconducting RF cavities will permit the beam energy to be increased to a value above the threshold for the process $e^+e^- \to W^+W^-$. The low event rates at the W pair threshold will make essential to optimize the scan to measure the threshold shape.

High Energy e^+e^- Cross Sections

The tree-level expression for the W-pair cross section is somewhat complex.[9] The inclusion of initial state radiation (as in equation (4.4)) and finite widths for the final state W bosons involves a four dimensional convolution of the tree-level expression. We therefore choose to present only the result of a Monte Carlo integration. The cross section for the process $e^+e^- \to W^+W^-$ is plotted in Figure 9 as a function of $E_b - M_W$ where E_b is the single beam energy. The mass and width of the W are assumed to be 80 GeV and 2.1 GeV, respectively. Note that three curves are plotted: the dashed curve is the basic tree-level cross section; the dashed-dotted curve is the cross section including the effect of initial state radiation; and the solid curve is the cross section including initial state radiation and the effect of a finite W width. The inclusion of initial state radiation reduces the size of the cross section. The finite W width produces non-zero cross section at energies below the nominal threshold at $E_b = M_W$.

The basic $e^+e^- \to f\bar{f}$ cross section for five quark and three lepton flavors increases from about 7 units of R at center-of-mass energies below the Z^0 pole to 10 units of R at energies above the Z^0 pole.[*] At $\sqrt{s} = 160$ GeV, the tree-level cross section is approximately 34 pb. Unfortunately, the initial state radiative corrections increase this number enormously. Although the photon structure functions decrease greatly as x is decreased from 1, the Z pole is sufficiently large that the convolution given in equation (4.4) is several times larger than the tree-level cross section. The process $e^+e^- \to \gamma Z^0$ therefore dominates the visible cross section at W-pair threshold. Using equation (4.4), we estimate the size of the visible cross section to be $\sim$150 pb at $\sqrt{s} = 160$ GeV.

$\underline{e^+e^- \to W^+W^-\ \text{Threshold Scan}}$

There are several different techniques that can be used to measure the W mass at LEP II. It is possible to extract M_W from the measured distributions of jet masses or lepton energies. These methods are are described in Reference 10. The technique that we'll discuss here is the measurement of the threshold behavior of the W pair cross section.

It is clear than the W mass can be extracted from the *step* in the cross section that is shown in Figure 9. Since there is a large background from ordinary processes, it is necessary to apply selection criteria to the data to improve the signal-to-noise ratio. The background processes produce mostly two- and three jet hadronic events or lepton pair events that are often highly boosted along the beam direction. The visible energy of the background is often small as compared with $\sqrt{s}$. The W-pair events appear most often as four-jet events ($\sim$44% of W-pairs) or as an energetic lepton and two jets ($\sim$44% of W-pairs). The authors of Reference 10 have studied a number of selection criteria to reduce the background cross section to less than $\sim$1 pb while retaining $\sim$75% of the four-jet and $\sim$45% of the lepton+two-jet events (we assume that τ leptons cannot be used and that one third of the remaining events are eliminated by the isolation cut used to suppress heavy flavor events). Assuming that the residual background is due to the large $\sqrt{s}$ continuum, the measured cross section would have the following form,

$$\sigma_{meas}(E_b) = \varepsilon\sigma_{ww}(E_b) + \frac{B}{(2E_b)^2},\qquad (4.22)$$

where: ε is the efficiency to identify a W-pair event ($\varepsilon \simeq 0.53$); $\sigma_{ww}(E_b)$ is the cross section plotted in Figure 9; and B is a constant that represents the residual background (which presumably scales as $1/s$).

$\underline{\text{Sensitivity Functions}}$

We can analyze the M_W and Γ_W sensitivity of a cross section scan of the W pair threshold by using the scanning theory that was discussed in the last section. Numerically

---

[*] The unit of R is the cross section for $e^+e^- \to \gamma^* \to \mu^+\mu^-$. Numerically, the cross section has the value $\sigma_R = 86.8$ nb-GeV$^2/s$.

differentiating the measured cross section (as defined in equation (4.22)), it is straightforward to calculate the sensitivity functions for M_W, Γ_W, and the background constant B. For the purpose of this exercise, we assume that $B = 1$ pb $\cdot (2M_W)^2$ or that the background cross section is 1 pb at W-pair threshold.

The sensitivity function $S(E_b, M_W)$ is plotted in Figure 10 as a function of $\epsilon_b = E_b - M_W$. Note that the maximum sensitivity occurs at $\epsilon_b \simeq 0.5$ GeV.

The sensitivity function $S(E_b, \Gamma_W)$ is shown in Figure 11 as a function of ϵ_b. As one would expect, it peaks just below the nominal threshold ($\epsilon_b = -1$ GeV) where the width-induced *tail* in the cross section is largest. The function $S(E_b, \Gamma_W)$ decreases rapidly as E_b is increased. It passes through zero near $\epsilon_b = 1$ GeV and plateaus above $\epsilon_b = 3$ GeV. The sensitivity in the plateau region is due to the reduction in the cross section caused by the finite width (see Figure 9). The maximum value of $|S(E_b, \Gamma_W)|$ is smaller than the maximum value of the mass sensitivity function by a factor of three. A good measurement of Γ_W will clearly require a substantial commitment of luminosity to a point of very small cross section. Note that the product $S(E_b, M_W) \cdot S(E_b, \Gamma_W)$ is an odd function about the point $\epsilon_b = 1$ GeV. In principle, the M_W-Γ_W correlation can be cancelled by measuring the cross section on both sides of this point. The functions $S(E_b, M_W)$ and $S(E_b, \Gamma_W)$ are not large in the region $\epsilon_b > 1$ GeV. The cancellation of the correlation therefore requires a substantial commitment of luminosity to a relatively insensitive region.

The function $S(E_b, B)$ is plotted as a function of ϵ_b in Figure 12. As one would expect, the background sensitivity is largest at small beam energy and decreases dramatically as E_b increases through the W pair threshold. Note that it is possible to cancel the B-Γ_W correlation but that it is not possible to cancel the B-M_W correlation.

<u>Scan Strategies</u>

It is clear that precise measurements of M_W and Γ_W require that LEP be operated in regions of small cross section. Since all other studies of the W-pair system require a large sample of data, there will be considerable pressure to operate the machine on the cross section plateau at the largest available energy. In order to estimate how precisely M_W and Γ_W could be measured in a 1-2 year run (500 pb^{-1}), we assume that 50% of the luminosity is dedicated to operating at the largest available energy (we assume that $\epsilon_b = 15$ GeV or $\sqrt{s} = 190$ GeV is achieved) and the remaining 50% is dedicated to operation in the threshold region.

It is instructive to first consider an extremely unrealistic scan scenario. We assume that we will measure only one parameter and that the other parameters are precisely known. In this case, we need only one scan point in the threshold region for a constrained fit. We choose to allocate the entire 250 pb^{-1} luminosity to operation at the most mass-sensitive point ($\epsilon_b = 0.5$ GeV) or at the most width-sensitive point ($\epsilon_b = -1$ GeV). Using equation (4.17) we estimate the precision of these measurements to be

$$\delta M_W = 92 \text{ MeV} \quad \text{or} \quad \delta \Gamma_W = 286 \text{ MeV}.$$

The M_W measurement would be a very desirable result. The Γ_W measurement is not competitive with the recent indirect determinations that). have been published by the CDF and UA2 collaborations,[11,12]

$$\Gamma_W = (0.85 \pm 0.08) \cdot \Gamma_Z = 2.19 \pm 0.20 \text{ GeV (CDF)}$$
$$\Gamma_W = (0.89 \pm 0.08) \cdot \Gamma_Z = 2.30 \pm 0.20 \text{ GeV (UA2)}.$$

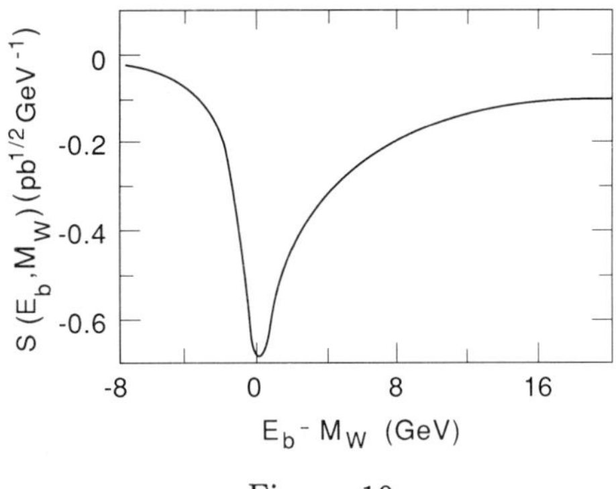

Figure 10

The sensitivity function for M_W as a function of the single beam energy about the W pair threshold $E_b - M_W$.

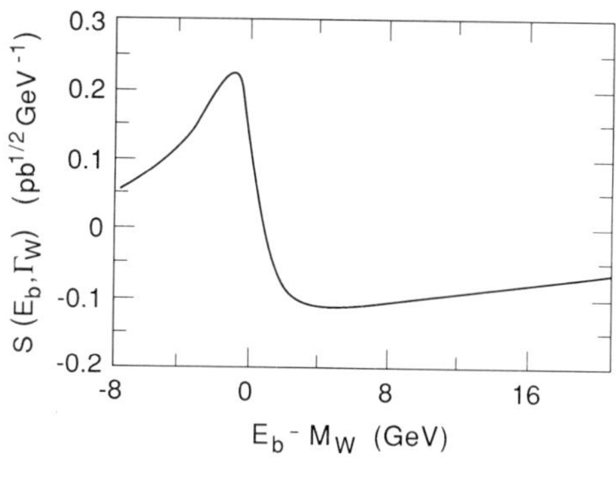

Figure 11

The sensitivity function for Γ_W as a function of the single beam energy about the W pair threshold $E_b - M_W$.

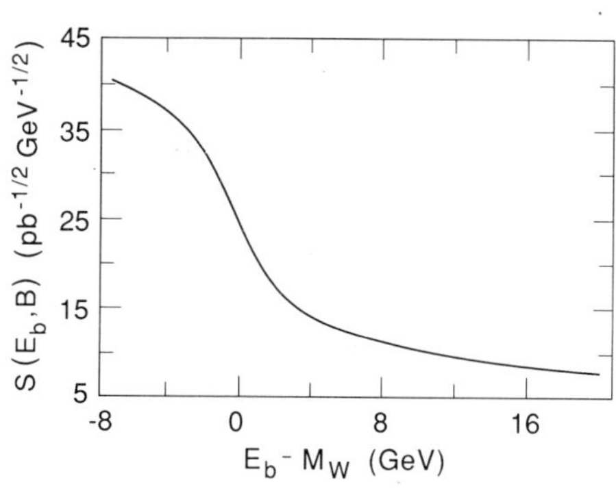

Figure 12

The sensitivity function for the background parameter B as a function of the single beam energy about the W pair threshold $E_b - M_W$.

Since the width cannot be measured to an interesting level, it is clearly unwise to design a scan to measure Γ_W. We therefore concentrate on the measurement of M_W.

A real measurement of M_W will require that the background constant B be varied as a fit parameter. Unfortunately, the B-M_W correlation cannot be canceled by a clever choice of scan points. It is therefore necessary to measure both parameters well.

The number of scan points is somewhat arbitrary. A minimum of three points are required to constrain the two parameter problem. The presence of a high energy point implies that only two points are needed in the threshold region. Equation (4.15) implies that several closely spaced points in a region of large sensitivity are equivalent to a single point in the same region. We can therefore analyze the optimization of the M_W measurement by considering a two-point threshold measurement.

An optimal scan must include an energy point in a region of large background sensitivity $|S(E_b, B)|$ and a point near the maximum of the mass sensitivity function $|S(E_b, M_W)|$. We choose the scan point energies to be $\epsilon_b = -5$ GeV and $\epsilon_b = 0.5$ GeV, respectively.[*] The apportionment of the available luminosity between the two points is a straightforward problem in one-dimensional optimization. We find that the error δM_W has a very broad minimum about the ratio of luminosities, $L(0.5 \text{ GeV})/L(-5 \text{ GeV}) \simeq 2/1$. If the luminosities of the -5 GeV and 0.5 GeV points are 85 pb^{-1} and 165 pb^{-1}, respectively, the minimum value of the error δM_W is approximately 155 MeV.

A two-point threshold scan is somewhat risky. It is safer to bracket the region of maximum M_W sensitivity with several scan points. We therefore construct an optimal four-point scan (a five-point measurement when the $\epsilon_b = 15$ GeV point is included) by assigning one third of the 165 pb^{-1} (55 pb^{-1}) to each of three points: $\epsilon_b = 0$ GeV, 0.5 GeV, and 1.0 GeV. It is instructive to compare this scan (Scan 1) with a slightly modified version. The modified version (Scan 2) is created by shifting the luminosity from the $\epsilon_b = 0$ GeV point to $\epsilon_b = -1$ GeV. We expect the second scan strategy to improve the width measurement at the expense of the mass measurement. Finally, we note that our modified scan strategy is similar to the scan strategy that was studied in Reference 10 (which we label Scan 3). The authors of Reference 10 assigned 100 pb^{-1} to each of the following five points: $\epsilon_b = -5$ GeV, -1 GeV, 0 GeV, 1 GeV, and 15 GeV.

Using equation (4.15) and the sensitivity functions, the performance of each scan scenario can be estimated. The expected number of detected events and the expected precisions δM_W, $\delta \Gamma_W$, and δB are listed in Table VIII for each of the three scan strategies. The presence of a high energy point in each strategy reduces the M_W-Γ_W correlation sufficiently that the M_W precision obtained from the three parameter fit is essentially identical to that obtained from a two-parameter fit.

As one might expect, the third scan strategy which allocates 400 pb^{-1} to the threshold measurement provides the most precise M_W measurement, $\delta M_W = 150$ MeV. *The M_W precision obtained from the optimized mass scan (Scan 1) is worse by 7%. Note however, that Scan 1 produces nearly 60% more events than does Scan 3.* Surprisingly, the second scan strategy provides a slightly better width measurement than does the third strategy. This occurs because the second scan produces a smaller B-Γ_W correlation than does the third scan strategy.

---

[*] Varying the energy of the second point about $\epsilon_b = 0.5$ GeV verifies that the B-M_W correlation does not shift the point of maximum M_W sensitivity.

The predicted results of three different five-point measurements of the W-pair threshold. Scan 1 is optimized for the measurement of M_W. Scan 2 is an attempt to improve the measurement of Γ_W. Scan 3 is identical to the threshold scan used in Reference 10. The results are presented for several assumptions about the level of residual background B and the W-pair detection efficiency.

Quantity	Scan 1	Scan 2	Scan 3
$L[\text{-}5\text{ GeV}]$ (pb^{-1})	85	85	100
$L[\text{-}1\text{ GeV}]$ (pb^{-1})	0	55	100
$L[0\text{ GeV}]$ (pb^{-1})	55	0	100
$L[0.5\text{ GeV}]$ (pb^{-1})	55	55	0
$L[1\text{ GeV}]$ (pb^{-1})	55	55	100
$L[15\text{ GeV}]$ (pb^{-1})	250	250	100
$B = 1.0\text{ pb}\cdot[2M_W]^2$ $\varepsilon_{ww} = 0.53$			
Number of Events	2951	2912	1863
δM_W (MeV)	160	176	150
$\delta\Gamma_W$ (MeV)	531	482	492
δB (pb $\cdot[2M_W]^2$)	0.12	0.12	0.12
$B = 0.5\text{ pb}\cdot[2M_W]^2$ $\varepsilon_{ww} = 0.53$			
Number of Events	2737	2698	1627
δM_W (MeV)	137	154	130
$\delta\Gamma_W$ (MeV)	508	450	448
δB (pb $\cdot[2M_W]^2$)	0.096	0.098	0.098
$B = 1.0\text{ pb}\cdot[2M_W]^2$ $\varepsilon_{ww} = 0.70$			
Number of Events	3760	3709	2309
δM_W (MeV)	130	144	123
$\delta\Gamma_W$ (MeV)	453	407	410
δB (pb $\cdot[2M_W]^2$)	0.12	0.13	0.13

It is clear from equation (4.22) that the functions $S(E_b, a_j)$ are sensitive to the level of residual background and to the W-pair detection efficiency. We investigate these effects by reducing the background constant to $B = 0.5 \text{ pb} \cdot (M_W)^2$ and by increasing the detection efficiency to $\varepsilon_{ww} = 0.70$. The results are listed in Table VIII. The error δM_W is improved by approximately 20 MeV in the case that the background is reduced by a factor of two.

The mass error is improved by approximately 30 MeV when the efficiency is increased. Note that the optimal luminosity ratio $L(0.5 \text{ GeV})/L(-5 \text{ GeV})$ is nominally sensitive to both effects. However, the optimal region is so broad that the use of a 2/1 ratio degrades the result by less than 1%.

<u>Systematic Errors</u>

The measurement of the W-pair threshold is affected by systematic uncertainties on the energy scale and cross section normalization. The energy scale uncertainty affects the M_W measurement directly. Assuming that the fractional error on the beam energy scale is constant, the uncertainty on M_W should be comparable to the one that applies to the M_Z measurement. By 1994, this uncertainty is expected to be ~ 20 MeV.

The sensitivity of the results given in Table VIII to normalization errors can be estimated from equation (4.18). Taking the first scan strategy as an example, we estimate that the uncertainties on the parameters are related to an overall normalization uncertainty $\delta\sigma/\sigma$ as follows,

$$\delta M_W = -2.26 \text{ GeV} \cdot \frac{\delta\sigma}{\sigma}$$

$$\delta\Gamma_W = -19.3 \text{ GeV} \cdot \frac{\delta\sigma}{\sigma}.$$

The normalization error must be controlled to the 3% level to avoid inflating the M_W error.

<u>Sensitivity to Assumptions</u>

Our analysis assumes that we have complete *a priori* knowledge of the W resonance parameters. Although the characteristic width in E_b space of the M_W-sensitive region is larger than the current uncertainty on M_W, our precision estimates are likely to be somewhat optimistic. It is possible to alter the results by $\lesssim 10\%$ by varying the resonance parameters over reasonable intervals.

<u>Conclusions</u>

Despite the uncertainties on the ultimate W-pair detection efficiency and residual background contamination, several conclusions can be drawn from this analysis:

1. The most sensitive scan region for the measurement of M_W is $\epsilon_b = 0\text{-}1$ GeV. *The mapping of the entire threshold shape would produce a less precise measurement.*

2. It is not possible to remove the correlation between the background parameter and M_W by a clever choice of scan point energies. This implies that a scan point of energy below the nominal threshold is quite important. If the energy is chosen to be $\epsilon_b = -5$ GeV ($E_b = 75$ GeV), an M_W-optimized scan strategy would allocate twice as much integrated luminosity to the M_W sensitive region as is allocated to the low energy point.

3. A measurement of M_W at the $\lesssim 160$ MeV level is possible with the dedication of a large integrated luminosity (250 pb^{-1}) and good control of the background contamination.

4. The measurement of Γ_W to an interesting level is difficult or impossible. It is probably unwise to attempt anything more than a cursory measurement.

5. The Search for New Particles

The Z^0 is the largest-mass neutral particle known to exist. Its couplings to fundamental particles are unambiguously determined from the quantum numbers of the particle in question. The strengths of these couplings are fairly uniform which implies that the Z^0 is remarkably democratic in its choice of final state. The branching ratio of the Z^0 into most hypothetical final states is typically larger than 1% unless it is suppressed by phase space or virtual intermediate states. The Z pole is therefore a good place to search for new particles.

The branching ratios of the Z^0 into hypothetical final states are large enough that the very modest Mark II data sample is adequate to perform new particle searches. A number of searches have been performed for new quarks and leptons,[13] supersymmetric particles,[14] and non-standard extensions to the Higgs sector.[15]

Since most of these searches have also been performed by the LEP collaborations, I will discuss the only search which is unique to the Mark II, the search for doubly charged Higgs bosons.

5.1. DOUBLY CHARGED HIGGS BOSONS

There are currently two popular scenarios that give rise to doubly charged Higgs bosons. The first, known as the Gelmini-Roncadelli model,[16−17] is a straightforward extension of the standard model to include a Majorana mass for the left-handed neutrino. The second scenario is the left-right symmetric extension of the standard model.[18] Before describing these two models, it is worth reviewing the mass generation scheme of the Standard Model.

5.2. THE STANDARD MODEL

The Standard Model describes the masses of all leptons in terms of a trilinear Lagrangian of the form,

$$\mathcal{L}_{\text{Dirac}} = f_D^\ell \bar{l}_L \Phi \ell_R + f_D^\nu \bar{l}_L (i\tau_2 \Phi^*)\nu_R + \text{h.c.} \tag{5.1}$$

where: f_D^ℓ, f_D^ν are dimensionless coupling constants;[*] Φ is the ordinary (isodoublet) Higgs field; ℓ_R is the right-handed charged-lepton field; τ_2 is the standard Pauli matrix; ν_R is the right-handed neutrino field; and where l_L is the left-handed doublet

$$l_L \equiv \begin{pmatrix} \nu \\ \ell - \end{pmatrix}_L .$$

---

[*] There is a pair of constants f_D^ℓ, f_D^ν for each lepton generation. For simplicity, generational labels are suppressed in this and the following expressions.

Note that the quantum numbers (I_w, Y) of the Higgs boson, $(\frac{1}{2}, -1)$, are equal to the quantum numbers of the bilinear product $\bar{e}_L e_R$, $(\frac{1}{2}, +1) \otimes (0, -2)$.

The actual mass terms are generated by the same spontaneous symmetry breaking that generates the gauge boson masses. The Higgs field is expanded about its non-zero minimum,

$$\Phi = e^{i\phi(x)} \begin{pmatrix} 0 \\ v + \eta \end{pmatrix} \tag{5.2}$$

where: $\phi(x)$ is a phase function; v is the vacuum expectation value of the Higgs field; and η is the physical Higgs field. Substituting equation (5.2) into equation (5.1) the usual Dirac mass terms for the charged and neutral leptons emerge

$$\mathcal{L}_{\text{Dirac}} = m_\ell(\bar{\ell}_L \ell_R + \bar{\ell}_R \ell_L) + m_\nu(\bar{\nu}_L \nu_R + \bar{\nu}_R \nu_L)$$

where the neutrino and lepton masses are $m_\nu = f_D^\nu v$ and $m_\ell = f_D^\ell v$, respectively.

5.3. The Gelmini-Roncadelli Model

The entire motivation of the Gelmini-Roncadelli model is to give the neutrino a mass without adding a right-handed (neutrino) field to the theory. The secret of doing this is to note that the charge conjugate of the left-handed lepton field doublet, l_L^c, is projected from the charge conjugate field $l^c = C\bar{l}^T$ (C is the charge conjugation matrix) by a *right-handed* projection operator,

$$l_L^c \equiv C\overline{\left[\frac{1}{2}(1 - \gamma_5)l\right]}^T = \frac{1}{2}(1 + \gamma_5)C\bar{l}^T = \frac{1}{2}(1 + \gamma_5)l^c.$$

It is important to emphasize that l_L^c creates left-handed fermions and destroys right-handed anti-fermions exactly as $\bar{l}_L$ does. Some authors like to confuse everything by labelling l_L^c with an R subscript. It is clear that the bilinear $\bar{l}_L^c l_L$ does not vanish and represents a kind of mass term. In fact, it represents the mass term for a self-conjugate or Majorana field.

To generate a Majorana mass term from the vacuum expectation value of a Higgs field, we note that the Higgs field must have the same quantum numbers as the bilinear $\bar{l}_L^c l_L$,

$$(\frac{1}{2}, -1) \otimes (\frac{1}{2}, -1) = (1, -2) \oplus (0, -2).$$

The bilinear must be coupled to a weak isotriplet or a weak isosinglet. The charge of the isosinglet must be $Q = I_3 + Y/2 = -1$. It therefore cannot be coupled to the neutral neutrino pair. On the other hand, the charges of the isotriplet Higgs are $Q = I_3 + Y/2 = 0, -1, -2$. The presence of a neutral member allows one to construct a mass term of the form

$$\mathcal{L}_{\text{Majorana}} = f_M \bar{l}_L^c \vec{H} \cdot (i\tau_2 \vec{\tau})l_L + \text{h.c.} \tag{5.3}$$

where: f_M is a dimensionless coupling constant; $\vec{H}$ are the three Higgs fields; and $\vec{\tau}$ are the three Pauli matrices.

The actual mass term is generated by giving the neutral member of the isotriplet a vacuum expectation value as follows

$$\langle \vec{\tau} \cdot \vec{H} \rangle = \left\langle \begin{pmatrix} H^- & \sqrt{2}H^{--} \\ \sqrt{2}H^\circ & -H^- \end{pmatrix} \right\rangle = \begin{pmatrix} 0 & 0 \\ v_T & 0 \end{pmatrix}. \tag{5.4}$$

Substituting equation (5.4) into equation (5.3) produces the desired Majorana mass term

$$\mathcal{L}_{\text{Majorana}} = M_\nu(\bar{\nu}_L^c \nu_L + \bar{\nu}_L \nu_L^c) \tag{5.5}$$

where the Majorana mass, M_ν, is defined as $M_\nu = f_M v_T$. Note that the vacuum expectation value v_T must be small to avoid disturbing the ρ parameter,

$$\rho \equiv \frac{M_W^2}{M_Z^2 \cos^2\theta_w} = \frac{v^2 + 2v_T^2}{v^2 + 4v_T^2}.$$

However, since we know that any left-handed neutrino is light as compared with the ordinary Higgs VEV (v is roughly 250 GeV), this condition is rather naturally satisfied.

Equation (5.3) also contains pieces that involve the coupling of the H^- to a lepton-neutrino pair (the H^- doesn't couple to quarks since $Y = -2$),

$$\mathcal{L}_{H^\pm} = -\frac{1}{\sqrt{2}}g_{\ell\ell}\left[(\bar{\nu}_L\ell_L^c + \bar{\ell}_L\nu_L^c)H^- + (\bar{\nu}_L^c\ell_L + \bar{\ell}_L^c\nu_L)H^+\right] \tag{5.6}$$

and the coupling of the H^{--} to a left-handed charged-lepton pair

$$\mathcal{L}_{H^{\pm\pm}} = -g_{\ell\ell}(\bar{\ell}_L\ell_L^c H^{--} + \bar{\ell}_L^c\ell_L H^{++}) \tag{5.7}$$

where $g_{\ell\ell} \equiv \sqrt{2}f_M$ is a dimensionless coupling constant. Note that we would expect to $g_{\ell\ell}$ to increase with neutrino mass. The coupling of the H^{--} to $\tau^-\tau^-$ *could* therefore be significantly larger than the couplings to $\mu^-\mu^-$ or e^-e^-. Although both the singly and the doubly charged Higgs bosons formally violate lepton flavor, only the doubly charged member visibly manifests the effect (because a light Majorana neutrino is virtually indistinguishable from a Dirac neutrino). Since the H^{--} couples only to charged lepton pairs, most existing searches for lepton flavor violation are fairly insensitive to doubly charged Higgs bosons. The existing limits are therefore relatively weak.[19-21]

5.4. Left-Right Symmetric Models

Another model that incorporates doubly charged Higgs bosons is the so-called Left-Right Symmetric model of Pati and Salam.[18] As its name implies, this model begins by treating both left- and right-handed fermions in a symmetric fashion. A right-handed weak isospin quantum number is added to the theory. All fermions are singlets of one isospin and doublets of the other. This permits one to discard the weak hypercharge quantum number and to replace it with a more physical one, $B - L$ ($B \equiv$ baryon number and

$L \equiv$ lepton number). The electric charge of each state is then given by the relationship

$$Q = I_3^L + I_3^R + \frac{1}{2}(B - L)$$

where I^L and I^R are the left- and right-handed weak isospins, respectively. The quantum numbers $(I^L, I^R, B - L)$ of the quark and lepton doublets are therefore:

$$\begin{pmatrix} u \\ d \end{pmatrix}_L \equiv (\tfrac{1}{2}, 0, \tfrac{1}{3}) \qquad \begin{pmatrix} \nu \\ \ell^- \end{pmatrix}_L \equiv (\tfrac{1}{2}, 0, -1)$$

$$\begin{pmatrix} u \\ d \end{pmatrix}_R \equiv (0, \tfrac{1}{2}, \tfrac{1}{3}) \qquad \begin{pmatrix} \nu \\ \ell^- \end{pmatrix}_R \equiv (0, \tfrac{1}{2}, -1)$$

The gauge group of this model is expanded to $SU(2)_L \otimes SU(2)_R \otimes U(1)_{B-L}$. Since we know that nature is left-handed in the low energy limit, the right-handed gauge symmetry must be broken at a significantly larger mass scale than its left-handed counterpart. The minimal Higgs sector that preserves the correct low energy phenomenology consists of a bidoublet Φ with quantum numbers $(\tfrac{1}{2}, \tfrac{1}{2}, 0)$ and two triplets, $\vec{H}_L$ and $\vec{H}_R$, with the quantum numbers $(1, 0, -2)$ and $(0, 1, -2)$, respectively. The vacuum expectation values of the Higgs sector are

$$\langle \Phi \rangle = \left\langle \begin{pmatrix} \phi_1^0 & \phi_1^+ \\ \phi_2^- & \phi_2^0 \end{pmatrix} \right\rangle = \begin{pmatrix} \kappa & 0 \\ 0 & \kappa' \end{pmatrix}$$

$$\langle \vec{\tau} \cdot \vec{H}_L \rangle = \left\langle \begin{pmatrix} H_L^- & \sqrt{2}H_L^{--} \\ \sqrt{2}H_L^o & -H_L^- \end{pmatrix} \right\rangle = \begin{pmatrix} 0 & 0 \\ v_L & 0 \end{pmatrix}$$

$$\langle \vec{\tau} \cdot \vec{H}_R \rangle = \left\langle \begin{pmatrix} H_R^- & \sqrt{2}H_R^{--} \\ \sqrt{2}H_R^o & -H_R^- \end{pmatrix} \right\rangle = \begin{pmatrix} 0 & 0 \\ v_R & 0 \end{pmatrix}.$$

The Standard model phenomenology is preserved by choosing the vacuum expectation values κ, κ', v_R, and v_L to have the following hierarchy

$$v_R \gg \left[\kappa^2 + \kappa'^2 \right]^{\frac{1}{2}} \gg v_L.$$

The Higgs sector now contains 20 degrees of freedom. Six degrees of freedom are used to provide mass for the $W_L^\pm$, Z_L^0, $W_R^\pm$, and Z_R^0 gauge bosons. The remaining 14 degrees of freedom contain four doubly-charged states, four singly-charged states, and six neutral states. Since the $\vec{H}_L$ and $\vec{H}_R$ triplets already have the quantum numbers of the bilinear combinations $\bar{l}_L^c l_L$ and $\bar{l}_R^c l_R$, it is quite natural to include the following Majorana mass term for the neutrino sector (in addition to the normal Dirac mass term)

$$\mathcal{L}_{\text{Majorana}} = f_M^L \bar{l}_L^c \vec{H}_L \cdot (i\tau_2\vec{\tau}) l_L + f_M^R \bar{l}_R^c \vec{H}_R \cdot (i\tau_2\vec{\tau}) l_R + \text{h.c.} \tag{5.8}$$

There are now three masses for each neutrino generation: a left-handed Majorana mass, $M_L = f_M^L v_L$; a Dirac mass, $m_D = f_D^\nu \kappa$; and a right-handed Majorana mass,

$M_R = f_M^R v_R$. Unless something very perverse is done with the coupling constants, the natural hierarchy of the masses is $M_L \ll m_D \ll M_R$. In fact, the most *natural* occurrence would be to have $m_D = m_\ell$. Making this assumption, the mass lagrangian for a neutrino generation can be written in the form*

$$\mathcal{L}_{\text{Neutrino mass}} = (\,\bar{\psi}_L \quad \bar{\psi}_R\,) \begin{pmatrix} M_L & \frac{1}{2}m_\ell \\ \frac{1}{2}m_\ell & M_R \end{pmatrix} \begin{pmatrix} \psi_L \\ \psi_R \end{pmatrix} \tag{5.9}$$

where the fields $\psi_L = \nu_L + \nu_L^c$ and $\psi_R = \nu_R + \nu_R^c$ are the left- and right-handed Majorana fields, respectively. The physical neutrino masses are given by the eigenvalues of equation (5.9),

$$m_1 \simeq \frac{m_\ell^2}{4M_R} \tag{5.10}$$

$$m_2 \simeq M_R.$$

There is a light, left-handed neutrino (presumably the *standard* one) and a very heavy, right-handed neutrino (the magnitude of M_R is roughly that of the right-handed W boson). The beauty (sic) of this model is that light, left-handed neutrinos are generated without *diddling* coupling constants. Since the mass of each light neutrino generation increases with m_ℓ^2, the coupling constants f_M^L and f_M^R do not necessarily increase with generation.

The same comment must therefore apply to the couplings of the doubly charged Higgs bosons to left- and right-handed charged-lepton pairs. Using equation (5.8), the $H^{\pm\pm} - \ell^\pm\ell^\pm$ couplings are

$$\mathcal{L}_{H^{\pm\pm}} = -\left[g_{\ell\ell}^L(\bar{\ell}_L \ell_L^c H_L^{--} + \bar{\ell}_L^c \ell_L H_L^{++}) + g_{\ell\ell}^R(\bar{\ell}_R \ell_R^c H_R^{--} + \bar{\ell}_R^c \ell_R H_R^{++}) \right] \tag{5.11}$$

where $g_{\ell\ell}^L = \sqrt{2} f_M^L$ and $g_{\ell\ell}^R = \sqrt{2} f_M^R$.

Unlike the Gelmini-Roncadelli model, there is No reason to expect that the $H^{\pm\pm}$ decays dominantly into $\tau^\pm\tau^\pm$ pairs.

5.5. Existing Limits

If we assume that the coupling of the doubly charged Higgs to lepton pairs is diagonal in lepton flavor, the best limits[19-21] on the existence of doubly-charged Higgs bosons are derived from the Bhabha scattering data of several PEP and PETRA experiments and from searches for conversion of muonium ($\mu^+ e^-$) into antimuonium ($\mu^- e^+$). Both processes involve the t-channel exchange of a doubly charged Higgs boson. These searches have the unfortunate property that they limit the ratio $g_{\ell\ell}/M_H$. Searches for real $H^{\pm\pm}$ production can extend the limits to very small values of the coupling constant for moderate mass values.

$\star$ The hierarchy $M_L \ll m_\ell \ll M_R$ is commonly known as the *Seesaw Mechanism*.

5.6. The Cross Section for $e^+e^- \to H^{++}H^{--}$

The coupling of the Z^0 to a pair of doubly charged Higgs bosons of either the Gelmini-Roncadelli type[17] or of the Left-Right Symmetric type[22] is given by the following expression,

$$\mathcal{L}_{Z^0 - H^{\pm\pm}} = -ie\left[\frac{I_3^L - Q_H \sin^2\theta_w}{\sin\theta_w \cos\theta_w}\right] Z^\alpha \phi_H^\dagger (\overrightarrow{\partial}_\alpha - \overleftarrow{\partial}_\alpha)\phi_H \tag{5.12}$$

where: e is the electric charge; Q_H is the charge of the Higgs boson; $\sin\theta_w$ is the sine of the electroweak mixing angle; Z^α is the Z^0 field; and ϕ_H is the Higgs field. Note that the quantum numbers (I_3^L, Q_H) are $(1, 2)$ for the Gelmini-Roncadelli Higgs and for the left-handed Higgs of the Left-Right Symmetric model. The right-handed Higgs has the quantum numbers $(0, 2)$.

Using equation (5.12) it is straightforward to calculate the cross section for the process $e^+e^- \to H^{++}H^{--}$,

$$\frac{d\sigma}{d\cos\theta} = \frac{9\pi}{M_Z^2} \frac{\Gamma_{HH}^\circ \Gamma_{ee} s}{(s - M_Z^2)^2 + s\Gamma_Z^2}\left[1 - \frac{4M_H^2}{s}\right]^{\frac{3}{2}} (1 - \cos^2\theta) \tag{5.13}$$

where: M_Z, Γ_Z are the mass and width of the Z^0, respectively; Γ_{ee} is the leptonic width of the Z^0; s is the square of the center-of-mass energy; and Γ_{HH}° is the partial width for the decay $Z^0 \to H^{++}H^{--}$ (unsuppressed by phase space),

$$\Gamma_{HH}^\circ = \frac{G_F M_Z^3}{6\pi\sqrt{2}}(I_3^L - 2\sin^2\theta_w)^2. \tag{5.14}$$

The ratio of the decay rates of the Z^0 into Higgs pairs and into a single neutrino species can therefore be written as

$$\frac{\Gamma(Z^0 \to H^{++}H^{--})}{\Gamma(Z^0 \to \nu\bar{\nu})} = 2(I_3^L - 2\sin^2\theta_w)^2\beta^3 \simeq \beta^3 \cdot \begin{cases} 0.57, & H_L^{\pm\pm}; \\ 0.43, & H_R^{\pm\pm} \end{cases}$$

where the Higgs velocity β is defined as

$$\beta \equiv \left[1 - \frac{4M_H^2}{M_Z^2}\right]^{\frac{1}{2}}.$$

The branching ratio of the Z^0 into Higgs pairs is roughly one half of the branching ratio for a neutrino species (multiplied by a β^3 factor).

5.7. THE DECAY OF THE DOUBLY CHARGED HIGGS BOSON

The conservation of quantum numbers greatly restricts the available decay modes
of triplet higgs bosons. A member of the triplet can decay into a lepton pair or into a
(perhaps virtual) W boson and another member of the Higgs triplet. To fully understand
the branching ratios of a doubly charged Higgs boson, we must know the hierarchy of
the triplet masses. In order to simplify the search, it is assumed that the doubly charged
member of the triplet can decay only into lepton pairs. The decay rate of a doubly charged
Higgs boson into a same-sign pair of leptons, $\Gamma_{\ell\ell}$, s given by the following expression,[17,21]

$$\Gamma_{\ell\ell} = \frac{g_{\ell\ell}^2}{8\pi} M_H \left[1 - \frac{2m_\ell^2}{M_H^2} \right] \left[1 - \frac{4m_\ell^2}{M_H^2} \right]^{1/2} \tag{5.15}$$

where m_ℓ is the lepton mass. The Higgs bosons are therefore short-lived (in an experimental sense) unless the coupling constants $g_{\ell\ell}$ are *very small* (less than 10^{-7}).

5.8. THE MARK II SEARCH

In order to place limits on the decay $Z^0 \to H^{++}H^{--}$, we must search for final states
of the form $l^+l^+\ell^-\ell^-$ where l and ℓ may or may not be the same flavor of lepton. Of the
six possible four-lepton final states, the most difficult to detect is the one consisting of four
τ leptons. The strategy of this analysis is to define a set of topological selection criteria
that can identify the four-τ final state with high efficiency. It is clear that such criteria
select four-lepton final states that contain two or more stable leptons with comparable or
larger efficiency.

Since 90% of all four-τ events decay into six or fewer charged particles, we consider
only those event candidates that contain six or fewer charged tracks that project into a
cylindrical volume of 2 cm radius and 6 cm length that is centered on the interaction
point of the SLC. In order to suppress two-photon events and badly accepted hadronic
final states, we require that the scalar sum of the track momenta be at least 10 GeV/c.

Isolated τ leptons appear as isolated single tracks or as low-mass clusters of tracks.
The 4-vectors of the charged tracks are thus subjected to a mass-based clustering algorithm. Initially, each track is defined to be a cluster. The pair of clusters with the smallest
invariant mass is merged if its mass is less than 2.0 GeV/c^2. The procedure is repeated
until all pairs of clusters have invariant masses larger than 2.0 GeV/c^2.

We expect most $H^{++}H^{--}$ events to appear as four-cluster events. There is a reason-
able probability, however, that a cluster occurs in one of the forward regions, $|\cos\theta| >$
0.80, and is not detected (approximately 30% of all events fall into this category). We
therefore require that each event candidate contain either three or four clusters of energy
larger than 1.0 GeV. The net charge of each cluster must be unity. The event must not
contain any clusters with charges larger than unity. The net event charge must be zero
for four-cluster candidates or unity for three-cluster candidates.

None of the events in the Mark II data sample pass the selection criteria. Using
a Monte Carlo simulation, we can predict the number of events that would have been
observed if doubly charged Higgs bosons were present in the data. The detection probabilities for electron and muon final states are essentially identical. The limit is therefore a
function of the Higgs boson to $\tau\tau$ branching ratio, B_τ, only. In the regions of parameter

space that the expected number of observed events is larger than 2.3 and 3.0, we can exclude the presence of the $H^{\pm\pm}$ with 90% and 95% confidence. In the short lifetime region, the intervals of M_H that are excluded with 90% confidence and with 95% confidence are listed in Table IX for several values of B_τ and I_3^L. The upper limits on M_H are due to the β^3 suppression of the cross section. The lower limits are due to the loss of efficiency as M_H becomes small. The efficiency function for stable leptons falls sharply at the cluster mass of 2.0 GeV/c^2. The 90% and 95% confidence limits occur quite close to this point. For $B_\tau = 0.5$, the number of events with four stable leptons is sufficient to exclude values of M_H down to the τ-pair threshold. The short lifetime constraint requires that there be at least one coupling constant in the region $g_{\ell\ell} \gtrsim 7.4 \times 10^{-7}/\sqrt{M_H}$ (M_H in GeV/c^2). This implies that the dominant coupling(s) be larger than $\sim 5 \times 10^{-7}$ in the small mass region and $\sim 1 \times 10^{-7}$ in the high mass region.

Table IX

The intervals of M_H that are excluded at 90% confidence and at 95% confidence for left-handed ($I_3^L = 1$) and right-handed ($I_3^L = 0$) doubly charged Higgs bosons. The excluded intervals are tabulated as a function of the τ branching ratio B_τ. They are valid in the region of coupling constant $g_{\ell\ell} \gtrsim 5 \times 10^{-7}$.

I_3^L	B_τ	90% Limit (GeV/c^2)	95% Limit (GeV/c^2)
0	1.0	$6.5 < M_H < 36.4$	$7.3 < M_H < 34.3$
0	0.5	$3.6 < M_H < 37.7$	$3.6 < M_H < 36.0$
0	0.0	$2.0 < M_H < 38.5$	$2.0 < M_H < 36.7$
1	1.0	$5.9 < M_H < 38.2$	$6.5 < M_H < 36.6$
1	0.5	$3.6 < M_H < 39.2$	$3.6 < M_H < 37.9$
1	0.0	$2.0 < M_H < 39.7$	$2.0 < M_H < 38.4$

The excluded regions overlap and significantly extend the existing small-coupling limits on M_H (which are independent of I_3^L).[21] The 90% confidence limit for $B_\tau = 0$ is extended from approximately 21.5 GeV/c^2 to 39.7 GeV/c^2 (38.5 GeV/c^2) for left-handed (right-handed) Higgs bosons. For $B_\tau = 1$, the 90% limit is extended from 14 GeV/c^2 to 38.2 GeV/c^2 (36.4 GeV/c^2) for left-handed (right-handed) Higgs bosons.

In order illustrate the dependence of the limits upon the mass and coupling constants, the least restrictive 90% limit ($B_\tau = 1$, $I_3^L = 0$) is plotted in $g_{\ell\ell} - M_H$ space in Figure 13 (the solid curve). Note that it extends to $g_{\ell\ell} = 7.2 \times 10^{-8}$. The limit is compared with two rather specific limits from Reference 21.

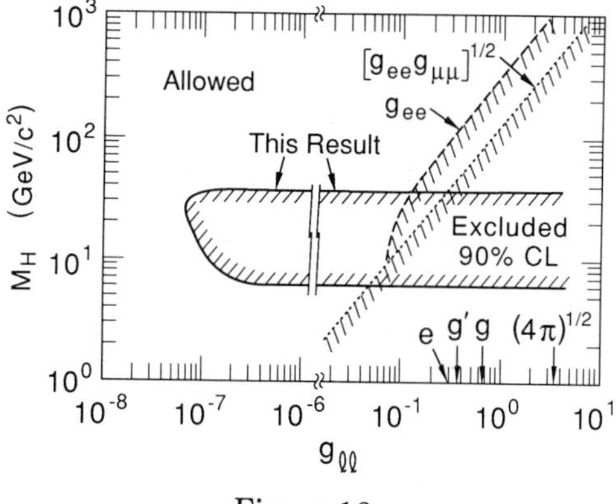

Figure 13

The 90% confidence contours of M_H versus the leptonic coupling strength $g_{\ell\ell}$ that are obtained from several processes. The excluded regions are indicated by the shaded side of each contour. The result of this search is shown as the solid contour (the limit is independent of lepton flavor). The limit[21] that is obtained from the limit on muonium to antimuonium conversion is shown as a dotted line ($\sqrt{g_{ee}g_{\mu\mu}}$ is plotted along the horizontal axis). The limit[21] that is obtained from the Bhabha scattering data of several PEP and PETRA experiments is shown as a dashed curve (g_{ee} is plotted along the horizontal axis). For reference, the sizes of the coupling constants g, g', and e are indicated. The strong coupling limit occurs at the value $\sqrt{4\pi}$.

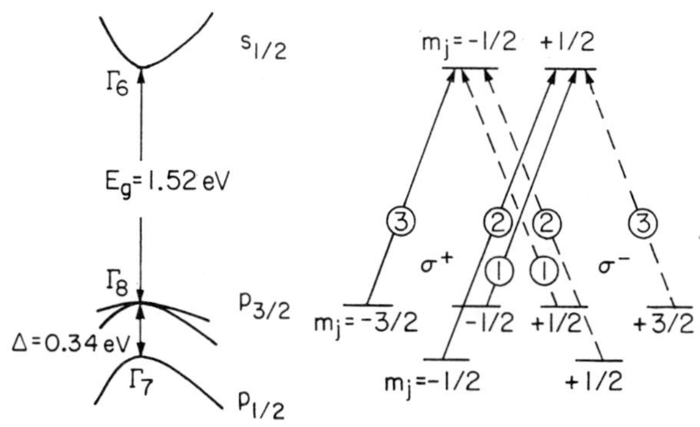

Figure 14

The band structure of GaAs near the bandgap minimum.[23] The energy levels of the states are shown on the right. Allowed transitions for the absorption of right (left) circularly polarized photons are shown as solid (dashed) arrows. The circled numbers indicate the relative transition rates.

6. Polarization Physics at the SLC

6.1. THE POLARIZED SLC

One of the advantages of linear colliders is that they are relatively straightforward to polarize. A layout of the polarized SLC is shown in Figure 1. The orientation of an electron spin vector is shown as the electron is transported from the electron gun to the interaction point.

A gallium arsenide based photon emission source produces pulses of up to 10^{11} longitudinally polarized electrons at repetition rates of up to 120 Hz. The electrons are then accelerated in the first sector of the linac. The beam pulse achieves an energy of 1.21 GeV as it arrives at the entrance of the LTR (Linac To Ring) transfer line.

The electrons must be stored in the North Damping Ring for one machine cycle (the cycle time is $\simeq$ 8 ms). A system consisting of the LTR bend magnets and a superconducting solenoid is used to rotate the spins into the vertical direction that is necessary for storage in the damping ring. After one machine cycle, the bunch is extracted and passed through another spin rotation system consisting of the bend magnets of the RTL (Ring-To-Linac) transfer line and two superconducting solenoids. The system is sufficiently flexible to provide essentially any spin orientation as the bunch reenters the linac at the beginning of sector 2.

The beam pulse is then accelerated to 46.5 GeV in the linac (due to synchrotron radiation losses, the energy at the end of the LINAC is $\sim$1 GeV larger than it is at the interaction point). To insure that the spin gymnastics in the damping ring have worked properly and to study many of the potential sources of depolarization, a Møller polarimeter is located at the end of the linac near the PEP injection line. This polarimeter is used primarily for diagnostic purposes.

The beam pulse is then transported through the north machine arc and the final focus section to the interaction point. At full energy, the spin vectors precess roughly 26 times. Vertical precession also occurs in the nonplanar arcs. Since longitudinal polarization is required at the interaction point, the total precession angle must be calculated for the exact machine energy and the polarization at the arc entrance must be adjusted appropriately.

After colliding with the unpolarized positron bunch, the electron beam is transported through the south final focus system where a Compton polarimeter is located. The beam continues to the south extraction line where a second Møller polarimeter is located. The bending magnets of the final focus and extraction line cause an additional spin precession of roughly 540° between the interaction point and the Møller target. Both polarimeters continuously monitor the beam polarization.

6.2. THE POLARIZED SOURCE

The SLC polarized electron source is based upon polarized photoemission from Gallium Arsenide (GaAs). The band structure of GaAs at the energy maximum of the valence band and energy minimum of the conduction band is shown in Figure 14. The band energy versus momentum is shown on the left-hand side and the energy level structure is shown on the right-hand side of the figure. The band gap of the material is $E_g = 1.52\ eV$.

At the minimum of the conduction band and the maximum of the valence band, the electron wavefunctions have S and P symmetry, respectively. Spin-orbit splitting causes the $P_{3/2}$ states to reside in energy above the $P_{1/2}$ states by an amount $\Delta = 0.34\ eV$. The absorption of single photons proceeds via an electric dipole transition. The selection rules for the absorption of right- and left-handed circularly polarized photons are $\Delta m_j = +1$ and $\Delta m_j = -1$, respectively. They are indicated by the solid and dashed arrows in Figure 14. Since the electric dipole operator changes the orbital angular momentum of the initial state by one unit, the spin of the electron remains unchanged in the process.

Let's consider what happens when a right-circularly polarized photon is incident upon a GaAs crystal. The photon direction is the only vector in the system. All angular momentum projections refer to the incident photon direction. If the photon energy E_γ is in the range $E_g \leq E_\gamma \leq E_g + \Delta$, then transitions can only occur from the $P_{3/2}$ states to the $S_{1/2}$ states. Specifically, the P state with $m_j = -3/2$ can make a transition to the S state with $m_j = -1/2$ and the P state with $m_j = -1/2$ can make a transition to the S state with $m_j = +1/2$. In the former case, the emitted electron has spin antiparallel to the incident photon direction (or parallel to its ejected direction). In the latter case, the spin of the emitted electron is parallel to the incident photon direction (antiparallel to its ejected direction). Due to Clebsch-Gordon coefficients (the P state with $m_j = -3/2$ is a pure spin state whereas state with $m_j = -1/2$ is not), the former transition is three times more likely than the latter. The relative transition rates are indicated by circled numbers in Figure 14. This implies that the absorption of a right circularly polarized photon produces a right-handed electron with a polarization

$$\mathcal{P} = \frac{3 - 1}{3 + 1} = 50\%.$$

Actually, all that's been shown so far is that polarized electrons can be pumped into the conduction band with a beam of circularly polarized photons. In order to make a polarized source, the electrons must leave the material. In normal GaAs, the energy gap from the bottom of the conduction band to the free electron state is approximately 2.5 electron volts. Even with a large applied electric field, pure GaAs is a poor photoemitter. The magic that is necessary to make it an efficient photoemitter is shown in Figure 15. The energy of the various bands is shown as a function of depth near the surface for several materials: pure GaAs, GaAs with a cesiated surface, and GaAs with a surface layer of Cs_2O. The energy of the free electron state is shown as E_∞. The addition of cesium to the surface causes the energy gap between the conduction band and the free electron state to decrease to zero. The addition of Cs_2O to the surface causes the gap to become negative! Quantum efficiencies as large as 5% have been observed for GaAs photocathodes that have been treated with Cs_2O (actually CsF is currently used instead).

Figure 16 shows the electron polarization that was measured[25] for several photocathodes as a function of the photon wavelength. The photocathodes are composed of pure gallium arsenide and several forms of gallium aluminum arsenide. The gallium aluminum arsenide is made by substituting aluminum atoms for fraction x of the arsenic atoms. At the wavelength of the SLC polarized light source, 715 nm, the pure GaAs sample produces a polarization of only 35%. The electron polarization will be improved to 42% by the use of $GaAl_{0.1}As_{0.9}$.

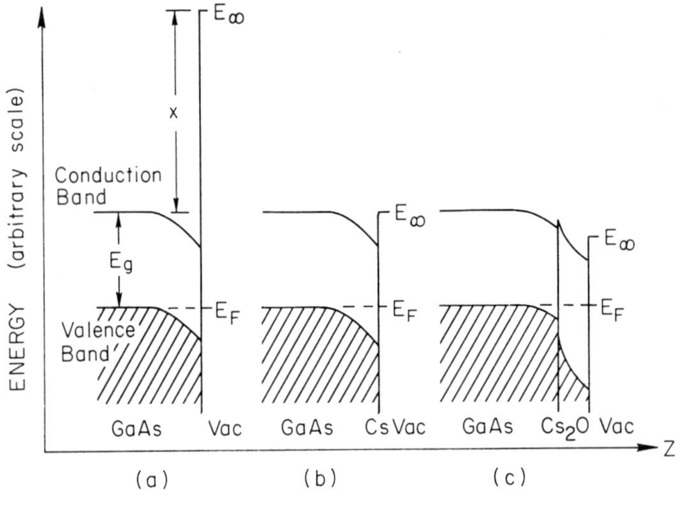

Figure 15

The band structure of Gallium Arsenide near its surface[23] for: (a) pure GaAs, (b) GaAs with a cesiated surface, and (c) GaAs with a layer of Cs_2O on its surface.

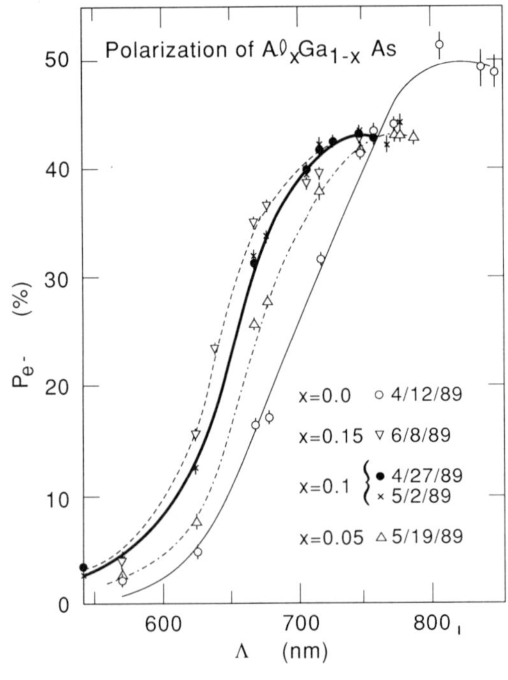

Figure 16

The polarization of electrons emitted from several GaAs photocathodes as functions of photon wavelength.[25] The cathodes consisted of pure gallium arsenide and several compositions of gallium aluminum arsenide.

6.3. The Spin Rotation System

The second major element of the polarized SLC is the spin rotation system. As was mentioned at the beginning of this chapter, the spin rotation system has two functions:

1. To rotate the (initially longitudinal) polarization vector of the electron bunch into the vertical direction for storage in the North Damping Ring.

2. To allow the orientation of the electron polarization vector to be controlled as the bunch reenters sector 2 of the linac. This is necessary to compensate for precession in the machine arcs.

A detailed representation of the north damping ring, the north LTR transfer line, and the north RTL transfer line is shown in Figure 2. The orientation of the polarization vector at various places is shown by the double arrow. The electron bunch arrives at the entrance to the LTR transfer line with an energy of 1.21 GeV. At this energy, the spins precess by 90° for each 32.8° that the electron trajectories are bent by a transverse magnetic field. The initial bend angle of the LTR has been chosen to be $5 \times 32.8°$. The longitudinal polarization of the beam emerging from sector 1 of the linac is therefore rotated into the horizontal direction. A superconducting solenoid of strength 6.34 T-m is introduced into the LTR optics after the first bend. The solenoid has only a small effect on the optics of the transport system but causes a rotation of the spin vector about the beam axis by 90°. The spins are therefore rotated into the vertical (either upward or downward) direction. After one machine cycle ($\simeq$ 8 ms), the electron bunch is extracted from the damping ring with a horizontal kicker magnet and passed through a second superconducting solenoid magnet. The horizontal bend magnets of the RTL transfer line then deflect the beam by an angle of $3 \times 32.8°$ before it reenters the linac at the beginning of sector 2. A third superconducting solenoid is introduced into the linac lattice just downstream of the reentry point. If the second (RTL) solenoid is adjusted to have the same strength as the first (LTR) solenoid has, the system will restore the longitudinal beam polarization. If it is not energized, the beam polarization will be vertical upon reentry into the linac. The third (linac) solenoid can then rotate the polarization vector to any transverse orientation. The combination of the two solenoids and the RTL bending magnets permits the selection of any orientation of the polarization vector.

6.4. Polarimetry at SLC

The polarization of the SLC electron beam will be monitored by three polarimeters. Two of the polarimeters are based upon polarized electron-electron scattering (Møller scattering) and one is based upon polarized electron-photon scattering (Compton scattering).

The Møller Polarimeters

The Møller polarimeters measure the elastic scattering of the polarized electron beam from the polarized atomic electrons in a magnetized foil target. The cross section for this process has the form (in the limit of zero electron mass),

$$\frac{d\sigma}{d\Omega} = \frac{\alpha^2}{s} \frac{(3 + \cos^2\theta)^2}{\sin^4\theta} \left\{ 1 - \mathcal{P}_z^1 \mathcal{P}_z^2 A_z(\theta) - \mathcal{P}_t^1 \mathcal{P}_t^2 A_t(\theta) \cos(2\phi - \phi_1 - \phi_2) \right\} \qquad (6.1)$$

where: s is the square of the total energy in the cm frame; θ is the cm frame scattering angle; ϕ is the azimuth of the scattered electron (the definition of $\phi = 0$ is arbitrary);

$\mathcal{P}_z^1$, $\mathcal{P}_z^2$ are the longitudinal polarizations of the beam and target, respectively; $\mathcal{P}_t^1$, $\mathcal{P}_t^2$ are the transverse polarizations of the beam and target, respectively; ϕ_1, ϕ_2 are the azimuths of the transverse polarization vectors; and $A_z(\theta)$ and $A_t(\theta)$ are the longitudinal and transverse asymmetry functions which are defined as

$$A_z(\theta) = \frac{(7 + \cos^2\theta)\sin^2\theta}{(3 + \cos^2\theta)^2}$$

$$A_t(\theta) = \frac{\sin^4\theta}{(3 + \cos^2\theta)^2}.$$

The differential cross section is the product of the unpolarized cross section and the sum of one and two polarization dependent terms. The first is the product of the longitudinal polarizations of the beam and target particles and the function $A_z(\theta)$. The second is the product of the transverse polarizations of the two electrons, an azimuthal factor, and the function $A_t(\theta)$. Both asymmetry functions are maximal for $90°$ scattering. The longitudinal asymmetry function becomes quite large ($A_z(90°) = 7/9$) whereas the transverse asymmetry function never exceeds $1/9$. The analyzing power of any polarimeter scales as the product of the unpolarized cross section and the square of the asymmetry. This combination is also largest at $\theta = 90°$ but has a rather broad maximum. Because the longitudinal and transverse asymmetries have maxima at the same scattering angle, it is quite straightforward to build three-axis polarimeters.

The polarimeters function by measuring the asymmetry in the scattering rate when either the beam or target polarizations are reversed,

$$A_{ee} \equiv \frac{\sigma(\mathcal{P}^1\mathcal{P}^2) - \sigma(-\mathcal{P}^1\mathcal{P}^2)}{\sigma(\mathcal{P}^1\mathcal{P}^2) + \sigma(-\mathcal{P}^1\mathcal{P}^2)} = -\mathcal{P}_z^1\mathcal{P}_z^2 A_z(\theta) - \mathcal{P}_t^1\mathcal{P}_t^2 A_t(\theta)\cos(2\phi - \phi_1 - \phi_2). \qquad (6.2)$$

The beam polarization $\mathcal{P}_x^1$ is inferred from the measured asymmetry A_{ee}, the measured target polarization $\mathcal{P}_x^2$, and the theoretical asymmetry $A_x(\theta)$ (where x is z or t).

The precision of the Møller polarimeters is limited by the uncertainties on the target polarization and on the subtraction of background from the radiative process $e^-N \rightarrow e^-N\gamma$ where N is a nucleus in the target material. We expect that a precision $\delta\mathcal{P}/\mathcal{P}$ of 5% is possible.

The Compton Polarimeter

The Compton polarimeter is the primary monitor of the polarization of the SLC electron beam. It measures the large asymmetry in the scattering of polarized electrons from circularly polarized laser photons. The light source is a frequency doubled Nd:YAG laser which produces 2.23 eV photons. The backscattered electrons are separated from the SLC beam by the first large dipole magnet of the SLC final focus system after it has passed through the interaction point.

The kinematical properties of the scattering of a high energy electron with an optical photon seem quite strange to those accustomed to working in reference frames that are nearer the center-of-mass frame. The energy of the electron is typically 10 orders of magnitude larger than that of the photon. It is clear that all final state particles are swept into the forward direction (along the incident electron direction). It is therefore convenient to define all angles with respect to the incident electron direction. The direction of the outgoing photon, θ_K, differs from the normal definition of the scattering angle by $180°$ (if the colliding e-γ are collinear). Let E, E', K, and K' be the incident electron energy, scattered electron energy, incident photon energy, and scattered photon energy, respectively. The maximum energy of the scattered photon K'_{max} and the minimum energy of the scattered electron E'_{min} can then be written as

$$K'_{max} = E(1 - y)$$
$$E'_{min} = Ey$$

where the parameter y is defined as

$$y \equiv \left(1 + \frac{4EK}{m^2}\right)^{-1}.$$

The emission angle of the scattered photon θ_K is related to the scattered photon energy by the following expression,

$$K' = K'_{max}\left[1 + y\left(\frac{E\theta_K}{m}\right)^2\right]^{-1}$$
$$= K'_{max} \cdot x$$

where the definition of x is obvious. The parameter x varies from unity at zero emission angle to zero at larger angles. The scale of the angular range is set by the angle for which the energy has been reduced by a factor of two. This occurs when $E\theta_K\sqrt{y}/m = 1$ or at the angle $\theta_K = m/E\sqrt{y}$. For the SLC Compton polarimeter operating with a 46 GeV beam, the value of the parameter y is 0.389. Therefore, the maximum photon energy is 28.1 GeV and the minimum electron energy is 17.9 GeV. The angle at which the photon energy has been decreased by a factor of two is 1.8×10^{-5} radians. The scattered electron and photon both remain along the beam direction.

The polarized cross section can be expressed in terms of the laboratory variables x, y, and the azimuth of the photon with respect to the electron transverse polarization ϕ as follows,[26]

$$\left(\frac{d^2\sigma}{dxd\phi}\right)_{Compton} = \left(\frac{d^2\sigma}{dxd\phi}\right)_{unpol}\left\{1 - \mathcal{P}^\gamma\left[\mathcal{P}^e_z A^{e\gamma}_z(x) + \mathcal{P}^e_t \cos\phi A^{e\gamma}_t(x)\right]\right\} \qquad (6.3)$$

where: the unpolarized cross section is defined as

$$\left(\frac{d^2\sigma}{dxd\phi}\right)_{unpol} = r_o^2 y\left\{\frac{x^2(1-y)^2}{1-x(1-y)} + 1 + \left[\frac{1-x(1+y)}{1-x(1-y)}\right]^2\right\};$$

$\mathcal{P}_z$, $\mathcal{P}_t$ are the longitudinal and transverse polarizations of the electron; $\mathcal{P}^\gamma$ is the *circular* polarization of the photon; and where the longitudinal and transverse asymmetries are defined as

$$A^{e\gamma}_z(x) = r_o^2 y\left[1 - x(1+y)\right]\left\{1 - \frac{1}{[1-x(1-y)]^2}\right\} \cdot \left(\frac{d^2\sigma}{dxd\phi}\right)^{-1}_{unpol}$$
$$A^{e\gamma}_t(x) = r_o^2 yx(1-y)\frac{[4xy(1-x)]^{1/2}}{1-x(1-y)} \cdot \left(\frac{d^2\sigma}{dxd\phi}\right)^{-1}_{unpol}$$

For $y = 0.389$, the unpolarized cross section is very large (several hundred millibarns) and peaked at $x = 1$. The longitudinal asymmetry has a maximum of 75% also at $x = 1$.

Note, however, that as x is decreased, $A_z^{e\gamma}$ decreases rapidly and becomes negative near $x = 0.72$. It reaches a minimum of -25% near $x = 0.47$ and returns to zero at $x = 0$. The transverse asymmetry is zero at both endpoints and reaches a maximum of 33% near $x = 0.75$.

The acceptance of the SLC polarimeter integrates over the entire azimuth. The polarimeter therefore measures only the longitudinal asymmetry. Using equation (6.2), the longitudinal beam polarization is measured. It appears that the polarimeter is capable of measuring the polarization with a precision of 1% to 2%.

6.5. Depolarization Effects

There are numerous possible sources of electron beam depolarization. None of them are expected to be serious. The following is a summary of the most important.

Depolarization in the Linac

The depolarization of a longitudinally polarized electron beam by the SLAC linac has been calculated to be very small.[27] This has been verified by several experiments.[28,24] The polarized SLC does differ in two respects from the old SLAC linac:

1. The electron bunches are much smaller than they were for the fixed target experiments. It was pointed out by W.K.H. Panofsky[29] that the intra-bunch fields could depolarize the bunch via an effect that is analogous to Thomas precession. More detailed calculations indicate that this effect causes less than a one percent depolarization of the beam.

2. The SLC must accelerate beams with transverse components of the polarization vector. This is not expected to be a problem, however, detailed calculations and experimental verification are still needed.

Depolarization in the Damping Ring

Since the helicity of each electron pulse is determined by a random number generator at the source, half of the electron pulses stored in the north damping ring will have their polarization vectors aligned with the guide field and half will have anti-aligned polarization vectors. The natural polarization of a storage ring by the Sokolov-Ternov effect[30] causes the spins to anti-align themselves with the guide field. This would cause the depolarization of the aligned bunches if they were stored in the ring for an appreciable fraction of a polarizing time. The polarizing time for the damping ring is approximately 15 minutes. Since the storage time of an electron bunch is only 8 milliseconds (at a 120 Hz repetition rate), this effect is negligible. Indeed, the short storage time (which is several damping times) implies that the only process that could cause a serious problem is resonant depolarization. The resonance condition is

$$\nu = N + I\nu_x + J\nu_y + K\nu_s$$

where: ν is the spin tune of the damping ring (the number of spin precessions per orbit); N, I, J, K are integers; ν_x and ν_y are the horizontal and vertical betatron tunes, respectively; and ν_s is the synchrotron tune of the damping ring. The SLC damping ring is designed to operate at an energy $E = 1.21$ GeV. The spin tune at this energy is given by the expression

$$\nu = \frac{g-2}{2} \cdot \frac{E}{m_e} = \frac{E}{440.65 \text{ MeV}} = 2.746$$

(where $(g-2)/2$ is anomalous magnetic moment of the electron). The horizontal and vertical betatron tunes are $\nu_x = 7.20$ and $\nu_y = 3.20$, respectively (the $\nu_x - \nu_y = 4$ coupling resonance is used to produce round beams). The synchrotron tune is very small ($\nu_s \simeq 0.04$). Therefore, the nearest spin depolarizing resonance occurs when $N, I, J = 6, 0, -1$ (the synchrotron tune is ignored since only relatively weak resonances are associated with it). The right hand side of the resonance equation is equal to 2.80 in this case. Since the natural width of this sideband resonance is expected to be less than 0.001, no serious resonant depolarization is expected.

Depolarization in the Arcs

The SLC arcs are fairly achromatic transport systems (they can transport a momentum interval $\Delta P/P = 5\%$). Since the total precession angle is a sensitive function of the beam energy, the finite energy spread of the beam ($\Delta P/P = 0.3\%$) causes a spread in the final spin directions of the electrons. The average longitudinal polarization at the interaction point is reduced by a factor 0.93.

Depolarization from Beam-Beam Interactions

Because the SLC beams are very small at the interaction point, each beam is subjected to very strong electromagnetic fields during the collision. These fields cause some depolarization of the electron bunch. The size of the effect is given by the expression

$$\Delta\theta_s = \frac{g-2}{2} \cdot \frac{E}{m_e} \cdot \theta_d$$

where: $\Delta\theta_s$ is the average precession angle of beam particles; E is the beam energy; and θ_d is the disruption angle of the beam. Since the disruption angle at SLC is roughly one milliradian, the average depolarization is less than one percent.

Systematic Effects

It is possible that the average beam polarization as measured by the two downstream polarimeters be different from the luminosity weighted average polarization. There are two possible causes for this effect.

1. The beam-beam interaction obviously changes the polarization before it arrives at the Compton and extraction line Møller polarimeters. The size of this effect is estimated to be less than 0.5%.

2. If the electron beam at the interaction point has a non-zero dispersion function, it is possible that a beam-beam targeting error could cause the luminosity weighted beam energy and polarization to differ from the average beam energy and polarization. The beam-beam deflection process allows the beam to be targeted to within a small fraction of the beam sizes. Therefore, even if the dispersion function at the interaction point were as large as 3 mm (which is quite large), the fractional deviation of the monitored polarization from the average one is less than two percent. If the dispersion function is the more normal 1 mm, this effect is a few tenths of one percent.

6.6. THE LEFT-RIGHT POLARIZATION ASYMMETRY

In order to understand the utility of a polarized electron beam, we must consider the cross section for the (longitudinally) polarized process $e^+ e^- \rightarrow f\bar{f}$. The beam polarizations, $\mathcal{P}^-$ and $\mathcal{P}^+$, are described in terms of a helicity basis ($\mathcal{P} = +1$ describes a right-handed beam, $\mathcal{P} = -1$ describes a left-handed beam). We can then write the tree-level cross section in the cm frame as follows,

$$\frac{d\sigma_f}{d\Omega} = \frac{\alpha^2 N_c^f}{64 s \sin^4 2\theta_w} \cdot \left\{ (1 - \mathcal{P}^+ \mathcal{P}^-)[\sigma_u^{\gamma Z} + \sigma_u^Z] + (\mathcal{P}^+ - \mathcal{P}^-)[\sigma_p^{\gamma Z} + \sigma_p^Z] \right\} \tag{6.4}$$

where: the unpolarized partial cross sections due to γZ interference and pure Z exchange are defined as,

$$\sigma_u^{\gamma Z} = -8Q_f \sin^2 2\theta_w \mathrm{Re}[\Gamma(s)]\left[(1 + \cos^2 \theta^*)v v_f + 2\cos\theta^* a a_f\right]$$

$$\sigma_u^Z = |\Gamma(s)|^2\left[(1 + \cos^2 \theta^*)(v^2 + a^2)(v_f^2 + a_f^2) + 8\cos\theta^* v a v_f a_f\right];$$

θ^* is the angle of the outgoing fermion relative to the incident electron; the polarized partial cross sections due to γZ interference and pure Z exchange are defined as,

$$\sigma_p^{\gamma Z} = 8Q_f \sin^2 2\theta_w \mathrm{Re}[\Gamma(s)]\left[(1 + \cos^2 \theta^*)a v_f + 2\cos\theta^* v a_f\right]$$

$$\sigma_p^Z = -|\Gamma(s)|^2\left[(1 + \cos^2 \theta^*)2 v a(v_f^2 + a_f^2) + 2\cos\theta^* (v^2 + a^2)2 v_f a_f\right];$$

the constant N_c^f is the color factor (3) for quark final states; $\Gamma(\hat{s}) = \hat{s}/(\hat{s} - M_Z^2 + i\Gamma_Z \hat{s}/M_Z)$ is the normalized Z propagator; and where the coupling constants without subscript, v and a, refer to the electron vector and axial-vector coupling constants. Note that we've assumed that the masses of all final state fermions are small as compared with $\sqrt{s}$ and that the unpolarized cross section for pure photon exchange is small as compared with the pure Z and interference terms.

The forward-backward asymmetries are defined to select the part of the $e^+ e^-$ cross section that is odd under spatial reflection. The left-right polarization asymmetry is designed to select the part of the cross section that is odd in difference of the beam polarizations $P^+ - P^-$. It is therefore useful to define a generalized beam polarization $\mathcal{P}_g$ that is proportional to $P^+ - P^-$ and has a convenient normalization,

$$\mathcal{P}_g \equiv \frac{\mathcal{P}^+ - \mathcal{P}^-}{1 - \mathcal{P}^+ \mathcal{P}^-}. \tag{6.5}$$

Note that $\mathcal{P}_g$ is positive whenever the electron beam is left-handed and/or the positron beam is right-handed. It is negative whenever the reverse is true. The generalized polarization becomes unity when either beam is completely polarized. The positron beam of the SLC is unpolarized. The generalized polarization therefore has the simple form, $\mathcal{P}_g = -\mathcal{P}^-$.

The left-right polarization asymmetry is defined as the ratio of the difference of the total Z^0 production rates with left-handed and right-handed beams to the total rate. This can be expressed more precisely as,

$$A_{LR} \equiv \frac{\sum_f \left\{ \int_{-x_f}^{x_f} dc\,\sigma_f(c, \mathcal{P}_g = +1) - \int_{-x_f}^{x_f} dc\,\sigma_f(c, \mathcal{P}_g = -1) \right\}}{\sum_f \left\{ \int_{-x_f}^{x_f} dc\,\sigma_f(c, \mathcal{P}_g = +1) + \int_{-x_f}^{x_f} dc\,\sigma_f(c, \mathcal{P}_g = -1) \right\}}, \tag{6.6}$$

where: $c \equiv \cos\theta^*$; $\sigma_f(c, \mathcal{P}_g)$ is shorthand for the differential cross section $d\sigma_f/d\Omega^*$; $\pm x_f$ are integration limits that depend upon fermion type; and where the sum is taken over all visible final state fermions except electrons (to exclude the t-channel scattering process). Note that the integrals must be taken over symmetric limits (which is a natural property of most e^+e^- detectors).

Substituting equation (6.4) (actually, the version of equation (6.4) with finite final state masses) into equation (6.6) it is straightforward to show that the left-right asymmetry takes the following form **on the Z^0 pole,**

$$A_{LR} = \frac{-2va \sum_f \int_{-x_f}^{x_f} dc[(v_f^2 + a_f^2)(1 + \beta_f^2 c^2) + (v_f^2 - a_f^2)(1 - \beta_f^2)]}{(v^2 + a^2) \sum_f \int_{-x_f}^{x_f} dc[(v_f^2 + a_f^2)(1 + \beta_f^2 c^2) + (v_f^2 - a_f^2)(1 - \beta_f^2)]},$$

where β_f is the velocity of the final state fermion in the $f\bar{f}$ center-of-mass frame. Cancelling the common factor, we recover a familiar expression,

$$A_{LR} = \frac{-2va}{v^2 + a^2} = \frac{2(1 - 4\sin^2\theta_w)}{1 + (1 - 4\sin^2\theta_w)^2}. \tag{6.7}$$

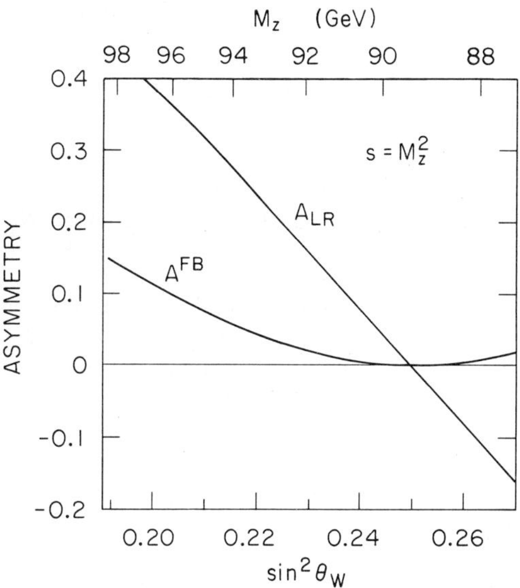

Figure 17

The left-right asymmetry A_{LR} is plotted as a function of $\sin^2\theta_w$ and M_Z (for some choice of m_t and m_h). The leptonic forward-backward asymmetry A^{FB} is shown for comparison.

A number of conclusions can be drawn from this derivation:

1. A_{LR} depends upon the Z^0-electron couplings alone. The dependence on the final state couplings cancels in the ratio.

2. A_{LR} is independent of the detector acceptance. This remains true even if each final state fermion is accepted differently.

3. A_{LR} is independent of final state mass effects (which would cause β_f to differ from unity).

4. All of the visible final states except the electron pairs can be used to measure A_{LR}. The measurement therefore utilizes about 96% of the visible decays. The various other Standard Model tests that are performed on the Z^0 pole make use of much smaller fractions of the event total ($\sim 4\%$ for the muonic forward-backward asymmetry, $\sim 0.9\%$ for the τ polarization measurement, and $\sim 4\%$ for the b-quark forward-backward asymmetry).

5. A_{LR} is very sensitive to the electroweak mixing parameter $\sin^2\theta_w$. This is shown graphically in Figure 17. Small changes in A_{LR} are related to changes in $\sin^2\theta_w$ by the following expression,

$$\delta A_{LR} \simeq -8\delta\sin^2\theta_w. \tag{6.8}$$

For $M_Z = 91.17$ GeV, the asymmetry is expected to be in the range 13%-15%.

<u>Radiative Corrections</u>

The left-right asymmetry has the property that it is insensitive to a large class of relatively uninteresting real and virtual radiative corrections and is very sensitive to an interesting set of virtual electroweak corrections. This behavior can be summarized as follows:

1. The left-right asymmetry is very insensitive to initial state radiative corrections. The emission of real photons by the incident electron and positron causes a smearing of the center-of-mass energy of the $f\bar{f}$ system ($\sqrt{\hat{s}}$). The left-right asymmetry is quite insensitive to small changes in $\sqrt{\hat{s}}$. The energy dependence of A_{LR} is compared with those of several forward backward asymmetries in Figure 18. The size of the initial state radiative correction to A_{LR} is calculated to be[32] $\delta A_{LR} \simeq 0.002$ (this is a 2% correction to the asymmetry). The uncertainty on the correction to A_{LR} is smaller by an order of magnitude.

2. The QCD corrections to the left-right asymmetry vanish entirely to all orders in the strong coupling constant α_s at the leading order in the electromagnetic coupling constant α. The leading QCD corrections to A_{LR} are the (extremely small) corrections to the weak vector boson box diagrams.

3. The theoretical uncertainty on A_{LR} is completely dominated by the uncertainty on the renormalization of the electromagnetic coupling constant to the Z^0 mass scale. The current value of this uncertainty is[33] $\delta A_{LR} \simeq 0.002$.

4. The left-right asymmetry is quite sensitive to virtual electroweak corrections and to the presence of new particles. The sensitivity of the asymmetry to the top quark mass (m_{top}) and the Higgs boson mass (m_{Higgs}) will be discussed in the last section of this document.

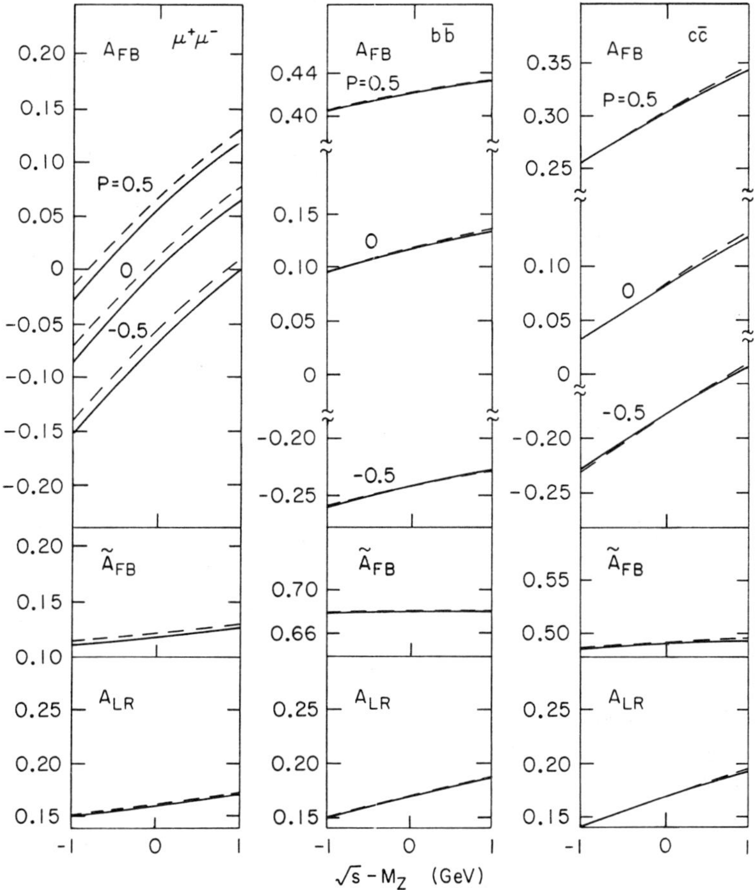

Figure 18

The forward-backward asymmetries for leptons, u-quarks, and d-quarks are plotted as a functions of the center-of-mass energy about the Z^0 pole. The asymmetries are also shown for the case that the incident beams are polarized. The energy dependence of the left-right asymmetry and an improved polarized forward-backward asymmetry $\tilde{A}^f_{FB}$ (from Reference 31) are also shown. The Z^0 mass is assumed to be 94 GeV. Note that the unimproved forward-backward asymmetries are much more sensitive to the center-of-mass energy than are the improved ones and the left-right asymmetries.

<u>Experimental Errors</u>

At the SLC, the measurement of A_{LR} is performed by randomly flipping the sign of the beam polarization on a pulse-to-pulse basis and by counting the number of Z^0 events that are produced from each state. The measured asymmetry, A_{LR}^{exp}, is related to the theoretical asymmetry, A_{LR}, by the following expression,

$$A_{LR}^{exp} \equiv \frac{N_Z(\mathcal{P}_g = +\mathcal{P}_0) - N_Z(\mathcal{P}_g = -\mathcal{P}_0)}{N_Z(\mathcal{P}_g = +\mathcal{P}_0) + N_Z(\mathcal{P}_g = -\mathcal{P}_0)} = \mathcal{P}_0 A_{LR}, \qquad (6.9)$$

where $\mathcal{P}_0$ is the magnitude of the beam polarization ($\mathcal{P}_0 \sim 0.40$), and $N_Z(\mathcal{P})$ is the number of Z^0 events logged with beam polarization $\mathcal{P}$. Since the left-handed and right-handed Z^0 cross sections are measured simultaneously, any systematic effects due to variations in detector livetime, luminosity, beam energy, beam position, etc., are cancelled in the ratio of the cross sections. This technique was used successfully to measure a very small polarized asymmetry ($\sim 10^{-5}$) in electron-deuteron scattering in 1978.[24] The dominant systematic error is expected to be the uncertainty on the beam polarization measurement. We expect that the SLC Compton polarimeter is capable of measuring the beam polarization with a precision of 1-2% ($\delta \mathcal{P}_0 / \mathcal{P}_0 = 1\text{-}2\%$).

There are a number of consistency checks that can be made with the SLC polarization hardware. It is possible to reverse the circular polarization optics of the electron source laser to search for systematic problems in that system. The polarity of the spin rotation system can be reversed to check for systematic problems in the damping rings. The polarization direction of each polarimeter target is reversible. The beam polarization can be measured separately with each target polarization direction (and must be consistent). Finally, the left-right asymmetry for small-angle Bhabha scattering is *very small* ($\sim 10^{-4}$). The luminosity monitors therefore provide an important check that the left-handed and right-handed luminosities are equal (the left-right asymmetry of the Bhabha signal must be consistent with zero).

Assuming that the dominant systematic error is the beam polarization uncertainty, the combined statistical and systematic uncertainty on A_{LR} is given by the following expression,

$$\delta A_{LR} = \left[A_{LR}^2 \left(\frac{\delta \mathcal{P}_0}{\mathcal{P}_0} \right)^2 + \frac{1 - (\mathcal{P}_0 A_{LR})^2}{\mathcal{P}_0^2 N_{tot}} \right]^{1/2}, \qquad (6.10)$$

where N_{tot} is the total number of Z^0 events. The expected precision of the A_{LR} measurement and the corresponding precision on $\sin^2 \theta_w$ are listed in Table X for several values of N_{tot}. Note that the statistical uncertainty dominates the total error in the region $N_{tot} \leq 10^6$. At $N_{tot} = 3 \times 10^6$, the statistical and systematic components are comparable.

The experimental confidence intervals that are presented in Table X are compared with the theoretical expectation for A_{LR} in Figures 19 and 20. The solid curves in Figure 19 enclose the 68.3% confidence region that is expected for $m_{Higgs} = 500$ GeV and m_{top} varying between 60 GeV and 240 GeV. The finite width of the region is due to a ± 20 MeV uncertainty on the Z^0 mass (we assume $M_Z = 91.17 \pm 0.02$ GeV). The solid curves in Figure 20 enclose the 68.3% confidence region that is expected for $m_{top} = 150$ GeV and m_{Higgs} varying from 100 GeV to 900 GeV. The size of the theoretical error on A_{LR} (± 0.002) is shown as the dotted vertical error bar in each figure. The sizes of the experimental 68.3% confidence intervals that correspond to the various values of N_{tot} are indicated by the solid vertical error bars. Since the τ polarization asymmetry is formally

Table X

The expected error on A_{LR} and $\sin^2\theta_w$ as a function of the number of Z^0 events. The left-right asymmetry is assumed to be $A_{LR} = 0.135$ (which is in the middle of the range that is expected for $M_Z = 91.17$ GeV). The beam polarization is assumed to be $\mathcal{P}_0 = 0.40$ and the precision of the polarization monitoring is assumed to be $\delta\mathcal{P}_0/\mathcal{P}_0 = 0.01$.

N_{tot}	δA_{LR}	$\delta\sin^2\theta_w$
100K	0.008	0.0010
300K	0.005	0.0006
1M	0.003	0.00035
3M	0.002	0.00025

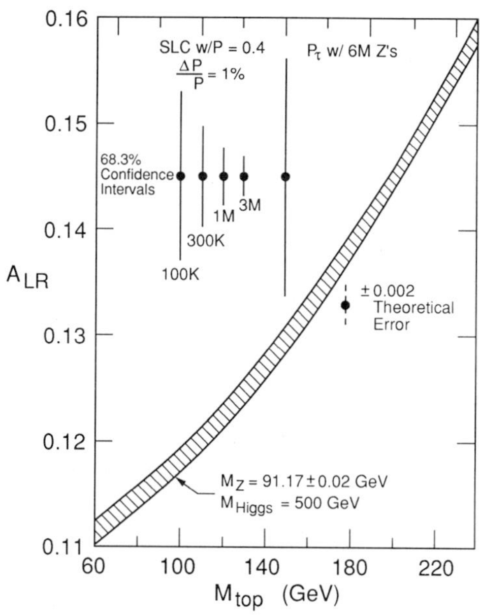

Figure 19

The left-right asymmetry as a function of the top quark mass (m_{top}). The Higgs boson mass (m_{Higgs}) is assumed to be 500 GeV. The solid curves enclose the 68.3% confidence region that is expected for a ± 20 MeV uncertainty on M_Z (we assume $M_Z = 91.17 \pm 0.02$ GeV) as m_{top} is varied from 60 GeV to 240 GeV. The dotted vertical error bar shows the size of the theoretical error (± 0.002) on A_{LR}. The sizes of the experimental 68.3% confidence intervals that are expected for the various values of N_{tot} are indicated by the solid vertical error bars. The confidence interval that is expected from a measurement of the τ polarization asymmetry with 6M Z^0 events is also shown.

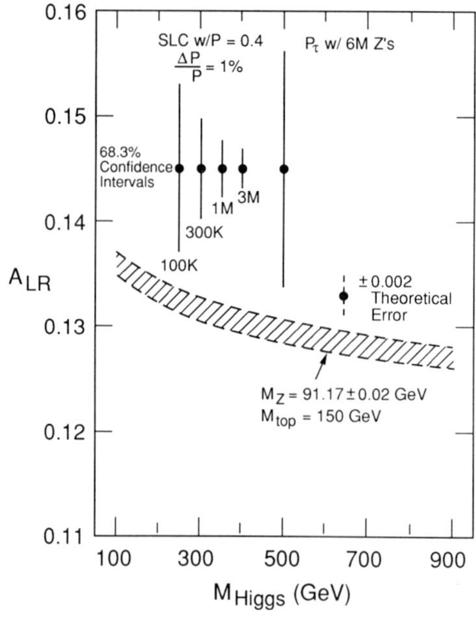

Figure 20.

The left-right asymmetry as a function of the Higgs boson mass. The top quark mass is assumed to be 150 GeV. The dashed curves enclose the 68.3% confidence region that is expected for a ±20 MeV uncertainty on M_Z (we assume $M_Z = 91.17 \pm 0.02$ GeV) as m_{Higgs} is varied from 100 GeV to 900 GeV. The dotted vertical error bar shows the size of the theoretical error (±0.002) on A_{LR}. The sizes of the experimental 68.3% confidence intervals that are expected for the various values of N_{tot} are indicated by the solid vertical error bars. The confidence interval that is expected from a measurement of the τ polarization asymmetry with 6M Z^0 events is also shown.

equivalent to A_{LR}, we plot the confidence region that is expected from a measurement with a 6M Z^0 sample. It is clear that A_{LR} is quite sensitive to m_{top}. A measurement with 300K Z^0 events constrains the top quark mass to a region of roughly $\delta m_{top} = \pm 17$ GeV which is comparable to a 100 MeV determination of M_W. The sensitivity to m_{Higgs} is clearly much smaller. A very high statistics measurement of A_{LR} could provide, at best, an indication of m_{Higgs}.

REFERENCES

1. See for example, K. Steffen, *Proceedings of the CERN Accelerator School*, CERN 85-19, November 1985.

2. E.L. Ginzton, W.W. Hansen, and W.R. Kennedy, *Rev. Sci. Instr.* **19**, 89 (1948).

3. W.A. Koska *et al.*, *Nucl. Instr. Meth.* **A286**, 32 (1990).

4. *SLC Performance in 1991*, M. Breidenbach *et al.*, SLAC internal report, Rev. January 1990.

5. E.R. Cohen and B.N. Taylor, *Rev. Mod. Phys.* **59**, 1121 (1987).

6. *Review of Particle Properties*, J.J. Hernandez et al., *Phys. Lett.* **B239**,1 (1990).

7. It would be impossible to list all of the contributors to this topic. A nice pedagogical review is given by M. Peskin, Proceedings of the 17th SLAC Summer Institute on Particle Physics: *Physics at the 100 GeV Mass Scale*, SLAC-Report-361, p 71.

8. G. Abrams et al., *Phys. Rev. Lett.* **63**, 2173 (1989).

9. T. Muta, R. Najima, and S. Wakaizumi, *Mod. Phys. Lett.* **A1**, 203 (1986).

10. J. Bijnens et al., *Proceedings of the ECFA Workshop on LEP 200*, edited by A. Böhm and W. Hoogland, CERN 87-08, June 1987, pg 85.

11. F. Abe et al., *Phys. Rev. Lett.* **62**, 154 (1990).

12. J. Alitti et al., *Zeit. Phys.* **47**, 11 (1990).

13. G.S. Abrams *et al.*, *Phys. Rev. Lett.* **63**, 2447 (1989); C.K. Jung, R. Van Kooten *et al.*, *Phys. Rev. Lett.* **64**, 1091 (1990); P.R. Burchat, M. King *et al.*, *Phys. Rev.* **D41**, 3542 (1990).

14. T. Barklow *et al.*, *Phys. Rev. Lett.* **64**, 2984 (1990).

15. M.L. Swartz *et al.*, *Phys. Rev. Lett.* **64**, 2877 (1990); S. Komamiya *et al.*, *Phys. Rev. Lett.* **64**, 2881 (1990).

16. G.B. Gelmini and M. Roncadelli, *Phys. Lett.* **99B**, 411 (1981).

17. V. Barger, H. Baer, W.Y. Keung, and R.J.N. Phillips, *Phys. Rev.* **D26**, 218 (1982).

18. J.C. Pati and A. Salam, *Phys. Rev.* **D10**, 275 (1974);
R.N. Mohapatra and J.C. Pati, *Phys. Rev.* **D11**, 566 and (1975);
R.N. Mohapatra and G. Senjanovic, Phys. Rev. **D12**, 1502 (1975), *Phys. Rev. Lett.* **44**, 912 (1980), and Phys. Rev. **D23**, 165 (1981);
R.N. Mohapatra and J.D. Vergados, *Phys. Rev. Lett.* **47**, 1713 (1981);
For a good review of left-right symmetric models, see R.N. Mohapatra, *Proceedings of the NATO Advanced Study Institute: Quarks, Leptons, and Beyond*, ed. H. Fritzch et al., Plenum Publishing Co. (New York, 1985), pg. 219.

19. D. Chang and W.-Y. Keung, *Phys. Rev. Lett.* **62**, 2583 (1989).

20. F. Hoogeveen, DESY 89-028, Feb. 1989.

21. M.L. Swartz, *Phys. Rev.* **D40**, 1521 (1989).

22. J.A. Grifols, A. Mendez, and G.A. Schuler, UAB-FT-196/88, February 1989.

23. D.T. Pierce and F. Meier, *Physical Review* **B13**, 5484 (1976); C.K. Sinclair, SLAC-PUB-3505, November 1984. Also published in *Proceedings of the 6th International Symposium on High Energy Spin Physics, Marseille France*, edited by J. Soffer, Les Editions de Physique, Les Ulis France, 1985.

24. C.Y. Prescott et al., *Phys. Lett.* **77B**, 347 (1978); and *Phys. Lett.* **84B**, 524 (1979).

25. T. Maruyama, private communication.

26. C.Y. Prescott, SLAC-TN-73-1, January 1973.

27. R.H. Helm and W.P. Lysenko, SLAC-TN-72-1 (March 1972) and M.J. Alguard et al., *Nuclear Instruments and Methods* **163**, 29 (1979).

28. P.S. Cooper et al., *Physical Review Letters* **34**, 1589 (1975); M. J. Alguard et al., *Physical Review Letters* **37**, 1258 and 1261 (1976); M. J. Alguard et al., *Physical Review Letters* **41**, 70 (1978); and W.B. Atwood et al., *Physical Review* **D18**, 2233 (1978).

29. W.K.H. Panofsky, SLAC-CN-361, (August 1987).

30. A.A. Sokolov and I.M. Ternov, *Soviet Physics-Dolkady* **8**, 1203 (1964).

31. A. Blondel, B.W. Lynn, F.M. Renard, and C. Verzegnassi, Montpellier preprint PM/87-14, March 1987.

32. D. Kennedy, B.W. Lynn, and C.J.C. Im, *Nucl. Phys.* **B321**, 83 (1989).

33. H. Burkhardt, F. Jegerlehner, G. Penso, and C. Verzegnassi, Vol. I *Polarization at LEP*, CERN 88-06, September 1988, pg 145.

PHYSICS FROM TRISTAN

Young-Kee Kim

Physics Department
University of Rochester
Rochester, N.Y. 14627

1. INTRODUCTION

The TRISTAN e^+e^- storage ring has delivered a total of about 60 pb^{-1} to each of the three general purpose solenoidal detectors: AMY, TOPAZ and VENUS. The center-of-mass energy range, between 50 and 64 GeV, is where interference effects between the virtual photon and Z^0 mediators of the annihilation process are largest. Experiments at TRISTAN study the electroweak interactions, QCD, and search for new phenomena. Among the reported results are: measurements of R; the observation of forward-backward charge asymmetries for $e^+e^- \rightarrow$ lepton pairs, $e^+e^- \rightarrow b\bar{b}$, and $e^+e^- \rightarrow q\bar{q}$; limits on lepton compositeness; determinations of $\Lambda_{\overline{MS}}$; measurements of various multihadron event properties; evidence for differences between quark and gluon jets; the demonstration of the running of α_s and the existence of the triple-gluon coupling; and searches for a variety of new particles. This report describes a few selected results from the AMY experiment.

2. FORWARD-BACKWARD CHARGE ASYMMETRIES

In the standard model, the process $e^+e^- \rightarrow f\bar{f}$ (here $f\bar{f}$ denotes a fermion-antifermion pair) can proceed via virtual photon and Z^0 intermediate states (see Figure 1).

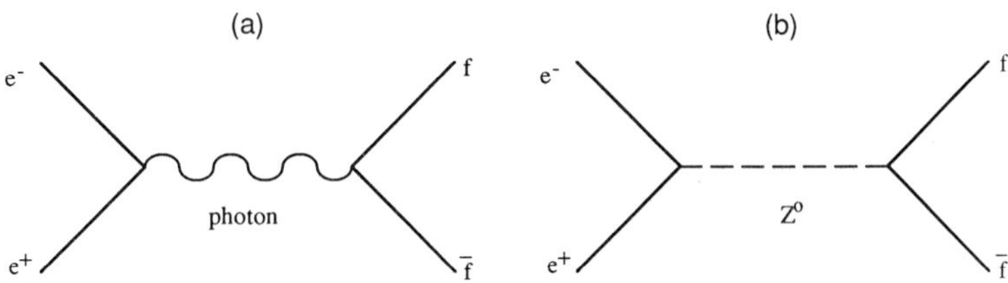

Fig. 1. Electromagnetic and weak contributions to the production of a fermion pair in the lowest order.

The coupling of the fermion f to the photon involves only the vector electromagnetic current, while that to the Z^0 involves weak couplings which have both vector and axial-vector components. A forward-backward charge asymmetry is induced by the interference between the axial-vector coupling to the Z^0 and the vector coupling to both the photon and the Z^0. At TRISTAN energies, $\sqrt{s}=50-64$ GeV, this asymmetry is substantial as indicated in Figure 2: about 20-50% for a μ (or τ) pair; about 30-60% for a u-type quark pair ($u\bar{u}, c\bar{c}$); about 50-60% for a d-type quark pair ($d\bar{d}, s\bar{s}, b\bar{b}$). The asymmetry of $e^+e^- \to q\bar{q}$, summed over all flavors, is also predicted to be nearly maximal in this energy range (see Figure 2). TRISTAN, therefore, provides excellent opportunities to test the predictions of the electroweak theory.

The electroweak prediction for the differential cross section of the fermion pair production ($e^+e^- \to f\bar{f}$) can be written in the form

$$\frac{d\sigma}{d\cos\theta} = \frac{\pi\alpha^2}{2s} R_f (1 + \cos^2\theta + \frac{8}{3} A_f \cos\theta), \tag{1}$$

where α is the fine structure constant and θ is the angle of the outgoing $f(\bar{f})$ direction with respect to the incoming $e^-(e^+)$ direction. R_f is the ratio of the cross section for $e^+e^- \to f\bar{f}$ to the lowest-order QED cross-section for $e^+e^- \to \mu^+\mu^-$ and A_f is the forward-backward charge asymmetry. These are given by

$$R_f = C_f \, [Q_f^2 - 8Q_f g_V^e g_V^f \mathrm{Re}(\chi) + 16(g_V^{e\,2} + g_A^{e\,2})(g_V^{f\,2} + g_A^{f\,2})|\chi|^2], \tag{2}$$

$$A_f = C_f \, [-6Q_f g_A^e g_A^f \mathrm{Re}(\chi) + 48 g_V^e g_V^f g_A^e g_A^f |\chi|^2] \, / \, R_f, \tag{3}$$

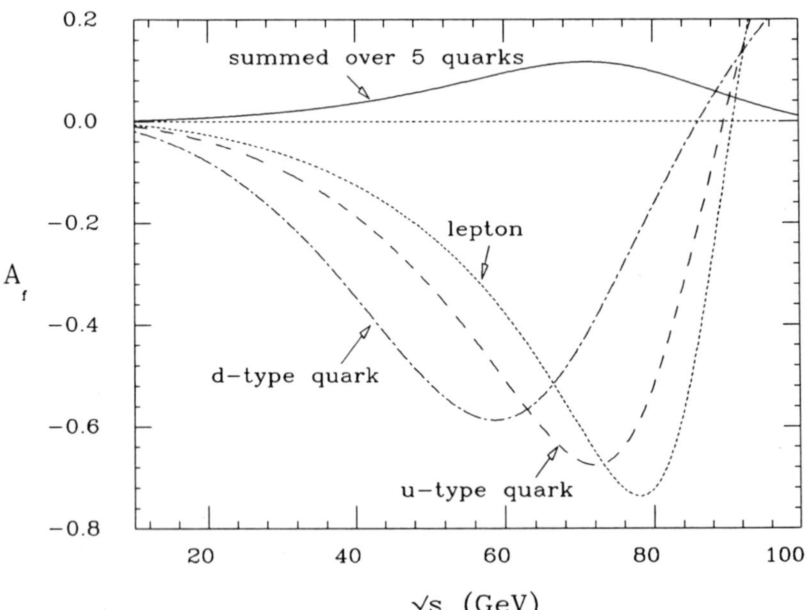

Fig. 2. Forward-backward asymmetry A_f of a fermion pair production in e^+e^- collisions, predicted by the standard model with $M_Z = 91.1$ GeV and $\sin^2\theta_W = 0.23$. The dotted, dashed, dot-dashed, and solid curve shows the asymmetry in μ (or τ), u-type quark, d-type quark production and the combined quark asymmetry, respectively.

where C_f is the number of colors — 3 for quarks and 1 for leptons, Q_f is the charge of the fermion, χ is the contribution from Z^0, given by

$$\chi = \frac{1}{16\sin^2\theta_W\cos^2\theta_W}\frac{s}{(s - M_Z^2 + i\Gamma_Z M_Z)}, \tag{4}$$

and g_V^f and g_A^f are the vector and axial-vector couplings to the fermion, respectively.

2.1 Purely Leptonic Processes $\left(e^+e^- \to \mu^+\mu^-,\ \tau^+\tau^-\right)$

Since the leptonic reactions, $e^+e^- \to \mu^+\mu^-$ and $e^+e^- \to \tau^+\tau^-$, have only point-like particles in both the initial and final state, they provide sensitive and unambiguous tests of the standard model.

Events of the type $e^+e^- \to \mu^+\mu^-$ are selected on the basis of its charged track topology of two back-to-back tracks with momenta consistent with beam energy. The energy deposits in the shower counter are required to be consistent with that from minimum ionizing particles, and the observed tracks must extrapolate to hits in the muon detection system.

The branching ratio for τ decay into a single charged particle is $\sim$83%; that for three charged particles is $\sim$17%. Thus, about two-thirds of the $e^+e^- \to \tau^+\tau^-$ events have two almost back-to-back charged tracks, with most of the remainder having a topology of one $vs.$ three charged tracks. Except cases where there are two tracks where both tracks are identified as either muons or electrons, these two topologies are accepted as $\tau^+\tau^-$ events.

The forward-backward charge asymmetry is extracted by fitting Equation (1) to the angular distribution, after taking into account of the detection efficiency and radiative corrections. The asymmetry for $\mu^+\mu^-$ and $\tau^+\tau^-$ is shown in Figure 3 (a) and (b),

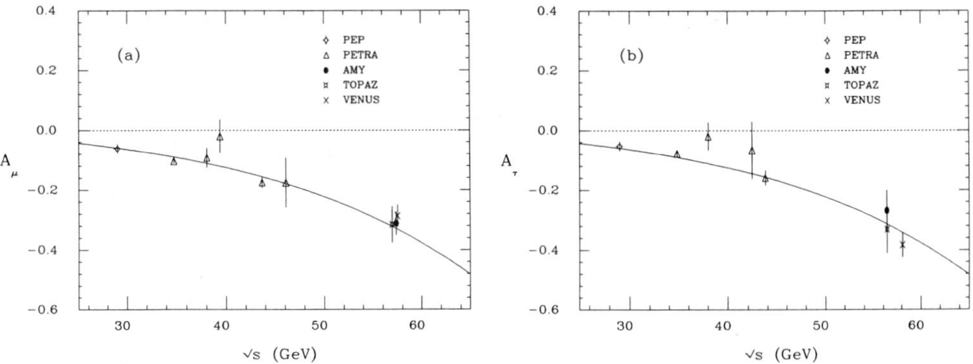

Fig. 3. Forward-backward asymmetry for (a) $e^+e^- \to \mu^+\mu^-$ and (b) $e^+e^- \to \tau^+\tau^-$ as a function of the center-of-mass energy. The solid-line curve is the standard-model prediction using M_Z=91.1 GeV and $\sin^2\theta_W$=0.23.

Table 1. Results for $< g_A^e g_A^\mu >$, $< g_A^e g_A^\tau >$, and $< g_A^\tau / g_A^\mu >$ measured at TRISTAN, together with results from PEP and PETRA, and the standard model predictions

	$< g_A^e g_A^\mu >$	$< g_A^e g_A^\tau >$	$< g_A^\tau / g_A^\mu >$
Standard Model	0.25	0.25	1.00
TRISTAN (AMY+TOPAZ+VENUS)	0.275 ± 0.013	0.238 ± 0.020	0.90 ± 0.08
PEP+PETRA	0.273 ± 0.015	0.215 ± 0.020	0.78 ± 0.09

respectively, together with results from other experiments. Using the lowest-order formula for A_l, i.e.

$$A_l = 6 g_A^e g_A^l \mathrm{Re}\chi, \tag{5}$$

with $\sin^2\theta_W = 0.23$ and $M_Z = 91.1$ GeV/c^2, and including measurements from the other experiments at TRISTAN, we obtain $< g_A^e g_A^\mu >=0.275 \pm 0.013$ and $< g_A^e g_A^\tau >=0.238 \pm 0.020$. We also determine the τ/μ axial coupling constant ratio to be $< g_A^\tau / g_A^\mu >=0.90 \pm 0.08$. These values are shown in Table 1 with the standard model predictions and the PEP and PETRA results. From these results, we can conclude that the TRISTAN results for the leptonic axial-vector couplings are in good agreement with the lower energy data and agree well with the standard model predictions.

The total cross sections for the $\mu^+\mu^-$ and $\tau^+\tau^-$ processes are also determined from the fit; these are shown in Figure 4 (a) and (b) with results from other experiments. It is observed that while R_τ is in good agreement with the prediction, R_μ is a bit lower than the prediction. Averaging over all TRISTAN experiments we find $R_\tau/R_\mu = 1.15 \pm 0.06$.

For the details of the analysis see refs. [1].

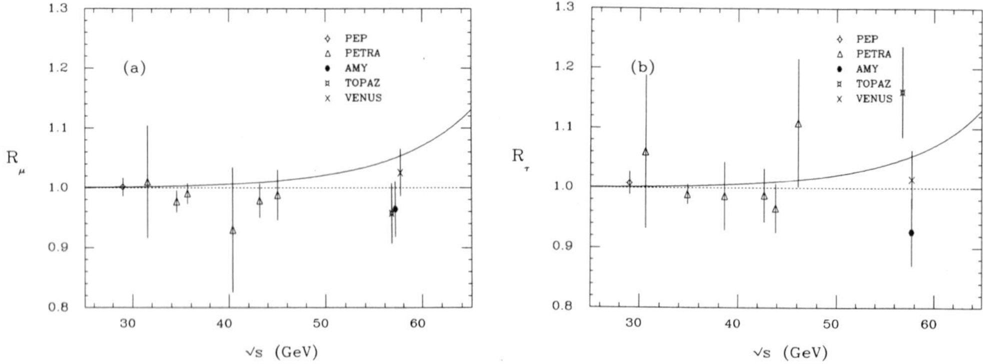

Fig. 4. R value for (a) $e^+e^- \rightarrow \mu^+\mu^-$ and (b) $e^+e^- \rightarrow \tau^+\tau^-$ as a function of the center-of-mass energy. The solid-line curve is the standard-model prediction using $M_Z = 91.1$ GeV and $\sin^2\theta_W = 0.23$.

2.2 $e^+e^- \rightarrow b\bar{b}$

The forward-backward charge asymmetry for the process $e^+e^- \rightarrow b\bar{b}$ is extracted from multi-hadronic annihilation events that contain muons. This inclusive muon data sample contains events of interest, namely $e^+e^- \rightarrow b\bar{b}$ with subsequent semi-leptonic decays — either from direct decays, $b \rightarrow \mu^-$ ($\bar{b} \rightarrow \mu^+$), or from cascade decays, $b \rightarrow c \rightarrow \mu^+$ ($\bar{b} \rightarrow \bar{c} \rightarrow \mu^-$) [2]. It also contains muons coming from $e^+e^- \rightarrow c\bar{c}$ followed by $c \rightarrow \mu^+$ ($\bar{c} \rightarrow \mu^-$), and hadron fakes which arise from punch-through of hadron showers in the hadron filter or from the decay-in-flight of $\pi^\pm$ and $K^\pm$ mesons.

The contributions from the $c\bar{c}$ production as well as those from hadron fakes are estimated by using a Monte Carlo simulation, where five flavors are generated according to the standard model using the LUND 6.3 event generator [3]. Figure 5 shows the distribution of muon p_t for the data together with the estimated contributions from $b\bar{b}$, $c\bar{c}$, and hadron fakes, where p_t is the transverse momentum of muons with respect to the event thrust axis. This figure indicates that we can maximize the fraction of b quark by applying a cut of muon p_t. A 33.3 pb^{-1} sample of multi-hadronic events containing muons with $p_t \geq 0.7$ GeV/c are used for this analysis [4].

The angular distribution, where the angle θ is defined as the direction of the thrust axis associated with $\mu^-(\mu^+)$ with respective to the incoming $e^-(e^+)$ direction, is shown in Figure 6. Here backgrounds from $c\bar{c}$ and hadron fakes have been subtracted and corrections are made for effects due to the definition of θ, the b quark detection efficiency, the effects of the cascade decay $b \rightarrow c \rightarrow \mu$, and the muon detection efficiency. No corrections are made for $B^0 - \bar{B}^0$ mixing.

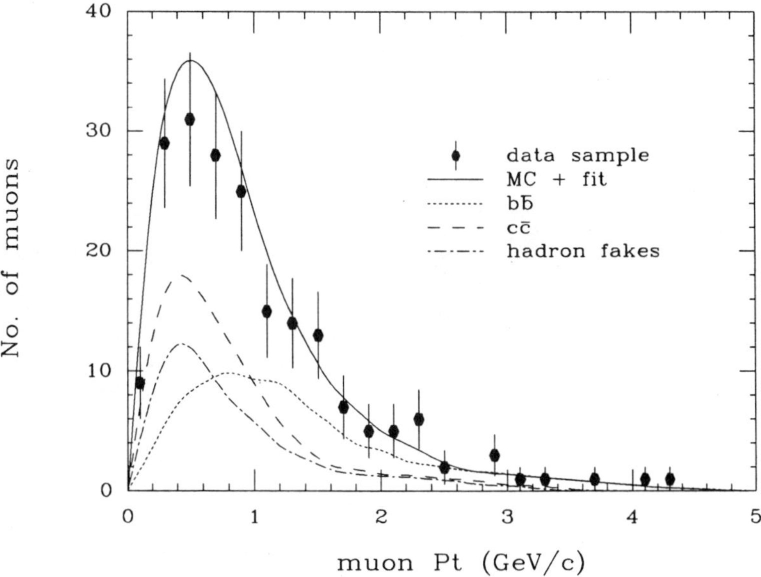

Fig. 5. The transverse momentum distribution of the muons with respect to the thrust axis for all multihadronic events including muons. Estimated contributions from $b\bar{b}$, $c\bar{c}$ and hadron fakes were determined from a Monte Carlo simulation.

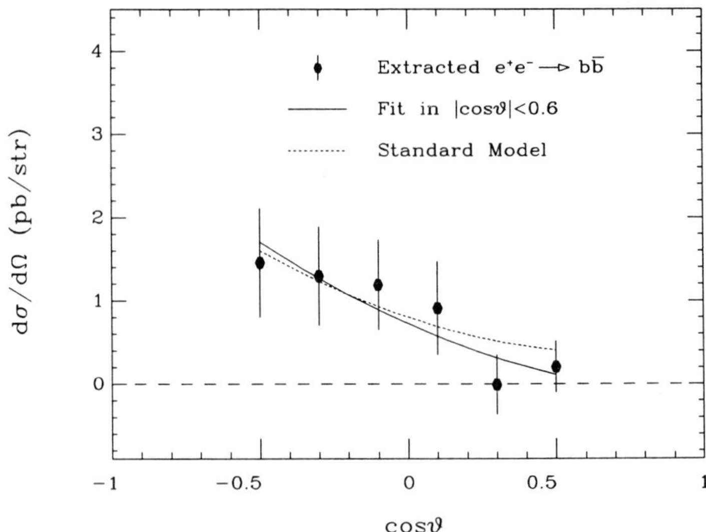

Fig. 6. $d\sigma/d\Omega$ of $e^+e^- \to b\bar{b}$ with results of the fit. The standard prediction is also shown.

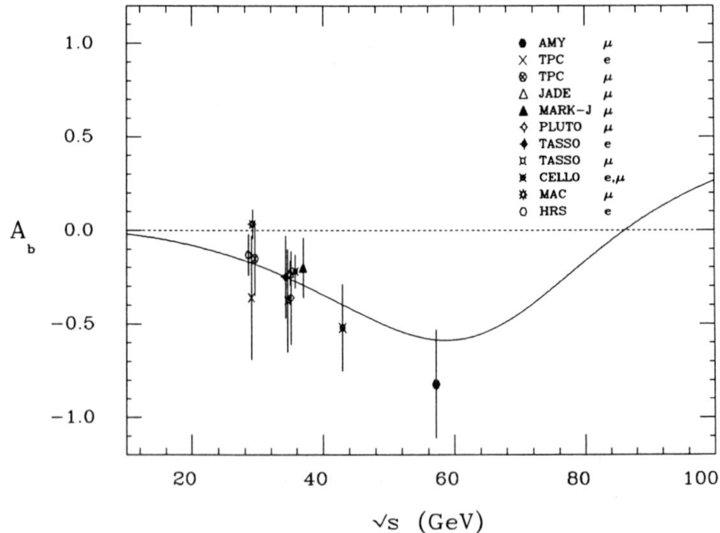

Fig. 7. The forward-backward charge asymmetry for $e^+e^- \to b\bar{b}$ as a function of the center-of-mass energy. The result of the AMY experiment at a mean $\sqrt{s}$=57.2 GeV is compared with previous measurements at lower energies. The solid-line curve is the standard-model prediction using M_Z=91.1 GeV and $\sin^2\theta_W$=0.23.

The results of a fit are $A_b=-0.82\pm0.25\pm0.14$ and $R_b=0.47\pm0.12\pm0.10$, where the errors are statistical and systematic, respectively. These observed results are consistent with the standard model predictions of $A_b=-0.58$ and $R_b=0.56$. Figure 7 compares our result for A_b with those from previous measurements [5][6][7], which are also not corrected for the $B^0-\bar{B}^0$ mixing. The measurements are consistent with the standard model prediction throughout the energy region explored so far.

$B^0-\bar{B}^0$ mixing can result in events where a μ^- is produced from an initial $\bar{b}$ quark, thereby confusing the quark identification. Consequently the asymmetry is reduced from the value A_b given by the standard model;

$$A_b^{obs} = (1 - 2\chi)A_b \tag{6}$$

where χ is the $B^0-\bar{B}^0$ mixing parameter defined by

$$\chi = \frac{\Gamma(B^0 \rightarrow \bar{B}^0 \rightarrow \bar{X})}{\Gamma(B^0 \rightarrow X \text{ or } \bar{X})} . \tag{7}$$

Using the 3:3:1 ratio for $(u\bar{u} : d\bar{d} : s\bar{s})$ quark pair production from the color fields and assuming that they are independent of the primary quark, we expect the production ratio for $B_u^+ : B_d^0 : B_s^0$ is to be about 3:3:1.[1] Charge conservation demands that mixing can only happen in the neutral meson system and, thus, χ_u must be zero. Assuming equal semi-leptonic branching ratios for all B mesons, the effective χ is then

$$\chi = 3/7\chi_d + 1/7\chi_s. \tag{8}$$

CLEO [8] measures $(\chi_d=0.123\pm0.048)$ and ARGUS [9] reports $(\chi_d=0.167\pm0.055\pm0.046)$. There are good theoretical reasons to believe that B_s^0 mixing is maximal [10] (i.e., $\chi_s=0.5$). We therefore expect $\chi=0.13\pm0.02$. Any experimental observation that significantly deviates from this value will indicate that the above assumptions are incorrect.

Using the measured asymmetry of $A_b^{obs} = -0.82\pm0.29$ and the standard model prediction of $A_b=-0.58$, we can set a 90% confidence level limit of $\chi < 0.20$. This is consistent with the above expectation. This result agrees with the 90% CL limits found by MARK II [11] $(\chi < 0.12)$ and by JADE [5] $(\chi < 0.13)$ but is in poor agreement with the 90% CL limit reported by MAC [6] $(\chi > 0.21)$.

2.3 $e^+e^- \rightarrow q\bar{q}$

Measurements of the forward-backward asymmetry for individual quark flavors are only feasible for heavy quarks. Alternatively, all five quark flavors can combined and the asymmetry in hadron production can be measured by distinguishing between negatively charged jets (from $d, s, b, \bar{u}, \bar{c}$) and positively charged jets (from $u, c, \bar{d}, \bar{s}, \bar{b}$). Defining θ to be the angle between the incoming e^- and the outgoing negatively charged jet, the hadronic asymmetry is given by

$$A_h = \sum_q{}' f_q A_q, \tag{9}$$

---

[1] Here, baryon productions are ignored.

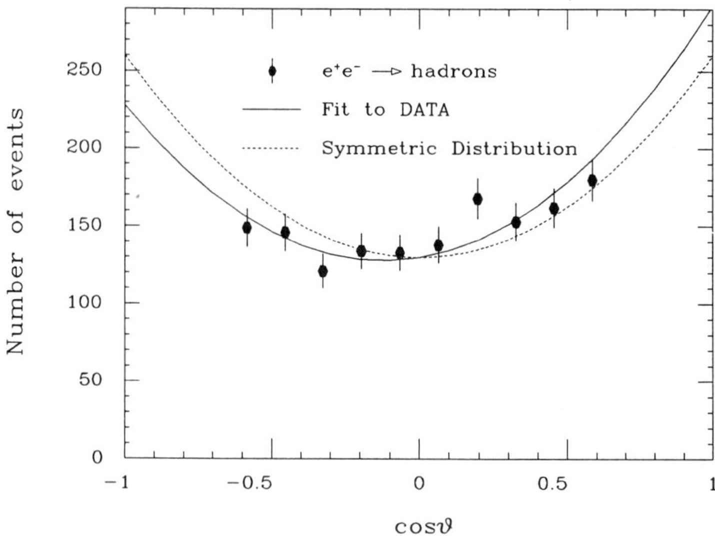

Fig. 8. Production angular distribution using rapidity weighting which combines the angular distributions of two u-type and three d-type quark production. The solid curve is a minimum χ^2 fit (χ^2=7.9/8 degrees of freedom) giving an asymmetry of 9.3%. The dotted curve is the corresponding symmetric distribution which has χ^2=17.2/9 degrees of freedom.

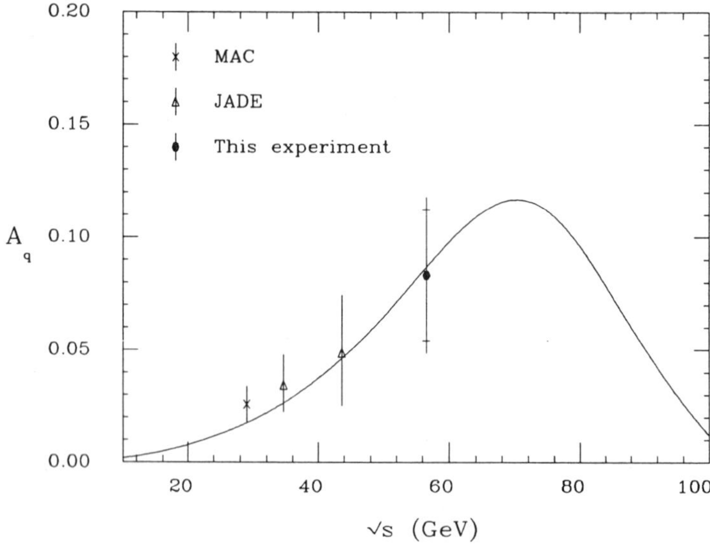

Fig. 9. Forward-backward asymmetry A_h of quark production in e^+e^- collisions compared with the standard-model values with M_Z=91 GeV and $\sin^2\theta_W$=0.23.

where $f_q = R_q/\sum_i R_i$, the fraction of q quarks in the sample. The primed summation symbol, $\sum'$, is used instead of $\sum$ to denote the fact that we must now use $-A_q$ for the positively charged $u-$ and $c-$quarks.

For this analysis [12], a 27.4 pb^{-1} sample of 3211 multihadronic events at $\sqrt{s}$=50$-$60.8 GeV ($<\sqrt{s}>$=56.6 GeV) is used. The particles in each event are divided into two hemispheres (jets) by the plane perpendicular to the thrust axis. The effective charge of a jet is defined as

$$Q = \sum_{j=1}^{n} Q_j \eta_j \Big/ \sum_{j=1}^{n} \eta_j, \tag{10}$$

where n is the number of particles in a jet and Q_j and η_j are the charge of particle j and its rapidity w.r.t. the thrust axis, respectively. The angular distribution, where the angle θ is defined as the direction of the thrust axis associated with the negatively charged jet with respective to the incoming e^- direction, is shown in Figure 8. A minimum χ^2 fit gives a measured asymmetry of A_h^M=(9.3 $\pm$ 3.1 $\pm$ 2.0)% (χ^2=7.9/3 degrees of freedom). The standard-model prediction at 56.6 GeV, with the quark-charge misidentification probabilities folded in, is A_h^M=(9.7 $\pm$ 0.6)% (the error is due to the uncertainty in the misidentification probabilities). The effect of $B^0-\bar{B}^0$ mixing would be to increase this predicted asymmetry by reducing the negative contribution from $b\bar{b}$. For example, a 20% mixing for B_d^0 and 100% mixing for B_s^0 changes the predicted asymmetry from A_h^M=9.7%, to A_h^M=10.9%, a change that is much less than our measurement uncertainty.

Assuming universality, the measured value for A_h^M corresponds to a quark axial-vector coupling constant (squared) of g_A^2=0.23$^{+0.29+0.12}_{-0.10-0.06}$ — the standard model predicts g_A^2=1/4. From this value for g_A^2, a "method-independent" asymmetry A_h=(8.3 $\pm$ 2.9 $\pm$ 1.9)% is unfolded. For comparison, the standard-model prediction at 56.6 GeV is A_h=8.7%. In Figure 9, A_h is compared to the standard-model prediction together with results from PEP and PETRA, calculated using the results for g_A from similar analyses by the MAC and JADE Collaborations. The higher-energy data at TRISTAN continue to agree well with the standard model for five quark flavors.

3. $\Lambda_{\overline{MS}}$ DETERMINATION IN NLL QCD APPROXIMATION

There are two main approaches used for applying perturbative expansions of QCD: the Matrix-Element (ME) approach and the QCD-cascade or Parton-Shower (PS) approach. In the ME approach, diagrams up to a fixed order in the QCD coupling strength, α_s, are calculated exactly. However, the possibility of summing all orders of perturbative diagrams motivated the PS approach to generating partons. In this approach, the leading-logarithm terms or the leading- and next-to-leading-logarithm terms in all orders are summed.

The ME approach allows one to determine the QCD scale parameter $\Lambda_{\overline{MS}}$ (where $\overline{MS}$ denotes the modified minimal-subtraction scheme) and provides a good approximation for large-angle parton generation such as those involved in three-jet events, using second-order ME calculations, which are the highest order ME calculations carried out to date. However, Monte Carlo event generators based on these calculations have shown large deviations from experiments for those cases where soft partons or low-transverse-momentum partons are important. On the other hand, in the Leading-Logarithm Approximation (LLA) based on PS models, where the leading term in each order is summed, it is possible to generate large number of partons and it provides a good approximation for the production of low-transverse-momentum gluons. However, the LLA evolution fails to give a correct production rate of QCD for large-angle parton generation such as the three-jet production rate. The LUND PS model cures this problem by correcting the first branching according to the first order ME result [3]. This has, in general, provided rather good agreement with experimental results over a wide energy range.

However, in the LLA PS method, the renormalization scheme for the QCD scale parameter, Λ, is undefined. In order to use the PS scheme to determine Λ in a definite scheme (we use $\Lambda_{\overline{MS}}$), we have to go beyond the LLA. This leads to the Next-to-Leading Logarithm approximation (NLL) where the leading and next-to-leading terms in each order are summed. The LLA expansion has the form

$$\sum_{n=0} c_n (\alpha_s \ln Q^2/Q_0^2)^n \tag{11}$$

and the NLL expansion has the form

$$\sum_{n=0} c_n (1 + d_n \alpha_s)(\alpha_s \ln Q^2/Q_0^2)^n, \tag{12}$$

where Q_0 is the minimum virtuality for parton branching. In the LLA expansion, α_s, which in turn depends on $\Lambda_{\overline{MS}}$, is tightly correlated with Q_0 and therefore the two parameters can not be independently determined; in the NLL expansion, α_s and Q_0 are not tightly correlated and $\Lambda_{\overline{MS}}$ can be fixed. In addition, the NLL PS model can generate partons with both small and large transverse momentum without resorting to a ME-generated correction to the first branching. Figure 10 (a) and (b) show diagrams contributing to the leading and next-to-leading vertices, respectively.

(a)   (b)

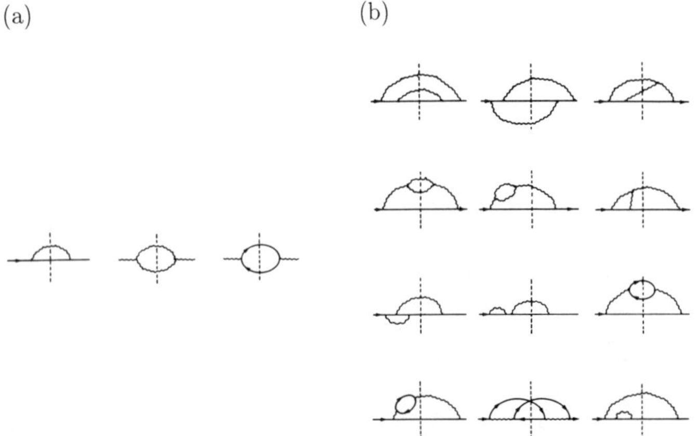

Fig. 10. Diagrams contributing to (a) the leading vertices (b) the next-to-leading vertices. The dotted vertical line indicates the final-state cut.

Based on theoretical calculations beyond the LLA [13][14], and including detailed considerations of the kinematics, Kato and Munehisa developed a program to generate events according to the prescription of the NLL Parton-Shower model [15]. This generator is combined with the LUND String-Fragmentation model [3] to generate hadronic events. We compare the general properties of multi-hadron final states observed in the AMY experiment with predictions of the NLL Parton-Shower model and measure $\Lambda_{\overline{MS}}$ based on this model using three different methods: event-shape variables, the three-jet ratio, and the Energy-Energy Correlation (EEC) [16].

3.1 Tuning the NLL Generator Using AMY Data

There are three NLL jet parameters: $\Lambda_{\overline{MS}}$, the QCD scale parameter; Q_0, the minimum virtuality; and δ, the parameter to determine the type of primary vertex, and three parameters in the LUND String-Fragmentation model: σ_q, the width in the transverse momentum distribution of primary hadrons; and the a and b parameters of the symmetric LUND fragmentation function.[2] Using 1191 multi-hadronic events at center-of-mass energies between 52 and 57 GeV, the NLL generator was tuned by varying the parameters $\Lambda_{\overline{MS}}$, σ_q and a. Here the parameters, b, δ, and Q_0, are fixed to be 0.9, 0.1, and 1.0 GeV, respectively.[3]

For tuning, we fit to the distributions, reported in Ref. [17], for: the scaled charged particle momentum $X_p=2p/\sqrt{s}$; the charged particle transverse momentum in and out of the event plane, p_T^{in} and p_T^{out}; and the invariant mass difference between two hemispheres by the plane perpendicular to the sphericity axis, $\Delta M^2 = |M_1{}^2 - M_2{}^2|/E_{vis}{}^2$. $\Lambda_{\overline{MS}}$ is sensitive to p_T^{in} and ΔM^2; σ_q is sensitive to p_T^{out}; and a is sensitive to X_p. A minimum value of $\chi^2=67.4$ is found at $a=0.73\pm0.07$, $\Lambda_{\overline{MS}}=0.31\pm0.02$ GeV, and $\sigma_q=0.41\pm0.01$ GeV.

Table 2 gives the χ^2 values that result from comparisons of the tuned NLL Parton-Shower, the LUND 6.3 LLA Parton-Shower, the LUND 6.2 ME, and the Webber LLA Parton-Shower models to all of the distributions given in Ref [17]. (For the latter three models we used the default values for the parameters.) The total χ^2 values for the LUND 6.3 LLA Parton-Shower ($\Sigma\chi^2=371$) and for the tuned NLL Parton-Shower ($\Sigma\chi^2=386$) models are almost the same.

We examined the effect of varying the Q_0 cutoff using values of $Q_0 = 1, \sqrt{2}, 2, 3,$ and 4 GeV. The parameters a, $\Lambda_{\overline{MS}}$ and σ_q are adjusted for each Q_0. Table 3 shows the total χ^2, together with the best values for a, $\Lambda_{\overline{MS}}$, and σ_q for each value of the Q_0

---

[2]The symmetric LUND fragmentation function is given as

$$f(z) = (1/z)(1 - z)^a exp(-bm_T^2/z),$$

where $z = (E + p)_{hadron}/(E + p)_{parton}$ is the longitudinal scaling variable and $m_T = \sqrt{m^2 + p_T^2}$ is the transverse mass of the hadron.

[3]Since the parameters a and b are strongly correlated, we fixed b and only allowed a to vary. Although some quantities that appear at intermediate levels of the generation depend on δ, the final observables do not vary much with moderate changes in δ [15]. The δ parameter, which smoothly connects the two-body and three-body vertices, is fixed at $\delta = 0.1$. We choose a small value for $Q_0=1.0$ GeV, near the limit of validity for the NLL approximation, in order to reduce the dependence on the hadronization model.

Table 2. Summary of χ^2 values for the comparison of different models with the AMY data.

Distribution	a) Lund PS	b) Lund ME	c) Webber	d) NLL	Number of points
Rapidity	52.3	309.2	25.6	26.6	10
X_p	13.7	197.1	18.2	20.0	10
p_L	18.4	135.5	17.0	13.6	10
p_T	30.3	175.1	29.5	48.2	10
p_T^2	19.5	112.5	19.3	20.4	8
p_T^{in}	26.5	97.0	23.4	32.5	10
p_T^{out}	5.6	239.9	33.3	5.6	6
$< p_T^{2in} >$	12.4	20.1	28.6	11.1	10
$< p_T^{2out} >$	6.8	115.4	53.3	12.2	8
Charged Particle Flow	63.5	324.7	21.5	24.7	15
Energy Flow	45.0	575.1	131.4	88.4	15
Thrust	11.3	37.6	22.6	9.8	8
Major Value	12.7	45.6	15.9	11.5	8
Minor Value	8.4	187.4	43.8	11.3	8
Oblateness	3.7	14.4	7.2	9.5	8
Sphericity	3.4	29.4	12.1	6.1	8
Aplanarity	2.8	48.4	25.5	6.9	6
Q_X	2.6	20.5	12.0	3.9	6
$Q_2 - Q_1$	1.7	1.5	5.6	2.7	8
M_{SL}^2/s	2.9	50.3	12.1	4.7	4
M_{BR}^2/s	13.4	26.6	15.5	7.0	9
$M_{SL}^2/s - M_{BR}^2/s$	14.8	19.8	27.1	9.3	9
Total	371	2783	601	386	194

Table 3. Summary of the tuned parameter values for each choice of the Q_0 cutoff in the NLL PS model. The "tuned χ^2" is the sum of χ^2 of the four distributions used for tuning; the "total χ^2" is the sum of χ^2 of for all 22 distributions listed in Table 2.

Parameter	Q_0				
	a) 1GeV	b) $\sqrt{2}$GeV	c) 2GeV	d) 3GeV	e) 4GeV
a	0.73	0.97	1.17	1.37	1.53
$\Lambda_{\overline{MS}}$	0.31	0.31	0.34	0.42	0.49
σ_q	0.41	0.41	0.42	0.42	0.43
tuned χ^2	67.4	65.6	73.4	79.8	178.5
total χ^2	386	552	566	756	1563

cutoff. The best value of the Q_0 cutoff is found to be 1 GeV. The value of $\Lambda_{\overline{MS}}$ varies by 0.11 GeV when the Q_0 cutoff varies from 1 GeV to 3 GeV. At the latter value, the total χ^2 doubles. We use this range of $\Lambda_{\overline{MS}}$ values as a preliminary estimate of the systematic error. Varying Q_0 over the range ($1 \leq Q_0 \leq 3$ GeV), results in a change of a from 0.66 to 1.4, and a change in σ_q from 0.40 to 0.43 GeV.

3.2 $\Lambda_{\overline{MS}}$ Determination from Three-Jet Ratio

Since a main source of three-jet events is $q\bar{q}g$ production, the fraction of three-jet events, R_3, is sensitive to the value of $\Lambda_{\overline{MS}}$. We define the jet by using the jet-clustering algorithm developed by the JADE group [18]. In this algorithm, the scaled mass-squared,

$$y_{ij} = m_{ij}^2 / E_{vis}^2,$$

where

$$m_{ij} = 2\epsilon_i\epsilon_j(1 - cos\theta_{ij}),$$

is calculated for each pair of particles in the event. Here ϵ_i is the energy of particle i, θ_{ij} is the opening angle between particles, i and j, and E_{vis} is the total visible energy. If the smallest of the y_{ij} values is less than a parameter, y_{cut}, the corresponding pair of particles is combined into a cluster by summing the four-momenta. This process is repeated until all the y_{ij} values exceed y_{cut}. The clusters remaining at this stage are identified as jets.

Figure 11 shows how the jet multiplicity fractions vary with y_{cut}, together with results from the NLL Parton-Shower generator for different values of $\Lambda_{\overline{MS}}$. Figure 12 shows $\Lambda_{\overline{MS}}$ determined from R_3 for different values of y_{cut}. Studies using NLL-Parton Shower model generated events indicate that above $y_{cut}=0.04$, the ratio of the three-jet ratio at the parton level and that at the hadron level is almost constant. At $y_{cut}=0.08$, $\Lambda_{\overline{MS}}=0.24\pm0.07$. Systematic uncertainties in determining $\Lambda_{\overline{MS}}$ arise mainly from variations with the choice of y_{cut} and correlations with the cutoff parameter Q_0. The fluctuations of $\Lambda_{\overline{MS}}$ over the range $0.04 \leq y_{cut} \leq 0.12$ indicate an uncertainty from the y_{cut} dependence of 0.034 GeV. The uncertainty arising from the Q_0 dependence $(1 \leq Q_0 \leq 3$ GeV$)$ and the a dependence $(0.66 < a < 1.4)$ are estimated to be 0.050 GeV, 0.015 GeV, respectively. The uncertainty from σ_q $(0.40 < \sigma_q < 0.43$ GeV$)$ is negligible. Adding all the above in quadrature, the total systematic error of $\Lambda_{\overline{MS}}$ amounts to 0.06 GeV.

3.3 $\Lambda_{\overline{MS}}$ Determination from the EEC

The Energy-Energy Correlation (EEC) was introduced as a method to extract the strong coupling constant α_s. The EEC is an energy-weighted angular correlation defined as

$$EEC(\theta) = \frac{1}{N_{event}} \sum_{events} \sum_{i,j} \frac{E_i}{E_{vis}} \frac{E_j}{E_{vis}} \delta(\theta - \theta_{ij}), \tag{13}$$

where N_{event} is the total number of events. While $q\bar{q}$ events dominate the EEC near $\theta=0°$ and $180°$, events with hard gluon radiation dominate the EEC in the intermediate angular region. Figure 13 shows the EEC for the corrected AMY data together with results from the NLL Parton-Shower generator. We obtain $\Lambda_{\overline{MS}}$ by comparing the integral of EEC over the angular range $\Theta \leq \theta \leq 180°$-$\Theta$ for the data and the model, for different values of $\Lambda_{\overline{MS}}$. Figure 14 shows the dependence of the resulting value of $\Lambda_{\overline{MS}}$ on the integration limit Θ. For smaller values of Θ, fragmentation effects and the contribution from two jet events are expected to be large. We avoid uncertainties

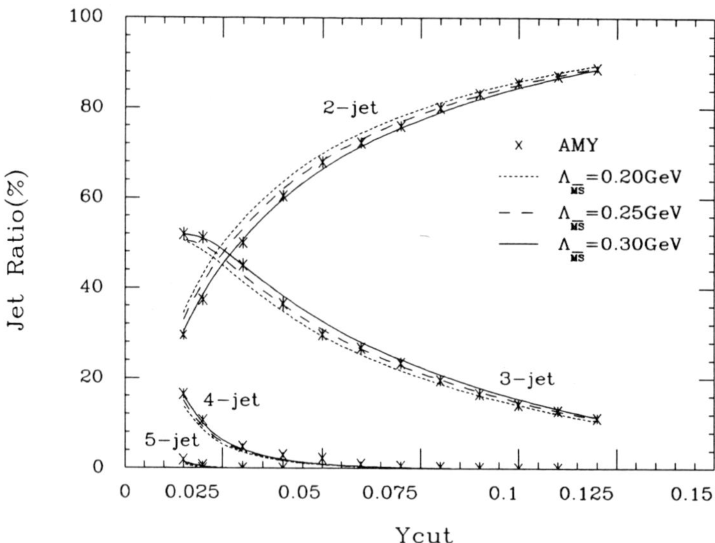

Fig. 11. The acceptance-corrected multi-jet production rates measured by AMY and predictions of the NLL Parton-Shower model for three different values of $\Lambda_{\overline{MS}}$.

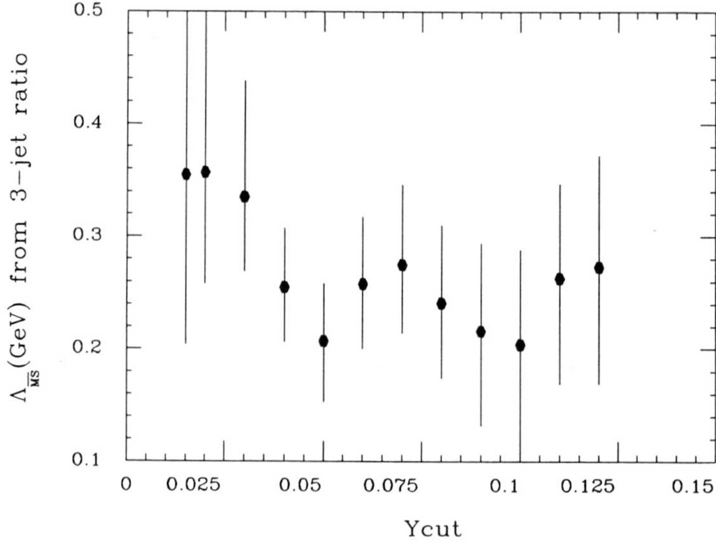

Fig. 12. The value of the QCD scale parameter $\Lambda_{\overline{MS}}$ determined from the three-jet ratio for different values of y_{cut}.

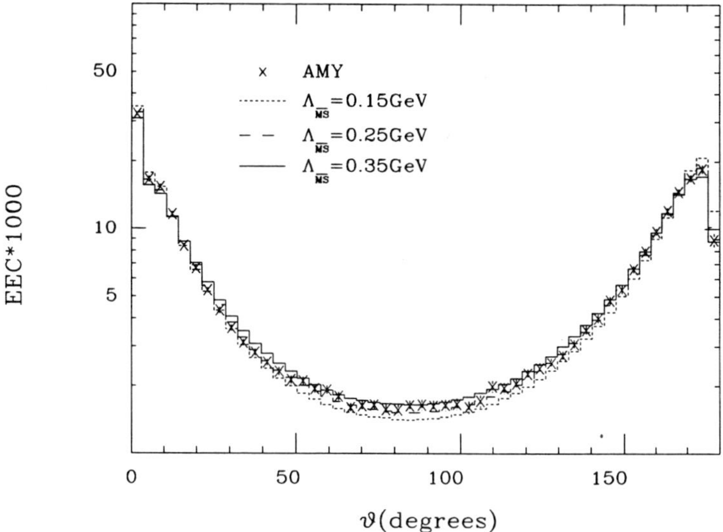

Fig. 13. The acceptance-corrected AMY results for the Energy-Energy Correlation (EEC) and predictions of the NLL Parton-Shower model for three different values of $\Lambda_{\overline{MS}}$.

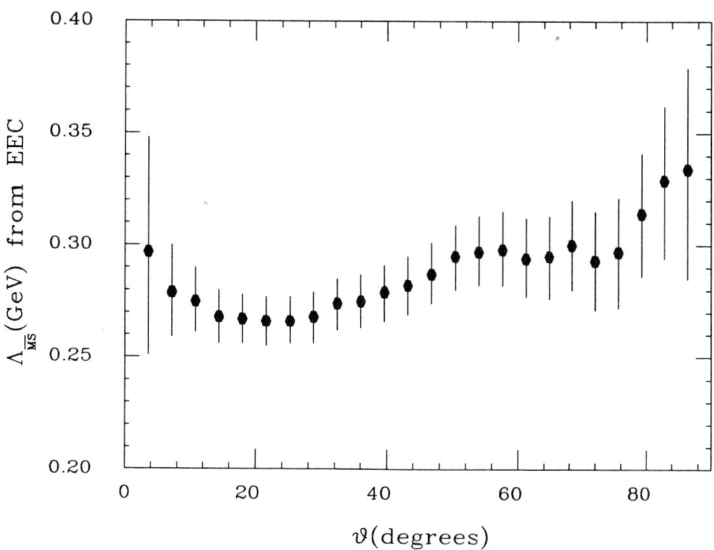

Fig. 14. The value of the QCD scale parameter $\Lambda_{\overline{MS}}$ determined from integrating the EEC between $\Theta \leq \theta \leq 180°\text{-}\Theta$, for different values of Θ.

Table 4. Summary of $\Lambda_{\overline{MS}}$ values, determined using the NLL PS model. The errors are statistical and systematic, respectively. Here, systematic errors are derived by varying parameters in the Monte Carlo generator.

Method	$\Lambda_{\overline{MS}}$
tunning data	$0.31 \pm 0.02 \pm 0.11$
three-jet ratio (y_{cut}=0.08)	$0.24 \pm 0.07 \pm 0.06$
energy-energy correlation	$0.30 \pm 0.02 \pm 0.13$

due to these effects by choosing the relatively large value of $\Theta=50^o$, with the result $\Lambda_{\overline{MS}}$=0.30±0.02. The systematic uncertainty arising from the dependence on the choice of Θ is estimated to be 0.039 GeV. The systematic uncertainties due to the Q_0 dependence ($1 \leq Q_0 \leq 3$ GeV), the a dependence ($0.66 < a < 1.4$) and the σ_q dependence ($0.40 < \sigma_q < 0.43$ GeV) are estimated to be 0.11 GeV, 0.075 GeV, and 0.011 GeV, respectively. Adding all of the above in quadrature gives a total systematic error for $\Lambda_{\overline{MS}}$ of 0.13 GeV.

The values of $\Lambda_{\overline{MS}}$ determined from these methods are summarized in Table 4.

4. DIFFERENCES BETWEEN QUARK AND GLUON JETS

In QCD, the color factor of gluons is 9/4 times that of quarks, leading one to expect significant differences in the hadronization of quarks and gluons [19][20][21][22] [23][24][25]. Calculations based on Leading-Logarithm approximations predict that, in the limit of infinite jet energy, the multiplicity of soft gluons in a gluon-initiated jet is 9/4 times that in a quark-initiated jet [20]. To the extent that the hadron multiplicity in an event follows that of the underlying partons, one expects that the particle multiplicity in gluon jets should be $\sim$ 9/4 times the particle multiplicity in quark jets, and, hence, that particles in gluon jets will tend to have lower momenta by roughly the same factor. Using similar arguments, it has also been shown that the angular widths of gluon and quark jets are related by $\delta_g = \delta_q^{4/9}$ [21] where δ is a half angle (in radians) of the cone in which a given fraction ϵ ($\epsilon < 1$) of jet energy is deposited, i.e., gluon jets should be wider than quark jets. However, there are effects that tend to dilute the factor of 9/4, for example, finite-energy corrections in α_s [23], and the effects of heavy-quark fragmentation [24]. Thus, the differences in particle multiplicity, width, and hardness of fragmentation may be less apparent than the naive expectations.

High-energy e^+e^- annihilations into $q\bar{q}g$ final states, which appear in the laboratory as $e^+e^- \rightarrow$ three jets of hadrons, have a simple underlying parton structure that makes them well-suited for comparisons of the hadronization process for quarks and gluons. Since the discovery of $e^+e^- \rightarrow$ three-jet events, they have been used for many experimental searches for differences between quark jets and gluon jets [26][27][28][29][30][31]. In this paper, we report on a comparison of various properties of samples of quark-enriched and gluon-enriched jets extracted from $e^+e^- \rightarrow$ three-jet events, observed in the AMY detector at center-of-mass energies between $\sqrt{s} = 50$ GeV and 60.8 GeV with the integrated luminosity of 27.6 pb^{-1}. This work with more details has been reported earlier [32][33]. These data are at higher energies than those of previous experiments and, thus, the finite-energy corrections and heavy-quark fragmentation effects, which reduce differences between quark and gluon jets, are expected to be less. In addition, experimentally, jets at higher energies are more distinct and better reflect the underlying parton configuration.

Jets are formed by means of the jet-clustering algorithm explained in the previous section. Three-jet events are selected by using $y_{cut} = (9\text{GeV})^2/s$ and additional selection cuts are applied to those events in order to get a clean data sample [33]. We determine the "calculated" energy of each jet, E_{cal}^{jet}, using energy-momentum conservation and the opening angles between the three jets. Here we neglect the jet's invariant mass. We categorize the jets in each event according to their calculated energy values; jet-1 and jet-2 are two highest energy jets and jet-3, the lowest energy jet. Since gluon radiation is a bremsstrahlung-like process, the gluon is typically emitted close to one of the primary quarks and is usually the lowest energy parton. Thus, it is expected that the jet-3 sample will be gluon enriched relative to the jet-1 and jet-2 samples, which are expected to be quark enriched. We obtain a second sample of quark-enriched jets from a set of two-jet events.

Since the gluon-enriched jet sample corresponds to the jets with the lowest value of E_{cal}^{jet} in each event, there is little energy overlap with jets in the quark-enriched sample over our range of center-of-mass energies. Thus, comparisons are best done using variables that have little variation with E_{cal}^{jet}. It has been shown that quarks and gluons do not hadronize independently in an event [34], which introduces ambiguities into the assignment of particles to jets. These, however, affect mainly the soft particles; the high momentum particles are produced nearly independently of inter-parton correlations [33]. In order to reduce the uncertainties due to soft particles, we prefer to rely on variables that are dominated by high momentum particles instead of variables that are uniformly weighted by all particles, such as n_{chg} and $< p_t >$. Specifically, we define the variables:

- the core-energy fraction, ξ ;

  the fraction of E_{vis}^{jet} that is contained in a cone of half angle θ_{cone} that is coaxial with the jet direction (see Figure 15 (a)):

$$\xi = \frac{\sum_i \epsilon_i \, \Theta(\theta_{cone} - \theta_i)}{E_{vis}^{jet}}, \tag{14}$$

  where ϵ_i and θ_i are energy and angle relative to the jet direction of particle i, θ_{cone} is $60^o/\sqrt{E_{cal}^{jet}}$ (E_{cal}^{jet} in GeV), and $\Theta(x)$ is a function which is 1 (0) if $x > 0$ ($x \leq 0$). The $\sqrt{E_{cal}^{jet}}$ denominator is motivated by the expectation that the widths of hadron jets decrease with the jet energy and leads to less energy dependence for ξ [25].

- the rapidity of the leading particle, η ;

  the rapidity of the most energetic particle in a jet, called the leading particle,

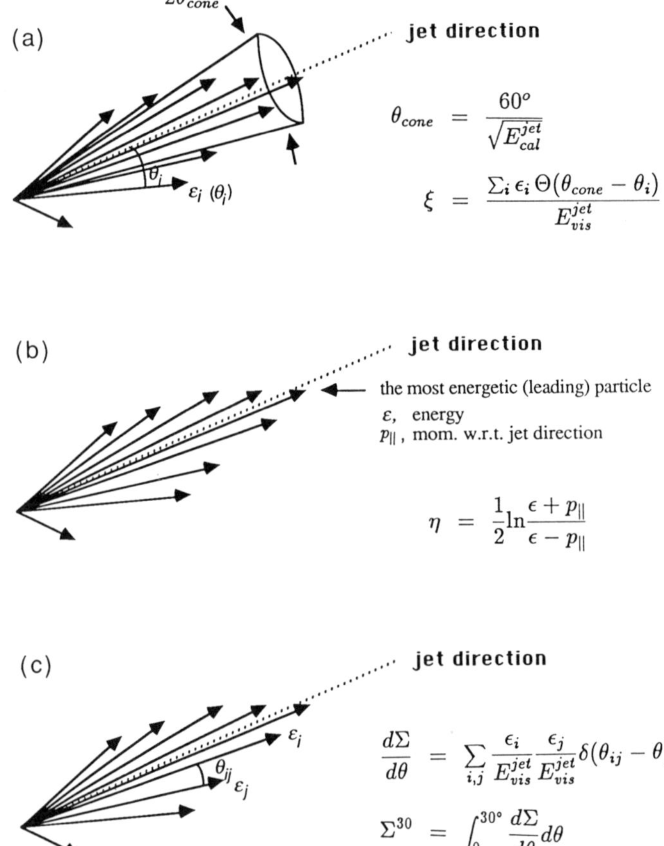

Fig. 15. Schematic descriptions of the variables used to compare quark- and gluon-jets: (a) the rapidity of leading particle η; (b) the core-energy fraction ξ; (c) the integrated Energy-Energy-Correlation Σ^{30}.

relative to the jet axis

$$\eta = \frac{1}{2}\ln\frac{\epsilon + p_{\parallel}}{\epsilon - p_{\parallel}}, \tag{15}$$

where ϵ is the leading particle's energy and $p_{\parallel}$ is its momentum component parallel to the jet direction (see Figure 15 (b)). For massless particles, $\eta = -\ln(\tan\frac{\theta}{2})$, where θ is the angle of the particle relative to jet direction.

- the integrated Energy-Energy Correlation, Σ^{30} ;

  the Energy-Energy Correlation EEC

$$EEC(\theta) = \frac{d\Sigma}{d\theta} = \frac{1}{N_{jet}}\sum_{jets}\sum_{i,j}\frac{\epsilon_i}{E_{vis}^{jet}}\frac{\epsilon_j}{E_{vis}^{jet}}\delta(\theta - \theta_{ij}) \tag{16}$$

is the energy weighted angular correlation, where N_{jet} is the number of jets, θ_{ij} is the opening angle between any pair of particles, i and j, in a jet and ϵ_i is the energy of particle i. We define the integrated EEC as the integral of the normalized EEC between $0°$ and $30°$ (see Figure 15 (c));

$$\Sigma^{30} = \int_0^{30°}\frac{d\Sigma}{d\theta}d\theta. \tag{17}$$

Note that $\Sigma^{180} = 1$, by definition.

Since ξ is determined by the energy flow in the core of a jet, η is determined from the most energetic particle in a jet, and Σ^{30} is an energy weighted variable, all of these variables are dominated by high momentum particles in different ways.

Since we are trying to compare properties of quarks and gluons, which are unobservable, we are forced to rely on theoretical models for guidance. Two different QCD-motivated Monte Carlo (MC) event generators are used: the LUND 6.2 Matrix Element model [3] with the independent fragmentation scheme of Hoyer et al. [35], subsequently called the $q = g$ model, and the LUND 6.3 Parton Shower model with the string fragmentation [3], subsequently called the PS model. In both cases, samples of generated events are passed through the detector simulation program and are subjected to the same three-jet analysis that is used for the data. In the $q = g$ model, the same algorithm is used to hadronize quarks and gluons and, thus, we don't expect any differences between the resulting jets. These events are used as a "control sample" to verify that the detector acceptance and our analysis procedures are not introducing artificial differences between quark and gluon jets.

In Figure 16 we show the mean value of the core-energy fraction ξ as a function of E_{cal}^{jet} for the jet-1,2 and jet-3 samples. This variable is not too sensitive to the jet energy. The data indicate that in quark jets the energy is concentrated near the jet axis, while in gluon jets it tends to be diffuse. The results for the $q = g$ MC event sample for the jet-1,2 and jet-3 samples, shown in the figure as solid lines, shows no significant discontinuity between the different jet samples; it agrees reasonably well with the jet-1,2 data points and lies considerably above those from the jet-3 sample. Included in the figure is the mean value of ξ for the jets from the two-jet data sample, which agrees well with the results from the jet-1,2 data sample.

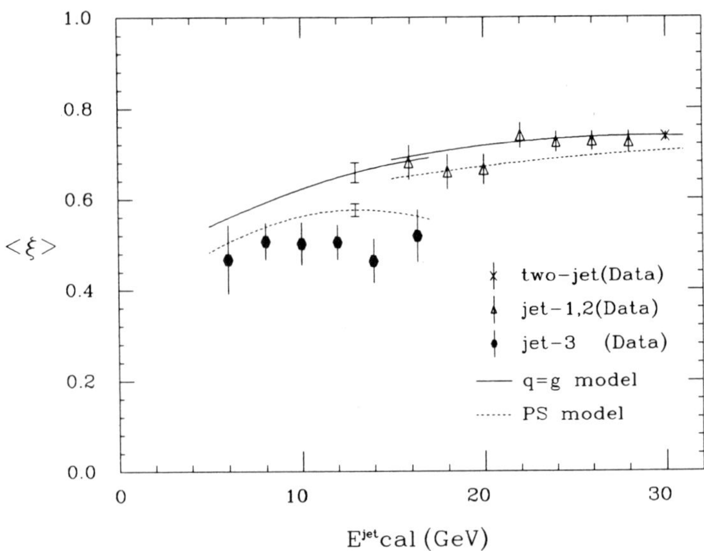

Fig. 16. The mean core-energy fraction $< \xi >$ as a function of the calculated jet energy, E_{cal}^{jet}. The solid (dashed) lines are the expectations from the ME + $q=g$ model (PS + string fragmentation model). The solid (open) points are for the gluon-enriched (quark-enriched) jet sample. The cross indicates the result from the two-jet events.

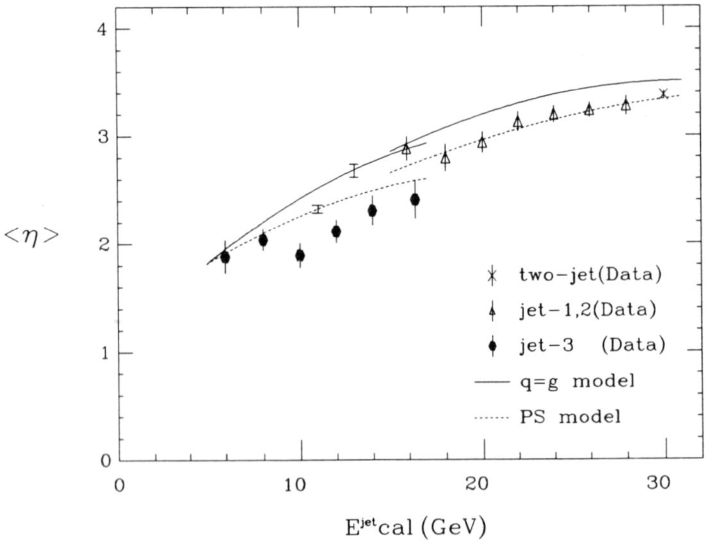

Fig. 17. The mean rapidity of the leading particle in the jet $< \eta >$ as a function of the calculated jet energy, E_{cal}^{jet}. The solid (dashed) lines are the expectations from the ME + $q=g$ model (PS + string fragmentation model).

Figure 17 shows the mean values of the leading particle's rapidity, η, which also indicate some distinction between the different jet samples. The leading particles tend to have a higher rapidity in quark jets than in gluon jets. Here, the results of the $q = g$ MC lie somewhat higher than the jet-1,2 data points but substantially overestimate those from jet-3.

The normalized EEC distribution is shown in Figure 18 (a) for the lowest energy portion of the jet-1,2 sample ($E_{cal}^{jet} \leq 19$ GeV; average=17.0 GeV) and the highest energy portion of the jet-3 sample ($E_{cal}^{jet} \geq 13$ GeV; average=14.7 GeV). The peak in the first bin is the particles' self correlation (i.e., when $i = j$). The $q = g$ MC event sample gives reasonable agreement with our data for jet-1,2 sample while the jet-3 distribution for the data sample is wider than that of $q = g$ MC sample. Figure 19 shows the mean values of Σ^{30}, the integral of the EEC distribution between $0°$ and $30°$, as a function of E_{cal}^{jet}, for the jet-1,2 and jet-3 samples. This also indicates some distinctions between the different jet samples.

The predictions of the PS model are shown as dashed lines in Figures 16, 17, 18, and 19. The agreement with the jet-1,2 data sample is reasonably good and the model's different treatment of quarks and gluons results in a different predicted behavior for the jet-3 sample. This difference, while evident in all four figures, is not as strong as the differences observed in the data.

In Figure 18 (b), we show distributions in ξ for the lowest energy portion of the jet-1,2 sample ($E_{cal}^{jet} \leq 19$ GeV; average=17.0 GeV) and the highest energy portion of the jet-3 sample ($E_{cal}^{jet} \geq 13$ GeV; average=14.7 GeV). The distributions for the jet-1,2 and jet-3 samples show a strikingly different character. The quark-enriched sample peaks at $\xi = 1$, corresponding to 100% of the visible energy in the core, while the gluon-enriched sample favors smaller values of ξ, corresponding to little of the visible energy in the core. The solid-line histograms are the results from the $q = g$ MC events. These give very similar distributions in both cases, showing reasonable agreement with the quark-enriched data sample ($\chi^2 = 11.9$ for 9 degrees-of-freedom) and clear disagreement with the gluon-enriched data sample ($\chi^2 = 49.6$). The PS model (dashed lines) predicts some distinction between the different jet samples although not as much as is observed in the data. The PS model gives good agreement with the jet-1,2 sample ($\chi^2 = 3.9$); the agreement with the jet-3 data is worse ($\chi^2 = 19.4$).

To check for possible systematic sources for the observed differences between the jet-1,2 and jet-3 samples, we made the comparison for a variety of selection criteria [33]. For example, if we increase y_{cut} to $(12\text{GeV})^2/s$, or eliminate the cut on $E_{vis}^{jet}/E_{cal}^{jet}$, or use only the charged tracks to compute ξ, η, and $d\Sigma/d\theta$, or use E_{vis}^{jet} rather than E_{cal}^{jet}, the observed differences persist.

5. BOSE-EINSTEIN CORRELATIONS [36][37][38]

Given a chaotic (thermal) source of pions, and requiring that the pion wave function be symmetric under the exchange of identical bosons, the correlation function between two like-signed pions is related to the Fourier transform of the pion source distribution,

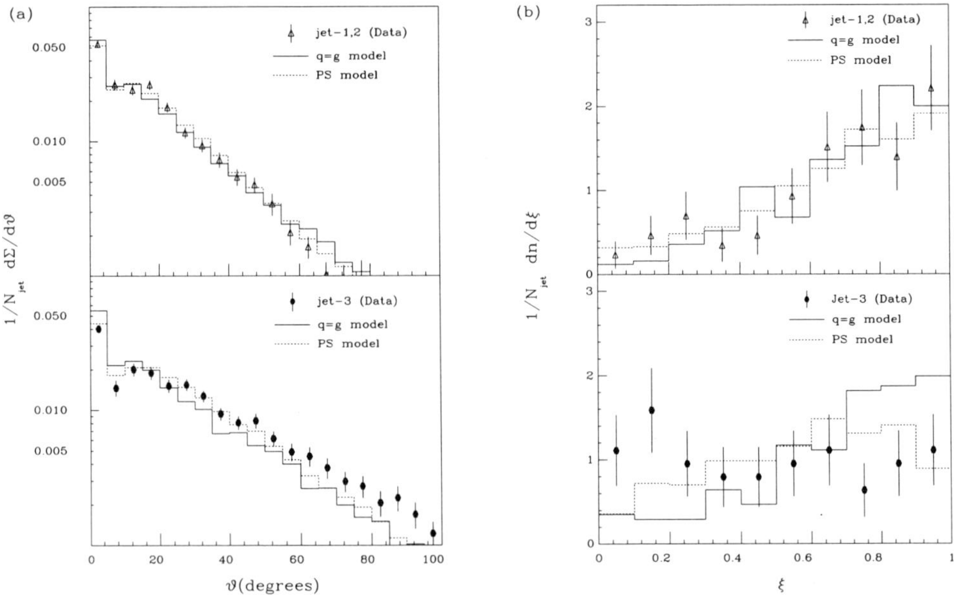

Fig. 18. The normalized distributions of (a) Energy-Energy Correlation $d\Sigma/d\theta$ and (b) core-energy fraction ξ for quark-enriched jets with $E_{cal}^{jet} < 19$ GeV (up) and gluon-enriched jets with $E_{cal}^{jet} > 13$ GeV (down). The solid (dashed) lines are the expectations from the ME + $q=g$ model (PS + string fragmentation model). All distributions are normalized to have unit integral.

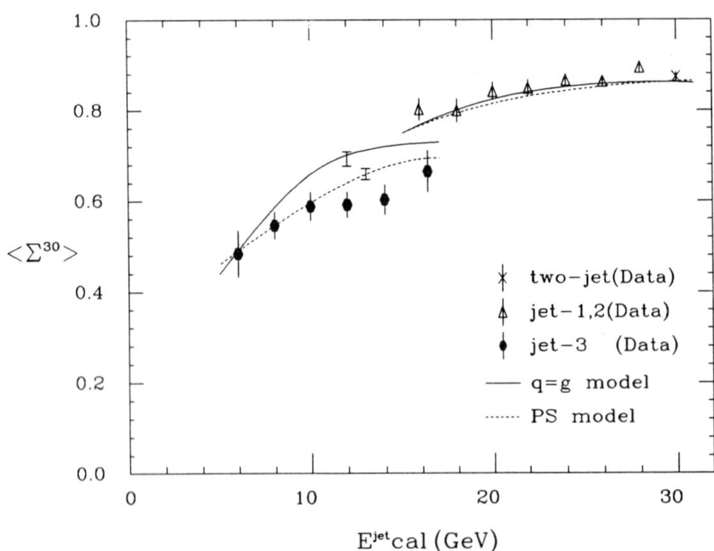

Fig. 19. The mean values of Σ^{30} as a function of the calculated jet energy, E_{cal}^{jet}. The solid (dashed) lines are the expectations from the ME + $q=g$ model (PS + string fragmentation model).

$\rho(\vec{k_1} - \vec{k_2})$, by:

$$C(\vec{k_1}, \vec{k_2}) = 1 + |\rho(\vec{k_1} - \vec{k_2})|^2 \qquad (18)$$

where $\vec{k_1}$ and $\vec{k_2}$ are the pion momentum. In order to extract this correlation function from data, it is customary to compare the distribution of like-signed pion pairs to that of unlike-signed pion pairs. This is due to the fact that the unlike-signed pair distribution contains much of the same physics as the like-signed distribution (i.e. phase space, momentum and angular distributions), but does not contain Bose-Einstein correlations. The ratio of the distributions, $R \equiv N_{like}/N_{unlike}$, is frequently parameterized in terms of a Gaussian function of the four-momentum difference, Q:

$$R(Q^2) = 1 + \lambda e^{-r_0^2 Q^2} \qquad (19)$$

$$Q^2(\vec{k_1}, \vec{k_2}) \equiv -(\vec{k_1} - \vec{k_2})^2 = M_{12}^2 - 4M_\pi^2 \qquad (20)$$

where M_{12} is the invariant mass of the two pion system. The parameters λ and r_0 are related to the strength of the enhancement and the size of the source, respectively, and are determined by fits to the data. Although this expression is only empirical, it has been shown to describe e^+e^- annihilation data well over a wide range of energies [39][40][41][42].

For each event, all possible pairs of charged tracks are formed and two histograms of the number of particle pairs are formed as a function of Q^2; one of like-signed pairs and one of unlike-signed paris. The ratio of these two distributions is the correlation function, after making corrections for differences not related to the Bose-Einstein symmetrization requirement of the like-signed pions, such as resonance decays, normalization, detector effects, and Coulomb interaction [43]. The ratio of like- to unlike-signed pairs as a function of Q^2, after applying all corrections, is shown in Figure 20. This distribution is parameterized by

$$R(Q^2) = N_0(1 + \lambda e^{-r_0^2 Q^2})(1 + \gamma Q^2) \qquad (21)$$

where N_0 is a normalization constant, the function $f_\pi(Q^2)$ is the polynomial fit to the proportion of like-signed pairs that are identical pions, and the term involving γ is used to take into account long-range correlations such as charge and energy conservation. A fit to this distribution yields parameters of $N_0 = 1.02 \pm 0.01$, $\lambda = 0.60 \pm 0.13 \pm 0.08$, $r_0 = 1.18 \pm 0.17 \pm 0.10$ fm, $\gamma = -0.04 \pm 0.03$ (GeV/c)$^{-2}$, and $\chi^2/\mathrm{NDF} = 92.5/96$. The detailed analysis is given in ref. [44].

The results of the fit parameters, λ and r_0, versus center-of-mass energy are shown in Figure 21 (a) and (b), respectively, together with other measurements in e^+e^- experiments [39][40][41][42]. Also indicated in this figure are the thresholds for charm and bottom production. The λ at TRISTAN energies is consistent with other results obtained above bottom threshold. The source size r_0 measured at TRISTAN seems larger compared with lower energy results which is $\sim$0.8 fm and independent of energy. However the discrepancy is only 2.2σ and not statistically significant.

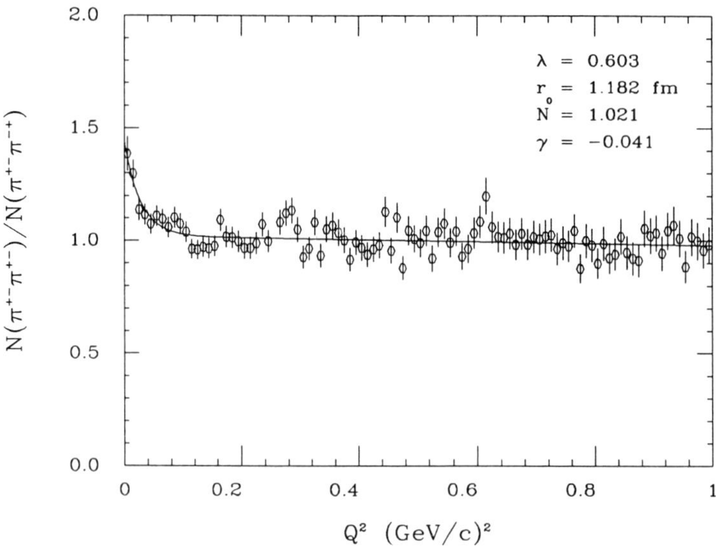

Fig. 20. Ratio of like- to unlike-signed particle pairs from the AMY data, after making all corrections.

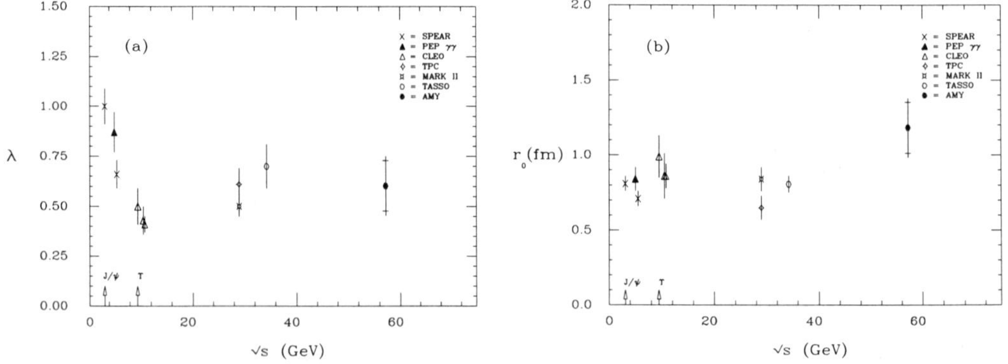

Fig. 21. The Bose-Einstein parameter (a) λ and (b) r_0 , shown versus center-of-mass energy, for a variety of experiments.

6. SUMMARY

The electroweak couplings in the processes $e^+e^- \to \mu^+\mu^-$, $\tau^+\tau^-$, $b\bar{b}$, and $q\bar{q}$ are in good agreement with the standard model although the production rate for $\mu^+\mu^-$ is lower than the expected value by about 2σ.

After tuning the parameters, the NLL Parton-Shower model agrees well with AMY data; the agreement is comparable to that of the LUND 6.3 LLA Parton-Shower model. The QCD scale parameter $\Lambda_{\overline{MS}}$ is derived in three ways; from tuning the NLL PS model to event shape distributions, from the three-jet event ratio, and from the integrated energy-energy correlation. The results are: $\Lambda_{\overline{MS}} = 0.31 \pm 0.02 \pm 0.11$ GeV (event-shape tuning); $\Lambda_{\overline{MS}} = 0.24 \pm 0.07 \pm 0.06$ GeV (three-jet ratio at $y_{cut} = 0.08$); $\Lambda_{\overline{MS}} = 0.30 \pm 0.02 \pm 0.13$ GeV ($\int_{50^\circ}^{130^\circ} EEC \, d\theta$).

We observe differences in the concentration of energy near the jet core, in the rapidity of the leading particles, and in the energy weighted angular correlation for quark and gluon jets. Studies with a QCD-motivated event generator that uses the same fragmentation procedure for quarks and gluons indicate that the observed differences are not introduced by the effects of detector acceptance or by our analysis procedures, but are reflections of basic differences in the fragmentation processes of the partons.

The Bose-Einstein correlation in $\pi\pi$ production have been observed. Results show correlations with a strength of $\lambda=0.60\pm0.13\pm0.08$, in agreement with lower energy results. The typical correlation size was determined to be $r_0=1.18\pm0.17\pm0.10$ fm, which is slightly larger than lower energy results.

ACKNOWLEDGEMENTS

I thank K. Abe, T. Kumita, S. Olsen, and H. Sagawa for help in preparing this summary report.

REFERENCES

[1] A. Bacala *et al.* (AMY Collaboration), Phys. Lett. **B218**, 112 (1989).

[2] The expression $b \to \mu^-$ is used for the process $B \to X\mu^-\bar{\nu}_\mu$, where B is the b-flavor hadron. The other expressions follow this notation.

[3] T. Sjöstrand, Computer Phys. Comm. **27**, 243 (1982); **28**, 229 (1983); **39**, 347 (1986);
M. Bengtsson and T. Sjöstrand, Computer Phys. Comm. **43**, 367 (1987).

[4] H. Sagawa *et al.* (AMY Collaboration), Phys. Rev. Lett. **63**, 2341 (1989);
J. Lim *et al.* (AMY Collaboration), Submitted to the 25th International Conference on High Energy Physics, Singapore (1990).

[5] W. Bartel *et al.* (JADE Collaboration), Phys. Lett. **B146**, 437 (1984).

[6] H. Band *et al.* (MAC Collaboration), Phys. Lett. **B218**, 369 (1989).

[7] M. Althoff *et al.* (TASSO collaboration), Phys. Lett. **B146**, 443 (1984);
F. Ould-Saada, DESY report 88-177 (1988);
C.R. Ng *et al.* (HRS collaboration), ANL-HEP-PR-88-11;
S.L. Wu, Nucl.Phys.**B** (Proc.Suppl.) 3 (1988);
C. Kiesling, in Proceedings of the 29th Moriond Conference, March,1989.

[8] M. Artuso *et al.* (CLEO Collaboration), Phys. Rev. Lett. **62**, 2233 (1989).

[9] H. Albrecht *et al.* (ARGUS Collaboration), Phys. Lett. **B192**, 245 (1987).

[10] K.R. Schubert, Proceedings: International Europhysics Conference on High Energy Physics, Uppsala(1987), p. 820.

[11] T. Schaad *et al.* (MARK II Collaboration), Phys. Lett. **B160**, 188 (1985).

[12] D. Stuart *et al.* (AMY Collaboration), Phys. Rev. Lett. **64**, 983 (1990).

[13] J. Kalinowski, K. Konishi, and T.R. Taylor, Nucl. Phys. **B181**, 221 (1981);
J. Kalinowski, K. Konishi, P.N. Scharbach, and T.R. Taylor, Nucl. Phys. **B181**, 253 (1981).

[14] J.F. Gunion and J. Kalinowski, Phys. Rev. **D29**, 1545 (1984);
J.F. Gunion, J.Kalinowski, and L. Szymanowski, Phys. Rev. **D32**, 2303 (1985).

[15] K. Kato and T. Munehisa, Mod. Phys. Lett. **A1**, 345 (1986);
Phys. Rev. **D36**, 61 (1987); Phys. Rev. **D39**, 156 (1989)

[16] H. Sagawa *et al.* (AMY Collaboration), Submitted to the 25th International Conference on High Energy Physics, Singapore (1990).

[17] Y.K. Li *et al.* (AMY Collaboration), Phys. ReV. **D41**, 2675 (1990).

[18] S. Bethke, LBL-25247, March 1988. Presented at XXII Rencontre de Moriond, March 1988.

[19] S.J. Brodsky, J.F. Gunion, Phys. Rev. Lett. **37**, 402 (1976).

[20] K. Konishi, A. Ukawa and G. Veneziano, Phys. Lett. **78B**, 243 (1978).

[21] K. Shizuya, S. Tye, Phys. Rev. Lett. **41**, 787 (1978);
M.B. Einhorn, B.G. Weeks, Nucl. Phys. **B146**, 445 (1978).

[22] K. Konishi, A. Ukawa and G. Veneziano, Phys. Lett. **80B**, 259 (1979).

[23] A.H. Mueller, Nucl. Phys. **B241**, 141 (1984);
A.H. Mueller, J.B. Gaffney, Nucl. Phys. **B250**, 109 (1985).

[24] G.Sh. Dzhaparidze, Z. Phys. **C32**, 59 (1986);
G.Sh. Dzhaparidze, IHEP preprint 86-100, Sepurkhov (1986).

[25] Yu.L. Dokshitzer, V.A. Khoze, S.I.Troyan, DESY Preprint 88-093 (1988).

[26] W. Bartel *et al.* (JADE Collaboration), Phys. Lett. **123B**, 460 (1983).

[27] A. Petersen *et al.* (MARK II Collaboration), Phys. Rev. Lett. **55**, 1954 (1985).

[28] R.J. Madaras *et al.* (TPC Collaboration), Rencontre de Moriond on Strong Interactions and Gauge Theories, Les Arcs (1986).

[29] Cello Collaboration, contributed paper at the International Symp. on Lepton and Photon Interactions at High Energies, Hamburg (1987).

[30] M. Derrick *et al.* (HRS Collaboration), Phys. Lett. **165B**, 449 (1985);
K. Sugano, Int. J. Mod. Phys. **A3**, 2249 (1988).

[31] W. Braunschweig *et al.* (TASSO Collaboration), Z. Phys. **C45**, 1 (1989).

[32] Y.K. Kim (AMY Collaboration), Phys. Rev. Lett. **63**, 1772 (1989).

[33] Y.K. Kim (AMY Collaboration), Ph.D. Thesis, University of Rochester (1990), unpublished.

[34] W. Bartel *et al.* (Jade Collaboration), Z. Phys. **C21** (1983);
Phys. Lett. **134B**, 275 (1984); **157B**, 340 (1985);
M. Aihara *et al.* (TPC Collaboration), Z. Phys. **C28**, 31 (1985);

M. Althoff *et al.* (TASSO Collaboration), Z. Phys. **C29**, 29 (1985).

[35] P. Hoyer, P. Osland, H.G. Sander, T.F. Walsh, P.M. Zerwas, Nucl. Phys. **B161**, 349 (1979).

[36] G. Goldhaber *et al.*, Phys. Rev. **120**, 300 (1960).

[37] A. Giovannini and G. Veneziano, Nucl. Phys. **B130**, 61 (1977).

[38] I. Juricic, Ph.D. thesis, Lawrence Berkeley Laboratory (1987).

[39] H. Aihara *et al.* (TPC Collaboration), Phys. Rev. **D31**, 996 (1985).

[40] P. Avery *et al.* (CLEO Collaboration), Phys. Rev. **D32**, 2294 (1985).

[41] M. Althoff *et al.* (TASSO Collaboration), Z. Phys. **C30**, 355 (1986).

[42] I. Juricic *et al.* (MARK II Collaboration), Phys. Rev. **D39**, 1 (1989).

[43] M. Gyulassy *et al.*, Phys. Rev. **C20**, 2267 (1979).

[44] R.C. Walker *et al.* (AMY Collaboration), Submitted to the 25th International Conference on High Energy Physics, Singapore (1990).

_W & Z_ PHYSICS AT $p\bar{p}$ COLLIDERS

R.J. Cashmore

Department of Nuclear Physics
Oxford

1. INTRODUCTION

Experiments at $p\bar{p}$ colliders have provided a wealth of information on _W_ and _Z_ particles. However with the advent of LEP detailed information on the properties (mass, width and decays) of the _Z_ from $p\bar{p}$ machines has been superceded. In this short summary the production properties of the _W_ and _Z_ are emphasised, together with some aspects of the decays of _W_s. _W_s as yet are not accessible at LEP and this is therefore the only source of systematic data.

2. PRODUCTION PROPERTIES OF THE _W_ AND _Z_

The _W_ and _Z_, interacting only electroweakly allow a systematic study of the production of these particles, comparatively untroubled by strong interactions (QCD) in the final states, and are thus a probe of the QCD parton production mechanisms. The principle production diagrams at leading order and next to leading order are shown in Fig 1. Recent calculations of these higher order diagrams[1] have reduced the uncertainty on the scale (M) used for α_s and the parton distributions in the calculations. Of course complete calculations (to all orders) would be scale independent but this has been a major step forward. These results are shown in Fig 2 and mean that the p_T distributions of the _W_s and _Z_s which are due to the higher order diagrams, are much more reliably predicted.

In Fig 3 the excellent agreement with the predictions for total cross-section is shown covering the energy range from the $Sp\bar{p}S$ to the Tevatron. Fig 4 shows a comparison of the calculations of Reno and Arnold[1] with preliminary CDF data[2]. The fractions of _W_s with different numbers of accompanying jets are shown in Fig 5 together with calculations[3,4] of higher order processes. The agreement is excellent.

As the quality and quantity of the data increases the production properties of the _W_s and _Z_s will allow a much more detailed measurement of parton distributions and processes at small x.

3. DECAY PROPERTIES OF _W_s AND _Z_s

The details of Z° decays and properties are the province of LEP. Here we concentrate on the _W_.

Z°Physics - Cargèse 1990, Edited by M. Lévy _et al._
Plenum Press, New York, 1991

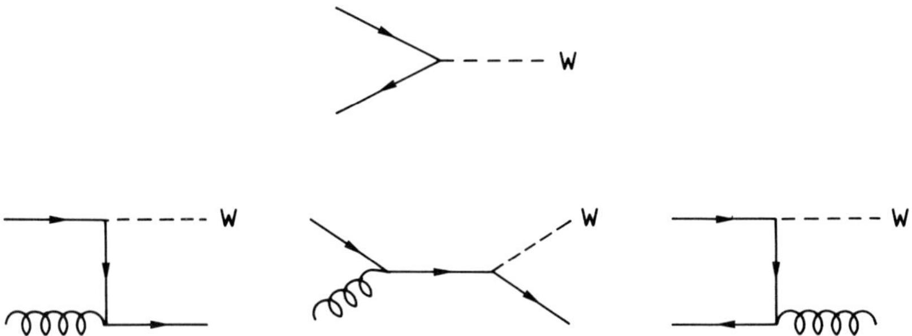

Fig 1. Leading Order and Next to Leading Order diagrams for W production.

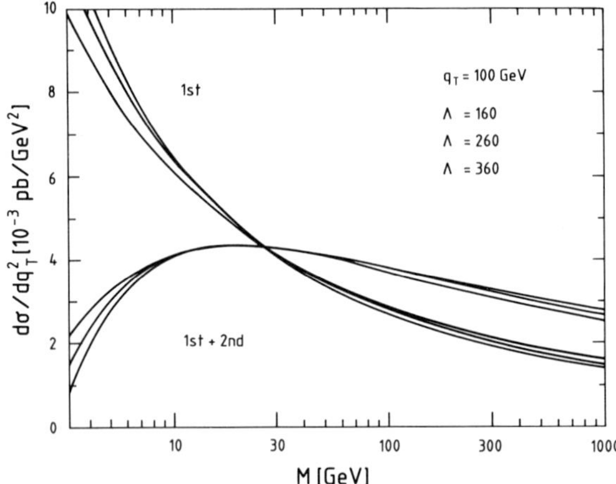

Fig 2. Leading Order (1st) and Next to Leading Order (2nd) contributions to W production with transverse momentum $q_T = 100$. M is the scale used for the calculation.

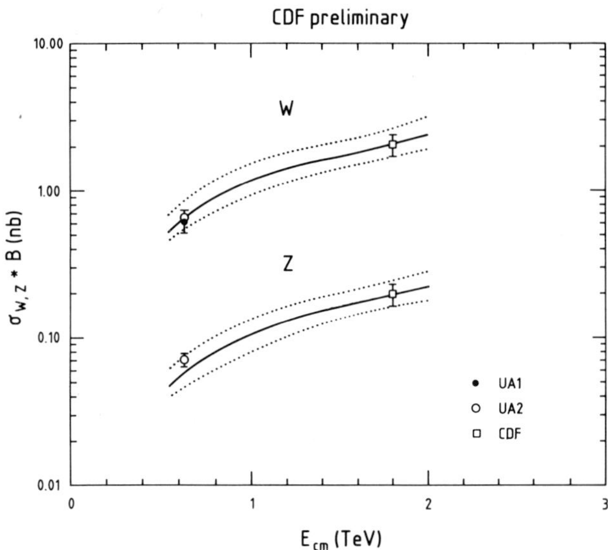

Fig 3. W and Z production at the $Sp\bar{p}S$ and the Tevatron.

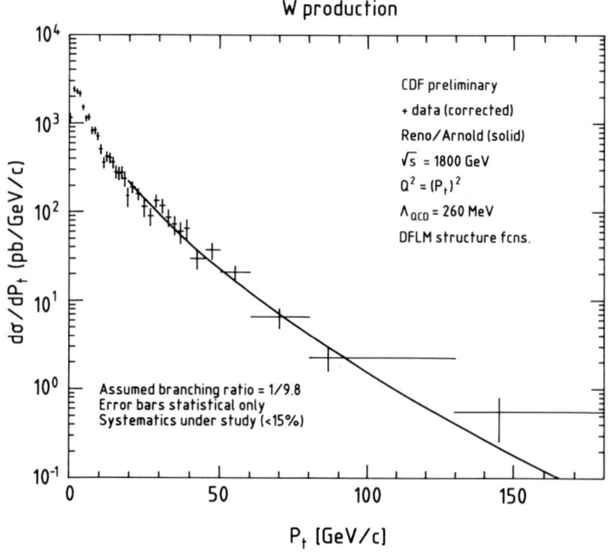

Fig 4. The p_T distribution of Ws compared with the calculations of Arnold and Reno.

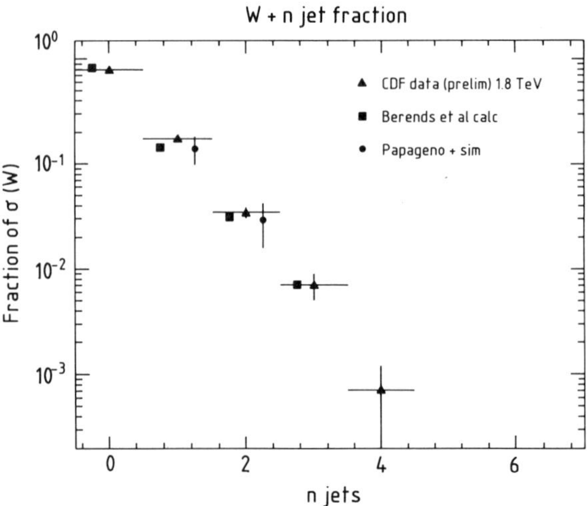

Fig 5. The number of jets accompanying W production, compared with various calculations.

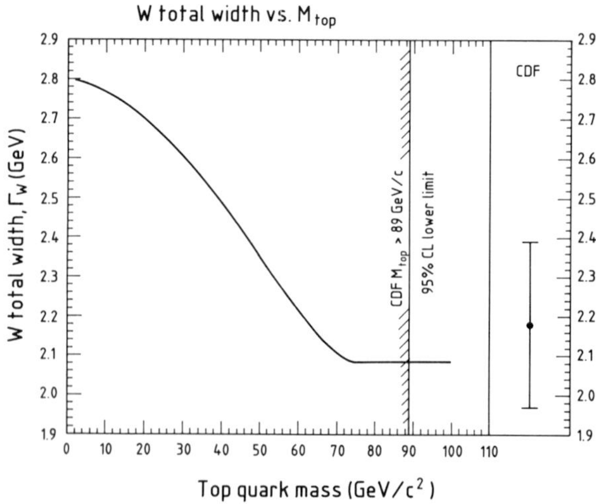

Fig 6. The Total Width of the W as a function of top mass compared with the experimental result.

Ws are observed primarily in their leptonic decay modes

$$W \to e\nu$$
$$W \to \mu\nu$$

by observation of an energetic lepton together with substantial missing E_T (transverse energy) associated with the neutrino. The signatures are very clear.

(i) Width of W

By measuring the quality

$$R = \frac{\sigma(W \to e\nu)}{\sigma(Z \to ee)} = \frac{\sigma(\bar{p}p \to Wx)}{\sigma(\bar{p}p \to Zx)} \cdot \frac{\Gamma(W \to e\nu)}{\Gamma(Z \to ee)} \cdot \frac{\Gamma(Z)}{\Gamma(W)}$$

it is possible to obtain a value for $\Gamma(W)$ by assuming

(a) the <u>relative</u> production cross-sections for W and Z, which are quite accurately predicted by QCD

(b) the measured value of $\frac{\Gamma(Z \to ee)}{\Gamma(Z \to)e}$ from LEP

(c) the theoretical prediction of $\Gamma(W \to e\nu)$.

A result from the CDF experiment[5] $\Gamma(W) = 2.18 \pm .021\,\mathrm{GeV}$ is shown in Fig 6 and compared with $\Gamma(W)$ as a function of top quark mass (for low top masses, $W \to t\bar{b}$ would be possible and hence increase the W total width). The agreement with the standard model is clear.

(ii) Mass of W

The W mass has been measured most accurately by the UA2[6] and CDF[7] experiments. It is principally determined from the transverse mass, M_T, distributions in $W \to e\nu$ and $W \to \mu\nu$

$$M_T = \sqrt{2E_T^e E_T^\nu(1 - cos\Delta\phi_{e\nu})}$$

where E_T^e is the electron transverse momentum
$\quad E_T^\nu$ is the neutrino transverse momentum inferred from $E_T^\nu = +\not{E}_T$ (missing E_T)
$\quad \Delta\phi_{e\nu}$ is the azimuthal angle between these quantities.

In Fig 7 we see a fit to the UA2 data. This measurement is non-trivial requiring careful understanding and calibration of the detectors, which are best appreciated in the publications from UA2 and CDF. The results obtained (using the LEP Z mass of $91.15 \pm .03$ GeV) are

$$\text{UA2} \quad M_W^e = 80.5 \pm 0.5$$
$$\text{CDF} \quad M_W^{e,\mu} = 79.9 \pm 0.4$$

In the future, with 10-100 times the statistics substantial improvements can be expected in the accuracy of this result.

4. ELECTROWEAK INTERACTIONS

The result for M_W, when combined with the results of LEP on the Z, can be used to probe the further ingredients of the Standard Model. In Fig 8 we see the expected dependence of M_W on the top quark mass and the Higgs mass. Clearly a top quark may

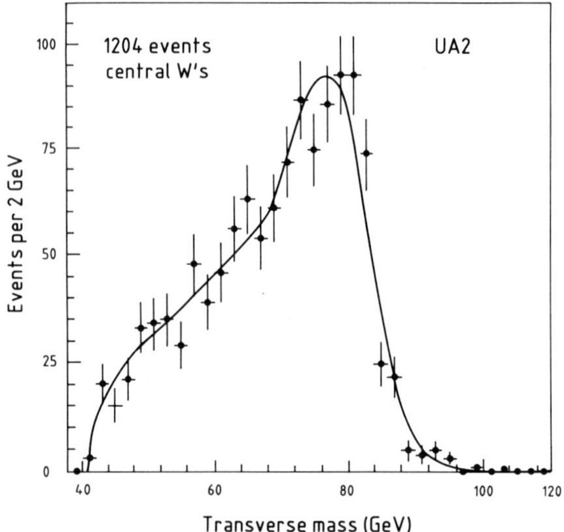

Fig 7. The transverse mass distribution for $W \rightarrow e\nu$ as measured and fitted by UA2.

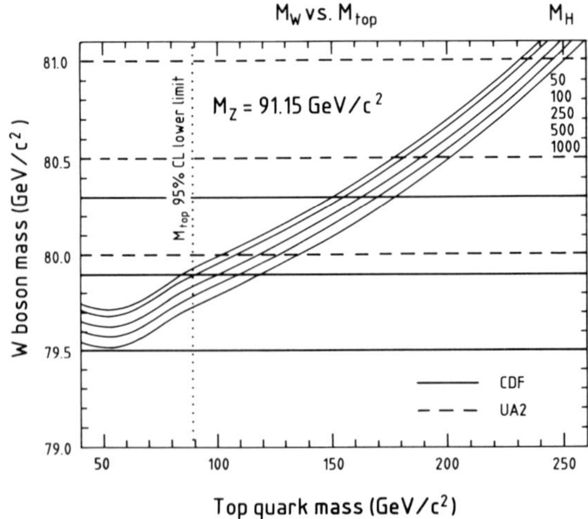

Fig 8. The W boson mass as a function of top quark and Higgs masses. The experimental results of CDF and UA2 are shown.

well be in the range of future tevatron runs, and eventually serve to further pin down the standard model.

5. BEYOND THE STANDARD MODEL, W' AND Z' SEARCHES

$Z' \to e^+e^-$ and $W' \to e\nu$ at high mass can be searched for to give evidence/limits on the production of these new bosons.

In CDF no events with $M(e^+e^-) > 200$ GeV are observed which gives a limit of

$$\sigma(Z')B(Z' \to e^+e^-) < 1.6pb \quad \text{at} \quad 95\% \quad \text{CL}$$

Assuming standard model couplings this implies

$$M_{Z'} > 380 \text{ GeV}$$

Similar studies of $e\nu$ show no events with $M_T > 150$ GeV and this implies

$$M_{W'} > 478 \text{ GeV}$$

6. SUMMARY

This is a very brief review of the results that can be obtained from $p\bar{p}$ colliders and only deals with Ws and Zs. Enormous progress has been made in QCD studies (jets, photons as well as Ws and Zs) and limits set for top quarks and SUSY particles, $\tilde{q}$ and $\tilde{g}$. Future progress in these latter areas can now only be made at $p\bar{p}$ machines in the next ~ 8 years. At the same time the ability to reconstruct B mesons opens up a whole new field of exciting physics. These results, together with large new data sample of particularly Ws, means that there is a clear programme of careful study of the standard model as well as clear potential for discovery at mass scales well in excess of 100 GeV.

ACKNOWLEDGEMENTS

It is a pleasure to acknowledge the help and advice of my many colleagues in particle physics, particularly those at CDF. The brevity of this paper does not do justice to all of their efforts. Finally it was a particular pleasure to be present at Cargèse and enjoy the atmosphere and hospitality. Such a phenomenological interface between theory and experiment is essential for progress in the field.

REFERENCES

1. P. Arnold & M. Reno. NP B319 (1989) 37.

2. T. Kamon, CDF collaboration, Proc. of 8th Topical Workshop on Proton-Antiproton Collider Physics, Castiglione, Italy. Sept 1989.

3. F. Berends, et al. PL B224 (1989), 237.

4. I. Hinchliffe, private communication.

5. CDF experiment. S. Errede. SLAC Summer Institute 1990.
   CDF collaboration, PRL 64 (1990) 152.

6. UA2 collaboration, PL B241 (1990) 241.

7. CDF experiment. S. Errede. SLAC Summer Institute 1990.

THEORY OF THE ELECTROWEAK INTERACTIONS

Luciano Maiani

Dipartimento di Fisica, Università di Roma "La Sapienza"
Sezione di Roma INFN
P.le A.Moro 2, 00185 Roma, Italy

Introduction

In 1979, Mary K. Gaillard and I lectured here in Cargese about Quarks and Leptons, that is, about the Standard Theory and beyond. When I was asked to come back again this year, the question came to mind: what has happened since?

The Standard Theory, in particular the unified theory of the Electroweak Interactions I am supposed to review, was dreamt in the fifties, shaped in the sixties and tested in the seventies. No doubt, the most important development of the last decade has been the opening of Intermediate Vector Boson physics, pioneered at the Cern collider, followed by the Tevatron and, since the last year, by SLC and LEP. Notably with LEP, a new phase has started, the high precision exploration of the 10^2 GeV energy-range, the natural mass scale of the Electroweak Theory.

It is also interesting to mention the things that did not happen ("the dog that did not bark", as Sherlock Holmes would say).

11 years ago, the t-quark mass limit was at about 15GeV. It has not been discovered, yet, and the mass limit is now around 90 GeV.

The low mass region for the Higgs boson has been cleared this summer by ALEPH, at LEP, and we know now that no Higgs particle exists below 45 GeV or so.

Most notably, proton decay has not been detected, to a level which raises serious doubts about the simplest Grand Unification scheme based on SU(5).

In parallel with the development of W and Z physics, a great amount of theoretical ingenuity has gone in understanding and organizing the electroweak higher order corrections, which are the second-generation test of the Electroweak Theory. Since the last year, we have been able to apply to the real data what was learned in theory .

Considerable progress has also taken place in the understanding of non-perturbative aspects of field theory. I am referring here to the massive development of lattice calculations, and also to the progress done in other areas, such as the $\lambda\phi^4$ theory (the triviality issue, numerical bounds to the Higgs boson mass, etc.) the Yukawa theory and the like.

These lectures are mainly focused on these two issues.

The main one, reported in Sect. 3, is Z-boson physics. In particular: the pattern of higher order electroweak corrrections to the Z boson physical properties and to the line-shape, the comparison of theory with LEP and collider data, the resulting limits to the t-quark mass. Also, a brief discussion is given of the consequences of the m_t limits for the weak mixing and CP violation parameters of K and B mesons.

The second issue, Sect. 4, concerns the theoretical bounds to the Higgs boson and to the t-quark masses, in particular those arising from the alleged triviality of non-asymptotically free theories, and the recent numerical, non-perturbative results, derived from lattice simulations.

The basic concepts of unified gauge theories, at the classical and quantum level, are illustrated in Sects. 1 and 2, respectively.

Obviously, many subjects have been left out, which perhaps would have been appropriate to add. This is partly because of my ignorance, e.g. about issues belonging to the "beyond the Standard Model" cathegory such as Technicolor, up to superstring-inspired phenomenology, and partly because of the lack of time, e.g. the phenomenological implications of supersymmetric models. I apologize to the reader for that.

1. GAUGE THEORIES: BASIC CONCEPTS

1.1 Locally invariant Lagrangians

Gauge theories, in particular non-abelian gauge theories[1], are discussed in many textbooks[2,3,4]. What follows is an abbreviated illustration, mainly to recall the basic concepts and establish the notation.

The construction of a gauge theory is based on the following elements:

a. a semisimple, compact group, G, with elements g, g'....
b. a set of matter fields, ψ, transforming in a definite way under **global**, i.e. space time independent, transformations of G:

$$\psi \rightarrow \psi' = U(g)\, \psi\, ;$$

$$U(g) = \exp\,(i\varepsilon^A T^A) \qquad\qquad (1.1.1)$$

ε^A are a set of continuous parameters which characterize the elements of G, T^A the representatives of the infinitesimal generators, normalized according to:

$$Tr(T^A T^B) = 2\,\delta^{AB}$$

A=1,2,..., N and a summation over repeated indices is understood. We will often suppress reference to the group element g and write U for U(g).

To obtain a renormalizable structure, matter fields must be restricted to spin 0 and 1/2. Otherwise, their composition in terms of multiplets of G is arbitrary;
c. a pre-gauge Lagrangian, $L_0(\psi, \partial_\mu \psi)$, invariant under the global transformations (1.1.1).

The corresponding gauge theory is obtained by requiring the dynamics to be invariant under **space-time dependent** transformations:

$$\psi \rightarrow \psi'(x) = U(x)\, \psi(x) \qquad\qquad (1.1.2)$$

with $U(x) = U(g(x)) = \exp\,[i\varepsilon^A(x)T^A]$.

To be able to compare fields belonging to different space-time points, it is necessary to introduce a set of vector fields (gauge fields or connections), one for each group generator, which we denote by A_μ^A. We shall assemble the fields A_μ^A into a single matrix, writing:

$$A_\mu = A_\mu^A \, T^A \tag{1.1.3}$$

In correspondence to the local gauge transformations (1.1.2), the matrix A_μ is required to transform according to:

$$A_\mu(x) \rightarrow A_\mu(x)' = U(x)\, A_\mu(x)\, U(x)^{-1} - i\, U(x)\, \partial_\mu U(x)^{-1} \tag{1.1.4}$$

Eq.(1.1.4) makes it possible to "parallel transport" the field $\psi(x)$ to the point $x + dx$. In fact, it is immediate to verify that the field:

$$\psi(x)_{\text{p.t.}} \equiv [1 - i\, A_\mu(x)\, dx^\mu]\psi(x)$$

transforms like $\psi(x+dx)$:

$$[\psi(x)_{\text{p.t}}]' = U(x+dx)\psi(x)_{\text{p.t}} + O(dx^2)$$

so that we can define the "covariant derivative" of ψ according to:

$$(D_\mu \psi)dx^\mu = \lim_{(dx \to 0)} \psi(x+dx) - \psi(x)_{\text{p.t}} = [\partial_\mu + i\, A_\mu(x)]\psi(x)dx^\mu \tag{1.1.5}$$

and:

$$D_\mu \psi(x)' = U(x) D_\mu \psi(x) \tag{1.1.6}$$

A locally invariant lagrangian is uniquely obtained from $L_0(\psi, \partial_\mu \psi)$ with the so-called "minimal substitution":

$$\partial_\mu \rightarrow D_\mu \tag{1.1.7}$$

$$L(\psi, \partial_\mu \psi, A_\mu) \equiv L_0(\psi, D_\mu \psi)$$

To obtain a non trivial dynamics for A_μ, we need terms quadratic in the derivatives, not present in eq.(1.1.7). To this aim, one introduces the Yang-Mills tensor:

$$G_{\mu\nu} \equiv G_{\mu\nu}^A \, T^A = \partial_\mu A_\nu - \partial_\nu A_\mu + i\, [A_\mu, A_\nu] \tag{1.1.8}$$

Analogously to the matter fields, $G_{\mu\nu}$ transforms linearly:

$$G_{\mu\nu}(x)' = U(x)\, G_{\mu\nu}(x)\, U(x)^{-1} \tag{1.1.9}$$

so we can construct with it the invariant Yang-Mills lagrangian:

$$L_{YM} = -\frac{1}{8g^2} \text{Tr}\,(G_{\mu\nu}G^{\mu\nu}) = -\frac{1}{4g^2}\, G_{\mu\nu}^A\, G^{\mu\nu A} \tag{1.1.10}$$

which contains terms quadratic in $\partial_\mu A_\nu$, as required. The total lagrangian is, in conclusion:

$$L_{tot} = L_{YM} + L_0(\psi, D_\mu \psi) \tag{1.1.11}$$

The constant g appears in eq.(1.1.10) as a normalization factor for L_{YM}. If we want the terms quadratic in A_μ^A to have the same form of the Maxwell lagrangian, i.e.:

$$-\frac{1}{4}(\partial_\mu A_\nu - \partial_\nu A_\mu)(\partial^\mu A^\nu - \partial^\nu A^\mu)$$

we must rescale the vector fields according to:

$$A_\mu^A \to g\, A_\mu^A$$

Correspondingly:

$$G_{\mu\nu} \to g\, G_{\mu\nu} = g\{ \partial_\mu A_\nu - \partial_\nu A_\mu + i\, g\, [A_\mu , A_\nu] \}$$

$$D_\mu \to \partial_\mu + i\, g\, A_\mu(x)$$

With the new normalization, g shows to be the universal coupling constant of the gauge fields with matter and with themselves, in the non-abelian case where the commutator term in $G_{\mu\nu}$ is non-vanishing.

The above equations are appropriate to the case where G is a simple group. When G contains mutually commuting factors, we can normalize independently the Yang-Mills lagrangians of the different factors. This leads to one independent coupling constant for each simple component (and for each U(1) factor). To account for this complication, we shall introduce, formally, one coupling g^A for each generator, with the understanding that all g^A within a simple component of G are equal among themselves. In this notation, for example, the covariant derivative is written according to:

$$D_\mu \to \partial_\mu + i\, g^A A_\mu^A(x)\, T^A \qquad (1.1.12)$$

(sum over A understood).

Ideally, we would like the pre-gauge lagrangian to be determined only by the kinematic properties of the matter fields, that is we would like L_0 to coincide with the free-particle Lagrangian. In this case, interactions would arise entirely from the principle of local gauge invariance, by the construction we have just illustrated, and would be of a completely geometrical origin. This is the case of QED, the interaction of electrons and photons, or QCD, the strong interaction of quarks and gluons.

However, life is not so simple for the electroweak interactions. In this case, L_0 **must** contain non trivial interactions of the scalar fields with themselves (ϕ^3 and ϕ^4 interactions) and with the spin 1/2 fields (Yukawa interactions), in order to account for the observed symmetry-breaking.

Table 1.1. - Local gauge symmetries of Particle interactions and the physical elements which led to their discovery.

Interaction	Symmetry	Physical Clue
Gravitation	Local coordinate transformations	Equivalence Principle, Einstein elevator
Electromagnetism	phase transformation	Maxwell-Dirac eqs. are invariant (Weyl, Pauli)
Weak	weak isospin SU(2)	Current-current interaction, CVC, PCAC
Strong	colour SU(3)	asymptotic freedom

Leaving aside the problem of the non-geometrical interactions, which, by the way, have not been observed thus far, the fact remains that all the observed interactions **are** described by a gauge principle, see Tab.1.1. It is remarkable how different among themselves have been the physical clues which, historically, have led to the discovery of the gauge structure of each particular interaction, see again Tab.1.1.

1.2. Equations of motion and current conservation

On the basis of eq.(1.1.7), the equations of motion of the matter fields in the complete theory are just the same as those obtained from L_0, normal derivatives being replaced by covariant derivatives (what else could it be?).

The equations for the gauge fields obtained by variation of (1.1.11), are:

$$\frac{1}{g^2} D_\mu G^{\mu\nu} = j^\nu \tag{1.2.1}$$

The covariant derivative appearing in (1.2.1) is the one appropriate to the transformation property specified in (1.1.9), that is:

$$D_\rho G^{\mu\nu} = \partial_\rho G^{\mu\nu} + i \; [A_\rho, G^{\mu\nu}] \tag{1.2.2}$$

and j^ν is the current associated with matter, in presence of the gauge fields:

$$j^\nu = j^{A\nu} \, T^A$$

$$j^{A\nu} = - \; \frac{\partial L_0}{\partial A_\nu{}^A} \tag{1.2.3}$$

$$= -i \; \frac{\partial L_0(\psi, D_\mu \psi)}{\partial D_\nu \psi} \, T^A \, \psi$$

For spinor fields, these currents coincide with the Noether currents associated to the global symmetry of L_0. For scalar fields, L_0 is quadratic in the field derivatives and the current contains additional (A_μ-dependent) terms, arising from the contact, seagull, interaction.

Eq.(1.2.1) is analogous to the Maxwell equations for the electromagnetic field. The analogy goes further since, also in the non-abelian case, one can prove the identity:

$$D_\mu D_\nu G^{\mu\nu} = 0$$

which leads to the "covariant conservation" of the matter current:

$$D_\mu j^\mu = 0 \tag{1.2.4}$$

Eq.(1.2.4) is of crucial importance and can be read in various, equivalent ways. On one hand, using (1.2.1), one obtains:

$$D_\mu j^\mu = \partial_\mu j^\mu + i \; [A_\mu, j^\mu] = \partial_\mu j^\mu + \frac{i}{g^2}[A_\mu, D_\rho G^{\rho\mu}] =$$

$$= \partial_\mu j^\mu + \frac{i}{g^2}[A_\mu, \partial_\rho G^{\rho\mu}] - \frac{1}{g^2} \; [A_\mu, [A_\rho, G^{\rho\mu}]] =$$

$$= \partial_\mu (j^\mu + \frac{i}{g^2}[A_\rho, G^{\mu\rho}]) - \frac{i}{2g^2}[\partial_\rho A_\mu - \partial_\rho A_\mu, G^{\rho\mu}] +$$

$$+ \frac{1}{2g^2}[[A_\rho,A_\mu],G^{\rho\mu}] =$$

$$= \partial_\mu J^\mu - \frac{i}{2g^2}[G_{\rho\mu},G^{\rho\mu}] = \partial_\mu J^\mu = 0$$

where we have made use of the Jacobi identity:

$$[A_\mu,[A_\rho,G^{\rho\mu}]] + [A_\rho,[G^{\rho\mu}, A_\mu]] + [G^{\rho\mu},[A_\mu,A_\rho]] = 0$$

Thus, eq.(1.2.4) implies the conservation of the total Noether current , **J**:

$$\mathbf{J}^\mu = j^\mu + \frac{i}{g^2}[A_\rho,G^{\mu\rho}] \tag{1.2.5}$$

associated with the invariance of the lagrangian (1.1.11) under global transformations. The matter current is not separately conserved, since charge can flow from matter into the gauge fields.

From another point of view, eq.(1.2.4) simply expresses the invariance of the total action under local gauge transformations.

Under a local gauge transformation with infinitesimal parameters $\varepsilon^A(x)$, one finds easily that:

$$\delta S = -\int d^4x \; \varepsilon^A(x) \; (D_\mu j^\mu)^A$$

In fact, using eqs. (1.1.1), (1.1.6), (1.1.9) and the matter equations of the motion, one finds:

$$\delta S = \int d^4x \; \varepsilon^A(x)$$

$$\{\frac{\partial L_0}{\partial\psi} iT^A \psi + \frac{\partial L_0}{\partial D_\nu\psi} iT^A D_\nu\psi +$$

$$- \frac{1}{2g^2} Tr (G_{\mu\nu} [iT^A, G^{\mu\nu}])\} =$$

$$= \int d^4x \; \varepsilon^A(x) \; \{D_\nu\frac{\partial L_0}{\partial D_\nu\psi} iT^A\psi + \frac{\partial L_0}{\partial D_\nu\psi} iT^A D_\nu\psi\} =$$

$$= -\int d^4x \; \varepsilon^A(x) \; (D_\nu j^\nu)^A$$

so that:

$$\frac{\delta S}{\delta\varepsilon^A(x)} = -(D_\nu j^\nu)^A \tag{1.2.6}$$

1.3 Spontaneous symmetry breaking

If there are scalar fields with the appropriate self-interaction, the gauge symmetry can be spontaneously broken even in the weak coupling limit[5].

Spontaneous symmetry breaking is indicated by a non-invariant ground, i.e. vacuum, state. We write the ϕ-dependent part of L_0 as:

$$L_0 = \frac{1}{2}\Sigma_i \, D_\mu\phi^i D^\mu\phi_i - V(\phi) + \dots \tag{1.3.1}$$

The ground state is, in general, determined by the absolute minimum of the energy. Classically[1] and assuming translation invariance, the vacuum field configuration thus corresponds to the absolute minimum of V:

$$\frac{\partial V}{\partial \phi} = 0 \quad ; \quad \frac{\partial^2 V}{\partial \phi^2} > 0$$

If at the minimum, ϕ_0:

$$T^A\phi_0 \neq 0 \tag{1.3.2}$$

for some group generator, G is said to be spontaneously broken[6] and the symmetry is reduced to the symmetry under the subgroup G' which leaves ϕ_0 invariant[2]. We denote by n the dimension of G'. When G' coincides with the identity transformation, i.e. n=0, the symmetry is completely broken.

Vector bosons corresponding to the broken generators acquire a non vanishing mass (Higgs mechanism[7]) and so do the chiral fermions coupled to ϕ. We discuss, in the following, a few characteristic features of spontaneously broken gauge theories.

i. Vector boson masses. The vector boson mass-matrix is determined by the covariant derivative term in eq.(1.3.1). It is convenient to use real fields for the components of ϕ and purely imaginary group generators, such that:

$$T^A = i\, t^A$$

with t^A real-antisymmetric matrices. Then, at the minimum, the quadratic term in A_μ in eq.(1.3.1) reads:

$$L_{mass} = \frac{1}{2}\Sigma_{AB}\, A_\mu^A\, M_{AB}^2\, A^{B\mu} \tag{1.3.3}$$

$$M_{AB}^2 = \Sigma_i\, (g^A t^A \phi_0)_i (g^B t^B \phi_0)_i = (v^A, v^B)$$

$$(v^A)_i = g^A (t^A \phi_0)_i \tag{1.3.4}$$

i denotes the index which labels the components of the scalar field. Fields with definite mass are linear combinations of the basic fields:

$$W_\mu^{(r)} = \Sigma_A\, c_A^{(r)}\, A_\mu^A$$

$$\Sigma_B\, M_{AB}^2\, c_B^{(r)} = (M^{(r)})^2 c_A^{(r)} \tag{1.3.5}$$

There is a one-to-one correspondence between massless eigenstates of M^2 and generators of the little group, G'. In fact, if Q is one of these generators:

$$Q\phi_0 = 0 \tag{1.3.6}$$

$$Q = \Sigma_A\, w_A T^A \tag{1.3.7}$$

the vector:

---

[1] For a discussion of the quantum corrections see, e.g. S. Coleman, ref.[5].

[2] The infinitesimal generators of G' annihilate ϕ_0. G' is also called the little group of ϕ_0.

$$c_A = \frac{1}{g^A}\, w_A \tag{1.3.8}$$

is a massless eigenvector of M^2:

$$\Sigma_B M^2_{AB} c_B = \Sigma_i (g^A t^A \phi_0)_i (Q\phi_0)_i = 0$$

Conversely, if c_A is a massless eigenvector, then:

$$0 = \Sigma_{AB}\, c_A M^2_{AB} c_B = \Sigma_i (Q\phi_0)_i (Q\phi_0)_i = 0$$

and eq.(1.3.6) holds, with:

$$Q = \Sigma_A (c_A g^A) t^A$$

In terms of the massive fields, the mass term (1.3.3) reads:

$$L_{mass} = \frac{1}{2} \Sigma_r (M^{(r)})^2\, W^{(r)}_\mu W^{(r)\mu} \tag{1.3.9}$$

with the sum restricted to the massive bosons only.

ii. Would-be-Goldstone fields. In the spontaneously broken situation, we can characterize part of the manifold formed by the Higgs field in a universal way, dictated by the group structure of G and G' (more precisely, by the structure of the quotient G/G').
Consider the vectors:

$$v^{(r)}_i = \Sigma_A c^{(r)}_A (v^A)_i = \Sigma_A c^{(r)}_A (g^A t^A \phi_0)_i \tag{1.3.10}$$

There are exactly (N-n) such vectors, the same number as massive vector bosons (for the massless eigenvectors of M^2 the combination (1.3.10) vanishes) and they are orthogonal to each other:

$$(v^{(r)}, v^{(s)}) = (M^{(r)})^2\, \delta_{rs}$$

Furthermore, we denote by $e^{(a)}$ ($a = 1, .., n$) the additional normalized and orthogonal vectors needed to obtain a complete, orthogonal basis.
In a small region around the vacuum configuration, ϕ_0, the Higgs field can be developed according to:

$$\phi(x) = \phi_0 + \phi'(x) \tag{1.3.11}$$

$$\phi'(x) = \Sigma_r \xi^{(r)}(x)\, \frac{v^{(r)}}{M^{(r)}} + \Sigma_a \sigma^{(a)}(x)\, e^{(a)} \tag{1.3.12}$$

The vectors $v^{(r)}$ span the manifold within which ϕ_0 moves under the action of the broken generators of G. Thus, the action of any such generator can produce a non-vanishing value of $\xi^{(r)}$ even if we started from vanishing components along the $v^{(r)}$'s, but it leaves $\sigma = 0$, if we so started. This means that, under the action of the broken generators, the fields $\xi^{(r)}$ transform in an inhomogeneous way:

$$\xi^{(r)}(x)' = \xi^{(r)}(x) + \varepsilon(x)\, M^{(r)} \tag{1.3.13}$$

while $\sigma^{(a)}$ transform homogeneously. On the other hand, both ξ's and σ's transform homogeneously under the action of the conserved generators. The latter leave ϕ_0 invariant, so that they cannot move away from zero either ξ or σ.

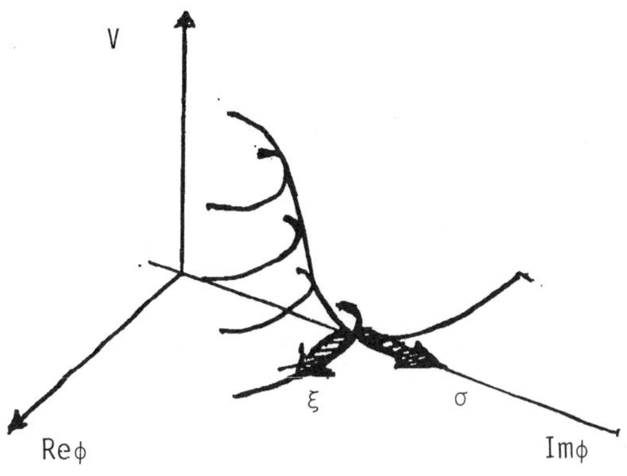

Fig. 1.1

The manifold of a complex scalar Higgs field. The potential, V, is invariant under U(1) phase transformations, represented by the rotations of the Reφ - Imφ plane. Lines of constant potential are reported. The manifold spanned by ξ, the Goldstone-boson field, lies along the minimum valley; also indicated is the manifold spanned by σ, in the in the orthogonal direction .

The situation is illustrated in Fig.1.1, for the familiar case of the complex Higgs field model.

Since it is invariant under G, the potential V in eq.(1.3.1) does not depend upon $\xi^{(r)}$. In particular, when developed in powers of the fields, V gives rise to a vanishing quadratic mass term for the $\xi^{(r)}$.

In the globally invariant theory described by the pre-gauge lagrangian, L_0, $\xi^{(r)}$ are the Goldstone, massless, fields associated to the spontaneous breaking[6] of G to G'. In the gauge theory we are considering, the ξ's are instead completely fictitious degrees of freedom. As shown by eq.(1.3.13), we may obtain everywhere ξ=0, with a space-time dependent gauge transformation. The field σ's, instead, represent true, dynamical, degrees of freedom. Their composition, mass and interactions depend from the particular representation we have choosen for φ, and from the specific form of the potential, V.

The fields ξ's and σ's are, usually, indicated as the would-be-Goldstone bosons and the physical Higgs fields of the gauge theory, respectively.

iii. Propagators in a renormalizable gauge. The gauge choice in which the would-be-Goldstone fields, $\xi^{(r)}$, are taken to vanish is called the unitary gauge.

It is often convenient to consider gauges where $\xi^{(r)}$ are non-vanishing, such as the Feynman-'t-Hooft gauge, to obtain superficially renormalizable propagators.

To derive the propagators[2], we consider the gauge-Higgs lagrangian:

$$L = \frac{1}{2}\Sigma_i \, D_\mu\phi^i D^\mu\phi_i - V(\phi) - \frac{1}{4}\Sigma_A \, W_{\mu\nu}^A \, W^{A\mu\nu} \qquad (1.3.14)$$

supplemented by a gauge fixing term of the form:

$$L_{gf} = -\frac{1}{2\lambda}\Sigma_A \, [\partial^\mu A_\mu^A + \lambda \, (\phi - \phi_0) \, g^A t^A \phi_0]^2 \qquad (1.3.15)$$

λ is an arbitrary parameter. We transform to the fields with definite mass and keep only quadratic terms, to obtain:

$$L + L_{gf} = \sum_r \{ -\tfrac{1}{4} W^{(r)}_{\mu\nu} W^{(r)\mu\nu} + \tfrac{1}{2} M^{(r)2} W^{(r)}_\mu W^{(r)\mu} + \tfrac{1}{2} \partial^\mu \xi^{(r)} \partial_\mu \xi^{(r)}$$

$$-M^{(r)} \partial^\mu \xi^{(r)} W^{(r)}_\mu - \tfrac{1}{2\lambda} [\partial^\mu W^{(r)}_\mu + \lambda M^{(r)} \xi^{(r)}]^2 \} -$$

$$- \tfrac{1}{4} \sum_a A^{(a)}_{\mu\nu} A^{(a)\mu\nu} - \tfrac{1}{2\lambda} \sum_a [\partial^\mu A^{(a)}_\mu]^2 + \sigma \text{ - dependent terms}$$

$$= \sum_r \{ -\tfrac{1}{4} [\partial_\mu W^{(r)}_\nu - \partial_\nu W^{(r)}_\mu]^2 - \tfrac{1}{2\lambda} [\partial^\mu W^{(a)}_\mu]^2 + \tfrac{1}{2} M^{(r)2} W^{(r)}_\mu W^{(r)\mu}$$

$$+ \tfrac{1}{2} \partial^\mu \xi^{(r)} \partial_\mu \xi^{(r)} - \tfrac{1}{2} \lambda M^{(r)2} \xi^{(r)2} \} +$$

$$\sum_a \{ -\tfrac{1}{4} A^{(a)}_{\mu\nu} A^{(a)\mu\nu} - \tfrac{1}{2\lambda} [\partial^\mu A^{(a)}_\mu]^2 \} + \sigma \text{ - dependent terms} \qquad (1.3.16)$$

The indices r and a go over the massive and massles vector fields, respectively, and we have omitted the model-dependent contribution of the physical Higgs fields.

The propagators obtained from eq.(1.3.16) are easily computed to be:

$$D^{(r)}_{\mu\nu} = \frac{1}{p^2 - M^{(r)2}} \left[-g_{\mu\nu} + \frac{(1-\lambda) p_\mu p_\nu}{p^2 - \lambda M^{(r)2}} \right] \qquad (1.3.17)$$

$$D^{(a)}_{\mu\nu} = \frac{1}{p^2} \left[-g_{\mu\nu} + \frac{(1-\lambda) p_\mu p_\nu}{p^2} \right] \qquad (1.3.18)$$

$$D^{(r)} = \frac{1}{p^2 - \lambda M^{(r)2}} \qquad (1.3.19)$$

for the massive or massless vector fields and for the would-be-Goldstone bosons, respectively. The propagators behave at infinity like p^{-2}, a renormalizable behaviour, but eqs.(1.3.17) and (1.3.18) contain a spurious pole, at $p^2 = \lambda M^{(r)2}$. As we discuss in Sect.2.2, the spurious poles in $W^{(r)}$ and $\xi^{(r)}$- exchange cancel each other because of current conservation, eq.(1.2.4). In the limit $\lambda \to \infty$, the ξ propagator vanishes and $D^{(r)}_{\mu\nu}$ takes the form:

$$D^{(r)}_{\mu\nu} = \frac{1}{p^2 - M^{(r)2}} \left[-g_{\mu\nu} + \frac{p_\mu p_\nu}{M^{(r)2}} \right] \qquad (1.3.20)$$

corresponding to the unitary-gauge propagator.

1.4 The Standard Electroweak Theory

The minimal gauge group of the observed electro-weak interactions is determined directly by the weak β-decays[8].

The observed characteristics of the processes:

$$\mu^+ \to \bar{\nu} + e^+ + \nu_e$$

$$P \to N + e^+ + \nu_e$$

are well described by the assumption that the e^+- ν_e pair is created by the operator[3]:

$$J_\mu = \frac{1}{2} \, \bar{\nu}_e \gamma_\mu (1-\gamma_5)e \; =$$

$$= \bar{\nu}_{eL} \, \gamma_\mu \, e_L \tag{1.4.1}$$

Defining the weak charge according to:

$$T = \int d^3x \, J_0(x,t) \tag{1.4.2}$$

the canonical equal-time anticommutation rules for the electron and neutrino fields lead to the commutation relations:

$$[T, e_L{}^\dagger] = \nu_{eL} \; ; \qquad\qquad [T, \nu_{eL}{}^\dagger] = 0$$

(we assume that currents are conserved, so that T is time-independent). Thus, T and $T^\dagger$ are the raising and lowering operators of the doublet:

$$l_L = \begin{pmatrix} \nu_{eL} \\ e_L \end{pmatrix} \tag{1.4.3}$$

The electromagnetic current and the electric charge have the form:

$$J_\mu{}^{em} = - \bar{e} \, \gamma_\mu e \; =$$

$$= - (\bar{e}_L \gamma_\mu e_L + \bar{e}_R \gamma_\mu e_R) \tag{1.4.5}$$

$$Q = - \int d^3x \, (e_L{}^\dagger e_L + e_R{}^\dagger e_R) \tag{1.4.6}$$

With only electrons and neutrinos T, $T^\dagger$ and Q do not form a closed algebra [4], since the commutator of T and $T^\dagger$ gives rise to an operator, T_3:

$$[T, T^\dagger] = 2 \, T_3 = \int d^3x \, (\nu_{eL}{}^\dagger \nu_{eL} - e_L{}^\dagger e_L) \tag{1.4.7}$$

different from Q.

However, the difference $Y = Q - T_3$ commutes with T, $T^\dagger$ and Q, as can be seen from its explicit expression:

$$Y = \int d^3x \, \{ -\frac{1}{2} \, (\nu_{eL}{}^\dagger \nu_{eL} + e_L{}^\dagger e_L) - e_R{}^\dagger e_R \} \tag{1.4.8}$$

$$Q = T_3 + Y \tag{1.4.9}$$

---

[3] We use the Bjorken-Drell γ-matrices and the metric $g_{\mu\nu} = \mathrm{diag}(-1,-1,-1,+1)$. Left and right-handed components of a Dirac field are projected by $(1-\gamma_5)/2$ and $(1+\gamma_5)/2$, respectively.

[4] this could be obtained by adding new, positively charged and neutral lepton fields as, e.g. in the Georgi-Glashow O(3) model, ref.[9].

The four charges: T, T†, T$_3$ and Y do make a closed, SU(2) x U(1), algebra, thus corresponding to the minimal algebra of the weak and electromagnetic charges.

Up and down left-handed quarks must also transform as a doublet, as implied by the universality of nuclear and muon beta decays. Together with the right-handed singlets and the e-v_e system, the u-d lefthanded doublet completes the first generation of quarks and leptons:

$$\begin{pmatrix} v_e \\ e_L \end{pmatrix} \qquad \begin{pmatrix} u_L \\ d_L \end{pmatrix} \qquad u_R, d_R$$

Quarks come in three colours and the exact SU(3)$_{colour}$ symmetry is the basis of the gauge theory of Strong Interactions. The quark weak hypercharge is determined again by the relation eq.(1.4.9), given that Q=2/3, -1/3 for up and down quarks respectively.

Muon and tau leptons are similarly accompanied by the charm-strange and by the top-beauty quark doublets, respectively:

2nd generation[10]:

$$\begin{pmatrix} v_{\mu L} \\ \mu_L \end{pmatrix} \qquad \begin{pmatrix} c_L \\ s_L \end{pmatrix} \qquad c_R, s_R$$

3rd generation[11,12,13]:

$$\begin{pmatrix} v_{\tau L} \\ \tau_L \end{pmatrix} \qquad \begin{pmatrix} t_L \\ b_L \end{pmatrix} \qquad t_R, b_R$$

The pre-gauge structure of the Standard Theory is completed by the complex, scalar Higgs doublet:

$$\phi = \begin{pmatrix} \phi^+ \\ \phi^0 \end{pmatrix} \tag{1.4.10}$$

Besides kinetic energy terms, the pre-gauge lagrangian, L_0, contains the Yukawa couplings of the fermion doublets and singlets to the Higgs field, of the form:

$$L_Y = \sum_{ij} (g_u)_{ij}\, \bar{Q}_L^i\, \phi_C\, u_R^j + \sum_{ij} (g_d)_{ij}\, \bar{Q}_L^i\, \phi\, d_R^j +$$

$$+ \sum_i (g_e)_i\, \bar{l}_L^i\, \phi^\dagger\, e_R^i + \text{hermitian conjugate} \tag{1.4.11}$$

and the scalar, renormalizable, self- interaction:

$$V(\phi) = \lambda(\phi^\dagger \phi - \eta^2)^2 \tag{1.4.12}$$

In (1.4.11), Q^i and l^i denote the left-handed doublets, u^i, d^i and e^i the right-handed singlets and i is a generation index The charge-conjugate, ϕ_C , of the doublet ϕ is defined according to:

$$\phi_C = i\tau_2\, \phi^*$$

The gauge group SU(3)xSU(2)xU(1) implies 3 different couplings, denoted by g_S, g and g' respectively and 12 vector fields. 8 gluons are associated to the Strong Interaction, SU(3) colour group. After spontaneous breaking due to the vacuum expectation value of ϕ^0, SU(2)xU(1) is reduced to the U(1)$_Q$ of the electric charge and 3 out of the 4 vector bosons of SU(2)xU(1) become massive.

In view of eqs.(1.3.8) and (1.4.9), the (normalized) photon field, expressed in terms of the gauge fields associated to T$_3$ and Y, is given by:

$$A_\mu = \frac{1}{\sqrt{g^{-2}+g'^{-2}}}\left(\frac{1}{g}W_{3\mu}+\frac{1}{g'}B_\mu\right) =$$

$$= \sin\theta\, W_{3\mu} + \cos\theta\, B_\mu$$

with:

$$\tan\theta = \frac{g'}{g} \tag{1.4.13}$$

while the massive neutral boson, Z, is given by the orthogonal combination:

$$Z_\mu = \cos\theta\, W_{3\mu} - \sin\theta\, B_\mu \tag{1.4.14}$$

The masses M_Z and M_W, computed according to eq.(1.3.3), are:

$$M_W^2 = \frac{1}{2}g^2\eta^2 \tag{1.4.15}$$

$$M_Z^2 = \frac{M_W^2}{\cos^2\theta} \tag{1.4.16}$$

where η is the vacuum expectation value of ϕ^0. Eq.(1.4.16) is a consequence of the weak isospin $=1/2$ of the Higgs doublet. In the Standard Theory, it can be used as an alternative definition of $\sin^2\theta$, see Sect.3.1.

According to the results of the previous Section, three out of the four components of the Higgs doublet correspond to fictitious, would-be-Goldstone fields. Explicitely, we may rewrite the fields in the doublet as:

$$\frac{1}{\sqrt{2}}\left(\xi^{(1)} + i\,\xi^{(2)}\right)$$

$$\eta + \frac{1}{\sqrt{2}}\left(\sigma + i\,\xi^{(3)}\right)$$

with σ the physical Higgs boson.

The top quark has not yet been observed, presumably because it is too heavy (see Sect.4.4), and the tau neutrino has not been detected, presumably because the ν_τ beams originating from τ decays are too feeble.

LEP data (Sect.4.4) as well as cosmological observations (as discussed by D.Schramm at this School) indicate the existence of only three types of light neutrinos. Extrapolating the remarkable lepton-quark symmetry observed thus far, we may conclude that fermionic matter is limited to the three generations listed above.

Right-handed neutrino fields could also be considered. At the present stage this would be quite immaterial as right-handed neutrinos would have vanishing SU(2) and U(1) charges, because of eq.(1.4.9), hence no gauge interaction at all. However, their existence would allow neutrinos to have a (lepton number conserving) non-vanishing mass, which could be of great physical relevance on cosmological scales and give rise to interesting phenomena at particle level, such as lepton flavour oscillations[14].

2. QUANTUM EFFECTS

In the previous Section we have emphasized properties already present in a gauge theory at the classical level. We discuss here the effects which show up at the quantum level.

Ghost fields and the possible arising of anomalies in current conservation are peculiar to gauge theories (chiral gauge theories, in the latter case). The breaking of scale invariance by quantum corrections and the resulting concepts of running coupling and dimensional transmutation are, instead, general features of any quantum field theory.

2.1 Unitarity and ghosts

Conservation of probability in quantum theory requires the scattering matrix, S-matrix, to be unitary:

$$S^\dagger S = SS^\dagger = 1 \qquad (2.1.1)$$

In terms of the reaction T-matrix:

$$S = 1 + iT$$

eq.(2.1.1) reads:

$$-i\,(T-T^\dagger) = T^\dagger\,T \qquad (2.1.2)$$

We take the diagonal matrix element of (2.1.2) in some state $|a>$, and define the scattering amplitude:

$$f(a \to n) = <n|T|a>$$

Eq.(2.1.2) leads then to the so-called optical theorem:

$$2\mathrm{Im}\,f(a \to a) = \sum_n |f(a \to n)|^2 \qquad (2.1.3)$$

which relates the forward scattering amplitude to the total cross-section.

The unitarity equations, (2.1.2) or (2.1.3), imply that the production of real intermediate states gives rise to singularities in the scattering amplitudes. Consider, for example, the forward scattering amplitude as a function of the center-of-mass energy. The contribution of a given intermediate state to its imaginary part is zero (not zero) if we are below (above) threshold, a clear sign of non-analiticity of the amplitude.

In perturbation theory, with real vertices, complex amplitudes result from the $i\varepsilon$ prescription in the denominators of intermediate (virtual) particle propagators:

$$2i\,\mathrm{Im}\,f = f - f^* = \lim_{\varepsilon \to 0}\,[f(+i\varepsilon)-f(-i\varepsilon)] \equiv \mathrm{Disc}(f) \qquad (2.1.4)$$

A non vanishing result indicates the presence of a cut, which, in turn, is related to the fact that particles in the intermediate state go on the mass-shell, in that particular kinematic region.

The above considerations lead directly to a very simple rule to compute the discontinuity associated with a particular intermediate state, known as the Cutkosky rule[15]. The discontinuity is obtained by the substitution:

$$\frac{1}{q^2-m^2} \to (-2\pi i)\,\delta^{(+)}(q^2-m^2) \qquad (2.1.5)$$

in all propagators corresponding to the particles of the intermediate state (the superscript in the delta-function indicates that one has to restrict to the positive energy solution, $q_0 > 0$).

Unitarity is not obvious in a quantum theory with constraints such as, for example, Quantum Electrodynamics or non-abelian gauge theories. The vector field is a four-component object, A_μ, but there are only two pysical states associated with it, the two transverse photons.

To illustrate the situation, we consider the scattering process:

$$e^+ e^- \to e^+ e^- \qquad\qquad (2.1.6)$$

and the discontinuity corresponding to the two-photon intermediate state, i.e. to the process:

$$e^+ e^- \to \gamma^A\, \gamma^B \to e^+ e^- \qquad\qquad (2.1.7)$$

(here γ denotes either a QED photon or a non-abelian Yang-Mills vector particle; we consider the non-spontaneously broken case, for simplicity). The amplitude corrisponding to the process:

$$e^+ e^- \to \gamma^A\, \gamma^B \qquad\qquad (2.1.8)$$

has the form:

$$M = M_{AB}^{\mu\nu}(q_1,q_2)\, \varepsilon_\mu^a(q_1)^* \, \varepsilon_\nu^b(q_2)^* \qquad\qquad (2.1.9)$$

where $q_{1,2}$ are the photon momenta, a and b polarization indices and:

$$\varepsilon_\mu^a(q_1) q_1{}^\mu = 0$$

The unitarity condition, eq.(2.1.3), requires that:

$$\mathrm{Disc}_{2\gamma}\, M(e^+ e^- \to e^+ e^-) =$$

$$\qquad\qquad (2.1.10)$$

$$= \int d\Omega\; \Sigma_{AB}\, M_{AB}^{\mu\nu}(q_1,q_2)\, [M_{AB}^{\rho\sigma}(q_1,q_2)]^* \{\Sigma_a\; \varepsilon_\mu^a(q_1)^* \, \varepsilon_\rho^a(q_1)\}\; \{\Sigma_b\, \varepsilon_\nu^b(q_2)^* \, \varepsilon_\sigma^b(q_2)\}$$

where:

$$d\Omega \propto \frac{d^3q_1\, d^3q_2}{4\omega_1\omega_2}\, \delta^{(4)}(q_1 + q_2 - P_{in}) =$$

$$= d^4q_1 d^4q_2\; \delta^{(+)}(q_1{}^2)\; \delta^{(+)}(q_2{}^2)\; \delta^{(4)}(q_1 + q_2 - P_{in})$$

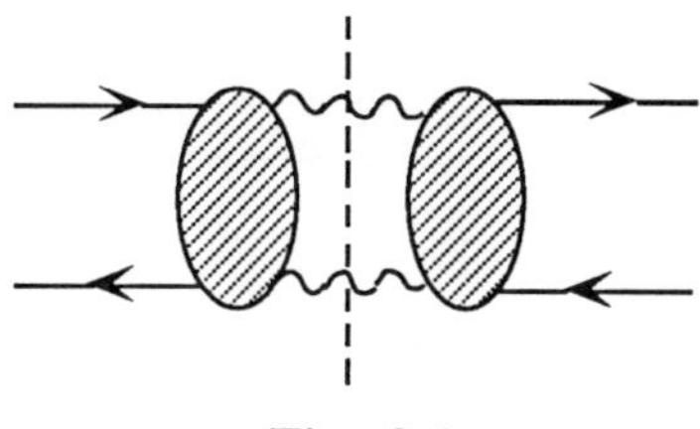

Fig. 2.1

Graphical representation of the two photon contribution to the discontinuity of the amplitude: $e^+ e^- \to e^+ e^-$. Lines cut by the vertical line correspond to on-shell particles.

The same result must be obtained if we compute process (2.1.6) from the diagram of Fig. 2.1 and deduce the discontinuity with the Cutkosky rule (2.1.5). In the Feynman gauge, the photon propagator is (Sect.1.3):

$$\frac{-g_{\mu\nu}}{q^2}$$

and the Cutkosky rule gives:

$$\text{Disc}_{2\gamma}\, M(e^+ e^- \to e^+ e^-) =$$

$$(2.1.11)$$

$$= \int d\Omega\, \Sigma_{AB}\, M^{\mu\nu}_{AB}(q_1,q_2)\, [M^{\rho\sigma}_{AB}(q_1,q_2)]^*\{-g_{\mu\rho}\}\,\{-g_{\nu\sigma}\}$$

Eqs.(2.1.10) and (2.1.11) illustrate the problem. For the two results to agree we must be allowed to replace the sum over polarization vectors with $-g_{\mu\nu}$, that is:

$$-g_{\mu\nu} \sim \{\Sigma_a\ \varepsilon^a_\mu(q_{1,2})^* \varepsilon^a_\nu(q_{1,2})\} \qquad (2.1.12)$$

when multiplied by the amplitudes of process (2.1.8), and for $q_1^2 = q_2^2 = 0$.

To see if (2.1.12) is correct, we explicitate the difference between the two sides of the equation, by introducing a four vector $s(q)$, longitudinal with respect to q and orthogonal to the polarization vectors:

$$-g_{\mu\rho} = \{\Sigma_a\ \varepsilon^a_\mu(q)\ [\varepsilon^a_\rho)(q)]^*\} + \frac{1}{2\omega}\ [q_\mu s(q)_\nu + q_\nu s(q)_\mu] \qquad (2.1.13)$$

The identity is easily verified by taking:

$$q^\mu = \omega\ (0,0,1;\ 1)$$

$$s^\mu = (0,0,\ -1;\ 1)$$

For unitarity to be obeyed, the additional term in (2.1.13) must give zero when inserted in eq.(2.1.11). In QED this is always so, since, by gauge invariance:

$$q_{1\mu}\, M^{\mu\nu}(q_1,q_2) = 0 \qquad (2.1.14)$$

for $q_1^2 = q_2^2 = 0$ (and external fermions on shell).

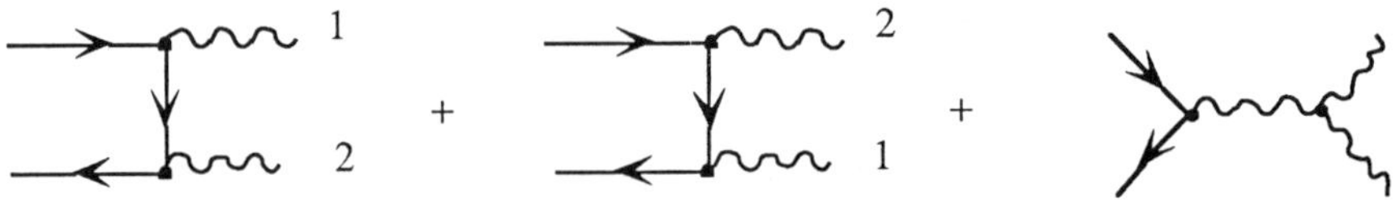

Fig. 2.2

Feynman diagrams for $e^+ e^-$ annihilation into two vector particles, in a non-abelian gauge theory.

The same condition is **not** obeyed in a non-abelian gauge theory. To lowest order, the amplitude for (2.1.8) is given by the diagrams of Fig.2.2 and an explicit calculation shows that:

$$q_{1\mu}\, M^{\mu\rho}_{AB}(q_1,q_2) =$$

$$(2.1.15)$$

$$= (ig^2)\ (g^{\rho\sigma}\, q_2^2 - q_2^\rho\, q_2^\sigma)\ \frac{-i\, g_{\rho\delta}}{(q_1+q_2)^2}\ i\, f_{ABC}\ \bar{u}(p')\, \gamma^\delta(\frac{\tau^C}{2})\, v(p)$$

The r.h.s. of eq.(2.1.15) is orthogonal to $q_{2\rho}$ (for any q_2) and to $\varepsilon_\rho^a(q_2)$ (for $q_2{}^2 = 0$), so that the difference between (2.1.10) and (2.1.11) is:

$$\Delta = \int d\Omega \sum_{AB}\{\frac{1}{2\omega_2}[q_{1\mu}\ M_{AB}^{\mu\nu}(q_1,q_2)\ s(q_2)_\nu]\ \frac{1}{2\omega_1}[s(q_1)_\rho M_{AB}^{\rho\sigma}(q_1,q_2)\ q_{2\sigma}]^*$$

$$+ \text{complex conjugate}\} \tag{2.1.16}$$

and, according to eq.(2.1.15):

$$\frac{1}{2\omega_2}[q_{1\mu}\ M_{AB}^{\mu\nu}(q_1,q_2)\ s(q_2)_\nu] =$$

$$\tag{2.1.17}$$

$$=(ig)\ (-i\ q_{2\sigma}\ f_{ABC})\ \frac{(-ig^{\sigma\nu})}{(q_1+q_2)^2}\ (ig)\ [\bar{u}(p')\ \gamma_\nu\ (\frac{\tau^C}{2})\ v(p)]$$

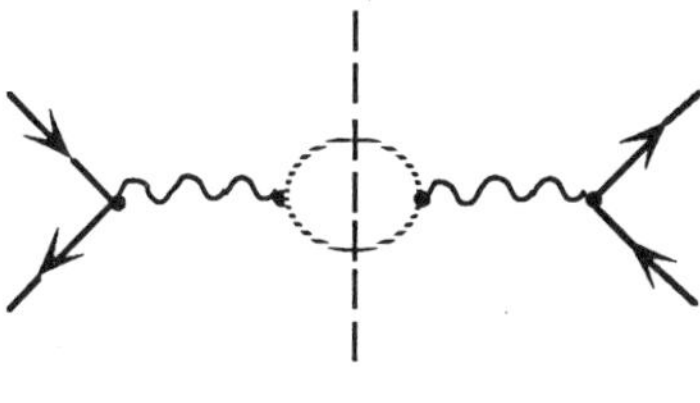

Fig. 2.3

Ghost loop in the amplitude for $e^+ e^- \to e^+ e^-$. The discontinuity shown in the figure compensates the extra-term in the discontinuity of Fig.2.1, present in the non-abelian case.

From (2.1.17) we see that the anomalous term, Δ, can itself be represented as the discontinuity of the one-loop diagram shown in Fig.2.3, where the dotted lines represent the propagation of scalar massless complex fields, c^A, with the same transformation properties under the gauge group as the vector fields and coupled according to:

$$L_{int} = gf_{ABC}\ \bar{c}^A\ \partial^\mu c^B\ A^C_\mu$$

Unitarity is restored if we add the gauge lagrangian a new term:

$$L_{ghost} = \ \partial^\mu\ \bar{c}^A\ \partial_\mu c^B + gf_{ABC}\ A^A_\mu\ \bar{c}^B\ \partial^\mu c^C$$

with the scalar massless multiplet quantized according to the **Fermi-Dirac statistics**. The minus sign associated with a loop of anticommuting fields makes so that the scalar-loop contribution cancels precisely the term Δ, arising from the gauge field loop.

This is Feynman's argument[16], showing the necessity of scalar ghost fields in order to quantize properly the Yang-Mills lagrangian. In the quantization of a massless spin-two field, the graviton, we would meet the very same phenomenon, ghosts being vector fields in this case[17].

To one-loop level, Feynman's is a complete argument. However, it is difficult to generalize it to all orders. This can be done following Faddev and Popov [ref.18], which have derived the ghost Lagrangian with functional integration methods, as shown, e.g. in ref.[19].

2.2 The Axial Anomaly

The axial anomaly reminds us that a quantum field theory can be defined only as the limit of a theory regularized with an ultraviolet cut-off, Λ. If the regularized theory violates some symmetry present at the classical level, the symmetry itself may not be recovered when the cutoff is removed, and even be completely lost in the full quantum theory.

The typical example is the anomalous non-conservation of the axial current:

$$J^5{}_\mu = \bar{e}\gamma_\mu\,\gamma_5\,e$$

in QED, discovered by Adler[20]. In the Pauli-Villars regularization, one introduces one (or more) fictitious, heavy, negative-metric, fermion(s), E, to cut-off the divergences in the electron loops. The total axial current, associated to the chiral transformations is then:

$$(J^5{}_\mu)_{reg} = \bar{e}\gamma_\mu\,\gamma_5\,e + \bar{E}\gamma_\mu\,\gamma_5\,E$$

and:

$$(\partial^\mu J^5{}_\mu)_{reg} = 2m\,i\,(\bar{e}\,\gamma_5\,e) + 2M\,i\,(\bar{E}\,\gamma_5\,E) \tag{2.2.1}$$

where M is the mass of the Pauli-Villars fermions. In the limit $M \to \infty$, the operator $i\bar{E}\gamma_5 E$ decouples from the physical states:

$$<\alpha|i\,\bar{E}\,\gamma_5\,E|\beta> \to 0$$

However, since it is multiplied by a divergent quantity in eq.(2.2.1), the matrix element of the divergence of the axial current may receive a finite (anomalous) contribution and may not vanish, even in the limit of a massless electron.

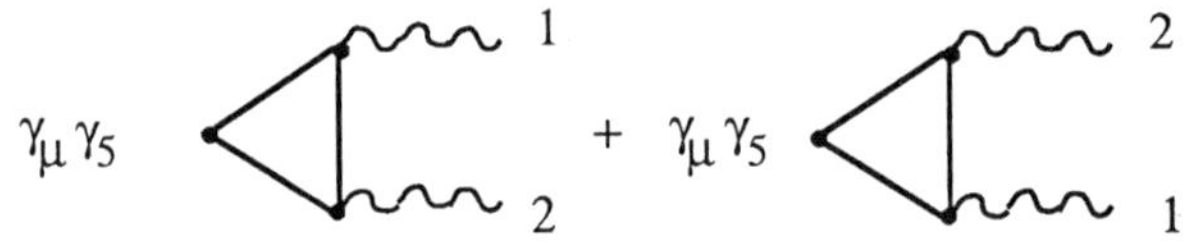

$\gamma_\mu\,\gamma_5$ 1   2 $+$ $\gamma_\mu\,\gamma_5$ 2   1

Fig. 2.4

Triangle diagrams giving rise to the abelian axial current anomaly.

This is precisely what happens in QED. Due to the triangle diagrams of Fig.2.4, the axial current is not conserved even for a massless electron, but it satisfies the anomalous equation:

$$\partial^\mu J^5{}_\mu = 2m\,i\,(\bar{e}\,\gamma_5\,e) + \frac{1}{16\pi^2}\,e^2 F_{\mu\nu}\,\tilde{F}^{\mu\nu} \tag{2.2.2}$$

with:

$$\tilde{F}^{\mu\nu} = \frac{1}{2}\,\varepsilon^{\mu\nu\rho\sigma}\,F_{\rho\sigma}$$

An analogous result is found in lattice regularized QCD. A theorem proved by Nielsen and Ninomiya[21] asserts that any lattice action with: i) local couplings, involving products of fields evaluated at finite distance from each other in units of the lattice spacing and ii) the correct number of fermion flavours, breaks necessarily the chiral symmetry at a finite value of the lattice spacing. As a consequence, the conservation equation for the quark axial current reads:

$$\nabla^\mu J^5_\mu = 2m\, i\, (\bar{q}\, \gamma_5\, q) + a\, X \tag{2.2.3}$$

were a denotes the lattice spacing and ∇^μ the discretized derivative. What happens now [ref.22] is that X has extra space derivatives, $X \propto (\bar{q}\gamma_5 \nabla^2 q)$, to compensate for the dimension of a. Its matrix elements between physical states have an ultraviolet linear divergence:

$$<\alpha|X|\beta> \propto \frac{1}{a}$$

and an anomalous contribution remains in the r.h.s. of eq.(2.2.3), in the continuum limit, $a\to 0$, so that:

$$\nabla^\mu J^5_\mu = 2m\, i\, (\bar{q}\, \gamma_5\, q) + \frac{(gs)^2}{64\pi^2}\, \mathrm{Tr}\, [G_{\mu\nu}\, \tilde{G}^{\mu\nu}] \tag{2.2.4}$$

where $G_{\mu\nu}$ is the gluon Yang-Mills tensor, see Sect. 1.1. Eq.(2.2.4) is the basis for the solution of the so-called axial U(1) problem[23].

Both in Pauli-Villars and lattice regularizations, vector currents are not anomalous (unlike chiral symmetry, a vector symmetry can be exact in the regularized theory). Correspondingly, gauge theories like QED or QCD, where gauge fields are coupled to[24] vector currents, are unaffected by these considerations.

On the other hand, in a chiral gauge theory, such as the Standard Electroweak Theory, the axial anomaly induces a non-conservation of the currents the gauge fields are coupled to[24] and gauge invariance is lost at the quantum level, see eq.(1.2.6).

The situation can be illustrated by a simple, abelian, example, quite representative of the general case.

We consider a theory with one fermion and one complex scalar fields, coupled to a U(1) gauge field. Only the left-handed component of the fermion (neutrino) field transforms under U(1). We take:

$$v_L' = e^{i\varepsilon(x)}\, v_L \qquad\qquad v_R' = v_R \tag{2.2.6}$$

$$\phi' = e^{i\varepsilon(x)}\, \phi \tag{2.2.7}$$

$$A_\mu' = A_\mu - \frac{1}{g}\partial_\mu\, \varepsilon \tag{2.2.8}$$

and the corresponding Lagrangian:

$$L = -\frac{1}{4g^2}\, F_{\mu\nu}F^{\mu\nu} + \bar{v}_L i \not{D} v_L + \bar{v}_R i \not{\partial}\, v_R + (\partial_\mu \phi)^\dagger(\partial^\mu \phi) -$$

$$\tag{2.2.9}$$

$$- V(\phi^\dagger\phi) - g_Y\, (\bar{v}_L\phi v_R + \bar{v}_R\phi^\dagger v_L)$$

When the symmetry is spontaneously broken, the vector and neutrino masses are:

$$M^2 = 2g^2\, \eta^2 \tag{2.2.10}$$

$$m = g_Y\, \eta \tag{2.2.11}$$

where η is the vacuum expectation value of ϕ. Following the discussion of Sect.1.3, the scalar field can be parametrized according to:

$$\phi = (\eta + \frac{\sigma}{\sqrt{2}}) \exp(i\frac{\xi}{\eta\sqrt{2}})$$

(2.2.12)

(σ is the Higgs boson and ξ the would-be-Goldstone boson). The total current, computed from eqs.(2.2.6) to (2.2.9), is:

$$J^\mu = j_f^\mu + j_s^\mu$$

$$j_f^\mu = \bar{\nu}\gamma_\mu\frac{(1-\gamma_5)}{2}\nu \equiv \bar{\nu}\gamma_\mu(g_V - g_A\gamma_5)\nu$$

$$j_s^\mu = i\,(\phi^\dagger D^\mu\phi - D^\mu\phi^\dagger\phi)$$

In the broken phase, j_s^μ is:

$$j_s^\mu = -\sqrt{2}\,\eta\partial_\mu\xi + \ldots$$

dots denoting terms of higher order in the fields.

At the tree level, the divergence of the total current is easily computed to be:

$$\partial_\mu J^\mu = -m\,i\,\bar{\nu}\gamma_5\nu - \sqrt{2}\,\eta(\partial^\mu\partial_\mu\xi) + \ldots =$$

(2.2.13)

$$= (-m + g_Y\,\eta)\,i\,\bar{\nu}\gamma_5\nu + \ldots$$

where we have used the equation of motion of ξ. Eqs.(2.2.13) and (2.2.11) show explicitly that J^μ is conserved also in the broken phase, in spite of the fact that the neutrino has become massive.

At the quantum level, current conservation is spoiled by the Adler-Bell-Jackiw anomaly, due to the same triangle diagram as in fig.2.4. The difference, with respect to QED, is that now the two gauge fields can also be emitted by axial currents. Following Bardeen[25], the anomalous conservation equation is, now:

$$\partial_\mu j_f^\mu = -m\,i\,\bar{\nu}\gamma_5\nu + g^2 C\,F_{\mu\nu}\tilde{F}^{\mu\nu}$$

(2.2.14)

$$C = \frac{1}{16\pi^2}\,g_A(\,g_V^2 + \frac{g_A^2}{3})$$

Fig. 2.5

In the presence of current non-conservation due to the axial anomaly, the unphysical pole in the amplitude of the first diagram is not cancelled by the amplitude of the second one, making the overall result gauge-dependent. Crossed diagrams are not shown.

The effect of the axial anomaly can be put into evidence by considering the annihilation amplitude of a neutrino pair into two vector bosons. Besides the usual pair–annihilation diagrams, which give rise to a gauge independent result, one must consider the diagrams of Fig.2.5. Using the propagators obtained in Sect.1.3, we can write schematically the amplitudes as follows:

$$M(\text{A-exchange}) = (-ig\, j_f^\mu\,) \frac{i}{q^2\text{-}M^2}\, [\, \text{-}g_{\mu\nu} + \frac{(1\text{-}\lambda)q_\mu q_\nu}{q^2\text{-}\lambda M^2}\,]\, (-ig\, <j_f^\nu>) \qquad (2.2.15)$$

$$M(\xi\text{-exchange}) = (-i\frac{g\gamma}{\sqrt{2}}\, j_5) \frac{i}{q^2\text{-}\lambda M^2}\, (-i\frac{g\gamma}{\sqrt{2}}\, <j_5>) \qquad (2.2.16)$$

where:

$$j_5 = i\, \bar{\nu}\gamma_5\nu$$

The first factor in the r.h. side of eqs.(2.2.15) and (2.2.16) represents the matrix element of the indicated current between the initial state and the vacuum, and the symbol $<\,>$ denotes the insertion of the current in the triangle diagram of Fig.2.5. Using the anomalous conservation equation (2.2.14), the A-exchange amplitude is rewritten as:

$$M(\text{A-exchange}) = (-ig\, j_f^\mu\,) \frac{-ig_{\mu\nu}}{q^2\text{-}M^2}\, (-ig\, <j_f^\nu>) +$$

$$\qquad (2.2.17)$$

$$+(j_5\,) \frac{i}{q^2\text{-}M^2}\, \frac{(1\text{-}\lambda)}{q^2\text{-}\lambda M^2}\, [g^2 m^2 <j_5> - g^2 m\, (g^2 C\, F_{\mu\nu}\, \tilde{F}^{\mu\nu})]$$

The second term in M(A-exchange) is λ-dependent, i.e. gauge-dependent. Its non-anomalous part, proportional to $<j_5>$, neatly combines with the ξ-exchange amplitude to yield a result which is λ (and gauge) independent, due to eqs.(2.2.10) and (2.2.11). The anomalous term, however, remains and it makes the full amplitude gauge-dependent, a meaningless result.

One may ask wether one could modify the action, adding to it appropriate terms, so as to restore gauge invariance. The answer is affirmative, in the spontaneously broken situation, as found by Wess and Zumino[26].

What is needed, is a furter interaction of the form:

$$L_{WZ} = g^2\frac{C}{\sqrt{2}\eta}\, \xi\, F_{\mu\nu}\, \tilde{F}^{\mu\nu} \qquad (2.2.18)$$

It is immediate to see that the corresponding addition to M(ξ-exchange) eliminates completely the λ-dependence of the full amplitude (we leave as an exercise to the reader to show that the final result coincides with the amplitude computed in the unitary gauge, $\lambda \to \infty$, without the Wess-Zumino term).

This result may be cast in more transparent form. Let us compute the gauge variation of the Wess-Zumino action. We find, from eq.(2.2.7):

$$\delta\xi(x) = \sqrt{2}\eta\, \varepsilon(x) \qquad (2.2.19)$$

so that:

$$\frac{\delta S_{WZ}}{\delta\varepsilon(x)} = g^2 C\, F_{\mu\nu}\, \tilde{F}^{\mu\nu}$$

On the other hand, for the action corresponding to our lagrangian, eq.(2.2.9), using eqs.(1.2.6) and (2.2.19), we see that:

$$\frac{\delta S}{\delta \varepsilon(x)} = - \partial_\nu J^\nu = -g^2 C \, F_{\mu\nu} \, F^{\mu\nu} \tag{2.2.20}$$

The total lagrangian:

$$L_{tot} = L + L_{WZ}$$

is therefore gauge invariant, so it is not surprising that it gives rise to a gauge invariant annhilation amplitude.

By power counting, the Wess-Zumino term is non-renormalizable. In a gauge theory with an anomalous fermion composition we are left with the choice between gauge variance and non-renormalizability. The first choice does not make sense, but the latter option is not to be excluded, a priori. A non-renormalizable interaction could be induced by fermions heavier than the mass scale we are considering. Turning the argument around, we can consider an anomaly free, renormalizable theory, and let to increase the mass of some fermion, M_Q, whose anomaly cancels the anomaly of the light fermions. The resulting low-energy effective theory will contain non-renormalizable interactions, precisely of the Wess-Zumino type, so as to be anomaly free[27]. Once we get to momentum scales larger than M_Q, of course, we "see" the elementary fermion and the original renormalizable theory comes back again.

We conclude by considering the formula for the axial anomaly in a general, non-abelian, theory. To this aim, we shift to a notation alternative, although equivalent, to the one used until now. We replace all right-handed fields, leptons and quarks, with the corresponding Charge-Conjugate left-handed fields, e.g. we replace the field e_R, which annihilates a right-handed electron and creates a left-handed positron, with the positron left-handed field, which annihilates a left-handed positron and creates a right-handed electron. In this way, the two fields e_R and $(e_C)_L$ are associated to the same set of physical states as e_L and e_R, with the advantage that all fermion fields have now the same chirality.

With this notation, there are only left-handed chiral currents which we denote by:

$$J^A_\mu = \bar{f}_L \, \gamma_\mu \, X^A \, f_L \tag{2.2.21}$$

with A running over the group generators and f_L a row vector including all fermion fields. Following again Bardeen's analysis, one sees that the anomaly is controlled by the three-index tensor:

$$D^{ABC} = \mathrm{Tr}(X^A \{X^B, X^C\}) \tag{2.2.22}$$

where the trace goes over all fermions.

In the Standard Theory, we can restrict to doublets and singlets under SU(2), and:

$$(2T_3)^2 = 1 \text{ , for SU(2) doublets}$$

$$= 0 \text{ , for singlets} \tag{2.2.23}$$

Furthermore, the identity:

$$\mathrm{Tr}(\tau^i \{\tau^j, \tau^k\}) = 0 \tag{2.2.24}$$

implies that the anomaly vanishes when all indices in ((2.2.22) belong to $SU(2)^1$.

To analyze the other cases, we replace the weak hypercharge, Y, with the electric charge, using the relation (1.4.9):

$$Q = T_3 + Y$$

We need to consider three cases.

i. Tr (Q^3). This quantity vanishes separately for leptons and quarks, since electrically charged particles are present always in conjugated pairs, e.g. e_L and e_{CL}.

ii. Tr $[(Q^2)T_3]$. We obtain:

$$\mathrm{Tr}\,[(Q^2)T_3] = \mathrm{Tr}\{[\,Y^2 + 2\,Y\,T_3 + (T_3)^2]\,T_3\} =$$

$$= 2\,\mathrm{Tr}[Y(T_3)^2] = \frac{1}{2}\,\mathrm{Tr}_{\mathrm{doub}}(Y) =$$

$$= \frac{1}{2}\,\mathrm{Tr}_{\mathrm{doub}}(Q)$$

where we have used (2.2.24), and $\mathrm{Tr}_{\mathrm{doub}}$ means that the trace is restricted to SU(2) doublets.

iii. Tr$[Q(T_3)^2]$. This is again proprtional to $\mathrm{Tr}_{\mathrm{doub}}(\,Q\,)$, because of eq.(2.2.24).

In conclusion, the presence of anomalies is determined by:

$$D = \mathrm{Tr}_{\mathrm{doub}}(\,Q\,) \tag{2.2.25}$$

For a single lepton or quark multiplet, e.g. the ν_e-e or the u-d doublets:

$$D_{\mathrm{lept}} = -1$$
$$D_{\mathrm{quark}} = 3\,(\frac{2}{3} - \frac{1}{3}) = +1$$

(the factor 3 comes from colour), so that each family of quarks and leptons has by itself a vanishing anomaly.

2.3 The breaking of scale invariance and the running coupling constant

Consider massless QCD with two flavours. The Lagrangian is:

$$L = \bar{q}i\,\slashed{D}q - \frac{1}{8}\,\mathrm{Tr}\,G_{\mu\nu}G^{\mu\nu} \tag{2.3.1}$$

(q=u, d), and it is believed to describe the strong interactions of non-strange hadrons, to a good approximation. The interaction is determined by the adimensional, bare coupling g_0, which appears in the covariant derivative and in the definition of $G_{\mu\nu}$. There are no dimensional parameters. As a consequence, the action is classically invariant under scale transformations:

$$x' = e^\lambda\,x$$

---

[1] In fact, this is true for any set of SU(2) multiplets, because any SU(2) representation, R, is equivalent to its complex conjugate, R*. From R=R* , it follows that $D^{ABC}(R) = D^{ABC}(R*)$. On the other hand, $D^{ABC}(R*) = -\,D^{ABC}(R)$ in general, since D contains an odd number of generators and therefore is odd under conjugation. Thus, $D^{ABC}(R) = 0$ for SU(2) and, in general, for any group with only real representations. In the usual terminology, SU(2) is anomaly free, albeit in a trivial way.

$$G_\mu(x')' = e^{-\lambda}\, G_\mu(x)$$

$$q(x')' = e^{-3/2\lambda}\, q(x) \tag{2.3.2}$$

$$S' = \int d^4x'\, L(q', G') = \int d^4x\, L(q, G) \tag{2.3.3}$$

The breaking of scale symmetry by quantum effects is another manifestation of the necessity of the ultraviolet cutoff, to regularize the theory before renormalization.

In fact, scale invariance of (2.3.1) is better be broken. Otherwise, we would not be able to account for any hadronic mass scale, e.g. the proton mass. An alternative possibility would be spontaneous scale symmetry breaking. However, this implies the existence of a massless Goldstone particle, the "dilaton", not observed in the hadron spectrum.

Let us suppose that we have indeed introduced an UV cut-off, for example by restricting the theory to a discrete lattice of points in 4-dimensional, Euclidean, space-time, with the lattice spacing:

$$a \sim \frac{1}{\Lambda}$$

The renormalized theory is obtained in the limit $\Lambda \to \infty$, with fixed physical quantities. For example, we may held fixed the renormalized coupling, g_R, defined as the value of the $\bar{q}qG$ vertex, at some value of the external momenta, μ. We cannot define g_R at $\mu=0$, because the massless theory is infrared-divergent there.

The presence of μ breaks scale invariance, even though, the value chosen for μ being entirely arbitrary, the observable quantities cannot really depend from it. Consider, for example, the cross-section ratio:

$$R(s) = \frac{\sigma(e^+e^- \to \text{hadrons})}{\sigma(e^+e^- \to \mu^+\mu^-)} \tag{2.3.4}$$

with s the center of mass energy squared. Since R is adimensional, it should not depend from s, in an exactly scale invariant theory. However, now we have μ at our disposal, so that:

$$R = f(\frac{s}{\mu^2}, g_R) \tag{2.3.5}$$

The fact that μ is arbitrary means that we must be able to compensate a change of μ with a change of g_R, so as to leave R invariant. In formulae, we obtain the renormalization group equation:

$$[\mu^2 \frac{\partial}{\partial\mu^2} + \beta(g_R) \frac{\partial}{\partial g_R}]\, f(\frac{s}{\mu^2}, g_R) = 0 \tag{2.3.6}$$

Note that the beta-function, $\beta(g_R)$, can depend only upon μ and g_R. Since it is adimensional, it is a function of g_R only.

The solution of (2.3.6) is well-known. We introduce the running coupling constant, $g(t; g_R)$, defined as the solution of the equation:

$$\frac{dg(t)}{dt} = \beta(g(t))$$

$$t = \ln(\frac{s}{\mu^2}) \tag{2.3.7}$$

with the boundary condition:

$$g(0) = g_R \tag{2.3.8}$$

Integration of (2.3.7) yields the equation:

$$\int_{g_R}^{g} \frac{dg'}{\beta(g')} = t \qquad (2.3.9)$$

form which we see that:

$$\frac{\partial g(t;g_R)}{\partial g_R} = \frac{\beta(g)}{\beta(g_R)}$$

The general solution of (2.3.6) is, then:

$$R(s) = f(\frac{s}{\mu^2}, g_R) = w[g(t;g_R)] \qquad (2.3.10)$$

The scale breaking implied by the existence of μ allows R to depend from s. However, this dependence may occur only through the running coupling, $g(t)$. Explicitly, in QCD:

$$R(s) = \Sigma \, Q^2 \, [1 + \frac{\alpha_S(t)}{\pi} + ...]$$

$$\alpha_S = \frac{g^2}{4\pi}$$

where dots indicate higher order terms in α_S.
In perturbation theory, with a finite cut-off Λ:

$$g_R = g_0 + \frac{b}{2}(g_0)^3 \ln(\frac{\Lambda^2}{\mu^2}) + \text{higher order terms} \qquad (2.3.11)$$

with b a calculable constant. Inverting the equation to lowest order we find:

$$g_0 = g_R - \frac{b}{2}(g_R)^3 \ln(\frac{\Lambda^2}{\mu^2}) + \text{higher order terms} \qquad (2.3.12)$$

We may find the beta-function by determining the variation of g_R needed to compensate a variation of μ^2 so as to leave g_0 invariant:

$$\beta(g_R) = -\mu^2 \frac{\partial}{\partial\mu^2} g_0 = - \frac{b}{2}(g_R)^3 \qquad (2.3.13)$$

The running copling thus obeys:

$$\frac{\partial g^2}{\partial t} = 2g \, \beta(g) = - bg^4$$

that is:

$$\frac{\partial}{\partial t} [\frac{1}{g(t)^2}] = b$$

which yields, finally:

$$\frac{1}{g(t)^2} = \frac{1}{g(t_0)^2} + b \ln(\frac{s}{s_0}) \tag{2.3.14}$$

for any value of $t = \ln(s/\mu^2)$ and $t_0 = \ln(s_0/\mu^2)$.

There are two different cases, according to the sign of b.
i. b>0. This is the case of non-abelian gauge theories, with not too many fermions, such as QCD, see below. We introduce a new quantity, Λ_{QCD} according to:

$$\frac{1}{g(t_0)^2} = b \ln(\frac{s_0}{\Lambda^2_{QCD}}) \tag{2.3.15}$$

Notice that, as shown by (2.3.15):

$$\frac{1}{g(t_0)^2} - b\ln(s_0) = \text{independent from } s_0$$

so Λ_{QCD} is independent from the particular value of s_0 we have chosen. Eq.(2.3.14) becomes now:

$$g(t)^2 = \frac{1}{b\ln(\frac{s}{\Lambda^2_{QCD}})} \tag{2.3.16}$$

The theory is asymptotically free[28], since g vanishes for $s\to\infty$. The dimensional parameter, Λ_{QCD}, has replaced the adimensional bare coupling constant, g_0. This "dimensional transmutation"[5] shows very clearly that scale invariance is broken and makes it possible for the Lagrangian (2.3.1) to give rise to a non vanishing proton mass: $M_P \propto \Lambda_{QCD}$, see next Section.
ii. b<0. This case applies to abelian gauge theories, such as QED, or to non-abelian gauge theories with a sufficently large number of matter (spin 0 or 1/2) multiplets. We have now:

$$\frac{1}{g(t)^2} = \frac{1}{g(t_0)^2} - |b| \ln(\frac{s}{s_0}) \tag{2.3.17}$$

and $g(t) > g(t_0)$ for $t > t_0$. At large energy, the interaction becomes inevitably strong. More about this situation in the next Section.

We close by giving the lowest order beta-function for the SU(N) gauge theory.
Fermions are described by left-handed fields (right-handed fermion fields are replaced by the corresponding left-handed antifermion fields, as in Sect. 2.2) and we consider the case of n multiplets in the fundamental, N, representation, and $\bar{n}$ in the complex conjugate one, $\bar{N}$. The gauge coupling g_N, for N>1, is defined from:

$$L_{int} = (g_N)\, \bar{\psi}_L\, \gamma_\mu\, \frac{\lambda^A}{2} \psi_L\, A^A_\mu \tag{2.3.18}$$

$$\text{Tr}(\lambda^A\lambda^B) = 2\,\delta^{AB}$$

while:

$$L_{int} = g_1 \bar{\psi}_L\, \gamma_\mu\, Y\psi_L\, A_\mu \tag{2.3.19}$$

for N=1. It is conventional to define:

$$\alpha_N = \frac{g_N^2}{4\pi}$$

$$\frac{\partial \alpha_N}{\partial t} = - b \, (\alpha_N)^2$$

then, one finds:

$$b = \frac{1}{3\pi} \left[\frac{11}{4} N - \frac{1}{4} (n + \bar{n}) \right] \qquad (N>1) \qquad\qquad (2.3.20)$$

$$b = \frac{1}{12\pi} \Sigma(-Y^2) \qquad\qquad (N=1) \qquad\qquad (2.3.21)$$

A scalar particle gives a contribution equal to 1/2 times that of a left-handed fermion of the same charge.

With these results at hand, one easily computes the beta functions of SU(3), SU(2) and $U(1)_Q$. With N_g generations of quark and leptons and one Higgs doublet (in parenthesis, we give separately the contribution of the fermions, Higgs and vector fields):

$$b_3 = \frac{1}{3\pi} \left[-N_g + 0 + \frac{33}{4} \right] \qquad\qquad (2.3.22)$$

$$b_2 = \frac{1}{3\pi} \left[-N_g - \frac{1}{8} + \frac{11}{2} \right] \qquad\qquad (2.3.23)$$

$$b_Q = \frac{1}{3\pi} \left[-\frac{8}{3} N_g - \frac{1}{4} + \frac{11}{2} \right] \qquad\qquad (2.3.24)$$

2.4 Critical value of g_0 and the removal of the UV cut-off

We consider here the relation between the bare coupling, g_0, and the UV cu-off, Λ, starting from the asymptotically free case.

According to eq.(2.3.12), we may regard g_0 as being approximately equal to the running coupling for $s \sim \Lambda^2$:

$$g_0^2 = \frac{1}{b \ln\left(\dfrac{\Lambda^2}{\Lambda^2_{QCD}}\right)} \qquad\qquad (2.4.1)$$

so that:

$$\Lambda^2 = \Lambda^2_{QCD} \, \exp\left(\frac{1}{b g_0^2}\right) \qquad\qquad (2.4.2)$$

This result is another manifestation of the asymptotic freedom: it shows that the cutoff is removed to infinity if we send g_0 to zero.

In lattice regularization, we start from the bare lagrangian (2.3.1), appropriately discretized[29] and expressed in term of the bare coupling, g_0. Furthermore, in numerical simulations, we obtain all quantities in units of the lattice spacing, a, i.e. pure numbers, depending from g_0 only. According to eq.(2.4.2), we are closer to the continuum limit the smaller g_0 is. Calling $m_{LATT}(g_0)$ any hadronic mass computed in the lattice, we write:

$$m_{LATT}(g_0) = m_{PHYS} \, a = \frac{m_{PHYS}}{\Lambda} = \frac{m_{PHYS}}{\Lambda_{QCD}} \exp\left(\frac{-1}{2 b g_0^2}\right) \qquad\qquad (2.4.3)$$

where m_{PHYS} is the physical mass.

Eq.(2.4.3) gives the scaling law obeyed by any lattice mass for sufficiently small g_0. Its approximate validity is the sign that we are close to the continuum limit in our calculation. For QCD with three colours, the scaling law (2.4.3) begins to be obeyed for the smallest values of g_0 which can be reached in the present calculations[30].

According to eq.(2.4.3), lattice masses vanish (and correlation lenghts diverge) when $g_0 \to 0$. The continuum limit is like a phase transition, $g_0=0$ being the critical value of the coupling constant, for asymptotically free lattice gauge theory.

The situation is completely different for non-asymptotically free theories, $b<0$. In this case, eq.(2.3.14) allows us to define an upper cut-off, Λ_{max}:

$$\frac{1}{g(t)^2} = |b|\ln(\frac{\Lambda^2_{max}}{s})$$

(2.4.4)

(Λ_{max} is again independent from the particular value of s we have chosen). Now:

$$\frac{1}{g_0^2} = |b|\ln(\frac{\Lambda^2_{max}}{\Lambda^2})$$

(2.4.5)

and, provided $g_0^2>0$, Λ cannot exceed Λ_{max}. Moreover, taking $s=\mu^2$, $g(0)=g_R$ in eq.(2.4.3), we see at once that:

$$\Lambda < \Lambda_{max} = \mu \exp(\frac{1}{2|b|g_R})$$

(2.4.6)

We can send the UV cutoff to infinity only in the free theory, $g_R=0$, or, equivalently, the continuum theory is free (trivial).

This phenomenon was discovered by Landau and collaborators in the fifties[31].

The argument presented here can be criticized on various grounds, all having to do with the fact that we have extrapolated the one-loop approximation, eq.(2.3.14), beyond its region of validity. In fact, the approximation in which we keep only the one-loop contributions to the beta-function in (2.3.13) corresponds, for the running coupling, to a resummation of all the leading terms, of the order:

$$(g_0^2)^n \, [\ln(s/\mu^2)]^n$$

(2.4.7)

and to neglect the next-to-leading ones, of order:

$$(g_0^2)^n \, [\ln(s/\mu^2)]^{n-1}$$

(2.4.8)

This is fine, even when the logarithm is so large that $(g_0^2)\ln(s/\mu^2)$ becomes of order unity, provided g_0 remains small. However, when Λ approaches Λ_{max} we see at once from (2.4.5) that g_0 does become of order unity and the non leading terms we have neglected can compete with the leading ones.

In spite of that, Landau's argument is supported by recent non-perturbative results found in the analogous case of the $\lambda\phi^4$ theory, see Sect.4, which indicate that non asymptotically free theories are indeed trivial in the continuum limit.

2.5. The gauge couplings of the Standard Theory

We summarize here the present knowledge about the gauge couplings of the Standard Theory.

Fig.2.6 gives a summary of the determinations of α_3, the SU(3) colour gauge coupling[32]. The suggested values of Λ_{QCD}, with 4 or 5 flavours ($\overline{MS}$-scheme) are:

$$\Lambda^{(4)}_{QCD} = (220\pm90)\text{MeV}$$

(2.5.1)

$$\Lambda^{(5)}_{QCD} = (140\pm60)\text{MeV}$$

(2.5.2)

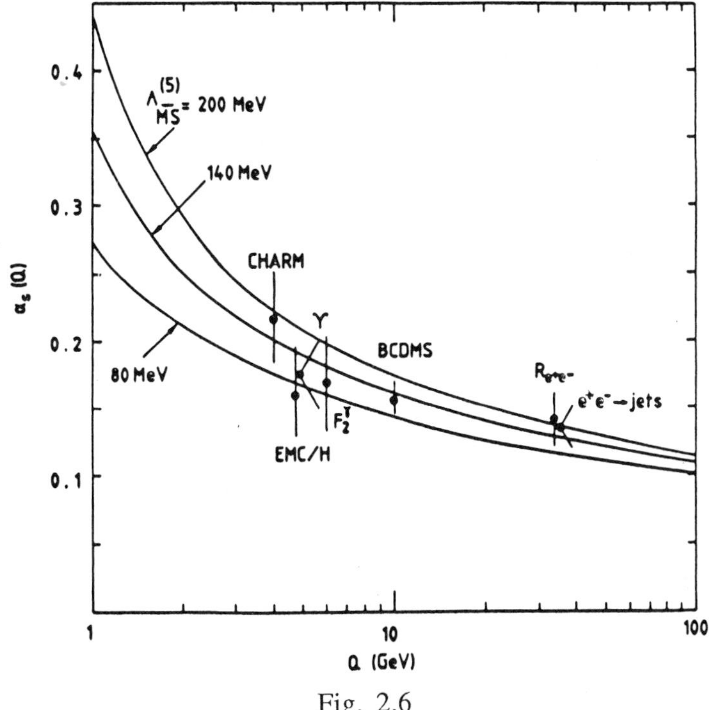

Fig. 2.6

The QCD running coupling constant, α_3, vs. $Q = \sqrt{q^2}$, determined from various processes. The figure is taken from ref.[32].

The corresponding value of α_3 , taking as reference point the Z mass, is:

$$\alpha_3(M_Z)= 0.12^{+0.01}_{-0.02} \tag{2.5.3}$$

This value has been confirmed by the LEP data on $Z\to$ jets[33]. The values quoted by the four collaborations are:

$0.127^{+0.006}_{-0.017}$	ALEPH
0.114 ± 0.012	DELPHI
0.125 ± 0.012	L3
0.116 ± 0.016	OPAL
0.120 ± 0.013	combined

To describe the electroweak couplings, we may use the fine structure function:

$$\frac{1}{\alpha} = 137.036$$

(which has a negligible error) and $\sin^2\theta$, defined according to eq.(1.4.16), namely:

$$\sin^2\theta = 1 - \frac{M_W^2}{M_Z^2} \qquad (2.5.4)$$

The SU(2)xU(1) gauge couplings are related to α and $\sin^2\theta$ according to:

$$\alpha_2 = \frac{\alpha}{\sin^2\theta}$$

$$\alpha_1 = \frac{\alpha}{\cos^2\theta}$$

We take again as reference point the Z mass, where:

$$\alpha(M_Z) = \frac{\alpha}{1-\epsilon_\alpha} \qquad (2.5.5)$$

with ϵ_α obtained from the total e^+e^- annihilation cross section:

$$\epsilon_\alpha = (6.01 \pm 0.04)\ 10^{-2}$$

$$\alpha(M_Z)^{-1} = 128.80 \pm 0.05 \qquad (2.5.6)$$

$\sin^2\theta$ can be taken from neutrino-Nucleon cross-sections or from the W-Z mass ratio, measured at $P\text{-}\bar{P}$ colliders. Both determinations are insensitive to the t-quark mass (which would appear through the higher order corrections, see Sect.3), the first because of an accidental cancellation, the second by its very definition. The most precise neutrino data come from deep-inelastic scattering on isoscalar nuclear targets. Langacker[34] quotes:

$$(\sin^2\theta)_{\nu\text{-}N} = 0.233 \pm 0.003 \pm 0.005 \qquad (2.5.7)$$

(the first error is experimental and the second theoretical). The UA2 and CDF determinations of the W/Z mass ratios[35], see Table 3.1, give:

$$(\sin^2\theta)_{W\text{-}Z} = 0.225 \pm 0.007 \quad . \qquad (2.5.8)$$

With a weighted average:

$$(\sin^2\theta)_{ave} = 0.229 \pm 0.005 \qquad (2.5.9)$$

Correspondingly, we obtain:

$$\alpha_2(M_Z)^{-1} = 29.5 \pm 0.6$$

$$\alpha_1(M_Z)^{-1} = 99.3 \pm 0.6$$

errors in α_1 and α_2 are, of course correlated, since the combination:

$$\frac{1}{\alpha_1} + \frac{1}{\alpha_2} = \frac{1}{\alpha}$$

has practically a vanishing uncertainty.

Figure 2.7 shows the running of the gauge couplings α_3, α_2, and $8/3\ \alpha$, from M_Z upward. The factor 8/3 is the appropriate normalization factor for the three couplings to coincide when SU(3)xSU(2)xU(1) is merged in the Grand Unified scheme of SU(5)[36]. We assume three generations, t-quark included, one Higgs doublet and no threshold for new physics between M_Z and the GUT scale.

There is a clear tendency for the three couplings to meet, but not quite exactly. The electromagnetic and strong couplings meet at:

$$M_{GUT} = (1\pm0.8)\ 10^{15} GeV \qquad\qquad (2.5.10)$$

a value which implies a rather large proton lifetime so as not to be in too flagrant conflict with the present bounds. From (2.5.10) one predicts [ref.37]:

$$\Gamma(P\rightarrow\pi^0\ e^+)^{-1} \sim 5\ 10^{32}\ sec \qquad\qquad (2.5.11)$$

with a large error, while, experimentally [ref.38]:

$$\Gamma(P\rightarrow\pi^0\ e^+)^{-1} > 6\ 10^{32}\ sec \qquad\qquad (2.5.12)$$

However, the fact that α_2 does not meet the other couplings at the same mass value makes so that extrapolating back from the crossing point of the two couplings at M_{GUT}, one predicts a much too large value for α_2 and therefore a much too small value of $\sin^2\theta$ at the Z mass scale:

$$(\sin^2\theta)_{GUT}=0.206\pm0.01 \qquad\qquad (2.5.13)$$

The same conclusion is found in reff.[34,39], where it is pointed out that a supersymmetric GUT would work much better, the extrapolation with three supersymmetric families, with a SUSY threshold at 10 TeV yielding a value of $\sin^2\theta$ in good agreement with the experimental value in (2.5.9).

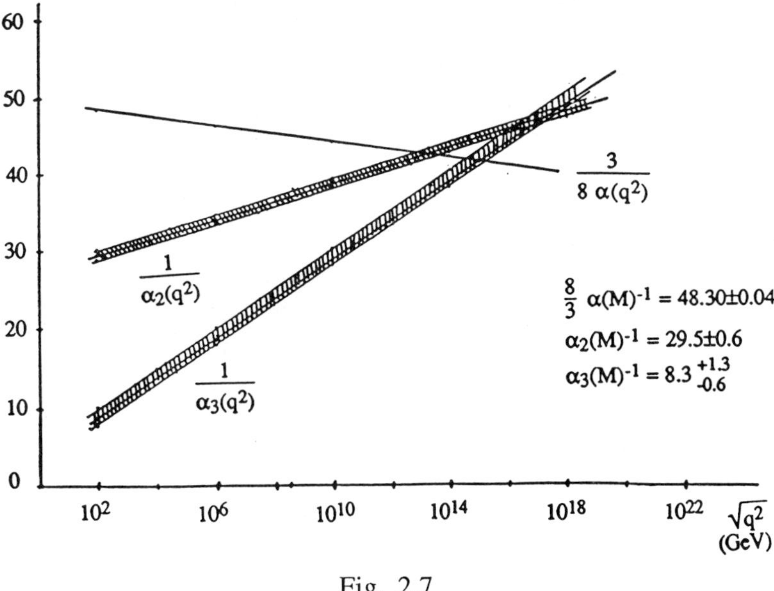

Fig. 2.7

Extrapolation of the $SU(3)_{color}xS(2)xU(1)$ gauge couplings to large momentum scales. The figure shows $[\alpha_3]^{-1}$, $[\alpha_2]^{-1}$ and $[8\alpha/3]^{-1}$ vs. $Q=\sqrt{q^2}$. Three fermion generations, with a t-quark mass of the order of M_Z, and one Higgs doublet are assumed. The normalization is chosen so that the three lines should meet at $Q = M_{GUT}$, in the simplest SU(5) Grand-Unified theory.

3. Z-PHYSICS

3.1 Renormalization of the Standard Theory

As we have seen in Sect.1.4, there are two independent gauge couplings for SU(2)xU(1), g and g'. To specify completely the theory, we must add the vacuum value of the Higgs field, η, which gives the mass-scale of the weak interactions, the Yukawa couplings of fermions to the Higgs field and the Higgs self-coupling.

We can replace some of the above quantities with equivalent, more convenient ones. For instance, Yukawa couplings can be replaced by the fermion mass and by the Cabibbo-Kobayashi-Maskawa mixing angles, and the Higgs self-coupling by the Higgs boson mass. Similarly, η can be replaced by the very well measured μ-decay Fermi constant, G_F, and one combination of g^2 and g'^2 replaced by the fine-structure constant, α. There remains a "last electroweak parameter", which we take, for the moment, to be determined by the ratio in (1.4.13), namely:

$$\frac{g'}{g} = \tan\theta$$

At the tree-level, the observable quantities are expressed as functions of these constants. Assuming all couplings but $\sin\theta$ to be known, we can determine the latter from the physical value of one observable, and then predict the value of all the others. Alternatively, we may obtain one value of $\sin\theta$ from each different observable. A test of the theory consists in showing that all such values of $\sin\theta$ coincide with each other, within errors. This is the plot frequently shown at conferences, with values of $\sin\theta$ from various sources reported on top of each other, to show how well they agree.

When higher order corrections[40] are included, the picture becomes more complicated.

First of all, we must distinguish between "bare" and "renormalized" couplings. We denote with c_0 and c the generic bare and renormalized coupling and write:

$$c = c_0 + \delta c$$

Introducing a finite UV cut-off, Λ, we compute the physical quantities as functions of the bare couplings and of the cut-off. Since the theory is renormalizable, finite expressions are obtained when observables are expressed in terms of the renormalized couplings, in the limit $\Lambda \to \infty$.

For instance, we may consider the one-loop corrected W mass[41-43]:

$$M_W^2 = M_{0W}^2 + \delta M_W^2$$

with:

$$M_{0W}^2 = \frac{\pi\alpha_0}{\sqrt{2}G_{F0}\sin^2\theta_0} \tag{3.1.1}$$

Replacing bare with renormalized couplings everywhere, we obtain ($s^2 \equiv \sin^2\theta$):

$$M_W^2 = \frac{\pi\alpha}{\sqrt{2}G_F\sin^2\theta}\,[1-$$

$$-\frac{\delta\alpha}{\alpha} + \frac{\delta G_F}{G_F} + \frac{\delta s^2}{s^2} + \frac{\delta M_W^2}{M_W^2}] \equiv \frac{(37.281\text{ GeV})^2}{\sin^2\theta}\,(1+\Delta r) \tag{3.1.2}$$

Each term in Δr is UV-divergent, but Δr itself is a finite quantity, well defined once the values of the renormalized couplings are specified in term of physical quantities.

Equations analogous to (3.1.2) apply to any other observable, X:

$$X = X_{tree}(1 + \Delta r_X)$$

If we use the tree-level expression to extract a value of $\sin^2\theta$ from X, different values of $\sin^2\theta$ would result from different observables, to one-loop accuracy. However, we may select one particular observable, $\bar{X}$, to yield a particular definition of $\sin^2\theta$ which we denote $(\sin^2\theta)_{\bar{X}}$, from:

$$\bar{X} = \bar{X}_{tree}$$

Unlike $\bar{X}$, all the other observables will have a non vanishing correction, in order for the same value, $(\sin^2\theta)_{\bar{X}}$, to be obtained from the value of the generic observable.

Of course, the choice of $\bar{X}$ is arbitrary. For each such choice, we have a different form of the one-loop correction Δr_X. What is invariant, and represents the physical prediction of the theory are the relations between any two different observables which are obtained after elimination of $(\sin^2\theta)_{\bar{X}}$.

The choice of a particular definition of $\sin^2\theta$ is thus inevitable, to make explicit calculations possible, and irrelevant, at least in principle, like the choice of a coordinate system.

In practice, it is of considerable importance to choose a definition of $\sin^2\theta$ which depends from precisely measured observables, and is theoretically clean. One such definition[43], adopted in the following, is given by eq.(1.4.16), namely:

$$\sin^2\theta = 1 - \frac{M_W^2}{M_Z^2} \tag{3.1.3}$$

We shall see below, eq.(3.2.3), another definition, more convenient in the case of a heavy t-quark.

Just to see how all this works in a particular case, we consider again the W mass, eq.(3.1.2). Since eq.(3.1.3) is obeyed by both bare and renormalized quantities, we obtain:

$$\frac{\delta s^2}{s^2} = -\frac{c^2}{s^2}\left[\frac{\delta M_W^2}{M_W^2} - \frac{\delta M_Z^2}{M_Z^2}\right] = -\frac{c^2}{s^2}\frac{\delta M_W^2 - c^2\,\delta M_Z^2}{M_W^2} \tag{3.1.4}$$

and eq.(3.1.2) becomes:

$$M_W^2 = \frac{\pi\alpha}{\sqrt{2}G_F\sin^2\theta}\,[1- \tag{3.1.5}$$

$$-\frac{\delta\alpha}{\alpha} + \frac{\delta G_F}{G_F} + \frac{\delta M_W^2}{M_W^2} - \frac{c^2}{s^2}\frac{\delta M_W^2 - c^2\,\delta M_Z^2}{M_W^2}\,]$$

The terms in the r.h.s. of this equation are determined by the one-loop corrections to the defining amplitude of the various quantities, e.g. the muon-decay amplitude for δG_F, the W and Z propagators for δM_W^2 or δM_Z^2, etc.. As we said before, the sum turns out to be finite. Repeating the same argument for the Z mass, the reader may easily verify that:

$$M_Z^2 = \frac{\pi\alpha}{\sqrt{2}G_F\sin^2\theta\cos^2\theta}\,(1+\Delta r) \tag{3.1.6}$$

with the same Δr as in eq.(3.1.3), in agreement with the definition (3.1.5).

Higher order corrections depend not only on $\sin^2\theta$, but also on the unknown t-quark and Higgs boson masses. Therefore, the value we obtain for $\sin^2\theta$ from a given observable depends, to some extent, on these quantities. The dependence from the Higgs boson mass turns out to be only logarithmic[44], but the dependence from m_t can be appreciable. The condition that the same value of $\sin^2\theta$ results from two different quantities can be used to obtain limitations on the value of m_t.

3.2 The approximate pattern of corrections

It is quite simple to identify the largest higher order corrections to electroweak amplitudes.

i. Large logs. The couplings α and G_F are both defined at an energy scale much smaller than the vector boson mass-scale, where $\sin^2\theta$ is defined. This mismatch makes so that quantities defined at the W or Z mass scale, e.g. the Z mass and width themselves, suffer large logarithmic corrections, in one and higher loop, of the order of:

$$(\alpha \ln\frac{M^2}{m_e^2})^n \; , \; (\alpha \ln\frac{M^2}{m_\mu^2})^n \qquad (3.2.1)$$

$(M\sim M_W\sim M_Z)$. These large logs can be estimated and resummed using renormalization group arguments[45]. The result is summarized by three simple rules:

a. α runs. The vacuum polarization due to light fermion loops makes so that α is replaced by the running constant introduced in Sect.2.5.

b. Amplitudes involving quarks have, of course, large logarithmic corrections determined by the running strong coupling constant, $\alpha_3(M)$.

c. G_F does not run. This result is a very simple consequence of the old result that μ-decay has UV-finite electromagnetic corrections. The argument goes as follows.

The difference between the effective G_F defined at the W mass-scale or at the muon mass-scale can be non-vanishing, to leading logarithmic order, because of the virtual photon-exchange corrections, the photon being the only vector particle coupled to leptons which survives below M_W. For the latter corrections, the W mass acts as an effective UV-cut-off so that we can determine corrections of the form (3.2.1) from the logarithmic divergences in the e.m. corrections to the local Fermi effective interaction.

The effective Fermi interaction mantains the V-A form when we bring it, with a Fierz rearrangement, in the so-called charge-retention form:

$$L_{eff} = \frac{G_F}{\sqrt{2}} \, \bar{e}\gamma_\lambda \, (1-\gamma_5)v_e \; \bar{v}_\mu\gamma^\lambda \, (1-\gamma_5)\mu =$$

$$= \frac{G_F}{\sqrt{2}} \, \bar{e}\gamma_\lambda \, (1-\gamma_5)\mu \; \bar{v}_\mu\gamma^\lambda \, (1-\gamma_5)v_e$$

In the latter form, photon-exchange corrections appear as a renormalization of the vector and axial $\mu \to e$ transition currents. By a well-known theorem[46], vector and axial currents suffer at most finite, i.e. cutoff independent, corrections so that, in conclusion, corrections of the form (3.2.1) between the Fermi couplings at the muon or at the W mass scale are absent.

ii. Large m_t corrections, I. The value of m_t affects the amplitudes involving light quarks and/or vector bosons only through vacuum polarization diagrams (with the exception of the $Z \to b\bar{b}$ amplitude which we consider next). For large values of m_t, the relevant parameter is[44]:

$$\delta\rho = \frac{3G_Fm_t^2}{8\sqrt{2}\pi^2} \sim 0.3 \; 10^{-2} \; (\frac{m_t}{100\,\text{GeV}})^2 \qquad (3.2.2)$$

We describe in more detail this effect in Appendix. In part, the correction can be reabsorbed into the definition of an effective $\sin^2\theta$[47]:

$$(\sin^2\theta)' \equiv s'^2 = s^2 + \delta\rho\ c^2$$

$$\hspace{2cm}(3.2.3)$$

$$c'^2 = (1-\delta\rho)\ c^2$$

Secondly, after amplitudes are expressed in terms of $(\sin^2\theta)'$, the large m_t effects give rise to a ρ-parameter different from unity:

$$\rho = 1+\delta\rho \hspace{2cm}(3.2.4)$$

that is they produce a different normalization of neutral versus charged current processes (as if there where Higgs scalar multiplets with weak isospin larger than 1/2).

ii. Large m_t corrections, II. Amplitudes involving external b quarks[48] have an additonal m_t dependence, since the t-quark appears also in vertex corrections. We consider the interesting case of the vertex corrections to the $Z\to b\bar{b}$ amplitude.

In the renormalizable Feynman-'t-Hooft gauge considered earlier, dominant corrections in the $m_t\to\infty$ limit are due to exchange of the would-be-Goldstone fields associated with the charged vector boson. The Yukawa coupling in the t-b-ξ vertex being proportional to m_t, the diagrams in Fig.3.1 give a contribution which is explicitly proportional to $\delta\rho$.

Individual terms in Fig.3.1 are logarithmically divergent, but the sum is finite, due to the same non-renormalization theorem of the vector and axial vector currents mentioned

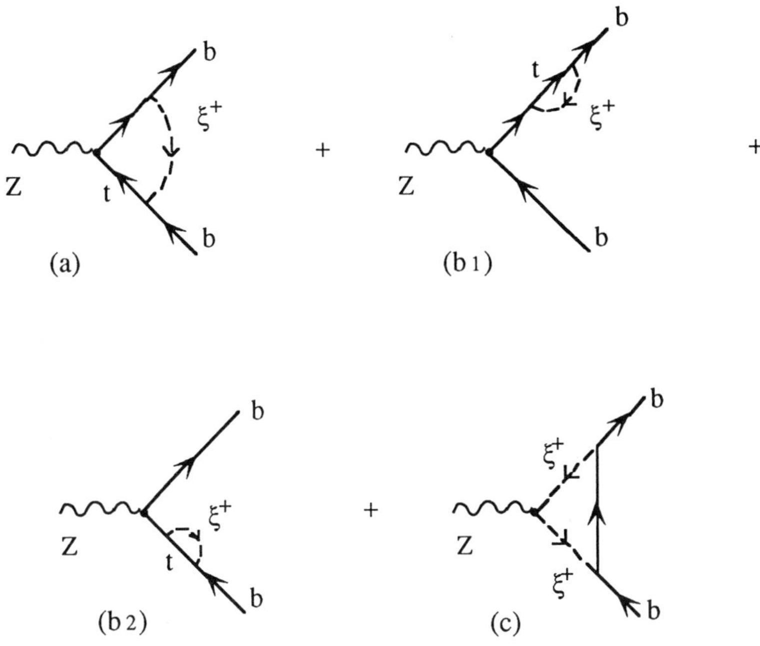

Fig. 3.1

Feynman diagrams for the leading correction to $Z\to b\bar{b}$, in the limit $m_t \to \infty$.

before[46]. We define the vector and axial vector couplings of the Z to any fermion f according to:

$$L_{eff} = \frac{g}{\cos\theta} Z^\mu \, \bar{f} \, \gamma_\mu \, [\frac{\tau_3}{2} \frac{(1-\gamma_5)}{2} - \sin^2\theta \, Q]f =$$

$$= \frac{g}{\cos\theta} Z^\mu \, \bar{f} \, \gamma_\mu \, [v_f(s) - a_f \gamma_5]f \tag{3.2.5}$$

Explicit calculation (see Appendix A) gives the result:

$$\delta v_b = -\delta a_b = \frac{1}{6} \delta\rho \tag{3.2.6}$$

The approximation in which we use tree-level amplitudes expressed in terms of the running α and of the effective $\sin^2\theta$ and ρ given by eqs.(3.2.3) and (3.2.4) is often called the "improved Born approximation". It reproduces adequately the electroweak higher order corrections up to an overall precision in the order of percent. For better accuracy, of few parts in 10^{-3}, the full one-loop corrections are needed.

As a first explicit example, we consider the W and Z mass again. The improved tree-level formulae are:

$$M_W^2 = \frac{\pi\alpha(M)}{\sqrt{2}G_F s'^2} \tag{3.2.7}$$

$$M_Z^2 = \frac{M_W^2}{\rho c'^2} = \frac{\pi\alpha(M)}{\sqrt{2}\rho G_F s'^2 c'^2} \tag{3.2.8}$$

with $M \sim M_W \sim M_Z$ and $\alpha(M)$ given by eq.(2.5.5).

Using eq.(3.2.3), we may, of course, come back to the previous definition of $\sin^2\theta$, bringing the masses into the to the "canonical" form (3.1.2) and (3.1.6), with:

$$\Delta r = \varepsilon_\alpha - \frac{c^2}{s^2} \delta\rho \tag{3.2.9}$$

Note that $\delta\rho$ works against the running α correction. As seen from eqs.(2.5.7) and (3.2.2), the two effects about cancel for $m_t \sim 250$ GeV.

The electromagnetic corrections embodied by ε_α are resummed to all orders if we bring Δr in the denominator of eqs.(3.1.2) and (3.1.6):

$$1 + \Delta r \rightarrow \frac{1}{1 - \Delta r} \tag{3.2.10}$$

The substitution (3.2.10) corresponds also to a resummation of the large-m_t corrections, but only a partial one, see ref.[49].

The other relevant example is the decay rate of the Z into a given fermion-antifermion pair. In tree-level, we have:

$$\Gamma(Z \rightarrow f \bar{f})_{tree} = \frac{1}{3} \frac{\alpha}{s^2 c^2} M_Z \, [v_f(s)^2 + a_f^2] \, N_f^0 =$$

$$\tag{3.2.11}$$

$$= \frac{\sqrt{2}}{3\pi} G_F M_Z^3 \, [v_f(s)^2 + a_f^2] \, N_f^0$$

($N_f^0 = 1$, 3 for leptons or quarks, respectively). In the improved approximation we find:

$$\Gamma(Z \to f\,\bar{f}) = \frac{\sqrt{2}}{3\pi} \frac{\alpha(M)}{\alpha} G_F M_Z^3 \rho \,[v_f(s')^2 + a_f^2]\, N_f \qquad (3.2.12)$$

with:

$$N_f = 1 \text{ (leptons)}; \qquad N_f = 3(1 + \frac{\alpha_S(M)}{\pi}) \text{ (quarks)}$$

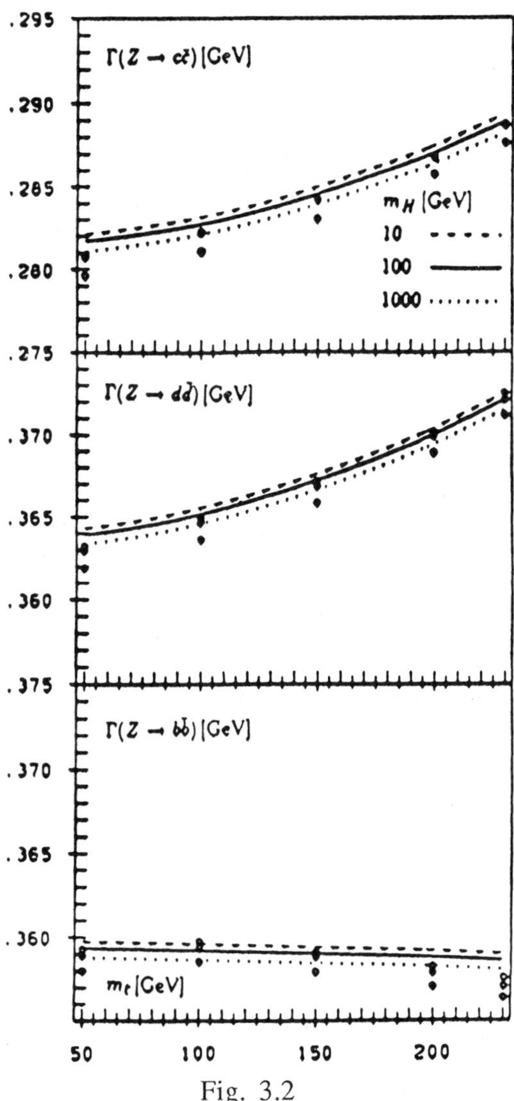

Fig. 3.2

One-loop corrected partial widths for $Z \to u\bar{u}$, $Z \to d\bar{d}$ and $Z \to b\bar{b}$, Kuhn and Zerwas, ref.[48].

The vector and axial couplings are defined as in eq.(3.2.5), for f$\neq$b, for beauty we have to add to v_b and a_b the further correction (3.2.6).

Alternatively, we can eliminate completely M_Z, to obtain:

$$\Gamma(Z \to f \bar{f}) = \frac{1}{3} \frac{v_f(s)^2 + a_f^2}{s^3 c^3} N_f \sqrt{\frac{\pi \alpha(M)^3}{\sqrt{2} G_F \rho}} \qquad (3.2.13)$$

One can use this equation to determine $\sin^2\theta$ from the experimental widths, to be comparedwith the one obained from M_Z.

The behaviour of Γ as function of the t-quark mass is illustrated in Fig.3.2, for up, down and beauty quarks. The difference between the last two widths is about 10 MeV, for m_t ~200 GeV, still too small to be detected. Its observation would allow us to distinguish the vertex correction, specific of the t-quark, from the $\delta\rho$ correction, which could arise from other sources.

3.3 The Z line-shape

To zeroth order, the energy dependent cross-section for $e^+e^- \to f$ ($f \neq e^+e^-$, the elastic cross section has an additional important contribution from photon exchange in the t-channel) takes the Breit-Wigner shape ($M_Z = M$):

$$\sigma_0(e^+e^- \to f) = \frac{12\pi}{M^2} \frac{\Gamma_e \Gamma_f}{(s-M^2)^2 + M^2\Gamma^2} \qquad (3.3.1)$$

The peak cross-section:

$$\sigma_0(f)_{peak} = \frac{12\pi}{M^2} \frac{\Gamma_e}{\Gamma} \frac{\Gamma_f}{\Gamma} \qquad (3.3.2)$$

has a simple interpretation. The first factor is essentially the maximum cross-section allowed by unitarity of the S-Matrix, the second and third ones are the probability for resonance formation from the initial state and for its decay into the final state, respectively. The shape of the line corresponds to the production of an unstable particle, with lifetime $1/\Gamma$, as discussed below. Thus, the Breit-Wigner cross-section is completely independent from the underlying theory which decribes the Z properties.

The peak position and width give directly M and Γ. The widths into the the visible channels, $\mu^+\mu^-$, $\tau^+\tau$, and multihadrons are obtained from the corresponding peak cross-sections, assuming lepton universality, $\Gamma_e = \Gamma_\mu = \Gamma_\tau \equiv \Gamma_l$. Comparison with the total width gives finally, by subtraction, the width into the invisible channels, i.e. neutrinos:

$$\Gamma_{inv} = \Gamma - \Gamma_{had} - 3\Gamma_l$$

The Breit-Wigner cross-section, within the line width:

$$\frac{s-M^2}{M^2} < \frac{\Gamma}{M} \qquad (3.3.3)$$

is of order unity, as can be seen from eq.(3.3.2): the numerator, $\Gamma_e\Gamma_f$, is of order α^2, like the cross-section outside the peak, but the denominator, Γ^2, is of the same order. However, altough formally of order unity, the Z branching ratio into a charged lepton pair is a small quantity, due to the many other channels available, so that, for muon or tau final states, the one photon exchange cross section:

$$\sigma_\gamma(e^+e^- \to \mu^+\mu^-) = \frac{4}{3}\pi \frac{\alpha^2(M)}{M^2} \sim 0.5 \; 10^{-2} \, \sigma_0(\mu^+\mu^-)_{peak}$$

is not entirely negligible, although formally of order α^2.

To describe adequately the Z line shape in e^+e^-, we need to take into account a number of corrections, which we describe briefly in the following.

i. External particle radiation. Bremmstrahlung from the initial particles is the largest source of corrections to the Breit-Wigner shape[50]. It can be accounted by folding the cross section σ_0, eq.(3.3.1), with a "structure function"[51,52] describing the probability that the initial electron or positron emit a photon wich carries away a fraction x of the initial momentum:

$$\sigma_{corr} = \int_0^{1-s_0/s} dx\ F(x,s)\ \sigma_0[s(1-x)] \tag{3.3.4}$$

s_0 is a threshold energy below which the final state does not exist or is not detected. For example, for $\mu^+\mu^-$, s_0 is the minimum energy for the muon pair to be detected, typically 1GeV.

For $s > M^2$, the initial state can go back to the top of the resonance by emitting one (or more) photons. As a consequence, the corrected cross section is larger **above** the peak than below, acquiring a characteristic asymmetric shape (for a detailed discussion, see the lectures at this School by F. Berends).

ii. Relativistic phase-space. Ignoring for simplicity 4-vector indices, the Z propagator can be represented as:

$$\Delta(q^2) = \frac{1}{q^2-M_0^2} + \frac{1}{q^2-M_0^2}\ \Pi(q^2)\ \frac{1}{q^2-M_0^2} + ... =$$

$$= \frac{1}{q^2-M_0^2-\Pi(q^2)} + ... \tag{3.3.5}$$

where $\Pi(q^2)$ represents the self-energy diagrams (one-particle irreducible with respect to Z-lines). In turn, we can expand the self-energy function, $\Pi(q^2)$, according to:

$$\Pi(q^2) = \Pi(M_0^2) + \Pi'(M_0^2)(q^2 - M_0^2) - i\ M_0\Gamma + ... \tag{3.3.6}$$

to obtain:

$$q^2-M_0^2-\Pi(q^2) = (1+\Pi')(q^2 - M^2 - i\ M\Gamma) + ... \tag{3.3.7}$$

with M, and Γ, the renormalized mass and width, given by :

$$M^2 = M_0^2 + \Pi(M^2) + ... \tag{3.3.8}$$

$$M\Gamma = -\ Im[\Pi(M^2)] + ... \tag{3.3.9}$$

(M^2 and M_0^2 can be identified, within Π, since their difference is of order α^2) so that, finally:

$$\Delta(q^2) = (\frac{1}{1+\Pi})\ \frac{1}{q^2-M^2-iM\Gamma} + ... \tag{3.3.10}$$

Using this form of $\Delta(q)$ in the amplitude of $e^+e^- \to f$, one obtains the cross-section in eq.(3.3.1) (this is how the Breit-Wigner formula is derived). We may obtain a more precise description[53] if we keep the full dependence from q^2 in $Im\Pi$, instead of fixing $q^2=M^2$. In turn, $Im[\Pi(q^2)]$ is obtained (by the Cutkoski rule, Sect.2.1) by cutting the $f\ \bar{f}$ virtual loop in the amplitude for:

$$Z \to f \, \tilde{f} \to Z$$

For massless fermion pairs we have, for dimensional reasons:

$$Im[\Pi(q^2)] \propto q^2 \tag{3.3.11}$$

Similarly, for a slightly off-shell Z, Γ_e and Γ_f in the numerator are proportional to $\sqrt{s}$, for massless $e^\pm$ and massless particles in f, so that we obtain the more precise expression:

$$\sigma_0(e^+e^- \to f) = \frac{12\pi}{M^2} \frac{\Gamma_e\Gamma_f}{(s-M^2)^2 + \frac{s^2}{M^2}\Gamma^2} \frac{s}{M^2} \tag{3.3.12}$$

The cross-section is maximum for:

$$\sqrt{s} = M - \frac{\Gamma^2}{4M} \tag{3.3.13}$$

The relativistic phase-space correction shifts the maximum of the cross-section downward, of about 17MeV for $\Gamma \sim 2.5$ GeV, a tiny effect, smaller but still comparable to the virtual effects discussed in the following item.

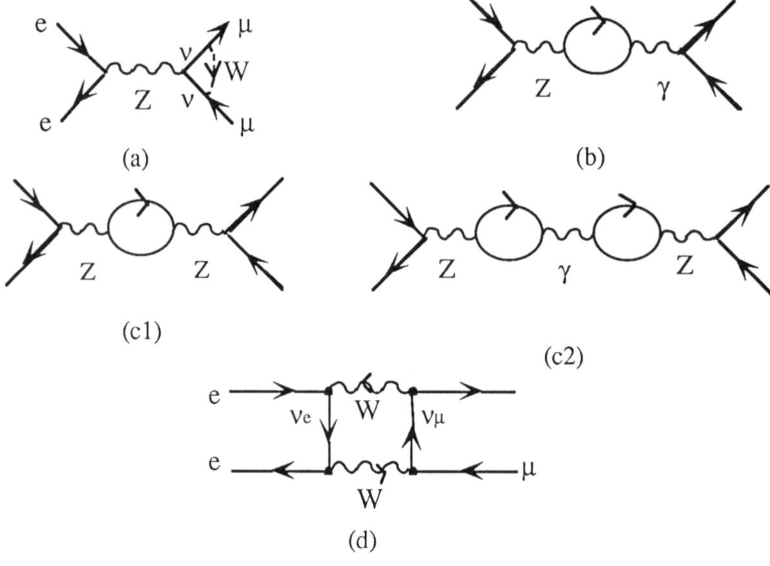

Fig. 3.3

Examples of one-loop corrections to the process $e^+ e^- \to f\tilde{f}$. a) and b) vertex and γ-Z mixing corrections, c_1) and c_2) self-energy corrections, d) box diagram. In the analysis of the Z line-shape it is convenient to define as one-particle irreducible those subdiagrams which do not have a Z-pole. Thus, the vertex correction in a) and the γ-Z mixing in b) are treated on the same footing as irreducible vertex corrections; similarly both c_1) and c_2) are considered irreducible self-energy corrections.

276

iii. Full electroweak corrections. Figs.3.3a to d give some examples of the first order virtual corrections[54].

To organize the corrections around the Z-pole, it is useful to introduce the notion of one-particle irreducible diagram with respect to the Z-line. For example, the vertex correction in Fig.3.3a **and** also that in Fig.3.3b[55], arising from one-loop γ-Z mixing, are to be considered as irreducible. Similarly, both self-energy diagrams in Fig.3.3c are irreducible. Reducible diagrams can be resummed so as to obtain the "skeleton diagram" of Fig.3.4. To this, we must add the one-photon exchange diagram (with the running α so as to take into account also the leading higher order corrections) and box diagrams, e.g. Fig.3.3d.

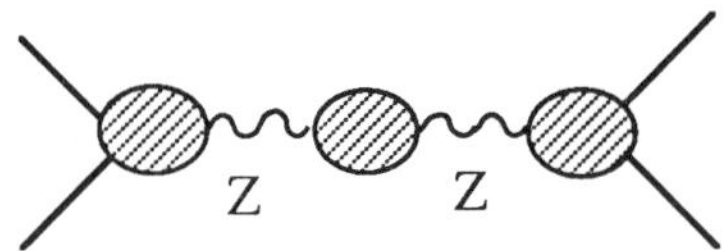

Fig. 3.4

Resonant skeleton diagram for $e^+ e^- \to f\bar{f}$. Dressed vertices include external fermion wave-function renormalization. The Z-propagator is fully dressed, including the one-Z-irreducible self-energy diagrams, like e.g. diagram c_2) in Fig 3.3.

The corrected cross-section is obtained in three steps. First, we square the skeleton diagram and add the interference with one-photon exchange and box diagrams. Secondly, we add the cross-section for the emission of radiation from the final state and the interference between initial and final state radiation. Finally, the cross section thus obtained, denoted by σ_{0w} in the following, is inserted into eq.(3.3.4) and folded with the initial state radiation to obtain the fully corrected cross-section.

Within the line-width, it is possible to describe σ_{0w} with a model-independent expression[56], that is to give the fully corrected cross-section in terms of a fixed number of physical parameters, which include the physical mass and partial widths, in a way which uniquely generalizes the Breit-Wigner formula. This is possible because the fractional deviation from the peak, eq.(3.3.3), is small, of order α, and can be used as an expansion parameter. To a finite order in α, e.g. to one-loop accuracy, only a finite number of powers in $(s-M^2)/M^2$ are required and, therefore, only a finite number of new, phenomenological parameters beyond those of the simple Breit-Wigner formula.

To one-loop order, the result is:

$$\sigma_0(e^+e^- \to f) = \frac{12\pi\Gamma_e\Gamma_f}{M^2} \frac{1}{|D(q^2)|^2} \left(\frac{s}{M^2} + R \frac{s-M^2}{M^2} + I \frac{\Gamma}{M} + ... \right) = \tag{3.3.14}$$

$$D(q^2) = s-M^2 + iM\Gamma \left(\frac{s}{M^2} + \varepsilon \frac{s-M^2}{M^2} + ... \right) \tag{3.3.15}$$

where dots denote higher orders in $(s-M^2)/M^2$. The constants I and ε depend upon the particle spectrum up the the Z mass, and have to be computed from the Standard Theory. M, Γ, Γ_e, Γ_f and R are instead free parameters, to be fitted from data.

We refer the reader to ref.[56] for details of the derivation and give only a sketch of the argument, restricting to the $\mu^+\mu^-$ case, for simplicity .

1. The amplitude corresponding to the resonant skeleton diagram, Fig.3.4, can be written as:

$$A_{res} = M_e^0(s)^* \, \Delta(s) M_\mu^0(s) \qquad (3.3.16)$$

The denominator of $\Delta(s)$ is developed around M^2. Since we want Γ to one loop accuracy, we have to keep one more term in $\text{Im}\Pi$ than we did in eq.(3.3.7), and we obtain the result:

$$\Delta(s)^{-1} = (1+\Pi')[\ s-M^2 +iM\Gamma\ (\frac{s}{M^2} +\varepsilon\ \frac{s-M^2}{M^2} +...) \ ] \qquad (3.3.17)$$

$\varepsilon=0$ corresponds to the scaling result (3.3.11). In general, if massive final states contribute to the Z-width:

$$1 + \varepsilon = [\frac{\partial \ln\Gamma(s)}{\partial \ln s}]_{s=M^2} \qquad (3.3.18)$$

(however, $\varepsilon \sim 2 \ 10^{-3}$ for the $b\bar{b}$ channel). As pointed out in ref.[53], we may rewrite:

$$s-M^2 +iM\Gamma\ (\frac{s}{M^2} +\varepsilon\ \frac{s-M^2}{M^2}\ ) \sim \eta\ (s-M'^2 +iM'\Gamma')$$

with:

$$\eta = 1+ i(1+\varepsilon)\frac{\Gamma}{M}; \qquad M'=M[1 - \frac{(1+\varepsilon)\Gamma^2}{2M^2}]; \qquad \Gamma'=\Gamma[\ 1 - \frac{(1+3\varepsilon)\Gamma^2}{2M^2}]$$

To first order in Γ/M ($\sim O(\alpha)$) η is a phase factor. It drops out from the square of A_{res} and it gives a higher order contribution to the interference of A_{res} with the one-photon and box amplitudes. This means that, to this order, it is impossible to disentangle the correction due to ε from a shift in the mass and the width of the Z. If we want to determine M and Γ, as defined by eqs.(3.3.8) and (3.3.9), we must know ε beforehand and correct for it explicitly in eq.(3.3.15).

2. We consider next the numerator of A_{res}, with the normalization constant arising from eq.(3.3.10) included:

$$N(s) = M_e(s)M_\mu(s) \qquad (3.3.19)$$

$$M_\mu (s) = \frac{M_\mu^0}{\sqrt{1+\Pi'}} \qquad (3.3.20)$$

and the same for $M_e(s)$.

At $s=M^2$, M_μ is the on-shell decay amplitude for Z-decay into a muon pair, inclusive of one-loop virtual corrections. When we square A_{res} and add the final state radiation, M_μ gives rise exactly to the factor Γ_μ of the Breit-Wigner numerator, interpreted now as the one-loop-corrected width for:

$$Z \to \mu^+\mu^- + \text{radiation} \qquad (3.3.21)$$

We can further subtract an appropriate term from the structure function in (3.3.4) and add it to $|M_e(M^2)|^2$, to obtain also the inclusive electron width, Γ_e, with a similar interpretation as (3.3.21).

For $s\neq M^2$, $N(s)$ is developed to first order in $(s-M^2)/M^2$. Higher than linear terms in this variable are of the same order of the two-loop corrections to $M_{e,\mu}(M^2)$, which have not been computed (and are not needed for the present experimental accuracy).

From the development of N(s) we get a first contribution, $R^{(1)}$, to the parameter R, in eq.(3.3.14). Note that the expansion in (3.3.14) is so defined that :

$$R^{(1)} + 1 = 2\mathrm{Re}[\frac{\partial \ln N(s)}{\partial \ln s}]_{s=M^2}$$

and $R^{(1)}$ vanishes to lowest order, which corresponds to $N(s) \propto \sqrt{s}$.

The largest contribution to R arises from the interference of the real part of A_{res} with the one-photon-exchange diagram. One finds a contribution formally of order unity (but in fact rather small, see below) which, in the leading approximation in the t-quark mass, is:

$$R_\gamma \sim [\frac{\pi\alpha(M)}{\sqrt{2}G_F(1+\delta\rho)M^2}] \frac{2Q_e Q_\mu v'_e v'_\mu}{(v'^2_e + a^2_e)(v'^2_\mu + a^2_\mu)} \qquad (3.3.22)$$

where $v'_{e,\mu}$ are the vector couplings defined as in eq.(3.2.5), but expressed in terms of $(\sin^2\theta)'$, eq.(3.2.3).

3. The interference of A_{res} with the other amplitudes gives also rise to the term $I\,\Gamma/M$ in eq.(3.3.14). At the peak, $\Delta(s)$ is imaginary and a term of this form arises from interference with the imaginary (real) part of the one-photon and box diagrams, when we take in A_{res} the real (imaginary) part of N. In all cases, the interference is determined by the particle spectrum below the Z mass. As shown by eq.(3.3.14), if we want to obtain the product $\Gamma_e\Gamma_\mu$ from the cross-section, we must correct for I, as we did for ε, computing it from the known particle spectrum.

The values of R and I computed in the Standard Theory are both quite small, of the order of few percent, wich corresponds to a very small correction to the (relativistic phase space) Breit-Wigner formula. This is an important result by itself: one-loop corrections in the Standard Theory affect the cross-section mostly through their influence on the values of M, Γ and $\Gamma_e\Gamma_f$. Theories with a different high energy particle content could lead to an appreciably different value of R.

It is clear from the derivation that the result (3.3.14) holds in any renormalizable theory with the same particle content as the Standard Theory below the Z mass, but with arbitrary structure at larger mass scales. The obvious advantage of eq.(3.3.14) is that we can obtain mass and widths of the Z from the experimental cross-section once and for all, independently from assumptions on the higher mass particle spectrum. This is true in the Standard theory itself, where physical mass and widths which are obtained by fitting the data with (3.3.14) are independent from any assumption on m_t or m_H.

In contrast, a universal parametrization is not possible **outside the line width**. In this case, the only parameters which describe the cross-section are the gauge, Yukawa and Higgs couplings of the fundamental theory, which bring in model-dependent effects related to the detailed particle and coupling structure of the theory.

3.4 Comparison with data

I can be brief here, since this subject is discussed at length by other lecturers at this School. We restrict first to the most recent LEP and P$\bar{\text{P}}$ data on the vector boson parameters, summarized in Tables 3.1 ref.[35a,b] and 3.2 ref.[57a,b,c,d].

Table 3.1. - CDF and UA2 results on M_W/M_Z. The first (second) error is statistical (systematic). Errors are combined in quadrature in the weighted average.

	M_W/M_Z	$\sin^2\theta_W$
CDF[35a]	0.8775±0.0047±0.0021	
UA2 [35b]	0.8831±0.0045±0.0026	
Average	0.8801 ±0.0037	0.225±0.007

Table 3.2. - Summary of LEP results . Errors reported in the first four lines are statistical only. Errors in the weighted average are obtained combining in quadrature the overall statistical error with a systematic error of 0.02 GeV, for M_Z, and a relative systematic error of 0.5 %, for the widths.

	M_Z	Γ_Z	Γ_e	Γ_h	$R = \dfrac{\Gamma_h}{\Gamma_e}$
	(GeV)	(GeV)	(MeV)	(GeV)	
ALEPH[57a]	91.193±0.016	2.497±0.031	84.3±1.3	1.754±0.027	20.85±0.35
DELPHI[57b]	91.188±0.013	2.476±0.026	83.7±1.0	1.756±0.023	20.8 ±0.6
L3[57c]	91.161±0.013	2.492±0.025	84.0±1.2	1.748±0.035	21.02±0.62
OPAL[57d]	91.174±0.011	2.505±0.020	83.6±1.0	1.778±0.026	21.72±0.49
Average	91.177±0.021	2.494±0.012	83.8±0.6	1.760±0.013	20.99±0.30

Curves (a), (b) and (c) in Fig.3.5 give[58] $\sin^2\theta$ as obtained from the Z mass, eq.(3.1.6), as functions of m_t and for three different values of the Higgs boson mass (40, 300, 800 GeV). The full one-loop calculation is used and $\sin^2\theta$ is defined as in eq.(3.1.3). The horizontal lines give the 1σ region of $\sin^2\theta$, as determined from the M_W/M_Z ratio measured by UA(1) and CDF in P$\bar{\text{P}}$ collisions, (2.5.8) (the latter determination is of course independent from m_t). As indicated by the figure, the two determinations are consistent, to 1σ, for:

$$80 \text{ GeV} < m_t < 210 \text{ GeV} \qquad\qquad (3.4.1)$$

Figs.3.6 to 3.8 show[58] $\sin^2\theta$ as obtained from the Z observed widths (averaged over the results of the four LEP collaborations), always as function of m_t. Again, the full one-loop calculation is used, with a fixed value of the Higgs boson mass, m_H=300GeV. The agreement with the previous value of $\sin^2\theta$ is startling. The total width gives the 1σ interval:

$$90 \text{ GeV} < m_t < 170 \text{ GeV} \qquad\qquad (3.4.2)$$

with an upper bound somewhat lower than the one in (3.4.1).

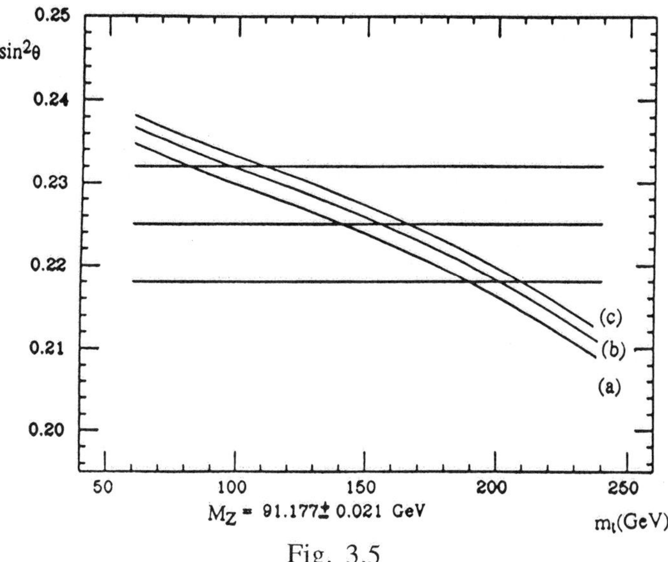

Fig. 3.5

The horizontal band represents the one-standard-deviation region of $\sin^2\theta$, defined as in eq.(3.1.3) of text, from the measured M_W/M_Z ratio. Weighted average of the CDF and UA(2) results, see Tab.3.1 and ref.[35a, b]. Curves (a), (b) and (c) give $\sin^2\theta$ vs.m_t, as obtained from the experimental value of M_Z, weighted average over the results of the four LEP experiments, Tab.3.2 and ref.[57a-d]. The assumed Higgs boson mass is: (a) 40GeV, (b) 300GeV, (c) 800GeV. Full one-loop corrections, ref.[58].

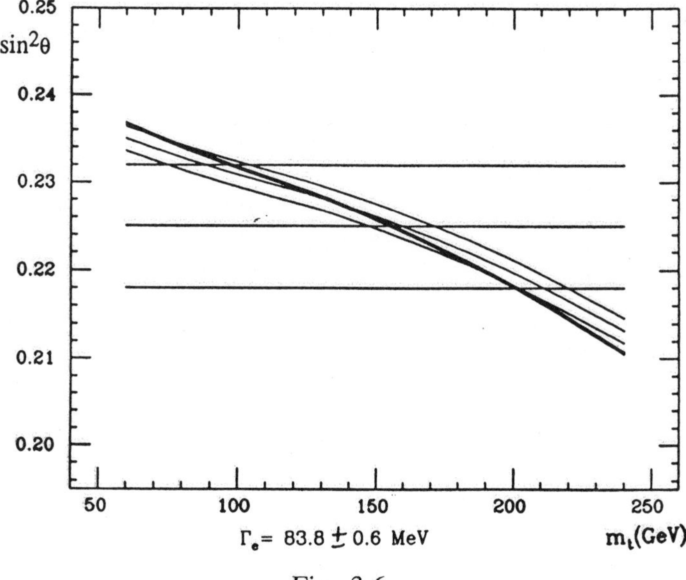

Fig. 3.6

Horizontal band as in Fig.3.1. The value of $\sin^2\theta$ vs. m_t from M_Z (thick curve) is compared to the one from the measured leptonic width, weighted average over the four LEP experiments, Tab.3.2 and ref.[57a-d]. m_H =300GeV assumed. Full one-loop corrections, ref.[58].

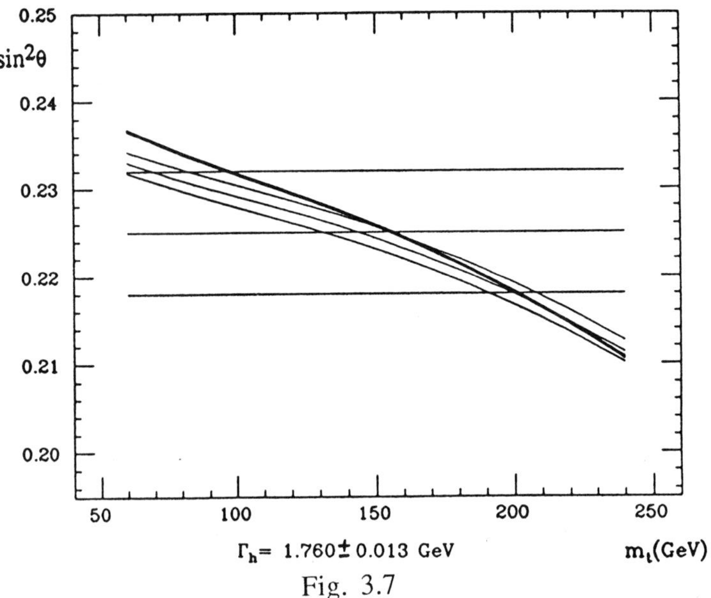

$$\Gamma_h = 1.760 \pm 0.013 \text{ GeV}$$

Fig. 3.7

Same as in Fig.3.6, for the hadronic width.

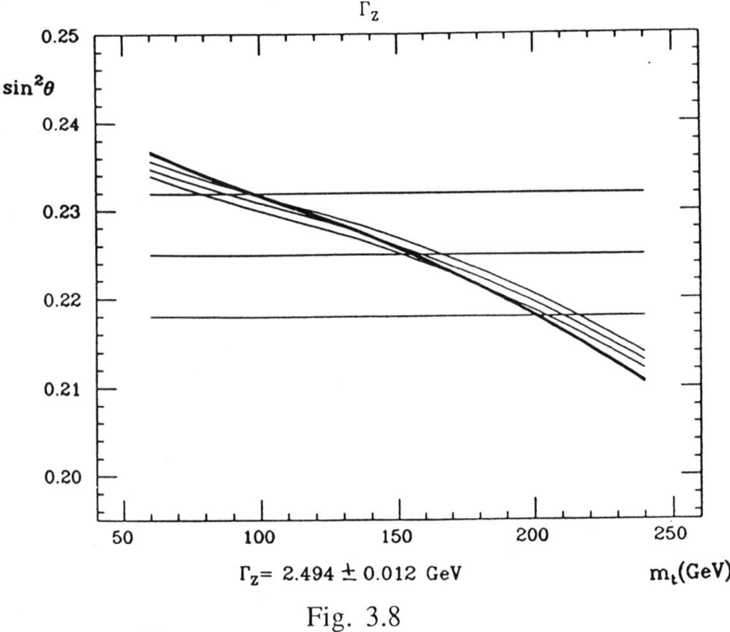

$$\Gamma_Z = 2.494 \pm 0.012 \text{ GeV}$$

Fig. 3.8

Same as in Fig.3.6, for the total width.

Quite similar results to ours have been obtained, independently, by other authors[34,59,60]. Using in addition the value of $\sin^2\theta$ determined from ν-N cross-section (which is quite independent from m_t because of an accidental cancellation of different corrections), more restrictive 1σ limitations to m_t have been given in ref.[59]:

$$m_t = 127 \,^{+24}_{-30} \text{ GeV} \tag{3.4.3}$$

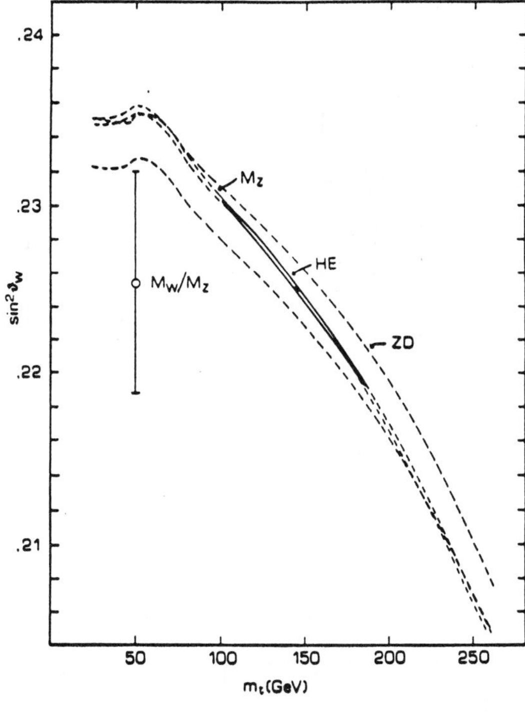

Fig. 3.9

$\sin^2\theta$ vs. m_t, all electroweak data, from ref.[59]. The restriction due to the Z decay widhts measured at LEP is indicated. The band marked ZD indicates the limits obtained from the Z decay widths.

see Figs.3.9, and in ref.[34], see Fig.3.10. The present precision is barely sensitive to the Higgs boson mass, and no significant restriction to m_H can be derived, see however ref.[59] and Fig.3.11.

Needless to say, the agreement of the total width with theory implies 3 neutrinos, with considerable accuracy, as discussed in other parts of this School (see also ref.[61]).

The improved Born approximation, described in Sect.4.2, proves to be quite adequate, at the present level of accuracy, particularly in the upper range of m_t. Using eq.(3.2.8) and the experimental value of M_Z, one finds:

$$\sin^2\theta = 0.230 \quad (m_t = 80 \text{ GeV})$$

$$\sin^2\theta = 0.224 \quad (m_t = 150 \text{ GeV})$$

$$\sin^2\theta = 0.218 \quad (m_t = 200 \text{ GeV}) \tag{3.4.4}$$

Comparing with the value of $\sin^2\theta$ from the W/Z mass ratio, eq.(2.5.8), we see that the upper limit to m_t comes out quite right, while the lower limit would be lowered or disappear at all.The upper limit comes down to about 150 GeV if we compare the results in (3.4.4) with the weigthed average of the v-N and W-Z determinations, eq.(2.5.9).

The reader may also easily verify that taking $\sin^2\theta$ from M_Z, via eq.(3.2.8), one obtains partial and total widths consistent with the weighted averages of Table 3.2, to 1σ, for m_t in the range:

$$m_t < 150 \text{ GeV (for } \Gamma_\mu) \tag{3.4.5}$$

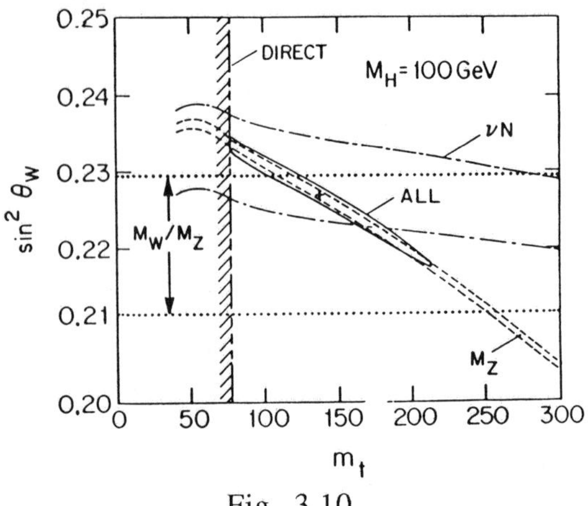

Fig. 3.10

$\sin^2\theta$ vs. m_t, all electroweak data, ref.[34].

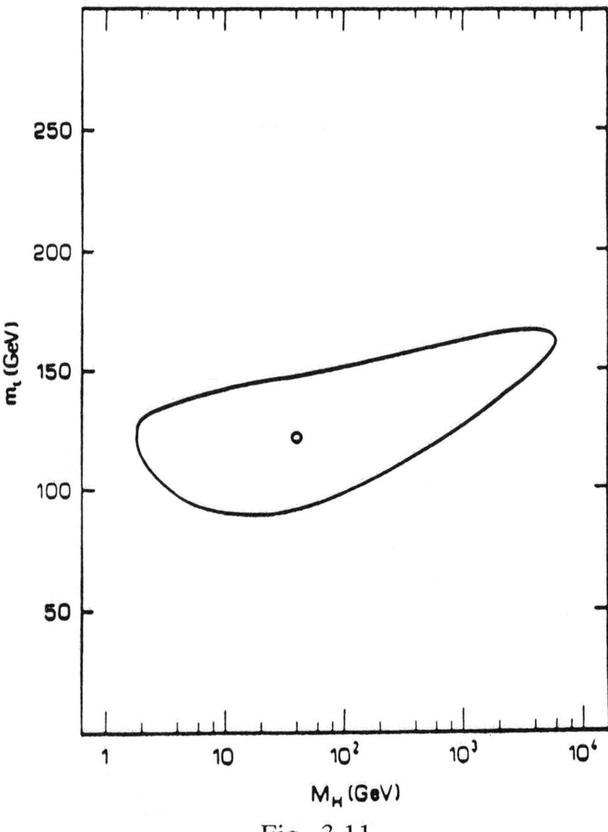

Fig. 3.11

1-σ contour in the m_t-m_H plane, from the fit to the electroweak data of ref.[59].

$$110 \text{ GeV} < m_t \quad \text{(for } \Gamma_{hadronic}) \tag{3.4.6}$$

$$m_t < 160 \text{ GeV} \quad \text{(for } \Gamma) \tag{3.4.7}$$

quite consistently with the more precise estimates reported above.

3.5 Top-quark mass and weak processes

Some weak amplitudes of light or beauty hadrons involve virtual t-quark exchange. One may ask how such effects compare with the range for m_t indicated by the electroweak corrections discussed in the previous Section.

In particular, we focus on the parameters ε and ε', describing CP-violation in K decays, and on the B_d-$\bar{B}_d$ mixing amplitude, x_d.

The neutral Kaon mass difference and K_L decay amplitudes such as:

$$K_L \to \mu^+ \mu^-, \gamma \gamma$$

do also depend from m_t, but are affected by long-distance, hard to control, strong interaction effects (see however ref.[62] for a recent analysis).

The question can be put as follows. Present experimental data on semileptonic decays (including the recent observation of charmless, inclusive, semileptonic decays of B-mesons, by the ARGUS[63] and CLEO[64] collaborations) determine all mixing angles of the Cabibbo-Kobayashi-Maskawa matrix, but the CP-violating phase, δ (we use the form of the CKM matrix given in Appendix B).

In principle, we could use ε to fix δ, x_d to determine m_t, and predict the ratio ε'/ε as well as the CP-violating amplitudes in beauty decays.

Things are more complicated, in practice, since the theoretical calculation of x_d involves the not-very-well known quantity f_B, the axial vector coupling of the pseudoscalar meson B. In this situation, we could use all three quantities, ε, ε'/ε, and x_d to determine δ, m_t and f_B (as proposed e.g. in ref.[65]). However, as we shall see, this is still too ambitious since the experimental value of ε'/ε is controversial and the theoretical expression for ε'/ε contains various ill-controlled strong interaction parameters[66].

A less ambitious, more realistic program is to take m_t in the range allowed by the electroweak data:

$$100 \text{ GeV} < m_t < 160 \text{ GeV} \tag{3.5.1}$$

determine δ and f_B from ε and x_d and see what predictions can be drawn for ε'/ε and for the CP-violating amplitudes in beauty decays . In addition, f_B can be compared with the results of theoretical calculations, e.g. lattice QCD or QCD sum rules.

The results of a recent analysis by our group[67] are illustrated in Fig.3.12, for the central value m_t=130 GeV, and in Figs. 3.13 and 3.14 for the extreme values, 100 and 160 GeV, respectively (for an independent analysis, see ref.[68]) . We can summarize the results as follows.

i. The value of $\sin\delta$ determined from ε decreases at the increase of m_t. For m_t>130 GeV, we find two distinct solutions, of about opposite sign for $\cos\delta$.

ii. In correspondence, we have two possibilities for f_B, as determined from B_d-$\bar{B}_d$ mixing: (a) f_B in the range 100-200 MeV and $\cos\delta$<0, (b) f_B in the range 200-300 and $\cos\delta$>0.

iii. Using lattice QCD results for the B-parameters of the various terms in the CP-violating hamiltonian (and some educated guess), one predicts rather small values for ε'/ε. The larger value corresponds to the $\cos\delta$>0 solution and is intermediate between the CERN and FNAL results. However, it is clear from the figure that we are unable, at present, to distinguish between the two possibilities (the solution with $\cos\delta$>0 and f_B large correponds to the one found in ref.[65]).

The result (ii) is easy to explain. x_d is dominated by box diagrams with t-quark exchange which give:

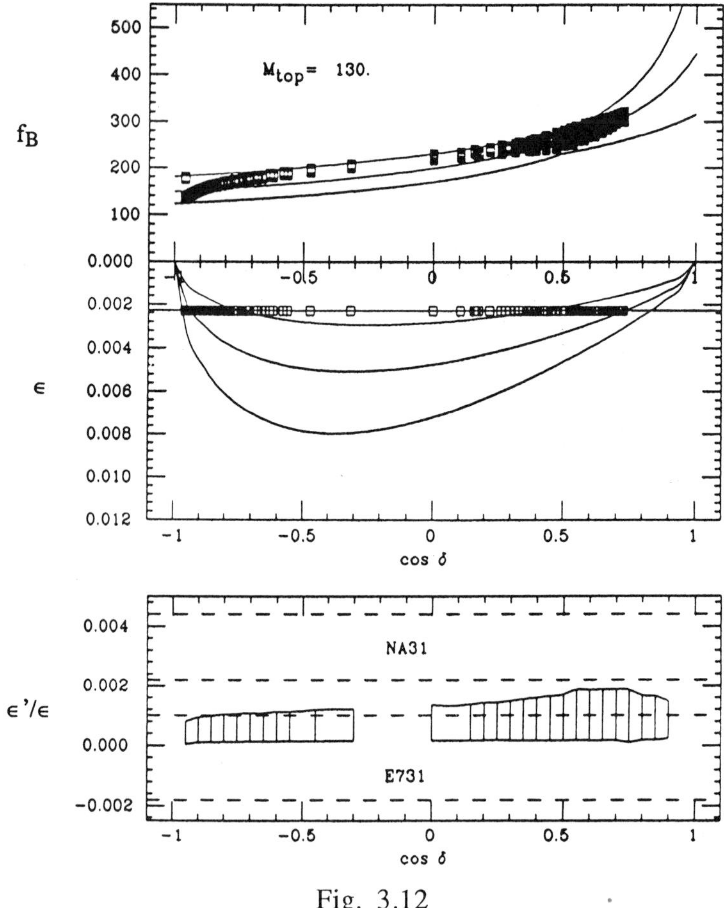

Fig. 3.12

m_t=130 GeV. Theoretical prediction of ε (middle section of the figure) and f_B (as derived from the B-$\bar{\text{B}}$ mixing parameter, x_d, upper section), vs. $\cos\delta$. Points indicate values compatible with both ε and x_d, with input parameters within 1-σ from central value. The allowed regions of ε'/ε are reported in the lowest section. A clear separation between two solutions: (a) $\cos\delta<0$; 100 MeV$<f_B<$200 MeV ; (b) $\cos\delta>0$; 200 MeV$<f_B<$300 MeV starts to appear. (b) is preferred by recent lattice QCD calculations and leads to a large CP-violating asymmetry in process (3.5.4) of text. Figure from ref.[67].

$$x_d \propto [\text{Re}(V_{td})f_B]^2 \, m_t^2 \, g(\frac{m_t^2}{M_W^2}) \tag{3.5.2}$$

g is a slowly varying function and, in Wolfenstein parametrization, $\text{Re}(V_{td})\propto (1-\rho\cos\delta)$ with $\rho\sim0.5$. For fixed m_t, $\cos\delta>0$, corresponding to V_{td} small, requires f_B large and viceversa, $\cos\delta<0$ and V_{td} large correspond to f_B small.

The result (ii) has two interesting connections.

A. The most recent lattice QCD calculations[69,70] indicate quite a large value for f_B. In these calculation, the b-quark is treated as a static (infinite mass) source. Neglecting $1/M_b$ corrections, one finds:

$$f_B \text{ (stat)} \sim 330\text{-}350\,\text{MeV} \tag{3.5.3}$$

a much larger value than one would have extrapolated with the asymptotic QCD scaling law, $f(M) \propto (\sqrt{M})^{-1}$, from the charm axial coupling, f_D which is predicted at about 180 MeV. Even accounting for a 25% reduction due to the $1/M_b$ corrections, estimated so as to fit both the charm and the static result, one would obtain a value of f_B in the upper range found previously (200-300 MeV) rather than the lower range, much used in the analyses of few years ago. If $m_t > 130$ GeV or so, lattice QCD results prefer definitely solution (b), with $\cos\delta > 0$.

In the competitive approach to strong interactions, namely QCD sum rules, one finds a value for f_D quite consistent with that from the lattice calculations. However, the most recent estimates[71] give $f_B < f_D$, opposite to the lattice result.

B. A glance at the unitarity triangle for the CKM matrix, Fig.3.15, shows that solution (b) corresponds to an acute, rather than obtuse, triangle. It is well known that the angles β and α shown in the same figure determine the CP-violating asymmetries (T is a tagging channel for B^0 or $\bar{B}^0$):

$$A_1 = \frac{N(B^0\bar{B}^0 \to T + J/\Psi + K_S) - N(B^0\bar{B}^0 \to \bar{T} + J/\Psi + K_S)}{\text{sum}} \qquad (\sin 2\beta) \qquad (3.5.4)$$

$$A_2 = \frac{N(B^0\bar{B}^0 \to T + \pi^+\pi^-) - N(B^0\bar{B}^0 \to \bar{T} + \pi^+\pi^-)}{\text{sum}} \qquad (\sin 2\alpha) \qquad (3.5.5)$$

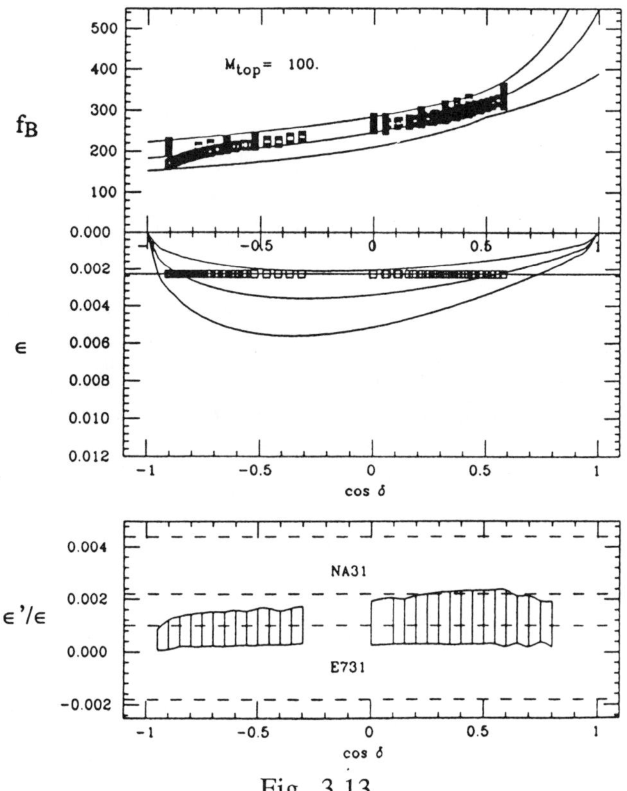

Fig. 3.13

Same as Fig.3.12, for $m_t=100$ GeV. At this value of the mass, the separation between the two solutions with $\cos\delta < 0$ and $\cos\delta > 0$ is disappearing.

Solution (a) gives the lower bound[68,72] $\sin 2\beta > 0.21$, solution (b) gives a much larger bound, of the order of 0.6.

For smaller values of m_t, the lattice result on f_B could be compatible even with $\cos\delta < 0$, and no definite statement can be made.

Thus, **if** m_t is not too small ($m_t > 130$ GeV) and **if** lattice QCD calculations of f_B are right, the CP-violating asymmetry in (3.5.4) should be considerably larger than previously expected. This can be quite an important effect. The lower bound $\sin 2\beta > 0.6$ is about a factor 3 larger than the lower bound for the other solution, which, in turn, implies a reduction of about a factor 10 in the luminosity needed to obtain a given statistical error in the asymmetry measurement [72].

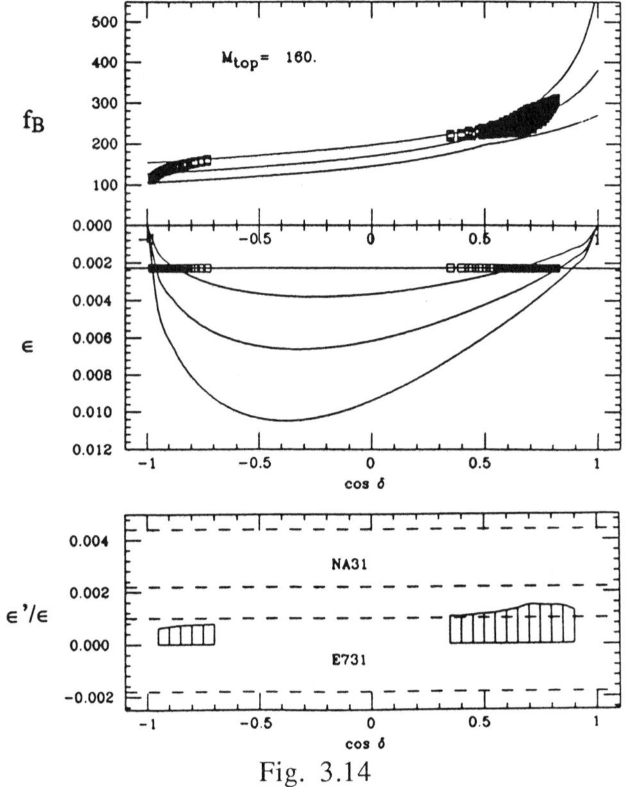

Fig. 3.14

Same as Fig.3.12, for m_t=160 GeV.

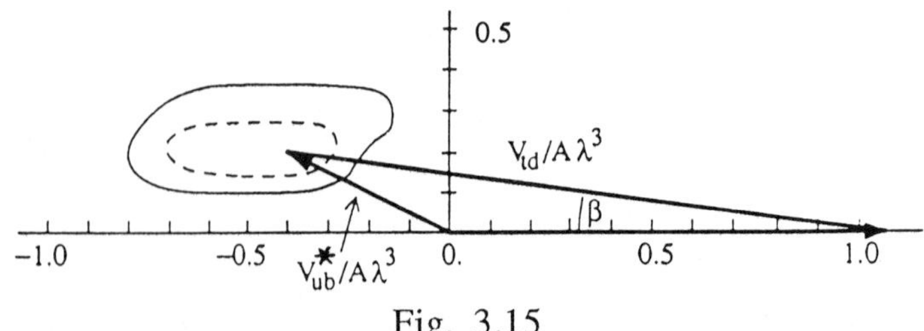

Fig. 3.15

Graphical representation of the unitarity relation: $V_{ub}+V_{td}^* - A\lambda^3 = 0$, see Appendix for notations. The triangle shown corresponds to the solution with $\cos\delta < 0$. For the other solution, $\cos\delta > 0$, the triangle is acute and, correspondingly, the lower bound to $\sin 2\beta$ is about a factor three larger. Figure from reff.[72,97]

4. THEORETICAL BOUNDS TO THE HIGGS BOSON AND THE T-QUARK MASS

The Higgs boson mass is related to the self-coupling λ, defined in (1.4.12), according to:

$$m_\sigma^2 = 4\lambda\eta^2 = \frac{\sqrt{2}\lambda}{G_F} \sim 4\lambda \, (174 \, \text{GeV})^2 \tag{4.0.1}$$

In perturbation theory, λ is arbitrary, so m_σ cannot be predicted. A similar result holds true for any fermion mass, in particular for the t-quark mass:

$$m_t^2 = g_t^2\eta^2 = \frac{g_t^2}{2\sqrt{2}G_F} \sim g_t^2 \, (174 \, \text{GeV})^2 \tag{4.0.2}$$

A useful reference point is given by the values of the coupling constants for which the amplitudes computed in the Born approximation violate unitarity[73,74]. In terms of masses, this happens for:

$$m_t \sim m_H \sim 1 \, \text{TeV} \tag{4.0.3}$$

On the other side, we have the experimental lower bounds to m_t and m_H, found by CDF and by the LEP collaborations[75]:

$$m_H > 45 \, \text{GeV}$$
$$(\text{CERN-LEP}) \tag{4.0.4}$$

$$m_t > 90 \, \text{GeV}$$
$$(\text{FNAL-CDF}) \tag{4.0.5}$$

Translated in terms of coupling constants, the above limitations read:

$$0.017 \, (\text{expt}) < \lambda < 8.2 \, (\text{unitarity limit})$$

$$0.52 \, (\text{expt}) < g_t < 5.7 \, (\text{unitarity limit})$$

Of course, the fact that the Born approximation violates unitarity is not a pathology by itself, it only signals that perturbation theory is inadequate, for such a large coupling constant.

The use of improved perturbation theory, that is the resummation of the leading logarithmic corrections of the form (2.4.7), shows that the theory is stable only inside a finite region in the λ-g_t plane. This leads to combined bounds to the Higgs boson and to the t-quark mass which will be discussed in this Section. Recent developments in the understanding of non-perturbative effects, within lattice theories, have given further support to this picure, and produced numerically significant bounds to the Higgs boson mass.

4.1 Triviality of the $\lambda\phi^4$ theory and upper bounds to m_H

We have already seen in Sect.2.4 that a non asymptotically free theory may posses only a trivial continuum limit. Let us first adapt the previous discussion to the case of a pure $\lambda\phi^4$ self-interaction (no gauge or Yukawa interactions).

Explicit calculation of the beta function leads to the renormalization group equation:

$$\mu\frac{\partial\lambda}{\partial\mu} = \frac{3}{2\pi^2}\lambda^2 \tag{4.1.1}$$

We take $\mu = \eta$ as the momentum scale where the renormalized coupling, λ_R, is defined, so that λ_R is related to the physical Higgs boson mass by eq.(4.0.1). Proceeding as in Sect.2.3, we obtain:

$$\frac{1}{\lambda_R} = \frac{1}{\lambda(\Lambda)} + \frac{3}{2\pi^2}\ln(\frac{\Lambda}{\eta}) \qquad (4.1.2)$$

The bare coupling, $\lambda(\Lambda)$, must be positive, in any case, which leads to the bound:

$$\frac{1}{\lambda_R} > \frac{3}{2\pi^2}\ln(\frac{\Lambda}{\eta}) \qquad (4.1.3)$$

Eq.(4.1.3) indicates again that the removal of the cut-off, $\Lambda\to\infty$, sends λ_R to zero. On the other hand, we know that λ_R cannot vanish, least the disappearence of the Higgs mechanism and the collapse of the Standard Theory. How do we have to interpret this phenomenon? Also, we expressed in Sect. 2.3 some criticism to the above result. Can we find evidence in favour or against, at the non-perturbative level? We address the two problems in sequence.

A direct consequence of the triviality phenomenon is that the Standard Theory must live with a, possibly large, but **finite cut-off**. Modifications have to take place at some finite energy scale, so as to cure the problem, represented otherwise by eq.(4.1.3). Furthermore, if all this is true, the physical Higgs boson mass must be within a definite upper bound, determined by the value of Λ (as firts pointed out by Maiani, Parisi and Petronzio[76]).
From eqs.(4.1.3) and (4.0.1), we obtain:

$$\frac{m_H}{174 GeV} < 2\pi \, [\frac{3}{2}\ln(\frac{\Lambda}{\eta})]^{-1/2} \qquad (4.1.4)$$

that is:

$$m_H < 144 \text{ GeV}, \; (\lambda_R < 0.17) \qquad \Lambda = 10^{19} \text{ GeV } (\sim M_{Planck})$$

$$m_H < 165 \text{ GeV}, \; (\lambda_R < 0.22) \qquad \Lambda = 10^{15} \text{ GeV } (\sim M_{GUT})$$

$$m_H < 675 \text{ GeV}, \; (\lambda_R < 3.8) \qquad \Lambda = 1 \text{ TeV} \qquad (4.1.5)$$

The numerical value of the bound depends from the assumed physical cut-off. However, it is remarkable that in the interesting range of Λ, i.e. Λ beteewn the present energies and the Planck mass, 10^{19} GeV, where quantum gravity will induce anyhow a change of regime in particle interactions, bounds are quite non trivial.
When the cut-off is large, of the order of M_{GUT} or M_{Planck}, λ_R is small. The running coupling remains small for most of the range between η and Λ, and the estimate of the upper bounde should be reliable. The last upper bound to m_H is, instead, a little stretched. In this case, λ_R is already close to the unitarity limity and non-perturbative methods are needed to cope with such a low value of the cut-off, see next Section.

The triviality problem has been addressed, at the non-perturbative level, by many authors[77,78,79]. The most advanced result has been found by Frølich[78], in the case of the $\lambda\phi^4$ on a cubic lattice (lattice spacing a) and nearest-neighbour interactions.
Frølich considers the 4-point correlation function, $u_4(x_1, x_2, x_3, x_4)$, in a euclidean d-dimensional space-time and for the non-spontaneously broken theory. On the basis of rigorous inequalities, he proves that, in the limit $a\to 0$:
i. $u_4 \to 0$, for $d \geq 5$;
ii. $u_4 \to 0$ in $d=4$ dimensions, if the field renormalization constant diverges in the limit $a\to 0$.
The proof is very technical and cannot be reproduced here.
The identical vanishing uf u_4 in euclidean space leads directly to the triviality of the $\lambda\phi^4$ theory, via the following chain of arguments.

First, by analytic continuation, one obtains the vanishing of the 4-point correlation function in Minkowski space-time, threfore the vanishing of the amplitude for (2 particles → 2 particles) scattering. The imaginary part of the latter amplitude is related by unitarity (Cutkoski rule, Sect.2.1) to the scattering probability for (2 particles→ n particles), which therefore is also vanishing. Finally, using crossing symmetry, we conclude that the general amplitude for (n particles→ m particles) has to vanish, i.e. that the S-matrix is unity and the theory is trivial.

It is very unlikely that the field renormalization constant remains finite for a→0. Also, triviality is a short-distance phenomenon (se Sect.2.4) and it should not matter if we are in the exact or in the spontaneously broken phase. Thus, Frølich result provides a very strong indication in favour of triviality in d=4 dimensions (triviality for d≥ 5, where the interaction is non renormalizable, is less surprising).

Triviality has been proven completely[80] for the O(N) symmetric $\lambda\phi^4$ interaction, in the N→∞limit.

The issue cannot be considered completely settled, as yet. If one changes the lattice action, e.g. by introducing next to nearest-neighbour interactions, the crucial inequalities disappear, and Frølich proof does not go through. Although unlikely, we cannot exclude that a more clever regularization may give rise to a non trivial interaction, in the continuum limit.

4.2 The 600 GeV bound

Reducing Λ from the GUT scale downwards, the upper bound to m_σ increases and one-loop perturbation theory becomes increasingly suspicious.

Dashen and Neuberger[81] have indicated how the largest possible upper bound can be determined with non-perturbative methods.

If we neglect gauge and Yukawa interactions, the Standard Theory reduces to the O(4) linear sigma model. This is seen by defining ϕ_i, $1=1,...,4$, according to:

$$\phi = \begin{pmatrix} \dfrac{\phi_1+i\phi_2}{\sqrt{2}} \\ \dfrac{\phi_3+i\phi_4}{\sqrt{2}} \end{pmatrix} \tag{4.2.1}$$

so that the potential in (1.4.12) becomes:

$$V = \frac{\lambda}{4} (\Sigma_i \, \phi_i\phi_i - \eta^2)^2 \tag{4.2.2}$$

In the spontaneously broken phase, we take $<\phi_4>=\eta \neq 0$ and all other components vanishing. $\phi_{1,2,3}$ are the Goldstone, massless, bosons and:

$$\phi_4 = \eta+\sigma \tag{4.2.3}$$

with σ the Higgs boson.

We can simulate the O(4) theory on a lattice, for example on a cubic lattice with nearest - neighbour interaction. Similarly to (4.2.2), the interaction is determined by two constants: the bare coupling, λ_0, and κ_0, equivalent to a bare mass term. For given values of λ_0 and κ_0, we can find η and the Higgs boson mass, m_H. Of course, the numerical simulation gives two pure numbers, to be interpreted as the corresponding quantities in unit of the lattice spacing, (ηa) and $(m_H a)$, $a\sim\Lambda^{-1}$. From their ratio we may obtain the Higgs boson/W mass ratio, to lowest order in the gauge interaction but exactly in the (regularized) O(4) interaction:

$$\frac{m_H}{M_W} = \frac{\sqrt{2}}{g} (\frac{m_H a}{\eta a}) +O(g) \tag{4.2.4}$$

By letting λ_0 and κ_0 to vary in all possible ways, we can determine the boundaries of the allowed region for η and m_H (in practice, the boundary is found by taking $\lambda_0 = \infty$, see eqs.(4.1.2) and(4.1.3), and varying κ_0).

In the past years, numerical simulations of O(4) have been performed by different groups[82].The results are shown in Fig.4.1, taken from ref.[79], where the ratio (4.2.4) is plotted versus the Higgs boson correlation lenght :

$$\xi = \frac{1}{m_H a} = \frac{\Lambda}{m_H} \qquad\qquad (4.2.5)$$

The allowed region lies **below** the experimental points, which thus represent the upper bound to the ratio (4.2.4), as function of the correlation lenght. For $\xi \to \infty$, the upper bound decreases slowly, to approach eventually the logarithmic behaviour implied by eq.(4.1.4). On the other side, the upper bound increases slowly, until m_H becomes of the order of the cut-off, $\xi \sim 1$. At this point all quantities start to be sensitive to the detailed form of the regularization, and we leave the continuum region. This happens when the ratio in (4.2.4) is about 8, which gives the absolute upper bound:

$$m_H < 8\, M_W \sim 650\,\text{GeV} \qquad\qquad (4.2.6)$$

corresponding to $\Lambda \sim 1.2$ TeV. This is well within reach of the presently planned proton colliders, SSC and LHC.

The numerical results agree with the analytic calculations[83a,b] which give the continuous curves in Fig. 4.1. More recent numerical calculations[84,85] have confirmed the result (4.2.6).

It is absolutely remarkable how well the non-perturbative upper bound agrees with the perturbative estimate in (4.1.5). The O(4) interaction does not really become strong and perturbation theory never fails!

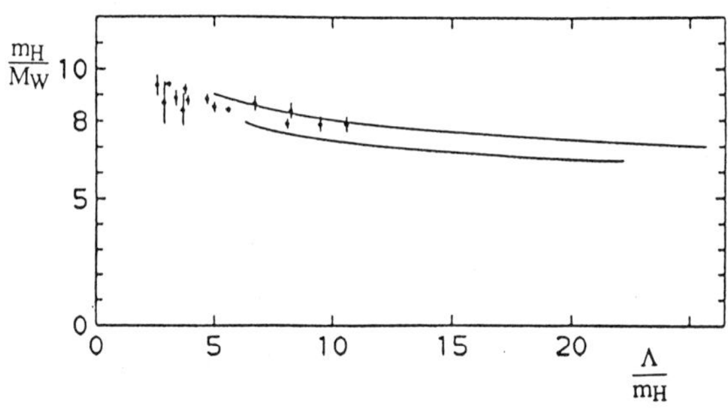

Fig. 4.1

Upper bound to the ratio m_H/M_W, as function of the correlation length: $\xi = \Lambda/m_H$. Points with errors are from MonteCarlo simulations, ref.[82]. Continuous curves from ref.[83a], upper, and ref.[83b], lower . Figure from ref.[79].

4.3 The effect of a heavy t-quark

It has not yet been possible to study systematically the effect of a heavy t-quark with lattice simulations, so we shall restrict ourselves to considerations based on renormalization group equations[86].

The coupled top-Higgs one-loop renormalization group equations are:

$$\mu\frac{\partial g_t^2}{\partial\mu} = \frac{9}{16\pi^2}\, g_t^4 \qquad\qquad (4.3.1)$$

$$\mu\frac{\partial\lambda}{\partial\mu} = \frac{3}{8\pi^2}\,(4\lambda^2 + 2\lambda\, g_t^2 - g_t^4) \qquad\qquad (4.3.2)$$

(for a more recent analysis, including two-loop effects, see Lindner, ref.[87]). The Yukawa interaction by itself is also non-asymptotically free, and we obtain, from the first equation, a bound entirely similar to that in eq.(4.1.3), namely:

$$\frac{1}{g_t^2} > \frac{9}{16\pi^2}\,\ln\frac{\Lambda}{\eta} \qquad\qquad (4.3.3)$$

The structure of the solutions can be approximately understood as follows. Eq.(4.3.2) admits two fixed points, where the r.h.s. vanishes, namely:

$$\lambda = \frac{g_t^2}{4}\,(-1\pm\sqrt{5}) \qquad\qquad (4.3.4)$$

The positive root is a repulsive fixed point, for $\mu\to\infty$. In fact, consider a fixed value of g_t. If λ is above the positive fixed point, the r.h.s. of eq.(4.3.2) is positive, and λ is pushed further above. Viceversa, if λ is below, the r.h.s. is negative , and λ is pushed further down. The other root, on the contrary, would be an attractive fixed point. However it is negative, and λ cannot become negative, least the theory becomes unstable, with an unbounded-from-below Hamiltonian.

The conclusion is that, for any value of g_t within the bound (4.3.3), λ has to be in a certain interval around the positive root in (4.3.4). If λ is too far above, it is pushed to $+\infty$ at a mass scale smaller than the physical cut-off, Λ. If it is too far below, it becomes negative before Λ.

The allowed interval shrinks, as g_t approaches its upper bound (4.3.3) because g_t varies more and more rapidly with $\ln\mu$ and so does λ. The interval reduces finally to a point, when g_t reaches the value (4.3.3).

The result of these considerations are reported in the diagram of Fig.4.2, taken from ref.[86] and computed for a GUT-like cut-off, $\Lambda\sim 10^{16}$ GeV. In correspondence to the value of m_t preferred by electroweak corrections (previous Section), one finds:

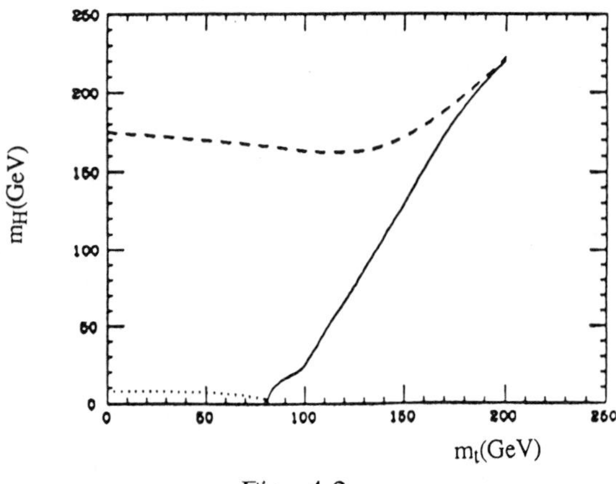

Fig. 4.2

Upper and lower bounds to m_H as function of m_t, with an ultraviolet cut-off $\Lambda\sim M_{GUT}\sim 10^{16}$ GeV, ref.[86]. Dots: Linde-Weinberg bound, dashes and solid: renormalization group. Three generations of quarks and leptons assumed.

$$80 \text{ GeV} \leq m_H \leq 160 \text{ GeV} \qquad (m_t = 130 \text{ GeV}) \qquad\qquad (4.3.5)$$

The lower side of the frontier of the allowed domain cuts the axis $m_H=0$ for $m_t \sim 80$ GeV. The lower bound to m_H for a lighter t-quark, is also indicated in the figure. It corresponds to the so-called Linde-Weinberg bound[88], and it arises from the requirement that the spontaneously broken vacuum state, $\eta \neq 0$, has indeed a lower energy density than the exactly symmetriuc state, $\eta=0$. The experimental lower bound (4.0.5) excludes already that we are on this branch, which is reported here for completness only.

The sensitivity of the allowed region to the value of Λ is indicated in Fig.4.3, taken from ref.[87].

Analogous considerations may be used to study the lower bound to m_H in minimal supersymmetric models, ref.[89,90].

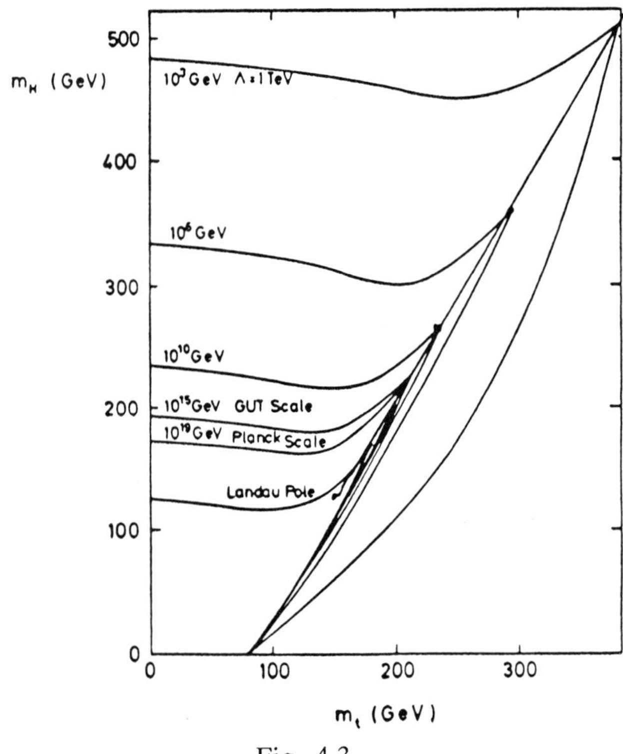

Fig. 4.3

Same as in Fig.4.2, for the indicated values of the ultraviolet cut-off, Λ., ref.[87]. Two-loop calculation, three generations of quarks and leptons assumed.

5. OUTLOOKS

The Standard Electroweak Theory has been extraordinarily successful in describing particle interactions, up to the energies thus far reached. In particular, LEP experiments have determined the physical properties of the Z-boson to a remarkable precision. Their results are in beautiful agreement with the theoretical expectation, provided the t-quark is really in a rather narrow mass range, approximately from 100 to 160 GeV.

Besides the t-quark, the Higgs particle is still missing, with a mass between the present experimental limit, about 45 GeV, and a theoretical limit of about 600 GeV.

It may seem that there is little room for new discoveries, with a dull future sitting in front of us. I do not share this opinion. Rather, I can see many areas which are ready for decisive progress. Let me close with a, most likely incomplete, list.

High energy side. The arguments presented in Sect. 4 leave little doubt that there is a physical cut-off energy, Λ, above which the Standard Theory undergoes deep modifications.

The worst case is the one in which the cut-off is provided by the gravitational interactions, with no new physics between us and $\Lambda \sim M_{Planck}$. This is the "desert" idea. Its signature would be the discovery of a standard Higgs boson and a standard t-quark, approximately within the boundaries illustrated in Fig.4.2.

This possibility, fortunately, meets with the well known problem of fine-tuning[91], which makes it completely unlikely. The problem is that the one-loop correction to the Higgs boson mass, computed with a finite cut-off, Λ, is, in general, not much smaller than the cut-off itself:

$$\delta\mu^2 \sim \frac{\alpha}{\pi}\Lambda^2 \tag{5.0.1}$$

It would take an extraordinary fine-tuning between the bare mass and the correction to have, at the same time:

$$\Lambda \sim M_{Planck} \sim 10^{19}\, \text{GeV} \text{ , and } \quad \mu^2 = \mu_o^2 + \delta\mu^2 \sim (G_F)^{-1} \sim (300\, \text{GeV})^2 \tag{5.0.2}$$

If there is no special symmetry to justify this cancellation, the cut-off must be low, such that $\delta\mu^2$ is at most of the order of $(G_F)^{-1}$, namely $\Lambda \sim$ TeV. This is the situation envisaged, for example, by t 'Hooft[92] in Cargese 1978. It corresponds to the option of a strongly interacting Higgs particle, or to a composite Higgs particle of the kind proposed by Technicolor theory[93], with new world of strongly interacting states in the critical TeV energy rangy.

However, there is a possible alternative, namely low-energy supersymmetry[94].

In a supersymmetric world, the corrections to the Higgs boson mass follow the same pattern as fermionic mass corrections. The latter are protected by chiral symmetry, and are typically of the order:

$$\delta m \sim \frac{\alpha}{\pi}m_0 \ln \left(\frac{\Lambda}{m_0}\right) \tag{5.0.3}$$

i.e. are naturally much smaller than the cut-off itself. In presence of the necessary supersymmetry breaking, the role of the cut-off in eq.(5.0.1) would be taken by the typical fermion-boson mass splitting:

$$\delta\mu^2 \sim \frac{\alpha}{\pi}\Delta M^2 \tag{5.0.4}$$

With $\Delta M \sim$ 1TeV we would be again in business, this time the critical energy of 1TeV representing the threshold for the production of the supersymmetric partners of the known particles. This is the elementary option: with supersymmetry, all particles, Higgs boson included, are allowed to be pointlike, up to a very large cut-off, of the order of M_{Planck}.

Thus, we see that we have really only two possibilities: the strongly interacting, most likely composite, Higgs particle, with new strong interactions, and the elementary Higgs particle, with no strong interactions at high energy, but a supersymmetry threshold[95].

In all cases, the TeV range is crucial, and this makes clearly a good case for the construction of new colliders in this energy region.

Supersymmetric models with a viable phenomenology exist, which is a good point in favour, while no completely consistent Technicolor model has been found, yet. However, Tecnhicolor has been analyzed, until now, by imitation of what we think are the salient features of QCD. Is this really true? or, rather, are we missing some unexpected non-perturbative behaviour, particularly in chiral gauge theories, that will make this a possible alternative? (think of the status of Yang-Mills theories in the fifties).

Intermediate energy. Contrary to the Electroweak Interactions, the really strong interactions that bind quarks inside the hadrons still defy a detailed theoretical understanding.

Beauty mesons and baryons, that is systems containing one very heavy quark may be very useful in this respect. The infinite mass limit for the beauty quark exists, in QCD, and leads to a quite simplified dynamics, although still fully non-perturbative[96].

In this direction, considerable theoretical progress can be envisaged in the coming years, particularly for lattice calculations in the static ($M\rightarrow\infty$ limit), with the possible inclusion of the next-to-leading, 1/M, corrections. This gives the exciting possibility of a systematic comparison of reliable theoretical calculations with B-particle spectroscopy and decays, such as can be obtained at a high-luminosity B-factory[97].

At a still lower energy scale, chiral dynamics can be tested in K decays, at the planned Kaon or Phi-factories.

CP-violation: Detection and study of CP-violation in B-decays is an essential step for a complete test of the electroweak theory. We still lack a basic understanding of CP-violation, and surprises are quite possible.

Very low energy: Solar neutrinos are the best example of frontier physics at very low energy. The gallium experiments, started at the Baksan and Gran Sasso Laboratories, are expected to give in quite a short time a crucial answer to the problem of the solar neutrino deficit, indicated since many years by Davies, and made more real by the recent Kamiokande data.

ACKNOWLEDGMENTS

I would like to acknowledge the stimulating discussions I had with lecturers and students, in Cargese, and with many other colleagues. In particular, I would like to thank R.Barbieri, F. Berends, A. Blondel, B. Borgia, A. Borrelli, M. Consoli, M. Diemoz, R.Eichler, F. Feruglio, V. Khoze, E. Longo, M. Lusignoli, G. Martinelli, A. Masiero, J. Steinberger. My thanks go also to the organizers, for their nice hospitality, for providing an exceptionally warm and clear sea and for not pressing me too much for the Lecture Notes. The technical assistance and skill of Mrs. A. Disilvestro is gratefully acknowledged.

REFERENCES

1.   C.N.Yang, R.L.Mills, Phys. Rev. <u>26</u> (1954) 191.
2.   J.C.Taylor, Gauge Theories of Weak Interactions, Cambridge Univ. Press, (Cambridge 1976).
3.   T.D.Lee, Particle Physics and Introduction to Field Theory, Harwood Academic Publisher (Chur, London, New York, 1981).
4.   L.B.Okun, Leptons and Quarks, Nort-Holland Physics Publishing (Amsterdam, 1982).
5.   See e.g. S.Coleman, Aspects of Symmetry, Cambridge Univ. Press (Cambridge, 1985).
6.   J.Goldstone, A.Salam, S.Weinberg, Phys. Rev.<u>127</u> (1962) 965.
7.   P.Higgs, Phys. Lett. <u>12</u> (1964) 132;
      F.Englert and R.Brout, Phys. Rev.Lett. <u>13</u> (1964) 321;
      T.W.Kibble, Phys. Rev. <u>155</u> (1967) 1554.
8.   S.L. Glashow, Nucl. Phys. <u>22</u> (1961) 579;
      S.Weinberg, Phys. Rev.Lett. <u>19</u> (1967) 1264;
      A.Salam, in Proc. 8th Nobel Symp., N.Svartholm ed. (Almqvist and Wicksell, Stockholm, 1968). The phenomenological foundations of the Standard Theory are discussed more in detail in ref.[3, 4]. See also:
      M.K.Gaillard, L.Maiani, Quarks and Leptons, Cargese 1979, ed. by R.Gastman et al..
9.   H.Georgi and S.L.Glashow, Phys.Rev.Lett. <u>28</u> (1972) 1494.

10. S.L. Glashow , J.Iliopoulos and L.Maiani, Phys. Rev. $\underline{D2}$ (1970) 1285.
11. M.Kobayashi and K.Maskawa, Progr. Theor. Phys. $\underline{49}$ (1973) 652.
12. M.Perl, Proc. of the 19th Int. Conf. on High-Energy Physics, Tokyo 1978, p.777.
13. Herb et al., Phys. Rev. Lett. $\underline{39}$ (1977) 252.
14. V.Gribov and B.M.Pontecorvo, Phys. Lett. $\underline{28B}$ (1969) 493. For a recent review see "Neutrino Physics", ed. by K.Winter, Cambridge Univ. Press, in publication.
15. R.J.Eden, P.V.Landshoff, D.I.Olive, J.C.Polkinghorn, The Analytic S-Matrix, Cambridge Univ. Press, 1966.
16. R.P.Feynman, Acta Physics Polonica $\underline{24}$ (1963) 697; R.P.Feynman in: Proc. of the Les Houches Summer School, 1976.
17. B.De Witt, Proc. of the Les Houches Summer School, 1983.
18. L.D.Faddeev, V.N.Popov, Phys. Letters $\underline{25}B$ (1967) 29.
19. E.S.Abers and B.W.Lee, Phys. Reports C $\underline{9}$ (1973) 1.
20. S.Adler, Phys. Rev. $\underline{177}$ (1969) 2426; J.S.Bell and R.Jackiw, Nuovo Cim. $\underline{60}$ (1969) 47; see also: R.Jackiw in: Lectures on Current Algebra and Its Applications, ed. by S.B.Treiman, R.Jackiw and D.J.Gross. (Princeton Univ. Press, N.J., 1972).
21. H.B.Nielsen and M.Ninomiya, Nucl. Phys. $\underline{B185}$ (1981) 20; E $\underline{B195}$ (1982) 541; $\underline{B193}$ (1981)173.
22. L.H.Karsten and J.Smit, Nucl. Phys. $\underline{B183}$ (1981) 103; M.Bochicchio, L.Maiani, G.Martinelli, G.C.Rossi and M.Testa, Nucl. Phys. $\underline{B262}$ (1985) 331.
23. S.Weinberg, Phys. Rev. D $\underline{11}$, 3583 (1975); J.Kogut and L.Susskind, Phys. Rev. D $\underline{11}$, 3594 (1976); G. 't-Hooft, Phys. Rev. Lett. $\underline{37}$ (1976) 8; Phys. Rev. D $\underline{14}$ (1976) 3432. See ref.[5] for a clear and detailed discussion.
24. C.Bouchiat, J.Iliopoulos and Ph.Meyer, Phys. Lett. $\underline{38B}$ (1972) 519; D.Gross and R.Jackiw, Phys.Rev. $\underline{D6}$ (1972) 477.
25. W.A.Bardeen, Phys. Rev. $\underline{184}$ (1969) 1848.
26. J.Wess and B.Zumino, Phys. Letters $\underline{37B}$ (1971) 95.
27. E.D'Hooker, E.Fahri, Nucl. Phys. $\underline{B248}$ (1984) 59.
28. D.J.Gross and F.Wilczeck, Phys. Rev. Lett. $\underline{30}$ (1973) 1343; D. Politzer, Phys. Rev.Lett. $\underline{30}$ (1973) 1346; G.'t-Hooft, unpublished. See also refs. [3, 4].
29. K.G.Wilson in New Phenomena in Subnuclear Physics, ed. by A.Zichichi (Plenum Press, New York, London, 1977); see also: M.Creutz, Quarks Gluons and Lattices, Cambridge Univ. Press (Cambridge, 1983).
30. For a review, see e.g. E.Marinari in: Lattice '88, ed. by A.S.Kronfeld and P.B. Mackenzie, Nucl. Phys. B (Proc. Suppl.) $\underline{9}$ (1989) 209.
31. L.D.Landau, A.A.Abrikosov and I.M.Khalatnikov, Dokl. Akad. Nauk USSR 95 (1954) 773, 1177; 96 (1954) 261; L.D.Landau and I.Pomeranchuk, Dokl. Akad. Nauk USSR 102 (1955) 489; L.D.Landau in: Niels Bohr and the development of physics, ed. W.Pauli (Pergamon Press, 1955).
32. G.Altarelli, Ann. Rev. Nucl. Part. Sci. $\underline{39}$ (1989) 357.
33. P.Abreu et al., DELPHI Collaboration, Phys. Lett. $\underline{B}$, to be published; B.Adeva et al., L3 Collaboration, Phys. Lett. $\underline{B248}$ (1990) 464. The results of the ALEPH and OPAL Collaborations have been reported in: Proc. of the Int. Conf. on High Energy Physics, Singapore, August 1990. See also the review by J.Steinberger, CERN-PPE/90-149. 12 October 1990.
34. P.Langacker in: Review of Particle Properties, 1990.
35. D.Froidevaux, Proc. of the International Conference on Neutrino Physics and Astrophysics, "Neutrino '90", CERN, June 1990. J.Alitti et al., UA2 Collab., Phys. Letters B241 (1990) 150.
36. S.L.Glashow, H.Georgi, Phys. Rev. Letters $\underline{32}$ (1974) 438.
37. See e.g. W.Marciano, Proc. Eighth Grand Unification Workshop, Syracuse, N.Y., April 1987, ed. by K.C.Wali, (World Scientific) p.185.

38. IMB, KAMIOKANDE and Frejus results, see e.g., Y.Totsuka, Rapporteur talk, Proc. 24th Conf. on High Energy Physics, Munich (Aug., 1988) p.282.

39. J.Ellis, S.Kelley, D.V.Nanopoulos, joint preprint CERN-TH-5773/90, CTP-TAMU-60/90, ACT-11, June 1990.

40. Earlier calculations of electroweak corrections in e^+e^- annihilation and neutrino scattering are found in:
G.Passarino and M.Veltman, Nucl. Phys. B160 (1979) 151;
M.Consoli, Nucl. Phys. B160 (1979) 208;
F.A.Berends, R.Kleiss and S.Jadach, Nucl. Phys. B202 (1982) 63;
A.Sirlin, Phys. Rev. D22 (1980) 971;
W.Marciano and A.Sirlin, Phys. Rev. D22 (1980) 2695;
D.Yu Bardin, P.Ch. Christova and O.M.Fedorenko, Nucl. Phys. B175 (1980) 435; B197 (1982) 1;
C.H.Llewellyn Smith and J.F.Wheather, Phys. Lett. 105B (1981) 486;
G.G.Ross, C.H.Llewellyn Smith and J.F.Wheather, Nucl. Phys. B177 (1981) 263.
A more recent source of information is: Z-physics at LEP1, CERN Yellow Report, 89-08 (Sept. 1989) eds. G.Altarelli, R.Kleiss and C.Verzegnassi.

41. M.Veltman, Phys. Lett. 91B (1980) 95;
F.Antonelli, M.Consoli and G.Corbò, Phys. Lett. 91B (1980) 90.

42. L.Maiani, Proc. of the Int. Conf. on Unified Theories and Experimental Tests, March 16-18, 1982, Venezia;
Z.Hioki, Progr. Theor. Phys. 68 (1982) 6;
M.Consoli, S.Lo Presti and L.Maiani, Nucl. Phys. B223 (1983) 474.

43. See the papers by A.Sirlin and W.Marciano quoted under ref.[40].

44. M.Veltman, Nucl. Phys. B123 (1977) 89.

45. F.Antonelli, L.Maiani, Nucl. Phys. B186 (1981) 269;
S.Dawson, J.S.Hagelin, L.Hall, Phys. Rev. D23 (1981) 2666;
E.Paschos, M. Wirbel, Nucl. Phys. B194 (1982) 189.

46. S.M.Berman, A.Sirlin, Am. Phys.(N.Y.) 20 (1962) 20. For a general proof of the non-renormalization theorem, see:
G.Preparata, W.Weisberger, Phys. Rev. 175 (1968) 1965.

47. The effective $\sin^2\theta$ was introduced by M.Consoli et al., quoted under ref.[42]. See also:
W.Hollik, preprint DESY 88-188 (1988);
D.C.Kennedy et al., Nucl. Phys. B321 (1989) 83;
D.C.Kennedy and B.W.Lynn, Nucl. Phys. B322 (1989) 1;
B.W.Lynn, Stanford report SU-ITP-867 (1989);
W.J.Marciano and A.Sirlin, Phys. Rev. Lett. 46 (1981) 163;
J.Sarantakos, W.J.Marciano and A.Sirlin, Nucl. Phys. B217 (1983) 84;
A.Sirlin, preprint CERN-TH. 5506/89 (1989).

48. A.A.Akhundov, D.Yu.Bardin and T.Riemann, Nucl. Phys. B276 (1986) 1. See also:
J.H.Kühn and P.M.Zerwas in: Z-Physics al LEP1, quoted in ref.[40].

49. M.Consoli, W.Hollik, F.Jegerlehner, CERN-TH. 5395/89 (1989). See also: M.Consoli and W.Hollik, Z-Physics at LEP1, quoted in ref.[40].

50. M.Greco, G.Pancheri and Y.Srivastava, Nucl. Phys. B171 (1980); B197 (1982) 543;
F.A.Berends, R.Kleiss and S.Jadach, Nucl. Phys. B202 (1982) 63;
M.Böhm and W.Hollik, Nucl. Phys. B204 (1982) 45; Z. Phys. C23 (1984) 31; Phys. Lett. B139 (1984) 213;
W.Wetzel, Nucl. Phys. B227 (1982) 1;
R.W.Brown, R.Decker and E.A.Paschos, Phys. Rev. Lett. 52 (1984) 1192;
B.W.Lynn and R.G.Stuart. Nucl. Phys. B253 (1985) 216.

51. E.A.Kuraev and V.S.Fadin, Sov. J. Nucl. Phys. 41 (1985) 446; G.Altarelli and G.Martinelli, in Physics at LEP, ed. J.Ellis and R.Peccei, CERN Yellow Report 86 (02), 1986, p.47.

52. F.A.Berends, G.Burgers and W.L.van Neerven, Phys. Lett. $\underline{B185}$ (1987) 395; Nucl. Phys. B297 (1988) 429; O.Nicrosini and L.Trentadue, Phys. Lett. $\underline{B196}$ (1987) 551; G.Burgers, in Polarization at LEP, ed. G.Alexander et al., CERN Yellow Report 88-06, 1988; R.N.Cahn, Phys. Rev. $\underline{D36}$ (1987) 2666. For more information, see F.Berends, this School.

53. D.Yu. Bardin, A.Leike, T.Reimann and M.Sachwitz, Phys. Lett. $\underline{B206}$ (1988) 539; F.A.Berends, G.Burgers, W.Hollik and W.L. van Neerven, Phys. Lett. $\underline{B203}$ (1988) 177.

54. Electroweak corrections to the line-shape,with reference to previous work, have been summarized by F.Berends in: Z-physics at LEP1, quoted under ref.[40], p.89.

55. G.Burgers, quoted under ref.[52].

56. A.Borrelli, M.Consoli, L.Maiani, R.Sisto, Nucl. Phys. $\underline{B333}$ (1990) 357.

57. a. D.Dechamp et al. ALEPH Collaboration, CERN-PPE/90-104;
b. P.Abreu et al. DELPHI Collaboration, CERN-PPE/90-119;
c. B.Adeva et al., L3 Collaboration, L3 preprint #8,9 and 17;
d. M.Z.Akrawy, OPAL Collaboration, OPAL Physics Note 90-16.

58. A.Borrelli, L.Maiani, R.Sisto, Phys. Lett. $\underline{B244}$ (1990) 117; Rendiconti dell'Accademia Nazionale dei Lincei, to be published.

59. J.Ellis, G.L.Fogli, CERN-TH 5817/90.

60. V.Barger, J.L.Hewett, T.G.Rizzo, Phys. Rev. Lett. $\underline{65}$ (1990) 1313.

61. B.Borgia, Electroweak Interactions at LEP, SLAC Summer Institute, Topical Conference, July 1990. Nota Interna n.962, Univ. di Roma I.

62. C.Q.Geng, J.N.Ng, Phys. Rev. $\underline{D41}$ (1990) 2351.

63. ARGUS Collaboration, H.Albrecht et al., Phys. Lett. $\underline{B234}$ (1990) 409.

64. CLEO Collaboration, R.Fulton et al., Phys. Rev. Lett. $\underline{64}$ (1990) 16.

65. J.Maalampi, M.Roos, Particle World $\underline{1}$ (1990) 148.

66. A lucid analysis is given in G.Buchalla, A.Buras, M.Harlander, MPI-PAE/PTh 63/89;

67. M.Lusignoli, L.Maiani, G.Martinelli, C.Reina, in preparation. The large range of variation for the kaon B-parameter considered in ref.[68] did not allow a clear separation of two solutions . Otherwise, we agree with the results of this paper.

68. C.Dib, I.Dunietz, F.Gilman, Y.Nir, Phys. Rev. D$\underline{41}$ (1990) 1522.

69. C.Allton et al., Nucl. Phys. $\underline{B349}$ (1991) 598.

70. C.Alexandrou et al., Wuppertal preprint, WU-B-90-18; C.Bernard et al., presented to the Lattice Workshop in Tallahasse, September 1990.

71. C.A.Dominguez, N.Paver, Phys. Letters B$\underline{246}$ (1990) 493; Phys. Letters $\underline{B197}$ (1987) 423.
For earlier calculations, see also:
S.Narison, Phys. Lett. $\underline{B198}$ (1987) 104;
A.Pich, Phys. Lett. $\underline{B206}$ (1988) 322.

72. See, e.g. R.A.Eichler, this School.

73. B.W.Lee, C.Quigg and H.B.Thacker, Phys. Rev. $\underline{D16}$ (1977) 1519.
A similar bound was derived earlier by:
D.A.Dicus, V.S.Mathur, Phys. Rev. $\underline{D7}$ (1973) 3111.

74 M.S.Chanowitz, M.A.Furman, I.Hinchliffe, Nucl. Phys. $\underline{B153}$ (1979) 402.

75. Reported in Proc. of the Int. Conf. on High Energy Physics, Singapore, August 1990.

76. L.Maiani, G.Parisi, R.Petronzio, Nucl. Phys. $\underline{B136}$ (1978) 115.

77. G.Parisi, Lecture notes of the 1973 Cargèse Summer School, Columbia Univ. preprint, 1974, unpublished;
J.Glimm, A.Jaffe, Phys. Rev. Lett. $\underline{33}$ (1974) 440;
J.Kogut, K.Wilson, Phys. Reports $\underline{12}$ (1975) 75;
R.Schrader, Commun. Math. Phys. $\underline{49}$ (1976) 131; 50 (1976) 97.

78. J.Fröhlich, Nucl. Phys. $\underline{B200}$ FS4 (1982) 281.

79. For a recent review, including numerical simulation results, see P.Hasenfratz in Lattice '88, quoted under ref.[30].

80. W.A.Bardeen, M.Moshe, Phys. Rev. $\underline{D28}$ (1983) 1372.

81. R.Dashen, H.Neuberger, Phys. Rev. Lett. $\underline{50}$ (1983) 1897.
82. J.Kuti, L.Lin and Y.Shen, Phys. Rev. Lett. $\underline{61}$ (1988) 678;
    A.Hasenfratz, K.Jansen, C.B.Lang, T.Neuhaus and H.Yoneyama, Phys. Lett. $\underline{B199}$ (1987) 531;
    G.Bhanot and K.Bitar, Phys. Rev. Lett. $\underline{61}$ (1988) 798.
83. a. P.Hasenfratz and J.Nager in: Nonperturbative methods in quantum field theory, ed. by Z.Horvath et al., p.89, Siofok (1986); Z.Phys. $\underline{C37}$ (1988) 477;
    b. M.Lüscher and P.Weisz, Nucl. Phys. $\underline{B290}$ [FS20] (1987) 25; Nucl. Phys. $\underline{B295}$ [FS21] (1988) 65; Phys. Lett. $\underline{B212}$ (1988) 472; Nucl. Phys. $\underline{B318}$ (1989) 705.
84. G.Bhanot, K.Bitar, U.Heller and H.Neuberger; FSU-SCRI-90-111 (RU-90-35) and FSU-SCRI-90-02 (RU-89-46); H.Neuberger, Phys. Lett. $\underline{B199}$ (1987) 5.
85. For a recent review, see J.Shigemitsu, Proc. of Lattice '90, Tallahasse, Nucl. Phys. B (Proc. Supp.) in preparation.
86. N.Cabibbo, L.Maiani, G.Parisi, R.Petronzio, Nucl. Phys. $\underline{B158}$, (1979) 296.
87. M.Lindner, Z. Phys. C-Particles and Fields $\underline{31}$ (1986) 295; M.Lindner, M.Sher, W.Zanglauner, Phys.Lett. $\underline{228B}$ (1989) 139.
88. A.Linde, JETP Letters $\underline{23}$ (1976) 64;
    S.Weinberg, Phys. Rev. Lett. $\underline{36}$ (1976) 294.
89. J.Ellis, G.Ridolfi and F.Zwirner, CERN preprint, TH 5946/90;
    H.Haber, unpublished.
90. R.Barbieri, M.Frigeni and F.Caravaglios, preprint, Universita' di Pisa, IFUP-TH 46/90.
91. E.Gildener, Phys. Rev. $\underline{D14}$, (1976) 1667.
92. G.t'Hooft, Proc. of the Cargèse Summer School, 1978.
93. S.Weinberg, Phys. Rev. $\underline{D13}$ (1976) 974; $\underline{D19}$ (1979) 1277;
    L.Susskind, Phys. Rev. $\underline{D20}$ (1979) 2615;
    S.Dimopoulos, L.Susskind, Nucl. Phys. $\underline{B155}$ (1979) 237.
94. The relevance of supersymmetry in this connection has been stressed by various authors:
    E.Gildener, S.Weinberg, Phys. Rev. $\underline{D13}$ (1976) 3333;
    L.Maiani, Proc. Summer School Gif-sur-Yvett, 1979 (IN2 P3, Paris, 1980);
    M.Veltman, Acta Physica Pol. $\underline{B12}$ (1981) 437;
    E.Witten, Nucl. Phys. $\underline{B188}$ (1981) 513;
    S.Dimopoulos, S.Raby, Nucl. Phys. $\underline{B199}$ (1981) 353.
95. For a review of supersymmetric models see:
    H.Nilles, Phys. Rep. $\underline{C10}$ (1984) 1;
    R.Barbieri, Riv. Nuovo Cimento $\underline{11}$ (1988).
    Experimental signatures at LEP are reviewed in: R.Barbieri, Z-Physics at LEP1, Vol.2 quoted under ref.[40]; M.Chen, C.Dionisi, M.Martinez and X.Tata, Phys. Rep. $\underline{159}$ (1988) 201.
96. M.B.Voloshin and M.A.Shifman, Sov. J. Nucl.Phys. $\underline{45}$ (1987) 292; Sov. J. Nucl. Phys. $\underline{47}$ (1988) 511;
    H.D.Politzer and M.B.Wise, Phys. Lett. $\underline{206B}$ (1988) 681; Phys. Lett. $\underline{208B}$ (1988) 504;
    N.Isgur and M.B.Wise, Phys. Lett. $\underline{232B}$ (1989) 113; CALT-68-1608;
    E.Eichten and Hill, Phys. Lett. $\underline{234B}$ (1990) 511;
    H.Georgi, Harvard University Preprint, HUTP-90/A002 (1990);
    A.Falk, H.Georgi, B.Grinstein and M.B.Wise, Harvard University Preprint HUTP-90/A011.
    A recent summary of this subject has been given by:
    J.D.Bjorken, Proc. of the Int. Conf. on High Energy Physics, Singapore, August 1990.
97. See e.g. the B-Factory Study in: CERN 90-02, T.Nakada (Editor).
98. L.Maiani, Proc. of the Int. Symposium on Lepton and Photon Interaction at High Energy, Hamburg, 1977.
99. H.Harari and M.Leurer, Phys. Lett. $\underline{B181}$ (1986) 123;
    H.Fritzsch and J.Plankl, Phys. Rev. $\underline{D35}$ (1987) 1732.
100. L.Wolfenstein, Phys. Rev. Lett. $\underline{51}$ (1983) 1945.

Appendix A

More about large m_t effects

i. We consider first the effect on M_W and M_Z of a large m_t, following ref.[42]. We start from the expression:

$$M_W^2 = \frac{\pi\alpha}{\sqrt{2}G_F \sin^2\theta}\, [1-$$

$$-\frac{\delta\alpha}{\alpha} + \frac{\delta G_F}{G_F} + \frac{\delta M_W^2}{M_W^2} - \frac{c^2}{s^2}\frac{\delta M_W^2 - c^2\,\delta M_Z^2}{M_W^2}] \tag{A.1}$$

We define the Z coupling according to:

$$L_{eff} = \frac{g}{\cos\theta}Z^\mu[J_\mu^3 - \sin^2\theta\, J_\mu^{em}] \tag{A.2}$$

Vacuum polarization amplitudes, with external vector fields i and j, arising from fermion loops are described by two form factors:

$$-i\,\Pi_{ij}^{\mu\nu}(q^2) = -i\,[g^{\mu\nu}A_{ij}(q^2)+ (g^{\mu\nu}q^2 - q^\mu q^\nu)F_{ij}(q^2)]$$

where $-i\,\Pi_{ij}^{\mu\nu}$ is the amplitude obtained from the Feynman diagram. Because of exact charge conservation:

$$A_{\gamma\gamma}(q^2) = A_{3\gamma}(q^2) = 0$$

Inserting the above amplitudes in the vector propagators, we find directly the following relations:

$$\frac{\delta\alpha}{\alpha} = -F_{\gamma\gamma}(0) = -e^2\,\bar{F}_{QQ}(0) \tag{A3}$$

$$\frac{\delta G_F}{G_F} = g^2\,\frac{\bar{A}_W(0)}{M_W^2} \tag{A4}$$

$$\frac{\delta M_W^2}{M_W^2} = -g^2[\,\frac{\bar{A}_W(M_W^2)}{M_W^2} + \bar{F}_W(M_W^2)] \tag{A5}$$

$$\frac{\delta M_Z^2}{M_Z^2} = -g^2\{\,\frac{\bar{A}_{33}(M_Z^2)}{M_W^2} + [\bar{F}_{33}(M_Z^2)-2\,s^2\,\bar{F}_{3Q}(M_Z^2)+ s^4\bar{F}_{QQ}(M_Z^2)] \tag{A6}$$

where, going to the new, barred, form factors we have separated explicitly the dependence from the coupling constants. Substituting in eq.(A1), and using the relation:

$$Q = T_3 + Y$$

we obtain:

$$M_W^2 = \frac{\pi\alpha}{\sqrt{2}G_F s^2}\,[1+\Delta r] \qquad (A7)$$

$$\Delta r = e^2[\,\bar{F}_{QQ}(0) - \bar{F}_{QQ}(M_Z^2)\,] + \frac{c^2}{s^2}\frac{g^2[\bar{A}_W(0)-\bar{A}_{33}(0)]}{M_W^2} + $$

$$+ (1-\frac{c^2}{s^2})\,\frac{g^2[\bar{A}_W(0)-\bar{A}_W(M_W^2)]}{M_W^2} + \frac{c^2}{s^2}\frac{g^2[\bar{A}_{33}(0)-\bar{A}_{33}(M_Z^2)]}{M_W^2}\,] + $$

$$+ \frac{c^2-s^2}{s^2}g^2[\bar{F}_W(M_W^2)-\bar{F}_{33}(M_Z^2)] + $$

$$+ 2g^2\bar{F}_{3Y}(M_Z^2) \qquad (A8)$$

Judging from the superficial degree of divergence, form factors of type F and A are logarithmically and quadratically divergent, respectively. However, vector and axial vector currents are partially conserved, which means that form factors of type A vanish in the limit of massless internal fermions. Thus, they are proportional to the internal fermion masses and only logarithmically divergent. Keeping this into account, we see from eq.(A8) that Δr is finite, although individual corrections in (A1) are divergent. Indeed, each contribution in (A8) is either the difference of two form factors with the same logarithmic divergence, or it

is finite, as in the case of $\bar{F}_{3Y}$.

When we let m_t to increase above $M_{W,Z}$ with m_b fixed, all form factors become constant, type A form factors being proportional to m_t^2. Restricting to leading terms, we find:

$$(\Delta r)_t = -\frac{c^2}{s^2}\,\delta\rho \qquad (A9)$$

$$\delta\rho = \frac{g^2[\bar{A}_{33}(0)-\bar{A}_W(0)]}{M_W^2} \qquad (A10)$$

Light fermions give a leading logarithmic correction to Δr, embodied by the first term in the r.h.s. of eq.(A8), which coincides with the correction we have denoted by ε_α in Sect.2.5. In conclusion, we find, in the leading approximation:

$$M_W^2 = \frac{\pi\alpha}{\sqrt{2}G_F s^2}\,[1 + (\Delta r)_{\text{light}} + (\Delta r)_t\,] = $$

$$= \frac{\pi\alpha}{\sqrt{2}G_F s^2}\,[1+\varepsilon_\alpha - \frac{c^2}{s^2}\,\delta\rho] = \frac{\pi\alpha(M)}{\sqrt{2}G_F s'^2} \qquad (A11)$$

with:

$$s'^2 = s^2 + c^2\,\delta\rho \qquad (A12)$$

Identification of the expression in (A10) with $\delta\rho$ is made more clear by noting that, on the basis of the normalization condition (3.13) and of eq.(A12), we find:

$$M_Z^2 = \frac{M_W^2}{c^2} = \frac{M_W^2}{c'^2(1+\delta\rho)} \tag{A13}$$

which is the relation which would arise in the general gauge theory, with Higgs fields not restricted to SU(2) doublets.

We leave to the reader the generalization of the above argument to the partial widths for $Z\to f\bar{f}$, with $f\neq b$, to obtain eq.(3.2.11) of text.

The explicit calculation of $\delta\rho$ goes as follows. From the vacuum polarization Feynman diagrams we obtain:

$$-i\,[\Pi_{33}^{\mu\nu}(0) - \Pi_W^{\mu\nu}(0)] = -ig^{\mu\nu}\frac{g^2[\bar{A}_{33}(0)-\bar{A}_W(0)]}{M_W^2} =$$

$$= -3(i\,g^2)\frac{(i^2)}{(2\pi)^4}\frac{1}{4}\int d^4k[F^{\mu\nu}(k,m_t,m_t)+F^{\mu\nu}(k,m_b,m_b)-2F^{\mu\nu}(k,m_t,m_b)]$$

The factor 3 comes from colour, the minus sign from Fermi statistics and:

$$F^{\mu\nu}(k,m_t,m_b) =$$

$$= \frac{1}{(k^2-m_t^2+i\varepsilon)(k^2-m_b^2+i\varepsilon)}\,\mathrm{Tr}[\gamma^\mu\frac{(1-\gamma_5)}{2}(\not{k}+m_t)\gamma^\nu\frac{(1-\gamma_5)}{2}(\not{k}+m_b)]$$

Computing traces and replacing $4k^\mu k^\nu$ with $g^{\mu\nu}k^2$ inside the integral, due to symmetric integration, we obtain:

$$\frac{g^2[\bar{A}_{33}(0)-\bar{A}_W(0)]}{M_W^2} = \frac{3g^2}{64\pi^4}\int i d^4k\frac{k^2(m_t^2-m_b^2)^2}{(k^2-m_t^2+i\varepsilon)^2(k^2-m_b^2+i\varepsilon)^2}$$

so that, after a Wick rotation ($dk_0 = idk_4$, $k^2 \to -k^2$) and elementary integrations, we obtain, for $m_t \gg m_b$:

$$\delta\rho = \frac{g^2[\bar{A}_{33}(0)-\bar{A}_W(0)]}{M_W^2} = \frac{3g^2}{64\pi^2 M_W^2}\,m_t^2 = \frac{3G_F m_t^2}{8\sqrt{2}\pi^2} \tag{A14}$$

ii. Additional, large-m_t corrections to $Z\to b\bar{b}$ arise, as explained in text, from the diagrams in Fig.3.1a to c, involving the exchange of charged would-be-Goldstone fields. Following Sect.1.4, we write the relevant coupling as:

$$L_{\xi\,tb} = -g_t\,[\xi\,\bar{b}t_R + \xi^\dagger\,\bar{t}b_L]$$

$$\xi = \frac{1}{\sqrt{2}}(\xi^{(1)} + i\,\xi^{(2)})$$

$$g_t^2 = g^2\frac{m_t^2}{2M_W^2} = 2\sqrt{2}G_F m_t^2 \tag{A15}$$

The Z coupling to $t\bar{t}$ and $b\bar{b}$ are taken from eq.(3.2.5) of text.

Vertex corrections from the diagrams of Fig.3.1 a to c can be written as:

$$\delta L^{(a)} = \frac{g}{\cos\theta}\, Z^\mu\, \bar{b}\, \gamma_\mu\, \frac{(1-\gamma_5)}{2} b\, (v_t V + a_t A)$$

$$\delta L^{(b)} = \frac{g}{\cos\theta}\, Z^\mu\, \bar{b}\, \gamma_\mu\, \frac{(1-\gamma_5)}{2} b\, (v_b - a_b)\, Z$$

$$\delta L^{(c)} = \frac{g}{\cos\theta}\, Z^\mu\, \bar{b}\, \gamma_\mu\, \frac{(1-\gamma_5)}{2} b\, v_\xi\, H \tag{A16}$$

with V, A, Z and H divergent constants. The $1-\gamma_5$ structure is required by the fact that only b_L is coupled, in the limit of a vanishing b-quark mass.

To the leading order we are considering, external particle mass and momenta are to be neglected with respect to m_t. The same constants V to A would then appear in the renormalization of the photon vertex at zero momentum transfer, which has to vanish because of exact charge conservation. After the appropriate substition of Z-couplings with electric charges, we obtain:

$$0 = e\,(Q_t V + Q_b Z + Q_\xi H)\,[A^\mu\, \bar{b}\, \gamma_\mu\, \frac{(1-\gamma_5)}{2} b]$$

that is (since $Q_\xi = Q_b - Q_t = -1$):

$$0 = Q_t(V+Z) - (H+Z) \tag{A17}$$

Q_t is arbitrary, so that both terms in parenthesis must vanish, and we have only two independent vertices, namely V and A, describing the insertion of a vector or an axial vector current in the diagram of Fig.3.1a. Substuting now in (A16) the expressions:

$$v_t = +\frac{1}{4} - \frac{2}{3}\sin^2\theta; \qquad v_b = -\frac{1}{4} + \frac{1}{3}\sin^2\theta; \qquad v_\xi = -\frac{1}{2} + \sin^2\theta;$$

$$a_t = -a_b = -\frac{1}{4}$$

we obtain:

$$\delta L = \frac{g}{\cos\theta}\, Z^\mu\, \bar{b}\, \gamma_\mu\, (\delta v - \delta a\, \gamma_5) b$$

$$\delta v = -\delta a = \frac{1}{2}[\, v_t V + a_t A + (v_b - a_b) Z + v_\xi H\,] =$$

$$= \frac{1}{8}(V-A) - \frac{1}{4}Z - \frac{1}{4}H = \frac{1}{8}(V-A)$$

The calculation is now elementary. The insertion of a $\gamma_\mu\,(1-\gamma_5)$ vertex in Fig.3.1a gives rise to the convergent integral:

$$\delta v \;=\; \frac{1}{8}\,(-ig_t)^2\,2\int\frac{d^4k}{(2\pi)^4}\,\frac{(i)^3 m_t^2}{k^2(k^2-m_t^2)^2}\;=\;$$

$$=\frac{g_t^2}{64\pi^2}\int_0^\infty dk_E^2\,\frac{m_t^2}{(k_E^2-m_t^2)^2}$$

We have neglected the W mass and all external momenta, and Wick-rotated the integration path as before. After integration, we find, finally, the result quoted in eq.(3.2.6) of text:

$$\delta v =- \delta a = \frac{1}{6}\,\delta\rho \tag{A18}$$

On the basis of eq.(3.2.6), we find:

$$\frac{\delta\Gamma}{\Gamma}=\frac{\Gamma(Z\to b\,\bar b)-\Gamma(Z\to d\,\bar d)}{\Gamma(Z\to b\,\bar b)}=\frac{2(v_b-a_b)\delta v}{v_b^2+a_b^2}=-\frac{16}{6}\,\frac{1-\frac{2}{3}s^2}{1+(1-\frac{4}{3}s^2)^2}\,\delta\rho$$

For $\Gamma(Z\to b\,\bar b)\sim 360$ MeV and $s^2\sim 0.23$, this gives $\delta\Gamma\sim 7$ MeV for $m_t = 200$GeV, not far from the more exact estimate of ref.[48], $\delta\Gamma\sim 10$ MeV, see Fig.3.2.

Appendix B

Parameters of the Cabibbo-Kobayashi-Maskawa matrix

We follow the parametrization suggested in ref.[98], with the CP-violating phase displaced into the upper right corner, as suggested in ref.[99]. We write the charged weak current as:

$$J_\mu = \bar q_u U\,\gamma_\mu\,(1-\gamma_5)q_d\,; \qquad q_u=(u,c,t);\; q_d=(d,s,b)$$

and the CKM matrix, U, as:

$$U=\begin{pmatrix} c_\beta\,c_\theta & c_\beta\,s_\theta & s_\beta\,e^{-i\delta}\\[4pt] -c_\gamma s_\theta - c_\theta s_\gamma s_\beta\,e^{+i\delta} & c_\gamma c_\theta - s_\gamma s_\theta s_\beta\,e^{+i\delta} & c_\beta\,s_\gamma\\[4pt] s_\theta s_\gamma - c_\gamma c_\theta s_\beta\,e^{+i\delta} & -c_\theta s_\gamma - c_\gamma s_\theta s_\beta\,e^{+i\delta} & c_\gamma\,c_\beta \end{pmatrix} \tag{B1}$$

θ is the Cabibbo angle, δ the CP violating phase.

For small angles, the Wolfenstein parametrization[100] is convenient. We put:

$$s_\theta = \lambda\,; \qquad |\,c_\beta\,s_\gamma\,| = A\,\lambda^2\,; \qquad |\,s_\beta\,| = \rho A\,\lambda^3 \tag{B2}$$

and develop to lowest significant order the matrix elements of U, to obtain:

$$U=\begin{pmatrix} 1-\frac{\lambda^2}{2} & \lambda & \rho A\,\lambda^3 e^{-i\delta}\\[4pt] -\lambda(1+\rho A^2\lambda^4 e^{+i\delta}) & 1-\frac{\lambda^2}{2} & A\,\lambda^2\\[4pt] A\lambda^3(1-\rho e^{+i\delta}) & -A\,\lambda^2 & 1 \end{pmatrix} \tag{B3}$$

The values used in our analysis[67] are (see ref.[72] for an extended discussion of the experimantal input data):

$$s_\theta = \lambda = 0.221 \pm 0.002;$$

$$|V_{cb}| = A \lambda^2 = 0.047 \pm 0.002;$$

$$|V_{ub}| = \rho A \lambda^3 = 0.005 \pm 0.001 \tag{B4}$$

that is:

$$A = 1.00 \pm 0.05; \qquad \rho = 0.50 \pm 0.14 \tag{B5}$$

Unitarity of the U matrix implies that:

$$0 = V_{ud}^* V_{ub} + V_{cd}^* V_{cb} + V_{td}^* V_{tb} \sim V_{ub} + V_{td}^* - A \lambda^3 \tag{B6}$$

Thus, the vectors V_{ub}, V_{td}^* and $-A \lambda^3$ form a closed triangle in the complex plane. This is explicitly born out by the parametrization in (B3) which gives, for these three quantities, the complex vectors: $\rho e^{-i\delta}$, $1 - \rho e^{-i\delta}$, -1, after dividing by $A \lambda^3$. The unitarity Bjorken's triangle is shown in Fig.3.15. The unitarity triangle is obtuse or acute for the solutions found in text, Sect.3.5, which correspond to $\cos\delta < 0$, or $\cos\delta > 0$.

ELECTROWEAK RADIATIVE CORRECTIONS FOR Z PHYSICS

F.A. Berends

Instituut-Lorentz, University of Leiden
P.O. Box 9506, 2300 RA Leiden
The Netherlands

1 INTRODUCTION

In these lectures an overview of electroweak radiative corrections (EWRC) for Z physics will be given.

Electroweak radiative corrections consist of two parts: QED or photonic corrections and weak corrections. The former are due to the real or virtual emission of photons and the weak corrections are the remaining virtual corrections. For convenience, we include the fermion loops in the latter category, although in pure QED one also has to calculate a vacuum polarization.

The QED corrections are in principle known but in practice sometimes hard to evaluate. They can be large and are experiment dependent. Theoretically one does not learn anything about new physics from them, but without an accurate knowledge one cannot interpret experimental results. On the other hand, the weak corrections differ from model to model. In the standard model (SM) they still depend on the unknown masses of the top quark and Higgs boson, so in the SM the experimental verification of the weak corrections gives us information on these masses or tells us that no mass values are compatible with the measurements. Clearly, the weak corrections are the most interesting ones from the point of view of physics. They are however small so that they can only be determined with an accurate knowledge of the QED corrections.

The lectures will focus on those radiative corrections which are most relevant for the present Z physics. The most relevant processes are

$$e^+(p_+) + e^-(p_-) \to \bar{f}(q_+) + f(q_-) \ . \tag{1.1}$$

Typical examples are mupair production

$$e^+(p_+) + e^-(p_-) \to \mu^+(q_+) + \mu^-(q_-) \tag{1.2}$$

and Bhabha scattering

$$e^+(p_+) + e^-(p_-) \to e^+(q_+) + e^-(q_-) \ . \tag{1.3}$$

In these reactions one studies amongst others the Z-line shape and forward-backward asymmetry. From the measurements one extracts the mass M_z , the total width Γ_z and

Z° Physics - Cargèse 1990, Edited by M. Lévy *et al.*
Plenum Press, New York, 1991

the partial widths of the Z. The Born cross sections in which these parameters occur
are simple. The radiative corrections, and again mainly the QED radiative corrections
distort considerably the simple Born predictions.

The emphasis of the lectures is on a discussion of potentially the most uncertain
part of EWRC, the QED corrections. In particular the question of the effect of experimental cuts and the necessity of higher order corrections will be addressed. This leaves
little room to explain the many technicalities of the weak corrections. Let me at least
list some of these points.

For doing calculations one needs Feynman rules and for that the SM Lagrangian.
For a more complete discussion I refer to some introductory lectures [1] and the quoted
literature there. Here I mention some steps one takes before one can start with the
actual calculations. The Lagrangian of the standard model consists of a sum of different
parts each containing some of the particles and their interactions. There is a part which
contains the $SU(2) \times U(1)$ gauge bosons $\mathcal{L}_G$ also containing two coupling constants g_2
and g_1 . Then there is a part containing the fermions and their interactions with gauge
bosons $\mathcal{L}_F$. The part $\mathcal{L}_H$ containing the Higgs sector contains a mass term for the Higgs
fields, characterized by a parameter μ and a quartic self interaction of the Higgs fields
with a coupling constant λ . Moreover the interaction with the gauge fields is present
in $\mathcal{L}_H$. Finally, another part $\mathcal{L}_Y$ contains Yukawa couplings between fermions and
Higgs fields. Coupling constants g_F determine the strength of the couplings. Thus the
Langrangian before quantization i.e. the classical $SU(2) \times U(1)$ invariant Lagrangian
can be written as

$$\mathcal{L}_{cl} = \mathcal{L}_G + \mathcal{L}_F + \mathcal{L}_H + \mathcal{L}_Y \; . \tag{1.4}$$

The self interaction in $\mathcal{L}_H$ leads to symmetry breaking and three gauge bosons become
massive as well as most of the fermions. Instead of the original parameters of the model
i.e. $g_1, g_2, \mu^2, \lambda, g_F$ one can also use e, M_W, M_Z, M_H, m_f . Thus one uses the physical
masses and the coupling constant of electromagnetic interactions. This shall be done
in the following.

In order to derive the Feynman rules one has to add two parts to the Lagrangian.
One part $\mathcal{L}_{GF}$ has to do with the necessity to choose a specific gauge in order to obtain
the propagators for the gauge bosons. This gauge fixing term makes it possible. One
then also has to introduce Faddeev-Popov ghosts by adding $\mathcal{L}_{FP}$ to the Lagrangian.
The total Lagrangian required for the derivation of Feynman rules becomes

$$\mathcal{L} = \mathcal{L}_{cl} + \mathcal{L}_{GF} + \mathcal{L}_{FP} \; . \tag{1.5}$$

For loop calculations the Feynman-'t Hooft gauge is usually taken, but for Born term
calculations the unitary gauge is more convenient. The former gets simpler gauge boson
propagators at the price of the introduction of ghosts and their diagrams. The latter
does not require ghosts. We give a few Feynman rules as an illustration. First some
propagators

$$\text{photon} \qquad \qquad \frac{-i g_{\mu\nu}}{q^2} \; , \tag{1.6}$$

$$\text{Vector boson} \qquad \qquad \frac{i}{q^2 - M_v^2} \left(-g_{\mu\nu} + \frac{q_\mu q_\nu}{M_v^2} \right) \; , \tag{1.7}$$

$$\text{Fermion} \qquad \qquad \frac{i(\not{q} + m_f)}{q^2 - m_f^2} \; . \tag{1.8}$$

Then some vertices

$$\gamma \bar{f} f \qquad -ieQ_f\gamma_\mu = -ieQ_f\left[\gamma_\mu\frac{1+\gamma_5}{2}+\gamma_\mu\frac{1-\gamma_5}{2}\right], \qquad (1.9)$$

$$Z\bar{f}f \qquad ie\gamma_\mu(v_f-a_f\gamma_5)=ie\left[g_f^+\gamma_\mu\frac{1+\gamma_5}{2}+g_f^-\gamma_\mu\frac{1-\gamma_5}{2}\right], \qquad (1.10)$$

$$We\nu \qquad i\,\frac{e}{\sqrt{2}s_w}\,\gamma_\mu\,\frac{(1-\gamma_5)}{2}, \qquad (1.11)$$

with

$$g_f^+ \;=\; v_f-a_f=\frac{-Q_fs_w^2}{s_wc_w},$$

$$g_f^- \;=\; v_f+a_f=\frac{I_3^f-Q_fs_w^2}{s_wc_w}, \qquad (1.12)$$

$$v_f \;=\; \frac{I_3^f-2Q_fs_w^2}{2s_wc_w},\; a_f=\frac{I_3^f}{2s_wc_w}. \qquad (1.13)$$

Here Q_f and I_3^f are the charge and weak isospin of the fermion. The sine and cosine of the weak mixing angle θ_w are denoted by s_w and c_w. The above vector boson propagator is in the unitary gauge. In the 't Hooft-Feynman gauge the $q_\mu q_\nu$ term is absent, but one needs Feynman diagrams with ghosts. The full set of Feynman rules can e.g. be found in ref [2].

For Born level calculations it is sufficient to know these Feynman rules. For loop calculations, where divergencies occur we need more techniques of field theory. They will be briefly mentioned later on.

The outline of the lectures is as follows. In section 2 we list the various helicity amplitudes for mupair production and Bhabha scattering and we consider their role in a number of experimental quantities. Section 3 discusses the weak corrections. The question of input parameters, renormalization schemes is addressed. The correction to the Z-width is treated and the effect of the weak radiative corrections on the helicity amplitudes is discussed. In section 4 QED corrections in $O(\alpha)$ are presented. By treating a simple example, the separation into virtual, soft and hard photon radiation is illustrated. Also the separation into initial and final state corrections is made. Collinear singularities show up. The choice between an analytical and Monte Carlo treatment is discussed. The next section gives $O(\alpha^2)$ corrections, the resummation of soft photons, and illustrates the difficulties a Monte Carlo approach has. Section 6 discusses the full line shape and some rules of thumb for it. Section 7 introduces the structure function method. For the total cross section the leading log (LL)corrections of section 5 are recovered. For differential cross sections the formalison leads to new LL calculations. Section 8 gives the results for A_{FB} in mupair production, whereas section 9 presents results for Bhabha scattering.

2   BORN HELICITY AMPLITUDES AND EXPERIMENTAL QUANTITIES

In Born approximation the diagrams contributing to

$$e^+(p_+,\lambda_+)+e^-(p_-,\lambda_-)\rightarrow\mu^+(q_+,\lambda'_+)+\mu^-(q_-,\lambda'_-) \qquad (2.1)$$

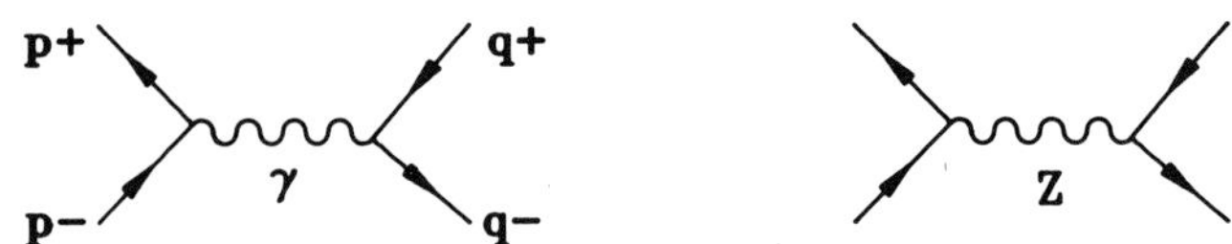

Fig. 2.1. The lowest order Feynman diagrams for mupair production.

come from γ and Z exchange (fig. 2.1) and $M(p_+\lambda_+; p_-\lambda_-; q_+\lambda'_+; q_-, \lambda'_-)$ denotes the amplitudes.

We find for instance, neglecting masses and using helicity spinors $u_\pm, v_\pm$

$$M(+-+-) = ie^2 \left(\frac{Q_e Q_f}{s} + \frac{g_e^- g_f^-}{s - M_z^2 + iM_z\Gamma_z} \right) \bar{u}_-(q_-)\gamma^\mu v_+(q_+)$$

$$\bar{v}_+(p_+)\gamma_\mu u_-(p_-) \ . \tag{2.2}$$

Here we introduce in the Z propagator a total Z width Γ_z which is strictly speaking an effect of weak corrections, to which we come back in the next section. One can verify that besides an irrelevant phase factor one has

$$M(+-+-) = 2ie^2 P(--) \, u \ ,$$

$$M(-++-) = 2ie^2 P(+-) \, t \ ,$$

$$M(+--+) = 2ie^2 P(-+) \, t \ , \tag{2.3}$$

$$M(-+-+) = 2ie^2 P(++) \, u \ .$$

Due to the neglect of the masses the helicities of the incoming particles are opposite and similarly of the outgoing particles. The function $P(\lambda_-, \lambda'_-)$ is the combination of propagators

$$P(\lambda, \lambda') = \frac{Q_e Q_f}{s} + \frac{g_e^\lambda g_f^{\lambda'}}{s - M_z^2 + iM_z\Gamma_z} \ . \tag{2.4}$$

Here we introduce the Mandelstam variables

$$s = (p_+ + p_-)^2 = 4E^2 \ , s' = (q_+ + q_-)^2 \ ,$$

$$t = (p_+ - q_+)^2 = \frac{-s}{2}(1 - \cos\theta) \ , t' = (p_- - q_-)^2 \ , \tag{2.5}$$

$$u = (p_+ - q_-)^2 = \frac{-s}{2}(1 + \cos\theta) \ , u' = (p_- - q_+)^2 \ ,$$

where E is the beam energy and θ the scattering angle. Since in the Born cross section no momentum is lost by radiation we have $s' = s, t' = t, u' = u$.

The differential cross section for one of the helicity combinations, e.g. $(-,-)$, is

$$\frac{d\sigma}{d\Omega} = \frac{4e^4|P|^2u^2}{(8\pi)^2 s} = \frac{\alpha^2}{s}|P|^2u^2 \ , \tag{2.6}$$

or

$$\frac{d\sigma}{dt} = \frac{2}{s}\frac{d\sigma}{dc} = \frac{4\pi\alpha^2}{s^2}|P|^2u^2 \ . \tag{2.7}$$

The unpolarized $\bar{f}f$ cross section reads

$$\frac{d\sigma}{d\Omega} = \frac{\alpha^2}{4s}\left[\left(|P(--)|^2 + |P(++)|^2\right)u^2 + \left(|P(+-)|^2 + |P(-+)|^2\right)t^2\right] \ . \tag{2.8}$$

From this the forward and backward cross sections can be defined

$$\sigma_F = \int_0^1 dc\frac{d\sigma}{dc} = 2\pi\frac{\alpha^2 s}{16}\left[\left(|P(--)|^2 + |P(++)|^2\right)\frac{7}{3} + \left(|P(+-)|^2 + |P(-+)|^2\right)\frac{1}{3}\right] ,$$
$$\tag{2.9}$$

$$\sigma_B = \int_{-1}^0 dc\frac{d\sigma}{dc} = 2\pi\frac{\alpha^2 s}{16}\left[\left(|P(--)|^2 + |P(++)|^2\right)\frac{1}{3} + \left(|P(+-)|^2 + |P(-+)|^2\right)\frac{7}{3}\right] ,$$
$$\tag{2.10}$$

$$\sigma_F - \sigma_B = \frac{\pi\alpha^2 s}{4}\left[\left(|P(--)|^2 + |P(++)|^2\right) - \left(|P(+-)|^2 + |P(-+)|^2\right)\right] , \tag{2.11}$$

$$\sigma_F + \sigma_B = \frac{\pi\alpha^2 s}{3}\left[|P(--)|^2 + |P(++)|^2 + |P(+-)|^2 + |P(-+)|^2\right] \ . \tag{2.12}$$

And from these the forward-backward asymmetry is constructed

$$A_{FB} = \frac{\sigma_F - \sigma_B}{\sigma_F + \sigma_B} = \frac{3}{4}\frac{|P(--)|^2 + |P(++)|^2 - |P(+-)|^2 - |P(-+)|^2}{|P(--)|^2 + |P(++)|^2 + |P(+-)|^2 + |P(-+)|^2} \ . \tag{2.13}$$

For left and right longitudinally polarized beams one can also define forward and backward cross sections and total cross sections

$$\sigma_F^L = \frac{\pi\alpha^2 s}{2}\left[|P(--)|^2\frac{7}{3} + |P(-+)|^2\frac{1}{3}\right] , \tag{2.14}$$

$$\sigma_F^R = \frac{\pi\alpha^2 s}{2}\left[|P(++)|^2\frac{7}{3} + |P(+-)|^2\frac{1}{3}\right] , \tag{2.15}$$

$$\sigma_B^L = \frac{\pi\alpha^2 s}{2}\left[|P(--)|^2\frac{1}{3} + |P(-+)|^2\frac{7}{3}\right] , \tag{2.16}$$

$$\sigma_B^R = \frac{\pi\alpha^2 s}{2}\left[|P(++)|^2\frac{1}{3} + |P(+-)|^2\frac{7}{3}\right] , \tag{2.17}$$

$$\sigma^L = \sigma_F^L + \sigma_B^L \ , \tag{2.18}$$

$$\sigma^R = \sigma_F^R + \sigma_B^R \ . \tag{2.19}$$

One can also measure the asymmetries

$$A_{LR} = \frac{\sigma^L - \sigma^R}{\sigma^L + \sigma^R} = \frac{|P(--)|^2 + |P(-+)|^2 - |P(++)|^2 - |P(+-)|^2}{|P(--)|^2 + |P(-+)|^2 + |P(++)|^2 + |P(+-)|^2} \ , \tag{2.20}$$

$$A_{FB}^{Pol} = \frac{\sigma_F^L - \sigma_B^L - (\sigma_F^R - \sigma_B^R)}{\sigma_F^L + \sigma_B^L + \sigma_F^R + \sigma_B^R}$$

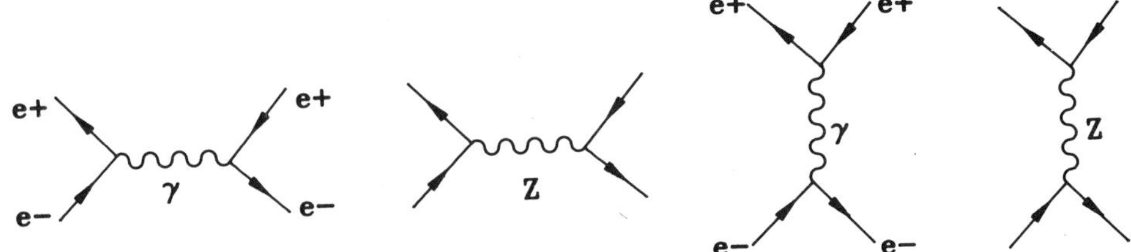

Fig. 2.2. The lowest order diagrams for Bhabha scattering.

$$= \frac{3}{4} \frac{|P(--)|^2 - |P(-+)|^2 - |P(++)|^2 + |P(+-)|^2}{|P(--)|^2 + |P(-+)|^2 + |P(++)|^2 + |P(+-)|^2} \cdot \tag{2.21}$$

By measuring the total cross section and all the above asymmetries one can extract every $|P(\lambda, \lambda')|^2$. From eq. (2.4) it is clear that it would mean a measurement of M_z, Γ_z and combinations of coupling constants.

For Bhabha scattering one has for the Born amplitudes four contributing diagrams (fig. 2.2)
Due to the t-channel diagrams there are more non vanishing helicity amplitudes. On the other hand only g_e^λ is present, which simplifies the expressions. We find

$$M(+ + + +) \;=\; M(- - - -) = 2ie^2 P'(+-)s \; ,$$

$$M(+ - + -) \;=\; 2ie^2 \left[P(--) + P'(--)\right] u \; ,$$

$$M(- + - +) \;=\; 2ie^2 \left[P(++) + P'(++)\right] u \; , \tag{2.22}$$

$$M(+ - - +) \;=\; M(- + + -) = 2ie^2 P(+-)t \; ,$$

where

$$P'(\lambda, \lambda') = \frac{Q_e^2}{t} + \frac{g_e^\lambda g_e^{\lambda'}}{t - M_z^2} \cdot \tag{2.23}$$

Thus the differential cross section becomes

$$\frac{d\sigma}{d\Omega} \;=\; \frac{\alpha^2}{4s} \Big[2s^2 |P'(+-)|^2 + 2t^2 |P(+-)|^2$$
$$+\; u^2 (|P(--) + P'(--)|^2 + |P(++) + P'(++)|^2) \Big] \; . \tag{2.24}$$

3 WEAK CORRECTIONS

3.1 Input parameters and renormalization scheme

As mentioned in the introduction loop calculations require additional techniques. For one-loop corrections one has to carry out integrals over $d^4\ell$, where ℓ is the momentum going around in the loop. These integrals often diverge and one likes to make them convergent by some regularization scheme. For gauge theories the so-called dimensional regularization [3] turns out to be very useful. The above mentioned integral is

generalized to an n-dimensional integral, where n can also become complex. Answers
for the integrals can be found depending on n or on $\varepsilon = 4 - n$. The one-loop results can
be studied around $\varepsilon \approx 0$. They often show poles ε^{-1} , reflecting the above mentioned
divergencies. Due to such a regularization scheme all the divergencies from various
loop diagrams contributing to a specific S-matrix element can be collected. It may be
that in the end no divergence remains, but usually a term proportional to ε^{-1} survives.

In order to give a meaning to the experimentally accessible matrix element, which
in the theory diverges, one needs a renormalization scheme. It makes use of redefinition
of the parameters of the theory like masses and coupling constants. For instance the
coupling e in the original Lagrangian is replaced by

$$e_0 = e + \delta e \tag{3.1}$$

with e the physical charge. Inserting e_0 in $\mathcal{L}$ will give the original Lagrangian in terms
of e and so called counterterms containing δe . These counterterms give rise to new
Feynman diagrams which have to be added to the S-matrix elements. When δe and
similar quantities δM^2 can be chosen in such a way that these S-matrix elements
become finite the theory is renormalizable. This is the case for the standard model.
Since there is a freedom of choice in the independent parameter set there exist different
renormalization schemes. On top of that one can also choose to renormalize the fields,
which makes it possible to obtain finite Green functions. One obtains for instance finite
expressions for certain vertices, or in other words for certain form factors. These finite
Green functions are the building blocks for many S-matrix elements, so one can obtain
results for many processes without going through the renormalization every time.

In the introduction the basic parameters of the theory were chosen to be e and
the physical masses of the participating particles. After one loop corrections these
parameters should not change their values. This means that the counterterms should
be chosen in such a way that

- Real parts of renormalized self energies vanish for $k^2 = \text{mass}^2$ i.e.

$$\mathbf{Re} \quad \overset{\mathbf{k}}{-\!\!-\!\!\bigcirc\!\!-\!\!-} \quad = 0 \text{ for } k^2 = M^2 \ , \tag{3.2}$$

$$\mathbf{k}\sim\!\!\sim\!\!\sim\!\!\bigotimes \quad = ie\gamma_\mu \text{ for } k^2 = 0 \text{ and on-mass shell electrons } . \tag{3.3}$$

Furthermore wave function renormalization constants can be introduced and corresponding counter terms. In a particular scheme[2], which will be used here for the
electroweak corrections the counterterms are chosen in such a way that one has the
additional relations

$$\overset{\gamma}{\sim\!\!\sim\!\!\bigcirc\!\!\sim\!\!\sim}{}_{\mathbf{Z}} \quad = 0 \text{ for } k^2 = 0 \ , \tag{3.4}$$

- residue of renormalized photon propagator and $I_3 = -1/2$ fermion

propagators equal one . $\tag{3.5}$

The above renormalization scheme is a generalization of the standard QED scheme,
which imposes (3.2), (3.3) and (3.5). Moreover the condition (3.4) makes it possible to
take over all known photonic corrections from QED.

Fig. 3.1. A fermion loop contribution
to the photon self-energy.

As an illustration of the above sketched procedures we give some examples. Consider a fermion loop contribution to the photon self-energy, also called vacuum polarization (see fig. 3.1) Inserting both photon propagators we obtain an expression D in which we introduce $\Sigma_{\gamma\gamma}$

$$D_{\alpha\beta} = -i\frac{g_{\alpha\mu}}{k^2}\left[-i\Sigma^{\mu\nu}_{\gamma\gamma}(k^2)\right]\left(-i\frac{g_{\nu\beta}}{k^2}\right) . \tag{3.6}$$

Carrying out the loop integration in n dimensions one obtains

$$\Sigma^{\mu\nu}_{\gamma\gamma} = \left(g^{\mu\nu} - \frac{k^\mu k^\nu}{k^2}\right)\Sigma_{\gamma\gamma}(k^2) \tag{3.7}$$

with

$$\Sigma_{\gamma\gamma} = \frac{\alpha}{4\pi}Q^2\left[\frac{4}{3}k^2\Delta + \frac{4}{3}(k^2 + 2m^2)F(k^2,m^2) - \frac{4}{9}k^2\right] , \tag{3.8}$$

$$\Delta = \frac{2}{\varepsilon} - \gamma - \ell n\left(\frac{m^2}{4\pi\mu^2}\right) . \tag{3.9}$$

The quantity Δ contains the divergence ε^{-1} , the Euler constant γ and a mass μ which is introduced in the dimensional regularization formalism.

The real part of F has the following behaviour

$$k^2 << m^2 \qquad \mathrm{Re}F(k^2,m^2) = \frac{1}{6}\frac{k^2}{m^2} . \tag{3.10}$$

The imaginary part has the form

$$\mathrm{Im}\ F(k^2,m^2) = \pi\sqrt{1 - \frac{4m^2}{k^2}}\theta(k^2 - 4m^2) . \tag{3.11}$$

It is clear that condition (3.2) is satisfied. Introducing

$$\Pi_{\gamma\gamma}(k^2) = \Sigma_{\gamma\gamma}(k^2)/k^2 \tag{3.12}$$

one sees that

$$\Pi_{\gamma\gamma}(0) = \frac{\alpha}{3\pi}Q^2\Delta . \tag{3.13}$$

Using the explicit form for F one derives

$$\hat{\Pi}_{\gamma\gamma}(k^2) = \Pi_{\gamma\gamma}(k^2) - \Pi_{\gamma\gamma}(0) = \frac{\alpha}{\pi}Q^2\left[\frac{8}{9} - \frac{a^2}{3} + a\left(\frac{1}{2} - \frac{a^2}{6}\right)\ell n\ b\right] \tag{3.14}$$

$$= \frac{\alpha}{\pi}Q^2\left[\frac{5}{9} - \frac{1}{3}\ell n\frac{k^2}{m^2}\right] \qquad k^2 >> m^2 , \tag{3.15}$$

where

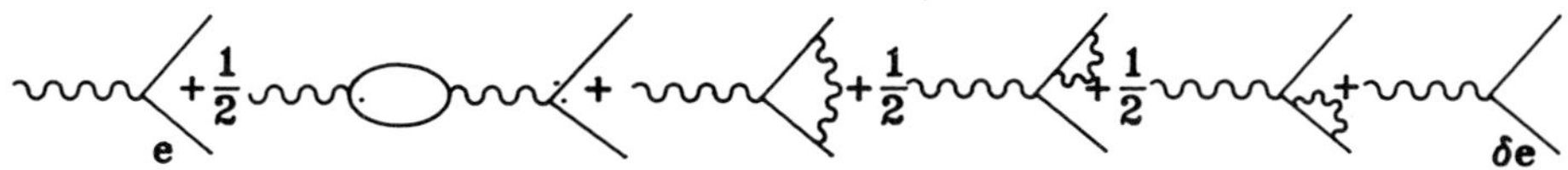

Fig. 3.2. The γee vertex and the order α corrections.

$$a = \left(1 - \frac{4m^2}{k^2}\right)^{1/2} \,, \tag{3.16}$$

$$b = \frac{1-a}{1+a} \,. \tag{3.17}$$

When the photon is coupled to a conserved current only the $g_{\mu\nu}$ term in (3.6) contributes. The usual photon propagator and the correction (3.6) together lead to a modified propagator

$$D_{\alpha\beta} = -i\frac{g_{\alpha\beta}}{k^2}\left[1 - \Pi_{\gamma\gamma}(k^2)\right] = -i\frac{g_{\alpha\beta}}{k^2}\frac{1}{1+\Pi_{\gamma\gamma}(k^2)} \,. \tag{3.18}$$

The residue of the $1/k^2$ pole is divergent due to the presence of Δ . In QED only fermion loops contribute to $\Sigma_{\gamma\gamma}$, so one obtains the above result with a summation over all fermions. We now illustrate how in QED this infinity is removed.

When the γee vertex is corrected up to order α one has to consider the following diagrams in fig. (3.2) The factors $1/2$ have to be included when a self-energy correction is inserted on the external lines. The infinities of diagrams 3,4 and 5 turn out to cancel. The resulting corrected vertex with on shell fermions in the limit $k^2 \to 0$ becomes

$$\text{Corrected vertex} = ie\gamma^\mu\frac{-i}{k^2}\left(1 - \frac{\Pi_{\gamma\gamma}}{2}(0) + \frac{\delta e}{e}\right) \,. \tag{3.19}$$

The renormalization condition (3.3) then fixes δe

$$\delta e = \frac{e}{2}\Pi_{\gamma\gamma}(0) \,. \tag{3.20}$$

As a result of this renormalization procedure one can use in QED for the corrected photon propagator the finite expression $\hat{\Pi}_{\gamma\gamma}(k^2)$ instead of $\Pi_{\gamma\gamma}(k^2)$ in eq. (3.18). The residue of the photon propagator now equals one, so condition (3.5) is also fulfilled. In the electroweak theory similar procedures can be carried out and in the renormalization scheme of ref [2] one can use for the renormalized self- energies and vertex correction finite expressions. Due to the large number of diagrams those expressions are often lengthy and we refer to the literature [2,4].

For later purposes we mention that the imaginary part of the Z self-energy from a fermion loop with mass m takes the form

$$\text{Im } \Sigma_{zz}(k^2) = \frac{\alpha}{\pi}\left[(v_f^2 - a_f^2)m^2 + \frac{1}{3}(v_f^2 + a_f^2)(k^2 - m^2)\right]\text{Im } F(k^2,m^2) \,. \tag{3.21}$$

For $v_f = 1$ and $a_f = 0$ this reduces to $\text{Im } \Sigma_{\gamma\gamma}$.

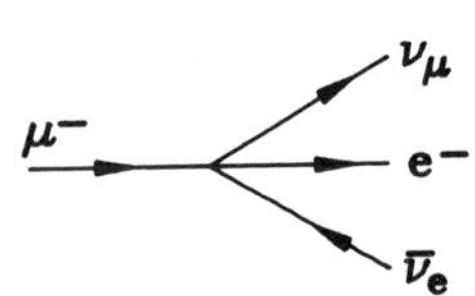

Fig. 3.3. Muon decay in the
Fermi theory.

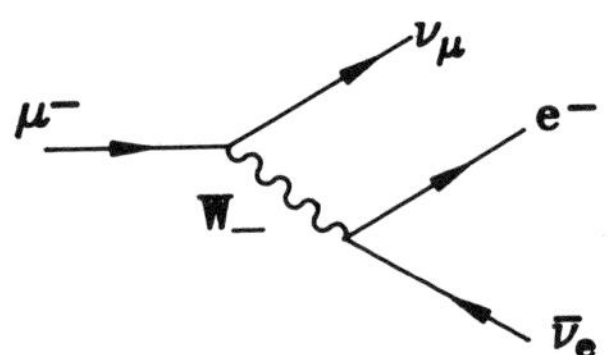

Fig. 3.4. Muon decay in the S.M.

We now turn to the question of the actual values of the input parameters. The fine structure constant $\alpha = e^2/4\pi$ is well known. When later on also QCD corrections are required α_s has to be taken from experiment. Lepton masses are known. The quark masses are chosen in such a way that the hadronic vacuum polarization is well reproduced. On one hand one can evaluate this quantity from a dispersion relation [5] which uses the measured total e^+e^- hadronic cross section as input. On the other hand, one can express this on the basis of QCD in terms of quark loop diagrams in which the masses figure. The actual masses used are (in GeV)

$$m_a = m_d = 0.041 \ , \ m_c = 1.5 \ , \ m_s = 0.15 \ , \ m_b = 4.5 \ . \tag{3.22}$$

These masses reproduce the dispersion relation result. The remaining masses are m_t , M_z , M_w and M_H .

The previous discussion used M_w as input. Although we shall formally continue to do so, we do not take M_w from a direct experiment but from an indirect measurement. Once M_w has been well determined directly one can omit the indirect information. Or rather confront it with the direct measurement. The indirect M_w determination uses the accurately known Fermi constant G_μ as obtained from the muon lifetime. In the Fermi theory the lifetime is expressed in G_μ using the diagram of fig. 3.3, whereas in the SM fig. 3.4 describes the process. The standard model predicts the lifetime in terms of the parameters of the theory. When one uses the Born term prediction of fig. 3.4 and relates this to the prediction of fig. 3.3 one finds

$$M_w^2 \sin^2 \theta_w = A \ , \tag{3.23}$$

with

$$A = \frac{\pi \alpha}{\sqrt{2} G_\mu} = (37.281 \ \text{GeV})^2 \tag{3.24}$$

and the definition of $\sin \theta_w$ is in terms of the vector boson masses

$$\sin^2 \theta_w = 1 - \frac{M_w^2}{M_z^2} \ . \tag{3.25}$$

Thus a choice of M_z would determine M_w through the eqs. (3.23) - (3.25) and one does not need M_w as a separate parameter.

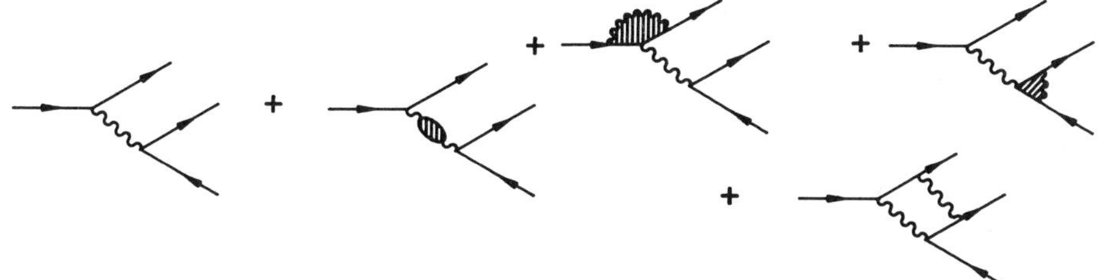

Fig. 3.5. The Born contribution and order α corrections to muon decay.

3.2 The quantity Δr

The Born term prediction of G_μ is however not accurate. When radiative corrections are included one gets an accurate relation

$$M_w^2 \sin^2 \theta_w = \frac{A}{1 - \Delta r} \, , \tag{3.26}$$

where Δr depends on the parameters of the theory. The quantity Δr receives contributions from self-energies, vertex corrections and box diagrams, as depicted symbolically in fig. 3.5

For specific M_z , m_t and M_H values one can solve for M_w . The sensitivity on m_t is greater than that on M_H for a fixed M_z as can be seen from fig. 3.6. The curves are based on the calculations of ref [4], which are accurate for $m_t \lesssim 150$ GeV. Above these values it is advisable to take m_t^4 effects into account. For the details we refer to [7]. As we saw in section 2 a Born cross section depends on the coupling constant $g_f^\pm$ and on Γ_z . These quantities depend on $\sin \theta_w$ which in turn depends on M_w and therefore on Δr . So the knowledge of Δr influences all predictions. The quantity Δr is a manifestation of a weak radiative correction. It is the most essential weak correction.

3.3 The partial and total widths

In lowest order the partial width of a Z decaying into a fermion pair is given by

$$\begin{aligned}
\Gamma^{(0)}_{z \to \bar{f}f} &= \frac{\alpha}{6} N_c M_z \sqrt{1 - \frac{4m_f^2}{M_z^2}} \left((g_f^-)^2 + (g_f^+)^2 \right. \\
&\quad + \left. \frac{m_f^2}{M_z^2} \left(6 g_f^- g_f^+ - (g_f^-)^2 - (g_f^+)^2 \right) \right) \, ,
\end{aligned} \tag{3.27}$$

where N_c represents the number of colours. At this point it is useful to notice that expressed in v_f and a_f one verifies

$$M_z \Gamma^{(0)}_{z \to \bar{f}f} = \operatorname{Im} \Sigma_{zz}(M_z^2) \, . \tag{3.28}$$

Making use of the tree level relation (3.23) and eqs. (3.24) and (3.25) one can also write

$$\Gamma^{(0)}_{z \to \bar{f}f} = N_c \frac{G_\mu M_z^3}{24\pi \sqrt{2}} \sqrt{1 - \frac{4m_f^2}{M_z^2}} \left(1 - \frac{4m_f^2}{M_z^2} + \left(2I_f^3 - 4Q_f s_w^2 \right)^2 \left(1 + \frac{2m_f^2}{M_z^2} \right) \right) . \tag{3.29}$$

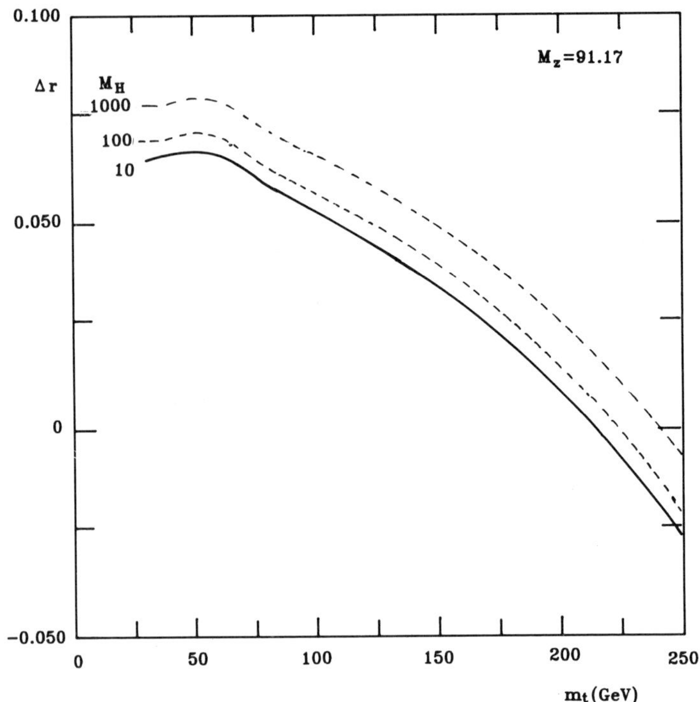

Fig. 3.6. The quantity Δr as function of m_t for three values of M_H .

For a fixed M_z , m_t and M_H one can determine from eqs. (3.24) - (3.26) M_w and s_w^2 . The two different representations of the width eqs. (3.27) and (3.29) give different values. The values would of course be the same when eq. (3.23) instead of (3.26) would be used to determine M_w and s_w^2 . Although the different Born expressions give different numerical values the addition of one-loop radiative corrections makes the corrected widths almost the same. We shall apply corrections to eq. (3.27) in the following.

The total width in lowest order is denoted by

$$\Gamma_z^{(0)} = \Sigma_f \Gamma_{z \to \bar{f}f}^{(0)} \ . \tag{3.30}$$

In the massless fermion case the lowest order formula can be considered as the sum of two widths: one for a decay into a negative helicity fermion and one for a positive helicity fermion

$$\Gamma_{z \to \bar{f}f}^{(0)} = \Gamma_- + \Gamma_+ = \frac{\alpha}{6} \, N_c \, M_z \, \left((g_f^-)^2 + (g_f^+)^2 \right) \ . \tag{3.31}$$

The first order corrections to $\Gamma_{z \to \bar{f}f}^{(0)}$ can be divided into four classes [6]:

1. Weak corrections

2. Photonic loop corrections and radiative decay

3. QCD corrections

4. Decay into three or more particles

318

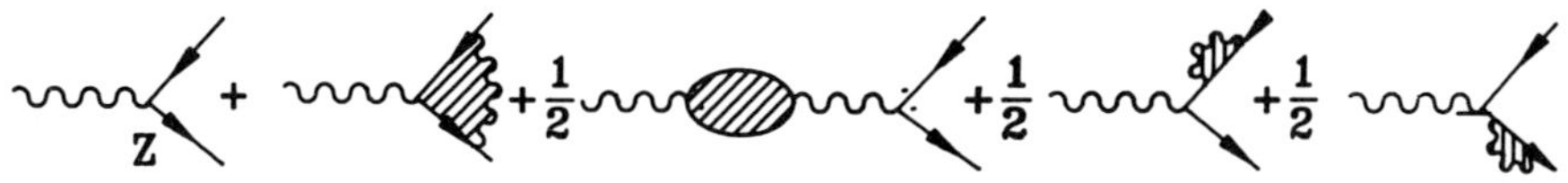

Fig. 3.7. The lowest order contribution and order α corrections to the Z width.

Fig. 3.8. The virtual and real gluon corrections to Z decay into quarks.

The one-loop weak corrections to the width (3.27) are considerable in the scheme we follow here. The contributions arise from vertex corrections and self-energy diagrams, as pictured in fig. 3.7. They contain all contributions except from photons. In particular the wave function renormalization of the Z is important. In another scheme one starts from eq. (3.29) and the one-loop corrections are tiny at least for $m_t < 120$ GeV . So for this m_t range eq. (3.29) is a good description of the corrected width. For larger m_t values a modified eq. (3.29) has been suggested [7].

The photonic correction is the same as the so-called final state QED correction for mupair production. The formula is given in section 4 and reads

$$\delta_{\text{QED}} = \frac{3\alpha}{4\pi}\, Q_f^2 \tag{3.32}$$

which for $Q_f = 1$ gives 0.17% . Although we postpone the general discussion of QED radiative corrections, we want to calculate the total width as well as we can and include therefore the correction (3.32).

The QCD correction for a Z decaying into massless fermions is similar to δ_{QED}

$$\delta_{\text{QCD}} = \frac{\alpha_s}{\pi}\, . \tag{3.33}$$

It arises from the QCD virtual diagrams and gluon emission (fig. 3.8). From a determination of δ_{QCD} at $s = (34 \text{ GeV})^2$

$$\delta_{\text{QCD}} = 0.047 \pm 0.009 \tag{3.34}$$

Table 3.1. Partial and total widths in MeV for $M_z = 91.17$ GeV, $\alpha_s = 0.12$ and various m_t, M_H (in GeV).

m_t	M_H	$\sin^2\theta_w$	$\Gamma_z^{(0)}$	Γ_z	$\Gamma_{z\to\nu\bar\nu}$	$\Gamma_{z\to e^+e^-}$	$\Gamma_{z\to u\bar u}$	$\Gamma_{z\to d\bar d}$	$\Gamma_{z\to b\bar b}$
	10	0.2300	2285	2480	165.9	83.2	295.7	381.7	378.6
90	100	0.2316	2270	2483	166.2	83.4	295.9	382.0	378.9
	1000	0.2347	2241	2475	165.9	83.2	294.6	380.6	377.5
	10	0.2254	2329	2488	166.3	83.5	297.2	383.3	378.2
130	100	0.2270	2313	2491	166.7	83.6	297.3	383.7	378.5
	1000	0.2303	2282	2483	166.4	83.4	296.1	382.3	377.2
	10.	0.2200	2384	2498	166.9	83.8	299.1	385.5	377.5
170	100	0.2217	2366	2501	167.3	84.0	299.2	385.9	377.9
	1000	0.2250	2333	2493	167.0	83.8	298.0	384.5	376.6
	10	0.2133	2456	2511	167.7	84.2	301.4	388.3	376.5
210	100	0.2150	2437	2514	168.1	84.3	301.6	388.7	377.0
	1000	0.2184	2400	2506	167.8	84.2	300.4	387.3	375.7

$\Gamma_z^{(0)}$ is the result from eq. (3.27). The other widths are the corrected ones.

one obtains from the running of the coupling constant

$$\delta_{\mathrm{QCD}}\left((92\ \mathrm{GeV})^2\right) = 0.040 \pm 0.007 \tag{3.35}$$

or $\alpha_s = 0.12$. For the $b\bar b$ channel a mass dependent QCD correction will be used [8]. It amounts to 0.045 ± 0.007 instead of (3.35). The quoted errors will introduce an error of 10 MeV in Γ_z .

Finally, Z decays into three or more particles also constitute corrections to the lowest order total width. Photon and gluon emission have already been dealt with. Of the other possible decays only

$$Z \to H\bar f f \tag{3.36}$$

could be of relevance but only for $M_H \leq 10$ GeV e.g. for $M_H = 10$, $m_t = 90$ (230) we find for the partial width (3.36) the value 5.3 (5.8) MeV .

Combining all corrections one finds the set of values for total and partial widths as given in table 3.1.

3.4   The EWRC for scattering amplitudes

We now discuss the weak radiative corrections to the helicity amplitudes of the previous section. One has three types of one-loop diagrams

1. Self-energy corrections for Z, γ exchange and $\gamma-Z$ transition diagrams

2. Vertex corrections to the electron and muon vertex

3. Box diagrams

Technically speaking the loop diagrams are performed after choosing a certain gauge. In this gauge ('t Hooft-Feynman gauge) it turns out that the box diagrams in the resonance region are negligible. They can therefore be omitted.

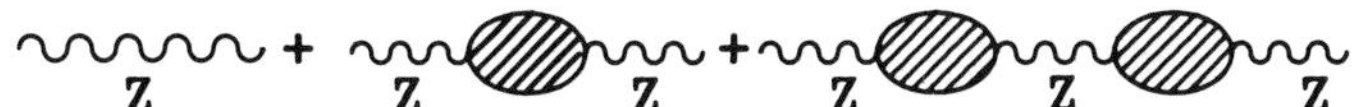

Fig. 3.9. The Z propagator and self-energy corrections.

The vertex corrections are taken into account but are less important than the self-energy corrections. For the self-energy corrections it makes a great difference whether one just adds the interference of a self-energy diagram with the Born amplitude or whether one sums a Dyson series. Consider the renormalized one particle irreducible transverse Z-self-energy $\Sigma_{zz}(s)$ and the sum of the diagrams of fig. 3.9 which gives the series

$$\frac{-i}{s - M_z^2} + \frac{-i}{s - M_z^2}\,(-i\Sigma_{zz})\,\frac{-i}{s - M_z^2} + \frac{-i}{s - M_z^2}\,(-i\Sigma_{zz})\,\frac{-i}{s - M_z^2}\,(-i\Sigma_{zz})\,\frac{-i}{s - M_z^2}$$

$$= \frac{-i}{s - M_z^2 + \Sigma_{zz}(s)} \, . \tag{3.37}$$

For a strict $O(\alpha)$ correction one would consider only the interference of the first and second term. We sum the series which means that the original Z-propagator $-i/(s - M_z^2)$ is replaced by the expression of eq. (3.37). In the region around $s = M_z^2$ we can expand $\Sigma_{zz}(s)$ and find

$$\Sigma_{zz}(s) = i\mathrm{Im}\,\Sigma_{zz}(M_z^2) + \Pi_{zz}(M_z^2)(s - M_z^2) \tag{3.38}$$

with

$$\Pi_{zz}(s) = \frac{\partial\,\mathrm{Re}\,\Sigma_{zz}(s)}{\partial s} \tag{3.39}$$

The $\mathrm{Re}\,\Sigma_{zz}(M_z^2)$ term vanishes due to the imposed renormalization condition. The propagator (3.37) can therefore be written as

$$\frac{1}{s - M_z^2 + \Sigma_{zz}(s)} \sim \frac{1}{s - M_z^2 + iM_z\Gamma_z}\,\frac{1}{1 + \Pi_{zz}(M_z^2)} \tag{3.40}$$

where

$$M_z\Gamma_z = \frac{\mathrm{Im}\,\Sigma_{zz}(M_z^2)}{1 + \Pi_{zz}(M_z^2)} \simeq \mathrm{Im}\,\Sigma_{zz}(M_z^2)(1 + O(\alpha^2) + \cdots) \, . \tag{3.41}$$

The first factor in (3.40) is the propagator used for the Born cross section in order to avoid a singularity for $s = M_z^2$. The occurring width is related to the imaginary part of the self-energy, as we already saw in order α in eq. (3.28). The Dyson series of fig. 3.9 leads to a new propagator for the Z. Similarly a Dyson series for the photon propagator would lead to the replacement

$$\frac{1}{s} \longrightarrow \frac{1}{s + \Sigma_{\gamma\gamma}(s)} \, . \tag{3.42}$$

When one considers the diagrams of fig. 2.1 with these new propagators (3.42) and (3.37) one notices that for $s = M_z^2$ the second diagram becomes of order α^0 . Both numerator and denominator are of order α . Outside of the resonance this is not the case. Therefore we want to know the imaginary part of Σ_{zz} up to second order in α . We shall calculate this by using the approximation

$$\text{Im } \Sigma_{zz}^{(2)}(s) = \frac{s}{M_z^2} \text{ Im } \Sigma_{zz}^{(2)}(M_z^2) \tag{3.43}$$

and relating the r.h.s. of eq. (3.43) to the first order correction to the width. This formula is correct at $s = M_z^2$ and incorporates some energy dependence.

Sofar we have not discussed the fact that there exists also a $\gamma-Z$ self-energy diagram. Its existence means that we have to redo the Dyson summation. The Dyson series has to be done for a 2×2 symmetric matrix with $\Sigma_{\gamma z}$ as off-diagonal matrix elements. When one carries this out the propagator replacements change. For the photon propagator one has

$$\frac{1}{s} \longrightarrow \frac{1}{s + \Sigma_\gamma(s)} , \tag{3.44}$$

with

$$\Sigma_\gamma(s) = \Sigma_{\gamma\gamma}(s) - \frac{\Sigma_{\gamma z}^2(s)}{s - M_z^2 + \Sigma_{zz}(s)} . \tag{3.45}$$

For the Z-propagator one makes the substitution

$$\frac{1}{s - M_z^2 + iM_z\Gamma_z} \longrightarrow \frac{1}{s - M_z^2 + \Sigma_z(s)} \tag{3.46}$$

where

$$\Sigma_z(s) = \Sigma_{zz}(s) - \frac{\Sigma_{\gamma z}^2}{s + \Sigma_{\gamma\gamma}(s)} . \tag{3.47}$$

Finally a propagator which represents a $\gamma-Z$ mixing has to be included

$$D_{\gamma z}(s) = \frac{-\Sigma_{\gamma z}(s)}{s[s - M_z^2 + \Sigma_z(s)]} . \tag{3.48}$$

The renormalization condition is modified so that for $s = M_z^2$ the real part of Σ_z vanishes.

It turns out that the energy dependence of $\Sigma_z(s)$ is very relevant: it shifts the peak position in the cross section with -35 MeV with respect to the cross section based on the original propagator in eq. (3.46). We shall come back to this below.

Summarizing, the weak corrections amount to a replacement of the coupling constants $g_f^\pm$ by form factors and by a replacement of propagators. Several computer programs have been made to evaluate cross sections with weak corrections [6]. The results of different groups agree. Here the program ZSHAPE is used for the weak corrections.

The total cross sections for mupair production and hadrons are listed in table 3.2 for $M_z = 91.1$ and $M_H = 100$ GeV. In the table the peak value of the cross section is given together with the values of $\sqrt{s}$ for which the maximum and half maxima positions are reached. The peak position is lower than M_z by about 17 MeV. The apparent width $\sqrt{s_+} - \sqrt{s_-}$ in mupair production is greater than Γ_z by about 15 MeV. These qualitative features will be discussed in the next subsection. It should be kept in mind that for $m_t = 170$ and 210 GeV the predictions in table 3.2 will somewhat change when m_t^4 terms in Δr are taken into account.

Table 3.2. Exact and approximate peak positions before QED corrections, but with weak corrections.

m_t	Γ_{tot}	Γ_e	Γ_f	$10^5 C_R$	$10^6 C_I$	σ_{max}	$\sqrt{s_{max}}$	$\sqrt{s_-}$	$\sqrt{s_+}$
90	2.476					2.001	91.083	89.847	92.337
		0.0830	0.0830	3.1282	0.712	2.001	91.083	89.847	92.337
130	2.484					2.002	91.083	89.843	92.341
		0.0833	0.0833	3.1473	0.835	2.000	91.083	89.843	92.341
170	2.494					2.004	91.083	89.838	92.346
		0.0836	0.0836	3.1714	1.009	1.999	91.083	89.838	92.346
210	2.507					2.007	91.083	89.831	92.352
		0.0840	0.0840	3.1995	1.225	1.996	91.083	89.832	92.353
90	2.476					41.465	91.084	89.855	92.332
		0.0830	1.729	65.125	41.71	41.457	91.084	89.855	92.332
130	2.484					41.484	91.084	89.851	92.336
		0.0833	1.735	65.546	45.45	41.456	91.084	89.851	92.336
170	2.494					41.518	91.084	89.846	92.342
		0.0836	1.742	66.081	50.35	41.450	91.084	89.846	92.342
210	2.507					41.577	91.084	89.840	92.348
		0.0840	1.751	66.730	56.71	41.428	91.084	89.840	92.349

The first line gives the result from the program ZSHAPE, while the entries on the second line have been calculated using eqs.(3.65) - (3.67). The upper half of the table is for mupairs, the lower half for hadrons. C_R and C_I are dimensionless, the unit for $\sigma_{\max}$ is nb, for the other quantities it is GeV.

3.5 Approximations to the corrected amplitudes

For the Born amplitudes (2.4) we found expressions depending on the coupling constants g_f^λ . We can replace these coupling constants by the partial helicity widths of eq. (3.31). In this way one finds

$$e^2 P(\lambda, \lambda') = \frac{e^2 Q_e Q_f}{s} + \frac{R_{\lambda_e \lambda_f}}{s - M_z^2 + i M_z \Gamma_z} \tag{3.49}$$

with

$$R_{\lambda_e \lambda_f} = \pm \lambda_e \lambda_f \frac{24\pi \Gamma_{\lambda_e}^{1/2} \Gamma_{\lambda_f}^{1/2}}{N_c^{1/2} M_z} \tag{3.50}$$

where the $\pm$ sign corresponds to $I_f^3 = \mp\frac{1}{2}$ of the final state. The total cross section in Born approximation eq. (2.12) takes the form

$$\sigma_0(s) = \frac{s N_c}{(s - M_z^2)^2 + M_z^2 \Gamma_z^2} \left[\frac{12\pi \Gamma_e \Gamma_f}{M_z^2 N_c} + \frac{I(s - M_z^2)}{s} \right] + \frac{4\pi Q_f^2 \alpha^2 N_c}{3s} \tag{3.51}$$

with a coefficient for the interference term

$$I = \pm \frac{4\pi Q_e Q_f \alpha}{N_c^{1/2} M_z} \left(\Gamma_+^{1/2}(e) - \Gamma_-^{1/2}(e) \right) \left(\Gamma_+^{1/2}(f) - \Gamma_-^{1/2}(f) \right) . \tag{3.52}$$

The first term in eq. (3.51) is the Breit-Wigner form for a spin 1 resonance, the last term is the pure QED cross section. The coefficient I is positive for realistic s_w^2 values and is smallest for mupair production.

The expression (3.51) is here obtained from the Born cross section and the elimination of the coupling constants by lowest order partial widths. After weak corrections the above formula can still be a good approximation to the corrected cross section. One then has to take the corrected total and partial widths, one incorporates the energy dependence of Im Σ_z by replacing $M_z^2\Gamma_z^2$ by $s^2\Gamma_z^2/M_z^2$ and one takes care of the photonic self-energy by replacing α by

$$\alpha(M_z^2) = \frac{\alpha}{1+\Pi_{\gamma\gamma}(M_z^2)} \;. \tag{3.53}$$

We end up with the approximate formula [9,6]

$$\begin{aligned}
\sigma(s) \; = \; & \left\{ \frac{s}{(s-M_z^2)^2 + s^2\Gamma_z^2/M_z^2} \left[\frac{12\pi\Gamma_e\Gamma_f}{M_z^2} + \frac{IN_c(s-M_z^2)}{s} \right] \right. \\
& + \; \left. \frac{4}{3}\pi Q_f^2 \frac{\alpha^2(M_z^2)N_c}{s} \right\} (1+\delta_{\mathrm{QCD}})(1+\delta_{\mathrm{QED}}) \;.
\end{aligned} \tag{3.54}$$

The partial widths do not contain anymore the QCD and QED corrections. These corrections have been added as overall correction. Also in I the quantity α is replaced by $\alpha(M_z^2)$. The approximate expression (3.54) describes the exact corrected cross section within 0.2% in the range $(M_z - \Gamma_z, M_z + \Gamma_z)$. The values for maxima and half maxima are within 1 MeV of the exact values.

The expression (3.54) is very useful as a parametrization of the resonance shape where in principle $\Gamma_e, \Gamma_f, \Gamma_z, M_z$ and I are parameters to be determined. More general paramatrizations can also be given [10]. They paramatrize also the occurrence of imaginary parts of form factors and photon self-energy. In the standard model the latter effects are negligible.

Also for a discussion of features of the resonance curve the expression (3.54) is useful. It is possible to rewrite the Breit-Wigner part with a constant width, when we introduce new variables [11]

$$\chi(s) = \frac{1}{s - M_z^2 + is\gamma} = \frac{1}{1+i\gamma}\,\frac{1}{s - \tilde{M}_z^2 + i\tilde{M}_z\tilde{\Gamma}_z} \tag{3.55}$$

with

$$\begin{aligned}
\gamma \; &= \; \Gamma_z/M_z \;, & (3.56)\\
\tilde{M}_z \; &= \; M_z/(1+\gamma^2)^{1/2} \;, & (3.57)\\
\tilde{\Gamma}_z \; &= \; \Gamma_z/(1+\gamma^2)^{1/2} \;. & (3.58)
\end{aligned}$$

Thus a compact form of (3.54) is

$$\sigma(s) = \frac{1}{(1-\tilde{M}_z^2/s)^2 + \tilde{M}_z^2\tilde{\Gamma}_z^2/s^2} \left\{ \frac{C_R}{s} + \frac{C_I}{s}\left(1 - \frac{M_z^2}{s}\right) \right\} + \frac{C_Q}{s} \;, \tag{3.59}$$

with

$$C_R = \frac{12\pi\Gamma_e\Gamma_f}{M_z^2(1+\gamma^2)} \;, \tag{3.60}$$

$$C_I = \mp\frac{4\pi Q_f\alpha(M_z^2)}{M_z(1+\gamma^2)}N_c^{1/2}\left(\Gamma_+^{1/2}(e) - \Gamma_-^{1/2}(e)\right)\left(\Gamma_+^{1/2}(f) - \Gamma_-^{1/2}(f)\right)(1+\delta_{\mathrm{QCD}})^{1/2} \;, \tag{3.61}$$

$$C_Q = \frac{4}{3}\pi Q_f^2 \alpha^2(M_z^2) N_c(1 + \delta_{\mathrm{QCD}}) \ . \tag{3.62}$$

Note that the widths Γ_f contain the QCD correction. The QED correction $1 + \delta_{\mathrm{QED}}$ has been omitted since it is very small anyhow, but can easily be included.

We consider M_z, Γ_z, C_R, C_I and C_Q as parameters and we want to express $\sigma_{\max}, \sqrt{s}_{\max}$ and $\sqrt{s}_\pm$ in terms of these. So we take s around M_z^2 . Specifically we write

$$s = \tilde{M}_z^2(1 - \gamma y) \tag{3.63}$$

where y is of $O(1)$ or smaller

$$y = y^{(0)} + y^{(1)} + y^{(2)} \tag{3.64}$$

with $y^{(i)}$ of order γ^i . The numerical values of the coefficients C indicate that $\frac{C_I}{C_R}\gamma^{-1}$ and $\frac{C_Q}{C_R}\gamma$ can be treated as of order 1. We now find [12]

$$\sigma_{\max} = \frac{C_R}{\tilde{M}_z^2\gamma^2}\left[1 + \frac{1}{4}\gamma^2 + \gamma^2(1 - \gamma^2)\frac{C_Q}{C_R}\right] , \tag{3.65}$$

$$\sqrt{s_{\max}} = M_z\left[1 - \frac{1}{4}\gamma^2 + \frac{1}{4}\gamma^2\frac{C_I}{C_R} - \frac{1}{4}\gamma^4\frac{C_Q}{C_R}\right] , \tag{3.66}$$

$$\sqrt{s_\pm} = M_z\left[1 - \frac{1}{8}\gamma^2 + \frac{1}{2}\gamma^2\frac{C_I}{C_R} - \frac{3}{4}\gamma^4\frac{C_Q}{C_R}\right]$$
$$\pm \frac{\Gamma_z}{2}\left[1 - \frac{5}{8}\gamma^2 + \gamma^2\frac{C_Q}{C_R} + \frac{1}{2}\left(\gamma^2\frac{C_Q}{C_R}\right)^2\right] . \tag{3.67}$$

The peak position is not for $\sqrt{s} = M_z$ but for $\sqrt{s} = M_z(1 - \frac{1}{4}\gamma^2) = \tilde{M}_z(1 + \frac{1}{4}\gamma^2)$ which is 17 MeV lower than M_z . Here we neglect the far smaller other terms in eq. (3.65). Would one use the original propagator in eq. (3.46), one would find the maximum for $\sqrt{s} = M_z(1 + \frac{1}{4}\gamma^2)$ i.e. 17 MeV higher than M_z . So the use of a propagator which takes the correct energy dependence of $\Sigma_z(s)$ into account gives a difference of about 34 MeV in the peak position. The apparent width $\sqrt{s}_+ - \sqrt{s}_-$ differs only significantly from Γ_z when C_Q is sizeable. This only is the case for mupair production. It means that the QED background is not entirely negligible. At the peak position it still contributes about 0.5% of the cross section for mupair production.

The approximate formulae (3.65) - (3.67) are compared with the exact calculation in table 3.2.

4   QED CORRECTIONS IN $O(\alpha)$

4.1   Introduction

In this section a discussion of order α corrections will be given. We distinguish virtual, soft and hard photon corrections. These can occur in the initial or final state and in an interference term. When the fermions were massless collinear divergencies would arise. Keeping the fermion masses in those situations gives large logarithms which often determine the size of the corrections.

The specific case of initial state corrections is studied in detail. It will illustrate analytical answers and the method of Monte Carlo simulation of radiative effects. It also demonstrates the large corrections to the line shape.

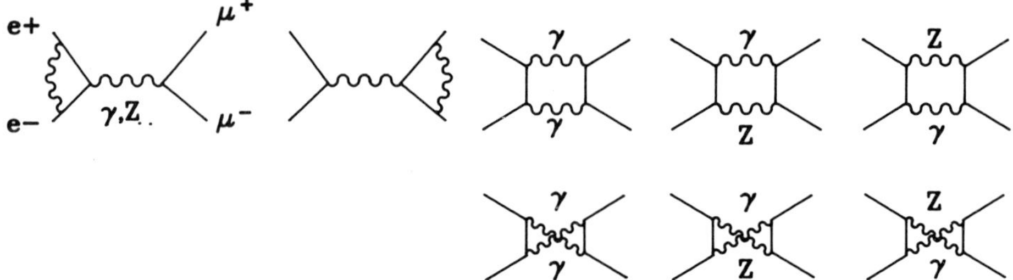

Fig. 4.1. The photonic one-loop corrections to mupair production.

4.2   Virtual corrections

We distinguish vertex corrections and box diagrams, as depicted in fig. 4.1 For small
fermion masses the vertex correction is the same for the γ_μ and $\gamma_\mu\gamma_5$ coupling. One
finds the expression $(s >> m^2)$

$$\delta_{vc} = \frac{\alpha}{\pi}\left\{-1 + \frac{\pi^2}{3} + \frac{3}{4}\ell n\,\frac{s}{m^2} - \frac{1}{4}\ell n^2\,\frac{s}{m^2} + \left(1 - \ell n\,\frac{s}{m^2}\right)\ell n\,\frac{m}{\lambda}\right\}\,. \tag{4.1}$$

The quantity δ_{vc} multiplies the lowest order matrix element. Here the mass m is either
the electron or muon mass and λ is a fictitious photon mass to regularize the infrared
divergence. The $\gamma\gamma$ box diagrams give amplitudes

$$M_{\gamma\gamma}(\lambda, \lambda') = Q_e Q_f\, M^0_\gamma(\lambda, \lambda')\,[V^{\gamma\gamma} + \lambda\lambda' A^{\gamma\gamma}] \tag{4.2}$$

whereas the γZ boxes give

$$M_{\gamma Z}(\lambda, \lambda') = Q_e Q_f\, M^0_z(\lambda, \lambda')\,[V^{\gamma z} + \lambda\lambda' A^{\gamma z}]\,. \tag{4.3}$$

Here the lowest order amplitude $M^0(\lambda, \lambda')$ is divided into the photon and Z propagator
parts (see eqs. (2.2) - (2.4)). As in section 2 the fermion masses are assumed to be
small with respect to the beam energy. A set of functions has been introduced

$$
\begin{aligned}
V^{\gamma\gamma} &= \frac{\alpha}{2\pi}\left[G(s,t) - G(s,u) + 2\ell n\left(\frac{\lambda^2}{-s - i\varepsilon}\right)\ell n\,\frac{t}{u}\right]\,, \\[2mm]
A^{\gamma\gamma} &= \frac{\alpha}{2\pi}\left[G(s,t) + G(s,u)\right]\,, \\[2mm]
G(s,t) &= \frac{s}{2(s+t)}\ell n\left(\frac{t}{s+i\varepsilon}\right) - \frac{s(s+2t)}{4(s+t)^2}\left[\ell n^2\left(\frac{t}{s+i\varepsilon}\right) + \pi^2\right]\,, \\[2mm]
V^{\gamma z} &= \frac{\alpha}{2\pi}\left[A(s,t) - A(s,u) + 2Li_2\left(1 + \frac{M^2}{t}\right) - 2Li_2\left(1 + \frac{M^2}{u}\right)\right. \\[2mm]
&\quad\left. + \; 4\ell n\left(\frac{M\lambda}{M^2 - s}\right)\ell n\,\frac{t}{u}\right]\,, \\[2mm]
A^{\gamma z} &= \frac{\alpha}{2\pi}\left[A(s,t) + A(s,u)\right]\,, \\[2mm]
A(s,t) &= \frac{s - M^2}{s+t}\left\{\ell n\left(\frac{t}{s - M^2}\right) + \frac{M^2}{s}\ell n\left(1 - \frac{s}{M^2}\right) + \frac{s + 2t + M^2}{s+t}\left[\vphantom{\frac{M^2}{M^2}}\right.\right. \\[2mm]
&\quad \left.\left.\ell n\left(\frac{-t}{M^2}\right)\ell n\left(\frac{M^2 - s}{M^2 + t}\right) + Li_2\left(\frac{s}{M^2}\right) - Li_2\left(\frac{-t}{M^2}\right)\right]\right\}\,,
\end{aligned}
\tag{4.4}
$$

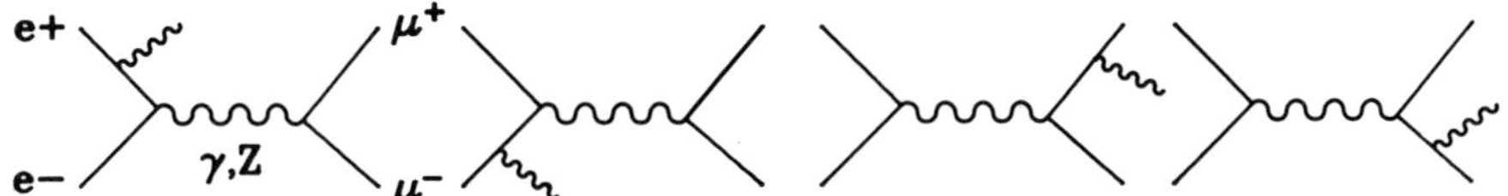

Fig. 4.2. The bremsstrahlung diagrams for mupair production.

$$M^2 = M_z^2 - i M_z \Gamma_z \ ,$$

$$Li_2(y) = -\int_0^y \frac{dx}{x} \ell n(1-x) \ .$$

The last function is called dilogarithm.

4.3 <u>Soft corrections</u>

Photon emission is described by the eight diagrams of fig. 4.2. Soft photon emission factorizes so that the bremsstrahlung matrix element squared can be written as

$$|M_\gamma|^2 = -e^2 \left(\frac{p_+}{p_+ \cdot k} - \frac{p_-}{p_- \cdot k} - \frac{q_+}{q_+ \cdot k} + \frac{q_-}{q_- \cdot k} \right)^2 |M^0|^2 \ . \tag{4.5}$$

The integration over an isotropic photon emission phase space, where photons can be emitted up to an energy $k_0 = k_m$ leads to a well known factor. In the calculation one again introduces a fictitious photon mass in order to avoid a divergent result. We divide the result into three parts, the initial state, final state part and the interference.

For the initial state radiation one finds the factor δ_s , such that the differential cross section with soft radiation becomes

$$\frac{d\sigma}{d\Omega} = \frac{d\sigma^0}{d\Omega} \delta_s \ , \tag{4.6}$$

where $d\sigma^0/d\Omega$ is the lowest order dressed (i.e. weak corrections included) cross section. The expression for δ_s reads

$$\delta_s = \frac{\alpha}{\pi} \left[\frac{-\pi^2}{3} + 2 \left(\ell n \, \frac{s}{m_e^2} - 1 \right) \ell n \, \frac{2k_m}{\lambda} + \ell n \, \frac{s}{m_e^2} - \frac{1}{2} \ell n^2 \, \frac{s}{m_e^2} \right] \ . \tag{4.7}$$

For final state radiation the same expression is found but with the electron mass replaced by the muon mass.

The interference term gives

$$\delta_s = \frac{2\alpha}{\pi} \left\{ 4\ell n \left(\tan \frac{\theta}{2} \right) \ell n \frac{2k_m}{\lambda} + 2\ell n^2 \left(\sin \frac{\theta}{2} \right) - 2\ell n^2 \left(\cos \frac{\theta}{2} \right) \right.$$

$$\left. - Li_2 \left(\sin^2 \frac{\theta}{2} \right) + Li_2 \left(\cos^2 \frac{\theta}{2} \right) \right\} \ . \tag{4.8}$$

In the following it will be useful to combine the effect of virtual and soft corrections of initial state radiation

$$\frac{d\sigma}{d\Omega} = \frac{d\sigma^0}{d\Omega}\,(1 + \delta_s + 2\delta_{vc})\,, \tag{4.9}$$

$$\delta_s + 2\delta_{vc} = \frac{2\alpha}{\pi}\left[\frac{\pi^2}{6} - 1 + \frac{3}{4}\,\ell n\,\frac{s}{m_e^2} + (\ell n\,\frac{s}{m_e^2} - 1)\ell n\,\frac{k_m}{E}\right]\,, \tag{4.10}$$

where E is the beam energy. For final state radiation $m_e \to m_\mu$. In this expression the photon mass cancels. The expression contains two large logarithms. The logarithm

$$L_e = \ell n\,\frac{s}{m_e^2} \tag{4.11}$$

and the logarithm

$$L_s = \ell n\,\frac{k_m}{E}\,. \tag{4.12}$$

The first logarithm is large at LEP energies, whereas the second can become large and negative for very small photon energies. Depending on k_m the quantity $\delta_s + 2\delta_{vc}$ can become positive or negative.

For a very stringent photon energy cut first order perturbation theory leads to a nonsensical result: the cross section becomes negative. The solution to this problem has been given long ago [13]. One should consider multiple soft photon emission and perform a resummation of the perturbation series. One then finds

$$\begin{aligned}
\frac{d\sigma}{d\Omega} &= \frac{d\sigma^0}{d\Omega}\,\delta^{v+s}\,\left(\frac{k_m}{E}\right)^\beta \\
&= \frac{d\sigma^0}{d\Omega}\,\delta^{v+s}\,e^{\beta\ell n\,\varepsilon}
\end{aligned} \tag{4.13}$$

where $\varepsilon = k_m/E$

$$\delta^{v+s} = 1 + \frac{\alpha}{\pi}\left(\frac{3}{2}\,L_e + \frac{\pi^2}{3} - 2\right)\,, \tag{4.14}$$

$$\beta = \frac{2\alpha}{\pi}\,(L_e - 1)\,. \tag{4.15}$$

Upon expansion in α we see that the correct $O(\alpha)$ correction is reproduced. Explicit order α^2 corrections will indicate that one indeed obtains a form where δ^{v+s} factorizes.

Although the above formulae are correct for soft photon emission $\varepsilon = k_m/E \ll 1$ one has also to consider hard photon corrections for which one has to perform explicit calculations. For initial state radiation we shall carry them out.

4.4 Hard photon corrections in general

The photon can have a helicity $\lambda_\gamma = \pm 1$ and it can be shown [14] that neglecting fermion masses the hard bremsstrahlung helicity amplitudes $M\left(\lambda_-, \lambda'_-, \lambda_\gamma\right)$ factorize as follows

$$M(+ - + - +) = 2ie^2 u G(- - +)\,,$$

$$M(+ - + - -) = 2ie^2 u' G(- - -)\,,$$

$$M(+--++) = 2ie^2 t G(-++) \,,$$

$$M(+--+-) = 2ie^2 t' G(-+-) \,,$$

$$M(-++-+) = 2ie^2 t' G(+-+) \,,$$

$$M(-++--) = 2ie^2 t G(+--) \,,$$

$$M(-+-++) = 2ie^2 u' G(+++) \,,$$

$$M(-+-+-) = 2ie^2 u G(++-) \,,$$

(4.16)

where

$$|G(\lambda_-, \lambda'_-, \pm)|^2 = -\frac{1}{2}\frac{s'}{s}\, v_p^2 |P(\lambda_-, \lambda'_-, s')|^2$$

$$-\frac{1}{2}\frac{s}{s'}\, v_q^2 |P(\lambda_-, \lambda'_-, s)|^2 + (v_p \cdot v_q) Re\left(P(\lambda_-, \lambda'_-, s') \cdot P^*(\lambda_-, \lambda'_-, s)\right)$$

$$\mp \frac{1}{2}\frac{(s-s')\varepsilon(p_+ p_- q_+ q_-)}{(p_+ \cdot k)(p_- \cdot k)(q_+ \cdot k)(q_- \cdot k)}\, Im\left(P(\lambda_-, \lambda'_-, s') P^*(\lambda_-, \lambda'_-, s)\right), \quad (4.17)$$

where

$$v_p = eQ_e \left(\frac{p_+}{p_+ \cdot k} - \frac{p_-}{p_- \cdot k}\right) \,, \tag{4.18}$$

$$v_q = eQ_f \left(\frac{q_+}{q_+ \cdot k} - \frac{q_-}{q_- \cdot k}\right) \,. \tag{4.19}$$

For the quantities $P(\lambda, \lambda')$ one can take the expressions corrected for weak loops i.e. the dressed expressions. The quantity $\varepsilon(p_+ p_- q_+ q_-)$ denotes the contraction between the Levi-Civita tensor and the four indicated momenta.

4.5 Initial state bremsstrahlung

Restricting ourselves to initial state radiation we have

$$|M|^2 = e^2 \frac{s'}{s} \frac{p_+ \cdot p_-}{p_+ \cdot k \, p_- \cdot k} 4e^4 \left\{ \left[|P(--s')|^2 + |P(++s')|^2\right](u^2 + u'^2) \right.$$
$$\left. + \left[|P(-+,s')|^2 + |P(+-,s')|^2\right](t^2 + t'^2) \right\} \,. \tag{4.20}$$

Consider the cms frame of the mupairs.
Thus

$$\vec{Q} = \vec{q}_+ + \vec{q}_- = 0 \tag{4.21}$$

and therefore

$$\vec{p}_+ + \vec{p}_- - \vec{k} = 0 \,. \tag{4.22}$$

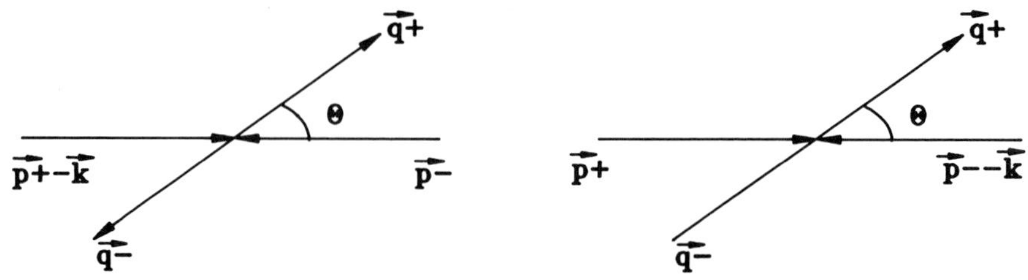

Fig. 4.3. Two different choices of z-axis in the mupair c.m.s.

In this frame one can consider the z-axis to be $-\vec{p}_-$ or $\vec{p}_+$ (fig. 4.3) and introduce $c = \cos\theta$ for the μ^+ . For the invariants we find

$$t' = -2p_- \cdot q_- = -2p_{-0}q_{-0}(1 - \cos\theta) = -p_{-0}\sqrt{s'}(1 - c) \; , \tag{4.23}$$

$$u' = -2p_- \cdot q_+ = -2p_{-0}q_{+0}(1 + \cos\theta) = -p_{-0}\sqrt{s'}(1 + c) \; , \tag{4.24}$$

$$t = -2p_+ \cdot q_+ = -p_{+0}\sqrt{s'}(1 - c) \; , \tag{4.25}$$

$$u = -2p_+ \cdot q_- = -p_{+0}\sqrt{s'}(1 + c) \; . \tag{4.26}$$

For t' and u' the $-\vec{p}_-$ axis is used and for t and u the $\vec{p}_+$ axis. In the non bremsstrahlung case we have $(s = s')$

$$t = t' = \frac{-s'}{2}(1 - c) \; , \tag{4.27}$$

$$u = u' = \frac{-s'}{2}(1 + c) \; , \tag{4.28}$$

and we call the matrixelement M^0 , where c is with respect to a specific z-axis, using the cms energy $\sqrt{s'}$ of the muon system.

We can calculate $p_{\pm 0}$ in the two pictures

$$p_{+0} + p_{-0} - k_0 = \sqrt{s'} \; , \tag{4.29}$$

$$\vec{p}_+ - \vec{k} = -\vec{p}_- \; , \tag{4.30}$$

$$(p_+ - k)^2 = \left(\sqrt{s'} - p_0\right)^2 - \vec{p}_-^{\,2} = s' - 2p_{-0}\sqrt{s'} \; , \tag{4.31}$$

$$p_{-0} = \frac{s' + 2p_+ \cdot k}{2\sqrt{s'}} \; . \tag{4.32}$$

In the second case

$$p_{+0} = \frac{s' + 2p_- \cdot k}{2\sqrt{s'}} \; . \tag{4.33}$$

We can now express the bremsstrahlung matrixelement M in the non-bremsstrahlung one M^0

$$|M|^2 = e^2 \frac{s'}{s} \frac{p_+ \cdot p_-}{(p_+ \cdot k)(p_- \cdot k)} \left\{ |M^0(\vec{p}_+)|^2 \left(\frac{s' + 2p_- \cdot k}{s'} \right)^2 + \right.$$

$$\left. |M^0(-\vec{p}_-)|^2 \left(\frac{s' + 2p_+ \cdot k}{s'} \right)^2 \right\} . \tag{4.34}$$

The z-axes in the mupair cms are respectively along $\vec{p}_+$ and $-\vec{p}_-$ for M^0 . The bremsstrahlung cross section can be written as

$$d\sigma = \frac{1}{2(2\pi)^3} \frac{s'k_0}{s} dk_0 d\Omega_\gamma (2\pi)^{-2} \delta(p_+ + p_- - k - q_+ - q_-)$$

$$\times \frac{1}{2^3 s'} \frac{1}{4} \sum |M|^2 \frac{d\vec{q}_+}{q_{+0}} \frac{d\vec{q}_-}{q_{-0}} . \tag{4.35}$$

The lowest order cross section is

$$d\sigma^0 = (2\pi)^{-2} \delta(p_+ + p_- - q_+ - q_-) \frac{1}{2^3 s'} \frac{1}{4} \sum |M^0|^2 \frac{d\vec{q}_+}{q_{+0}} \frac{d\vec{q}_-}{q_{-0}} . \tag{4.36}$$

Here we take $\vec{p}_+$ as z-axis or equivalently $-\vec{p}_-$. The bremsstrahlung cross section now takes the form

$$\frac{d\sigma}{dk d\Omega_\gamma d\Omega_\mu} = \frac{e^2}{2(2\pi)^3} \frac{k_0}{2s} \frac{1}{(p_+ \cdot k)(p_- \cdot k)} \left[(s' + 2p_- \cdot k)^2 \frac{d\sigma^0}{d\Omega_\mu}(\vec{p}_+) \right.$$

$$\left. + (s' + 2p_+ \cdot k)^2 \frac{d\sigma^0}{d\Omega_\mu}(-\vec{p}_-) \right] . \tag{4.37}$$

The lowest order cross sections are in the muon cms and z-axes are $\vec{p}_+$ or $-\vec{p}_-$. The bremsstrahlung cross section is a sum of two factorized terms. However the $\vec{p}_+$ or $-\vec{p}_-$ axis are not anymore the same axes. Written more compactly one has [15]

$$\frac{d\sigma}{dk_0 d\Omega_\gamma d\Omega_\mu} = \frac{\alpha}{4\pi^2 s} \left[g_+ \frac{d\sigma^0}{d\Omega_\mu}(\vec{p}_+) + g_- \frac{d\sigma^0}{d\Omega_\mu}(-\vec{p}_-) \right] , \tag{4.38}$$

where $dk_0 d\Omega_\gamma$ are taken in the lab. frame, $d\Omega_\mu$ in the mupair cms. The factors $g_\pm$ follow from above. Including those electron mass terms which can still constribute one has

$$g_\pm = \frac{-m^2 k_0 s'}{(p_\mp \cdot k)^2} + \frac{(s' + 2p_\mp \cdot k)^2}{2(p_+ \cdot k)(p_- \cdot k)} k_0 . \tag{4.39}$$

That the first term can contribute we see as follows. In non-collinear situations $p_\pm \cdot k \sim Ek_0$ such that we have

$$\frac{m^2}{(p_\pm \cdot k)^2} \sim \frac{m^2}{E^2 k_0^2} \quad \text{and} \quad \frac{s'}{(p_+ \cdot k)(p_- \cdot k)} \sim \frac{E^2}{E^2 k_0^2} .$$

So the first term is negligible. In a collinear situation

$$p_+ \cdot k = k_0(E - |\vec{p}_+|) \equiv k_0 |\vec{p}_+|(e - 1) \simeq \frac{k_0 m^2}{2E}$$

and we have

$$\frac{m^2}{(p_+ \cdot k)^2} \sim \frac{E^2}{k_0^2 m^2} \quad \text{and} \quad \frac{s'}{(p_+ \cdot k)(p_- \cdot k)} = \frac{E^2}{k_0^2 m^2} .$$

So one has to keep the $m^2/(p_+ \cdot k)^2$ terms in collinear situations. All other terms with m^2 in the numerator can be neglected.

From the hard bremsstrahlung cross section one can derive $d\sigma/dk_0$ or $d\sigma/ds'$, the energy loss spectrum. One performs the following steps.

1. Integration over $d\Omega_\mu$. The result is not dependent on the chosen z-axis in the mupair cms. One obtains a formula first introduced by Bonneau and Martin [16]

$$\frac{d\sigma}{dk_0 d\Omega_\gamma} = \frac{\alpha}{4\pi^2 s} \, (g_+ + g_-) \, \sigma^0_{tot}(s') \tag{4.40}$$

with

$$s' = (q_+ + q_-)^2 = s\left(1 - \frac{k_0}{E}\right) . \tag{4.41}$$

2. Integration over $d\Omega_\gamma$. Since we have

$$\begin{aligned}
g_+ + g_- &= \frac{-4m^2 s'}{sk_0}\left[\frac{1}{(e - \cos\theta_\gamma)^2} + \frac{1}{(e + \cos\theta_\gamma)^2}\right] - 4k_0 \\
&+ \frac{s}{k_0}\left(1 + \left(\frac{s'}{s}\right)^2\right)\left[\frac{1}{e - \cos\theta_\gamma} + \frac{1}{e + \cos\theta_\gamma}\right]
\end{aligned} \tag{4.42}$$

with

$$e = \frac{E}{|\vec{p}_+|} \tag{4.43}$$

integration gives

$$\begin{aligned}
\int d\,\cos\theta_\gamma (g_+ + g_-) &= \frac{-4m_e^2 s'}{sk}\,\frac{s}{m_e^2} - 8k_0 + \frac{s}{k_0}\left(1 + \left(\frac{s'}{s}\right)^2\right) 2\ell n\,\frac{s}{m_e^2} \\
&= \frac{2s}{k_0}\left(1 + \left(\frac{s'}{s}\right)^2\right)\left(\ell n\,\frac{s}{m_e^2} - 1\right) .
\end{aligned} \tag{4.44}$$

The final azimuthal angle integration gives a factor of 2π. So the end result for the spectrum is

$$\frac{d\sigma}{dk_0} = \frac{\alpha}{\pi}\,\sigma(s')\left(\ell n\,\frac{s}{m_e^2} - 1\right)\left(1 + \left(\frac{s'}{s}\right)^2\right)\frac{1}{k_0} . \tag{4.45}$$

Three remarks are in order. In the first place the spectrum peaks as $1/k_0$ i.e. for soft photon emission. When we integrate over it and have k_m as minimum energy we get a $\ell n\, k_m$ term which cancels against a similar term in the soft photon emission part. The second remark concerns other possible peaks in the photon spectrum. When $s > M_z^2$ there is of course a peak for

$$\frac{k_0}{E} = 1 - \frac{M_z^2}{s} . \tag{4.46}$$

Another peak arises for very hard photons due to the $1/s'$ behaviour of the photon propagator in $\sigma(s')$ (eq. (2.12)). The third remark concerns the large logarithm, which occurs

$$L_e = \ell n\,\frac{s}{m_e^2} . \tag{4.47}$$

It orgininates from the collinear peaking. When $m_e = 0$ the peak would diverge. The integral

$$J = \int_a^b d \, \cos \theta_\gamma \, \frac{1}{e - \cos \theta_\gamma} = \ell n \, \frac{e - a}{e - b} \tag{4.48}$$

becomes large when $b \to 1$, since then $e - 1 = \frac{2m^2}{s}$. The peaking occurs when the photon is emitted along the e^+ or e^- direction.

When one would consider only one photon exchange $\sigma(s') = \frac{4\pi\alpha^2}{3s'}$ one can calculate the total correction [17]

$$\sigma(k_0 > k_m) + \sigma(k_0 < k_m) = \sigma^0(1 + \delta_T) \,, \tag{4.49}$$

with

$$\delta_T = \frac{2\alpha}{\pi} \left[\frac{1}{2}\ell n \, \frac{s}{m_e^2}\ell n \, \frac{s}{m_\mu^2} - \frac{7}{12}\ell n \, \frac{s}{m_e^2} - \frac{1}{2}\ell n \, \frac{s}{m_\mu^2} + \frac{\pi^2}{6} + \frac{1}{3} \right] \,, \tag{4.50}$$

where m_μ is the muon mass. The correction δ_T is large at LEP energies ($\delta_T = 0.69$) . For later use we rewrite the spectrum by introducing $z = s'/s$ or $k_0 = (1 - z)E$ such that

$$
\begin{aligned}
\frac{d\sigma}{ds'} \;=\; & \frac{\sigma^0(s')}{s} \left\{ \delta(1 - z) + \frac{\alpha}{\pi} \, \delta(1 - z) \left[2(L - 1)\ell n \, \varepsilon + \frac{3}{2}L \right. \right. \\
& \left. + \;\; 2\zeta(2) - 2 \right] + \theta(1 - z - \varepsilon) \, \frac{\alpha(L - 1)}{\pi} \, \frac{1 + z^2}{1 - z} \right\}
\end{aligned}
\tag{4.51}
$$

where $\varepsilon = k_m/E$, $L = L_e$ and $\zeta(2) = \pi^2/6$. In this spectrum the soft and virtual photon correction is included as a δ function.

4.6   Collinear approximation

When one would neglect masses everywhere except to regulate poles $(p_\pm \cdot k)^{-1}$ we obtain

$$g_\pm = \frac{(s' + 2p_\mp \cdot k)^2}{2(p_+ \cdot k)(p_- \cdot k)} \, k_0 \, . \tag{4.52}$$

When one approximates these expressions by the respective collinear poles and residues one obtains for $\vec{k} \| \vec{p}_+$

$$s' + 2p_\mp k = s \text{ or } s' \tag{4.53}$$

$$2p_- \cdot k = s - s' \tag{4.54}$$

and

$$g_+ = \frac{s^2 k_0}{(s - s')(p_+ \cdot k)} \, , \quad g_- = \frac{s'^2 k_0}{(s - s')(p_+ \cdot k)} \, . \tag{4.55}$$

And similarly for $\vec{k} \| \vec{p}_-$

$$g_+ = \frac{s'^2 k_0}{(s - s')(p_- \cdot k)} \quad g_- = \frac{s^2 k_0}{(s - s')(p_- \cdot k)} \, . \tag{4.56}$$

In the collinear approximation the quantities $g_\pm$ are a sum of pole terms

$$g_+ = \frac{1}{4E} \left(\frac{s^2}{p_+ \cdot k} + \frac{s'^{\,2}}{p_- \cdot k} \right) , \tag{4.57}$$

$$g_- = \frac{1}{4E} \left(\frac{s'^{\,2}}{p_+ \cdot k} + \frac{s^2}{p_- \cdot k} \right) . \tag{4.58}$$

Using these forms of $g_\pm$ one finds for the spectrum

$$\frac{d\sigma}{dk_0} = \frac{\alpha}{\pi}\, \sigma^0(s') L_e \left(1 + \left(\frac{s'}{s} \right)^2 \right) \frac{1}{k_0} . \tag{4.59}$$

In other words, only the leading log term of eq. (4.45) is obtained. One can also show that in the virtual and soft corrections a collinear approximation gives the terms proportional to L_e .

4.7 Final state radiation

Similarly one can show that for final state radiation one gets the spectrum

$$\frac{d\sigma}{dk_0} = \sigma^0(s) \frac{\alpha}{\pi} \left[(L_\mu - 1) + \ell n\, \frac{s'}{s} \right] \left(1 + \left(\frac{s'}{s} \right)^2 \right) \frac{1}{k_0} . \tag{4.60}$$

The leading log is

$$L' = L_\mu + \ell n\, \frac{s'}{s} = \ell n\, \frac{s'}{m_\mu^2} . \tag{4.61}$$

In a collinear approximation L_μ instead of $L_\mu - 1$ would have been obtained.

When one calculates the full order α correction due to final state radiation one finds

$$\sigma(k_0 > k_m) = \sigma_0 \left[\frac{2\alpha}{\pi}(L_\mu - 1) \left(\ell n\, \frac{E}{k_m} - \frac{3}{4} \right) + \frac{\alpha}{\pi} \left(-\frac{1}{3}\pi^2 + \frac{5}{4} \right) \right] . \tag{4.62}$$

Combined with $\sigma(k_0 < k_m)$ one finds

$$\sigma_{\text{tot}} = \sigma_0 \left(1 + \frac{3\alpha}{4\pi} \right) . \tag{4.63}$$

So from final state radiation one gets a tiny correction, not containing any large logarithm. This is a result of the KLN theorem [18]: the collinear singularities cancel. This does not apply to initial state radiation as the result (4.50) for the one photon exchange cross section explicitly shows.

4.8 Radiative corrections with experimental cuts:
an event generator versus a semi-analytical approach

In eqs. (4.49) and (4.50) an analytical result is given for a radiative correction to an experimental quantity. In this case it is the total cross section. The integration over the hard photon phase space could be carried out completely.

In general this cannot easily be done. One often has to impose experimental cuts which make the boundaries in phase space involved. Therefore the analytical integration becomes too tedious. In that case numerical methods are applied. We want to illustrate two approaches. Consider the differential cross section for mupair production. Reasonable conditions to select the mupair events are

i The muons should be back to back within a certain maximum acollinearity angle ζ (typically $10^o - 20^o$)

ii The muons should have a sizeable fraction of the beam energy $q_{\pm 0} > E_0$ where E_0 is typically $0.5E$.

We shall refer to these cuts as canonical cuts. They were already introduced for the first hard photon correction calculations in e^+e^- collisions [19]. The conditions are such that there will be anyhow a part in phase space in which a photon can be emitted isotropically, say up to an energy k_1 . One can then write the expression for the order α corrected differential cross section as follows

$$
\frac{d\sigma}{d\Omega_\mu} = \frac{d\sigma^0}{d\Omega_\mu}[1 + \delta_{\mathrm{T}}(E,\theta,\zeta,E_0)] \tag{4.64}
$$

$$
= \frac{d\sigma^0}{d\Omega_\mu}[1 + \delta_v + \delta_s(k_1)]
$$

$$
+ \int_0^{k_1} dk_0 d\Omega_\gamma \left(\frac{d\sigma^B}{d\Omega_\mu d\Omega_\gamma dk_0} - \frac{d\sigma^s}{d\Omega_\mu d\Omega_\gamma dk_0} \right)
$$

$$
+ \int_{k_1 < k_0 < k_2} dk_0 d\Omega_\gamma \frac{d\sigma^B}{d\Omega_\mu d\Omega_\gamma dk_0} . \tag{4.65}
$$

Here the exact bremsstrahlung cross section $d\sigma^B$ and the approximate soft one $d\sigma^s$ are introduced. The latter follows from eq. (4.5). The cross sections $d\sigma^0, d\sigma^s$ and $d\sigma^B$ can be taken including the most important weak corrections. The total correction δ_{T} depends on the applied cuts. The evaluation is done with an analytical term using virtual and soft corrections δ_v and δ_s . Since k_1 is not necessarily so small that the soft expression can be used it is corrected in the second term. The hard bremsstrahlung which is not emitted isotropically is represented by the third term in eq. (4.65), where k_2 is the maximum photon energy. The infrared cancellation (i.e. the λ dependence) is automatically done in the first term. The second term is a rather straightforward numerical integration since the peaking structures of both terms cancel. It is the third term where the collinear peakings cause problems. The numerical integration has to be performed with great care [19].

We call this approach the semianalytical approach. It gives the exact $O(\alpha)$ correction with certain specified cuts. The advantage is the high accuracy which can be obtained. The draw-back is that one has no freedom to impose cuts on different variables without changing the program.

Another approach has this freedom [17,15]. It uses a Monte Carlo simulation of the reactions

$$
e^+e^- \rightarrow \mu^+\mu^- \tag{4.66}
$$

and

$$
e^+e^- \rightarrow \mu^+\mu^-\gamma . \tag{4.67}
$$

The distinction between the two reactions is made through a maximum soft photon energy k_1 . This quantity has to be chosen small such that the presence of a soft photon in (4.66) is experimentally hardly observable. In practice it means that a program generates N events i.e. N_1 sets of momenta $\{q_+, q_-\}$ and N_2 sets

$\{q_+, q_-, k\}$ with $N = N_1 + N_2$. The momenta $\{q_+, q_-\}$ have a distribution governed by $d\sigma^0/d\Omega_\mu [1 + \delta_v + \delta_s(k_1)]$ whereas the momenta $\{q_+, q_-, k\}$ have a distribution $d\sigma^B/d\Omega_\mu d\Omega_\gamma dk_0$. For small enough k_1 the former distribution is a good approximation.

The ratio N_1/N_2 should satisfy

$$\frac{N_1}{N_2} = \frac{\sigma(k_0 < k_1)}{\sigma(k_0 > k_1)} \, , \tag{4.68}$$

where $\sigma(k_0 \lessgtr k_1)$ denote total cross sections. In fact

$$\sigma_{\rm T}(E) = \sigma(k_0 < k_1) + \sigma(k_0 > k_1) \tag{4.69}$$

is the known total cross section for mupair production including the order α correction i.e. the sum of the total cross sections of reactions (4.66) and (4.67). From the N generated events one can obtain the previous correction $\delta_{\rm T}$ of eq. (4.64) by selecting all $N_{\rm sel}$ events where μ_+ has a scattering angle θ, φ in an interval $\Delta\Omega_\mu$ and for which the acollinearity and energy requirements are satisfied. One now has

$$\frac{d\sigma}{d\Omega_\mu}\Delta\Omega_\mu = \frac{N_{\rm sel}}{N} \, \sigma_{\rm T}(E) \tag{4.70}$$

and since $d\sigma^0/d\Omega_\mu$ is explicitly known the correction $\delta_{\rm T}(E, \theta, \zeta, E_0)$ can be obtained. One can however obtain more general corrections $\delta_{\rm T}$ with any type of cuts since one can impose these to the simulated N events. This general purpose ability of the event generator is a great advantage.

There is however also a disadvantage to which we now turn. Although k_1 has to be small it cannot be chosen too small since then $\sigma(k_0 < k_1)$ becomes negative. This is in contrast to the semianalytical approach, where the second term in eq. (4.65) is present and compensates for an incorrect approximation. Here the k_1 dependence drops out, whereas in an event generator there is a dependence on k_1 . As long as k_1 can be chosen small enough this dependence is no problem since the soft photon bremsstrahlung expression is then a very good approximation to the real hard bremsstrahlung.

The technical implementation of the Monte Carlo simulation can easily be sketched in the case of initial state radiation. In the general case the procedure is similar [17,20].

The event generation is done in the following steps

1. Choose a soft or hard event according to the prohabilities $\sigma(k_0 < k_1)/\sigma_{\rm T}$ and $\sigma(k_0 > k_1)/\sigma_{\rm T}$. These probabilities are known from eq. (4.9) and the integral over expression (4.45).

2. When the event is soft the scattering angle θ is generated with the distribution (4.9). The azimuthal angle φ_γ is generated uniformly in $(0, 2\pi)$.

3. When the event is hard the photon energy is generated using the photon spectrum (4.45).

4. The azimuthal angle φ_γ is generated uniformly whereas θ_γ is generated with distribution (4.40).

5. Since the four momentum k is known the functions $g_\pm$ are known. One now chooses as z-axis the $\vec{p}_+$ or $-\vec{p}_-$ direction with relative probability g_+/g_- . In the muon c.m.s. the scattering angle θ and azimuthal angle φ are generated with respect to the chosen z-axis. So one knows the four momenta q_+, q_- in the muon c.m.s. system, which are then transformed to the lab system.

In the above procedure every variable is generated with a single distribution. When the distribution strongly peaks special strategies are used. For the photon spectrum a histogram can be set up when the program is started. The bins have different widths but the same area. One then chooses a bin at random and k_0 inside this bin. For the $\cos\theta_\gamma$ distribution the integral (4.48) with $a = -1$ and b arbitrary is helpful. One chooses at random a fraction of the total integral and then solves for b i.e. a $\cos\theta_\gamma$ value.

5 HIGHER ORDER QED CORRECTIONS

When one considers initial state corrections second order in α, one has the following contributions [21].

1. Second order corrections to the electron vertex, so one obtains an $O(\alpha^2)$ virtual correction to the process

$$e^+e^- \to \mu^+\mu^- \ . \tag{5.1}$$

   Diagrams are listed in fig. 5.1. In all diagrams the dashed line stands for a Z and γ decaying into a mupair.

2. Virtual corrections to the process

$$e^+e^- \to \mu^+\mu^-\gamma \ . \tag{5.2}$$

   When the photon is soft, one just obtains the standard soft factor times an $O(\alpha)$ corrected process (5.1). When the photon is hard new loop diagrams have to be evaluated, see fig. 5.2.

3. The double bremsstrahlung process

$$e^+e^- \to \mu^+\mu^-\gamma\gamma \tag{5.3}$$

   with diagrams listed in fig. 5.3.

In order α^2 there are however more mechanisms that produce a mupair. Thus we add the process

$$e^+e^- \to \mu^+\mu^-e^+e^- \ . \tag{5.4}$$

Instead of an e^+e^- pair in general an $\bar{f}f$ can be emitted. The mechanisms producing (5.4) have different structures, so we list them separately.

A Initial state radiation of a virtual photon decaying into a fermion pair (fig. 5.4).

B Bhabha scattering with the emission of a mupair (fig. 5.5).

C The interference between A and B.

D Multiperipheral production of an $\bar{f}f$ pair (fig. 5.6).

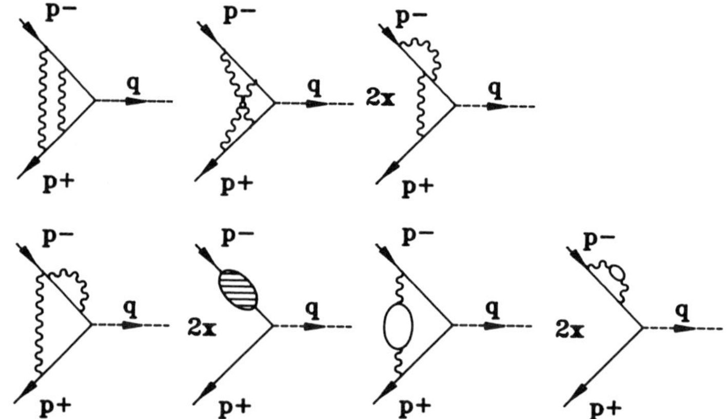

Fig. 5.1. The initial state $O(\alpha^2)$ corrections to mupair production.

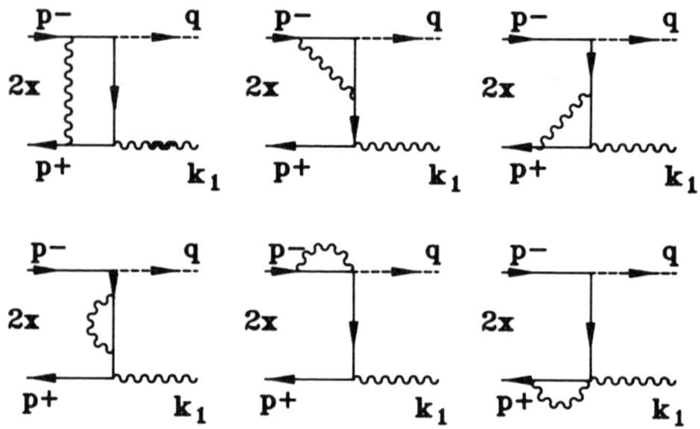

Fig. 5.2. The $O(\alpha)$ corrections to radiative mupair production.

We can calculate from all of these processes $d\sigma/ds'$. The relative importance can be seen in fig. 5.7. The highest curve is due to the combination of processes (5.2) and (5.3). It is clear that it is dominant. When one is satisfied with a description of the Z peak with an accuracy of 0.2% the inclusion of the above photonic corrections is adequate and contributions A, B, C can be neglected. When we assume that experimentally the multiperipheral events can be removed, which can anyhow be done by a cut on s' , we can also omit process D.

Should one want to increase the precision the process (5.4) through mechanism A should be included. One will then notice that in evaluating the total mupair cross section a cancellation occurs between LL terms arising from reaction (5.4) with $\bar{f}f$ emission and the fermion loop insertion diagram in fig. 5.1 (i.e. the last diagrams). Therefore when considering photonic corrections the last diagram in fig. 5.1 will be omitted.

So we now turn to $O(\alpha^2)$ photonic corrections. We give the result for $d\sigma/ds'$ from an exact calculation [21]. Later on we discuss how the LL terms can also be obtained in a different way. In fact they have been initially obtained in that way. The expression

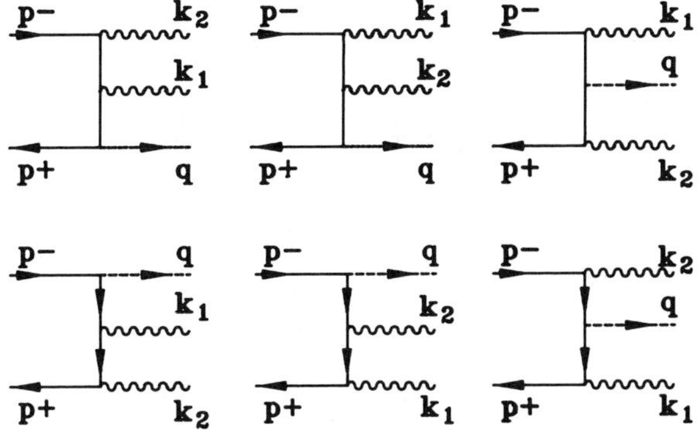

Fig. 5.3. Mupair production with double bremsstrahlung.

resulting from the exact calculation is

$$
\begin{aligned}
\frac{d\sigma}{ds'} =\ & \frac{1}{s}\sigma^0(s')\,\Big\{\delta(1-z) \\
& +\ \delta(1-z)\Big[\beta\ell n\ \varepsilon + \delta_1^{v+s}\Big] + \theta(1-z-\varepsilon)\Big[\beta\frac{1}{1-z}+\delta_1^H\Big] \\
& +\ \delta(1-z)\Big[\frac{\beta^2}{2}\ell n^2\ \varepsilon + \beta\ell n\ \varepsilon\delta_1^{v+s}+\delta_2^{v+s}\Big] \\
& +\ \theta(1-z-\varepsilon)\Big[\beta^2\frac{\ell n(1-z)}{1-z}+\beta\frac{1}{1-z}\delta_1^{v+s}+\delta_2^H\Big]\Big\}
\end{aligned}
\tag{5.5}
$$

where

$$
\beta = \frac{2\alpha}{\pi}(L-1)
\tag{5.6}
$$

with

$$
L = L_e\ ,
$$

$$
\delta_1^{v+s} = \frac{\alpha}{\pi}\left(\frac{3}{2}L + 2\zeta(2)-2\right)\ ,
\tag{5.7}
$$

$$
\delta_1^H = -\frac{\alpha}{\pi}(L-1)(1+z)\ ,
\tag{5.8}
$$

$$
\begin{aligned}
\delta_2^{v+s} =\ & \left(\frac{\alpha}{\pi}\right)^2\Big[\left(\frac{9}{8}-2\zeta(2)\right)L^2 + \left(-\frac{45}{16}+\frac{11}{2}\zeta(2)+3\zeta(3)\right)L \\
& -\ \frac{6}{5}\zeta(2)^2 - \frac{9}{2}\zeta(3) - 6\zeta(2)\ell n\ 2 + \frac{3}{8}\zeta(2) + \frac{19}{4}\Big]\ ,
\end{aligned}
\tag{5.9}
$$

$$
\begin{aligned}
\delta_2^H =\ & \left(\frac{\alpha}{\pi}\right)\Big\{X - (1+z)\Big[2\,\ell n(1-z)(L-1)^2 \\
& +\ (L-1)(\frac{3}{2}L+2\zeta(2)-2)\Big]\Big\}\ ,
\end{aligned}
\tag{5.10}
$$

339

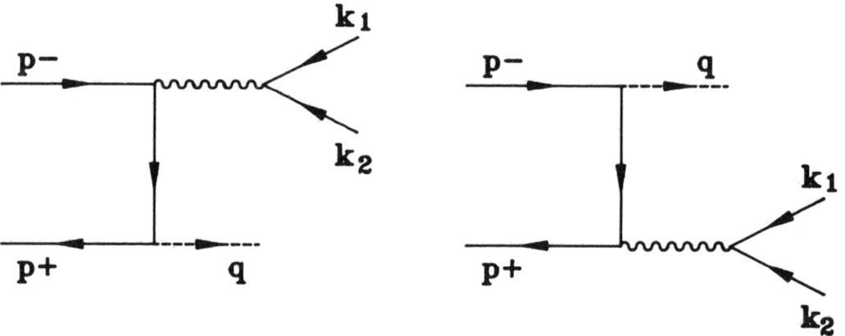

Fig. 5.4. Mupair production with initial state radiation of a fermion pair.

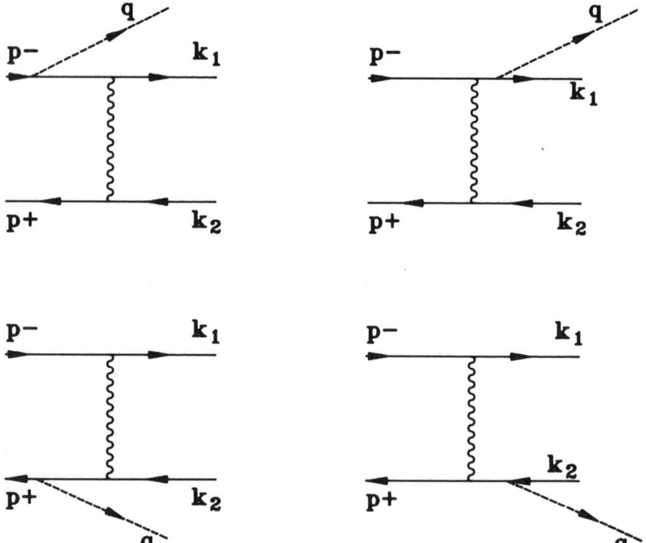

Fig. 5.5. Bhabha scattering with emission of a mupair.

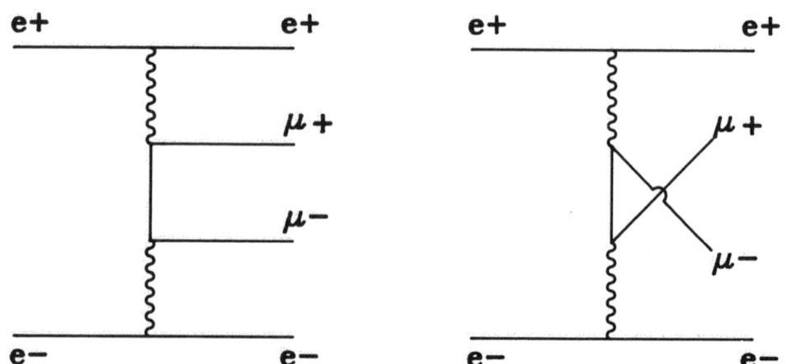

Fig. 5.6. Two-photon production of a mupair.

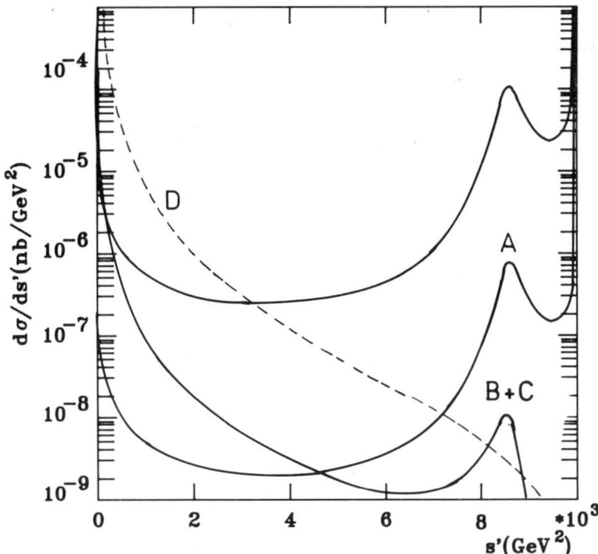

Fig. 5.7. The distributions $d\sigma/ds'$ for $\sqrt{s} = 100$ GeV arising from various sources.

$$
\begin{aligned}
X \;=\;& \left(-\frac{1+z^2}{1-z}\,\ell n\ z + (1+z)\frac{1}{2}\ell n\ z + z - 1\right) L^2 \\[4pt]
+\;& \left[\frac{1+z^2}{1-z}\left(Li_2(1-z) + \ell n\ z\,\ell n\ (1-z) + \frac{7}{2}\ell n\ z - \frac{1}{2}\ell n^2\ z\right)\right. \\[4pt]
+\;& \left. (1+z)\frac{1}{4}\ell n^2\ z - \ell n\ z + \frac{7}{2} - 3z\right] L \\[4pt]
+\;& \frac{1+z^2}{1-z}\left(-\frac{1}{6}\ell n^3\ z + \frac{1}{2}\ell n\ z\,Li_2(1-z) + \frac{1}{2}\ell n^2\ z\,\ell n(1-z)\right. \\[4pt]
-\;& \left.\frac{3}{2}Li_2(1-z) - \frac{3}{2}\ell n\ z\,\ell n(1-z) + \zeta(2)\ell n\ z - \frac{17}{6}\ell n\ z - \ell n^2\ z\right) \\[4pt]
+\;& (1+z)\left(\frac{3}{2}Li_3(1-z) - 2S_{1,2}(1-z) - \ell n(1-z)Li_2(1-z) - \frac{1}{2}\right) \\[4pt]
-\;& \frac{1}{4}(1-5z)\ell n^2(1-z) + \frac{1}{2}(1-7z)\ell n\ z\,\ell n(1-z) - \frac{25}{6}z\,Li_2(1-z) \\[4pt]
+\;& (-1 + \frac{13}{3}z)\zeta(2) + \left(\frac{3}{2} - z\right)\ell n(1-z) + \frac{1}{6}(11+10z)\ell n\ z \\[4pt]
+\;& \frac{2}{(1-z)^2}\,\ell n^2\ z - \frac{25}{11}z\ell n^2\ z - \frac{2}{3}\,\frac{z}{1-z} \\[4pt]
& \left(1 + \frac{2}{1-z}\,\ell n\ z + \frac{1}{(1-z)^2}\,\ell n^2\ z\right) .
\end{aligned}
\tag{5.11}
$$

In these definitions the polylogarithms $Li_n(x)$ and $S_{n,p}(x)$ have been introduced (cf. refs. [22] and [23]) and the Riemann zeta function $\zeta(2) = \pi^2/6$ and $\zeta(3) \approx 1.202$. These formulae are due to explicit calculations with Feynman diagrams. It can be seen that we have terms of order $\alpha L/\pi$, α/π and $(\alpha L/\pi)^2$, $(\alpha/\pi)^2 L$ and $(\alpha/\pi)^2$. We call them leading log (LL), subleading log terms and non-log terms. It should be noted that the leading log terms can be obtained in a relatively straightforward way from the structure function approach of section 7. From this formula it is clear that the $ln\ \varepsilon$ terms cancel against terms from the hard bremsstrahlung part. Moreover that part of the bremsstrahlung leading to the $ln\ \varepsilon$ terms can be resummed leading to

$$\frac{d\sigma}{ds'} = \frac{1}{s}\sigma^0(s') \left[\beta(1 - z)^{\beta-1}\delta^{v+s} + \delta^H \right] \tag{5.12}$$

with

$$\delta^{v+s} = 1 + \delta_1^{v+s} + \delta_2^{v+s} , \tag{5.13}$$
$$\delta^H = \delta_1^H + \delta_2^H . \tag{5.14}$$

The explicit $O(\alpha^2)$ calculation is reproduced when eq. (5.12) is expanded. We confirm the anticipated form of eq. (4.13). The advantage of the resummed distribution is twofold : the soft photon effects are included to all orders and the distribution can be integrated. Thus we obtain for the total cross section

$$\sigma(s) = \int ds' \frac{d\sigma}{ds'} = \int_{z_{\min}}^{1} dz\ \sigma^0(sz) \left[\beta(1 - z)^{\beta-1}\delta^{v+s} + \delta^H \right] \tag{5.15}$$

$$= \int_{z_{\min}}^{1} dz\ \sigma^0(sz)G(z) .$$

The lower bound is $z_{\min} = \frac{s'_{\min}}{s}$. The lowest value is $z_{\min} = \frac{4m_\mu^2}{s}$. We call the function $G(z)$ the flux function. The above formula will be used in the next section to obtain the Z line shape.

One may wonder whether a second order event generator i.e. one which generates events of the reactions (5.1) - (5.3) can be made. Again the problem is that soft photon emission should get a positive probability, which requires a not too small ε . On the other hand the cross section near the Z-peak varies drastically which requires a small ε . For initial state radiation alone still a reasonable ε value can be found and an event generator can be made. If one could handle multiple soft photon emission one could use the resummed formula as a starting point. When one generates s' one has to determine whether this is due to a single photon or more photons and then use the single or multiple photon distributions to generate the momenta. In particular when multiple photon distributions are required approximations should be made. There are efforts in this direction [24], which are reasonably succesful for initial state radiation. The full treatment with interference between initial and final state radiation has not yet been given. For the line shape without experimental cuts this does not matter, since eq. (5.15) can be used. When experimental quantities with cuts are measured an event generator including multiple photon emission from initital and final state would be ideal. In the absence one could try to perform semianalytical calculations incorporating canonical cuts. The structure function method makes this possible. In this method one can treat the $O(\alpha)$ corrections exactly, while $O(\alpha^2)$ corrections are taken in the LL approximation and soft photons are resummed, to all orders. This is the strategy we shall follow in section 7. First we want to discuss the results for the Z-line shape using the resummed expression (5.15).

6 THE Z LINE SHAPE

6.1 Introduction

We want to predict the Z line shape with such an accuracy that the peak and half peak positions are correct within 10 MeV and the overall normalization is given within 0.3% .

With this aimed accuracy it is sufficient to adopt the weak corrections from section 3 i.e. to use a "dressed" Born amplitude. On this $\sigma_{ew}(s)$ we then apply purely photonic corrections from the initial state. So we neglect lepton pair production from the initial state, the interference between initial and final state corrections and final state corrections. The latter could be easily added since the overall normalization is just multiplied by a factor $1 + \frac{3\alpha}{4\pi}Q_f^2$, which gives for lepton pairs the largest factor i.e. 1.0017 .

In this section we don't apply any cuts. When one wants to do so the convolution becomes more involved as we will show in the next section.

6.2 Exact numerical results

The calculation of the line shape is based on

$$\sigma(s) = \int\limits_{z_{\min}}^{1} dz \; \sigma_{ew}(sz)G(z) \; . \tag{6.1}$$

where $G(z)$ is given by (5.15) . For $z_{\min} = \frac{s'_{\min}}{s}$ we take $\sqrt{s'_{\min}} = 2m_\mu$ for mupairs and 10 GeV for quark pairs. The numerical results are presented in tables 6.1 - 6.2, whereas the corresponding cross sections before the photonic corrections can be found in table 3.2. Further results on specific $\bar{f}f$ channels can be found in ref [6].

Comparing the two set of tables we notice the following features: a reduction ρ of the peak height, shifts $\Delta\sqrt{s}_{\max}$, $\Delta\sqrt{s}_{-}$ and $\Delta\sqrt{s}_{+}$ of the peak and half peak positions. For hadrons we find approximately

$$\rho \;=\; 0.738 \pm 0.001 \; , \tag{6.2}$$
$$\Delta\sqrt{s}_{\max} \;=\; 108 \pm 2 \text{ MeV} \; , \tag{6.3}$$
$$\Delta\sqrt{s}_{-} \;=\; 58 \pm 1 \text{ MeV} \; , \tag{6.4}$$
$$\Delta\sqrt{s}_{+} \;=\; 415 \pm 5 \text{ MeV} \; , \tag{6.5}$$

and for mupairs

$$\rho \;=\; 0.744 \pm 0.001 \; , \tag{6.6}$$
$$\Delta\sqrt{s}_{\max} \;=\; 108 \pm 2 \text{ MeV} \; , \tag{6.7}$$
$$\Delta\sqrt{s}_{s} \;=\; 48 \pm 1 \text{ MeV} \; , \tag{6.8}$$
$$\Delta\sqrt{s}_{+} \;=\; 429 \pm 5 \text{ MeV} \; . \tag{6.9}$$

The indicated errors represent the spread of the numbers in the tables with varying s_w^2 .

The reduction in peak height is considerable. For mupairs the reduction is slightly smaller. The reason is that the QED background term increases through radiative corrections whereas the Z-peak itself decreases. In mupair production the QED background is much more important than in hadron production. The apparent width increases considerably and due to the QED background more for mupair production

Table 6.1. Exact and approximate peak positions after QED corrections for muons.

m_t	$\sin^2\vartheta_W$	Γ_Z	σ_{max}	$\sqrt{s_{max}}$	$\sqrt{s_-}$	$\sqrt{s_+}$
90	0.232	2.476	1.488	91.191	89.895	92.766
			1.490	91.191	89.897	92.760
			1.494	91.188	89.895	92.760
130	0.228	2.484	1.489	91.191	89.891	92.772
			1.490	91.191	89.893	92.765
			1.493	91.188	89.891	92.766
170	0.222	2.494	1.491	91.191	89.886	92.778
			1.490	91.192	89.888	92.773
			1.493	91.188	89.886	92.773
210	0.216	2.507	1.494	91.192	89.880	92.787
			1.488	91.192	89.882	92.781
			1.492	91.189	89.880	92.781

The first line gives the result from the program ZSHAPE, while the entries on the second line have been calculated using eqs. (6.23)-(6.25). The Z and Higgs masses are kept fixed at 91.1 and 100 GeV.

Table 6.2. Exact and approximate peak positions after QED corrections for hadrons.

m_t	$\sin^2\vartheta_W$	Γ_Z	σ_{max}	$\sqrt{s_{max}}$	$\sqrt{s_-}$	$\sqrt{s_+}$
90	0.232	2.476	30.597	91.192	89.913	92.747
			30.689	91.192	89.912	92.743
			30.760	91.188	89.909	92.742
130	0.228	2.484	30.620	91.192	89.909	92.753
			30.698	91.192	89.909	92.748
			30.769	91.188	89.905	92.747
170	0.222	2.494	30.658	91.193	89.904	92.760
			30.705	91.193	89.904	92.755
			30.777	91.188	89.900	92.754
210	0.216	2.507	30.717	91.193	89.898	92.769
			30.704	91.193	89.898	92.764
			30.777	91.189	89.894	92.762

The same conventions as in table 6.1.

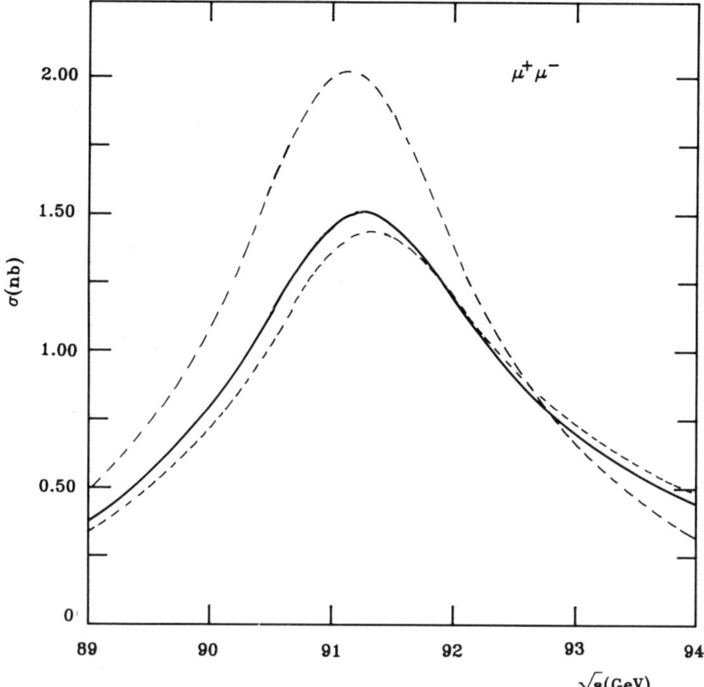

Fig. 6.1. The mupair cross section without and with QED $O(\alpha)$ (small-dashed line) and $O(\alpha^2)$ (solid line) corrections.

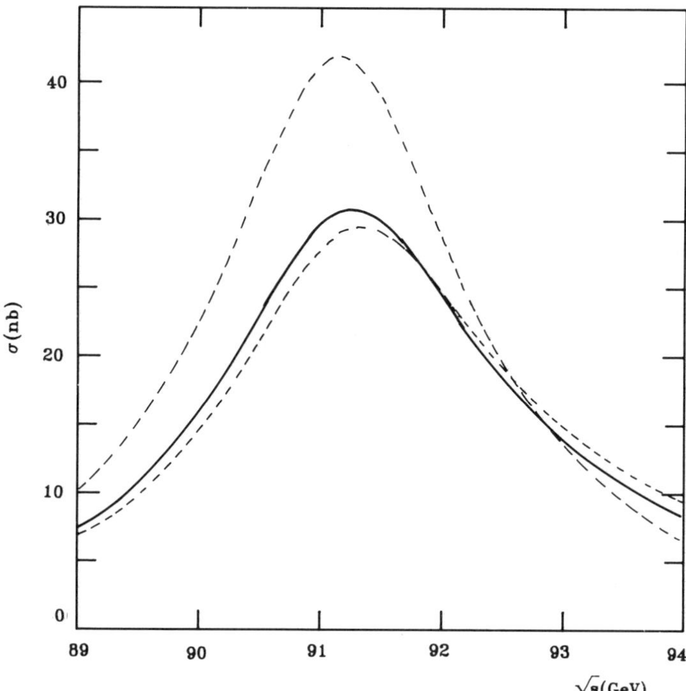

Fig. 6.2. The hadron cross section without and with QED corrections.

than for quark pair production. The change from $\sigma_{ew}(s)$ to $\sigma(s)$ through the radiative corrections can also be seen in figs. 6.1 and 6.2. The largest cross section is σ_{ew}, whereas the solid line represents the $O(\alpha^2)$ exponentiated corrected cross section from eq. (6.1). A pure $O(\alpha)$ correction is not sufficient (short-dashed line).

Although a numerical integration of (6.1) provides these results it would be useful to have some qualitative analytical results which make the above features understandable. This is possible by making a number of approximations to eq. (6.1) .

The first one is to use for $\sigma_{ew}(s')$ expression (3.59). As second approximation we use for $G(z)$ a first order exponentiated form i.e.

$$G(z) = \beta(1 - z)^{\beta-1}\delta^{v+s} + \delta^H \tag{6.10}$$

with

$$\delta^{v+s} = 1 + \delta_1^{v+s} , \tag{6.11}$$
$$\delta^H = \delta_1^H . \tag{6.12}$$

For the convolution of the pure QED term the exponentiation is omitted and the first order correction is used. The third approximation has to do with the integration region of the convolution. Using the integration variable $x = 1 - z$ the integration region becomes $(0, 1 - z_{\min})$. The approximation is that this integration region is extended to $(0, \infty)$ for the exponentiated part of G and to $(0,1)$ for the δ^H part.

We can express the convolution (6.1) in a number of integrals [25]

$$J_\beta = \beta \int\limits_0^\infty dx \, \frac{x^{\beta-1}}{x^2 - 2\eta x \cos \zeta + \eta^2} = \eta^{\beta-2}\phi(\cos \zeta, \beta) , \tag{6.13}$$

$$\phi(\cos \zeta, \beta) = \pi\beta \, \frac{\sin[(1 - \beta)\zeta]}{\sin \pi\beta \sin \zeta} , \tag{6.14}$$

with

$$\eta = \sqrt{a^2 + b^2} , \quad \cos \zeta = a/\eta , \tag{6.15}$$
$$a = \tilde{M}_z^2/s - 1 , \quad b = \tilde{M}_z\tilde{\Gamma}_z/s , \tag{6.16}$$
$$J_1 = \int\limits_0^1 dx \, \frac{2 - x}{(x + a)^2 + b^2} = \frac{2 + a}{b} A - \frac{B}{2} , \tag{6.17}$$
$$J_2 = \int\limits_0^1 dx \, \frac{x(2 - x)}{(x + a)^2 + b^2} = -\frac{a^2 + 2a - b^2}{b} A + (1 + a)B - 1 , \tag{6.18}$$

with

$$A = \arctan \frac{a + 1}{b} - \arctan \frac{a}{b} , \tag{6.19}$$

$$B = \ell n \, \frac{2a + 1 + a^2 + b^2}{a^2 + b^2} . \tag{6.20}$$

The resulting total cross section reads

$$\sigma(s) = \left(\frac{C_R + C_I}{s} - \frac{\tilde{M}_z^2 + \tilde{\Gamma}_z^2}{s^2}C_I\right)\left(\eta^{\beta-2}\varphi(\cos \zeta, \beta)\delta^{v+s} - \frac{\beta}{2}J_1\right)$$
$$- \frac{C_R + C_I}{s}\left(\frac{\beta\eta^{\beta-1}}{\beta+1}\varphi(\cos \zeta, \beta + 1)\delta^{v+s} - \frac{\beta}{2}J_2\right) + \frac{C_Q'}{s} , \tag{6.21}$$

where

$$C'_Q = C_Q \left[1 + \frac{\alpha}{\pi} L_e \left(2\ell n(1 - \frac{s_m}{s}) - \ell n \, \frac{s_m}{s} + \frac{1}{2} + \frac{s_m}{s} \right) \right.$$

$$\left. + \, \frac{\alpha}{\pi} \left(\frac{\pi^2}{3} - 1 - 2\ell n(1 - \frac{s_m}{s}) + \ell n \, \frac{s_m}{s} - \frac{s_m}{s} \right) \right] , \tag{6.22}$$

and $s_m = s'_{\min}$.

Again $\sigma(s)$ can be considered in the region $s = M_z^2$ as we did in section 3.5. Keeping the numerically most important terms in the expansions one finds [12]

$$\sigma_{\max} = \frac{C_R}{\tilde{M}_z^2 \gamma^2} \, \gamma^\beta \delta^{v+s} \left\{ 1 + \frac{5}{48} \pi^2 \beta^2 - \frac{3}{4} \pi \beta \gamma + \frac{\gamma^{(2-\beta)} C'_Q}{C_R \delta^{v+s}} \right\} , \tag{6.23}$$

$$\sqrt{s}_{\max} = M_z(1 - \frac{1}{4}\gamma^2 + \frac{1}{8}\pi\beta\gamma) , \tag{6.24}$$

$$\sqrt{s}_\pm = M_z \left[1 + \frac{1}{4}\pi\beta\gamma \left(1 + \beta - \frac{5}{4}\gamma + \frac{\gamma^{(2-\beta)} C'_Q}{C_R \delta^{v+s}} \right) \right]$$

$$\pm \frac{\Gamma_z}{2} \left[1 + (1 + \beta + \frac{1}{4}\pi\beta) \left(\frac{1}{4}\pi\beta + \frac{\gamma^{(2-\beta)} C'_Q}{C_R \delta^{v+s}} \right) + \frac{\beta}{2} \left(1 + \frac{\pi\beta}{4} \right) \ell n \, 2 \right] . \tag{6.25}$$

These formulae are a very good approximation to the exact calculations. The peak height is correct within 0.5%, $\sqrt{s}_{\max}$, $\sqrt{s}_-$ within 1 MeV whereas $\sqrt{s}_+$ is correct within 5 MeV. One can see this from the tables 6.1 and 6.2. One can even invert these relations. Measuring these characteristics of the resonance curves, one determines

$$\gamma_{\text{obs}} = \frac{\sqrt{s}_+ - \sqrt{s}_-}{\sqrt{s}_{\max}} , \tag{6.26}$$

$$\beta_{\text{obs}} = \frac{2\alpha}{\pi} \left(\ell n \, \frac{s_{\max}}{m_e^2} - 1 \right) . \tag{6.27}$$

Then the measured mass and width of the Z follow from the formulae

$$M_z^{\text{exp}} = \sqrt{s}_{\max} \left[1 + \frac{1}{4}\gamma_{\text{obs}}^2 - \frac{1}{8}\pi\beta_{\text{obs}}\gamma_{\text{obs}} \right] , \tag{6.28}$$

$$\Gamma_z^{\text{exp}} = \left(\sqrt{s}_+ - \sqrt{s}_- \right) \left[1 - \left(\frac{1}{4}\pi\beta_{\text{obs}} + \frac{1}{2}\beta_{\text{obs}} \, \ell n \, 2 + \frac{C'_Q}{s_{\max}\sigma_{\max} - C'_Q} \right) \right.$$

$$\times \left. \left(1 + \beta_{\text{obs}} - \frac{1}{2}\beta_{\text{obs}} \, \ell n \, 2 - \frac{C'_Q}{s_{\max}\sigma_{\max}} \right) + \frac{1}{2}\beta_{\text{obs}}^2 \, \ell n \, 2 \right] . \tag{6.29}$$

Finally one can also extract the hadronic and leptonic widths from the observed quantities when one assumes $\Gamma_e = \Gamma_\mu$:

$$\frac{\Gamma_\mu}{\Gamma_h} = \frac{\sigma_{\max,\mu}}{\sigma_{\max,h}} \left(1 - \frac{C'_{Q,\mu}}{s_{\max,\mu}\sigma_{\max,\mu}} \right) , \tag{6.30}$$

$$\Gamma_\mu^2 = \frac{(s_{\max,\mu})^2 \sigma_{\max,\mu}}{12\pi \delta^{v+s}} \left(\frac{\Gamma_{z,\mu}^{\text{exp}}}{M_{z,\mu}^{\text{exp}}} \right)^{2-\beta_{\text{obs}}} \left[1 - \frac{C'_{Q,\mu}}{s_{\max,\mu}\sigma_{\max,\mu}} \right] . \tag{6.31}$$

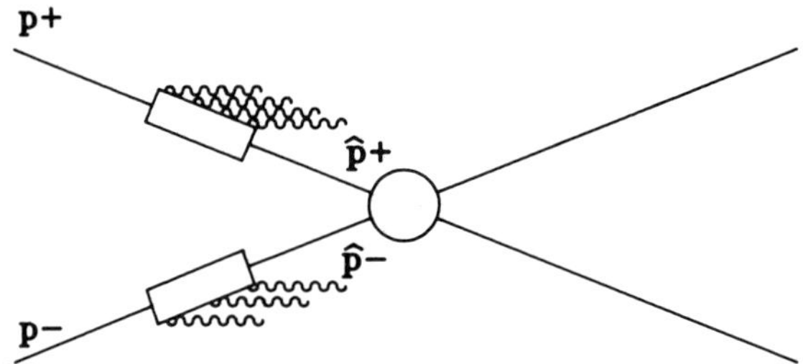

Fig. 7.1. The structure function method view of
initial state photon emission.

7   THE STRUCTURE FUNCTION METHOD

7.1   Total cross sections

In the structure function method we apply techniques developed for QCD. It enables
one to obtain stepwise leading log and subleading log terms. We shall first consider
a total cross section calculation e.g. for e^+e^- annihilation into muon pairs [26-29,21].
We consider only initial state photonic corrections, which is sufficient according to the
findings of section 5.

The master formula based on mass factorization [30] reads for the total mupair
cross section

$$\sigma_{\text{tot}}(s) = \int_0^1 dx_1 \int_0^1 dx_2 \, \Gamma_{ee}(x_1) \, \Gamma_{ee}(x_2) \, \hat{\sigma}_{\text{tot}}(x_1 x_2 s) \ . \tag{7.1}$$

Note that in $\hat{\sigma}_{\text{tot}}$ a stepfunction $\theta(x_1 x_2 s - 4m_\mu^2)$ is absorbed. Pictorially the formula is
represented in fig. 7.1. The electron and positron emit photons and scatter afterwards
to produce a mupair. The multiple photon emission reduces both momenta so that in
the "hard scattering" process the incoming momenta are

$$\hat{p}_+ \;=\; x_1 p_+ \ , \tag{7.2}$$

$$\hat{p}_- \;=\; x_2 p_- \ . \tag{7.3}$$

The approximation in the formula (7.1) is that the emission is collinear, which is natural
since we are interested in the LL terms. The behaviour of the electron and positron
are the same, so there is only one type of splitting function $\Gamma_{ee}(x)$. This function is
universal and can be written as the following series

$$\Gamma_{ee}(x) = \delta(1 - x) + \sum_{n=1}^{\infty} \left(\frac{\alpha}{\pi}\right)^n \Gamma_{ee}^{(n)} \ , \tag{7.4}$$

where

$$\Gamma_{ee}^{(n)} = \sum_{i=1}^{n} b_{n_i} L_e^i \ . \tag{7.5}$$

So Γ_{ee} is a series in α and the coefficient of the α^n term i.e. $\Gamma_{ee}^{(n)}$ can be expanded in the
large logarithms L_e . The highest power is n . Taking directly L_e in eq. (7.5) means
that we have chosen the scale s in the large logarithm. This looks reasonable, but also
follows from the explicit calculations in sections 4 and 5. Also the hard scattering cross
section can be expanded

$$\hat{\sigma} = \sum_{n=0}^{\infty} \left(\frac{\alpha}{\pi}\right)^n \hat{\sigma}^{(n)} . \tag{7.6}$$

Here $\hat{\sigma}^{(0)}$ is the Born cross section, $\hat{\sigma}^{(1)}$ the first order correction to it, but with the large logarithms removed i.e. it is the non log $O(\alpha)$ corrected cross section.

So in principle eq. (7.1) contains LL terms but also all subleading log terms. The explicit form of $\Gamma_{ee}(x)$ up to $O(\alpha^2)$ reads

$$\begin{aligned}
\Gamma_{ee}(x) &= \delta(1-x) + \frac{\alpha}{2\pi} P_{ee}^{(0)}(x) L_e \\
&+ \frac{1}{2}\left(\frac{\alpha}{2\pi}\right)^2 \left[\left(P_{ee}^{(0)} \otimes P_{ee}^{(0)}\right)(x) L_e^2 + P_{ee}^{(1)}(x) L_e\right]
\end{aligned} \tag{7.7}$$

where $P_{ee}^{(i)}(x)$ are the Altarelli-Parisi splitting functions [31] and the convolution symbol $\otimes$ is defined by

$$(f \otimes g)(x) = \int_0^1 dx_1 \int_0^1 dx_2 \, \delta(x - x_1 x_2) f(x_1) g(x_2) . \tag{7.8}$$

The splitting function $P_{ee}^{(0)}$ is given by

$$P_{ee}^{(0)} = \delta(1-x)(\frac{3}{2} + 2\ln \varepsilon) + \theta(1 - x - \varepsilon) \frac{1+x^2}{1-x} \tag{7.9}$$

and therefore

$$\begin{aligned}
\frac{1}{2}\left(P_{ee}^{(0)} \otimes P_{ee}^{(0)}\right)(x) &= \delta(1-x)\left(2\ln^2\varepsilon + 3\ln \varepsilon + \frac{9}{8} - 2\zeta(2)\right) . \\
&+ \theta(1-x-\varepsilon)\left\{\frac{1+x^2}{1-x}\left[2\ln(1-x) - \ln x + \frac{3}{2}\right] + \frac{1}{2}(1+x)\ln x - (1-x)\right\} (7.10)
\end{aligned}$$

Also $P_{ee}^{(1)}$ is explicitly known in the literature [32], but the higher order ones $P^{(i)}(x)$ have not yet been evaluated. Note that

$$\int dx \, \Gamma_{ee}(x) = 1 . \tag{7.11}$$

For a pure LL calculation of $\sigma_{\text{tot}}(s)$ up to $O(\alpha^2)$ one needs the Born cross section $\hat{\sigma}^{(0)}$, $P_{ee}^{(0)}$ and $P_{ee}^{(0)} \otimes P_{ee}^{(0)}$. For the $O(\alpha^3)$ one needs in addition $P_{ee}^{(0)} \otimes P_{ee}^{(0)} \otimes P_{ee}^{(0)}$ and so on. For the subleading logs up to $O(\alpha^2)$ one has terms of order $\frac{\alpha L}{\pi}$, $\frac{\alpha}{\pi}$, $\left(\frac{\alpha L}{\pi}\right)^2$ and $\left(\frac{\alpha}{\pi}\right)^2 L$ for which one needs also $\hat{\sigma}^{(1)}$ and $P_{ee}^{(1)}$.

Here we shall indicate that eq. (7.1) indeed reproduces the LL terms of the explicit calculation i.e. of eq. (5.5). One can rewrite eq. (7.1) as

$$\sigma_{\text{tot}}(s) = \int_{4m_\mu^2/s}^1 dz \, \phi_{\text{tot}}(z)\hat{\sigma}_{\text{tot}}(sz) \tag{7.12}$$

with

$$\begin{aligned}
\phi_{\text{tot}}(z) &= \int dx_1 \int dx_2 \, \delta(x_1 x_2 - z)\Gamma_{ee}(x_1)\Gamma_{ee}(x_2) \\
&= \Gamma_{ee} \otimes \Gamma_{ee}(z) .
\end{aligned} \tag{7.13}$$

Insertion of Γ_{ee} gives

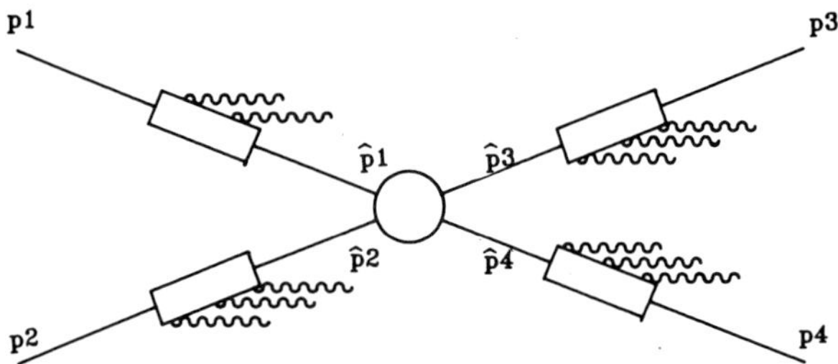

Fig. 7.2. The structure function method view of photon emission.

$$\phi_{\text{tot}}(z) = \delta(1-z) + \frac{\alpha}{\pi}P^{(0)}_{ee}(z)L_e + \frac{1}{2}\left(\frac{\alpha}{\pi}\right)^2 P^{(0)}_{ee} \otimes P^{(0)}_{ee}(z)L_e^2 \ . \tag{7.14}$$

With the explicit expression (7.9) one sees that indeed the LL terms of eq. (5.5) are reproduced. One can substantially improve on the LL calculation by replacing L by L -1 . In the soft photon part one can show that due to the $m^2/(p_{\pm}\cdot k)^2$ terms one should obtain L -1 . Since the $\ell n\,\varepsilon$ terms should cancel against integration of the hard photon part also L -1 terms should occur there at least in the terms containing $\frac{1}{1-z}$. Again one can resum the soft photon terms obtaining the LL version of eq. (5.12).

Also the subleading logs can be calculated from eq. (7.1) and are found to agree [21] with the explicit calculations from Feynman diagrams i.e. with the result of eqs. (5.5). These comparisons give also confidence that we use eq. (7.5) with the correct scale i.e. with

$$L_e = \ell n\,\frac{s}{m_e^2}\ . \tag{7.15}$$

7.2 Differential cross sections

We consider photonic corrections to a differential cross section [33] for mupair production or Bhabha scattering. The cross section can be differential in energies and angles. For convenience we shall use kinematical invariants. In the end we can still insert laboratory variables like energies and angles. So the conditions for the validity of the KLN theorem are not necessarily fulfilled, since we do not always integrate over all final states. In this case one should start as general as possible and allow for photon emission from the final state. For the LL calculation one has a more general picture than fig. 7.1. The situation is depicted in fig. 7.2. We now denote the incoming momenta by p_1, p_2 and the momenta of the detected particles (electrons or muons) by p_3, p_4 . Due to the radiation there is no energy momentum conservation and one has the following independent kinematical invariants, characterizing the process

$$\begin{aligned}
s &= (p_1 + p_2)^2 \quad , \quad s' = (p_3 + p_4)^2 \ , \\
t_1 &= (p_1 - p_3)^2 = -E_3\sqrt{s}(1 - \cos\theta_3) \ , \\
u_1 &= (p_2 - p_3)^2 = -E_3\sqrt{s}(1 + \cos\theta_3) \ , \\
t_2 &= (p_2 - p_4)^2 = -E_4\sqrt{s}(1 + \cos\theta_4) \ , \\
u_2 &= (p_1 - p_4)^2 = -E_4\sqrt{s}(1 - \cos\theta_4) \ .
\end{aligned} \tag{7.16}$$

The hard scattering process is in our LL calculation represented by the Born cross section, where the incoming momenta are

$$\hat{p}_1 \;=\; x_1 p_1 \; , \tag{7.17}$$
$$\hat{p}_2 \;=\; x_2 p_2 \; ,$$

and the outgoing momenta

$$\hat{p}_3 \;=\; \frac{1}{x_3} p_3 \; ,$$
$$\hat{p}_4 \;=\; \frac{1}{x_4} p_4 \; . \tag{7.18}$$

For the hard scattering process the cross section depends on the reduced invariants

$$\hat{s} \;=\; x_1 x_2 s \; ,$$
$$\hat{t}_1 \;=\; \frac{x_1}{x_3} t_1 \; , \qquad\qquad \hat{u}_1 = \frac{x_2}{x_3} u_1 \; , \tag{7.19}$$
$$\hat{t}_2 \;=\; \frac{x_2}{x_4} t_2 \; , \qquad\qquad \hat{u}_2 = \frac{x_1}{x_4} u_2 \; .$$

The Born cross section can be written as the multidifferential cross section

$$\frac{d^4\sigma}{dt_1 du_1 dt_2 du_2} = \frac{1}{64\pi s^2}\,|M^0|^2 \delta(s + t_1 + u_1)\delta(t_1 - t_2)\delta(u_1 - u_2) \tag{7.20}$$

where $\frac{1}{4}|M^0|^2$ is the Born spin averaged matrix element squared, depending on s and t_1 . Of course one can also introduce the Born cross sections

$$\frac{d^2\sigma}{dt_1 du_1} = \frac{1}{64\pi s^2}|M^0|^2 \delta(s + t_1 + u_1) \; , \tag{7.21}$$

$$\frac{d\sigma}{dt_1} = \frac{1}{64\pi s^2}|M^0|^2 \; , \tag{7.22}$$

$$\sigma_{\text{tot}}(s) = \int dt_1\, \frac{d\sigma}{dt_1} \; . \tag{7.23}$$

The master formula for the LL calculation for differential cross sections is

$$\frac{s^4 d^4\sigma}{dt_1 du_1 dt_2 du_2} \;=\; \int\limits_0^1 \frac{dx_1}{x_1^2} \int\limits_0^1 \frac{dx_2}{x_2^2} \int\limits_0^1 \frac{dx_3}{x_3^2} \int\limits_0^1 \frac{dx_4}{x_4^2} \Gamma(x_1)\Gamma(x_2)$$
$$D(x_3)D(x_4)\, \frac{\hat{s}^4 d^4\hat{\sigma}}{d\hat{t}_1 d\hat{u}_1 d\hat{t}_2 d\hat{u}_2} \; . \tag{7.24}$$

Besides the splitting functions $\Gamma(x) = \Gamma_{ee}(x)$ also fragmentation functions $D(x) = D_{ee}(x)$ are introduced. The latter are in LL the same as $\Gamma_{ee}(x)$, except that the scale Q^2 may be different. In general one has

$$\mathrm{L} \;=\; \ell n\, \frac{Q^2}{m^2} \tag{7.25}$$

where for the initial state radiation $\mathrm{L} = \mathrm{L}_e$ for a total cross section calculation. For the final state radiation we come back to the question of the correct L.

From the master formula (7.24) one can derive other formulae. When we only consider particle 3 to be detected, we can integrate over t_2 and u_2 and find

$$\frac{s^2 d^2\sigma}{dt_1 du_1} = \int \frac{dx_1}{x_1} \int \frac{dx_2}{x_2} \int \frac{dx_3}{x_3^2} \int dx_4 \ \Gamma(x_1)\Gamma(x_2)D(x_3)D(x_4)\ \hat{s}^2 \ \frac{d^2\hat{\sigma}}{d\hat{t}_1 d\hat{u}_1}, \quad (7.26)$$

$$\frac{s^2 d\sigma}{dt_1 du_1} = \int \frac{dx_1}{x_1} \int \frac{dx_2}{x_2} \int \frac{dx_3}{x_3^2} \ \Gamma(x_1)\Gamma(x_2)D(x_3)\ \hat{s}^2 \ \frac{d^2\hat{\sigma}}{d\hat{t}_1 \ d\hat{u}_1} \ . \quad (7.27)$$

Since the x_4 dependence is only in $D(x_4)$ we can carry out the integration by using eq. (7.11). Integration over u_1 gives

$$s \frac{d\sigma}{dt_1} = \int dx_1 \ \int \frac{dx_2}{x_2} \ \int \frac{dx_3}{x_3} \ \Gamma(x_1)\Gamma(x_2)D(x_3)\ \hat{s} \ \frac{d\sigma}{d\hat{t}_1} \ . \quad (7.28)$$

Finally integration over t_1 gives the total cross section. The x_3 integration can be done with eq. (7.11)

$$\sigma_{\text{tot}}(s) = \int dx_1 \ \int dx_2 \ \Gamma(x_1)\Gamma(x_2)\sigma_{\text{tot}}(\hat{s}) \ . \quad (7.29)$$

One can also select initial state or final state radiation effects by switching off one of them i.e. by taking $\Gamma(x) = \delta(1-x)$ or $D(x) = \delta(1-x)$ .We discuss two examples and compare to the results of section 4.

For instance, when one considers final state radiation one obtains from (7.26)

$$\frac{s^2 d^2\sigma}{dt_1 du_1} = \int \frac{dx_3}{x_3^2} \ \int dx_4 \ D(x_3)D(x_4)\ \hat{s}^2 \ \frac{d^2\hat{\sigma}}{d\hat{t}_1 d\hat{u}_1} \ . \quad (7.30)$$

For the total cross section one obtains

$$\sigma(s) = \int dx_3 \ \int dx_4 \ D(x_3)D(x_4)\sigma(s) \ , \quad (7.31)$$

which gives an identity when eq. (7.11) is applied which is a result of the KLN theorem. One can use (7.31) to derive the mupair invariant mass (s') spectrum by introducing $z = s'/s$ and rewriting (7.31)

$$\begin{aligned}
\sigma(s) &= \sigma(s) \int dz \ \delta(z - x_3 x_4) \int dx_3 \ \int dx_4 \ D(x_3)D(x_4) \\
&= \frac{\sigma(s)}{s} \int ds' \ D \otimes D\left(\frac{s'}{s}\right) \ .
\end{aligned} \quad (7.32)$$

For the hard bremsstrahlung part this gives

$$\frac{d\sigma}{ds'} = \frac{\sigma(s)}{s} \frac{\alpha}{\pi} \ \text{L} \ \frac{1+z^2}{1-z} \ , \quad (7.33)$$

or

$$\frac{d\sigma}{dk_0} = \frac{\sigma(s)}{s} \frac{\alpha}{\pi} \text{L} \left(1 + \left(\frac{s'}{s}\right)^2\right) \frac{1}{k_0} \ . \quad (7.34)$$

Comparison with the explicit calculation shows that the scale is s' i.e. $\text{L} = \text{L}'$, with L' given by (4.61).

When we take initial state radiation alone in eq. (7.24) and moreover only from the e^+ we get

$$\frac{s^4 d^4\sigma}{dt_1 du_1 dt_2 du_2} = \int_0^1 \frac{dx_1}{x_1^2} \ \Gamma(x_1) \ \frac{\hat{s}^4 d^4\hat{\sigma}}{d\hat{t}_1 d\hat{u}_1 d\hat{t}_2 du_2} \ . \quad (7.35)$$

In $O(\alpha)$ the hard bremsstrahlung part can easily be compared with eq. (4.35). In the collinear limit eq. (4.55) holds and the $d\Omega_\gamma$ integration can be carried out, while in $|M^0|^2$ the axes are kept fixed. Since the two particle phase space can be rewritten

$$
\begin{aligned}
dPS_2 &= \frac{1}{(2\pi)^2}\, \delta(\hat{p}_+ + p_- - q_+ - q_-)\, \frac{d\vec{q}_+}{2q_{+0}}\, \frac{d\vec{q}_{-0}}{2q_{-0}} \\
&= \frac{1}{8\pi\hat{s}}\, d\hat{t}_1 du_1 dt_2 d\hat{u}_2 \delta(\hat{s} + \hat{t}_1 + u_1)\delta(\hat{t}_1 - t_2)\delta(u_1 - \hat{u}_2)
\end{aligned}
\tag{7.36}
$$

we can write the result into

$$
\begin{aligned}
d^5\sigma &= \frac{\alpha}{2\pi}\, L_e \frac{1 + x_1^2}{x_1(1 - x_1)}\, dx_1\, \frac{1}{16\pi s\hat{s}}\, \frac{1}{4}\, \Sigma|M_0(s')|^2\delta(\hat{s} + t_1 + u_1) \\
&\quad \delta(\hat{t}_1 - t_2)\delta(u_1 - \hat{u}_2)d\hat{t}_1 du_1 dt_2 d\hat{u}_2 \ .
\end{aligned}
\tag{7.37}
$$

With eq. (7.20) we obtain the multi-differential cross section

$$
\frac{d^4\sigma}{dt_1 du_1 dt_2 du_2} = \frac{\alpha}{2\pi}\, L_e \int dx_1\, \frac{1 + x_1^2}{1 - x_1}\, x_1^2\, \frac{d^4\sigma}{d\hat{t}_1 du_1 dt_2 d\hat{u}_2} \ ,
\tag{7.38}
$$

which is the same as (7.34) for $\Gamma(x_1) = P_{ee}^H(x_1)$, i.e. the bremsstrahlung part. When we insert $\Gamma(x)$ up to $O(\alpha^2)$ we get of course an answer which goes beyond the $O(\alpha)$ calculation of section 4.

When cuts on angles and energies of the produced particles 3 and 4 are imposed we have to use eq. (7.24) as starting point. One then performs t_i and u_i integrations compatible with the imposed cuts. When only particle 3 is detected the starting point is eq. (7.27). A total cross section compatible with cuts on the angle and energy of particle 3 is obtained by a suitable t_1, u_1 integration. For the line shape in Bhabha scattering which involves anyhow angle and energy cuts eq. (7.24) will be the starting point. For the forward-backward asymmetry A_{FB} in mupairs it depends on the definition used for A_{FB}. In case the definition uses only information on particle 3 one can use eq. (7.27). When both particles 3 and 4 figure in the definition again eq. (7.24) is indispensable. It is clear that both eqs. (7.24) and (7.27) are unlikely to reduce to a form like eq. (7.12) when limited momenta ranges for particles 3 and 4 are considered. Nevertheless one sometimes finds eq. (7.12) applied to differential cross sections, which is clearly incorrect.

Let us illustrate this for the relatively simple example of the forward-backward asymmetry without cuts and where the definition only involves the detection of particle 3 in the forward or backward hemisphere.

Using eqs. (7.16), (7.21) and (7.27) we obtain the differential cross section

$$
\begin{aligned}
\frac{d^2\sigma}{dc_3 dE_3} &= 2E_3 s\, \frac{d^2\sigma}{dt_1 du_1} \\
&= \frac{1}{32\pi s} \int \frac{dx_1}{x_1} \int \frac{dx_2}{x_2} \int \frac{dx_3}{x_3}\, \Gamma(x_1)\Gamma(x_2)D(x_3)|\hat{M}^0(\hat{s}, \hat{t}_1, \hat{u}_1)|^2 \\
&\quad E_3\delta(x_1 x_2 x_3 s - E_3\sqrt{s}[x_1(1 - c_3) + x_2(1 + c_3)]) \ .
\end{aligned}
\tag{7.39}
$$

Integration over E_3 fixes E_3 and gives

$$
\hat{t}_1 = \frac{-x_1^2 x_2 s(1 - c_3)}{x_1(1 - c_3) + x_2(1 + c_3)} \ , \quad \hat{u}_1 = \frac{-x_1 x_2^2 s(1 + c_3)}{x_1(1 - c_3) + x_2(1 + c_3)} \ .
\tag{7.40}
$$

Since the only remaining x_3 dependence is in $D(x_3)$ the x_3 integration can also be carried out and we end with

$$\frac{d\sigma}{dc_3} = \frac{1}{32\pi s} \int dx_1 \int dx_2 \, \Gamma(x_1)\Gamma(x_2) \, \frac{|M^0(\hat{s},\hat{t})|^2}{[x_1(1-c_3)+x_2(1+c_3)]^2} \, . \tag{7.41}$$

Defining the forward and backward cross sections as in eqs. (2.9) and (2.10) and using eq. (2.3), we can write for the difference

$$
\begin{aligned}
\sigma_{FB}(s) &= \sigma_F - \sigma_B = \int_0^1 dc_3 \left[\frac{d\sigma}{dc_3}(c_3) - \frac{d\sigma}{dc_3}(-c_3) \right] \\
&= 2\pi\alpha^2 s \int_0^1 dc_3 \int_0^1 dx_1 \int_0^1 dx_2 \, \Gamma(x_1)\Gamma(x_2) \, \frac{x_1^2 x_2^2 [x_2(1+c_3)-x_1(1-c_3)]}{[x_1(1-c_3)+x_2(1+c_3)]^3} \\
&\quad \left[|P(--)|^2 + |P(++)|^2 - |P(+-)|^2 - |P(-+)|^2 \right] \, . \tag{7.42}
\end{aligned}
$$

Carrying out the c_3 integration and using the Born term expression (2.11) for σ_{FB} we obtain

$$\sigma_{FB}(s) = \int dx_1 \int dx_2 \, \Gamma(x_1)\Gamma(x_2) \, \frac{4x_1 x_2}{(x_1+x_2)^2} \, \sigma_{FB}(x_1 x_2 s) \, . \tag{7.43}$$

Introducing a forward-backward flux

$$\frac{\phi_{FB}(z)}{(1+z)^2} = \int dx_1 \int dx_2 \, \frac{\delta(x_1 x_2 - z)}{(x_1+x_2)^2} \, \Gamma(x_1)\Gamma(x_2) \tag{7.44}$$

we obtain up to $O(\alpha^2)$

$$
\begin{aligned}
\phi_{FB} &= \delta(1-z) + \frac{\alpha}{\pi} \, P_{ee}^{(0)} \, \mathrm{L} \, + \frac{1}{2}\left(\frac{\alpha}{\pi}\right)^2 P_{ee}^{(0)} \otimes P_{ee}^{(0)}(z) \, \mathrm{L}^2 \\
&\quad + \left(\frac{\alpha}{2\pi}\right)^2 \int dx_1 \int dx_2 \, \delta(x_1 x_2 - z) \left[\left(\frac{1+z}{x_1+x_2}\right)^2 - 1 \right] P_{ee}^H(x_1) P_{ee}^H(x_2) \, \mathrm{L}^2 \, . \tag{7.45}
\end{aligned}
$$

The expression is the same as for $\phi_{\mathrm{tot}}(z)$ up to $O(\alpha^2)$ except for the last term. The last term only contributes for hard photons. So we can write up to $O(\alpha^2)$

$$\phi_{FB} = \phi_{\mathrm{tot}} + \tilde{\phi}_{FB} \tag{7.46}$$

where an explicit evaluation gives

$$
\begin{aligned}
\tilde{\phi}_{FB} &= \left(\frac{\alpha}{2\pi}\right)^2 \mathrm{L}_e^2 \left\{ \frac{(1-z)^3}{2z} + \frac{(1-z)^3}{\sqrt{z}} \left[\arctan\left(\frac{1}{\sqrt{z}}\right) - \arctan(\sqrt{z}) \right] \right. \\
&\quad \left. - \, (1+z)\ell n \, z + 2(1-z) \right\} \, . \tag{7.47}
\end{aligned}
$$

This is a pure LL result. Since we know that mass terms occur which replace L_e by $\mathrm{L}_e - 1$, we can improve on the LL result by making this substitution. Moreover in ϕ_{tot} we can resum the soft photons.

Summarizing we obtain the A_{FB} the expression

$$A_{FB} = \frac{\sigma_{FB}}{\sigma_{\mathrm{tot}}} = \frac{1}{\sigma_{\mathrm{tot}}} \int dz \, \phi_{FB}(z) \frac{4z}{(1+z)^2} \, \sigma_{FB}(zs) \, . \tag{7.48}$$

Here we see clearly the difference between the convolution for a total cross section i.e. eq. (7.12) and one for restricted cross sections. One cannot convolute σ_{FB} with ϕ_{tot} as is sometimes done in the literature. One not only neglects the kinematical factor $4z/(1+z)^2$ but also forgets $\tilde{\phi}_{FB}$. The numerical differences which these incorrect calculations give are fortunately not dramatic in the resonance region. This is due to the fact that soft photons give a dominant contribution and for soft photons the kinematical factor can be set equal to one and $\tilde{\phi}_{FB}$ equal to zero.

In the case one measures both particles 3 and 4 and applies kinematical cuts one has to start with eq. (7.24). The kinematical constraints have to be translated into restrictions on the invariants. In general semi-analytical results like eq. (7.48) are not anymore possible. Numerical integrations with certain boundaries are feasible. We now outline the procedure in such cases.

Starting from the master formula (7.24) integrations over u_1, t_2, u_2 can be carried out using the δ-functions in the Born cross section (7.20) which occurs in the integrand. The resulting differential cross section is

$$\frac{d\sigma}{dt_1} = \frac{1}{64\pi s^2} \int \frac{dx_1}{x_1} \int \frac{dx_2}{x_2^2} \int \frac{dx_3}{x_3} \int dx_4 \; \Gamma(x_1)\Gamma(x_2)D(x_3)D(x_4)\Sigma|\hat{M}^0|^2 \; . \quad (7.49)$$

A total cross section with cuts is obtained by integration over the allowed $t_1 = \frac{x_3}{x_1}\hat{t}_1$ range

$$\sigma = \frac{1}{64\pi s^2} \int \frac{dx_1}{x_1^2} \int \frac{dx_2}{x_2^2} \int d\hat{t}_1 \int dx_3 \int dx_4 \; \Gamma(x_1)\Gamma(x_2)D(x_3)D(x_4)\Sigma|\hat{M}^0|^2. \quad (7.50)$$

The integrations for x_i run in the interval $(x_{i\,\mathrm{min}}, 1)$ and for $\hat{t}_1$ from $\hat{t}_{1\,\mathrm{min}}$ to $\hat{t}_{1\,\mathrm{max}}$. Suppose we demand the following energy and angular cuts

$$E_3 \; > \; \hat{E} + \frac{1}{2}\sqrt{s} \; , \tag{7.51}$$

$$E_4 \; > \; \hat{E} - \frac{1}{2}\sqrt{s} \; , \tag{7.52}$$

$$\theta_{m+} \; < \; \theta_3 < \theta_{M+} \; , \tag{7.53}$$

$$\theta_{m-} \; < \; \pi - \theta_4 < \theta_{M-} \; , \tag{7.54}$$

In terms of $c_i = \cos\theta_i$, $c_{m\pm} = \cos\theta_{M\pm}$ and $c_{M\pm} = \cos\theta_{m\pm}$ we obtain

$$c_{m+} \; < \; c_3 < c_{M+} \; , \tag{7.55}$$

$$c_{m-} \; < \; -c_4 < c_{M-} \; , \tag{7.56}$$

or

$$-c_{M-} \; < \; c_4 < -c_{m-} \; . \tag{7.57}$$

The reason of the introduction of $\theta_4' = \pi - \theta_4$ is that its role with respect to $\vec{p}_2$ is the same as that of θ_3 with respect to $\vec{p}_1$.

In order to translate these constraints into the integration boundaries it is convenient to see the effect of relations (7.51) - (7.54) on the hard scattering process in its cms system. Using $\frac{1}{2}\sqrt{s} = E$ as unit we have in the laboratory system a total 4-momentum $(\hat{E}, \hat{p})$ with (z-axis along $\vec{p}_1$)

$$\hat{E} \; = \; x_1 + x_2 \; , \tag{7.58}$$

$$\hat{p} \; = \; (0, 0, x_1 - x_2) \; , \tag{7.59}$$

and an invariant mass

$$M^2 = 4x_1 x_2 \; . \tag{7.60}$$

The hard scattering process gives particles with momenta $\hat{p}_3$ and $\hat{p}_4$. Switching off for the moment the fragmentation, we have $p_3 = \hat{p}_3, p_4 = \hat{p}_4$. The hard scattering process has the kinematics of a particle with mass M decaying into particles with momenta p_3, p_4 . In the laboratory system this particle is moving and in the cms of the hard scattering process it is at rest. Denoting a 4-vector in the lab. system by q and in the cms system by $\tilde{q}$ we have a boost relating them

$$
\begin{aligned}
\tilde{q}_x &= q_x , \ \tilde{q}_y = q_y , \\
\tilde{q}_z &= \frac{1}{M}\left(\hat{E}q_z + \hat{p}_z q_0\right) , \\
\tilde{q}_0 &= \frac{1}{M}\left(\hat{E}q_0 + \hat{p}_z q_z\right) .
\end{aligned}
\tag{7.61}
$$

The momenta $\tilde{p}_3, \tilde{p}_4$ into which the particle M decays can be characterized by the angle θ with respect to $\hat{\vec{p}}_1 (s_\theta = \sin\theta, c_\theta = \cos\theta)$

$$
\begin{aligned}
\tilde{p}_3 &= \frac{M}{2}\left(1, s_\theta, 0, c_\theta\right) , \\
\tilde{p}_4 &= \frac{M}{2}\left(1, -s_\theta, 0, -c_\theta\right) .
\end{aligned}
\tag{7.62}
$$

The constraints (7.51) - (7.54) can be translated with eq. (7.61) into constraints on c_θ , which in turn give restrictions on

$$
\hat{t}_1 = -2\hat{p}_1 \cdot \hat{p}_3 = -\frac{M^2}{2}(1 - c_\theta) .
\tag{7.63}
$$

For the $\hat{t}_1$ integration range we find

$$
\hat{t}_{1,min} =
\begin{cases}
\max\left(\dfrac{-x_1 x_2 s}{1 + \dfrac{x_1}{x_2}\dfrac{1 + c_{m-}}{1 - c_{m-}}} , \ \dfrac{-x_1 x_2 s}{1 + \dfrac{x_2}{x_1}\dfrac{1 + c_{m+}}{1 - c_{m+}}} , \ -x_2 s\dfrac{x_1 - \hat{E}_0^+}{1 - x_2/x_1}\right) & \text{for } x_1 > x_2 \\[3em]
\max\left(\dfrac{-x_1 x_2 s}{1 + \dfrac{x_1}{x_2}\dfrac{1 + c_{m-}}{1 - c_{m-}}} , \ \dfrac{-x_1 x_2 s}{1 + \dfrac{x_2}{x_1}\dfrac{1 + c_{m+}}{1 - c_{m+}}} , \ -x_1 s\dfrac{x_2 - \hat{E}_0^-}{1 - x_1/x_2}\right) & \text{for } x_1 < x_2
\end{cases}
,
$$

$$
\hat{t}_{1,max} =
\begin{cases}
\min\left(\dfrac{-x_1 x_2 s}{1 + \dfrac{x_1}{x_2}\dfrac{1 + c_{M-}}{1 - c_{M-}}} , \ \dfrac{-x_1 x_2 s}{1 + \dfrac{x_2}{x_1}\dfrac{1 + c_{M+}}{1 - c_{M+}}} , \ -x_1 s\dfrac{x_2 - \hat{E}_0^-}{1 - x_1/x_2}\right) & \text{for } x_1 > x_2 \\[3em]
\min\left(\dfrac{-x_1 x_2 s}{1 + \dfrac{x_1}{x_2}\dfrac{1 + c_{M-}}{1 - c_{M-}}} , \ \dfrac{-x_1 x_2 s}{1 + \dfrac{x_2}{x_1}\dfrac{1 + c_{M+}}{1 - c_{M+}}} , \ -x_2 s\dfrac{x_1 - \hat{E}_0^+}{1 - x_2/x_1}\right) & \text{for } x_1 < x_2
\end{cases}
.
$$

$$
\tag{7.64}
$$

From the condition

$$
\hat{t}_{1\min} \leq \hat{t}_{1\max}
\tag{7.65}
$$

one obtains amongst others

$$
\begin{aligned}
\frac{x_1}{x_2} &> \sqrt{\frac{1 + c_{m+}}{1 - c_{m+}}\frac{1 - c_{M-}}{1 + c_{M-}}} , \\
\frac{x_2}{x_1} &> \sqrt{\frac{1 - c_{M-}}{1 + c_{M+}}\frac{1 + c_{m-}}{1 - c_{m-}}} , \\
x_1 + x_2 &> \hat{E}_0^+ + \hat{E}_0^- .
\end{aligned}
\tag{7.66}
$$

There are in fact more restrictions on x_1, x_2 following from (7.65). Since the integrations over x_1, x_2 have to be carried out numerically it is more convenient to take x_1, x_2 values between (0,1) and satisfying (7.66). When also condition (7.65) is fulfilled the pair (x_1, x_2) is an acceptable integration point.

The boundaries on x_3 and x_4 depend also on $\hat{t}_1$. A fixed $\hat{t}_1$ determines c_θ and therefore $\hat{E}_3, \hat{E}_4$. The latter should obey eqs. (7.51) and (7.52). One then finds

$$x_{3\min} = \frac{\hat{E}_0^+}{x_1 - \dfrac{\hat{t}_1}{s}\dfrac{1 - x_1/x_2}{x_1}} \ , \tag{7.67}$$

$$x_{4\min} = \frac{\hat{E}_0^-}{x_2 + \dfrac{\hat{t}_1}{s}\dfrac{1 - x_1/x_2}{x_1}} \ . \tag{7.68}$$

Having established the integration boundaries in eq. (7.50) it is worthwile to perform as many integrations analytically as is possible. The x_3, x_4 integrations are just a separate factor. In $O(\alpha)$ they can be carried out easily and the result depends on $\hat{t}_1$ through eqs. (7.67) and (7.68). Combined with $\Sigma|\hat{M}^0|^2$ the $\hat{t}_1$ integration can still be carried out analytically such that a numerical x_1, x_2 integration remains. For the $O(\alpha^2)$ final state correction an analytical result can also be obtained. Insertion of $x_{3\ \min}$ and $x_{4\ \min}$ gives a complicated formula which combined with $\Sigma|\hat{M}^0|^2$ is too involved for an analytical evaluation. It is however reasonable to omit initial state corrections on the $O(\alpha^2)$ final state corrections. In that case $x_1 = x_2 = 1$ in eqs. (7.67) and (7.68) and the $O(\alpha^2)$ final state correction simplifies. The only remaining integration is over $\hat{t}_1$ and can be carried out analytically. So except for the $O(\alpha^2)$ final state correction the x_1, x_2 integration takes the form

$$\begin{aligned}
\sigma &= \int dx_1 \int dx_2 \ \Gamma(x_1)\Gamma(x_2)f(x_1, x_2) \\
&= \int dx_1 \int dx_2 \ \Gamma(x_1)\Gamma(x_2)\left[f(x_1 x_2, 1) + f(x_1, x_2) - f(x_1 x_2, 1)\right] \ ,
\end{aligned} \tag{7.69}$$

where $f(x_1, x_2)$ incorporates eq. (7.65) by means of a θ-function. When we impose symmetric conditions on both particles $f(x_1, x_2)$ is symmetric and we see that for either $x_1 \rightarrow$ or $x_2 \rightarrow 1$ only the first integrand survives. In other words, the combination of second and third terms in the integrands represent only hard photon contributions, whereas the $f(x_1 x_2, 1)$ term contains soft and hard photons. It should be noted that the part of the cross section depending on $|P(\lambda, \lambda')|^2$ naturally depends on $x_1 x_2$ and not on x_1, x_2 separately. The kinematical constraints give the additional separate x_1, x_2 dependence. By using eq. (7.69) the part which includes soft photons now has only $x_1 x_2$ dependence which makes it possible to resum soft photons in analogy with previous occasions. First we introduce $\varphi_{\text{tot}}(z)$ and write for the first part of eq. (7.69)

$$\int dx_1 \int dx_2 \ \Gamma(x_1)\Gamma(x_2)f(x_1 x_2, 1) = \int_0^1 dz \ \varphi_{\text{tot}}(z)f(z, 1) \ . \tag{7.70}$$

In φ_{tot} we can replace L by L-1 and resum soft photons. When the kinematical constraints are not symmetric $f(x_1, x_2)$ has to be divided into a symmetric and antisymmetric part. For the former the procedure can be repeated, the latter does not contribute since the x_1, x_2 integration is symmetric. It should be noted that a condition on the acollinearity angle ζ between the detected particles 3 and 4 can also be incorporated by a constraint on $\hat{t}_1$. We refer to the literature [34].

In summary, a total cross section with cuts leads to an expression for symmetric $f(x_1, x_2)$:

$$\sigma = \int\limits_0^1 dz\, \varphi_{\text{tot}}(z) f(z, 1) + \int\limits_0^1 dx_1 \int\limits_0^1 dx_2 \left[f(x_1, x_2) - f(x_1 x_2, 1) \right]$$
$$+ \quad O(\alpha^2) \text{ final state corrections .} \tag{7.71}$$

The integrals in eq. (7.71) have to be carried out numerically.

7.3   The non log order α terms

For two reasons we want to know the exact non-log order α correction. The first reason is that we want to have a high precision evaluation of radiative corrections. For that a precise $O(\alpha)$ correction has to be known. The second reason has to do with the choice of scale in the LL calculation. The scale is chosen such that the LL part is really the dominant part of a correction in each order of α . For the total cross section explicit $O(\alpha^2)$ calculations indicated that L_e arises as the natural large logarithm. For differential cross sections we don't have explicit exact $O(\alpha^2)$ calculations neither without cuts nor with cuts. When we have a numerical evaluation in $O(\alpha)$ of the LL and non-log correction we can verify whether the LL term is really the dominant one.

In order to evaluate the exact $O(\alpha)$ correction one can use a numerical integration of eq. (4.65) incorporating the imposed cuts. Since the bremsstrahlung cross section σ^B has an involved peaking structure the numerical integration is non trivial. There is however another approach possible since we already know the LL term in $O(\alpha)$. It is possible to write down explicitly the hard bremmstrahlungs cross section which leads to the LL result. It is a certain collinear approximation σ^{BC} to the exact cross section σ^B . The non-log term in order α can now be obtained from the integration of

$$\frac{d\sigma^{\text{NL}}}{d\Omega_\mu} = \frac{d\sigma^0}{d\Omega} \left[1 + \delta_v^{\text{NL}} + \delta_s^{\text{NL}}(k_1) \right]$$
$$+ \int\limits_{k_1}^{k_2} dk_0 d\Omega_\gamma \left(\frac{d\sigma^B}{d\Omega_\mu d\Omega_\gamma dk_0} - \frac{d\sigma^{BC}}{d\Omega_\mu d\Omega_\gamma dk_0} \right) . \tag{7.72}$$

Here the non-log terms of the virtual and soft corrections are taken. This now includes box diagrams and soft photon interference terms. The value k_1 is chosen so small that the soft photon approximation is reliable and that we therefore can omit a term like the second one in eq. (4.65). The hard photon part has an integrand which is the difference between two cross sections with similar peaking structures. This integration can be carried out by a numerical multidimensional Monte Carlo integration procedure. Although time consuming a satisfactory precision can be obtained. A computerprogram ALIBABA has been made which performs the evaluation of eqs. (7.71) and (7.72) both for Bhabhascattering and mupair production.

8   THE FORWARD-BACKWARD ASYMMETRY A_{FB}

In this section the methods of section 7 will be applied to A_{FB} . The LL corrections up to order α^2 and higher order soft photon resummation is applied to a Born cross section which contains the weak corrections of section 3. The non-log order α term is calculated as in section 7.3, but includes in the virtual corrections also weak boxes.

Table 8.1. The LL and non-log first order corrections to the forward and backward cross sections for mupair production.

$\sqrt{s}$	Forward				Backward			
	'Born'	LL $\mathcal{O}(\alpha)$	$\mathcal{O}(\alpha)$	$\mathcal{O}(\alpha^2)$	'Born'	LL $\mathcal{O}(\alpha)$	$\mathcal{O}(\alpha)$	$\mathcal{O}(\alpha^2)$
88.17	109.9	77.1	77.9(1)	84.2(1)	188.7	140.0	140.8(1)	148.7(3)
		73.4	73.6(1)	79.4(1)		136.3	136.6(1)	144.1(1)
89.17	234.0	154.3	155.7(1)	169.9(2)	330.6	230.8	232.3(1)	248.5(2)
		150.4	150.9(1)	164.7(1)		226.9	227.1(2)	243.5(1)
90.17	567.7	358.6	361.8(2)	398.3(2)	662.6	440.3	444.0(3)	479.9(3)
		354.5	356.2(2)	392.0(2)		436.1	437.6(3)	474.1(2)
91.17	1014.9	689.2	695.5(3)	741.4(5)	983.3	704.1	710.4(6)	744.3(5)
		684.3	687.6(4)	733.6(4)		699.3	701.5(4)	737.0(6)
92.17	668.1	583.1	587.6(3)	578.7(3)	540.5	518.4	522.6(4)	504.1(2)
		577.7	580.8(3)	571.9(4)		513.0	514.1(2)	497.1(2)
93.17	338.3	371.0	373.3(2)	352.9(2)	230.0	300.2	301.9(3)	277.7(2)
		365.3	367.0(1)	346.9(1)		294.6	295.2(1)	271.7(1)
94.17	195.8	251.9	253.2(2)	235.9(1)	112.7	190.8	191.6(1)	173.4(2)
		246.3	247.4(1)	230.1(1)		185.1	185.7(1)	167.5(1)

The column under the heading LL $\mathcal{O}(\alpha)$ contains the cross sections with first order LL corrections, the column with $\mathcal{O}(\alpha)$ lists the fully corrected first order result. The column with $\mathcal{O}(\alpha^2)$ includes on top of that the second order LL corrections plus soft photon exponentiation. The numbers between brackets indicate the estimated numerical error in the last digit. For each energy the first line displays the results for definition 1 (see text) and the second line for definition 2. All cross sections are in picobarn, energies in GeV.

We present results for two definitions of A_{FB} which differ because σ_F and σ_B can be defined in different ways. Introducing

$$c_+ = \cos \angle(e^+, \bar{f}) = \cos \angle(\vec{p}_1, \vec{p}_3) ,$$

$$c_- = \cos \angle(e^-, f) = \cos \angle(\vec{p}_2, \vec{p}_4) ,$$

$$\tag{8.1}$$

we have as first definition

$$\sigma_{F,1} \; : \; c_+ > 0 \quad , \quad c_- \text{ arbitrary} ,$$

$$\sigma_{B,1} \; : \; c_+ < 0 \quad , \quad c_- \text{ arbitrary} ,$$

$$\tag{8.2}$$

and as second one

$$\sigma_{F,2} \; : \; c_\pm > 0 \quad , \quad \sigma_{B,2} \; : \; c_\pm < 0 . \tag{8.3}$$

The first definition involves only the detection of particle 3 and the second of both particles, which every experiment will anyhow do. For stringent acollinearity cuts the definitions become the same, which is certainly the case for the Born approximation.

In addition to the angular ranges which define A_{FB} energy cuts (7.51) and (7.52) with $\hat{E}^+ = \hat{E}^-$ can be taken into account. Also an acollinearity condition

$$\zeta < \zeta_{\max} \tag{8.4}$$

can be imposed.

Table 8.2. The forward-backward asymmetry for mupair production without any cuts.

$\sqrt{s}$	'Born'	$\mathcal{O}(\alpha)$		ZFITTER	ALIBABA	
		def. 1	def. 2		def. 1	def. 2
88.17	-0.264	-0.288	-0.300	-0.279	-0.277	- 0.289
88.67	-0.218	-0.244	-0.252	-0.235	-0.235	- 0.242
89.17	-0.172	-0.198	-0.202	-0.189	-0.188	- 0.193
89.67	-0.125	-0.150	-0.152	-0.141	-0.140	- 0.144
90.17	-0.077	-0.102	-0.103	-0.093	-0.093	- 0.095
90.67	-0.031	-0.053	-0.055	-0.046	-0.045	- 0.046
91.17	0.016	-0.011	-0.010	-0.002	-0.002	- 0.002
91.67	0.061	0.028	0.028	0.037	0.037	0.037
92.17	0.106	0.059	0.061	0.069	0.069	0.070
92.67	0.149	0.085	0.087	0.096	0.097	0.099
93.17	0.191	0.106	0.108	0.118	0.119	0.122
93.67	0.231	0.124	0.127	0.137	0.138	0.141
94.17	0.269	0.138	0.142	0.152	0.153	0.157
94.67	0.306	0.152	0.157	0.165	0.166	0.171

The rightmost two columns show the result for the two definitions of the asymmetry, corrected at $\mathcal{O}(\alpha^2)$ (plus exponentiation).

Table 8.3. The forward-backward asymmetry for mupair production using an energy cut of $\frac{1}{4}\sqrt{s}$ on both final state particles and an acollinearity cut of 10°.

$\sqrt{s}$	'Born'	$\mathcal{O}(\alpha)$		ZFITTER	ALIBABA	
		def. 1	def. 2		def. 1	def. 2
88.17	-0.264	-0.295	-0.296	-0.289	-0.285	- 0.286
88.67	-0.218	-0.248	-0.248	-0.240	-0.238	- 0.239
89.17	-0.172	-0.199	-0.200	-0.191	-0.190	- 0.190
89.67	-0.125	-0.149	-0.150	-0.142	-0.141	- 0.141
90.17	-0.077	-0.100	-0.101	-0.093	-0.093	- 0.093
90.67	-0.031	-0.053	-0.054	-0.045	-0.046	- 0.045
91.17	0.016	-0.010	-0.009	-0.001	-0.002	- 0.002
91.67	0.061	0.028	0.029	0.038	0.038	0.037
92.17	0.106	0.059	0.060	0.071	0.071	0.071
92.67	0.149	0.086	0.086	0.099	0.098	0.099
93.17	0.191	0.106	0.108	0.122	0.121	0.123
93.67	0.231	0.125	0.127	0.142	0.141	0.143
94.17	0.269	0.142	0.143	0.159	0.158	0.160
94.67	0.306	0.157	0.159	0.174	0.173	0.175

The rightmost two columns show the result for the two definitions of the asymmetry, corrected at $\mathcal{O}(\alpha^2)$ (plus exponentiation).

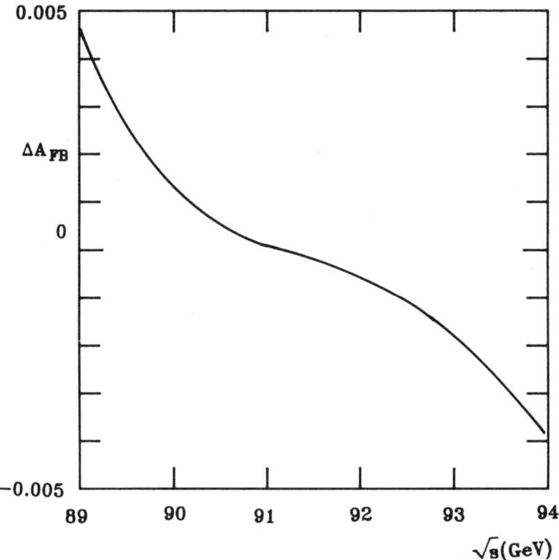

Fig. 8.1. The difference between the two forward-backward asymmetries.

For mupair production we have carried out the approach of section 7 [35]. In it the scale s and therefore $L = L_e$ have been used. Its choice is justified by the LL $O(\alpha)$ and complete $O(\alpha)$ results as are given in table 8.1. Results without cuts are given in table 8.2 whereas table 8.3 includes cuts. From the numbers it is clear that $O(\alpha^2)$ corrections are necessary in view of an aimed experimental accuracy $\delta A_{FB} = 0.0035$. Both tables include also results of a program ZFITTER, which is an extension of the work of ref. [36]. This calculation uses definition 1 for A_{FB} and considers some $O(\alpha^2)$ corrections. Within the above experimental accuracy there is no difference between the two calculations. In fig. 8.1 the difference between the two definitions is plotted. From the tables it is clear that the acollinearity requirement decreases the difference.

9 BHABHA SCATTERING

9.1 Introduction

It is well known that the Bhabha cross section diverges in the forward direction. An angular cut on one of the particles e.g. the positron is certainly required. One could then start from eq. (7.24) and evaluate a total cross section with an angular cut. We want to demonstrate that such a cross section gets an enormous correction. To that end we consider only the pure t channel one photon exchange contribution to the cross section. So take

$$|M(s,t)|^2 = 256\pi^2\alpha^2\frac{s^2}{t_1^2}\theta(s - 4m_e^2) \ . \tag{9.1}$$

From eq. (7.41) one finds the $O(\alpha)$ LL correction

$$\frac{d\sigma}{dc_3} = \frac{8\pi\alpha^2}{s}\frac{1}{(1-c_3)^2}\left[1 + \frac{\alpha}{2\pi}\,L_e\left(4\ell n(1 - x_{\min})\right)\right.$$

Table 9.1. Peak height and positions for Bhabha
scattering with $M_Z = 91.17$ GeV.

$\sqrt{s_{max}}$	$\sqrt{s_-}$	$\sqrt{s_+}$	$\sqrt{s_+} - \sqrt{s_-}$	σ_{max}(pb)
91.16	89.91	92.41	2.50	1318.
90.96	89.32	92.17	2.85	1495.
91.14	89.30	92.77	3.47	1094.
91.06	89.28	92.57	3.29	1136.

The first line corresponds to only s-channel contributions at Born level, including weak corrections. Lines 2-4 correspond to lines 1-3 of figure 9.1, so the second line in this table is for Born s- and t-channel, including pure weak corrections, the third line includes first order LL QED corrections and the last line has second order exponentiated LL QED corrections. All positions are in GeV.

$$ - \quad \log x_{\min} + \frac{1}{2} + \frac{1}{x_{\min}} + x_{\min} + \frac{1}{2}x^2_{\min} \Big) \Big] \; . \tag{9.2}$$

Since $x_{\min}$ is the lower bound on the integration variables x_1 and x_2 it is of order $m_e/\sqrt{s}$ or smaller. So the corrections will diverge at least as $\sqrt{s}/m_e$. This dramatic behaviour is avoided when the other particle, the electron, also is required to have a non vanishing angle with the electron beam. So theoretically there is a reason to require cuts on both particles. Experimentally this is done anyhow. Thus we have to start with eq. (7.50).

9.2   Results for Bhabha scattering

The precision of the numerical evaluation of a corrected Bhabha cross section is the same as for σ_F and σ_B of the previous section. We apply symmetric angular and energy cuts i.e.

$$E_{3,4} > E_0 \; , \tag{9.3}$$

$$|\cos\theta_{3,4}| < \cos\theta_m \; . \tag{9.4}$$

In particular two sets of values are used

$$1 : \qquad \theta_m = 42° \; , \qquad\qquad E_0 = 10 \text{ GeV} \; , \tag{9.5}$$

$$2 : \qquad \theta_m = 10° \; , \qquad\qquad E_0 = 10 \text{ GeV} \; . \tag{9.6}$$

All calculations are done with $M_z = 91.17$, $M_H = 100$ and $m_t = 100$ GeV .

When one considers LL calculations in $O(\alpha)$ and $O(\alpha^2)$, the latter with resummed soft photons one gets the behaviour as in figs. 9.1 and 9.2 for conditions (9.5) and (9.6) respectively. In the latter the t-channel contribution is far larger than in the former case. The solid line is the $O(\alpha^2)$ exponentiated result, the short-dashed line the LL $O(\alpha)$ result.

The shape can again be characterized by $\sqrt{s}_{\max}, \sqrt{s}_-, \sqrt{s}_+$. For the case of fig. 9.1 this has been done in table 9.1. For the Born cross section with weak corrections also

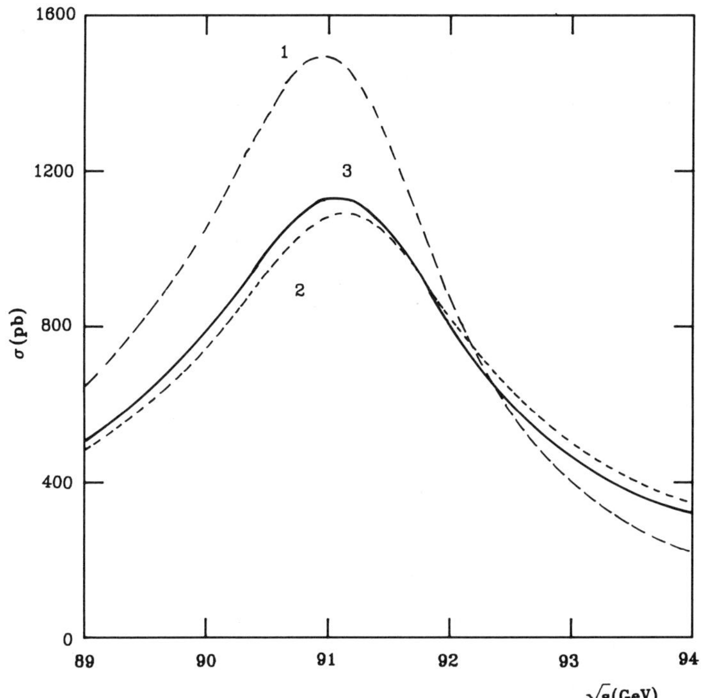

Fig. 9.1. The Bhabha cross section for $\theta_m = 42°$.

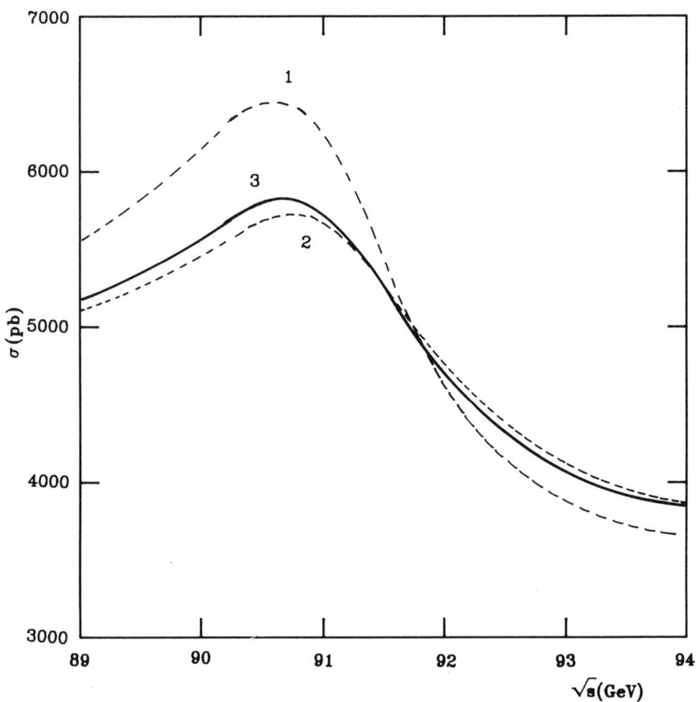

Fig. 9.2. The Bhabha cross section for $\theta_m = 10°$.

Table 9.2. Results for the total cross section (in picobarn) using an acollinearity cut, compared with BABAMC.

ζ_{max}	BABAMC	LL $\mathcal{O}(\alpha)$	$\mathcal{O}(\alpha)$	$\mathcal{O}(\alpha^2)$	$\mathcal{O}(\alpha^2)$+exp.
180°	5750.±5.	5744.	5749.±7.	5838.±7.	5865.±7.
20°	5619.±5.	5633.	5629.±7.	5725.±7.	5753.±7.
10°	5505.±5.	5534.	5506.±7.	5611.±7.	5640.±7.
5°	5329.±5.	5380.	5336.±8.	5461.±8.	5492.±8.
180°	1127.±2.	1120.	1127.±1.	1180.±1.	1179.±1.
20°	1108.±2.	1112.	1109.±1.	1163.±1.	1162.±1.
10°	1087.±2.	1104.	1089.±1.	1144.±1.	1143.±1.
5°	1056.±2.	1090.	1057.±1.	1115.±1.	1114.±1.

All entries are calculated at $\sqrt{s} = 91.1$ GeV. The upper half of the table is for an angular cut of 10°, the lower half for an angular cut of 42°.

the pure s-channel case is given for comparison. The necessity of corrections higher than order α clearly follows from the differences between the third and fourth lines.

Sofar we have not presented any results for the non-log order α correction. An idea of these can be obtained from table 9.2. In this table both LL order α and the complete order α corrections are listed for different acollinearity cuts. It is seen that the LL is dominant which justifies the choice of scale adopted in the calculations. Also in this table the results of an $O(\alpha)$ event generator BABAMC [37] are listed so that the $O(\alpha)$ results of the presented method can be checked. Within the errors of the numerical integration there is good agreement. For completeness, it should be noted that the weak corrections in the event generator do not include a Dyson series summation. For this table the dressed Born cross section in the structure function approach has been corrected for weak interactions in the same way as in the event generator. It is only here for this comparison that we deviate from the approach sketched in section 3.

REFERENCES

[1] W. Hollik, Lectures CERN-JINR School of Physics 1989, CERN-TH. 5661/90.
L. Maiani, Lectures in these Proceedings.

[2] M. Böhm, W. Hollik and H. Spiesberger, Fortschr. Phys. 34 (1986) 687.

[3] G. 't Hooft and M. Veltman, Nucl. Phys. B44 (1972) 189.

[4] W. Hollik, Fortschr. Phys. 38 (1990) 165.

[5] H. Burkhardt, F. Jegerlehner, G. Penso and C. Verzegnassi, Z. Phys. C43 (1989) 497.

[6] For more details, cf D. Bardin et al. in "Z Physics at LEP 1", CERN 89-08, Vol. 1, p. 89, eds. G. Altarelli, R. Kleiss and C. Verzegnassi.

[7] M. Consoli, W. Hollik and F. Jegerlehner, ibidem p. 7.

[8] T.H. Chang, K.J.F. Gaemers and W.L. van Neerven, Nucl. Phys. B202 (1982) 407.

[9] F.A. Berends in Radiative Corrections for e^+e^- Collisions, Ringberg Castle, 1989, p. 55, ed. J.H. Kühn, Springer Berlin.

[10] A. Borelli, M. Consoli, L. Maiani, R. Sisto, CERN-TH. 5541/89.

[11] D. Bardin, A. Leike, T. Riemann and M. Sachwitz, Phys. Lett. B206 (1988) 539.

[12] W. Beenakker, F.A. Berends and S.C. van der Marck, Z. Phys. C46 (1990) 687.

[13] F. Bloch and A. Nordsieck, Phys. Rev. 52 (1937) 54;
D.R. Yennie, S.C. Frautschi and H. Suura, Ann. Phys. (N.Y.) 13 (1961) 379.

[14] F.A. Berends, R. Kleiss, P. de Causmaecker, R. Gastmans, W. Troost and T.T. Wu, Nucl. Phys. B206 (1982) 61.

[15] F.A. Berends and R. Kleiss, Nucl. Phys. B178 (1981) 141 and B260 (1985) 32.

[16] G. Bonneau and F. Martin, Nucl. Phys. B27 (1971) 381.

[17] F.A. Berends and R. Kleiss, Nucl. Phys. B177 (1981) 237.

[18] N. Nakanishi, Progr. Theor. Phys. 19 (1958) 159;
T. Kinoshita, J. Math. Phys. 3 (1962) 650;
T.D. Lee and M. Nauenberg, Phys. Rev. 133 (1964) B 1549.

[19] F.A. Berends, K.J.F. Gaemers and R. Gastmans, Nucl. Phys. B57 (1973) 381.

[20] F.A. Berends, R. Kleiss and S. Jadach, Nucl. Phys. B202 (1982) 63; Comp. Phys. Comm. 29 (1983) 183.

[21] F.A. Berends, W.L. van Neerven and G.J.H. Burgers, Nucl. Phys. B297 (1988) 429.

[22] R. Barbieri, J.A. Mignaco and E. Remiddi, Nuovo Cim. 11A (1972) 824, 865.

[23] A. Devoto and D.W. Duke, Riv. Nuovo Cim. 7 no. 6 (1984).

[24] S. Jadach and B.F.L. Ward, Phys. Rev. D38 (1988) 2897.

[25] R.N. Cahn, Phys. Rev. D36 (1987) 2666.

[26] E.A. Kuraev and V.S. Fadin, Sov. J. Nucl. Phys. 41 (1985) 466.

[27] G. Altarelli and G. Martinelli, CERN 86-02 (1986) 47.

[28] D. Nicrosini and L. Trentadue, Phys. Lett. B196 (1987) 551.

[29] V.S. Fadin and V.S. Khoze, Novosibirsk preprint 87-157.

[30] J.C. Collins and G. Sterman, Nucl. Phys. B185 (1982) 172.
J.C. Collins, D.E. Soper and G. Sterman, Phys. Lett. 134B (1984) 263; Nucl. Phys. B261 (1985) 104.
G.T. Bodwin, Phys. Rev. D31 2616 and references therein.

[31] V.N. Gribov and L.N. Lipatov, Sov. J. Nucl. Phys. 15 (1972) 438; 675.
G. Altarelli and G. Parisi, Nucl. Phys. B126 (1977) 298.

[32] E.G. Floratos, D.A. Ross and C.T. Sachrajda, Nucl. Phys. B129 (1977) 66; Erratum Nucl. Phys. B139 (1978) 545.
A. González-Arroyo, C. López and F.J. Ynduraïn, Nucl. Phys. B153 (1979) 161.
A. González-Arroyo and C. López, Nucl. Phys. B166 (1980) 429.
E.G. Floratos, C. Kounnas and R. Lacaze, Phys. Lett. 98B (1981) 89; 285.

[33] W. Beenakker, F.A. Berends and W.L. van Neerven, in Radiative Corrections for e^+e^- Collisions, Ringberg Castle, 1989, p. 3, ed. J.H. Kühn, Springer, Berlin.

[34] W. Beenakker, F.A. Berends and S.C. van der Marck, Nucl. Phys. B (in print).

[35] W. Beenakker, F.A. Berends and S.C. van der Marck, Phys. Lett. B (in print).

[36] D. Bardin, M. Bilenky, A. Chizov, A. Sazonov, O. Fedorenko, T. Riemann and M. Sachwitz, Berlin-Zeuthen PHE 89-19 preprint.

[37] M. Böhm, A. Denner and W. Hollik, Nucl. Phys. B304 (1988) 687;
F.A. Berends, R. Kleiss and W. Hollik, Nucl. Phys. B304 (1988) 712.

QCD AND JETS AT LEP

Torbjörn Sjöstrand

CERN/TH
CH-1211 Geneva 23
Switzerland

ABSTRACT

The various aspects of our understanding of QCD in the context of LEP are discussed. Perturbative QCD may be described either in a matrix element approach or in a parton shower one, each with its advantages and drawbacks. The nonperturbative hadronization is not understood from first principles, but is modelled with string, cluster or independent fragmentation. Existing data from PETRA, PEP, TRISTAN and, most recently, LEP has been used to test and constrain the models that have been developed. The plan of these lecture is as follows:

- 1. Introduction
- 2. Perturbative QCD
    - 2.1. Matrix Elements
    - 2.2. Parton Showers
    - 2.3. Testing QCD
- 3. Fragmentation Models
    - 3.1. String Fragmentation
    - 3.2. Independent Fragmentation
    - 3.3. Cluster Fragmentation
    - 3.4. Other Fragmentation Approaches
    - 3.5. Particles and Their Decays
- 4. Our Experimental Knowledge
    - 4.1. Event Characterization Methods
    - 4.2. Event Shapes
    - 4.3. Multiplicities and Correlations
    - 4.4. Jet Type Separation
    - 4.5. String and Coherence Phenomena
- 5. Summary and Outlook

1 Introduction

LEP has brought spectacular confirmation of the basic validity of the standard model, both in the electroweak and in the strong sector. Whereas tests of the electroweak model typically aim at very high precision, the QCD aspects are rather less under control. In particular, the fragmentation process has not yet been understood from first principles, but only in terms of vaguely QCD-inspired models, with many issues unsolved. Even in the large momentum transfer regime, where perturbative calculations can be used to describe jet production, the strong coupling constant α_S is large enough that yet uncalculated higher order corrections could well shift current results significantly. The status is reflected in these lectures, which contain a mixed bag of success stories and not yet resolved issues, but it should always be kept in mind that the underlying theory of QCD is not in question.

QCD is a non-Abelian gauge theory, with a well-defined Lagrangian, see e.g. [1]. Quarks appear in a colour triplet representation, and gluons in a colour octet one. There are three fundamental vertices in the theory: a quark-(antiquark-)gluon vertex, a three-gluon vertex and a four-gluon vertex. While the first is the analogue of the standard QED coupling of a charged particle to a photon, the latter two are absent in QED and reflect the non-Abelian structure of QCD. The four-gluon vertex appears only in higher orders, and does not have any major experimental consequences; for the renormalizability of the theory its presence is essential, however.

The theory contains one common coupling parameter, the strong coupling constant α_S. In the lectures of Maiani [2] it is discussed how the renormalization group leads to a running of α_S with the momentum transfer scale Q^2. To first order, this is expressed by the relation

$$\alpha_S(Q^2) = \frac{12\pi}{(33 - 2n_f)\ln(Q^2/\Lambda^2)},$$ (1)

where n_f gives the effective number of quarks, for e^+e^- annihilation normally taken to be 5 at the Z^0 scale, and Λ is a free dimensional parameter of the theory.

At short distances, which by the uncertainty principle corresponds to large momentum transfer scales, α_S is a reasonably small number. This is the region of 'asymptotic freedom', where quarks and gluons behave as almost free particles, and standard perturbation theory can be used to calculate their interactions. At large distances, on the other hand, the effective Q^2 scale is small and α_S becomes very large. Therefore the perturbative treatment in terms of quarks and gluons breaks down. Although it has not been fully proven mathematically, lattice QCD calculations give strong support to the concept of 'confinement', i.e. that quarks never appear as free particles, but are always confined inside colour singlet hadrons. The transition process, whereby the quarks and gluons turn into a set of hadrons, is called fragmentation or hadronization.

The structure of a typical multihadronic event in e^+e^- annihilation is shown in Fig. 1. A few introductory paragraphs, based on this figure, follow. A number of complications are here swept under the carpet, and only covered (if at all) in the subsequent, more detailed discussion.

In a first phase, an e^+e^- pair annihilates into a virtual γ/Z^0 state, which in its decay produces a primary quark-antiquark pair $q\bar{q}$. Before the annihilation, initial state QED bremsstrahlung may occur, so that the mass of the hadronic final state is reduced from the naive value. For the precision needed for Z^0 line shape studies, also higher order (loop) corrections to the basic graph are important. These aspects are covered in the lectures of Berends [3].

In the second phase, the initial $q\bar{q}$ pair may radiate gluons g, which in their turn

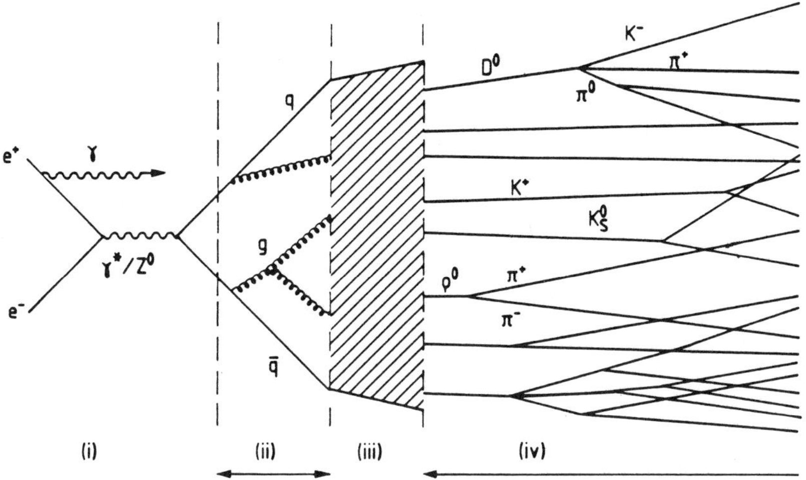

Figure 1. Schematic illustration of an e^+e^- annihilation event.

may radiate. While the primary $q\bar{q}$ production is mainly given by electroweak perturbation theory, strong perturbation theory must be used to describe this second stage. The strong coupling constant being larger than the electroweak ones, the degree of accuracy is less, in particular for soft parton emission.

In the third phase, the coloured partons fragment into a number of colourless hadrons. Although we believe this process to be given by QCD, it is not perturbatively calculable, and therefore it is an area where phenomenological models have to be invoked.

In a fourth phase, unstable hadrons decay into the experimentally observable particles. This includes everything from $\pi^0 \to \gamma\gamma$ decays to long decay chains of charm and bottom hadrons. Whereas the qualitative features of these decays usually are well known, little quantitative understanding exists. Instead the main input here comes from experimentally determined branching ratios.

Given the complexity of the problem, purely analytical techniques are of limited usefulness for LEP physics studies (however, see the lecture of Khoze [4] for a different point of view). Instead the Monte Carlo simulation of complete hadronic events constitutes one of the main tools for improving our understanding of QCD at LEP. The use of Monte Carlo methods, i.e. the selection of variables according to rules which contain random numbers, is well suited to describe a Nature which is random in itself, and in addition allows the subdivision of a complex task into more manageable subtasks. In order to put the original theory or model on a computer, some compromises may be necessary, but on the other hand the Monte Carlo event generator approach makes it necessary to specify a number of details that often are dismissed cavalierly in purely analytical studies. It should still be recognized that the Monte Carlo approach, based as it is on a probabilistic description, in the end cannot be used to describe all subtle quantum mechanical effects [5], and that therefore analytical studies at times may be the only viable alternative.

A detailed review of existing QCD event generators is found in [5], from which part of the material in these lectures is taken, and to which we refer for all program details. The list of programs studied in the report is given in Table 1. The contents of these programs is classified according to the phases listed above, which also pretty much

Table 1: Physics components included in the programs covered in the LEP physics yellow book [5]. A '•' signifies that a task is handled in the program, a '+' that it is obtained by calling on another program (of the ones listed here), and a '-' that it is not considered at all. For some programs, the classification may not always be trivial, so this table should only be taken as a first indication. The symbols contain no grading on the quality of a program for a given task.

Program (comment)	Hard inter.	Initial γ rad.	Parton shower (or matrix el.)	Fragmentation	Decays
ARIADNE	+	+	•	+	+
CALTECH-II	•	-	•	•	•
COJETS	•	•	•	•	•
DPSJET (non-QCD)	-	-	•	•	-
EPOS (non-QCD)	•	-	•	•	•
EURODEC	-	-	-	•	•
HERWIG	•	-	•	•	•
JETSET ('Lund')	•	•	•	•	•
NLLJET	•	•	•	+	+
PARJET	•	-	•	•	•
TIPTOP (heavy ferm.)	•	•	-	+	+

represent the general 'flow chart' of the programs. However, the hard interaction and initial state γ radiation aspects typically only correspond to less than 10% of the code (and the precision is nowhere near that of dedicated line shape programs) — most of the emphasis is put on the perturbative and fragmentation sections of the programs. Not all programs in Table 1 are equally ambitious, and in these lectures only a few will appear explicitly. Thus, essentially all LEP studies presented so far have been based on JETSET ('the Lund Monte Carlo') or HERWIG ('Marchesini-Webber').

The rapid variation of the electroweak cross-section around the Z^0 peak necessitates special attention to fine details, like the effects of multiple soft photon initial state radiation. Programs written for use at lower energies (PETRA and PEP, say) can therefore in general not be taken over for LEP studies. In QCD the situation is quite different. The only significant consequences of the Z^0 peak for QCD studies are a somewhat changed primary quark composition and a different forward-backward quark production asymmetry. For the rest, QCD physics extrapolates smoothly from lower energies, such that the accumulated experience from PETRA, PEP and TRISTAN could be directly applied for LEP predictions. The LEP QCD event generators therefore are pretty much the same as those which have already seen heavy use at lower energies.

2 Perturbative QCD

The modelling of perturbative QCD is, together with the fragmentation modelling, the central objective of QCD studies. While the hard electroweak interaction provides a description of the production of a primary $q\bar{q}$ pair, the perturbative QCD description sets out to describe the emergence of multijet events. As the CM energy is increased, hard QCD emission is increasingly important, relative to fragmentation, in determining the event structure. At LEP, three- and four-jet event structures abound.

Two traditional approaches exist to the modelling of perturbative QCD. One is the matrix element method, in which Feynman diagrams are calculated, order by order. In principle, this is the correct approach, which takes into account exact kinematics, and the full interference and helicity structure. The only problem is that calculations become increasingly difficult in higher orders, in particular for the loop graphs. The calculations have therefore only been carried out, in full, up to $\mathcal{O}(\alpha_S^2)$. We have indirect but strong evidence that, in fact, the emission of multiple soft gluons plays a significant rôle in building up the event structure at LEP energies, and this sets a limit to the applicability of matrix elements. However, the perturbative expansion by itself is more well behaved at LEP than at lower energies, due to the smaller α_S value, so inclusive measurements should yield more reliable results, e.g. in terms of α_S measurements [6].

The second possible approach is the parton shower one. Here an arbitrary number of branchings of one parton into two (or more) may be put together, to yield a description of multijet events, with no explicit upper limit on the number of partons involved. This is possible since the full matrix element expressions are not used, but only approximations derived by simplifying the kinematics, and the interference and helicity structure. Parton showers therefore are expected to give a good description of the substructure of jets, but in principle the shower approach has limited predictive power for the rate of well-separated jets (i.e. the 2/3/4/5-jet composition). In practice, shower programs may be patched up to describe the hard gluon emission region reasonably well. Nevertheless, the shower description is not optimal for absolute α_S determinations.

Thus the two approaches in many respects are complementary, and both have found use at LEP. However, because of its simplicity and flexibility of use, the parton shower option is generally the first choice, while the matrix elements one is mainly used for α_S determinations and three-gluon vertex studies. Obviously, the ultimate goal would be to have an approach where the best aspects of the two worlds are harmoniously married.

2.1   Matrix Elements

For the discussion in this section, we will use the word 'jet' also for properties on the partonic level, almost interchangeably with the word 'parton'. However, while a 'parton' is any quark or gluon appearing in the process description, a 'jet' is one or several nearby partons, lumped together according to some jet resolution criterion (see below). The 'true' number of partons in an event is thus an ill-defined concept (and may well be infinite), while the number of jets is unique (for a given jet definition).

2.1.1   First order matrix elements

The Born process $e^+e^- \rightarrow q\bar{q}$ is modified in first order QCD by the probability for the q or $\bar{q}$ to radiate a gluon, i.e. by the process $e^+e^- \rightarrow q\bar{q}g$, Fig. 2a. The matrix element is conveniently given in terms of scaled energy variables in the CM frame of the event, $x_1 = 2E_q/E_{CM}$, $x_2 = 2E_{\bar{q}}/E_{CM}$, and $x_3 = 2E_g/E_{CM}$, i.e. $x_1 + x_2 + x_3 = 2$.

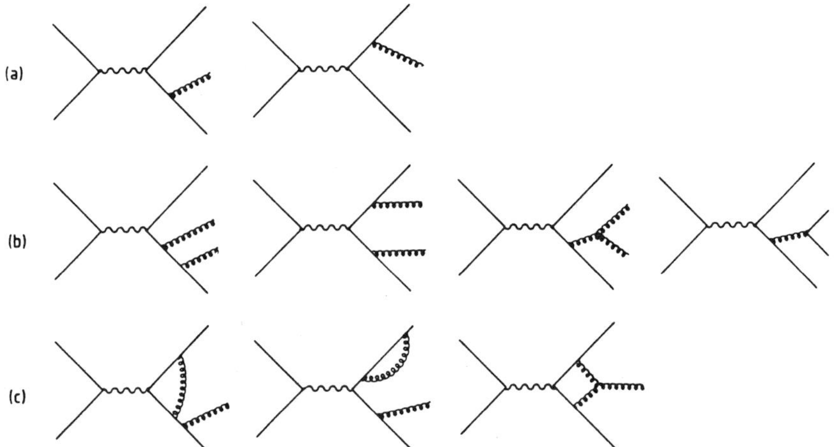

Figure 2. Feynman graphs for three- and four-jet production.
(a) The two graphs which contribute to three-jet production in first order.
(b) A few of the graphs which contribute to four-jet production (the ones left out can be obtained by symmetry).
(c) A few of the loop (vertex and propagator) graphs which contribute to three-jet production in second order.

For massless quarks the matrix element reads [7]

$$\frac{1}{\sigma_0}\frac{d\sigma}{dx_1 dx_2} = \frac{\alpha_S}{2\pi} C_F \frac{x_1^2 + x_2^2}{(1-x_1)(1-x_2)}, \qquad (2)$$

where σ_0 is the lowest order cross-section, $C_F = 4/3$ is the appropriate colour factor, and the kinematically allowed region is $0 \leq x_i \leq 1, i = 1, 2, 3$. The x_k variable for parton k is related to the invariant mass m_{ij} of the remaining two partons i and j by

$$y_{ij} \equiv \frac{m_{ij}^2}{E_{CM}^2} = 1 - x_k. \qquad (3)$$

In order to separate two-jets from three-jets, it is useful to introduce jet resolution parameters, either (ϵ, δ) or y.

- (ϵ, δ) : a three-parton configuration is called a two-jet event if $\min(x_i) < \epsilon$ or if $\min(\theta_{ij}) < \delta$ (where θ_{ij} is the angle between partons i and j in the CM frame of the event).
- y : a three-parton configuration is called a two-jet event if
$\min(y_{ij}) = \min(m_{ij}^2/E_{CM}^2) < y$.

In the discussion which follows, we will mainly refer to the y cut case, since this is the simpler one.

The cross-section in eq. (2) diverges for $x_1 \to 1$ or $x_2 \to 1$ but, when first order propagator and vertex corrections are included, a corresponding singularity with opposite sign appears in the $q\bar{q}$ cross-section, so that the total continuum hadronic cross-section is finite, $\sigma_{tot}/\sigma_0 = 1 + \alpha_S/\pi$. In analytical studies, the average value of any well-behaved quantity Q can therefore be calculated as

$$\langle Q \rangle = \frac{1}{\sigma_{tot}} \lim_{y \to 0} \left(Q(2parton)\sigma_{2parton}(y) + \int_{y_{ij} > y} Q(x_1, x_2)\frac{d\sigma_{3parton}}{dx_1 dx_2} dx_1 dx_2 \right), \qquad (4)$$

where any explicit y dependence disappears in the limit $y \to 0$. One should note that the hadronic Z^0 width receives a corresponding QCD correction factor as does σ_{tot} above; the net result is a reduction of the Z^0 peak cross-section.

In Monte Carlo programs, it is not possible to work with a negative total two-jet rate, and so it is necessary to introduce a fixed non-vanishing y cut-off in the three-jet phase-space. Experimentally, there is evidence for the need of a low y cut-off, i.e. a large three-jet rate. (Already at PETRA, this was apparent from the behaviour of the Energy-Energy Correlation Asymmetry [8].) For LEP applications, the recommended value is $y = 0.01$, which is about as far down as one can go and still retain a positive two-jet rate. With $\alpha_S = 0.12$, in full second order QCD, the $2:3:4$ jet composition is then approximately $11\% : 77\% : 12\%$.

A fixed cutoff in y corresponds to a minimum mass between any two jets that grows linearly with CM energy. For PETRA/PEP a cutoff at $y = 0.01$ (or slightly above, so as to keep a positive two-jet rate even for the somewhat higher α_S value at lower energies) corresponds to a mass separation between jets of 3.5 GeV, at LEP it corresponds to 9 GeV. The nonperturbative fragmentation is expected to take over below a mass scale of around 1 GeV, independent of CM energy. As the energy is increased, there is therefore a widening soft gluon phase space region which cannot be simulated in a standard matrix elements based event generator. One may try to sum up the soft gluon effects into an effective energy-dependent fragmentation description, but this approach is fairly ugly and must eventually break down once sufficiently high energies are reached. As we shall see later, the Z^0 region is just at the borderline.

2.1.2 Second order matrix elements

Two new event types are added in second order QCD, $e^+e^- \to q\bar{q}gg$ and $e^+e^- \to q\bar{q}q'\bar{q}'$, Fig. 2b. Of the 12% four-jet rate quoted above, 11.5% is $q\bar{q}gg$ and only 0.5% $q\bar{q}q'\bar{q}'$.

The four-jet cross-section has been calculated by several groups [9,10,11,12], which agree on the result. The formulae are too lengthy to be quoted here. In one of the calculations [9], quark masses were explicitly included. The original calculations were for the pure γ exchange case; recently it has been pointed out [13] that an additional contribution to the $e^+e^- \to q\bar{q}q'\bar{q}'$ cross-section arises from the axial part of the Z^0. This term is not included in any QCD program so far, but fortunately it is finite and small.

As for the first order, a full second order calculation consists both of real parton emission terms and of vertex and propagator corrections, Fig. 2c. These modify the three-jet and two-jet cross-sections. Although there was some initial confusion, everybody eventually agreed on the size of the loop corrections [11,14,15]. In analytic calculations, the procedure of eq. (4), suitably expanded, can therefore be used unambiguously for any well-behaved variable.

For Monte Carlo event simulation, it is again necessary to impose some jet resolution criteria. This means that four-parton events which fail the four-jet cuts should be reassigned either to the three-jet or to the two-jet event class. It is this area which has caused quite a lot of confusion in the past [16]. The reasons for the confusion are well understood by now, but this does not mean we have reached agreement on a unique procedure to resolve the issue. Most likely, agreement will never be reached, since indeed there are ambiguous points in the procedure, related to uncertainties on the theoretical side. This is illustrated in the following paragraphs.

For the y-cut case, any two partons with an invariant mass $m_{ij}^2 < yE_{cm}^2$ should be recombined into one. If the four-momenta are simply added, Fig. 3, the sum

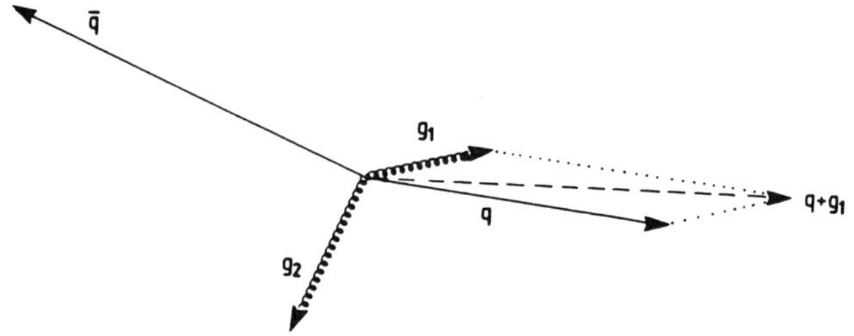

Figure 3. A four-parton event, with two nearby partons that are to be recombined. Construction of momentum sum $q+g_2$ is shown, but that gives a composite parton with non-zero mass.

will correspond to a parton with a positive mass, namely the original m_{ij}. The loop corrections are given in terms of final massless partons, however. In order to perform the (partial) cancellation between the four-parton real and the three-parton virtual contributions, it is therefore necessary to get rid of the bothersome mass. Several procedures have been proposed; the following two are probably the most frequently used ones.

- The $\overline{p}$ recombination scheme: keep the constructed three-momentum sum $\overline{p}_{ij} = \overline{p}_i + \overline{p}_j$ (in the CM frame of the event), and redefine the energy of the recombined parton as being $E_{ij} = |\overline{p}_i + \overline{p}_j|$. Since $E_{ij} < E_i + E_j$, the total CM energy of the event has been reduced, which can be compensated by rescaling all the four-momenta in the event by a common factor.

- The E recombination scheme: require $E_{ij} = E_i + E_j$. The three variables x_1, x_2 and x_3 are then easily obtained, but the three-momenta have to be modified in a non-trivial manner to keep momentum conserved.

The E scheme obviously gives more energy to the recombined jet than does the $\overline{p}$ one. Since the typical situation is that the recombined jet still is the lowest-energy one (two soft gluons recombined into a medium soft one), the E scheme gives more three-jetlike topologies than the $\overline{p}$ one, and so has a larger (i.e. more positive) second order correction to the three-jet rate.

The total number of different recombination schemes proposed is fairly large, and so is the list of different parametrizations of the second order three-jet rate. Fig. 4 shows a comparison of some of these parametrizations [5]. (Further recent comparisons between schemes may be found in [6,17].) While the GKS curve is known to be wrong (in the sense that it is based on a number of analytical approximations we now know are unacceptably crude), there is enough of a spread between the other ones. Experimental α_S determinations are typically based on the shape of the thrust distribution, and are thus particularly sensitive to the region $T < 0.85$. Here, however, four-jets give a non-negligible contribution (actually the ratio becomes infinite at $T = 2/3$, since three-jet phase-space vanishes at that point, while four-jets may have $T < 2/3$). The net effect is thus that the uncertainty in the value of the thrust distribution is nowhere larger than $\pm 10\%$, based on the generators above. This gives some feeling for what 'theory systematic error' need to be assigned to any α_S determination, just from the matrix element point of view. Additional uncertainties come from higher orders, from fragmentation, from experimental errors, and so on.

If the situation is complicated enough for the y cut, it is even worse for the (ϵ, δ) alternative. Separate recipes need to be specified for what to do when a parton fails the ϵ cut, when a pair fails the δ cut, and when both happen in the same event. In particular, for the ϵ cut, one could either discard the soft parton outright, or choose

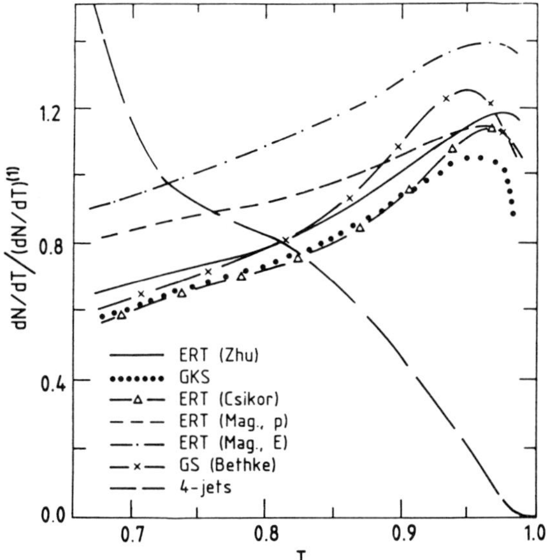

Figure 4. Full second order three-jet thrust distribution, normalized to the first order one, for $\alpha_S = 0.12$ and $y = 0.01$. Also shown is the four-jet thrust distribution, normalized the same way.

between several possible recombination algorithms. The spread in effective three-jet rate, i.e. in terms of α_S values found for given data, has therefore always been larger in the (ϵ, δ) alternative. None of the algorithms in use today are based on (ϵ, δ) cuts.

In a complete second order description, consistency requires the use of the second order expression for α_S. Several alternative forms exist, which only differ by terms nominally of order α_S^3. The standard proposed by the Particle Data Group [18] is

$$\alpha_S(Q^2) = \frac{12\pi}{(33 - 2n_f)\ln(Q^2/\Lambda^2)} \left[1 - 6\frac{153 - 19n_f}{(33 - 2n_f)^2}\frac{\ln(\ln(Q^2/\Lambda^2))}{\ln(Q^2/\Lambda^2)} \right]. \tag{5}$$

Further, the total continuum hadronic cross-section is [19]

$$\frac{\sigma_{tot}}{\sigma_0} = 1 + \frac{\alpha_S}{\pi} + (1.986 - 0.115n_f)\left(\frac{\alpha_S}{\pi}\right)^2. \tag{6}$$

It should be mentioned that σ_{tot} was calculated to third order a few years ago [20]. This calculation gave a surprisingly large third order term. The authors have now found an error in that calculation, but have not yet provided a corrected result. Currently, σ_{tot} should therefore be considered to be known only to second order.

2.1.3   Optimized perturbation theory

The second order corrections to the three-jet rate are large. It is therefore not unreasonable to expect large third order corrections to the four-jet rate. Indeed, the experimental four-jet rate is much larger than second order predicts (also when fragmentation effects have been taken into account), if α_S is determined based on the three-jet rate [8,21,22,23].

The only consistent way to resolve this issue is to go ahead and calculate the full next order. This is a tough task, however, so people have looked at possible shortcuts. For example, one can try to minimize the higher order contributions by a suitable choice of the renormalization scale — 'optimized perturbation theory'. This is equivalent to a different choice for the Q^2 scale in α_S, a scale which is not unambiguous anyway. Indeed the standard value $Q^2 = s = E_{CM}^2$ $(= M_Z^2$ at LEP) is larger than the natural physical scale of gluon emission in events, given that most gluons are fairly soft. One could therefore pick another scale, $Q'^2 = fs$, with $f < 1$. Since α_S is increased by a reduction of the Q^2 scale, the $\mathcal{O}(\alpha_S)$ three-jet rate would then be increased, and so would the number of four-parton events, including those which collapse into three-jet ones. The loop corrections depend on the Q^2 scale, however, and compensate the changes above by giving a larger negative contribution to the three-jet rate.

The transformation to a different scale can be done easily as follows. Suppose that the three-jet rate R_3 (differential or integrated over some region of phase space) is given in a certain renormalization scheme and for a scale $Q^2 = s$

$$R_3 = r_1 \alpha_S + r_2 \alpha_S^2 + \mathcal{O}(\alpha_S^3), \tag{7}$$

with

$$\alpha_S \equiv \alpha_S(Q^2) = \frac{12\pi}{(33 - 2n_f)\ln(Q^2/\Lambda^2)} = \frac{1}{b\ln(Q^2/\Lambda^2)} \tag{8}$$

(in the formulae in this section we use the first order α_S expression for simplicity; to our required level of approximation an inclusion of the full second order expressions changes nothing in the following argumentation).

When the coupling is chosen at a different scale, $Q'^2 = fQ^2$, this corresponds to a change of the α_S value according to

$$\begin{aligned}
\alpha_S' &\equiv \alpha_S(Q'^2) = \frac{1}{b\ln(fQ^2/\Lambda^2)} = \frac{1}{b\left[\ln(Q^2/\Lambda^2) + \ln f\right]} \\
&= \frac{1}{b\ln(Q^2/\Lambda^2)} \frac{1}{1 + \frac{\ln f}{\ln(Q^2/\Lambda^2)}} = \alpha_S\left(1 - \frac{\ln f}{\ln(Q^2/\Lambda^2)} + \cdots\right) \\
&= \alpha_S\left(1 - b\alpha_S \ln f + \cdots\right) = \alpha_S - b\alpha_S^2 \ln f + \mathcal{O}(\alpha_S^3). \tag{9}
\end{aligned}$$

The expression for the three-jet rate reads

$$\begin{aligned}
R_3' &= r_1' \alpha_S' + r_2' \alpha_S'^2 + \mathcal{O}(\alpha_S'^3) = r_1'(\alpha_S - b\alpha_S^2 \ln f + \cdots) + r_2'(\alpha_S + \cdots)^2 + \mathcal{O}(\alpha_S^3) \\
&= r_1' \alpha_S + (r_2' - r_1' b\ln f)\alpha_S^2 + \mathcal{O}(\alpha_S^3) \tag{10}
\end{aligned}$$

The concept of the renormalization group asserts that, if a physical quantity like the three-jet rate R_3 is calculated to infinite order, no dependence can remain on the original choice of the expansion parameter α_S. If calculated to infinite order, one must therefore have $R_3' \equiv R_3$. Thus, in particular, if R_3 and R_3' are given in terms of power series in the same α_S parameter, the coefficients must be the same, order by order (since formally α_S can be varied continuously, independently of the f choice, by a change of the CM energy). Comparing eqs. (7) and (10) one concludes that

$$\begin{aligned}
r_1' &= r_1, \\
r_2' &= r_2 + r_1 b\ln f, \tag{11}
\end{aligned}$$

i.e.

$$R'_3 = r_1\alpha'_S + (r_2 + r_1 b \ln f)\alpha'^2_S + \mathcal{O}(\alpha'^3_S). \tag{12}$$

The net effect is to make the physical three-jet rate less dependent on the choice of scale Q'^2, i.e. f, in α_S: if f is made smaller (keeping Λ fix), the effective parameter α'_S becomes larger, but simultaneously r'_2 is made smaller, such that R'_3 is not changed very much. That R'_3 is not completely independent of f is only a consequence of the approximations made in relating α'_S to α_S; the differences are formally of $\mathcal{O}(\alpha_S^3)$, but may become large when f is varied over a large range. When fitting to an experimentally measured three-jet rate, such variations are reflected in a necessity to use slightly different Λ values for different f:s.

Since only the Born term has been calculated for four-jets, the four-jet rate is directly proportional to α_S^2. As f is decreased (again for fixed Λ), the increase in α'_S therefore directly leads to an enhanced four-jet rate. It is exactly this feature that makes it possible to solve the mystery of the lacking four-jets in the older standard matrix element implementations, which used $f = 1$.

While the experimental motivation for considering optimized perturbation theory in e^+e^- came from the problems with the four-jet rate, in fact a number of theorists have earlier proposed recipes for selecting which scale to use for a process, if the one loop corrections to the Born term cross-section are already available. These recipes go by names such as PMS (principle of minimal sensitivity) [24], FAC (fastest apparent convergence) [25], or BLM [26]. While different in detail, they have in common that, the larger the relative size of the second order term (i.e. the ratio r_2/r_1), the smaller the preferred scale. Therefore these schemes all strongly suggest the correct scale to be smaller than the naive $Q^2 = M_Z^2$ one. Also on physical grounds it can be argued that, reasonably, the scale for the emission of a gluon should be related to the kinematics of this emission, and thus be smaller than M_Z (a more precise statement will come in the parton shower section). However, only a full higher order calculation would in the end tell whether a particular choice is favoured or not.

A number of α_S determinations have now been presented by the LEP groups [27]. With $f = 1$, typically one obtains $\alpha_S(M_Z^2) \approx 0.12$, or $\Lambda \approx 0.2$ GeV. If f is varied over a wide range, down to $f = 0.001$, say, Λ is found to be somewhere between 0.1 and 0.2 GeV. Plugging this Λ range into eq. (5), the LEP collaborations quote $\alpha_S(M_Z^2) \approx 0.11$—$0.12$, and conclude that this value is rather independent of the choice of f. Unfortunately, the claimed constancy of α_S may lead the unwary reader astray. For a proper understanding, it is important to make a distinction between $\alpha_S(fM_Z^2)$ and $\alpha_S(M_Z^2)$ in studies where an $f < 1$ is used. The $\alpha_S(fM_Z^2)$ value is α_S as actually extracted from the experimental data, e.g. in terms of jet rates or the Energy-Energy Correlation Asymmetry. This α_S typically is larger the smaller the f chosen, and may vary significantly: with $f = 0.001$ and $\Lambda = 0.1$ GeV, $\alpha_S(fM_Z^2) \approx 0.20$. It is only if the widely different $\alpha_S(fM_Z^2)$ values, determined for a wide range of f values, are evolved up to a common scale, M_Z^2, that the small $\alpha_S(M_Z^2)$ range noted above is obtained. The small range thus simply reflects the stability of the Λ value under variations of the choice of optimized scale. This aspect would not readily be appreciated by just quoting $\Lambda \approx 0.1$–0.2 GeV: since Λ only enters logarithmically into α_S, and therefore into any experimental observable, a factor 2 uncertainty would look more than it really is. In summary: the experimentally determined α_S value strongly depends on what optimized scale is picked, but the Λ value that is extracted remains fairly stable, as reflected in the small range of α_S values at the common scale M_Z^2.

Actually it should come as no surprise that the use of optimized perturbation theory is related to the use of a higher $\alpha_S(fM_Z^2)$ value, at least not if we remember the original motivation of pushing up the theoretical four-jet rate to the experimental level. Since this rate, to the order we have available, is directly proportional to α_S^2 and nothing else, obviously we would not be helped if optimized perturbation theory did not allow us to make α_S bigger (compared to $f = 1$), while still keeping the three-jet rate unchanged. Put this way, optimized perturbation theory could also be considered as a consistent scheme to decouple the descriptions of three-jet and four-jet production: in general, for an α_S adjusted to get the four-jet rate right, it is afterwards possible to find an f such that also the three-jet rate is well described. It is amusing to note that a pure first order fit to LEP data would give $\alpha_S \approx 0.18$, which then is reduced, for the three-jet rate, by second order contributions. It is therefore not a vain hope that, once third order corrections are included fully, the α_S needed for the four-jet rate will drop to around 0.12.

The appearance of an α_S value as high as 0.20 should not lead anybody to fear that QCD corrections to the total hadronic width of the Z^0 have a corresponding uncertainty: as was the case with the three-jet rate above, as soon as an $f < 1$ is used, such that the first order contribution to the width is increased, a compensating term automatically appears in the second order contribution, such that the total hadronic width is fairly insensitive to the choice of Q^2 scale. Let us work this out in a general case.

Consider two measures C and D, each with its expansion in terms of $\alpha_S = \alpha_S(Q^2)$,

$$
\begin{aligned}
C &= c_0 + c_1\alpha_S + c_2\alpha_S^2, \\
D &= d_0 + d_1\alpha_S + d_2\alpha_S^2.
\end{aligned}
\tag{13}
$$

The same derivation that gave eq. (12) now gives

$$
\begin{aligned}
C' &= c_0 + c_1(\alpha_S' + b\ln f\alpha_S'^2) + c_2\alpha_S'^2, \\
D' &= d_0 + d_1(\alpha_S' + b\ln f\alpha_S'^2) + d_2\alpha_S'^2,
\end{aligned}
\tag{14}
$$

at another scale $Q'^2 = fQ^2$. If the quantity C is used for defining the α_S value as a function of the f scale then, by definition, $C' = C$. This implies that

$$
\alpha_S' + b\ln f\alpha_S'^2 = \alpha_S - \frac{c_2}{c_1}(\alpha_S'^2 - \alpha_S^2).
\tag{15}
$$

If this is plugged into the expression for D', one obtains

$$
D' = D + d_1\left(\frac{d_2}{d_1} - \frac{c_2}{c_1}\right)\left(\alpha_S'^2 - \alpha_S^2\right).
\tag{16}
$$

Note that any change in D appears only in second order in α_S.

The QCD corrections to the Z^0 hadronic width, which have the same form as the corrections to the continuum hadronic cross-sections, eq. (6), are characterized by a small second order coefficient, while typical event measures have a much larger (positive) ratio of second to first order coefficients. Therefore the tendency would be towards a smaller Γ_Z if optimized perturbation theory is used, with α_S' determined by event measures such as thrust or the multi-jet rate at the same optimized Q'^2 scale as is to be used for the hadronic width. However, it should be noted that the different ratios of second to first order coefficients for Γ_Z and typical event measures also implies different theoretically preferred optimized scales. For Γ_Z the scale should therefore be closer to M_Z^2, i.e. it is not motivated to consider values down to $f = 0.001$, as has been done in jet rate studies. Further, while eq. (16) gives a transparent description of the variation of an experimental quantity with the f parameter, it may not be the most convenient recipe for experimental studies.

Given the current large use of optimized scales, it is maybe appropriate to end this section with a warning. The second order corrections to the three-jet rate (i.e. the coefficient r_2) is positive in most regions of the three-jet phase space, but not everywhere. One region where it is negative is when the q and $\bar{q}$ are fairly close in phase space: since the q and $\bar{q}$ are in a relative colour octet (together with the colour octet gluon they have to make up a colour singlet), the colour force between them is repulsive, which translates into a negative loop correction [28]. Another region is where the gluon is very close either to the q or $\bar{q}$, and the first loop correction is partly related to the eventual buildup of a Sudakov form factor (see section 2.2.2). While regions with a negative differential first+second order three-jet rate can exist already for the simple choice $Q^2 = E_{CM}^2$, they become larger as smaller scales are used, simply because the relative importance of the second order terms to the three-jet rate becomes larger. Thus the total theoretical three-jet rates computed for small f values are positive only because kinematical variables have been integrated out, so that positive and negative regions have cancelled — a not very satisfactory state of affairs. This points the way towards further potentially necessary 'improvements' of the current treatment of three-jets, such as exponentiation of some of the terms into a Sudakov form factor (which, as a by-product, would also make it possible to use smaller y cut values).

2.1.4 Higher order matrix elements

A few years ago, three groups calculated the five-jet Born cross-sections [29,30]. This is certainly impressive, and could be useful for specific applications. So far, these formulae have not found wide spread use, for a number of reasons. First, the formulae are lengthy; generation of unweighted events is therefore slow. Second, the actual rate of five-jet production is small, even when increased by the use of an optimized scale as described above. Third, inclusion of five-jets will not help at all for α_S determinations so long as loop corrections to the same order are not available.

The techniques developed not only make it possible to calculate five-jet matrix elements, but also processes of the type $q\bar{q}(ng), n \geq 0$, i.e. with an arbitrary number of gluons [30]. Matrix element evaluation is slow for many gluons, but simpler approximate expressions exist [31], which preserve the full pole structure, and numerically seem to agree with the exact results on the 10% level. These could come in very handy for topological studies.

2.2 Parton Showers

The parton shower picture is derived within the framework of the leading logarithm approximation, the LLA. In this picture, only the leading terms in the perturbative expansion are kept and, where need be, resummed. Subleading corrections, which are down in order by factors of $\ln Q^2$ or $\ln z$ ($\ln(1-z)$), or by powers of $1/Q^2$, are thus neglected. Different schemes have been devised for taking into account some subleading corrections; they appear under names like MLLA (M for modified) or NLLA (N for next-to-) [32]. The overall theoretical picture is rather encouraging: there is reason to believe that neglected subleading effects are small, and the predictive power of this approach is increasing year by year.

Phenomenologically, the main reason for the LLA success is the ability to formulate LLA in terms of a probabilistic picture, suitable for event generation. The approach is based on simplifications in kinematical variables, however, so the predictive power for hard, wide-angle parton emission remains limited.

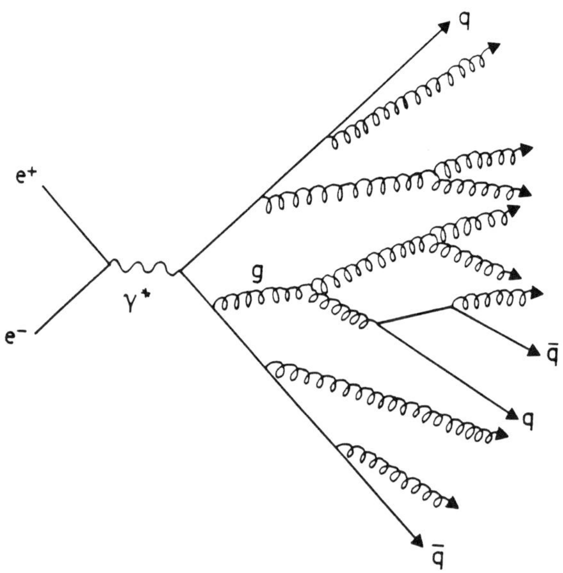

Figure 5. Schematic picture of parton shower evolution in e^+e^- events.

Much of the theory and the analytical predictions are described in the lecture of Khoze [4]. The emphasis here will therefore be on the basic concepts and the Monte Carlo applications.

2.2.1 The evolution equations

Most parton shower algorithms are based on an iterative use of the basic branchings $q \to qg$, $g \to gg$, and $g \to q\bar{q}$, Fig. 5. A probabilistic picture is used to describe these branchings, as follows.

The probability $\mathcal{P}$ that a branching $a \to bc$ will take place during a small change $dt = dQ^2_{evol}/Q^2_{evol}$ of the evolution parameter $t = \ln(Q^2_{evol}/\Lambda^2)$ is given by the Altarelli-Parisi equations [33]

$$\frac{d\mathcal{P}_{a \to bc}}{dt} = \int dz \frac{\alpha_S(Q^2)}{2\pi} P_{a \to bc}(z).$$ (17)

For gluons it is necessary to sum over all allowed final state flavour combinations b and c to obtain the total branching probability. The $P_{a \to bc}(z)$ are the Altarelli-Parisi splitting kernels

$$\begin{aligned}
P_{q \to qg}(z) &= C_F \frac{1+z^2}{1-z}, \\
P_{g \to gg}(z) &= N_C \frac{(1-z(1-z))^2}{z(1-z)}, \\
P_{g \to q\bar{q}}(z) &= T_R(z^2 + (1-z)^2),
\end{aligned}$$ (18)

with $C_F = 4/3$, $N_C = 3$, and $T_R = n_f/2$, i.e. T_R receives a contribution of $1/2$ for each allowed $q\bar{q}$ flavour. The z variable specifies the sharing of four-momentum between the daughters, with daughter b taking fraction z and c taking $1 - z$. Usually the first order α_S is used, eq. (1); as we shall see, the Q^2 scale of α_S need not agree with the evolution scale Q^2_{evol}.

Persons familiar with analytical calculations may wonder why the '+ prescriptions' and $\delta(1-z)$ terms of the splitting kernels in eq. (18) are missing. These complications fulfil the task of ensuring flavour and energy conservation in the analytical equations. The corresponding problem is solved trivially in Monte Carlo programs, where the shower evolution is traced in detail, and flavour and four-momentum are conserved at each branching. The legacy left is the need to introduce a cut-off on the range of the z integral in the Altarelli-Parisi equations, so as to avoid the singular regions corresponding to excessive production of very soft gluons. Typically, this is achieved by introducing an effective (fictitious) gluon mass (in programs usually denoted by Q_0 or $Q_0/2$).

Also note that $P_{g \to gg}(z)$ is given here with a factor N_C in front, while it is sometimes shown with $2N_C$. The difference of a factor of two comes from either considering the number of gluons that branch or the number of gluons that are produced.

2.2.2 The Sudakov form factor

Starting at the maximum allowed virtuality t_{max} for parton a, the t parameter may be successively degraded. This does not mean that an individual parton runs through a range of t values: each parton in the end is associated with a fixed t value, and the evolution procedure is just a way of picking that value. It is only the ensemble of partons in many events that evolve continuously with t, cf. the concept of structure functions. The probability that no branching occurs during a small range of t values, δt, is given by $(1 - \delta t\, d\mathcal{P}/dt)$. When summed over many small intervals, the no-emission probability exponentiates

$$\mathcal{P}_{no-emission}(t_{max}, t) = \exp\left(-\int_t^{t_{max}} dt' \frac{d\mathcal{P}_{a \to bc}}{dt'}\right).\tag{19}$$

Thus the actual probability for a branching of a given t is the naive probability, eq. (17), multiplied by the probability that a branching has not already taken place, eq. (19). This is nothing but the exponential decay law of radioactive decays, with a t-dependent decay probability.

It is customary to introduce the Sudakov form factor, i.e. the probability that a parton starting from a maximum virtuality t will reach the fixed lower cut-off t_{min} (related to the effective gluon mass Q_0) without branching

$$S_a(t) = \exp\left(-\int_{t_{min}}^{t} dt' \int_{z_{min}(t')}^{z_{max}(t')} dz \frac{\alpha_S(Q'^2)}{2\pi} P_{a \to bc}(z)\right).\tag{20}$$

The no-emission probability above is then just $S_a(t_{max})/S_a(t)$.

Once the branching of parton a has been selected, the products b and c may be allowed to branch in their turn, and so on, giving a treelike structure. The branching of a given parton is stopped whenever the evolution parameter is below t_{min}.

2.2.3 Coherence

Very valuable input for model builders is provided by the theoretical studies of corrections beyond leading log, like coherence effects [34,28]. The latter come in two kinds.

- The intrajet coherence phenomenon is responsible for a decrease of the amount of soft gluon emission inside jets. It has been shown that an ordering in terms of a decreasing emission angle takes into account the bulk of soft gluon interference effects. Algorithms which contain angular ordering are loosely said to produce coherent showers, while those without generate conventional ones.

- The interjet coherence phenomenon, responsible for the flow of particles in between jets, with constructive or destructive interference depending on colour configuration ('colour drag phenomena'), cf. [28]. This form of coherence is not just a direct consequence of the ordering of (polar) emission angles mentioned above, but also requires that azimuthal angles of branchings be properly distributed.

Another effect beyond the predictability of leading logs concerns the scale choice in α_S. Here theoretical studies of loop corrections [32] strongly suggest the use of a scale $Q^2 = z(1 - z)m_a^2 \approx p_T^2$, i.e. the scale is set by the transverse momentum of a branching, rather than by the mass of the decaying parton, as might naively have been expected.

2.2.4 Shower programs

A wide selection of shower algorithms have been developed, see [5], which mainly differ in the interpretation of the variables t (or Q^2_{evol}), Q^2 and z. Many of the variations formally are of a subleading character, and therefore are not constrained by theoretical leading log analyses, while others imply quite different physics, at least at higher energies.

While z generically expresses the sharing of energy and momentum between the two daughters, the exact definition varies from program to program. It is possible to define z in terms of energy E, in terms of light-cone momentum $E + p_L$, in terms of $E + |\overline{p}|$, or in terms of some other combination of energy and momentum. The definition chosen affects the allowed region of z values, i.e. the $z_{min}(t)$ and $z_{max}(t)$ functions of eq. (20). The introduction of angular ordering actually constrains the allowed z range, compared to the kinematically allowed one. Since the details of the z interpretation are most significant at low z values, angular ordering as a 'fringe benefit' introduces a reduced dependence on the z choice made.

One should note that none of the z definitions given above are fully Lorentz covariant, and hence neither are most shower algorithms. From a theoretical point of view, this is related to the need to fix a gauge before the branching probabilities can be evaluated (it is e.g. possible to have an algorithm where only one of the initial two partons radiate). The general belief is that the breaking of Lorentz covariance here is a rather minor point, with no experimentally observable consequences ...

The best known coherent shower algorithm is probably the Marchesini-Webber (HERWIG) one [35], where angular ordering is built in from the onset. Here $Q^2_{evol} = E^2\xi$, with $\xi \approx 1 - \cos\theta$. E is the energy of the branching parton and θ the opening angle between its decay products. Since E is fix, a decrease of Q^2_{evol} is equivalent to a decrease of the opening angle in the branching. Once the decay angular variable ξ_a is known, the maximum Q^2_{evol} scale of the two daughters is now given by $E_b^2\xi_a$ and $E_c^2\xi_a$, respectively, because of the requirement of angular ordering. These daughters may be degraded in their turn, to find $\xi_b < \xi_a$ and $\xi_c < \xi_a$, etc.

The Marchesini-Webber algorithm is different from others in several respects. One is that parton masses are not defined during the evolution stage. Only afterwards are parton masses constructed from the final on-shell partons backwards to the shower initiator, by using the opening angle variable, and only after that is the actual kinematics of the shower found. Most other programs construct the kinematics in parallel with the shower evolution proper. Another difference is that the Marchesini-Webber algorithm makes use of an evolution variable which explicitly involves angles, and which thereby automatically incorporates angular ordering.

In many other algorithms, such as the one implemented in JETSET [36], the evolution variable is defined to be $Q^2_{evol} = m^2_a$, i.e. the mass of the decaying parton. This means that angular ordering is not included automatically, but must be imposed as an additional constraint on the combination of m^2 and z values which are allowed for a particular branching, given the m^2 and z values of the preceding branching.

Since showers are so important for phenomenological studies at higher energies, a steady evolution is taking place in the field. Three main approaches can be distinguished for this improvement. One is to refine the standard (modified) leading log picture as such, by including further corrections and effects. Another is to go beyond leading log, and to include also higher order effects. The third is to strive for an alternative (but equivalent) formulation of the shower process, where a number of non-trivial effects are included from the beginning. A few words about each of these.

One way to improve on the basic picture is to include non-isotropic azimuthal angles. This is particularly well explored by HERWIG, in which the following three sources of anisotropy are included:

- Gluon polarization, leading to a correlation between the production and decay planes of a gluon.
- Gluon polarization, also giving correlations between non-adjacent branchings, i.e. of the 'Bell inequality' type [37].
- Soft gluon interference, related to the fact that the soft gluon emission probability is actually obtained by summing emission amplitudes from a number of separate hard partons.

Another possible improvement, found e.g. in JETSET, is to constrain the first branchings of the shower to agree with the explicit three-jet matrix element form. This is an attempt to modify the shower formalism in the region where the kinematical approximations involved are known to be least reliable, while still preserving concepts of the LLA such as the Sudakov form factor and the $Q^2 \approx p^2_T$ argument in α_S.

As the name suggests, NLLjet [38] is a program which tries to go beyond leading order, to Next-to-Leading-Logs. The most apparent consequence is the introduction of $1 \to 3$ parton branchings: $q \to qgg, q \to qq'\overline{q}', g \to ggg$, and $g \to gq\overline{q}$. Also the ordinary $1 \to 2$ branchings are modified, in analogy with the second order three-jet matrix element modifications. Since loop graphs are explicitly involved in calculating the corrections, the Q^2 scale definition is under better control. From that point of view, NLLjet should be superior for Λ determinations. Unfortunately, as for leading logs, the whole approach is really only valid for collinear kinematics. A number of additional assumptions are needed for extrapolations to the interesting region of well separated jets.

An alternative to parton shower algorithms is the dipole formulation of ARIADNE [39], suggested by the Leningrad group [40], and studied in detail by the Lund people. The picture is based on identifying the string pieces between partons with colour dipoles or colour antennae, so that the emission of a gluon corresponds to the breaking of a dipole into two. This breaking is simple and well-defined in the rest frame of a dipole, and yet it automatically includes angular ordering and non-trivial azimuthal effects when the boost back to the overall CM frame is taken into account. The picture has a number of appealing features, e.g. an explicitly Lorentz covariant formulation, and may well come closer to describing nature as it really is than do the other algorithms. This does not mean it is a unique recipe. While angular ordering is built in, the transverse momentum of a dipole branching is a priori not required to be smaller than the p_T of the branching in which it was produced; the model is rather sensitive to what is assumed on this count. Additional degrees of freedom in branchings come from the angular orientation of daughter dipoles with respect to the mother one.

2.3 Testing QCD

The theory of QCD seems to be in excellent shape. This in part reflects on the quality of QCD, but also on the difficulty to construct a consistent alternative that is not already excluded by data. We also know of no good way to slightly 'deform' QCD, such that only subtle differences would remain in experimental observables. This in contrast to the electroweak sector, where e.g. the introduction of a Z' could be used to shift slightly the predictions for the ordinary Z.

Two toy models have been used frequently, just so as to have something to compare with QCD. One is a scalar gluon model, i.e. where the gluon has spin 0 rather than 1. This leads to observable differences already on the three-jet level [7,41]:

$$\frac{1}{\sigma_0}\frac{d\sigma}{dx_1 dx_2} \propto \frac{x_3^2}{(1-x_1)(1-x_2)}, \tag{21}$$

to be contrasted with eq. (2). PETRA three-jet data were used to exclude this possibility [42].

A somewhat more interesting alternative is the Abelian gluon model, where the gluon has spin 1 but does not carry any colour, i.e. where QCD is pretty much like ordinary QED. There are a number of reasons why this model is already counterindicated, ranging from the absence of confinement and the incorrect running of α_S [43] in the model, to the contradiction with Υ decay [44] and $p\bar{p}$ [45] data. The differential three-jet rate is the same as for ordinary QCD (modulo the different running of α_S), so tests based on angular distributions of jets have to be performed by looking at four-jet events. The changes here can be summarized as follows.

- The colour factor C_F, related to the bremsstrahlung of a gluon off a quark, is changed from 4/3 in ordinary QCD to 1 in the Abelian model; effectively this change is absorbed in the necessity to tune α_S to data anyway. The rate of four-jets from double gluon bremsstrahlung therefore remains unchanged.
- The colour factor N_C is changed from 3 to 0; the three-gluon vertex $g \to gg$ is thus absent in the Abelian theory, just like there is no vertex $\gamma \to \gamma\gamma$ in QED.
- The colour factor T_R is changed from $n_f/2$ to $3n_f$, i.e. the branchings $g \to q\bar{q}$ are greatly enhanced. For a $y = 0.01$ cutoff, the $q\bar{q}q'\bar{q}'$ fraction of all four-jets is increased from 4.7% in QCD to 31% in the Abelian model, something which could be directly observable if flavour tagging were available.

A more conventional method to study the differences is to look at angular distributions among the four-jets. Several angles have been defined [46] to gauge the expected differences. As one example, in the production of $q\bar{q}g^*$, the g^* is polarized in the plane of the three-jet event, and this is reflected in its subsequent splitting: in $g^* \to gg$ the azimuthal angle distribution of g around the g^* direction slightly favours splittings in the event plane, while splittings out of the event plane are rather strongly favoured for $g^* \to q'\bar{q}'$. In ordinary QCD the latter process is infrequent, so the net result is close to zero. In the Abelian model the $g^* \to gg$ branchings are absent, while the $g^* \to q'\bar{q}'$ ones are strongly enhanced; thus out-of-the-event-plane splittings are strongly favoured. One trivial consequence is that, while the $q\bar{q}q'\bar{q}'$ fraction of four-jets is around 30% in the Abelian model with a y cutoff, it is more like 50% if instead a cutoff is used based on the aplanarity of events. (In some of the past literature, these two numbers have been mixed up, leading to unnecessary confusion and, at times, to an incorrect simulation of the Abelian model.)

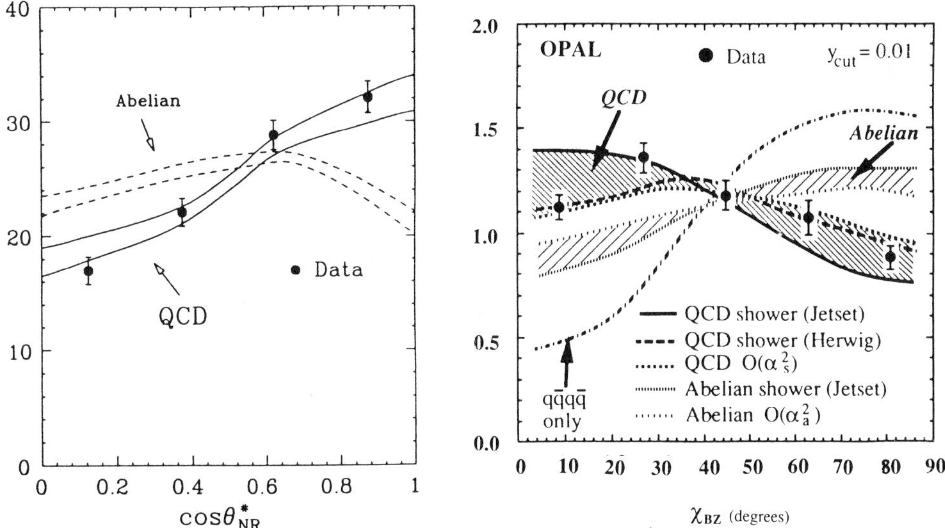

Figure 6. Examples of angular distributions in two of the four-jet angles used to distinguish between QCD and the Abelian model. Left the NR^* variable measured with L3 [48], right the BZ one of OPAL [47].

Results have been presented from OPAL [47] and L3 [48], which both show that the Abelian model is excluded by current LEP data, based on the angular distributions in four-jet events, Fig. 6. However, the most interesting study is that of DELPHI [49], which attempts a two-dimensional analysis in two different angles at the same time, and which uses this to determine the ratio of Casimir operators N_C/C_F and T_R/C_F. This method thus can be used to check on the validity of QCD without any specific assumptions about alternative models — of course, as a by-product, again the Abelian model is excluded. Similar studies have also been made at TRISTAN [50], unfortunately with too low statistics to be as conclusive.

An independent indication for the validity of QCD comes from the running of α_S, Fig. 7, which is based on a study of the variation of the three-jet rates within some fixed y cuts, using the JADE clustering algorithm. The observed running agrees very well with QCD predictions. Thus, several of the fundamental properties of perturbative QCD have been tested at LEP, with good results.

3 Fragmentation

The fragmentation process has yet to be understood from first principles, starting from the QCD Lagrangian. This has left the way clear for the development of a number of different phenomenological models. Being models, none of them can lay claims to being 'correct'. The best one can aim for is a good representation of existing data, plus a predictive power for properties not yet studied or results at higher energies.

All existing models are of a probabilistic and iterative nature. This means that the fragmentation process as a whole is described in terms of one (or a few) simple underlying branchings, of the type jet $\rightarrow$ hadron + remainder-jet, string $\rightarrow$ hadron + remainder-string, cluster $\rightarrow$ hadron + hadron, or cluster $\rightarrow$ cluster + cluster. At each branching, probabilistic rules are given for the production of new flavours, and for the sharing of energy and momentum between the products.

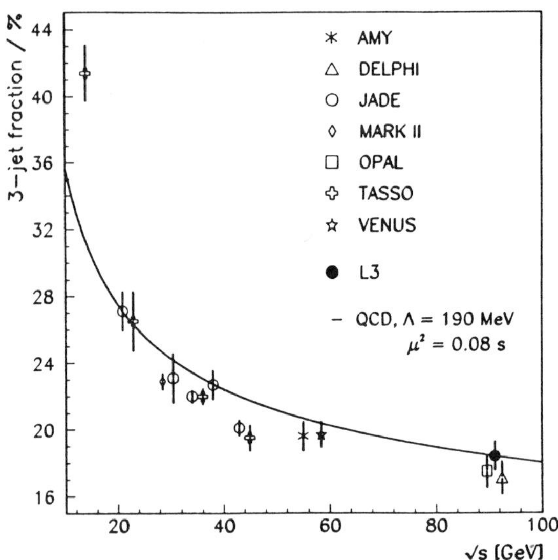

Figure 7. The three-jet fraction as a function of CM energy [23]. The three-jet definition is in terms of the JADE cluster algorithm [21], which is supposed to be insensitive to fragmentation effects (at least for CM energies above 20 GeV). Therefore a fixed α_S should have lead to a constant three-jet fraction, while the observed decrease with CM energy is in good agreement with QCD expectations.

Three main schools are usually distinguished, string fragmentation (SF), independent fragmentation (IF) and cluster fragmentation (CF) [51]. These need not be mutually exclusive; it is possible to have models which contain both cluster and string aspects, or models which interpolate between independent and string fragmentation.

While the evolution of fragmentation models was rapid in the early eighties, no really new algorithms have been introduced in the last five years, and only a modest amount of refinement of the existing approaches has been performed. New concepts, like local parton-hadron duality, and new experimental features, like intermittency, have recently led to a resurgence of fragmentation studies outside the framework of the existing programs. While very interesting, these studies have not yet led to fully-fledged programs of the kind needed to describe LEP events in full, and maybe they never will. A few examples of new ideas will appear later in this report, as a reminder for users to keep an open mind, given that nobody knows the ultimate truth.

3.1 String Fragmentation

The first example of a string fragmentation (SF) scheme was given by Artru and Mennessier [52]. With the elaborate string model developed by the Lund group in the years around 1980 [53], SF became more or less synonymous with the Lund model. However, also other SF or SF-inspired models exist.

Figure 8. A uniform colour flux tube stretched between a q and a $\bar{q}$ endpoint — one possibility of visualizing the linear confinement property.

3.1.1 The string concept

While nonperturbative QCD is not solved, lattice QCD studies lend support to a linear confinement picture (in the absence of dynamical quarks), i.e. the energy stored in the colour dipole field between a charge and anticharge increases linearly with the separation between the charges, if the short-distance Coulomb term is neglected. This is quite different from the behaviour in QED, and is related to the presence of a three-gluon vertex in QCD. The details are not yet well understood, however.

The assumption of linear confinement provides the starting point for the string model, most easily illustrated for the production of a back-to-back $q\bar{q}$ jet pair. As the partons move apart, the physical picture is that of a colour flux tube (or maybe colour vortex line) being stretched between the q and the $\bar{q}$, Fig. 8. The transverse dimensions of the tube are of typical hadronic sizes, roughly 1 fm. If the tube is assumed to be uniform along its length, this automatically leads to a confinement picture with a linearly rising potential. In order to obtain a Lorentz covariant and causal description of the energy flow due to this linear confinement, the most straightforward way is to use the dynamics of the massless relativistic string with no transverse degrees of freedom [54]. The mathematical, one-dimensional string can be thought of as parametrizing the position of the axis of a cylindrically symmetric flux tube. From hadron spectroscopy the string constant, i.e. the amount of energy per unit length, is deduced to be $\kappa \approx 1$ GeV/fm. The expression 'massless' relativistic string is somewhat of a misnomer: κ effectively corresponds to a 'mass density' along the string.

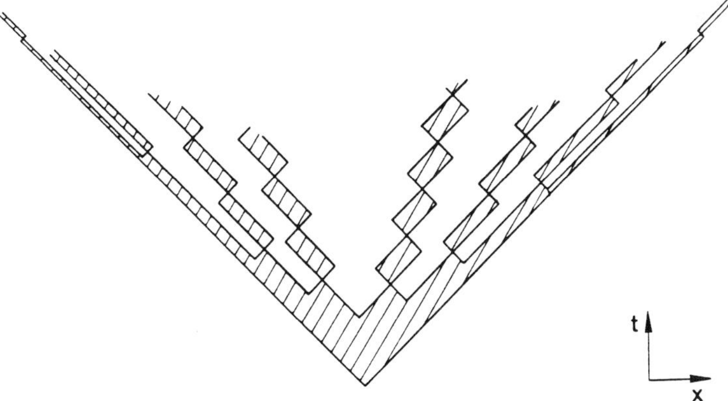

Figure 9. Breaking of a string in the Lund approach, into particles with discrete masses. Time is running upwards, and the longitudinal space dimension is horizontal. Massless quarks are moving along the light cone, corresponding to diagonal lines in the diagram. Hatched areas indicate regions of nonvanishing colour field. The quark from one break combines with the antiquark from the adjacent one to form a meson, in this figure represented by a sequence of rectangles (traced out by the 'yo-yo' motion of the quark and antiquark endpoints).

3.1.2 Fragmentation of a simple string

Let us now turn to the fragmentation process, and for simplicity consider a $q\bar{q}$ two-jet
event. As the q and $\bar{q}$ move apart, the potential energy stored in the string increases,
and the string may break by the production of a new $q'\bar{q}'$ pair, so that the system splits
into two colour singlet systems $q\bar{q}'$ and $q'\bar{q}$. If the invariant mass of either of these
string pieces is large enough, further breaks may occur. In the Lund string model,
the string break-up process is assumed to proceed until only on-mass-shell hadrons
remain, each hadron corresponding to a small piece of string, Fig. 9. Alternatively,
string breaking may be stopped at a typical mass scale of one or a few GeV, where
the string pieces are associated with clusters (see section 3.3).

In order to generate the quark-antiquark pairs $q'\bar{q}'$ which lead to string break-ups,
the Lund model invokes the idea of quantum mechanical tunnelling. In terms of the
transverse mass m_T of the q', the tunnelling probability (i.e. the probability that the
$q'\bar{q}'$ will appear) is given by

$$\exp\left(-\frac{\pi m_T^2}{\kappa}\right) = \exp\left(-\frac{\pi m^2}{\kappa}\right)\exp\left(-\frac{\pi p_T^2}{\kappa}\right). \tag{22}$$

The factorization of the transverse momentum and the mass terms leads to a flavour-
independent Gaussian spectrum for the p_T of $q'\bar{q}'$ pairs. Since the string is assumed
to have no transverse excitations, this p_T is locally compensated between the quark
and the antiquark of the pair. The total p_T of a hadron is made up out of the
p_T contributions from the quark and antiquark that together form the hadron. In a
perturbative QCD framework, a hard scattering is associated with gluon radiation, and
further contributions to what is naively called fragmentation p_T comes from unresolved
radiation. This is used as an explanation why the experimental $\langle p_T \rangle$ is somewhat
higher than obtained with the formula above.

The formula also implies a suppression of heavy quark production $u : d : s : c \approx$
$1 : 1 : 0.3 : 10^{-11}$. Charm and heavier quarks hence are not expected to be produced
in the soft fragmentation. Since the predicted flavour suppressions are in terms of
quark masses, which are notoriously difficult to assign (should it be current algebra,
or constituent, or maybe something in between?), the suppression of $s\bar{s}$ production is
left as a free parameter in the Lund program. At least qualitatively, the experimental
value agrees with theoretical prejudice.

When the quark and antiquark from two adjacent string breakings are combined
to form a meson, it is necessary to invoke an algorithm to choose between the different
possibilities allowed, notably to choose between pseudoscalar and vector mesons. Here
the string model is not particularly predictive. Qualitatively one expects a $1 : 3$ ratio,
from counting the number of spin states, multiplied by some wave function normalization factor, which should disfavour heavier states. Again, the relative composition
is left as a free parameter in Monte Carlo implementations.

A tunnelling mechanism can also be used to explain the production of baryons.
This is still a poorly understood area. In the simpest possible approach, a diquark
in a colour antitriplet state is just treated like an ordinary antiquark, such that a
string can break either by quark-antiquark or antidiquark-diquark pair production.
The production probabilities are then given by the effective diquark masses assumed,
plus simple flavour Clebsch-Gordan coefficients of the baryon wavefunctions. In this
approach, the baryon and antibaryon are produced next to each other, and share
(at least) two quark flavours. A more complex scenario is the 'popcorn' one, where
diquarks as such do not exist, but rather quark-antiquark pairs are produced one after
the other. Part of the time, this scenario gives back an effective diquark picture,
but in addition configurations are possible where one or more mesons are produced

in between the baryon and antibaryon, and where therefore these two are no longer required to be as strongly correlated in flavour content.

In general, the different string breaks are causally disconnected, see Fig. 9. This means that it is possible to describe the breaks in any convenient order, e.g. from the quark end inwards. One therefore is led to write down an iterative scheme for the fragmentation, as follows. Assume an initial quark q moving out along the $+z$ axis, with the antiquark going out in the opposite direction, and with total light cone momentum $W_+ = E + p_L$, where $p_L = p_z$ is the longitudinal momentum along the jet axis. By the production of a $q_1\bar{q}_1$ pair, a meson $q\bar{q}_1$ is produced, leaving behind an unpaired quark q_1. The sharing of the W_+ is given by some probability distribution $f(z_1)$, where z_1 is the fraction taken by the hadron, with $1 - z_1$ left for the remainder-jet q_1. A second pair $q_2\bar{q}_2$ may now be produced, to give a new meson $q_1\bar{q}_2$, which takes a fraction z_2 of the remaining light cone energy and momentum, $(1 - z_1)W_+$, etc. This process may be iterated until all energy is used up, with some modifications close to the $\bar{q}$ end of the string to make total energy and momentum come out right. By the relation $(E + p_L)(E - p_L) = m_T^2 = m^2 + p_T^2$, the peeling off of W_+ in fact also implies a reduction in the available $W_- = E - p_L$; it is necessary to conserve both.

The choice of starting the fragmentation from the quark end is arbitrary, however. A fragmentation process described in terms of starting at the $\bar{q}$ end of the system and fragmenting towards the q end should be equivalent. This 'left-right' symmetry constrains the allowed shape of fragmentation functions $f(z)$, where z is once again the fraction of $E + p_L$ along the jet axis chosen, i.e. $p_L = \pm p_z$ for quark/antiquark. Under some simplifying assumptions, the left-right symmetric fragmentation function takes the form

$$f(z) \propto \frac{1}{z}(1 - z)^a \exp(-bm_T^2/z), \tag{23}$$

with the two free parameters a and b.

If the objects produced during string break-up are not on-shell hadrons, as is the case e.g. in the Artru-Mennessier model, the logical recipe is to have a constant probability for the string to break per unit of invariant space-time area swept out by the string. This gives an exponential area decay law, if one remembers that a string can not break in the forwards light cone of an earlier break, i.e. when it has 'already' broken. Further, left-right symmetry is ensured. The specification of adjacent break-ups in the two-dimensional sheet of the string can be translated into a simultaneous probability distribution in the mass-squared and the z of the produced cluster:

$$\frac{d\mathcal{P}}{dm^2 dz} \propto \frac{1}{z} \exp(-bm^2/z), \tag{24}$$

with b a free parameter proportional to the probability of break-up per invariant unit of string area. The mass-spectrum can be obtained by integrating out z, and is proportional to $\mathrm{E}_1(bm^2)$, i.e. it possesses a logarithmic divergence at small masses (in programs often removed by a minimum cluster mass requirement). Once the m^2 is given, the fragmentation function can be read off eq. (24), i.e. it is exactly the same as was obtained in the discrete mass case, eq. (23), except that $a = 0$. The similarity in the final result is all the more surprising, since a discrete mass spectrum implies that string breaks are allowed only along one-dimensional hyperbolae, rather than inside two-dimensional areas.

3.1.3 Fragmentation of a multiparton system

If several partons are moving apart from a common origin, the details of the string drawing become more complicated. For a $q\bar{q}g$ event, a string is stretched from the q

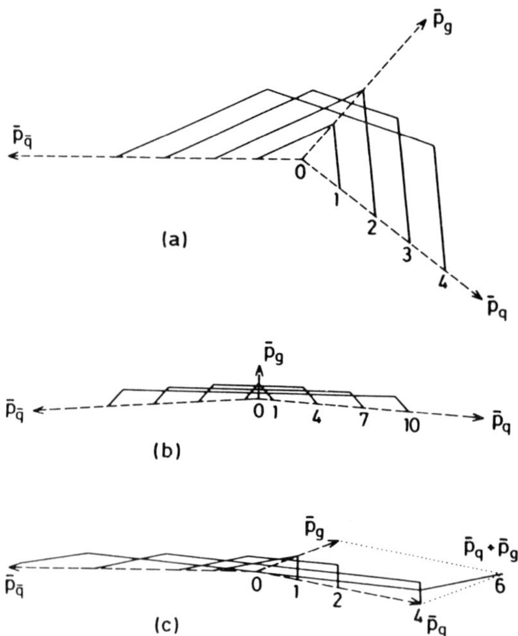

Figure 10. The string drawing for (a) an ordinary three-jet event, (b) a three-jet event with a soft gluon, and (c) a three-jet event with a collinear gluon. Dashed lines give the momenta (and hence the trajectories) of the partons. Full lines give the string shape at different times, with numbers representing time in some suitable scale.

end via the g to the $\bar{q}$ end, i.e. the gluon is a kink on the string, carrying energy and momentum, Fig. 10a. As a consequence, the gluon has two string pieces attached, and the ratio of gluon/quark string force is two, a number which can be compared with the ratio of colour charge Casimir operators, $N_C/C_F = 2/(1 - 1/N_C^2) = 9/4$. In this, as in other respects, the string model can be viewed as a variant of QCD where the number of colours N_C is not 3 but infinite. Note that the factor 2 above does not depend on the kinematical configuration: a smaller opening angle between two partons corresponds to a smaller string length drawn out per unit time, but also to an increased transverse velocity of the string piece, which gives an exactly compensating boost factor in the energy density per unit string length.

Consider the string motion in a $q\bar{q}g$ event. In the string piece between the g and the q ($\bar{q}$), g four-momentum is flowing towards the q ($\bar{q}$) end and q ($\bar{q}$) four-momentum towards the g end. Such packets of energy and momentum are called 'genes' [54]. When the gluon has lost all its energy, the g four-momentum continues moving away from the middle (i.e. where the gluon used to be), and instead a third string region is formed there, consisting of inflowing q and $\bar{q}$ four-momentum, Fig. 10a. If this third region would only appear at a time later than the typical time scale for fragmentation, it could not affect the sharing of energy between different particles. This is true in the limit of high energy, well separated partons.

For a small gluon energy, on the other hand, the third string region appears early, and the overall drawing of the string becomes fairly two-jetlike, Fig. 10b. In the limit of vanishing gluon energy, the two initial string regions collapse to naught, and the ordinary two-jet event is recovered. Also for a collinear gluon, i.e. θ_{qg} (or $\theta_{\bar{q}g}$) small, the stretching becomes two-jetlike, Fig. 10c. These properties of the string motion

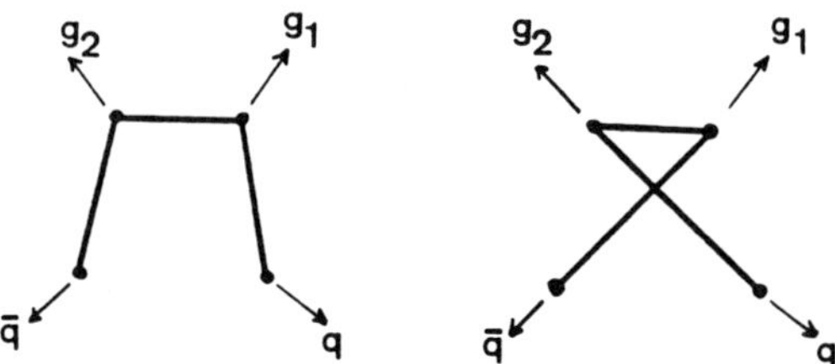

Figure 11. The two different ways of stretching a string in a $q\overline{q}gg$ event.

are the reason why the string fragmentation scheme is 'infrared safe' with respect to soft or collinear gluon emission.

In an event with several gluons, these still will appear as kinks on the string between the q and $\overline{q}$ ends. With several gluons present, the full string evolution may become rather complicated, but the basic principles remain the same as outlined for three-jet events. In particular two nearby partons (with matching colours) always collectively draw out the string.

For four-jet events (or events with more than four jets) which are generated using matrix elements, there are several possible topologies for the ordering of partons along the string. This is illustrated in Fig. 11 for $q\overline{q}gg$ events. A knowledge of quark and gluon colours, obtained by perturbation theory, would uniquely specify the stretching of the string [55], so long as the two gluons do not have the same colour. The probability for the latter is down in magnitude by a factor $1/N_C^2$, where $N_C = 3$ is the number of colours. Perturbative QCD gives no answer on how to handle these situations, but recipes have to be included in event generators.

In leading log shower programs, where only $1 \to 2$ branchings are included, the rules for colour flow in branchings are well-defined.
- $q \to qg$: the original colour of the quark is taken by the gluon, and a new colour-anticolour pair is shared by quark and gluon.
- $g \to gg$: the original colour is inherited by one gluon and the anticolour by the other, again a new colour-anticolour pair is shared.
- $g \to q\overline{q}$: the original colour goes to the quark and the anticolour to the antiquark; this means that the original string going through the gluon is split into two.

If one goes beyond this approximation, as in NLLjet, the same problems will arise as in the four-jet matrix elements, and similar solutions have to be found.

Whichever scenario is considered, matrix elements or showers, there is a tacit assumption that soft gluon exchanges between partons will not mess up the original colour assignment. Both theoretical and experimental arguments can be raised as to why this might be a fair approximation, but eventually this assumption is something which needs more rigorous testing [56].

3.2   Independent Fragmentation

The independent fragmentation (IF) approach dates back to the early seventies [57], and gained widespread popularity with the Field-Feynman paper [58]. Subsequently, IF was the basis for two programs widely used in the early PETRA/PEP days, the Hoyer *et al.* [59] and the Ali *et al.* [60] programs. These programs have not been updated since their conception, and therefore are not well suited for LEP physics studies. Anyway, as will be seen, the IF concept very much is in disrepute these days, and what little use there is can be fully covered by the IF options available in JETSET.

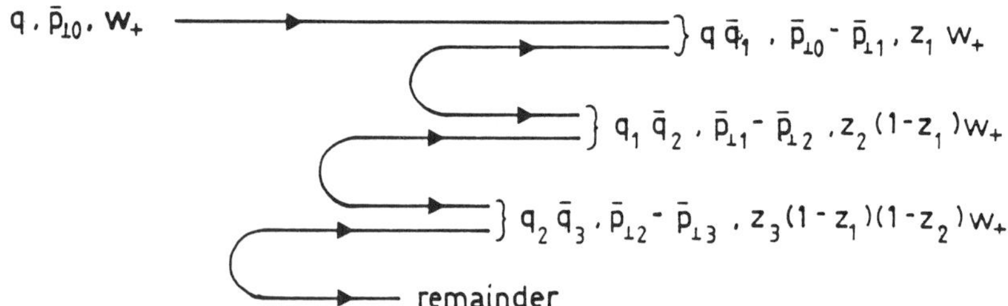

Figure 12. The iterative ansatz for flavour, transverse momentum, and light-cone energy-momentum ($W_+ = E + p_L$) fraction.

3.2.1 Single jet fragmentation

As in the case of SF, the fragmentation of a jet is described iteratively, see Fig. 12. From an original quark jet q, hadrons are split off one by one, leaving behind new remainder-jets with scaled-down energies. The function $f(z)$, which describes how big a fraction z of the remaining energy is taken by the hadron, is assumed to be the same at each step, i.e. independent of remaining energy. No special philosophy constrains the choice of $f(z)$, so several different proposals exist in the literature. If z is interpreted as the fraction of the jet $E + p_L$, this leads to a flat central rapidity plateau dn/dy for a large initial energy.

The normal z interpretation means that a choice of a z value close to 0 corresponds to a particle moving backwards, i.e. with $p_L < 0$. It makes sense to allow only the production of particles with $p_L > 0$, but to explicitly constrain z accordingly would destroy longitudinal invariance. The most straightforward way out is to allow all z values but discard hadrons with $p_L < 0$. Flavour, transverse momentum and $E + p_L$ carried by these hadrons are 'lost' for the forward jet. The average energy of the final jet comes out roughly right this way, with a spread of 1 - 2 GeV around the mean. The jet longitudinal momentum is decreased, however, since the jet acquires an effective mass during the fragmentation procedure. For a two-jet event this is as it should be, at least on average, because also the momentum of the compensating opposite-side parton is decreased.

Each splitting corresponds to the production of a $q'\bar{q}'$ pair. The relative composition of these pairs is parametrized as in the string model, with corresponding possibilities to extend the model also to baryon production. Also the way in which the q_1 from one break is combined with the $\bar{q}_2$ from an adjacent break to form the different possible hadrons is the same, as is the way the transverse momenta from the constituent quarks are added to give the transverse momentum of the hadron. The main differerence, at this point, is that the string model is based on a specific underlying space-time picture of the fragmentation process, while the iterative approach is introduced *ad hoc* in independent fragmentation.

Contrary to the string model, however, within the IF framework there is no unique recipe for how gluon jet fragmentation should be handled. One possibility is to treat it exactly like a quark jet, with the initial quark flavour chosen at random among $u, \overline{u}, d, \overline{d}, s$ and $\overline{s}$, including the ordinary s quark suppression factor. Since the gluon is supposed to fragment more softly than a quark jet, the fragmentation function may be chosen independently. Another common option is to split the g jet into a pair of parallel q and $\overline{q}$ ones, sharing the energy, e.g. according to the Altarelli-Parisi splitting function in eq. (18). The fragmentation function could still be chosen independently, if so desired. Further, in either case the fragmentation p_T could be chosen to have a different mean.

3.2.2 Fragmentation of a jet system

The concept of IF inevitably leads to the total flavour, energy and momentum not being exactly conserved in the fragmentation process proper. At the end of the generation, special algorithms are therefore used to patch this up.

Usually little attention is given to flavour conservation. Typically, that aspect is solved by reassigning the flavour content of centrally produced particles, without changing their three-momenta, in such a way that the net number of each flavour is vanishing.

Several different schemes for energy and momentum conservation have been devised. One [59] is to conserve transverse momentum locally within each jet, so that the final momentum vector of a jet is always parallel with that of the corresponding parton. Then longitudinal momenta may be rescaled separately for particles within each jet, such that the ratio of rescaled jet momentum to initial parton momentum is the same in all jets. Since the initial partons had net vanishing three-momentum, so do now the hadrons. The rescaling factors may be chosen such that also energy comes out right. In another common approach [60] a generated event is boosted to the frame where the total hadronic momentum is vanishing. After that, energy conservation can be obtained by rescaling all particle three-momenta by a common factor.

Depending on what choice is made, the implementation of momentum conservation may either force three-jet events to become even more three-jetlike, or to become more two-jetlike. As a consequence, the α_S value needed to describe data can be rather different. The difference decreases with increasing CM energy, but could still be 20% – 30% at LEP energies. The discarding of independent fragmentation models is the main reason why the range of published α_S values is so much smaller at LEP energies than it was at PETRA, not the higher energies or the refined choice of observables.

A serious conceptual weakness of the IF framework is the issue of Lorentz invariance. The outcome of the fragmentation procedure depends on the coordinate frame chosen, a problem circumvented by requiring fragmentation always to be carried out in the CM frame. This is a consistent procedure for two-jet events, but only a technical trick for multijets.

It should be noted, however, that a Lorentz covariant generalization of the independent fragmentation model exists, in which separate 'gluon-type' and 'quark-type' strings are used, the Montvay scheme [61]. For a three-jet event, the three string pieces are joined at a junction. The motion of this junction is determined by the vector sum of the string tensions acting on it. In particular, it is always possible to boost an event to a frame where this junction is at rest. In this frame, much of the standard naive IF picture holds for the fragmentation of the three jets; additionally, a correct treatment would automatically give flavour, energy and momentum conservation. Unfortunately, while the scheme is perfectly valid also with several gluon jets present, it is then very complicated to write an event generator based on it, and nobody has ever done so.

Another weakness of IF is the issue of collinear divergences. In a parton shower picture, where a quark or gluon is expected to branch into several reasonably collimated partons, the independent fragmentation of one single parton or of a bunch of collinear ones gives quite different outcomes, e.g. with a much larger hadron multiplicity in the latter case. Results therefore are sensitive to the choice of shower cutoff scale Q_0: any working set of fragmentation functions would have to be retuned significantly if Q_0 were changed.

As we will see in section 4.5, actually IF fails to describe the 'string effect'. This means that also from an experimental point of view, IF is not a particularly interesting model anymore. Attempts are occasionally made to revive IF [62], but these are not particularly convincing.

3.3 Cluster Fragmentation

While the Artru-Mennessier model [52] contained clusters, dedicated cluster models were first developed at CALTECH [63]. Today, cluster models are found in HERWIG and in the CALTECH-II program [64]. The main difference between existing cluster fragmentation (CF) schemes is the extent to which string fragmentation ideas are incorporated.

- In one extreme, a parton shower picture is used to produce a partonic configuration. At the end of the shower evolution, remaining gluons are split forcibly into $q\bar{q}$ pairs. With colour explicitly kept track of, the quark of one splitting may be combined with the antiquark from an adjacent one to form a colourless cluster, Fig. 13. These clusters subsequently decay into the final hadrons. This is, more or less, the HERWIG strategy.

- In the other extreme, parton showers or matrix elements may be used to generate a partonic configuration, with strings stretched in between the partons, as described in the SF section. These strings then fragment into clusters, which again decay into the final hadrons. Fig. 14 illustrates this approach, which is found in CALTECH-II.

3.3.1 The cluster concept

The concept of cluster fragmentation offers the great promise of a simple, local and universal description of hadronization. Gone are the long, ordered fragmentation chains present both in SF and IF. In their place appear simple clusters, which are assumed to be the basic units from which the hadrons are produced. A cluster is ideally only characterized by its total mass and total flavour content, i.e. unlike a string it does not possess an internal structure. If the shower evolution and/or string breaks are chosen such that most clusters have a mass of a few GeV, the cluster mass spectrum may be thought of as a superposition of fairly broad (i.e. short-lived) resonances. Phase-space aspects may then be expected to dominate the decay properties. This applies both for the selection of decay channels, and for the kinematics of the decay. Thus a decay is assumed to be isotropic in the rest frame of the cluster. This gives a compact description with few parameters. In particular, the separate longitudinal and transverse momentum fragmentation descriptions in SF and IF are here replaced by a unified framework, wherein parton showers and cluster phase-space decays give the full momentum distribution. As we shall see in the following paragraphs, there are some complications, however.

Ideally, parton shower evolution should in itself give a cluster mass spectrum

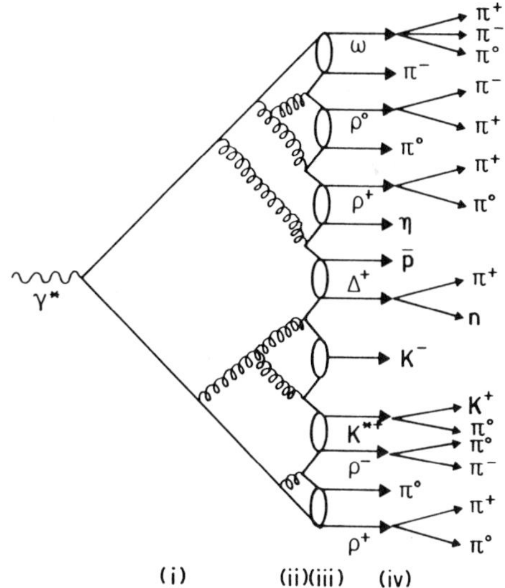

Figure 13. One cluster fragmentation scenario: (i) shower evolution, (ii) forced $g \to q\bar{q}$ branchings, (iii) cluster formation, and (iv) cluster decay. Occasionally, a cluster is associated with a single particle.

strongly damped at masses above a few GeV, so that two-body decays of clusters would give a sufficient description. This was the hope in the early days of CF, a hope founded on the concept of 'preconfinement' [65]. Unfortunately, despite the preconfinement property, there is no known way of avoiding a rather large spread of cluster masses. In programs, it is therefore necessary to introduce the possibility for a high-mass cluster to produce more than two hadrons. This is typically done by allowing branchings cluster $\to$ cluster + hadron or cluster $\to$ cluster + cluster.

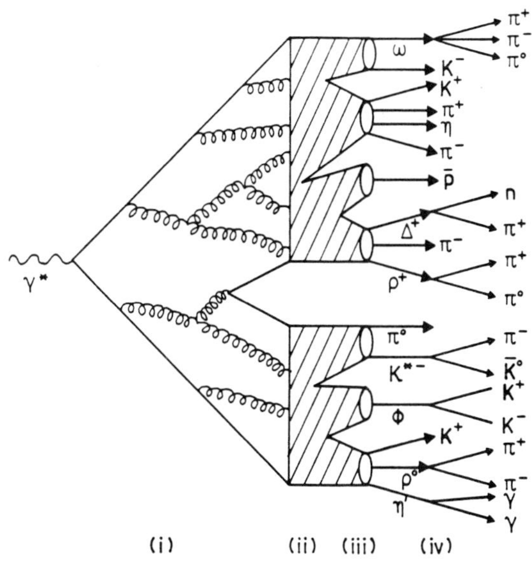

Figure 14. Another cluster fragmentation scenario: (i) shower evolution, (ii) string evolution and break-up into (iii) clusters, (iv) cluster decay into a varying number of particles.

3.3.2 Cluster formation and decay

Flavours are generated at several different stages. First, at the branchings $g \to q\bar{q}$, alternatively at the string breaks, i.e. when the clusters are formed. In CALTECH-II the relative probabilities appear as explicit parameters, while they are given by the parton mass assignments in HERWIG. In the original Webber-Marchesini model, only quark-antiquark pairs (or gluons) were created at branchings. This leads to some problems in the description of baryon production, see section 4.3.4. The present model therefore also includes, as an option, the possibility that a gluon may branch into a diquark-antidiquark pair. This process is turned on below some scale in the shower evolution, with an arbitrary strength relative to ordinary $q\bar{q}$ production.

A second stage of flavour production occurs when larger clusters decay into smaller ones. Typically, this means that a cluster $q_1\bar{q}_2$ breaks, by the production of an intermediate $q_3\bar{q}_3$ pair, into clusters $q_1\bar{q}_3$ and $q_3\bar{q}_2$. One of the two may, but need not, be directly associated with a hadron. In general, the symbol q may here represent either a quark or an antidiquark.

The third stage of flavour production is when a cluster decays into two hadrons. The flavour flow is as above, i.e. a new $q_3\bar{q}_3$ pair splits the old cluster in the middle. Again quark and diquark production is allowed, with relative probability dictated by phase-space alone.

The phase-space assumption means that each allowed cluster decay channel is assigned a weight proportional to the density of states, $(2s_1 + 1)(2s_2 + 1)(2p^*/m)$. Here s_1 and s_2 are the spins of the two hadrons produced, and p^* the common momentum of the products in the rest frame of the decaying cluster (with mass m). The weight gives the relative probability of the choice being retained; in case of rejection a new q_3 flavour is selected and the procedure repeated.

It should be noted that the new q_3 quark flavour is not associated with any dynamical properties, such as a mass or, for diquarks, a total spin. It is only the properties of the final, 'observable' particles that can influence the relative production rate. Further, the 'fragmentation' transverse momentum is determined by the average energy release in cluster decay, and in subsequent resonance decays, as opposed to the extra parameter needed in SF or IF.

In the decay of a large cluster into two, the kinematics is usually handled anisotropically, along the 'string' direction. The same kind of phase-space weight may still be used, provided clusters are assigned suitable spins and a cluster mass spectrum weight is folded in.

If all clusters are to decay into at least two particles, the probability of producing a single particle carrying a large fraction of the total jet energy is severely underestimated. Therefore cluster programs usually contain a mechanism, so that a sufficiently light cluster is assumed to collapse into a single particle. Four-momentum is shuffled to or from nearby clusters, so as to achieve overall energy and momentum conservation.

3.4 Other Fragmentation Approaches

As should be amply clear from the discussions above, the study of fragmentation is not a closed chapter, with one simple framework that does it all. In addition, none of the models discussed are able to describe all experimental features. This means that a broad spectrum of alternatives should be pursued. A few alternatives are described in [5]. They include the following ones.

- Local Parton-Hadron Duality (LPHD) [28]. In this approach, the fragmentation step is done away with altogether: by continuing the parton shower approach down to very small virtualities, a situation is reached where partons can be identified on a one-to-one basis with the final observable hadrons. How flavours and flavour correlations should be introduced in this approach is not addressed, but all kinds of energy and multiplicity correlations can be predicted analytically, and successfully [4].

- Intermittency models. Given the large interest in intermittency, a number of toy models have been developed to describe non-Poissonian particle production processes, which would give rise to additional fluctuations compared to the standard fragmentation approaches [66].

- Topological models. In a model proposed by Ellis and Kowalski, baryons are produced at topological defects [67]. These defects appear when chiral symmetry is broken during the fragmentation phase, and adjacent domains (corresponding to the final mesons) have a mismatch in chiral direction. A main prediction is a larger rate of baryon production at LEP energies than in conventional models.

- Non-QCD models. It may be interesting to see whether it is possible to construct models which do not make any explicit reference to standard QCD, and yet can be made to describe data. A few different suggestions are here available, where the original fragmenting object is an excited hadron with the full CM energy of the event, which cascades over a sequence of other excited hadron states down into the final light hadrons [68]. In this kind of approach, there is no reference at all to a quark/gluon phase at early times of the fragmentation.

3.5 Particles and Their Decays

A large fraction of the particles produced by fragmentation are actually unstable, and subsequently decay into the observable stable (or almost stable) ones. It is therefore important to include all particles with their proper mass distributions and decay properties. In fact, although involving little of deep physics, this is less trivial than it may sound: much information is available in the Review of Particle Properties [18], but there is also very much missing. If one goes beyond the lowest-lying pseudoscalar or vector meson multiplets, then many multiplet members are poorly known. One still gets by, since the contribution from higher multiplets usually is assumed to be small, or at least to give small effects in the observable hadron distributions.

Many resonances are very short-lived, so that a simple Breit-Wigner description of the mass distribution is not realistic. However, to include the correct shape is messy, since then the production and the decay of a particle become intertwined. Therefore usually simplified approaches are introduced, which work fine for general event shapes, but are insufficient for detailed studies of resonance production.

Further, the decay treatment involves a guessing at branching ratios when these are poorly known, like for B mesons. Even when a large fraction of the decays are known, like for the D mesons, the decay channels used in a program have to add up to 100%, i.e. extra plausible decay channels have to be added by hand. Also with the decay products known, it is often not sufficient just to distribute them according to phase space. An obvious example is weak semileptonic decays, where the produced electron or muon is well measured experimentally, and an accurate modelling is required e.g. to separate contributions from b and c quarks for quark tagging.

Finally, the effects of particle polarization often are neglected: particles are assumed to decay isotropically. In τ decays we know this is not correct, and there are also other cases where polarization effects might be important.

4 Experimental Knowledge

There are several reviews of results in e^+e^- physics at lower energies [69,51], and a summary of LEP QCD results is found in the lecture of Ward [27]. In this section therefore the objective is to illuminate the interplay between model building and experimental tests. For this purpose, a few examples will be picked.

4.1 Event Characterization Methods

Since the typical LEP hadronic event contains 20 charged particles and as many neutral ones, it is convenient to characterize an event in terms of collective variables. These variables may be classified as event measures, cluster topologies, one-particle inclusive distributions, two-particle inclusive correlations, etc.

The list of one-particle inclusive distributions include transverse and longitudinal momentum spectra, rapidity distributions, etc. The most well-known two-particle inclusive distribution is probably the energy-energy correlation [70]

$$EEC(\theta) = \sum_{i<j} \frac{2E_i E_j}{E_{CM}^2} \delta(\theta - \theta_{ij}),\tag{25}$$

and its asymmetry $EECA(\theta) = EEC(\pi - \theta) - EEC(\theta)$.

4.1.1 Thrust and Sphericity

One frequently used family of event measures is the thrust one [71]. Thrust T is defined as

$$T = \max_{|\overline{n}|=1} \frac{\sum_i |\overline{p}_i \overline{n}|}{\sum_i |\overline{p}_i|},\tag{26}$$

where the index i runs over all the particles of the event, and where the axis $\overline{n}$ for which maximum is obtained is called the thrust axis. Ideal two-jet events have $T = 1$, while perfectly spherical events have $T = 1/2$. Major is defined the same was as thrust, but with the major axis constrained to be orthogonal to the thrust one. The minor axis is chosen orthogonal both to the thrust and to the major ones, and the minor measure is defined as the scaled sum of projected momenta along this axis, just like for thrust and major. The difference between major and minor is called oblateness.

Another frequently used family is the sphericity one [72]. Here one defines a tensor

$$S^{ab} = \frac{\sum_i p_i^a p_i^b}{\sum_i p_i^2},\tag{27}$$

with $a, b = 1, 2, 3$ corresponding to the x, y and z directions in space. From the three eigenvalues $\lambda_1 > \lambda_2 > \lambda_3$, with sum unity, one may construct sphericity $S = 3(\lambda_2 + \lambda_3)/2$ and aplanarity $A = 3\lambda_3/2$. The former quantity runs between 0 and 1, with 0 corresponding to ideal two-jets and 1 to spherical events. The eigenvectors of the tensor can be used to define an event axis and an event plane.

Both the thrust and the sphericity family have their advantages and disadvantages [73]. There is no simple fast recipe to find the correct thrust axis; either one has to pick a slow algorithm or an approximate one. Further, the behaviour of the thrust axis can be pathological: a slight change of the momenta in an event can flip the axis to an altogether different position. The sphericity axis is easy to find, but the sphericity measure itself suffers from the disadvantage of being quadratic in momenta. This means that, if one single particle is split into two collinear ones, the sphericity measure and sphericity axis is changed, i.e. sphericity is sensitive to irrelevant details of parton branchings and particle decays. Therefore the limiting procedure of eq. (4) does not work, i.e., unlike thrust, sphericity cannot be calculated perturbatively.

Considering these known facts, it is maybe surprising that so little attention has been given to using better, or at least alternative, methods, in particular since such are already known. The most obvious is to use a linearized version of the sphericity method, i.e. to replace the sphericity tensor above by [74]

$$S_{lin}^{ab} = \frac{\sum_i p_i^a p_i^b / |\overline{p}_i|}{\sum_i |\overline{p}_i|}, \tag{28}$$

i.e. where one power of $|\overline{p}_i|$ has been removed from both numerator and denominator. This measure can be diagonalized just like the sphericity one, and eigenvectors are easy to find. Yet it is linear in momentum, and therefore infrared safe and perturbatively calculable. What little experience we have also indicates that event axes found this way are every bit as good, however that should be defined, as either sphericity or thrust axes.

4.1.2 Cluster algorithms

The usage of cluster algorithms has become more and more widespread. Today, the JADE algorithm [21] is the most popular one. In this algorithm a distance measure between clusters i and j is defined by

$$y_{ij} = \frac{2E_i E_j (1 - \cos \theta_{ij})}{E_{vis}^2}, \tag{29}$$

where E_{vis} is the total visible energy of the event. (The usage of E_{vis} in the denominator, rather than e.g. E_{CM}, tends to make the measure less sensitive to detector acceptance corrections.) The algorithm works as follows. Initially there is one cluster for each of the particles of the event. The y_{ij} value is calculated for each cluster pair, and the pair with the smallest y_{ij} is joined into one single cluster, with four-momentum given by adding the constituent four-momenta. The event now has one cluster less. Once again the pair with smallest y_{ij} is found, a new joining is made, etc. The recombination procedure is iterated until all distances $y_{ij} > y_{cut}$, where y_{cut} is a predetermined jet resolution power. The larger the y_{cut} scale, the smaller the final number of clusters.

The JADE algorithm has several advantages. The y_{cut} scale is very closely related to the y resolution parameter defined in the matrix element description (section 2.1.1). Specifically, if y_{cut} is chosen equal to y, the distribution in the number of clusters found very closely matches the generated parton distribution. This gives a direct possibility to test perturbative QCD with a minimum of worry about nonperturbative effects.

The successes of the JADE algorithm are legio. It is maybe therefore necessary to issue a general warning against blind faith in it.

- The algorithm is not Lorentz covariant, contrary to what is sometimes claimed. It is well known (since ten years or so) that an algorithm which uses the true invariant mass, i.e. $(E_i + E_j)^2 - (\overline{p}_i + \overline{p}_j)^2$ rather than $2E_i E_j (1 - \cos \theta_{ij})$, is

unstable: such an algorithm will start the clustering where most particles are
to be found, i.e. in the center of the event, and give quite funny clustering
assignments. The JADE algorithm breaks Lorentz covariance by favouring clustering of faster particles over that of central ones, and thus gives more reasonable
results. Premature joinings of central particles still remains one of the major
problems of the algorithm.

- The algorithm is not insensitive to fragmentation effects. It is true that, if string
  fragmentation is used, the number of clusters on the parton and on the hadron
  level agree well, at least for reasonably large CM energies. However, if independent fragmentation were to be used, quite large fragmentation corrections would
  appear. Thus, if the situation today looks much simpler than it did in the old
  PETRA/PEP days, it is not so much because we have hit on more fragmentation model independent measures, but rather because a number of fragmentation
  models taken seriously then are no longer in use today.

- While the JADE algorithm does a good job of finding the number of partons, it is
  not optimal for finding jet directions or energies. One reason is that the distance
  measure allows the joining of particles which go in quite different directions, if
  only they are soft. Another is that two particles, once joined, can never be
  reassigned to separate final clusters. Thus, in the angular region between two
  jets, a set of low-energy particles could be joined at an early stage, and then go
  en masse to one of the jets.

It may be interesting to compare with another cluster algorithm, such as the one
found in JETSET [75]. Here the distance measure is

$$d_{ij} = \frac{2p_i^2 p_j^2 (1 - \cos\theta_{ij})}{(p_i + p_j)^2}. \tag{30}$$

For two nearby particles ($\sin\theta_{ij} \approx 2\sin(\theta_{ij}/2)$) the d_{ij} measure has a simple geometrical interpretation as the p_T^2 of each of the two particles with respect to their vector
sum. The underlying physics idea is to define a jet as a collection of particles with
a limited transverse momentum with respect to the jet direction, rather than as a
collection of particles with a small invariant mass.

The algorithm works the same way as the JADE one, in that the number of clusters
is successively reduced by one, until all cluster pairs are separated by a distance above
some cutoff value. However, after each joining, all particles of the event are reassigned
to the cluster to which they are closest, again in terms of the d_{ij} measure. If some
low-momentum cluster is formed at an early stage, it can thus very well later be split
between two high energy jets.

The net result of these two differences is that the JETSET algorithm gives narrower
jets, which better agrees with the naive visual impression of jet structure than the
JADE algorithm clusters do, and that jet directions are better reconstructed. For
four-jet (triple gluon vertex) studies it is therefore likely to be the better algorithm.
However, the number of clusters reconstructed agrees less well with the number of
partons above some given y cut, the algorithm is even less Lorentz covariant than the
JADE one, and it is every bit as much fragmentation model dependent. It is therefore
not a panacea — today we do not have *one* algorithm that is best for everything!

4.2   Event Shapes

A number of event shape studies at LEP have already been presented [76]; two examples of the distributions obtained are given in Fig. 15. Generally, very good agreement

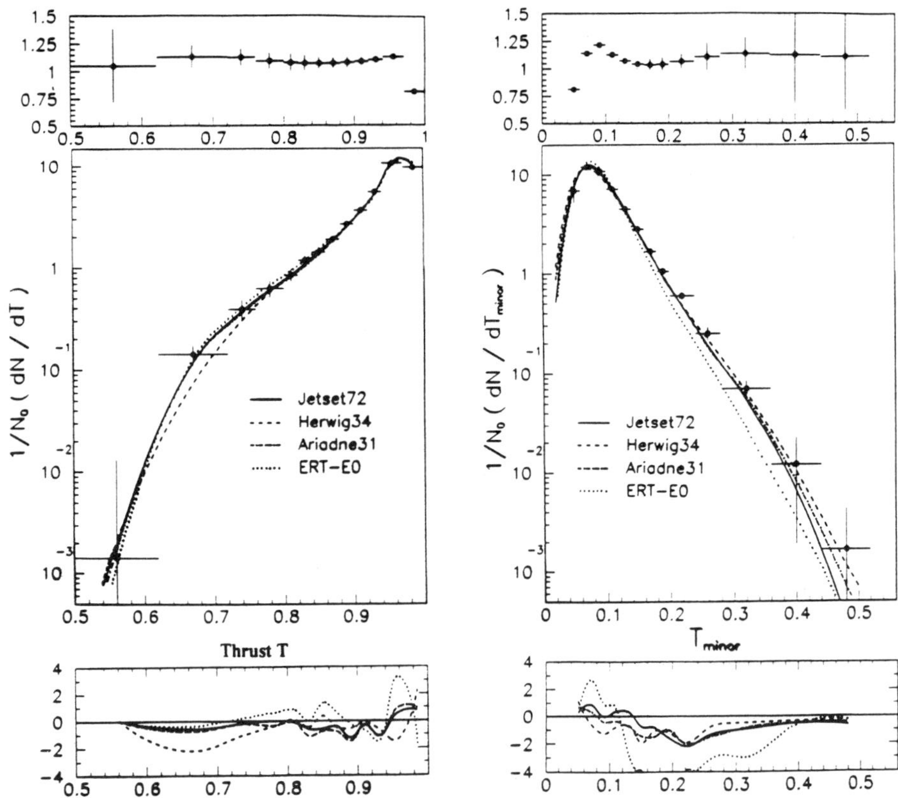

Figure 15. The thrust and minor distributions observed by OPAL, compared with the results of a few models [76]. Top: correction factors applied to the data for unfolding of detector effects. Bottom: difference between model predictions and data, in units of standard deviation.

is obtained between the data and models based on parton showers plus string or cluster fragmentation, be that JETSET, HERWIG or ARIADNE. It is not even necessary to tune the parameters of the programs to LEP data; already the values determined at around 30 GeV [77,78] give quite good descriptions.

A closer look reveals some interesting patterns. All programs provide a better description of LEP data than they do of PETRA/PEP data. Since the parton shower aspects are more dominant at larger energies, relative to fragmentation effects, this would indicate that the shower is fairly well simulated in all the programs studied. Despite large superficial differences, the JETSET, HERWIG and ARIADNE shower algorithms are pretty much based on the same kind of approach to the parton production process, and this approach indeed seems very successful — good news for experimentalists who need extrapolations to higher energies.

Conversely, the fact that agreement is worse at lower energies tends to indicate that fragmentation aspects are not as well modelled as are the parton shower ones. Here we also see a marked difference between JETSET and HERWIG, with the latter providing a clearly worse description at PETRA/PEP. In other words, most likely the cluster fragmentation model of HERWIG is inferior to the string one of JETSET.

Matrix element based models, if tuned at lower energies, fail miserably to describe LEP data. Particle multiplicities are too small and jets too narrow; one way or another this is reflected in distribution after distribution. These shortcomings can all be traced

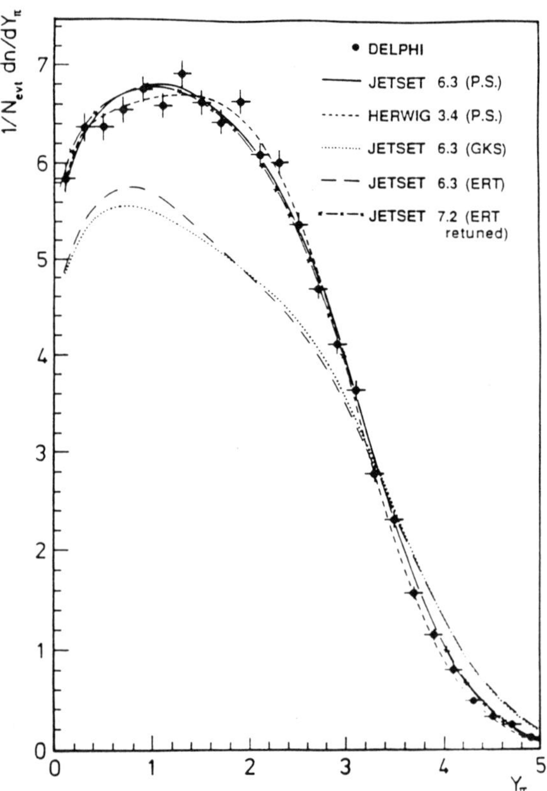

Figure 16. The rapidity distribution of charged particles (assumed to be all pions) with respect to the thrust axis, for shower programs (P.S.), matrix element programs extrapolated from lower energies (GKS and ERT), and for retuned matrix elements (ERT retuned), from DELPHI [76].

to the inability to account for multiple (semi)soft gluon emission, because of the need to keep sensical two-, three- and four-jet rates, i.e. to use cutoffs $y \geq 0.01$. The failure of the matrix element approach at LEP was predicted beforehand [77], and fits well in with our current picture of perturbative QCD, where parton showering is assumed to go on down to mass scales of around 1 GeV, much below acceptable matrix element cutoffs.

Granted that no energy-independent parametrization is possible in the matrix element approach, one may wonder if it is possible to tune parameters separately for each energy. In particular, could one make the fragmentation function softer and the fragmentation transverse momentum larger as the CM energy is increased, i.e. try to include the emission of soft gluons into some effective fragmentation parameters, in such a way that good agreement could still be obtained with LEP data? When this approach was tried years ago, at PETRA/PEP/TRISTAN energies, the answer was no: already there the matrix element approach was clearly inferior to the parton shower one [77,78,79]. The outcome at LEP was therefore never in doubt.

However, a new aspect has entered the game in the last few years: optimized perturbation theory. Many of the problems of the past were related to the lack of four-jet events in the matrix element implementations. When today we allow ourselves the additional freedom of increasing the four-jet rate by the use of a lower Q^2 scale in α_S, in fact it is possible to achieve quite good agreement between data and matrix elements even at LEP energies [80], see e.g. Fig. 16. This is rather a disappointment, since it indicates that any attempt to look at jet structures beyond four-jets will have to cope with an overwhelming fragmentation background. One of the tasks of the future will be to find measures where the tuned-up matrix element approach fails, and from there to gain increased understanding of soft gluon emission.

4.3 Multiplicities and Correlations

4.3.1 Multiplicity distributions

The average charged multiplicity at LEP agrees well with QCD extrapolations from lower energies [76,81], both for analytical formulae and for parton shower based programs. However, as already noted, a matrix element approach with fixed fragmentation parameters gives too slow an increase of multiplicity with energy.

Also the multiplicity distribution itself is well described in models. If plotted in terms of $z = n/\langle n \rangle$ (multiplicity over average multiplicity), the distribution is independent of CM energy, to a good degree of accuracy — KNO scaling [82]. This is less trivial than it may sound. It has been shown that, at asymptotic energies, the parton shower gives rise to an approximate negative binomial parton multiplicity distribution which KNO scales [83]. At currently explored energies the parton distribution is narrower, and does not KNO scale with energy. However, here fragmentation effects induce an additional broadening of the multiplicity distribution. Only the sum of the two contributions gives approximate KNO scaling over a wide range of energies [84]. In the framework of the parton shower plus string fragmentation model, the crossover between the fragmentation dominated and the shower dominated region is at around 50 GeV.

4.3.2 Particle composition

So far, no LEP results have been presented on the particle composition of events. From PETRA/PEP energies, a reasonably complete experimental picture has emerged [85,18], which thus remains to be tested. Noteworthy is that none of the existing fragmentation models does more than a passable job in describing the baryon composition, Fig. 17. JETSET does least bad, but contains quite a few parameters. CALTECH-II is about as wasteful with parameters, and still does not fit. The cluster decay of HERWIG is the only economical model, since the particle composition is given by the showering history and simple phase space considerations, but this program does worst.

Given the current state of affairs, there have been some attempts to find alternative descriptions. In the UCLA variation of the Lund model [86], the probability to produce different hadrons is determined by the integral of the Lund symmetric fragmentation function over all z values, i.e. by a monotonically decreasing function of the hadron mass, times Clebsch-Gordan coefficients of the quark content. Quark and diquark masses do not enter the game. This gives a description of comparable quality to the standard Lund one, but with significantly fewer parameters.

Disregarding the problems with the absolute normalizations, the momentum spectra of particles are reasonably well described. However, a closer look reveals problems e.g. in the p/π ratio as a function of particle momentum [87]; this distribution is not quite well described in either existing model.

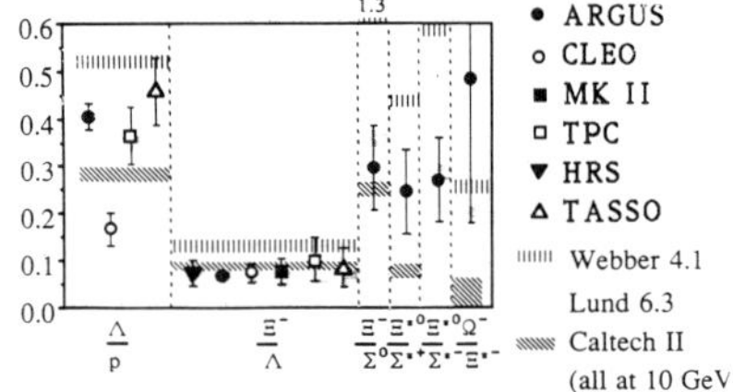

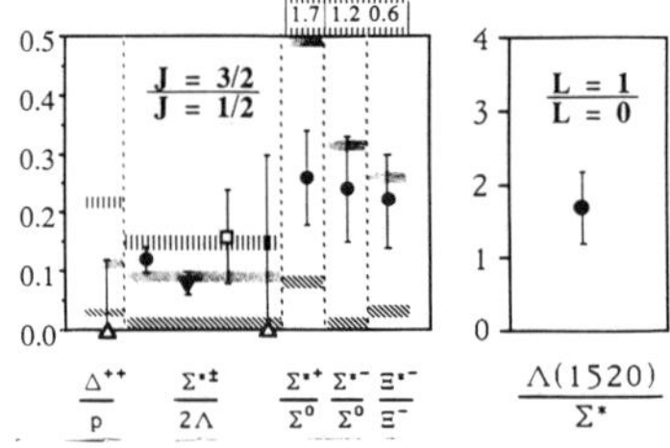

Figure 17. Compilation of baryon production ratios at 10 — 35 GeV, compared with Webber (HERWIG precursor), Lund (JETSET) and CALTECH-II predictions at 10 GeV, for baryons with varying (a) strangeness content, and (b) angular momentum [85].

4.3.3 Intermittency

One of the most discussed topics in recent years has been that of 'intermittency' [88,66,5], a term borrowed from the theory of turbulence. In high energy physics, it is used to denote multiplicity fluctuations which become stronger the smaller the regions of phase space considered. To distinguish trivial and non-trivial fluctuations, one introduced factorial moments [88]. If, e.g., the total rapidity range Y is divided into M bins, each with a width $\delta y = Y/M$, then the i:th factorial moment is defined as

$$F_i(\delta y) = \frac{M^{i-1}}{\langle N \rangle^i} \left\langle \sum_{m=1}^{M} n_m(n_m - 1) \cdots (n_m - i + 1) \right\rangle, \tag{31}$$

where n_m is the multiplicity in bin m, N is the total multiplicity and $\langle \cdots \rangle$ indicates average over all events in the sample. Slightly different averaging procedures are possible, but they all share the same basic properties. The factoral moments are constructed such that they are identically equal to unity for a longitudinal phase space model with purely Poissonian multiplicity fluctuations. The signal for 'true intermittency' is that the factorial moments keep on increasing as M is made larger, according to a power law behaviour, $F_i \propto (Y/\delta y)^{f_i} = M^{f_i}$.

However, there are also sources of 'false intermittency', where an approximate power law behaviour may be observed over a range of δy values, but where asymptotically F_i goes to a constant:

- Resonance production.
- Non-flat rapidity distributions, from edge effects, event axis determinations, particle misidentifications, etc.
- Jet production.
- Bose-Einstein effects.

(The list above does not include purely experimental problems, e.g. that a track with a kink may be reconstructed as two separate nearby tracks.) The first two effects typically contribute mainly for $\delta y > 1$, while the last one is believed to give only minor contributions in current studies. The main effect therefore is jet production: if a jet is produced at a given rapidity, its fragmentation will give rise to several hadrons close to this rapidity. Since the typical transverse width of a jet is at around 1 GeV, factorial moments will increase down to scales of $\delta y \approx (1 \text{ GeV})/p_{Tjet}$, where p_{Tjet} is the tranverse momentum of the jet with respect to the event axis. The amount of 'false intermittency' increases with the CM energy, since the phase space for gluon emission becomes larger.

It is a general misconception that an intermittency signal from jets has to be related to cascading, specifically to parton shower evolution. Even if only three-jet events were produced, i.e. events with no intermittency on the parton level, the fragmentation process would give particles at nearby rapidities, and hence rising factorial moments. In fact, the introduction of a parton shower picture rather leads to a broadening of jet profiles and thus to a reduction of factorial moments compared to the three-jet case [89].

Signals for intermittency were first observed in hadronic collisions. The first indirect evidence in e^+e^- came from an analysis of HRS data [90]. Later the TASSO analysis showed that programs like JETSET and HERWIG, when tuned to describe other data, cannot describe the factorial moments [91]. The disagreement is both in the variation as a function of δy and in the absolute level. In particular, the data points keep on rising for very small bin sizes, where the models tend to flatten out, see Fig. 18.

At LEP energies, so far only DELPHI has presented an intermittency analysis [92]. Contrary to TASSO, good agreement is observed between models and data, both in absolute level and in a tendency for F_i distributions to flatten out at small bin sizes, Fig. 18. If the analysis is extended to a simultaneous binning in both rapidity and azimuth, the factorial moments do not seem to flatten out, but again data and Monte Carlo follow suit.

The discrepancy between TASSO and DELPHI does not seem to be just a matter of having experiments at different energies, since recently CELLO has presented data taken at the same energy as TASSO, which disagrees with TASSO but agrees with model predictions [93]. One cannot exclude the possibility that the TASSO analysis is flawed, so the conservative judgement would be that intermittency in e^+e^- still remains to be convincingly demonstrated.

Finally, a personal comment on this controversial issue. People in the field have developed a terminology of their own, which still is evolving rapidly. A year ago, a factorial moments distribution had to have a power law rise at asymptotically small bin sizes to be called intermittent. By now it has been shown that a distribution, which follows a power law in three dimensions, may have projections on one dimension

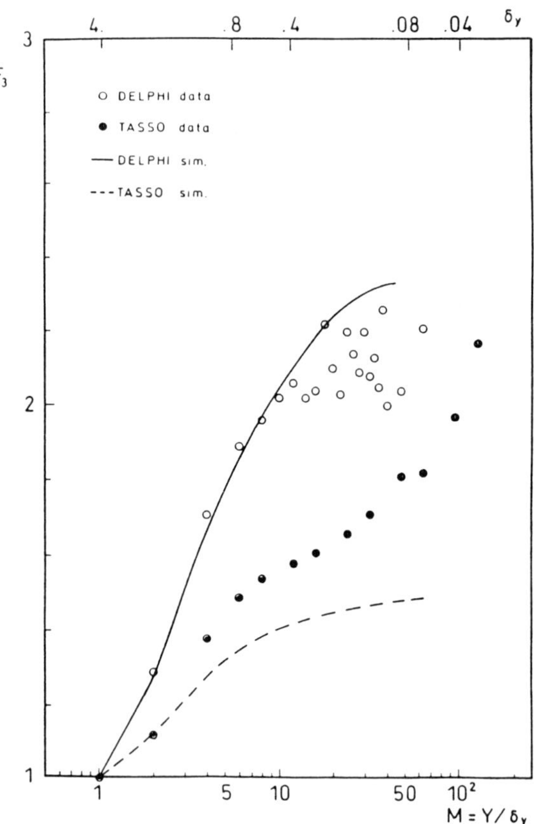

Figure 18. The third factorial moment as a function of the number of bins, TASSO 35 GeV and DELPHI 91 GeV data (circles) compared with JETSET shower + string fragmentation program at the same energies (lines) [92].

which flatten out at small bin sizes. Therefore any factorial moments distribution which shows an increase over some significant region of δy values, like the DELPHI ones do, today automatically is called intermittent. However, *if* the LEP data can be perfectly well described by the standard parton shower plus fragmentation pictures, then to use the intermittency language is rather uneconomical and imprecise, however 'photogenique' the term may be. It is the belief of this author that e^+e^- annihilation may not be the right process in which to look for 'true' intermittency, whatever that is, but that the challenge is to be found in the understanding of hadron collision data.

4.3.4 Baryon pair correlations

Flavour correlations have not yet been studied at LEP. At lower energies, nontrival results have been obtained. A particularly useful probe is baryon correlations, since baryons are rare enough that usually only one baryon-antibaryon pair is present in an event, and baryons are also less affected by resonance decays.

One example is the TPC/2γ study of the distribution in the opening angle θ between the event axis and the internal axis of the $p\bar{p}$ pair, the latter axis defined by boosting to the rest frame of the $p\bar{p}$ pair [94,85]. In a cluster fragmentation picture, where clusters are not assumed to have any net baryon number, a flat distribution is expected in $\cos\theta$, slightly modified by events with several baryon-antibaryon pairs.

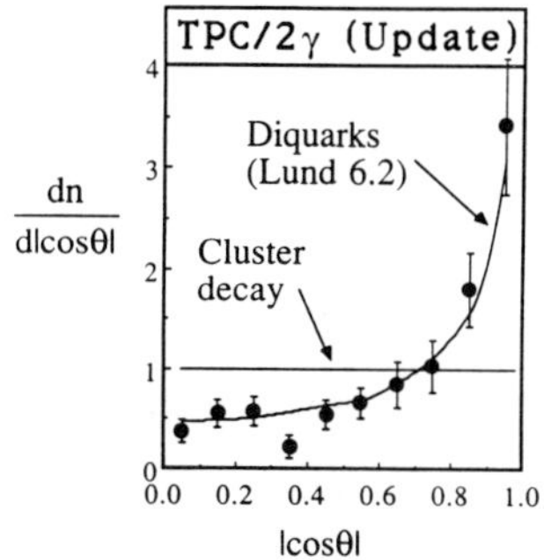

Figure 19. Distribution in $|\cos\theta|$, with θ the angle between the event axis and the $p\bar{p}$ axis in the rest frame of the pair; TPC/2γ data compared with isotropic cluster decay and JETSET baryon production [94,85].

Instead the data show a strong peaking at $\cos\theta = \pm 1$, Fig. 19, as predicted in the Lund model, where the baryon and antibaryon are pulled apart along the string direction.

Further, correlation studies have indicated that when baryons are produced, they do not predominantly appear as nearest neighbours in rank (i.e. sharing a diquark-antidiquark pair), but rather are separated by at least one intermediate meson most of the time.

4.4 Jet Type Separation

4.4.1 Heavy vs. light quark jets

Not all quark jets are expected to be the same. Among the light u, d, and s quarks, differences are expected to reflect the different charge and strangeness content. If quark and antiquark jets are averaged over, and if π/K mesons are not separated, such differences are minor, however.

The situation is more interesting for the heavier c and b quark jets. Here it is well established that the heavy flavour jet deposits a major fraction of its energy into the leading charm or bottom hadron, and that this fraction is larger the heavier the quark is. This is most succinctly expressed by the Bjorken formula [95]

$$\langle z \rangle \approx 1 - \frac{1\,\text{GeV}}{m_Q}, \tag{32}$$

where m_Q is the heavy flavour quark mass and $\langle z \rangle$ is the fraction of total jet energy (or momentum) taken by the heavy flavour hadron. The philosophy underlying this formula is that heavy quarks are not expected to be significantly decelerated during the fragmentation process.

Many explicit fragmentation functions have been based on this principle, the most famous being the Peterson *et al.* one [96],

$$f(z) \propto \left[z \left(1 - \frac{1}{z} - \frac{\epsilon_Q}{1-z} \right)^2 \right]^{-1}, \tag{33}$$

which is based on arguments in old-fashioned perturbation theory for the process $Q \to H(= Q\bar{q}) + q$. Here one expects $\epsilon_Q = (m_0/m_Q)^2$, with m_0 some reference scale related to light hadrons.

The Lund symmetric fragmentation function, eq. (23), is the only publicly propounded function which breaks the Bjorken relation. Instead the asymptotic behaviour here is given by

$$\langle z \rangle \approx 1 - \frac{1+a}{bm_H^2}, \tag{34}$$

with a and b the same parameters as used for ordinary light hadrons, and m_H the mass of the heavy hadron. Compared to the Bjorken approach, differences are not large for charm hadrons, while the Lund picture predicts a harder fragmentation function for B mesons.

The discussion above only covers the nonperturbative fragmentation aspects. In addition, it is necessary to take into account the effects of gluon emission, e.g. in the framework of one of the standard parton shower programs (for a recent analytical evaluation, see [97]). The gluon emission effects increase with increasing CM energy, and lead to a continuous softening of the observable fragmentation function.

At PETRA energies, data were not precise enough to really distinguish between the Bjorken-Peterson and the Lund philosophies, although some slight preference could be given to the former [98]. The ratio of $b\bar{b}$ to $c\bar{c}$ events is considerably more favourable at LEP, and so experiments have been able rapidly to present some first results on the average z value of B hadrons (indirectly, by studying lepton spectra) [99]. These values agree well with extrapolations from lower energies of the Peterson function [5], but are significantly softer than the predictions obtained with the Lund fragmentation function.

This means that some of the assumptions of the Lund approach have to be re-examined, at least for the case of heavy flavours. Other string-based fragmentation functions have been proposed, in particlar by Bowler [100] (derived in the framework of the Artru-Mennessier model, see section 3.1.2):

$$f(z) \propto \frac{1}{z^{1+bm_Q^2}} \exp(-bm_T^2/z). \tag{35}$$

(It would be possible to motivate the introduction of a $(1-z)^a$ term, to bring the form even closer to the standard Lund one [64].) There is some hope that this could provide a viable string alternative to the Peterson function. Obviously, once the actual shape of the fragmentation function is measured well, rather than just $\langle z \rangle$, it should be possible to differentiate better between the alternatives.

4.4.2 Gluon vs. quark jets

The gluon has a larger colour charge than a quark has. In perturbative QCD this is reflected in a larger probability for a gluon to radiate: referring back to the Altarelli-Parisi splitting kernels of eq. (18), the ratio of gluon to quark bremsstrahlung probability is roughly $N_C/C_F = 9/4$. The difference in colour charge should also be reflected in the nonperturbative treatment, as in the Lund string model, where a gluon is attached to two string pieces but a quark only to one. Altogether, there is therefore strong theoretical support to the idea that gluon jets ought to be softer than quark ones.

A number of factors act to obscure the theoretical picture: the naive factor of two multiplicity difference between quark and gluon jets is only expected at very large energies; presence of heavy quarks tend to soften the average quark jet, while comparisons with gluons should properly be made for light quarks only; it is difficult to know which is the gluon jet in a three-jet event; etc. It is therefore maybe not surprising that it has turned out to be very difficult to establish such differences. A review of the situation is given in the lecture of Kim [101]. With the larger jet energies and huge statistics available at LEP, this is obviously one area where one would expect the situation soon to clarify.

4.5 String and Coherence Phenomena

In a three-jet event, the Lund model is based on having a string stretched from the quark via the gluon to the antiquark, Fig. 10. The string piece between the quark and the gluon has a transverse motion out along a direction intermediate to the quark and gluon directions. The particles which are produced when the string piece breaks therefore receive a Lorentz boost, such that slow particles systematically are shifted slightly away from the origin. A corresponding boost in a direction intermediate to the gluon and antiquark directions is required for the string piece spanned by these two partons. Since there is not a string piece spanned directly between the quark and antiquark, no particles are produced in between these two partons, except by 'leakage' from the other two regions, by transverse momentum fluctuations and particle decays. In the Lund string picture, there is therefore a direct prediction that the region between the quark and antiquark directions should be significantly less populated than the two other regions between jets [102]. This contrasts with the behaviour in the independent fragmentation framework, where fragmentation takes place symmetrically around each of the three jet directions, and therefore none of the three regions between jets occupies a special position.

Comparisons with data have tended to favour the Lund scenario, and disfavour the independent fragmentation one [103]. One example of this kind of studies is shown in Fig. 20. The effects that are experimentally observed are actually much smaller than the ones predicted on the Monte Carlo level — it is difficult to know which jet is the gluon one, and therefore the true effect is reduced by the influence of events where the gluon is misidentified. So far, essentially all studies have been based on the assumption that the jet with lowest energy is the gluon one, which typically is true only 60% of the time. Obviously this is one area where the much higher statistics promised by LEP would help; one possibility would be to tag the quark and antiquark of an event by the presence of prompt leptons from semileptonic charm and bottom decays. Even so, the experimental evidence is fairly convincing, and few models manage to describe the data.

The Leningrad group has shown that the 'string effect' appears as a natural consequence of coherence phenomena in the parton shower evolution [40,4]. In lowest order, this may be viewed as follows. Start out with a quark, an antiquark and a gluon, all three with approximately the same energy, and let the three partons act as antennae that emit soft gluons in a semiclassical pattern. Due to interference effects between the colour charges of the three partons, there is then a surplus of radiation in the q–g and g–$\bar{q}$ regions, and a depletion in the q–$\bar{q}$ one. If a term proportional to $1/N_C$ (i.e. a colour suppressed term) is dropped, the two remaining terms may be interpreted as simple qg and $\bar{q}g$ dipole radiation, boosted from the the qg and $\bar{q}g$ rest frames into the overall $q\bar{q}g$ CM frame.

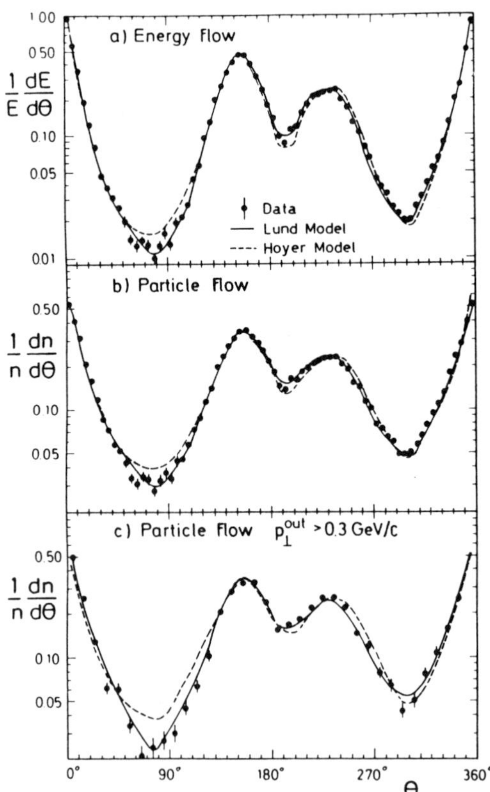

Figure 20. Energy and particle flow in three-jet events, with the rightmost valley likely to correspond to the q–$\bar{q}$ angular region. JADE data compared with the Lund string model and the Hoyer independent fragmentation one [103].

The scenario above literally repeats the explanation given in the string model, with the important difference that, where the string picture is based on purely nonperturbative deliberations, the colour antennae picture is purely perturbative. Despite the sharing of a common ideological basis, namely the importance of the colour flow in events, the two explanations are not describing quite the same physics. The perturbative and nonperturbative scenarios can be distinguished by more detailed tests, tests which can be carried out by comparing data at LEP with those from lower energies. Some examples can be found in [104], but here we will present another one [105].

Consider, for simplicity, a symmetric three-jet event. Denote by E_{qg} $(= E_{\bar{q}g})$ the energy flow in the middle of the angular region between the q and g jet directions (e.g. integrated over the central $0.4 \cdot 120° = 48°$ of this region), and by $E_{q\bar{q}}$ the corresponding energy flow between the q and $\bar{q}$ jet directions. In the string picture, E_{qg} and $E_{q\bar{q}}$ are essentially independent of the CM energy: as the energy is increased, more particles are produced along the three jet directions (at large momenta), but the center of the string pieces between the jets remain the same. In the perturbative framework, on the other hand, it is the *fraction* of energy radiated into gluons between the jets that stays the same, modulo the running of α_S and issues related to the cutoff of the shower evolution at low virtualities. One therefore expects $E_{qg}, E_{q\bar{q}} \propto \alpha_S E_{CM}$. In forming the ratio $E_{qg}/E_{q\bar{q}}$, as has traditionally been done, the energy dependence is factored out, so that both approaches predict a constant ratio as function of CM energy.

410

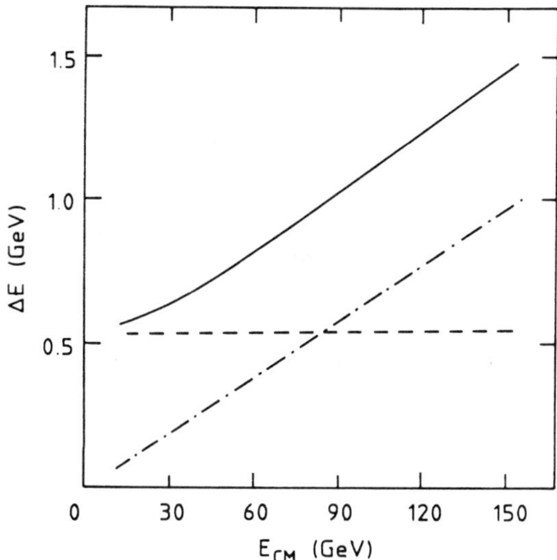

Figure 21. The difference ΔE of energy flows in the q–g and q–$\bar{q}$ angular ranges of symmetric three-jet events. In this figure only u and d quarks are included, and α_S is chosen fix, $= 0.20$. No jet misidentification or other experimental problems are included. The dashed line shows string fragmentation of three-jets only, the dashed-dotted line the energy flow on the parton level in a parton shower approach, and the full curve the hadronic energy flow when parton shower evolution is followed by string fragmentation. All curves have been obtained with JETSET.

The energy difference, $\Delta E = E_{qg} - E_{q\bar{q}}$, however, preserves the contrast between the two approaches. This is illustrated in Fig 21. Furthermore, it allows for a world in which not all is 'black and white', but where both perturbative and nonperturbative contributions coexist. In fact, in the full JETSET simulation, which includes both parton shower evolution and string fragmentation, it turns out that the two contributions add up more or less incoherently. It would therefore be motivated to fit data at different energies to a form

$$\Delta E(E_{CM}) = E_{qg}(E_{CM}) - E_{q\bar{q}}(E_{CM}) = c_1 + c_2\alpha_S E_{CM}. \tag{36}$$

The presence of both these terms in the data would thus demonstrate the complementary nature of the perturbative and the nonperturbative contributions. Obviously, perturbative effects are guaranteed to win out eventually, but LEP is still at around the crossover point (if the JETSET simulations are to be believed), and so is ideally placed to address this kind of issues. Needless to say, the experimental problems are considerable, in particular since gluon jet misidentifications significantly reduce observable effects, and since it is necessary to compare data obtained with different detectors operating at different energies. If nothing else, the discussion above provides a 'gedanken experiment', to help understand the distinction between perturbative and nonperturbative effects better.

5   Summary and Outlook

The perturbative aspects of QCD are fairly well under control. The running of α_S and the structure of the fundamental couplings of QCD (with the exception of the four-gluon one) have already been successfully tested at LEP (with some backing from

PETRA/PEP/TRISTAN data). In the light of the issue of optimized Q^2 scales, one may have different opinions about how well α_S is really determined, but this does not affect how well physical quantities can be predicted.

The matrix element approach has made a remarkable comeback. For a long time it was considered dead as a viable alternative for the description of event shapes at LEP energies. In particular, the lack of four-jet events was a serious shortcoming. The wholehearted adoption of optimized perturbation theory has at least solved that problem. If fragmentation parameters are retuned, i.e. if soft gluon effects are bundled together with the nonperturbative fragmentation description, it is even possible to get quite acceptable descriptions of the generic event properties studied so far.

We should not be swept off our feet by these successes, however. The estimate of higher order effects by the use of optimized Q^2 scales is no substitute for a full-length actual calculation of, in particular, the loop corrections to the four-jet rate. Furthermore, the need for quite a significant retuning of fragmentation parameters when going from PETRA/PEP to LEP energies shows that the effects of soft gluons, which are not included in the matrix element treatment, are large.

The alternative perturbative description is the parton shower one. Originally, this approach had no particular claim to accurate predictions, but with time the level of ambition has increased. We may reasonably expect further progress in the future, although the pace may be slower than in the past. One particular advantage of the parton shower approach is that we expect all model parameters to be independent of the CM energy. This indeed seems to be the case experimentally, which gives us confidence in predicting event shapes at energies beyond what has been observed so far.

On the fragmentation side, there used to be three competing models. Of these, independent fragmentation no longer is a serious alternative, but occasionally still is used as a contrast to the other two. Cluster fragmentation in principle is the model with least free parameters, but the minimal forms of the cluster model have not been very successful in accounting for the data. String fragmentation, on the other hand, contains many parameters, but also gives quite a good description of most data. In fact, at LEP, only one spectacular failure has been registered so far: the bottom fragmentation function is distinctly softer than predicted. Much of the evolution in the cluster models has been attempts to make them more stringlike. It is not unreasonable that 'the truth' lies somewhere in between the two extremes but, judging by our current experience, rather closer to the string approach.

If parton shower models still show a healthy learning curve upwards, the same cannot really be said about fragmentation models. Very little visible progress has been made in the last five years. Specifically, we see no 'fourth alternative' on the horizon, which could take up competition with our currently known three approaches. Further, the task of connecting up these models with a more solid understanding of nonperturbative QCD still lies ahead of us.

The study of QCD at LEP, perturbative and nonperturbative, is off to a good start. The 'shopping list', which is now being worked on, includes items like:

- Perturbative QCD.
    - The running of α_S and other scaling violations.
    - Studies of the fundamental vertices of QCD (specifically the triple-gluon vertex).
    - Improved understanding of the region of validity of optimized perturbation theory.
    - Exponentiation of matrix elements (i.e. introduction of Sudakov form factors).

- Coherence effects (mainly perturbative, but with some nonperturbative ingredients).
    - Momentum and rapidity distributions.
    - Particle and energy flow in three-jet events (including flavour dependence).
  Further examples may be found in the lecture of Khoze [4].
- Differences between jet types.
    - Gluon versus quark jets.
    - Heavy flavour fragmentation functions.
    - Flavour tagging (also for studies of electroweak couplings).
- Multiplicities and correlations.
    - Increase of average multiplicity with energy.
    - Shape of multiplicity distribution, in the event as a whole, and in rapidity windows.
    - Probability of having a very low multiplicity in a jet.
    - Intermittency, in one and several dimensions.
    - Bose-Einstein effects.
- Flavour properties.
    - Particle composition and its consistency with results at lower energies.
    - Rate of rare baryon production.
    - Baryon-antibaryon correlations.
    - Importance of tensor mesons and other excited states.
- Prompt photon production.
    - As probe of the primary flavour composition.
    - As probe of parton showering.

It would be unrealistic to assume that a revolutionary change of our understanding of strong interactions will come out of LEP studies. QCD is already too well entrenched for that. This should not unduly discourage us. Many aspects of QCD are still poorly understood, in particular in the semisoft and soft regions. Here LEP will be an excellent laboratory for increasing our understanding over the years to come.

Acknowledgements

Thanks to the organizers for having arranged two wonderful weeks of physics and sun. Also thanks to S. Bethke and V. Khoze for helpful comments on parts of the manuscript.

References

[1] see e.g. G. Altarelli, Phys. Rep. **81** (1982) 1

[2] L. Maiani, lectures in this volume

[3] F. Berends, lectures in this volume

[4] V. Khoze, lecture in this volume

[5] T. Sjöstrand *et al.*, 'QCD Generators for LEP', *in* eds. G. Altarelli, R. Kleiss, C. Verzegnassi, CERN 89-08, volume 3, p. 143

[6] Z. Kunszt *et al.*, *ibid.*, volume 1, p. 373

[7] J. Ellis, M.K. Gaillard, G.G. Ross, Nucl. Phys. **B111** (1976) 253

[8] T. Sjöstrand, Z. Physik **C26** (1984) 93

[9] A. Ali, J.G. Körner, Z. Kunszt, E. Pietarinen, G. Kramer, G. Schierholz, J. Willrodt, Nucl. Phys. **B167** (1980) 454

[10] K.J.F. Gaemers, J.A.M. Vermaseren, Z. Physik **C7** (1980) 81

[11] R.K. Ellis, D.A. Ross, A.E. Terrano, Nucl. Phys. **B178** (1981) 421

[12] D. Danckaert, P. De Causmaecker, R. Gastmans, W. Troost, T.T. Wu, Phys. Lett. **114B** (1982) 203

[13] B.A. Kniehl, J.H. Kühn, Phys. Lett. **B224** (1989) 229

[14] J.A.M. Vermaseren, K.J.F. Gaemers, S.J. Oldham, Nucl. Phys. **B187** (1981) 301

[15] K. Fabricius, G. Kramer, G. Schierholz, I. Schmitt, Z. Physik **C11** (1982) 315

[16] Z. Kunszt, Phys. Lett. **99B** (1981) 429, **107B** (1981) 123;
T.D. Gottschalk, Phys. Lett. **109B** (1982) 331;
R.-y. Zhu, MIT Ph.D. thesis (1983);
A. Ali, F. Barreiro, Nucl. Phys. **B236** (1984) 269;
F. Gutbrod, G. Kramer, G. Schierholz, Z. Physik **C21** (1984) 235;
T.D. Gottschalk, M.P. Shatz, Phys. Lett. **150B** (1985) 451, CALT-68-1172 (1984);
G. Kramer, B. Lampe, Fortschr. Phys. **37** (1989) 161;
F. Gutbrod, G. Kramer, G. Rudolph, G. Schierholz, Z. Physik **C35** (1987) 543

[17] G. Kramer, N. Magnussen, DESY 90-080

[18] Particle Data Group, J.J. Hernández *et al.*, Phys. Lett. **B239** (1990) 1

[19] M. Dine, J. Sapirstein, Phys. Rev. Lett. **43** (1979) 668;
K.G. Chetyrkin *et al.*, Phys. Lett. **85B** (1979) 277;
W. Celmaster, R.J. Gonsalves, Phys. Rev. Lett. **44** (1980) 560

[20] S.G. Gorishny, A.L. Kataev, S.A. Larin, Phys. Lett. **B212** (1988) 238

[21] JADE Collaboration, W. Bartel *et al.*, Z. Physik **C33** (1986) 23

[22] JADE Collaboration, S. Bethke *et al.*, Phys. Lett. **B213** (1988) 235;
TASSO Collaboration, W. Braunschweig *et al.*, Phys. Lett. **B214** (1988) 286;
Mark II Collaboration, S. Bethke *et al.*, Z. Physik **C43** (1989) 325;
AMY Collaboration, I.H. Park *et al.*, Phys. Rev. Lett. **62** (1989) 1713;
VENUS Collaboration, K. Abe *et al.*, Phys. Lett. **B240** (1990) 232

[23] OPAL Collaboration, M.Z. Akrawy *et al.*, Phys. Lett. **B235** (1990) 389, CERN-PPE/90-143;
DELPHI Collaboration, P. Abreu *et al.*, Phys. Lett. **B247** (1990) 167;
L3 Collaboration, B. Adeva *et al.*, L3 Preprint #011 (1990)

[24] P.M. Stevenson, Phys. Rev. **D23** (1981) 2916

[25] G. Grunberg, Phys. Lett. **95B** (1980) 70

[26] S.J. Brodsky, G.P. Lepage, P.B.Mackenzie, Phys. Rev. **D28** (1983) 228

[27] D. Ward, lecture in this volume

[28] Yu.L. Dokshitzer, V.A. Khoze, S.I. Troyan, Sov. J. Nucl. Phys. (Yad. Fiz.) **47** (1988) 1384, *in* Perturbative QCD, ed. A.H. Mueller (World Scientific, Singapore, 1989), p. 241

[29] K. Hagiwara, D. Zeppenfeld, Nucl. Phys. **B313** (1989) 560;
N.K. Falck, D. Graudenz, G. Kramer, Phys. Lett. **B220** (1989) 299, Nucl. Phys. **B328** (1989) 317

[30] F.A. Berends, W.T. Giele, H. Kuijf, Nucl. Phys. **B321** (1989) 39

[31] C. J. Maxwell, Phys. Lett. **B225** (1989) 425

[32] A.H. Mueller, Nucl. Phys. **B213** (1983) 85, **B241** (1984) 141;
Yu.L. Dokshitzer, S.I. Troyan, Leningrad preprint LNPI-922 (1984);
A. Bassetto, M. Ciafaloni, G. Marchesini, Phys. Rep. **100** (1983) 201;
B.R. Webber, Ann. Rev. Nucl. Part. Sci. **36** (1986) 253

[33] G. Altarelli, G. Parisi, Nucl. Phys. **B126** (1977) 298

[34] A.H. Mueller, Phys. Lett. **B104** (1981) 161;
B.I. Ermolaev, V.S. Fadin, JETP Lett. **33** (1981) 269

[35] G. Marchesini, B.R. Webber, Nucl. Phys. **B238** (1984) 1;
B.R. Webber, Nucl. Phys. **B238** (1984) 492;
G. Marchesini, B.R. Webber, Nucl. Phys. **B310** (1988) 461

[36] M. Bengtsson, T. Sjöstrand, Phys. Lett. **B185** (1987) 435, Nucl. Phys. **B289** (1987) 810

[37] J.C. Collins, Nucl. Phys. **B304** (1988) 794;
I.G. Knowles, Nucl. Phys. **B310** (1988) 571

[38] K. Kato, T. Munehisa, Phys. Rev. **D39** (1989) 156, CERN-TH.5719/90

[39] G. Gustafson, Phys. Lett. **B175** (1986) 453;
G. Gustafson, U. Pettersson, Nucl. Phys. **B306** (1988) 746;
L. Lönnblad, Lund preprint LU TP 89-10 (1989)

[40] Ya.I. Azimov, Yu.L. Dokshitzer, V.A. Khoze, S.I. Troyan, Phys. Lett. **B165** (1985) 147, Yad. Fiz. **43** (1986) 149

[41] J. Ellis, I. Karliner, Nucl. Phys. **B148** (1979) 141

[42] TASSO Collaboration, R. Brandelik *et al.*, Phys. Lett. **97B** (1980) 453

[43] S. Bethke, LBL-28112 (1989)

[44] T.F. Walsh, P.M. Zerwas, Phys. Lett. **93B** (1980) 53

[45] R. Cashmore, lectures in this volume

[46] J.G. Körner, G. Schierholz, J. Willrodt, Nucl. Phys. **B185** (1981) 365;
O. Nachtmann, A. Reiter, Z. Phys. **C16** (1982) 45;
M. Bengtsson, P.M. Zerwas, Phys. Lett. **B208** (1988) 306;
M. Bengtsson, Z. Phys. **C42** (1989) 75;
S. Bethke, A. Ricker, P.M. Zerwas, Aachen preprint PITHA 90/14 (1990)

[47] OPAL Collaboration, M.Z. Akrawy *et al.*, CERN-PPE/90-97 (1990)

[48] L3 Collaboration, B. Adeva *et al.*, Phys. Lett. **B248** (1990) 227

[49] DELPHI Collaboration, contribution to the 25th International Conference on High Energy Physics, Singapore, 2–8 August, 1990;
H. Müller, parallel session talk at this conference

[50] AMY Collaboration, I. H. Park *et al.*, Phys. Rev. Lett. **62** (1989) 1713;
VENUS Collaboration, K. Abe *et al.*, KEK Preprint 90-62 (1990)

[51] T. Sjöstrand, Int. J. Mod. Phys. **A3** (1988) 751

[52] X. Artru, G. Mennessier, Nucl. Phys. **B70** (1974) 93

[53] B. Andersson, G. Gustafson, G. Ingelman, T. Sjöstrand, Phys. Rep. **97** (1983) 31

[54] X. Artru, Phys. Rep. **97** (1983) 147

[55] G. Gustafson, Z. Physik **C15** (1982) 155

[56] G. Gustafson, U. Pettersson, P. Zerwas, Phys. Lett. **B209** (1988) 90

[57] A. Krzywicki, B. Petersson, Phys. Rev. **D6** (1972) 924;
J. Finkelstein, R.D. Peccei, Phys. Rev. **D6** (1972) 2606;
F. Niedermayer, Nucl. Phys. **B79** (1974) 355;
A. Casher, J. Kogut, L. Susskind, Phys. Rev. **D10** (1974) 732

[58] R.D. Field, R.P. Feynman, Nucl. Phys. **B136** (1978) 1

[59] P. Hoyer, P. Osland, H.G. Sander, T.F. Walsh, P.M. Zerwas, Nucl. Phys. **B161** (1979) 349

[60] A. Ali, J.G. Körner, G. Kramer, J. Willrodt, Nucl. Phys. **B168** (1980) 409;
A. Ali, E. Pietarinen, G. Kramer, J. Willrodt, Phys. Lett. **B93** (1980) 155

[61] I. Montvay, Phys. Lett. **84B** (1979) 331

[62] P. Biddulph, G. Thompson, Computer Physics Commun. **54** (1989) 13;
G. Ballocchi, R. Odorico, CERN-EP/89-162

[63] G.C. Fox, S. Wolfram, Nucl. Phys. **B168** (1980) 285;
R.D. Field, S. Wolfram, Nucl. Phys. **B213** (1983) 65

[64] T.D. Gottschalk, D.A. Morris, Nucl. Phys. **B288** (1987) 729;
D.A.Morris, Caltech Ph.D. thesis, CALT-68-1440 (1987)

[65] D. Amati, G. Veneziano, Phys. Lett. **83B** (1979) 87;
G. Marchesini, L. Trentadue, G. Veneziano, Nucl. Phys. **B181** (1981) 335

[66] W. Kittel, R. Peschanski, Nucl. Phys. B (Proc. Suppl.) **16** (1990) 445;
A. Bialas, CERN-TH.5791/90; and references therein

[67] J. Ellis, H. Kowalski, Phys. Lett. **B214** (1988) 161;
J. Ellis, M. Karliner, H. Kowalski, Phys. Lett. **B235** (1990) 341

[68] L. Angelini, L. Nitti, M. Pellicoro, G. Preparata, G. Valenti, Computer Physics Commun. **34** (1985) 371;
W. Ochs, Z. Physik **C23** (1984) 131, **C34** (1987) 397

[69] P. Mättig, Phys. Rep. **177** (1989) 141;
High Energy Electron Positron Physics, eds. A. Ali, P. Söding (World Scientific, Singapore, 1988)

[70] C. Basham, L. Brown, S. Ellis, S. Love, Phys. Rev. Lett. **41** (1978) 1585

[71] S. Brandt, Ch. Peyrou, R. Sosnowski, A. Wroblewski, Phys. Lett. **12** (1964) 57;
E. Fahri, Phys. Rev. Lett. **39** (1977) 1587

[72] J.D. Bjorken, S.J. Brodsky, Phys. Rev. **D1** (1970) 1416

[73] S. Brandt, H.D. Dahmen, Z. Physik **C1** (1979) 61

[74] G. Parisi, Phys. Lett. **74B** (1978) 65;
J.F. Donoghue, F.E. Low, S.Y. Pi, Phys. Rev. **D20** (1979) 2759

[75] T. Sjöstrand, Computer Physics Commun. **28** (1983) 229

[76] ALEPH Collaboration, D. Decamp *et al.*, Phys. Lett. **B234** (1990) 209;
DELPHI Collaboration, P. Aarnio *et al.*, Phys. Lett. **B240** (1990) 271;
OPAL Collaboration, M.Z. Akrawy *et al.*, Z. Physik **C47** (1990) 505

[77] Mark II Collaboration, A. Petersen *et al.*, Phys. Rev. **D37** (1988) 1

[78] TASSO Collaboration, W. Braunschweig *et al.*, Z. Physik **C41** (1988) 359, **C47** (1990) 187

[79] AMY Collaboration, Y.K. Li *et al.*, Phys. Rev. **D41** (1990) 2675

[80] W. de Boer, H. Fürstenau, J.H. Köhne, Karlsruhe preprint IEKP-KA/90-4 (1990)

[81] DELPHI Collaboration, P. Abreu *et al.*, CERN-PPE/90-117

[82] Z. Koba, H.B. Nielsen, P. Olesen, Nucl. Phys. **B40** (1972) 317

[83] E.D. Malaza, B.R.Webber, Nucl. Phys. **B267** (1986) 702;
E.D. Malaza, Z. Physik **C31** (1986) 143

[84] B. Andersson, P. Dahlqvist, G. Gustafson, Z. Physik **C44** (1989) 455

[85] W. Hofmann, Ann. Rev. Nucl. Part. Sci. **38** (1988) 279

[86] C.D. Buchanan, S.-B. Chun, Phys. Rev. Lett. **59** (1987) 1997, *in* Proceedings of the KEK Topical Conference on e^+e^- Collision Physics, eds. Y. Shimizu, F. Takasaki (KEK, Tsukuba, 1990), p. 384

[87] TPC/2γ Collaboration, H. Aihara *et al.*, Phys. Rev. Lett. **61** (1988) 1263;
G.D. Cowan, Ph.D. thesis, LBL-24715 (1988)

[88] A. Bialas, R. Peschanski, Nucl. Phys. **B273** (1986) 703, **B308** (1988) 857

[89] T. Sjöstrand, *in* Multiparticle Production, eds. R. Hwa, G. Pancheri, Y. Srivastava (World Scientific, Singapore, 1989), p. 327

[90] B. Buschbeck, P. Lipa, R. Peschanski, Phys. Lett. **B215** (1988) 788

[91] TASSO Collaboration, W. Braunschweig *et al.*, Phys. Lett. **B231** (1989) 548

[92] DELPHI Collaboration, P. Abreu *et al.*, Phys. Lett. **B247** (1990) 137;
A. De Angelis, CERN-PPE/90-129

[93] CELLO Collaboration, H.J. Behrend *et al.*, DESY 90-114

[94] TPC/2γ collaboration, H. Aihara *et al.*, Phys. Rev. Lett. **55** (1985) 1047

[95] J.D. Bjorken, Phys. Rev. **D17** (1978) 171;
M. Suzuki, Phys. Lett. **71B** (1977) 139;
V.A. Khoze, Ya.I. Azimov, L.L. Frankfurt, *in* Proceedings of the Conference on High Energy Physics, Tbilisi 1976

[96] C. Peterson, D. Schlatter, I. Schmitt, P. Zerwas, Phys. Rev. **D27** (1983) 105

[97] B. Mele, P. Nason, Phys. Lett. **B245** (1990) 635

[98] S. Bethke, Z. Physik **C29** (1985) 175, and references therein;
J. Chrin, Z. Physik **C36** (1987) 163, and references therein;
M. Zimmer, Heidelberg Ph.D. thesis (1989)

[99] L3 Collaboration, B. Adeva *et al.*, Phys. Lett. **B241** (1990) 416;
ALEPH Collaboration, D. Decamp *et al.*, Phys. Lett. **B244** (1990) 551

[100] M.G. Bowler, Z. Physik **C11** (1981) 169

[101] Y.-K. Kim, lecture in this volume

[102] B. Andersson, G. Gustafson, T. Sjöstrand, Phys. Lett. **94B** (1980) 211

[103] A. Petersen (JADE Collaboration), *in* Elementary Constituents and Hadronic Structure, ed. J. Tran Thanh Van (Editions Frontières, Dreux, 1980), p. 505;
JADE Collaboration, W. Bartel *et al.*, Phys. Lett. **B101** (1981) 129, **B134** (1984) 275, Z. Physik **C21** (1983) 37;
TPC/2γ Collaboration, H. Aihara *et al.*, Z. Physik **C28** (1985) 31;
TASSO Collaboration, M. Althoff *et al.*, Z. Physik **C29** (1985) 29

[104] V. Khoze, L. Lönnblad, Phys. Lett. **B241** (1990) 123

[105] T. Sjöstrand, presented at Leningrad QCD workshop, October 1989 (unpublished)

COLOUR-COHERENCE PHYSICS AT THE Z^0

V.A. Khoze*

CERN, Geneva, Switzerland

and

Leningrad Nuclear Physics Institute

Gatchina, Leningrad, USSR

Abstract

We review the recent results on the applications of the analytical parton-shower approach to QCD jet physics in $Z^0 \to$ hadrons. We concentrate on the studies at the Z^0 resonance of the colour-coherence phenomena, reflecting the collective nature of multihadron production.

1 INTRODUCTION

The existence of spectacular jets of hadrons in hard processes is one of the most striking successes of Quantum Chromodynamics (QCD), the SU(3) gauge theory of coloured quarks and gluons.

Jet-like events were first seen definitely in e^+e^- collisions at SPEAR in 1975 [1]. Five years later a gluon was discovered experimentally in three-jet events at PETRA [2]. These pioneering results opened up an important new area for experimental tests of both perturbative and non-perturbative QCD.

In the last decade hadron-jet physics has been intensively studied at both e^+e^- and hadron accelerators; see, for example, lectures given by Ward [3] and Cashmore [4] at this School. It will be one of the main problems of investigation for the e^+e^-, $pp(p\bar{p})$, and ep colliders of the future.

e^+e^- annihilation into hadrons proves to be a particularly wonderful laboratory for detailed experimental tests of QCD. A vast amount of data on multihadron production has been accumulated in the experiments at PETRA, PEP, and TRISTAN. With the start of SLC and LEP activity a wealth of new experimental results has become available (see Refs. [3],[5]).

Experiments at the Z^0 resonance provide substantial advantages compared with the previous e^+e^- data: higher energy, very high statistics, abundance of the heavy-flavour (bottom) production. In addition, these experiments have, in general, better detector resolutions and efficiencies than those at lower energies.

The first experimental studies of the hadronic events from Z^0 decays [3] appear to be very useful for detailed tests of perturbative QCD and for reducing our domain of ignorance on the physics of confinement. The data show that hadronic event characteristics calculated at the parton level agree very well with the measured ones. This supports the hypothesis of local parton–hadron duality (LPHD) [6],[7]. Such local duality is naturally connected with the 'pre-confinement' properties [8] of the QCD parton cascades.

On the basis of this hypothesis, a detailed analysis of perturbative QCD characteristics of data on hard collisions is possible (for reviews, see Ref. [9]). These studies are important not only for testing various aspects of QCD and for the design and analysis of experiments. The characteristic QCD features

---

*Supported in part by the World Laboratory ELOISATRON Project

of jet-like states could also provide us with a valuable additional tool, in helping to extract and to study the manifestation of New Physics.

It would be impossible to cover completely even this specialized subject of QCD jet physics at the Z^0 resonance. So I must apologize in advance for being selective in the topics discussed. In this lecture I shall review some recent results of the analytical perturbative approach (PA) [9]–[14] to jet physics in $Z^0 \to$ hadrons. The emphasis will be put on the manifestations at the Z^0 of colour-coherence phenomena.

For excellent recent reviews, covering other topics in QCD jet dynamics, see Refs. [15]–[20].

A topic closely related to the subject of this lecture is a comparison of PA with QCD-inspired Monte Carlo algorithmic models. The reader can find the presentation of the recent status of these schemes and of their development in the lectures given by Sjöstrand at this School [5] (for a review, see Ref. [16]).

Monte Carlo simulations based on a QCD parton-shower mechanism describe the existing data very successfully and are becoming better and better at building in realistic fragmentation and proper QCD evolution.

One of the approaches for the improvement of models is the most adequate incorporation of the colour coherence, reflecting the quantum mechanics of QCD. However, note that the very possibility of absorbing the interference effects into the probabilistic scheme is far from obvious, and the use of Monte Carlo generators for the evolution of the multiparton system proves to be of limited value in principle (for details, see Refs. [21],[22]). For example, under special conditions some subtle collective phenomena, breaking the standard probabilistic picture, may even become dominant and observable; see Section 6.

The outline of the lecture is as follows: In Section 1 an introduction is given. In Section 2 the parton-cascade mechanism of multihadron production is discussed. In Section 3 a guide to colour coherence is given. In Section 4 a modified leading logarithmic approximation is put forward. In Section 5 inclusive particle spectra and intrajet coherence are discussed. In Section 6 the Z^0 resonance as a laboratory to study interjet collective effects is studied. In Section 7 prospects for future experiments are given. Finally, in Section 8 the conclusions are presented.

2 PARTON-CASCADE MECHANISM OF MULTIHADRON PRODUCTION

The QCD picture of multihadron production has gone through a qualitative transformation in the last few years, reaching a mature level of sophistication, see Refs. [9]–[20] and references therein.

The milestones on the road to the present understanding of QCD jet physics were:

i) the preconfinement idea of Amati and Veneziano [8];

ii) the growth of particle multiplicity due to QCD showering [23],[24];

iii) KNO scaling in QCD jets [25];

iv) intrajet coherence and strong angular ordering (AO) in parton cascades [26],[27];

v) the modified leading logarithmic approximation (MLLA) and strict AO [6],[28];

vi) collective phenomena, string-effect [29], and interjet coherence [30].

Today the members[1] of the international QCD community (Columbia University– Leningrad– Novosibirsk–Italy–Cambridge $\equiv$ CLIC) are able to make testable quantitative predictions for jet characteristics, with controllable accuracy in terms of analytical perturbative calculations.

The key idea is to invoke the parton-shower picture where one views the evolution of a jet as a sequence of parton branchings. The main concept is to reorganize the perturbative expansion in such a way that its zero-order approximation is systematic and involves an arbitrary number of produced particles. This approximation can be achieved through an iteration of basic A $\to$ B + C branchings. In principle, it should be possible to include higher corrections in order to systematically improve the accuracy.

The approximation mentioned above is the MLLA, which takes care of both double-logarithmic (DL) and single-logarithmic (SL) effects in a systematic way [6],[28].

---

[1]The names can be found among the authors of Refs. [9]–[15] and in the references therein.

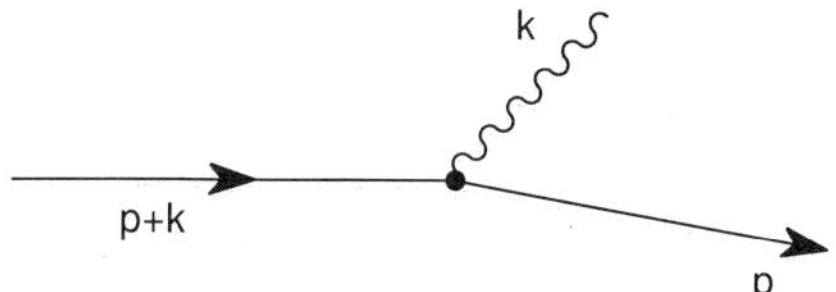

Figure 1. Kinematics of gluon emission.

In constructing the MLLA, one should pay special attention to maintaining the probabilistic interpretation of jet evolution. The existence of such an interpretation is far from trivial in the problems connected with the description of soft-particle distributions ($x \ll 1$). Here, interference contributions play an important role and prove to be unavoidable, unlike the case of the problems dealt with near $x \sim 1$. Nevertheless, it appears to be possible, by choosing an appropriate evolution parameter (jet opening angle θ_0) and accounting for the specific angular dependence of soft-emission probabilities, to maintain a probabilistic interpretation of the dynamics of soft-parton cascades.

Experimental data have convincingly demonstrated the dominant role of the perturbative stage of jet evolution and have confirmed the QCD bremsstrahlung nature of multiple hadroproduction. According to the LPHD concept [6],[7], the conversion of partons into hadrons occurs at a low virtuality scale, independent of the scale of the primary hard process, and involves only low-momentum transfer, leading to a close similarity between parton and hadron distributions.

The PA attempts to describe the global features of the hadronic systems, such as the mean multiplicities and multiplicity distributions, angular patterns of energy and multiplicity flow, inclusive energy spectra and correlations of particles, etc., without making any reference at all to a fragmentation scheme. The non-perturbative effects are reduced to normalizing constants relating hadronic characteristics to partonic ones. Since there are only a few parameters to vary in connecting PA results with experiment, these predictions are simply testable. Thus, the main equation of PA reads

$$PA = MLLA + LPHD. \tag{1}$$

At the Z^0 a quark, with $E \sim m_Z/2$, starting from the annihilation time $t^{\text{ann}} \sim 1/E \sim 5 \times 10^{-3}$ fm/c and up to its hadronization time $t^{\text{had}} \sim ER^2 \sim 10^2$ fm/c ($R \sim 1$ fm stands for the hadronization scale), behaves as if it were a true coloured object radiating gluons perturbatively without taking any care about its future confinement.

Since multiple QCD bremsstrahlung plays the key role in multihadron production, let us first discuss the basic QCD process, namely the gluon emission off a quark with momentum p, as shown in Fig. 1. The differential spectrum is given by the well-known formula

$$\mathrm{d}w_{\mathrm{q}}^{\mathrm{g}} = \frac{\alpha_{\mathrm{s}}(k_{\perp}^2)}{4\pi} 2C_{\mathrm{F}} \left[1 + \left(1 - \frac{k}{p} \right)^2 \right] \frac{\mathrm{d}k}{k} \frac{\mathrm{d}k_{\perp}^2}{k_{\perp}^2} , \tag{2}$$

$$C_{\mathrm{F}} = \frac{(N_{\mathrm{c}}^2 - 1)}{2N_{\mathrm{c}}} = \frac{4}{3} .$$

The effective coupling here runs with the gluon transverse momentum $k_{\perp}$, which comes from higher-order corrections to the Born probability.

Notice two important properties of the spectrum (2): i) the broad logarithmic distribution over transverse momentum $k_{\perp}$ (typical for a theory with dimensionless coupling); ii) the broad logarithmic energy distribution (specific for theories with massless vector particles).

At large emission angle and large energy one would get an extra gluon jet with a small probability.

$$\text{Multijet topology: } k_{\perp} \sim k \sim E \to w \sim \frac{\alpha_{\mathrm{s}}}{\pi} \ll 1 .$$

At the same time, the bulk of radiation (quasicollinear and/or soft partons) will not lead to the appearance of additional visible jets in an event but will instead populate the original jet with secondary partons influencing the particle multiplicity and other jet characteristics.

$$\text{Intrajet activity: } k_{\perp} \ll k \ll E \to w \sim \alpha_{\mathrm{s}} \log^2 E \sim 1 .$$

This DL q $\to$ qg process, together with two other basic parton splittings (DL g $\to$ gg radiation and SL decay g $\to$ q$\bar{\mathrm{q}}$), forms parton cascades.

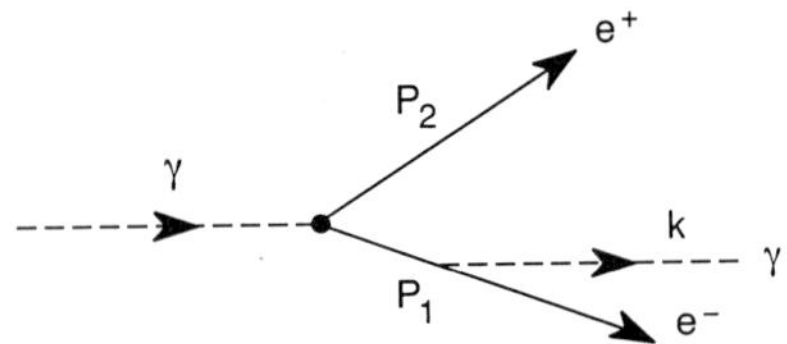

Figure 2. Bremsstrahlung radiation of a photon with momentum k after e^+e^- pair production.

To answer the question of how offspring partons influence the hadronic yield, one has to find out first what the condition is for a gluon to behave as an independent coloured object. Let us estimate the formation time, t_g^{form}, the time interval needed for a gluon to be radiated. Using the uncertainty relation to estimate the 'lifetime' of the virtual $(p+k)$ state, one finds:

$$t_g^{form} \sim \frac{E}{(p+k)^2} \simeq \frac{E}{kE\theta^2} \simeq \frac{k}{k_\perp^2} .$$ (3)

Comparing Eq. (3) with the hadronization time of a gluon

$$t_g^{had} \sim kR^2 ,$$ (4)

one concludes that it is the restriction

$$k_\perp > R^{-1} = \text{a few hundred MeV}$$ (5)

which guarantees the applicability of the quark–gluon language, i.e. of the perturbative QCD.

Most dangerous (from the perturbative point of view) is the radiation of gluons with $k_\perp \sim R^{-1}$ [$\alpha_s(k_\perp^2)$ 1]. These partons can hardly be treated as gluons at all, since owing to Eqs. [3],[4] they are forced to hadronize just immediately after being formed. The real strong interaction comes into play here, which results in the famous *hadronic plateau* of the old parton picture. Its height, according to Eq. (2), could be estimated qualitatively as

$$dN = \left[\int_{k_\perp \sim R^{-1}} \frac{dk_\perp^2}{k_\perp^2} 4C_F \frac{\alpha_s(k_\perp^2)}{4\pi} \right] \frac{dk}{k} = \text{const} \frac{dk}{k} .$$ (6)

Gluons with parametrically large transverse momenta $k_\perp \gg R^{-1}$ will perturbatively emit, in turn, new offsprings forming cascades. If such a gluon forms its own hadronic plateau, consisting of hadrons with energies $R^{-1} \leq E_h \leq k$, one would expect that the resulting spectrum would be strongly populated with the softest hadrons: $E_h \sim m_h$. However, this common wisdom was proved to be wrong at the beginning of the eighties, when *Quantum-Mechanical Coherence* had been rediscovered in the QCD context [26],[27].

The basic consequence of QCD coherence for the physics of *intrajet* cascades is AO [26],[27], which strongly reduces the multiplication of soft particles in parton cascades (see subsection 3.1).

Because of the AO, the structure of the partonic system representing the jet evolution can be treated as a tree of independent parton branchings into sequentially shrinking angular cones. The direct consequence of the AO was the prediction of the so-called *hump-backed* shape of QCD particle spectra xdN/dx [31],[32].

Not the softest particles with momentum fractions $x \sim m_h/E_j$, but those with intermediate energies $x \sim (m_h/E_j)^c (c \simeq 0.65)$ multiply most effectively in QCD cascades (see Sections 4 and 5).

3 A GUIDE TO COLOUR COHERENCE

Our aim here is to provide the reader with an elementary introduction to the basic ideas of coherence phenomena, see Ref. [21].

Coherence problems are basic to any gauge theory. Roughly speaking, there are two types of coherence effects which occur in QCD jet dynamics.

The first manifestation of coherence is the AO of the sequential parton decays [26],[27]. Angular ordering accounts for the appropriate dependence of the soft radiation on the prehistory of parton-jet development

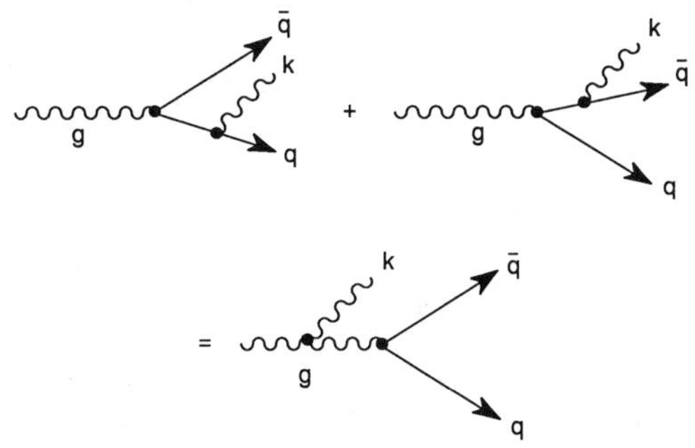

Figure 3. Wide-angle emission of a soft gluon with momentum k, off q and $\bar{\text{q}}$, acts as if the emission came off the parent gluon g imagined to be on shell.

Coherence of the second type deals with the angular structure of particle flows when three or more partons are involved in a hard process. Here the particle angular distributions depend on the geometry and colour topology of the whole jet ensemble giving rise to *QCD radiophysics of jets* (see Refs. [13],[21],[33]).

3.1   Angular ordering in parton cascades

To elucidate the physical origin of the AO let us consider a simple model of a jet cascade, namely the radiation pattern of soft photons produced by a relativistic e^+e^- pair in a QED shower (see Fig. 2).

The question is, To what extent do the e^+ and e^- independently emit γ's? To answer this question, let us estimate the formation time of the γ radiation from, say, e^- leg. According to Eq. (3) one has

$$t_{\text{form}} \approx \frac{1}{k\Theta_{\gamma e^-}^2} = \frac{\lambda_\perp}{\Theta_{\gamma e^-}} \tag{7}$$

($\lambda_\perp$ is the transverse wavelength of the radiated photon).

During this time the e^+e^- pair separate, transversely, by a distance

$$\rho_\perp^{e^+e^-} \approx \Theta_{e^+e^-}\, t_{\text{form}} \approx \lambda_\perp \frac{\Theta_{e^+e^-}}{\Theta_{\gamma e^-}} \ . \tag{8}$$

One concludes that for large-angle photon emissions,

$$\Theta_{\gamma e^-} \approx \Theta_{\gamma e^+} \gg \Theta_{e^+e^-} \ ,$$

the separation of two emitters, e^+ and e^-, proves to be smaller than $\lambda_\perp$. In this case the emitted photon cannot resolve the internal structure of the e^+e^- pair and probes only its total electric charge, which is zero. Thus for $\Theta_{\gamma e} \gg \Theta_{e^+e^-}$ we expect photon emission to be strongly suppressed[2]. The e^+ and e^- can be said to emit γ's independently only at $\rho_\perp^{e^+e^-} \gg \gamma_\perp$, that is when

$$\Theta_{\gamma e^+} \text{ or } \Theta_{\gamma e^-} \le \Theta_{e^+e^-} \ .$$

A similar physical picture appears for QCD parton cascades, where soft-gluon radiation is governed by the conserved *colour* current. The only difference is that the coherent radiation of soft gluons by an unresolved pair of quarks (or gluons) is no longer zero but the radiation acts *as if* it were emitted from the parent gluon *imagined* to be on shell, as is illustrated in Fig. 3. This property is universal and holds true for the soft radiation accompanying q $\to$ qg and g $\to$ gg splittings as well. A remarkable fact is that one gets all leading double and single logarithmic effects correctly, for angular averaged observables, by allowing the gluon emission, independently, off line q when $\Theta_{kq} \le \Theta_{q\bar{q}}$, off line $\bar{\text{q}}$ when $\Theta_{k\bar{q}} \le \Theta_{q\bar{q}}$, and off the parent line g, when $\Theta_{kg} \ge \Theta_{q\bar{q}}$ (see Fig. 3).

This observation furnishes the core idea of the Marchesini–Webber model [35], the first Monte Carlo simulation that included intrajet coherence effects.

---

[2]This phenomenon has been well known in cosmic-ray physics since the middle of the fifties—the so-called 'Chudakov effect' [34].

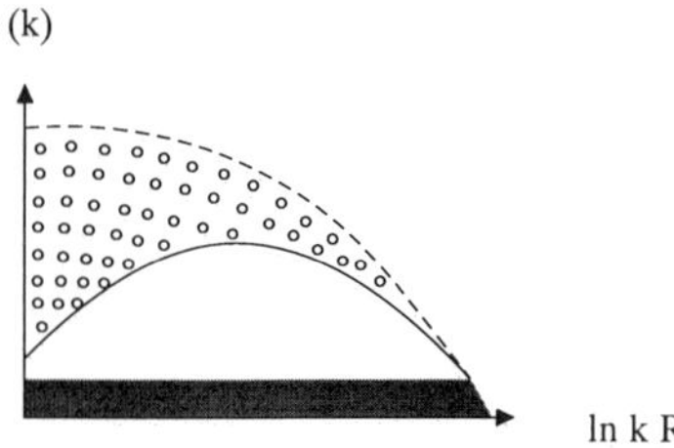

Figure 4. The effect of colour coherence on particle energy spectrum $\rho(k) = dn/d\ln k$. Dotted area corresponds to the contribution which is removed when turning from the incoherent model (*dashed*) to the coherent one (*solid*). Shaded area shows the old-fashioned plateau (without taking into account bremsstrahlung).

The yield of soft, wider-angle gluons remains unchanged when the number of particles inside the multiparton bunch increases, since such gluons are unable to resolve separate colour sources inside the parton jet. The soft gluon behaves as a classical probe, testing the colour charge of the jet as a whole, i.e. that of the original parton, initiating the jet. In this sense the QCD colour coherence can be said to suppress the soft·radiation at large angles. As a result, to describe the jet evolution in terms of independent sequential parton splittings one has to impose the AO condition—a uniform decrease of successive opening angles in the cascade.

The AO occurs not only for the time-like jet evolution but also for the space-like parton cascades determining the target fragmentation in deep-inelastic scattering and the structure of initial-state radiation in the Drell–Yan and large-$p_\perp$ scattering processes etc., see, for example, Refs. [13],[21].

3.2 Suppression of the soft intrajet radiation, hump-backed spectrum

The depletion of emission of soft particles inside a jet (hump-backed plateau in the inclusive energy spectrum) is one of the most striking predictions of perturbative QCD. It follows from the AO of the parton cascade in going from greater to lesser virtuality and is a direct manifestation of coherence. This can be understood on kinematical grounds as being the result of two conflicting tendencies: on the one hand, owing to the restriction $k_\perp > 1/R$ [see Eq. (5)] a soft particle is 'forced out' at large emission angle $\Theta > 1/kR$, and, on the other hand, the allowed decaying angle, after a few successive branchings, is shrunk to small values.

Let us illustrate the influence of the intrajet coherence on particle spectra with the help of the *toy model*[7], based on first-order QCD.

We start with an old-fashioned plateau $\rho(k) \equiv dn/d\ln k + \text{const}$ of hadrons with limited transverse momenta $k\Theta = k_\perp \sim R^{-1}$ for a quark jet with energy E (see shaded area in Fig. 4). Taking into account a gluon with energy ε and emission angle Θ_0, let us use the DL expression for the radiation probability

$$dw_g \propto \alpha_s \frac{d\varepsilon}{\varepsilon} \frac{d\Theta_0}{\Theta_0} \vartheta(\varepsilon\Theta_0 - R^{-1}) \,. \tag{9}$$

The step function ϑ restricts here the transverse momentum $p_\perp \approx \varepsilon\Theta_0 > R^{-1}$.

How does the gluon contribute to the particle yield? From the orthodox parton model one might expect the gluon to give rise to its own plateau of particles with

$$R^{-1} < k < \varepsilon$$

and limited transverse momenta with respect to the gluon: $k\Theta' \sim R^{-1}$.

Now let us verify that the coherence leads to the following reduction of this additional plateau:

$$(R\Theta_0)^{-1} < k < \varepsilon \,. \tag{10}$$

The distribution of particles from the gluon jet can be represented symbolically as

$$dN = \frac{dk}{k} \frac{d\Theta'}{\Theta'} \delta(k\Theta' - R^{-1}) \,. \tag{11}$$

424

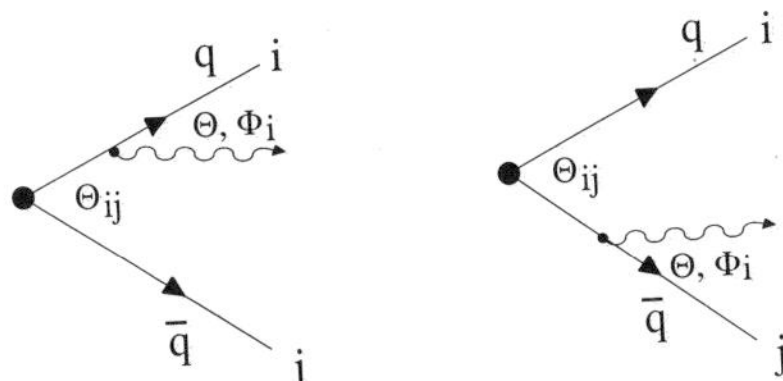

Figure 5. Soft-gluon emission off a hard colourless $q\bar{q}$ pair.

This expression can be thought of as a DL spectrum of bremsstrahlung from a gluon (ε, Θ_0) 'projected' onto the domain of the most intensive radiation. As follows from the AO in cascade, the offspring particles are independently emitted by the gluon only inside the cone with the opening angle $\Theta' < \Theta_0$. Applying this inequality to Eq. (11), one obtains restriction (10).

Finally, one finds

$$
N = \int\limits^{E} \frac{\mathrm{d}k}{k} \int\limits^{1} \frac{\mathrm{d}\Theta}{\Theta} \delta(k\Theta - R^{-1}) + \alpha_s \int\limits^{E} \frac{\mathrm{d}\varepsilon}{\varepsilon} \int\limits^{1} \frac{\mathrm{d}\Theta_0}{\Theta_0} \vartheta(\varepsilon\Theta_0 - R^{-1})
$$
$$
\times \int\limits^{\varepsilon} \frac{\mathrm{d}k}{k} \int\limits^{\Theta_{\max}} \frac{\mathrm{d}\Theta'}{\Theta'} \delta(k\Theta' - R^{-1}) .
\tag{12}
$$

The first term of this schematic expression stands for the background quark plateau, the second one is constructed from the gluon emission (9) and fragmentation (11). The parameter $\Theta_{\max}$ encodes the difference between the coherent and incoherent cases:

$$
\Theta_{\max}^{\mathrm{incoh}} = 1 ,
$$
$$
\Theta_{\max}^{\mathrm{coh}} = \Theta_0 .
$$

Using Eq. (12) one obtains for the particle energy spectrum $\rho(k)$

$$
\rho^{\mathrm{incoh}} = 1 + \frac{\alpha_s}{2}(\ln^2 ER - \ln^2 kR) ,
$$
$$
\rho^{\mathrm{coh}} = 1 + \alpha_s \ln \frac{E}{k} \ln kR .
\tag{13}
$$

The additional gluon-initiated multiplicity $\int \mathrm{d}\ln k[\rho(k) - 1]$ appears to be twice as large for the incoherent case. So the coherence substantially depletes the soft part of energy spectra, giving rise to a hump (see Fig. 4). The maximum is increasing with energy and peaks at $k \sim \sqrt{E}$.

3.3 Radiation pattern for the 'quark–antiquark' antenna

Let us briefly discuss soft emission associated with a colour-singlet $q\bar{q}$ pair. This radiation pattern is interesting not only because of the process $e^+e^- \to q\bar{q}$. Neglecting the terms of the order of $1/N_c^2$, one can represent the radiation in the case of the complex hard-parton system as a sum of terms in which each external quark line is uniquely connected to an external antiquark line of the same colour ($q\bar{q}$ antennae).

In the lowest order the soft-gluon distribution takes the familiar form (classical currents) (for notations see Fig. 5):

$$
\mathrm{d}W_{q\bar{q}} = -\frac{\mathrm{d}^3k}{(2\pi)^2 k} \alpha_s C_F \left(\frac{p_i}{p_i k} - \frac{p_j}{p_j k} \right)^2 = \frac{\mathrm{d}k}{k} \mathrm{d}\Omega_{\vec{n}} \frac{2C_F \alpha_s}{(2\pi)^2} (\widehat{ij}) = \frac{\mathrm{d}k}{k} \mathrm{d}\Omega_{\vec{n}} \frac{\alpha_s}{(2\pi)^2} W^{q\bar{q}}(\vec{n}) .
\tag{14}
$$

Here

$$
(\widehat{ij}) = \frac{a_{ij}}{a_i a_j} , \quad a_{ij} = (1 - \vec{n}_i \vec{n}_j) , \quad a_i = (1 - \vec{n}_i \vec{n}_j) ,
\tag{15}
$$

where $\vec{n}_i$ and $\vec{n}_j$ denote the directions of the q and $\bar{q}$ momenta respectively; $\vec{n}$ is the direction of the emitted gluon.

Let us call the distribution $(\widehat{ij})$, describing the radiation pattern of the colourless $q\bar{q}$ pair, the $q\bar{q}$

antenna. The antenna $(\hat{ij})$ may be represented in the form

$$(\hat{ij}) = P_{ij} + P_{ji} \, ,$$

where

$$P_{ij} = \frac{1}{2}\left[\frac{1}{a_i} + \frac{a_{ij} - a_i}{a_i a_j}\right] \, . \tag{16}$$

One can refer to the two terms in the square brackets in Eq. (16) as the *incoherent* and the *interference* terms. At fixed Θ_i, the incoherent term is independent of the azimuthal angle ϕ_i. The interference term depends on ϕ_i through the angle Θ_j. If the soft gluon lies within the cone, described by i and j, the interference is positive. If the gluon lies outside this cone the interference is negative. This term is largest when the soft gluon lies in the plane defined by i and j, $\phi_i = \phi_j$. The point about splitting the radiation pattern into two terms $P_{ij}(P_{ji})$ is that only the former (latter) has the pole at $\Theta_i = 0(\Theta_j = 0)$, so this term can be treated as 'belonging to' quark $i(j)$.

After integration over ϕ_i it turns out that the total contributions of the incoherent and interference terms are equal, so that after azimuthal averaging we find

$$\langle P_{ij}\rangle_{\phi_i} \equiv \int \frac{\mathrm{d}\phi_i}{2\pi} P_{ij} = \frac{1}{a_i}\vartheta(\cos\Theta_i - \cos\Theta_{ij}) \, . \tag{17}$$

In other words, $\langle P_{ij}\rangle_{\phi_i}$ is just the incoherent radiation P_i^{incoh} off a quark i, effectively confined to the cone

$$\cos\Theta_i > \cos\Theta_{ij} \, , \quad \Theta_i < \Theta_{ij} \, . \tag{18}$$

For the case of a multiparton system with n hard emitters ($i = 1...n$) the full soft radiation pattern may be replaced by

$$P(\Omega) \Rightarrow P_{\mathrm{AO}} \equiv \sum_i^n P_i^{\mathrm{incoh}}\vartheta(-\Theta_i + \Theta_{ij}) \, ,$$

which looks like a sum of independent emission probabilities.

Such a recipe was used, for example, to construct the Monte Carlo program HERWIG [11].

Let us make a comment concerning gluon emission from a heavy-quark pair ('$Q\bar{Q}$ antenna'), for details see Refs. [13],[36]. In such a case, the heavy-quark masses substantially reduce the phase space available for branching. As a result, after azimuthal averaging the emission of a soft gluon by a heavy quark Q of velocity β_i effectively takes place only in the following 'screened cone' [36], cf. Eq. (18),

$$\beta_j \, \cos\Theta_{ij} < \cos\,\Theta_i < \cos\,\Theta_i^{\mathrm{min}} \simeq \beta_i \, , \tag{19}$$

where β_j is the velocity of the parton, which is colour connected to the emitting parton.

3.4 Wave nature of drag phenomena

From the standpoint of the QCD community, the drag (or string) effect in the $q\bar{q}g$ events of e^+e^- annihilation is one of the best examples of the manifestation of the interjet colour coherence.

Let us recall that originally the string effect was predicted as the result of a Lorentz boost exerted by a gluon on the *non-perturbative* string stretched between quarks, see Refs. [29],[37]. But, in *perturbative* language, this phenomenon is a direct consequence of interference among the gluon waves radiated from the $q\bar{q}g$ composite emitter [30].

In Section 6 we shall discuss this coherence effect in more detail, but here our purpose is simply to illustrate the basic ideas.

So far, the most spectacular experimental test of the drag phenomenon is the comparison of associated hadron production in $q\bar{q}g$ events with that of $q\bar{q}\gamma$ events, with the g and γ having similar kinematics [38]. In the plane of the three jets, counting the photon as a jet, one finds a suppression of associated hadrons in the region between the q and $\bar{q}$ in $q\bar{q}g$ events as compared to $q\bar{q}\gamma$ events.

It might seem strange, at first sight, that inclusion of an additional 'emitter' (coloured gluon replacing a colourless photon) results in less QCD radiation in some direction. This is a typical quantum-mechanical interference effect, and one can illustrate its physical origin with the help of the 'QED' model, where quarks are replaced by electrons and the gluon by a collinear e^+e^+ pair, as shown in Fig. 6a. The $q\bar{q}\gamma$ event is illustrated in this model by Fig. 6b.

The corresponding radiation patterns are [cf. Eq. (14)]:

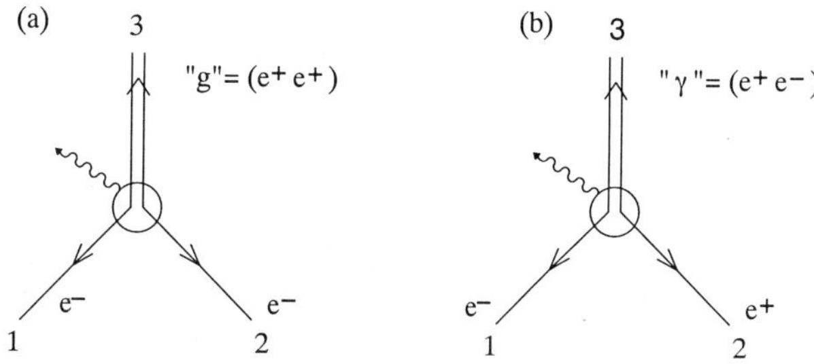

Figure 6. A 'QED' model illustrating the drag effect. The gluon is represented as having double electric charge. The photon is replaced by a collinear e^+e^- pair.

$$dW_{(q\bar{q}\gamma)} = \frac{dk}{k}\, d\Omega_{\vec{n}}\, \frac{\alpha}{2\pi^2}\, (\hat{12}) \,, \tag{20}$$

$$\begin{aligned} dW_{(q\bar{q}g)} &= -\frac{dk}{k} d\Omega_{\vec{n}}\, \frac{\alpha}{4\pi^2} \left(\frac{p_1}{p_1 k} + \frac{p_2}{p_2 k} - \frac{2p_3}{p_3 k} \right)^2 \\ &= \frac{dk}{k} d\Omega_{\vec{n}}\, \frac{\alpha}{2\pi^2} \left(2[(\hat{13}) + (\hat{23})] - (\hat{12}) \right) \,. \end{aligned} \tag{21}$$

For notations see Eq. (15). Note that this pattern mimics the QCD $q\bar{q}g$ sample at $N_c = \sqrt{2}$, cf. Eq. (70).

Let us turn our attention to the negative contribution of the antenna $(\hat{12})$, connected with the 'repulsion' of electrons.

For qualitative analysis it is convenient to rewrite the classical current expression for the $(q\bar{q}g)$ emitter (21) in terms of three-dimensional quantities

$$W^{(q\bar{q}g)}(\vec{n}) = - \left(\frac{p_1}{p_1 k} + \frac{p_2}{p_2 k} - \frac{2p_3}{p_3 k} \right)^2 = (\vec{A} \times \vec{n})^2 \,, \tag{22}$$

where

$$\vec{A} = \left(\frac{\vec{n}_1}{a_1} + \frac{\vec{n}_2}{a_2} - \frac{2\vec{n}_3}{a_3} \right) \,. \tag{23}$$

One can easily see that for a symmetric configuration $(a_{13} = a_{23})$ there is no radiation emitted directly opposite to the double charge: $\vec{A} \parallel \vec{n}$. This can be understood as the vanishing of the electric field midway between the equal charges (see Fig. 6a).

For the QCD case, soft-gluon radiation in this direction is non-zero but appears to be suppressed owing to the same physical reason: replacing γ by a gluon 'recharges' q and $\bar{q}$ to almost oposite colour charges. The resulting ratio of multiplicity flows in symmetric $q\bar{q}\gamma$ and $q\bar{q}g$ events looks as follows:

$$\frac{dN_\gamma/d\vec{n}}{dN_g/d\vec{n}} = \frac{2(N_c^2 - 1)}{N_c^2 - 2} = \frac{16}{7} \,. \tag{24}$$

Note that the destructive interference is strong enough to make even the most kinematically unfavourable direction transversal to the event plane better populated as compared to the $q\bar{q}$ valley. In the case of the threefold symmetric ('Mercedes'-like) $q\bar{q}g$ events the particle flow reads

$$\left(\frac{dN_\perp/d\vec{n}}{dN_{q\bar{q}}/d\vec{n}} \right)_g = \frac{2N_c^2 - 1}{2(N_c^2 - 2)} = \frac{17}{14} \,. \tag{25a}$$

Note that in the case of $q\bar{q}\gamma$ event this ratio is

$$\left(\frac{dN_\perp/d\vec{n}}{dN_{q\bar{q}}/d\vec{n}} \right)_\gamma = \frac{1}{4} \,. \tag{25b}$$

Owing to the constructive interference there is a surplus of radiation in the qg and g$\bar{q}$ valleys: for the 'Mercedes'-like $q\bar{q}g$ configuration midway between the parton directions, one gets

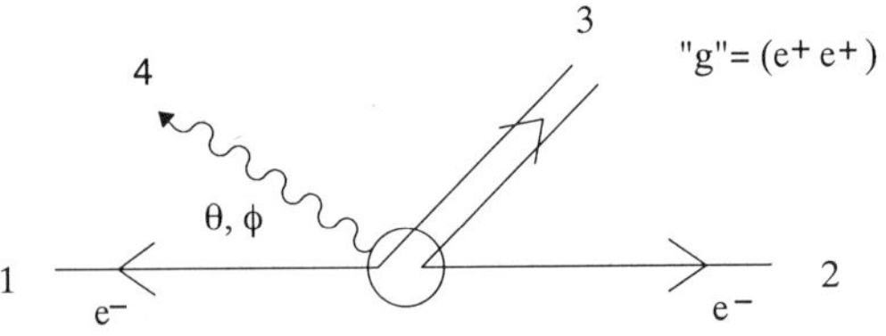

Figure 7. A 'QED' model illustrating two-soft-gluon emission in $e^+e^- \to q\bar{q}$.

$$\frac{dN_{qg}}{dN_{q\bar{q}}} = \frac{5N_c^2 - 1}{2N_c^2 - 4} = \frac{22}{7} \ . \tag{26}$$

For the jetty final hadronic states PA predictions for the ratios of particle flows similar to Eqs. (24)–(26) should remain correct, since non-perturbative hadronization effects should cancel, at least at high energies. Studying the coherent phenomena one gets plenty of infrared stable predictions deeply rooted in the basic structure of non-Abelian gauge theory.

The experimental observations of the above coherent phenomena can be said to test very detailed features of QCD colour flows.

3.5 Characteristic properties of the two-particle correlation function

In Refs. [14] it was suggested that one should analyse the ratio of energy-multiplicity correlations, which proves to be especially sensitive to colour-flow structure in jet formation.

In the case of e^+e^- annihilation the basic quantity is the gluon–gluon correlation function characterizing the emission of two soft gluons by a hard $q\bar{q}$ antenna:

$$C^{q\bar{q}}(\vec{n}_3, \vec{n}_4) = \frac{W^{q\bar{q}}(\vec{n}_3, \vec{n}_4)}{W^{q\bar{q}}(\vec{n}_3)W^{q\bar{q}}(\vec{n}_4)} \ . \tag{27}$$

Here $W^{q\bar{q}}(\vec{n}_i)$ describes the radiation pattern for the $q\bar{q}$ antenna ($i = 3,4$), see Eq. (14). $W^{q\bar{q}}(\vec{n}_3, \vec{n}_4)$ represents the angular distribution of two soft gluons with $E_1 \simeq E_2 \gg E_3 \gg E_4$. E_1 (E_2) is the quark (antiquark) energy, and $\vec{n}_1 \simeq -\vec{n}_2$.

To illustrate the specific features of the correlation function $C^{q\bar{q}}$ let us invoke once more the 'QED' model used in the previous subsection, see Fig. 7.

Consider, for example, the configuration when the gluons are emitted at the same polar angles relative to the direction $\vec{n}_1$ ($\Theta_3 = \Theta_4$) but in the back-to-back azimuthal directions ($\phi = \pi$). Using the three-dimensional representation for the $q\bar{q}$—g— radiation pattern [see Eqs. (22) and (23)] one can easily see that in this case $\vec{A}(\vec{n}_3) \parallel \vec{n} \equiv \vec{n}_4$ and the correlation function $C^{q\bar{q}}$ vanishes. This is a result of destructive interference, which is of the same magnitude as the drag effect in a symmetric $q\bar{q}$—g— configuration.

In orthogonal azimuthal directions ($\phi = \pi/2$) the gluons become uncorrelated, $C^{q\bar{q}} = 1$.

Thus gluon–gluon correlations in the orthogonal and back-to-back azimuthal directions demonstrate the same interference phenomena as the drag effect.

The correlation function $C^{q\bar{q}}$ for the $q\bar{q}$—g— case can be presented in the form

$$C^{q\bar{q}}(\vec{n}_3, \phi) = 1 + 2 \ \cos\phi \ \frac{\sin\Theta_3 \sin\Theta_4}{a_{34}} \ , \tag{28}$$

where

$$a_{34} \ \frac{1}{\sin\Theta_3 \ \sin\Theta_4} = \cosh\eta_{34} - \cos\phi \ . \tag{29}$$

Here $\eta_{34} = \eta_3 - \eta_4$, where η_3 and η_4 are gluon pseudorapidities, $\phi = \phi_3 - \phi_4$, and the corresponding azimuthal angles ϕ_3 and ϕ_4 are relative to the direction $\vec{n}_1 \simeq -\vec{n}_2$:

$$\cosh\eta_{34} \ \sin\Theta_3 \ \sin\Theta_4 = 1 - \cos\Theta_3 \ \cos\Theta_4 \ . \tag{30}$$

---

[3]This section is intended for the advanced reader. On first reading it is possible to skip it without a loss of continuity. For details see Refs. [9],[10].

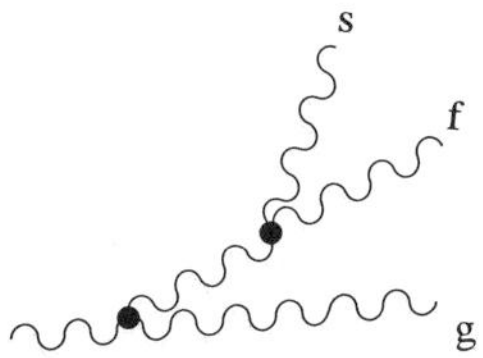

Figure 8. Genealogy of the gluon cascade.

4 MODIFIED LEADING LOGARITHMIC APPROXIMATION [3]

As has been mentioned above, the strong AO provides one with the basis for the probabilistic interpretation of soft-gluon cascades in the DL approximation:

$$k_\mathrm{s} \ll k_\mathrm{f} \ll k_\mathrm{g} , \quad \Theta_\mathrm{sf} \ll \Theta_\mathrm{fg} , \tag{31}$$

where the subscripts denote the cascade genealogy (first introduced by V. Fadin): 'grandpa', 'father', and 'son', as shown in Fig. 8.

The DL approximation happens to be too rough for making reasonable predictions even for asymptotically high energies. For example, it completely ignores the energy-momentum balance (recoil effects). Therefore, it overestimates parton multiplicities in cascading processes, the characteristic energy of the most actively multiplying partons (the position of the maximum of the particle spectra), etc.

Quantitatively, in the DL case one keeps track of the $\sim \sqrt{\alpha_\mathrm{s}}$ term in the anomalous dimension γ, disregarding $O(\alpha_\mathrm{s})$ contributions which give rise to the essential pre-exponential energy-dependent factors.

Thus to keep a quantitative control over parton branching processes one is forced to take into full account non-leading SL effects.

When constructing the probabilistic scheme, taking into account both DL and essential SL effects, one is faced with a dramatic growth of the number of *interference* contributions which must be analysed and interpreted. The interference graphs contain soft-gluon lines, connecting harder partons of quite different generations. Meanwhile, the very idea of the classical *shower picture* implies that the structure of elementary-parton decays, i.e. the 'blocks' for building up the parton cascade, should depend on just the nearest 'fore-fathers' of a considered parton. Thus the possibility of absorbing all essential interference terms into the local probabilistic scheme looks far from being obvious.

Therefore, it is even more striking that such a scheme not only exists but appears to be *a posteriori* a simple almost trivial generalization of the standard LLA scheme. That is the reason to refer to this approximation [6],[28] as the *modified* LLA (MLLA).

It is important to note that beyond the MLLA a probabilistic picture of the parton-cascade evolution simply does not exist (the so-called 'colour monsters': $1/N_\mathrm{c}^2$ suppressed soft contributions, $\Delta\gamma \sim \alpha_\mathrm{s}^2$ [6],[21]).[4]

In dealing with the subleading corrections to DL asymptotics it is helpful to invoke the technique of Generating Functionals (GF), which is perfectly suited to the description of intrajet cascades, see, for example, Ref. [25].

The structure of GF can be expressed symbolically as

$$Z = C(\alpha_\mathrm{s}(t))\exp\left\{ \int^t \gamma(\alpha_\mathrm{a}(t'))\mathrm{d}t' \right\} . \tag{32}$$

Owing to AO the evolution parameter ('time' t) *is* connected here with the jet opening angle $\mathrm{d}t = \mathrm{d}\Theta/\Theta$. One can easily see that the 'time' derivative of Eq. (32) produces the factor $\gamma(\alpha_\mathrm{s}(t))$. It is natural to call the symbolic quantities γ and C, the *anomalous dimension* and the *the coefficient function*, respectively.

The exponent of the integrated anomalous dimension γ incorporates the Markov chains of sequential angular ordered parton decays. The regular coefficient function C, being free of collinear (or 'mass')

---

[4]Another interesting example is connected with the interjet collective phenomena that could be reproduced in a probabilistic way only in the large N_c limit [21] (see also Section 6).

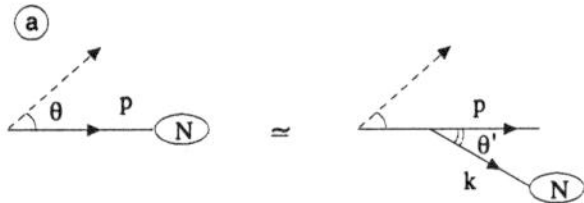

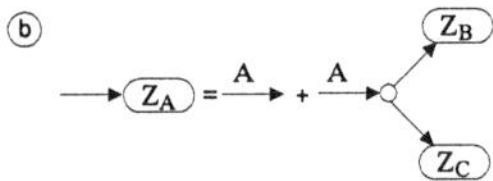

Figure 9. a) Schematic illustration of Eq. (35). b) Schematic illustration of Eq. (41).

singularities, could be said to describe wide-angle partonic configurations, i.e. *multijet contributions* to the evolution of the system.

Successive terms of symbolic series for $\gamma(\alpha_s)$,

$$\gamma = \sqrt{\alpha_s} + \alpha_s + \alpha_s^{3/2} + \alpha_s^2 + \dots , \tag{33}$$

correspond to the increasing accuracy of the description of elementary partonic decays at parametrically small angles $\Theta_{ij} \ll 1$ and thus of the jet evolution. Iterating the coefficient function

$$C = 1 + \sqrt{\alpha_s} + \alpha_s + \dots , \tag{34}$$

one accounts for the ensembles of increasing numbers of such jets with large relative angles $\Theta_{ij} \sim 1$.

To estimate the 'magnitude' of γ let us look at the simple DL evolution equation for parton multiplicity depending on the product of the energy and the opening angle of a jet

$$N(p\Theta) \approx \int^{\Theta} \frac{d\Theta'}{\Theta'} \left[\int^{p} \frac{dk}{k} 4N_c \frac{\alpha_s}{2\pi} \right] N(k\Theta') , \tag{35}$$

as illustrated by (Fig 9a).

Comparing this equation with Eq. (32) one can easily see that the expression in square brackets in Eq. (35) represents the anomalous dimension. Since both the left-hand side and the right-hand side of Eq. (35) contain multiplicity factors of the same order of magnitude, the two logarithmic integrations have to compensate α_s

$$\int dt' \int \frac{dk}{k} \alpha_s \sim 1 .$$

Therefore γ can be estimated as

$$\gamma^{DL}(\alpha_s) = \int \frac{dk}{k} \alpha_s = \alpha_s \ell \sim \sqrt{\alpha_s} . \tag{36}$$

Here we have denoted by ℓ the logarithmic integral over the gluon momentum fraction, which contributes effectively as

$$\int \frac{dk}{k} = \int \frac{dz}{z} = \ell \sim \alpha_s^{-1/2} .$$

Integrating $\gamma(\alpha_s)$ estimated by Eq. (36) in Eq. (32) one comes to the characteristic exponent $\exp (c\sqrt{\ln E})$, which describes the DL behaviour of the multiplicity growth. A non-leading correction to γ, namely $\Delta\gamma \sim \alpha_s$, causes a significant energy dependence $\exp [c_1 \ln \ln (E/\Lambda)] \propto \alpha_s^{-c_1}$ as well. To describe this correctly one has to analyse the following subleading effects: influence of the $\ln (1/z)$ dependence of the running coupling argument; $g \rightarrow q\bar{q}$, $q \rightarrow qg$, and $g \rightarrow gg$ splittings with hard momenta $z \sim 1$; exact angular integration in the $g \rightarrow ggg$ (or $q \rightarrow qgg$) 'double-soft' emission.

Taking these effects into account in the MLLA one gets symbolically

$$\gamma^{MLLA}(\alpha_s) = \sqrt{\alpha_s} + \alpha_s . \tag{37}$$

The MLLA parton-decay probabilities look as follows:

$$dw_A^{BC} = \frac{\alpha_s(k_\perp^2)}{2\pi} \Phi_A^{BC}(z) V(\vec{n}) dz \frac{d\Omega}{8\pi} \ , \tag{38}$$

$$V_{f(g)}^s(\vec{n}) = \frac{a_{sg} + a_{fg} - a_{sf}}{a_{sf}\, a_{sg}} \ , \quad a_{ik} = 1 - \cos(\vec{n}_i \vec{n}_k) \ , \tag{39}$$

where subscripts refer to the soft-gluon family as before.

Equation (38) takes into full account the SL effects. Here the Gribov–Lipatov–Altarelli–Parisi splitting functions Φ_A^{BC} include both soft-gluon emission and terms corresponding to a loss of energy logs in A $\to$ B + C decays. The exact angular kernel $V(\vec{n})$, depending on the directions of momenta of partons of three sequential generations, replaces the rough *strong* AO. To see that this angular factor is nothing but the *exact* AO, the reader is advised to check the nice property of the V-kernel [cf. Eq. (17)]:

$$\langle V_{f(g)}^s(\vec{n}) \rangle_{\text{azimuth average}} = \int\limits_0^{2\pi} \frac{d\phi}{2\pi} V_{f(g)}^s(\vec{n}) = \frac{2}{a_{sf}} \Theta(a_{fg} - a_{sf}) \ . \tag{40}$$

This means that the decay probability integrated over the azimuth of 'son' around 'father' results in the logarithmic Θ-distribution inside the parent cone $\Theta_{sf} \le \theta_{fg}$ and vanishes outside. It is important to emphasize that the 'V-scheme', Eq. (38), proves to eliminate not only A $\to$ A + g$'$ + g$''$ ($\Delta\gamma = \alpha_s$) but also A $\to$ A + g$'$ + g$''$ + g$'''$ ($\Delta\gamma = \alpha_s^{3/2}$) elementary splitting processes, factorizing them completely into the chains of two-parton decays.

When studying the global characteristics of parton systems, such as mean multiplicities and multiplicity fluctuations, energy particle spectra and correlations, etc., one is allowed to replace the full angular kernel $V(\vec{n})$ in Eq. (38) by its *azimuth averaged* analogue of Eq. (40). In this case the exact AO makes it possible to construct a simple evolution equation for jet GFs.

The system of two coupled equations for the quark (Z_F) and gluon (Z_G) functionals reads (A, B, C = F, G) (see Fig. 9b):

$$Z_A(E,\Theta;u(k)) = e^{-w_A(E\Theta)}u_A(k_0 = E) + \frac{1}{2}\sum_{B,C}\int^\Theta \frac{d\Theta'}{\Theta'}\int_0^1 dz\, e^{-w_A(E\Theta)+w_A(E\Theta')} \times$$

$$\times \frac{\alpha_s(k_\perp^2)}{2\pi}\Phi_A^{BC}(z)Z_B(zE,\Theta';u)Z_C((1-z)E,\Theta';u) \ , \tag{41}$$

where $u(k)$ is the probing function.

The first term on the right-hand side corresponds to the form-factor damped situation when the A-jet with energy E and opening angle Θ consists of the parent parton only. The integral term describes the first splitting A$\to$ B + C with the angle Θ' between the products. The exponential factor ensures that this decay is the first one indeed: it is the probability of emitting *nothing* in the angular interval between Θ and $\Theta' \le \Theta$. The last two factors account for the further evolution of the subjets produced having smaller energies and Θ' as the opening angles.

The MLLA form factors are the following:

$$w_F = \int^\Theta \frac{d\Theta'}{\Theta'}\int_0^1 dz\, \frac{\alpha_s(k_\perp^2)}{2\pi}\Phi_F^F(z) \ , \quad (F = q, \bar{q}) \ , \tag{42}$$

$$w_G = \int^\Theta \frac{d\Theta'}{\Theta'}\int_0^1 dz\, \frac{\alpha_s(k_\perp^2)}{2\pi}\left[\frac{1}{2}\Phi_G^G(z) + n_f\Phi_G^F(z)\right] \ . \tag{43}$$

Collinear and soft singularities presented in Eqs. (41)–(43) may be regularized by imposing the usual perturbative restriction, which we shall write here as

$$k_\perp \simeq Ez(1-z)\Theta' > Q_0 \ . \tag{44}$$

Differentiating the product $Z_A(\Theta)\exp[w_A(E\Theta)]$ over Θ and using Eqs. (42) and (43) one arrives at the master equation, which is actively exploited for studying properties of QCD jets, see Refs. [6],[10],[21]:

$$\frac{\mathrm{d}}{\mathrm{d}\ln\Theta}Z_{\mathrm{A}}(E,\Theta) = \frac{1}{2}\sum_{\mathrm{B,C}}\int_0^1 \mathrm{d}z\,\frac{\alpha_{\mathrm{s}}(k_\perp^2)}{2\pi}\Phi_{\mathrm{A}}^{\mathrm{BC}}(z)$$

$$\times\left\{Z_{\mathrm{B}}(zE,\Theta)Z_{\mathrm{C}}((1-z)E,\Theta) - Z_{\mathrm{A}}(E,\Theta)\right\} \tag{45}$$

Equation (45) accumulates information about *azimuth averaged* jet characteristics in MLLA. Taking the nth variational derivative of GF over the probing function $u(k_i)$ near the 'point' $u = 0$ one gets the *exclusive* n-parton cross-sections. Expansion of Z_{A} at $u = 1$ generates *inclusive* parton distributions and correlations.

To conclude this section let us present one of the most practically important and pedagogically instructive results of the MLLA technique, namely, the expression for the inclusive energy spectrum of partons B in the A-jet (for details, see Refs. [6],[21]).

The evolution equation for particle spectra following directly from Eq. (45) reads:

$$\frac{\mathrm{d}}{\mathrm{d}\ln\Theta}x\bar{D}_{\mathrm{A}}^{\mathrm{B}}(x,Y,\lambda) = \sum_{\mathrm{C}=q,\bar{q},g}\int_0^1 \frac{\mathrm{d}z}{z}\,\frac{\alpha_{\mathrm{s}}(k_\perp^2)}{2\pi}\Phi_{\mathrm{A}}^{\mathrm{C}}(z)\left[\frac{x}{z}\bar{D}_{\mathrm{C}}^{\mathrm{B}}\left(\frac{x}{z},\frac{\ln Ez\Theta}{Q_0},\lambda\right)\right], \tag{46}$$

where

$$Y = \ln\left(\frac{E\Theta}{Q_0}\right) = \ell + z\,, \quad \ell = \ln\frac{E}{k} = \ln\frac{1}{x}\,,$$

$$y = \ln\left(\frac{k\Theta}{Q_0}\right)\,, \quad \lambda = \ln\frac{Q_0}{\Lambda}\,. \tag{47}$$

Here E and Θ are the energy and the opening angle of a jet A,

$$k_\perp \simeq z(1-z)E\Theta \geq Q_0\,.$$

Through the chain of transformations one comes to the following equation for the leading contributions $D^+(\omega,Y,\lambda)$ to the momentum representation of the spectra

$$\left(\omega + \frac{\mathrm{d}}{\mathrm{d}Y}\right)\frac{\mathrm{d}}{\mathrm{d}Y}D^+ = 4N_{\mathrm{c}}\frac{\alpha_{\mathrm{s}}}{2\pi}D^+ - a\left(\omega + \frac{\mathrm{d}}{\mathrm{d}Y}\right)D^+\,. \tag{48}$$

Here $D(\omega,Y,\lambda)$ are the Mellin-transformed distributions

$$D(\omega,Y) = \int_0^1 \frac{\mathrm{d}x}{x}x^\omega[x\bar{D}(x,Y)] = \int_0^\infty \mathrm{d}\ell\,e^{-\omega\ell}D(\ell,Y)\,, \tag{49}$$

$$a = \frac{11}{3}N_{\mathrm{c}} + \frac{2n_{\mathrm{f}}}{3N_{\mathrm{c}}^2}\,. \tag{50}$$

Introducing the anomalous dimension as follows

$$D^+(\omega,Y) = D^+(\omega,Y_0)\exp\int_{Y_0}^Y \mathrm{d}y\,\gamma[\omega,\alpha_{\mathrm{s}}(y)]\,, \tag{51}$$

one finally arrives at the MLLA expression

$$\gamma = \gamma^{\mathrm{DL}} + \frac{\alpha_{\mathrm{s}}}{2\pi}\left[-\frac{a}{2}\left(1 + \frac{\omega}{\sqrt{\omega^2 + 4\gamma_0^2}}\right) + b\frac{\gamma_0^2}{\omega^2 + 4\gamma_0^2}\right] + O(\alpha_{\mathrm{s}}^{3/2})\,, \tag{52}$$

where

$$\gamma_0^2 = \frac{4N_{\mathrm{c}}\alpha_{\mathrm{s}}}{2\pi}\,, \quad b = \frac{11N_{\mathrm{c}}}{3} - \frac{2n_{\mathrm{f}}}{3} \tag{53}$$

$$\gamma^{\mathrm{DL}}(\omega,\alpha_{\mathrm{s}}) = \frac{1}{2}\left(-\omega + \sqrt{\omega^2 + 4\gamma_0^2}\right) \tag{54}$$

corresponds to the DL approximation.

One can solve the differential equation (48) explicitly, reducing it to the confluent hypergeometric equation for the product $\alpha_{\mathrm{s}}(Z)D$. Taking into account initial conditions, the solution in the physical region $\ell \leq Y$ reads:

$$x\bar{D}_{\mathrm{q}}^{\mathrm{g}}(x, Y, \lambda) = \frac{4C_{\mathrm{F}}(Y + \lambda)}{bB(B + 1)} \int\limits_{\varepsilon - \mathrm{i}\infty}^{\varepsilon + \mathrm{i}\infty} \frac{\mathrm{d}\omega}{2\pi\mathrm{i}} x^{-\omega} \, \Phi[-A + B + 1, B + 2, -\omega(Y + \lambda)] \, K(\omega, \lambda) ,$$

$$K(\omega, \lambda) = \frac{\Gamma(A)}{\Gamma(B)}(\omega\lambda)^B \, \Psi(A, B + 1, \omega\lambda) . \tag{55}$$

Φ and Ψ are the standard degenerate hypergeometric functions.

Here we have used the notations

$$A = \frac{4N_{\mathrm{c}}}{b\omega} , \quad B = \frac{a}{b} . \tag{56}$$

The MLLA expression (55) is formally expected to be valid at ($\Theta \sim 1$):

$$1 \gg x > Q_0/E , \quad k \gg Q_0, \Lambda .$$

Note that the dimensional quantity Q_0 regularizes collinear divergencies. This quantity represents the minimal value of the relative $k_\perp$ of decay products in jet evolution. Q_0 also puts a bound on parton energies $E_{\mathrm{p}} = xE \geq k_\perp \geq Q_0$, thus playing the role of the 'effective mass' of a parton.

The restored x-spectrum of partons exhibits the above-mentioned 'hump-backed' structure with a maximum at particle energies asymptotically approaching $\sqrt{E_{\mathrm{jet}}/\Theta}$. At $\Theta \sim 1, Q_0 = \Lambda$,

$$\ln\left(\frac{1}{x_0}\right) = \frac{1}{2}Y\left(1 + B\sqrt{\frac{b}{4N_{\mathrm{c}}Y}}\right) + O(1) \tag{57}$$

(width $\sigma \sim Y^{3/4}$).

Taking $\omega = 0$ as the moment in Eq. (55) one obtains the analytical expression for the jet-parton multiplicity through modified Bessel functions with the MLLA asymptotic

$$\ln\left(N^{\mathrm{g}}\right) \simeq \sqrt{\frac{32N_{\mathrm{c}}\pi}{\alpha_{\mathrm{s}}(E)}} \frac{1}{b} + \left(\frac{B}{2} - \frac{1}{4}\right) \ln \alpha_{\mathrm{s}}(E) + O(1) . \tag{58}$$

Note that a change of the scale parameter Λ by a factor of $O(1)$ would correspond to a correction of $O(\sqrt{\alpha_{\mathrm{s}}})$ in multiplicity N [21]. Such corrections are subleading to the MLLA and therefore this parameter cannot be compared directly with the values obtained from other experiments.

The first term in Eq. (58) is reduced by a factor of $1/\sqrt{2}$, relative to the incoherent case. This reduction is a direct result of the destructive interference of soft-gluon emission.

5 INCLUSIVE PARTICLE SPECTRA AND INTRAJET CO-HERENCE (GOOD NEWS)

At sufficiently high energies, events in hard processes should possess the clear geometry that reflects the topology of the participating partons. The space-energy portrait represents a natural partonmeter for registering the kinematics of the energetic partons. While the *hard component* of a hadron system (a few hadrons with $z \sim 1$) determines the partonic skeleton, the *soft component* (the other hadrons with $z \ll 1$) forms the bulk of the multiplicity. This component is concentrated inside the bremsstrahlung cones of jets. The opening angle θ_0 of each cone is bounded by the nearest other jet. Even though the bremsstrahlung cones of the neighbouring jets strongly overlap, the resulting total multiplicity can be presented as the additive sum of the contributions of the individual jets.

One can study the properties of an individual quark jet when measuring the different inclusive distributions in the process $e^+e^- \rightarrow$ hadrons. In spite of the great importance of coherence phenomena, the notion of an isolated jet makes sense if one does not deal with the azimuthal effects, but considers only multiplicities, energy spectra and correlations, etc.

The inclusive energy spectrum for e^+e^- annihilation is the sum of two q-jet contributions:

$$\frac{1}{\sigma} \frac{\mathrm{d}\sigma}{\mathrm{d}(\ln 1/x)} = 2\bar{D}_{\mathrm{q}}^{\mathrm{h}}(\ell, Y)|_{\Theta \simeq 1} . \tag{59}$$

Here $x = E_{\mathrm{h}}/E, E = W/2$, where W is the total c.m. energy.

To link hadronic spectra in jets with the calculated parton distribution let us recall that the Q_0

value in Eq. (55) sets a formal boundary between two stages of jet evolution: the perturbative one and then the phase of non-perturbative transition into hadrons. If the theory of hadroproduction should exist, the final result would be independent of the quantity Q_0. As a matter of fact, for large enough Q_0 the number of partons produced at recent energies is certainly small. So, one is forced to apply some *ad hoc* hadronization model describing the multihadron production as the evolution 'below Q_0' of a partonic system with large invariant masses of parton pairs. Unfortunately, an experimental verification of such results looks more like a tuning of parameters of a phenomenological model rather than a real test of QCD predictions.

However, by an intent look at Eq. (55) an opportunity to make a model-independent prediction may be found. The inclusive distribution of *hadrons* could be represented by the formula similar to Eq. (55) where the K-factor is replaced by the product

$$K^{\mathrm{h}}(\omega) = K(\omega, \lambda) C^{\mathrm{h}}(\omega, \lambda; m_{\mathrm{h}}, J^{\mathrm{PC}}) , \tag{60}$$

where $C^{\mathrm{h}}(\omega)$ is a Mellin-transformed parton fragmentation function $C^{\mathrm{h}}(z)$.

In the kinematical region of relatively soft particles, which we are mainly interested in, the essential values of ω under the integral are small, $\omega \ll 1$ [near the peak they are parametrically small, $\omega_0 \sim \sqrt{\alpha_{\mathrm{rms}}(Y)}$].

To understand how hadronization affects the spectrum shape one needs to know the behaviour of $K^{\mathrm{h}}(\omega)$ at $\omega \to 0$.

Let us consider two different possibilities [6],[7]:

i) The singular behaviour $K^{\mathrm{h}}(\omega) \sim 1/\omega + \ldots$ corresponds to the physical picture where each coloured parton produced hadrons with a plateau-like energy distribution: $C^{\mathrm{h}}(z) \propto 1/z$. In such a case the dip at small x, which is characteristic for parton spectra, would never manifest itself in hadron distributions.

ii) The regular behaviour $K \to$ const. corresponds to a *local* in the phase-space blanching and the hadronization of partons. In this case hadron and parton spectra prove to be similar.

It is perhaps surprising to see the x-dependence of $x\bar{D}^h(x)$ being given completely by means of the perturbative evolution. Non-perturbative effects can smear the distributions over a limited interval in $\ln 1/x$. This is, however, the higher-order effect from the MLLA point of view. Thus, the overall normalization factor remains the only arbitrary parameter. It may be fixed, for example, by fitting the average multiplicity (the value of Λ could be determined phenomenologically from comparison of the PA predictions for spectra with experiment).

So, in such a case, hadron distributions become similar to the parton ones, and LPHD is realized.

Whether or not present jet energies are sufficiently large for LPHD to be applied is, of course, a question to experiment. So far experimental data (see, for example, Refs. [39]–[42]) have demonstrated that the LPHD works unexpectedly well. A strong support of the PA ideas came recently from the results of the OPAL Collaboration at the Z^0 (see Ref. [40] and discussion below).

Before coming to the comparison with data let us emphasize an important property of the parton spectrum, given by Eq. (55). Namely, at very high energies, when the typical value of the product $\omega\lambda \sim \lambda/\sqrt{Y} \ll 1$, the *shape* of the spectrum appears to be insensitive to the value of Q_0:

$$K(\omega, \lambda) \approx \frac{2}{\Gamma(B)} \left(\frac{Z_0}{2}\right)^B K_B(Z_0) = \mathrm{const}\ (\omega), \quad Z_0 = \sqrt{\frac{16 N_{\mathrm{c}} \lambda}{b}}, \quad \lambda = \ln \frac{Q_0}{\Lambda} . \tag{61}$$

Thus when the bremsstrahlung cascade is developed enough, the shape of the resulting energy distribution of particles becomes insensitive to the processes occurring at the last steps of evolution (at $k_\perp \sim Q_0$). This observation may serve to justify an attempt to provide the developed cascade not with an expensive increase of total energy E but with a decrease of Q_0, thus enhancing the responsibility of perturbative QCD for jet evolution at recent energies, see Refs. [7].

At $\lambda = 0$ the expression (55) simplifies noticeably, since in this case $K(\omega, Q_0) = 1$ at all ω (the so-called 'limiting' spectrum). Decreasing λ means extending the responsibility of the perturbative stage for the jet dynamics beyond its formal range of applicability. In some sense the limiting choice $Q_0 \simeq \Lambda$ is a specific attempt to model the confinement.

Already the first comparisons [7] of the $\pi^\pm$ data at PETRA/PEP energies with the limiting MLLA spectrum have demonstrated a quite satisfactory agreement. The effective value of Λ appeared to be close to m_π. It has also been noticed that the relative yields of hadrons of different species (p, K, ...) can be reasonably simulated simply by cutting the development of the parton cascade, leading to the production of a hadron h, by the value $Q_0 = m_{\mathrm{h}}$ (preserving the same value of Λ). This procedure would also influence the shape of energy spectra of massive hadrons, stiffening them via the increase of the minimal energy fraction $x = k/E_{\text{jet}} > k_\perp/E_{\text{jet}} \geq Q_0/E_{\text{jet}} = x_{\min}$ available for the relevant partons. As a consequence, the position of the hump in the energy spectra becomes higher as the mass of the hadrons increases.

The conjecture [7] about an increase of $(Q_0)_{\text{eff}}$ with hadron mass has been supported by experiment (see Refs. [39]).

When using the limiting spectrum for comparison with the data on different types of hadrons, the effective value of Λ is expected to increase with the hadron mass [14].

In the limiting case $(Q_0 = \Lambda)$ the MLLA spectrum (55) can be presented in a form [6]

$$
\bar{D}_{\text{q}}(\ell, Y) = \frac{4C_{\text{F}}}{b}\Gamma(B) \int_{-\pi/2}^{\pi/2} \frac{d\tau}{\pi} \, e^{-B\alpha} \left[\frac{\cosh\alpha + (1-2\zeta)\sinh\alpha}{(4N/b)Y(\alpha/\sinh\alpha)} \right]^{\frac{B}{2}}
$$

$$
\times \, I_{\text{B}}\left(\sqrt{\frac{16N_{\text{c}}}{b}Y\frac{\alpha}{\sinh\alpha}[\cosh\alpha + (1-2\zeta)\sinh\alpha]} \right) , \tag{62}
$$

which is especially convenient for numerical integration.

Here $Y = \ln(E/\Lambda)$ and $\alpha = \alpha_0 + i\tau$ with

$$
\tanh\alpha_0 = 2\zeta - 1 , \quad \zeta = \frac{Y - \ell}{Y} . \tag{63}
$$

I_{B} is the modified Bessel function of order B.

Taking into account the next-to-MLLA effects leads to some change in the shape of spectra in q and g jets. The spectrum in the g jet shifts to lower x.

One may find analytically the asymptotic shape of the limiting distribution not too far from its peak by saddle-point evaluation of the integral (55)

$$
\bar{D}_{\text{q}}(\ell, Y) = N_q(E, \Lambda) \left(\frac{C}{\pi Y^{3/2}} \right)^{1/2} \exp\left[\frac{-c(\ell - \ln 1/x_0)^2}{Y^{3/2}} \right] ,
$$

$$
\ell = \ln\frac{1}{x} , \quad C = \sqrt{\frac{36N_{\text{c}}}{b}} . \tag{64}
$$

It has a broad Gaussian shape with the peak position given by Eq. (57). It grows rather slowly with the jet energy E

$$
\frac{E \, dE_0}{E_0 \, dE} = \frac{1}{2} - \sqrt{\frac{b}{64N_{\text{c}}Y}} , \quad E_0 = x_0 E . \tag{65}
$$

A sizeable shift in $(\ln 1/x_0)$, of the order of 0.5 units, has been predicted in going from PETRA to Z^0 energy $[(\ln 1/x_0)_{Z^0} = 3.6]$

In Refs. [12] has been derived a transparent next-to-leading-order expression for the $\ln 1/x$ distribution in the soft region in the form of a distorted Gaussian

$$
\bar{D}(\ell, Y) \simeq \frac{N(E, \Lambda)}{\sigma 2\pi}\exp\left[\frac{1}{8}k - \frac{1}{2}s\delta - \frac{1}{4}(2+k)\delta^2 + \frac{1}{6}s\delta^3 + \frac{1}{24}k\delta^4 \right] \tag{66}
$$

with

$$
\delta = (\ell - \bar{\ell})/\sigma , \quad \bar{\ell} \equiv \langle \ell \rangle ,
$$

$$
\bar{\ell} = \ln\left(\frac{1}{x_0} \right) + O(1) .
$$

Parameters σ, s, and k are the width, skewness, and kurtosis of the $(\ln 1/x)$ distribution [12]:

$$\sigma = \sqrt{\frac{Y}{3}} d^{1/4} \left(1 - \frac{b}{64} d^{-1/2} \right) + O\left(Y^{-1/4} \right) ,$$

$$s = -\frac{a}{16} \sqrt{\frac{3}{Y}} d^{-1/4} + O\left(Y^{-5/4} \right) ,$$

$$k = -\frac{27}{5Y} \left(d^{1/2} - \frac{b}{24} \right) + O\left(Y^{-3/2} \right) ,$$

$$d = \frac{bY}{48} = \frac{\pi}{24\alpha_{\mathrm{s}}(E)} . \tag{67}$$

One finds that s and k vanish asymptotically, while the so-called higher-order reduced cumulants of the distribution are $O(\alpha_{\mathrm{s}})$ or smaller and therefore should be neglected in the MLLA case.

Let us recall here that the limiting distribution (62), including $O(\sqrt{\alpha_{\mathrm{s}}})$ corrections (after phenomenological fixing of Λ), is expected to be valid for relativistic particles in a region $\ln 1/x \gg 1$. It means that at high energies the range of its validity is parametrically $\Delta\ell \sim Y$.

The Gaussian-like formulae describe spectra only in the region not so far from the peak, and are formally applicable in the range $\Delta\ell \sim \sigma \sim Y^{3/4}$.

Although the underlying MLLA physics is the same in both cases, one may expect (as some compensation for a rather cumbersome form) that the limiting spectrum at high energies provides us with somewhat more information[5].

As has already been mentioned, even before the start of the activity at the Z^0, the experimental data supported PA predictions. This nice tradition was continued by the ALEPH data [42] on the x distribution of LEP. Being replotted on a $\ln 1/x$ scale these data demonstrated a rather good agreement with the perturbative predictions[6], see also Ref. [42].

Below we shall concentrate on the new OPAL data on the inclusive momentum distribution of charged particles in e^+e^- collisions at the Z^0 energy. Being finely binned at small x, thse data provide a convincing test of the analytical QCD predictions. To study the predicted spectacular energy evolution of the spectra the OPAL analysis included data at lower energies from the recent TASSO paper [41].

Figure 10 shows the OPAL data on the $\ln (1/x_p)$ distribution ($x_p = 2p/W$, p being the momentum of each hadron[7]) together with the predictions of analytical formulae (62), (64), and (66). When confronted with the data, formula (62) was multiplied by the overall normalization factor $K(Y)$, connecting the numbers of hadrons and partons. Note that for the OPAL data the systematic error in the absolute normalization is 5%. The relative systematic error between the different x_p values is much smaller than 5%. LPHD-induced, this factor should be universal at very high energies. Dependence on Y effectively encodes the influence of the asymptotically vanishing perturbative effects as well as the possible deviations from the PA picture.

In good accordance with the theory, the measured spectrum has a broad maximum around $\ln (1/x_0) \simeq 3.6$. A fit performed using the limiting formula (62) around the maximum (in the interval $2.5 < \ln 1/x < 4.5$) gives the values:

$$K(m_Z) = 1.28 \pm 0.01 , \quad \Lambda = 0.253 \pm 0.030 \text{ GeV} . \tag{68}$$

From Fig. 10, one can easily see that in all three cases the region around the maximum is well described. This agreement of the shape and the peak position of the measured hadron distribution with the analytical result derived for parton distribution can be considered as a strong argument in favour of the LPHD–MLLA picture of multihadron production.

The deviation from the predictions, based on the relativistic treatment in the region of non-relativistic momenta at small x_p, is surely not unexpected.

Not surprisingly, the Gaussian-like formulae break down at larger x_p. The agreement with the limiting spectrum (62) in the finite-x region looks more challenging. The question immediately arises, Is it really a coincidence?

At the present level of understanding the answer is as follows (see footnotes[4],[6]): the agreement at finite x is *accidental* but *natural*.

---

[5] I am very grateful to Yu.L. Dokshitzer for his clarifying comments on this and many other problems. I hope that sooner or later the results of the enjoyable discussions with him and S.I. Troyan will be published elsewhere. See also Ref. [10].

[6] This observation was made by W. Hofmann and the author immediately after the ALEPH results appeared.

[7] Recall that the analytical formulae do not discriminate between momentum and energy and are valid only for relativistic particles. The mass effects are included, for example, in Monte Carlo algorithms.

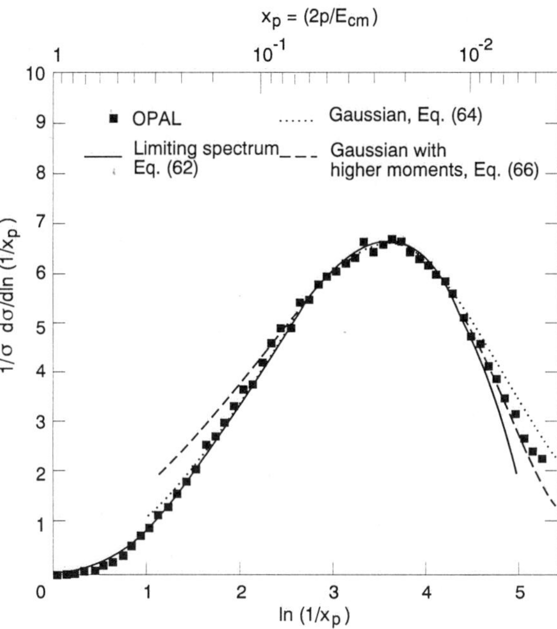

Figure 10. Single-particle inclusive $\ln(1/x_p)$ distribution at $W = 91\,\mathrm{GeV}$, compared with analytical formulae [40].

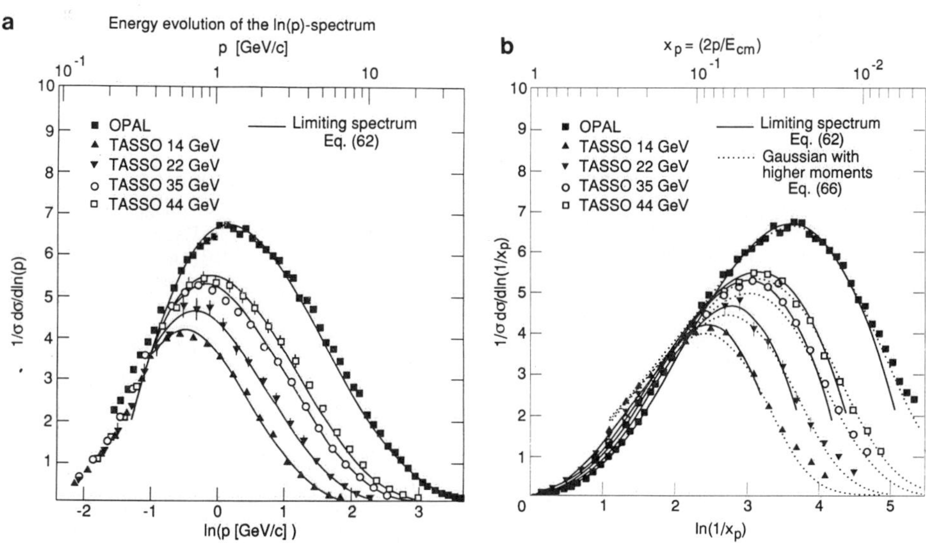

Figure 11. a) Single-particle inclusive $\ln(1/x_p)$ distributions at $W = 91, 44, 35, 22,$ and 14 GeV, [40]; b) ln p distributions at $W = 91, 44, 35, 22,$ and 14 GeV [40].

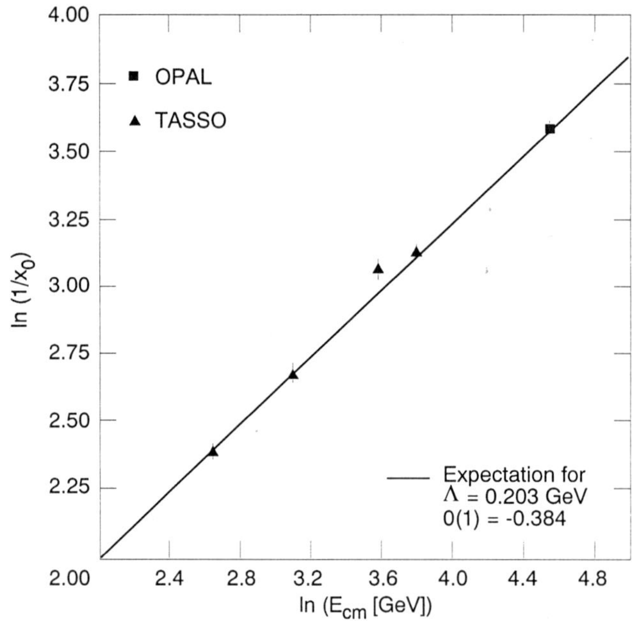

Figure 12. Energy evolution of the maximum of $(\ln 1/x_\mathrm{p})$ compared with the analytical formula (57).

Table 1. Normalization factors $K(Y)$ at different energies W
· (the errors do not include systematic effects) [40]

W (GeV)	91	44	35	22	14
$K(Y)$	1.28 ± 0.01	1.31 ± 0.01	1.36 ± 0.01	1.38 ± 0.02	1.46 ± 0.02

To elucidate the point, the reader is reminded that at $x \ll 1$ the essential values of ω under the integral in Eq. (55) are small $(\omega \sim 1/\sqrt{Y})$. The anomalous dimension $\gamma(\omega, \alpha_\mathrm{s}) = [\alpha_\mathrm{s}(Y)/2\pi]/\nu_+(\omega)$ is singular at $\omega \ll 1$ [see Eq. (52)], the trajectory $\nu_+(\omega)$ is $\nu_+(\omega) = (4N_\mathrm{c}/\omega) - a + O(\omega)$. This approximation, however, proves to simulate reasonably well the behaviour of γ also at $\omega \sim 1$, the region responsible for $x \sim 1\,(\ell \sim 1)$.

The formal extrapolation to $\omega = 1$ gives $\nu_+(1) = 12 - (101/9) = 7/9$, not far from the exact value $\nu_+ = 0$, following from the momentum conservation. Moreover, at larger $\omega > 1$, ν_+ becomes negative (as the true γ does), thus imitating the scaling violation in the fragmentation region, $x \to 1$.

The combination of the OPAL measurements with the results obtained by the TASSO Collaboration at energies between 14 and 44 GeV is shown in Figs. 11a and 11b. Figure 11a also displays the predictions of the analytical formulae (62) (full line) and (66) (dotted line). In Fig. 11b the ln p distribution is compared only with formula (62).

The Λ parameter was fixed at the value obtained at the Z^0 [see Eq. (68)]. The normalization factors $K(Y)$ were determined for each energy separately from a fit to a region around the peak (see Table 1).

From Figs. 11 one finds that the analytical MLLA formulae provide an excellent description of all the energy evolution of spectra. This result wonderfully confirms the QCD cascading picture of the multiple hadroproduction. As can be seen from Fig. 11a, the height of the hump rises with increasing energy, reflecting the increase of multiplicity. Especially spectacular is the energy evolution of the peak position ln $1/x_0$, which is independent of the normalization factors. It is displayed in Fig. 12 together with the MLLA prediction [Eq. (57)] for values of Λ and $O(1)$ obtained from the fit of Eq. (64) to the OPAL spectrum. The confirmation of this basic prediction of the cascading picture is really exciting.

The ln p distributions, shown in Fig. 11b, give evidence in favour of the scaling behaviour at low-p values (especially for the TASSO data). This observation provided convincing confirmation of the basic idea of QCD coherence that the low-momentum particles (hadrons!) do not multiply (see also Refs. [7], [43]). To come to a final conclusion on this important problem we need a better understanding of the relative normalization between the TASSO and OPAL data in the low-p region.

438

Data at low p also demonstrate that with increasing energy the peak of the spectrum becomes more separated from the region where the influence of hadron masses is important.

A rigorous reader could, nevertheless, be unsatisfied with the one-sided interpretation of the new data on particle spectra solely in favour of the PA picture. The point is that the data show a systematic increase of the normalization factor $K(Y)$ with decreasing c.m. energy from 91 to 14 (22) GeV by 14% (8%) (recall that the systematic error in normalization is $\pm 7\%$).

Before expanding on the intriguing problem of the possible influence of the hadronization stage, one should better understand the non-leading perturbative corrections to the limiting spectra [see footnote[6)]].

The formulae [Eqs. (55), (62)] for the MLLA spectra correspond only to the parton–gluon distributions. However, in the formation of the final hadrons the soft-sea $q\bar{q}$ pairs could participate as well. Taking them into account may result in an additional multiplicative factor $[1 + O(1/\sqrt{Y})]$, which decreases with c.m. energy. Some energy dependence of $K^{\mathrm{ch}}(Y)$ could be connected with the fact that $\overline{D}^{\mathrm{ch}}$ is a mixture of different particles, the spectra of which prove to be not exactly similar.

In order to understand this and some other problems of the LPHD picture, the measurements of inclusive spectra of identified hadrons (π, K, p, ...) are important. They should also provide a test of the conjecture [7] about the connection of $(Q_0)_{\mathrm{eff}}$ with the hadron mass.

Note that the inclusive momentum distributions of charged particles (Fig. 11a) can be described as the sum of MLLA spectra for π's, K's, and p's, assuming $Q_0 = m_h$ and $\Lambda \simeq m_\pi$, see Refs. [7], [21]. This implies the natural explanation that the effective Λ, as determined from the data, is $\Lambda \simeq 0.253$ GeV, larger than the value $\Lambda \simeq m_\pi$ found in Refs. [7] from fitting only $\pi^\pm$ spectra.

To conclude this section let us mention that one can study the energy dependence of the inclusive distribution $\overline{D}_{\mathrm{q}}(\ell, Y)$ by using the data at the Z^0 peak only, see Refs. [9], [14], [44]. This can be done by measuring the single inclusive distribution $\overline{D}_{\mathrm{q}}^{\theta}(\ell, Y)$ obtained by counting only the hadrons which enter into a cone of aperture θ around the jet axis. Particles entering into this cone come from the QCD cascade of a primordial parton with virtuality of the order of $E_{\mathrm{eff}} = E \sin \theta/2$. Therefore the inclusive distribution $\overline{D}_{\mathrm{q}}^{\theta}(\ell, Y)$ obtained within this cone corresponds to $\overline{D}_{\mathrm{q}}(\ell, Y_{\mathrm{eff}})$. More precisely, for small x, one should find that

$$\overline{D}_{\mathrm{q}}^{\theta}(\ell, Y) = \overline{D}_{\mathrm{q}}^{\theta}(\ell, Y_{\mathrm{eff}}) + [\overline{D}_{\mathrm{q}}(\ell, Y) - \overline{D}(\ell, \overline{Y})] , \tag{69}$$

where

$$Y_{\mathrm{eff}} = \ln \frac{E \sin \theta/2}{\Lambda}, \ \ \overline{Y} = \ln \frac{E \cos \theta/2}{\Lambda} \ \ (0 < \theta < \pi/2) .$$

The term in square brackets accounts for partons which are emitted in the *backward* jet. Notice that for $\theta = \pi/2$ we obtain the full distribution. Of course the value of E_{eff} should be reasonably high for perturbation theory to be applicable, so one should not take θ too small.

6 THE Z^0 RESONANCE AS A LABORATORY TO STUDY INTERJET COLLECTIVE EFFECTS

6.1 Radiophysics of particle flows

In the framework of LPHD the source of multihadron production in QCD jets is gluon bremsstrahlung, so one should expect that the produced hadrons are a consequence of the colour dynamics at small distances. Therefore, the detailed features of the parton-shower system, such as the flow of colour quantum numbers, influence significantly the distribution of colour-singlet hadrons in the final state.

Such a phenomenon has been first observed in the experiments (see Refs. [45]) studying the angular flows of hadrons in three-jet ($q\bar{q}g$) events from e^+e^- annihilation, the string [29] (or drag [30]) effect. The PETRA/PEP data have strongly supported the predicted drag of the interjet particles in the direction of the gluon jet (net destructive interference in the region between the q and the $\bar{q}$). They demonstrated that the wide-angle particles really do not belong to any particular jet, but have emission properties dependent on the overall jet ensemble.

Detailed studies of the string-like collective effects are of importance for the high-energy reactions. These effects are interesting, not only in their own right, as a test of QCD. They could also be valuable in helping to distinguish new physics signals from the conventional QCD backgrounds.

The Z^0 resonance seems to be an ideal laboratory to explore QCD collective phenomena in three-jet events for the following reasons.

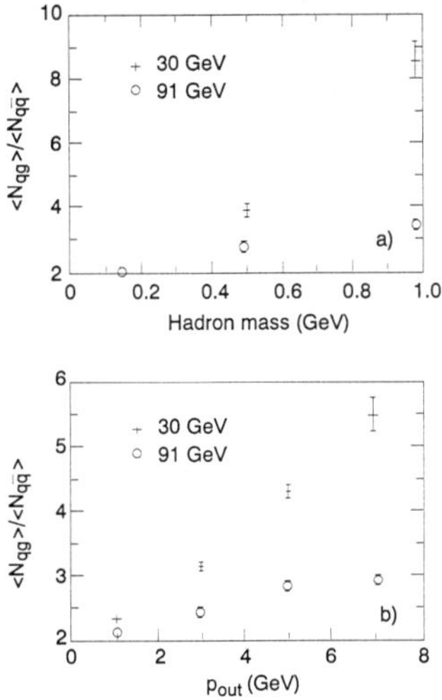

Figure 13. The ratio of particle flow between q and g jets to that between q and q̄ jets, for Mercedes-like configurations (using 60° sectors) [47]: a) mass dependence of the string effect at $W = 30$ and 91 GeV; b) p_{out} dependence of the string effect at $W = 30$ and 91 GeV.

i) Jets are more energetic and collimated than at lower energies and the gluon jet is better separated (see Ref. [46]).

ii) An essential advantage is the tagging of heavy-quark jets. It allows one not only to isolate gluon jets but also to study many subtle properties of interjet particle flows (see Refs. [21], [18]).

iii) At SLC/LEP energies one can convincingly discriminate between the Lund (string) [29] and the PA [30] interpretation of the celebrated string effect (see Refs. [47]).

Already the first experimental studies [46] of three-jet events from Z^0 decays look very promising.

To give a QCD explanation of the drag phenomenon we present the antenna pattern for soft-gluon emission from a hard, massless $q\bar{q}g$ system (see Refs. [30], cf. subsection 3.4). Denoting the quark, antiquark, and the hard- and soft-gluon directions by $\vec{n}_1, \vec{n}_2, \vec{n}_3$, and $\vec{n}_4$, respectively, we have

$$W^{q\bar{q}g}(\vec{n}_4) \propto N_{\mathrm{c}} \left(\frac{a_{13}}{a_{14}a_{43}} + \frac{a_{23}}{a_{24}a_{43}} \right) - \frac{1}{N_{\mathrm{c}}} \frac{a_{12}}{a_{14}a_{42}} , \qquad (70)$$

where $W^{q\bar{q}g}(\vec{n}_4)$ represents the angular distribution of the soft gluon and $a_{ij} = 1 - \vec{n}_i \cdot \vec{n}_j$. This may be compared with the pattern for a quark–antiquark system, with a hard photon replacing the gluon jet 3 to permit the directions $\vec{n}_1$ and $\vec{n}_2$ to remain the same:

$$W^{q\bar{q}\gamma}(\vec{n}_4) = W^{q\bar{q}}(\vec{n}_4) \propto 2C_{\mathrm{F}} \frac{a_{12}}{a_{14}a_{42}} , \qquad (71)$$

Let us emphasize here that, owing to colour coherence, the radiation of a secondary soft gluon $\vec{n}_4 \, (W \gg E_3 \gg E_4)$ at angles larger than the characteristic angular size of each parton jet proves to be insensitive to the jet internal structure: g_4 is emitted by a colour current, which is conserved when a jet splits.

Thus, the colour-coherence phenomena strongly affect the total three-dimensional shape of particle flows in three-jet events, practically excluding the very possibility of representing it as a sum of three parton contributions (for a detailed analysis see Ref. [21]).

Note that if a term proportional to $1/N_{\mathrm{c}}$ in Eq. (70) is dropped, the two remaining terms may be interpreted as the sum of two independent $(1\bar{3})$ and $(2\bar{3})$ antenna patterns, boosted from their respective rest frames into the overall $q\bar{q}g$ c.m.s. This scenario literally repeats the explanation given in the Lund string model [29].

Taking, for illustration, the symmetric $q\bar{q}g$ or $q\bar{q}\gamma$ configuration, the ratio of emission in the direction opposite to the hard gluon or photon, $\vec{n}_4 = -\vec{n}_3$, in the two processes is $W^{q\bar{q}g}/W^{q\bar{q}\gamma} = 7/16$ [see

Table 1. Ratios r_1 and r_2 as a function of energy W [47]

W (GeV)		30	45	75	91	750
Dipole cascade	r_1	2.74	2.65	2.43	2.38	2.13
+ string fragmentation	r_2	2.52	2.38	2.12	1.97	1.74
String fragmentation	r_1	2.69	2.94	3.01	2.99	2.89
only	r_2	2.69	2.94	2.99	2.97	2.90

Eq. (24)], showing the destructive interference in this region. This depletion was confirmed both by comparison of hadron densities of the jets [45] and by comparison of the $q\bar{q}g$ and $q\bar{q}\gamma$ final states [38].

It will be of importance to perform a comprehensive analysis at LEP energies, where, for example, the subasymptotic corrections are less essential.

A discriminative test between the string picture and PA is provided by the dependence of the multiplicity flow on the particle mass m_h or p_{out} (momentum out of the event plane).

The previous data [45] demonstrated that the string effect is enhanced by increasing m_h or p_{out}. If interpreted in terms of a Lorentz boosted string of the canonical Lund model, this enhancement will be of the same strength at all c.m. energies, given fixed angles between jets. In contrast, in the PA scenario, the effect is determined by the number of soft gluons emitted in the different interjet valleys. Therefore, at high enough energies no enhancement is expected for the subsamples with large m_h or p_{out}. While the subasymptotic effects may play an important role at low energies, ultimately the perturbative picture will win out at higher energies.

For illustration, we present in Fig. 13 the energy evolution of the particle-flow ratios in the ARIADNE program [49], which contains both soft-gluon emission and string fragmentation. One can easily see that the PA regime here becomes dominant at Z^0 energies.

6.2 Correlations of interjet particle flows

Studies of particle correlations in jets should provide an especially sensitive test of the connection between observed hadron distributions and the colour structure of an underlying hard process, see Refs. [14], [21].

For example, an interesting manifestation of the QCD wave nature of hadronic flows arises from studying the double-inclusive correlations of interjet flows in $e^+e^- \to q\bar{q}g$ events. The point is that here one faces such tiny effects as the mutual influence of different $q\bar{q}$ antennae. As a consequence $(+, - \equiv q, \bar{q}; 1 \equiv g)$:

$$r_2 = \frac{\mathrm{d}^2 N}{\mathrm{d}\Omega_{(1+)}\mathrm{d}\Omega_{(1-)}} \bigg/ \frac{\mathrm{d}^2 N}{\mathrm{d}\Omega_{(+-)}\mathrm{d}\Omega_{(1-)}} < r_1 = \frac{\mathrm{d}N}{\mathrm{d}\Omega_{(1+)}} \bigg/ \frac{\mathrm{d}N}{\mathrm{d}\Omega_{(+-)}} , \tag{72}$$

(the elaborated analytical formulae may be found in Ref. [21]). Such colour-screening effects cannot be mimicked by the canonical Lund string model. But the 'dipole formulation' of the Lund Monte Carlo [48] reproduces them [21]. To quantify the correlation effects one can compare the ratios of particle flows, projected onto the event plane, in the case of Mercedes-like symmetric events.

Table 2 illustrates the energy dependence of the ratios r_1 and r_2, calculated using the ARIADNE program [49]. One can easily see that when the dipole cascade is used, r_2 is clearly smaller than r_1 already for $W = 30$ GeV, while $r_2 \simeq r_1$ over the whole energy range in the case when only string fragmentation is used.

Of fundamental importance is the azimuthal asymmetry of QCD jets. Recall that the treatment of the structure of final states given by the string picture qualitatively reproduced the QCD radiation pattern only up to $O(1/N_c^2)$ corrections (the large N_c limit), see Ref. [30] and discussion below Eq. (71).

However, under specific conditions $O(1/N_c^2)$ terms become sizeable or even dominant.

The simplest example is given by the azimuthal asymmetry of a quark jet in the events $e^+e^- \to q_+\bar{q}_-g_1$. The azimuthal distribution of particles produced inside a cone of opening half-angle θ_0 may be characterized by an asymmetry parameter:

$$A(\theta_0) = \frac{N_{\to g}(\theta < \theta_0) - N_{\to q}(\theta < \theta_0)}{N_{\text{tot}}(\theta < \theta_0)} = \frac{(\Delta N)_{\text{as}}}{N_{\text{tot}}} . \tag{73}$$

For parametrically small θ_0,

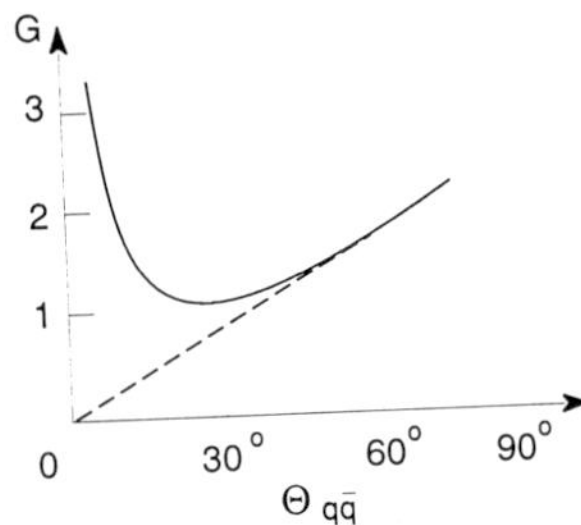

Figure 14. Comparison of the QCD (solid) and large-N_c limit (dashed) predictions for the G factor, see Eq. (74).

$$A(\theta_0) \; \simeq \; \frac{2\theta_0}{\pi} \, G \sqrt{4N_c \frac{\alpha_s(E\theta_0)}{2\pi}} \; , \tag{74}$$

$$G \; = \; \frac{N_c}{2C_F} \, \cot \frac{\theta_{+1}}{2} + \frac{1}{2N_c C_F} \, \cot \frac{\theta_{+-}}{2} \; .$$

The first colour-flavoured term in Eq. (74) describes the asymmetry due to the 'boosted string' connecting the q and g directions. The corresponding asymmetry vanishes with an increase of θ_{+1} as the string straightens. Here, however, the second term enters into the game, forcing the asymmetry to increase anew, as shown in Fig. 14. This behaviour might be interpreted as an additional repulsion between particles from q and q̄ jets.

For a symmetric configuration

$$\theta_{+1} = \theta_{-1} = \pi - \theta_{+-}/2 \; ,$$

the colour-suppressed term in Eq. (23) prevails when

$$\theta_{+1} \leq 2 \arctan(1/N_c) \approx 37° \; . \tag{75}$$

To realize this effect, one has to select qq̄g events with special kinematics, in which case the hard gluon moves in the opposite direction to the quasi-collinear qq̄ pair.

The observation of the drag (string) effect has beautifully demonstrated the connection between colour and hadronic flows. However, the 'string' analysis of three-jet events suffers from some inherent difficulties and weaknesses that one would prefer to avoid if possible. First of all, the necessity of selecting a three-jet event sample reduces the statistics and may introduce biases into the observed hadronic flow. The need to define jet directions introduces a dependence on the jet-finding algorithm. Discrimination between quark and gluon jets, on the basis of their relative energies, reduces the effect and prevents the use of symmetrical jet configurations.

Of course, the interjet collective phenomena could be better analysed if one could identify the quark jets. This can be done in the case of the heavy-flavour production events, in which one can tag the heavy-quark decay, see Refs. [18], [21].

Recently a new method has been proposed [14] to reveal the connection between observed hadron distributions and the colour structure of an underlying hard process. The method is related to the 'string effect' analysis but has the advantage of not requiring any special event selection or jet finding. It involves measuring a ratio of energy-multiplicity correlations which is especially sensitive to colour flows in jet formation. This quantity is infrared stable and can be calculated completely perturbatively.

To illustrate the new method, consider the emission of *two* soft gluons by a hard quark–antiquark system:

$$W^{q\bar{q}}(\vec{n}_3, \vec{n}_4) \propto 2C_F N_c \left(\frac{a_{12}}{a_{13}a_{34}a_{42}} + \frac{a_{12}}{a_{23}a_{34}a_{41}} - \frac{1}{N_c^2} \frac{a_{12}^2}{a_{13}a_{32}a_{14}a_{42}} \right) \tag{76}$$

(cf. subsection 3.5). Dividing by the product of the single-gluon distributions gives the gluon–gluon correlation function

$$C^{q\bar{q}}(\eta_{34}, \phi) = \frac{W^{q\bar{q}}(\vec{n}_3, \vec{n}_4)}{W^{q\bar{q}}(\vec{n}_3)W^{q\bar{q}}(\vec{n}_4)} = 1 + \frac{N_c}{2C_F} \left(\frac{\cos\phi}{\cosh\eta_{34} - \cos\phi} \right) \tag{77}$$

(for the definition of η_{34} and ϕ see subsection 3.5).

Equation (77) provides an infrared-finite measure of the correlation between colour flows in the directions (η_3, ϕ_3) and (η_4, ϕ_4). According to the local duality hypothesis, it can be applied directly to hadronic flows. Considering, for example, the flows in the orthogonal ($\phi = \pi/2$) and back-to-back ($\phi = \pi$) azimuthal directions, we have

$$C^{q\bar{q}}(\eta_{34}, \pi/2) = 1, \quad C^{q\bar{q}}(0, \pi) = \frac{N_c^2 - 2}{2(N_c^2 - 1)} = \frac{7}{16} \ . \tag{78}$$

This implies that for $\phi = \pi/2$ the hadronic multiplicities are uncorrelated, whilst for $\phi = \pi$ there is destructive interference, of the same magnitude as the string effect in symmetric jets [Eq. (24)]. Thus measurements of hadronic-flow correlations in the orthogonal and back-to-back azimuthal directions should demonstrate the same type of colour coherence as the string effect, without requiring the selection of a three-jet event sample.

To estimate the effects of finite energy corrections and hadronization [11] the Monte Carlo program HERWIG was used. The results clearly demonstrated the destructive interference at both the parton and hadron levels. The effects of hadronization are small at these energies. However, the $O(\sqrt{\alpha_S})$ corrections look rather significant. It would be interesting to compute the corrections to the correlation function analytically.

An analysis of colour flow along the above lines may be performed for other correlations. Another example is the study of azimuthal-asymmetry effects in e^+e^- events with a hard photon ($q\bar{q}\gamma$). It is interesting to observe that in correlations with an additional multiplicity flow ($q\bar{q}g\gamma$), the asymmetry depends on $1/N_c^2$ corrections, which are typically neglected in Monte Carlo simulations.

7 PROSPECTS FOR FUTURE COLOUR-COHERENCE STUDIES AT THE Z^0

We present here a brief 'shopping list' for an experimental investigation of hadron-jet phenomena, which should reveal the colour structure of QCD and the local duality features. We concentrated mainly on the two topics: i) hadron distributions in QCD jets; ii) QCD collective effects in interjet hadronic flows (for discussions of other topics see, for example, Refs. [14], [21], [22]).

a) Testing the various aspects of the MLLA–LPHD picture urgently requires measurements of spectra of different identified hadrons: $\pi^\pm, \pi^0, \mathrm{K}, \mathrm{p}, \dots$.

b) It will be interesting to separate contributions to particle spectra caused by the light quarks and by the heavy ones, for example, by tagging heavy-quark jets.

c) By measuring spectra of particles restricted to lie within the particular opening angles with respect to the jet [44] one can explore the specific features of the parton multiplication processes and, in particular, sharpen the influence of angular ordering on the particle branching.

d) Multiplicity distributions (KNO picture) in QCD jets are predicted analytically. It will be of interest to analyse these distributions in detail in the LEP energy region.

e) One of the most interesting tasks seems to be the measurement of the two-particle inclusive distribution of hadrons (for the DL result see Refs. [31], for calculation of the next-to-leading corrections see Refs. [12]):

$$R(\ell_1, \ell_2, Y) \equiv \frac{\overline{D}^{(2)}(\ell_1, \ell_2, Y)}{\overline{D}(\ell_1, Y)\overline{D}(\ell_2, Y)} \ , \tag{79}$$

with

$$Y = \ln \frac{E}{\Lambda}, \quad \ell_i = \ln \frac{1}{x_i} \quad (i = 1, 2) \ .$$

A preference is expected for $x_1 \simeq x_2$ and for both to be small. The range of these correlations is of the order of Y, so they extend over long distances in the $(\ln x_1, \ln x_2)$ plane. Thus they should be easily distinguishable from correlations due to hadronization, which would be expected to have ranges of the order of $\ln (Q_\mathrm{h}/\Lambda)$, where the hadronization scale Q_h is at most a few GeV.

f) To establish clear connections between theory and experiment, it is preferable [44] to work with inclusive quantities which are defined on the basis of simple jet characteristics such as energy or multiplicity flow, rather than on the basis of a given number of jets having specific directions, energies, momenta, masses, etc.

There is a direct correspondence between the jet direction and the energy-flow direction, so that we may naturally study the shapes of jets and any characteristic of the hadronic system produced in a hard interaction by introducing inclusive correlations among energy flows and multiplicity flows. In this case one does not need to apply event-selection procedures or jet-finding algorithms.

g) The detailed tests of colour-coherence effects and their discrimination against non-perturbative dynamics require comprehensive studies of the total three-dimensional pattern of particle flows in three-jet events. Comparison with analytical results (accounting for both interjet and intrajet coherence phenomena) should make it possible to distinguish reliably the PA predictions from fragmentation schemes. Of special interest here are the energy dependence of the multiplicity flow and its dependence on m_h and p_{out}.

h) One can study (for example by jet identification) the double-inclusive correlations r_2 of the interjet flows, where the quantum mechanical collective nature of radiation manifests itself.

i) The ratio of energy-multiplicity correlations (77) looks very promising. This method has the advantage of not requiring any special event selection or jet finding. Since finite energy corrections depend, in general, on the particle masses, it would be useful to study these corrections for different hadron species.

j) By studying the radiation pattern of a tagged quark jet in $q\bar{q}g$ events, one can measure the azimuthal asymmetry of a jet. An interesting vista on this problem is connected with the $Z^0 \to c\bar{c}g$ events. To investigate the role of $1/N_c^2$ interference terms, one has to select $c\bar{c}g$ events in which the hard gluon moves in the direction opposite to the quasi-collinear $c\bar{c}$ pair, and to measure the asymmetry parameter $A(\theta_0)$ at $\theta_0 \geq 3\text{--}4°$, see Ref. [21] for details. Of course, before the different behaviour in QCD prediction and its large N_c limit can be studied, the asymmetry itself must be investigated experimentally.

k) Of special importance are the particle distributions inside jets initiated by heavy quarks, see Refs. [18], [21] for details. One of the prospective ways of fixing jet direction arises from the vertex detection.

In particular, one can study here distributions of particles inside fixed cones, energy, and angular spectra. Of special interest is the observation of the particle depletion in the forward region and the search for the manifestations of non-perturbative dynamics.

One of the interesting characteristics is the spectrum $\overline{D}_Q^h(x, W)$ of associated light particles in the heavy-quark jet. In the regime $W \gg m_Q \gg \Lambda$ (within the MLLA accuracy) this spectrum is given by [50]:

$$\overline{D}_Q^h(x, W) = \overline{D}_q^h(x, W) - \overline{D}_q^h\left(\frac{x}{\langle z \rangle}, m_Q \sqrt{e}\right) , \qquad (80)$$

where $\overline{D}_q^h(x, W)$ is the usual hadron inclusive spectrum in e^+e^- collisions, and $\langle z \rangle$ is the averaged scaled energy of the heavy quark.

8   CONCLUSION

With the start of activity at the Z^0 peak a wealth of experimental data on QCD jets has become available for detailed tests of PA and for reducing our domain of ignorance on the physics of confinement. These experimental studies of jets show that the gross features of hadronic event shapes calculated at the partonic level agree well with the measured ones. This strongly supports the hypothesis of 'soft confinement', which leads to the LPHD picture. Starting from this hypothesis it is possible within the PA picture to obtain the number of detailed predictions for hard collisions.

Hadroproduction studies within the MLLA–LPHD scheme seem to be far from being exhausted. Future experiments, in particular at the Z^0 resonance, definitely need CLIC-like PA activity. They should provide convincing tests of both perturbative and non-perturbative aspects of QCD.

The main aims of this lecture were: i) to move the experimentalists closer to the QCD–CLIC picture of multihadron production, ii) to help them to overcome some prejudices stemming from the past, iii) to present a 'shopping list' for colour-coherence studies at the Z^0.

To conclude this lecture, let me try to respond to the argument of the devil's advocate: Why, at the present high level of credibility and maturity of QCD, should we explore so thoroughly one of its inherent properties — the quantum mechanical coherence?

The answer could be as follows. The main purpose is not the proof of colour coherence: it would be inexcusable (and somewhat expensive) to test quantum mechanics at modern accelerators. Of real importance is how the coherence reveals itself in hadron spectra, i.e. how confinement disturbs the perturbative picture of multihadron production.

One should also take into account that the spectacular collective effects in the jetty final states could provide a valuable additional tool, helping to extract and to study manifestations of New Physics.

Finally, let us mention that the most adequate incorporation of the coherence phenomena becomes now an important approach for the development of the QCD-inspired Monte Carlo algorithms.

9 Acknowledgements

It is a pleasure for me to thank the organizers of this excellent Institute for their kind invitation to and hospitality in Cargese. I wish to thank Ya.I. Azimov, Yu.L. Dokshitzer, G. Marchesini, A. Mueller, S. Troyan and B. Webber for their pleasant collaboration.

I am very much indebted to Yu.L. Dokshitzer, M. Jacob, H. Kreutzmann, P. Mättig and T. Sjöstrand for fruitful discussions.

References

[1] R.F. Schwitters et al., Phys. Rev. Lett. **35** (1975) 1320;
G.G. Hanson et al., Phys. Rev. Lett. **35** (1975) 1609.

[2] R. Brandelik et al., TASSO Collaboration, Phys. Lett. **86B** (1979) 243;
D.P. Barber et al., MARK J Collaboration, Phys. Rev. Lett. **43** (1979) 830;
C. Berger et al., PLUTO Collaboration, Phys. Lett. **86B** (1979) 418;
W. Bartel et al., JADE Collaboration, Phys. Lett. **91B** (1980) 142.

[3] D. Ward, these Proceedings.

[4] R. Cashmore, these Proceedings.

[5] T. Sjöstrand, these Proceedings.

[6] Yu.L. Dokshitzer and S.I. Troyan, Proc. 19th Winter School of the LNPI, Vol. 1, p. 144; preprint LNPI–922 (1984).

[7] Ya.I. Azimov et al., Z. Phys. **C27** (1985) 65 and **31** (1986) 213.

[8] D. Amati and G. Veneziano, Phys. Lett. **83B** (1979) 87;
G. Marchesini, L. Trentadue and G. Veneziano, Nucl. Phys. **B181** (1981) 335.

[9] A. Bassetto, M. Ciafaloni and G. Marchesini, Phys. Rep. **C100** (1983) 201;
Yu.L. Dokshitzer, V.A. Khoze, A.H. Mueller and S.I. Troyan, Rev. Mod. Phys. **60** (1988) 373;
Yu.L. Dokshitzer, V.A. Khoze and S.I. Troyan, *in* Perturbative QCD, ed. A.H. Mueller (World Scientific, Singapore, 1989), p. 241.

[10] Yu.L. Dokshitzer, Talks given at the International Schools of Subnuclear Physics, Erice, 1989, 1990.

[11] G. Marchesini and B.R. Webber, Nucl. Phys. **B310** (1988) 461.

[12] C.P. Fong and B.R. Webber, Phys. Lett. **229B** (1989) 289 and **249B** (1990) 255; Cavendish–HEP–90/7 (1990).

[13] Yu.L. Dokshitzer, V.A. Khoze and S.I. Troyan, in Proc. 6th Int. Conf. on Physics in Collision, ed. M. Derrick (World Scientific, Singapore, 1987).

[14] Yu.L. Dokshitzer, V.A. Khoze, G. Marchesini and B.R. Webber, Phys. Lett. **245** (1990) 243; preprint CERN–TH.5738/90 (1990).

[15] R.K. Ellis, Proc. 24th Int. Conf. on High Energy Physics, eds. R. Kotthaus and J. Kühn (Springer-Verlag, Berlin, 1989), p. 48.

[16] B. Bambah et al., QCD generators for LEP, Proc. 1989 Workshop on Z Physics at LEP 1, eds. G. Altarelli, R. Kleiss and C. Verzegnassi, CERN 89–08 (1989), Vol. 3 p. 143.

[17] Z. Kunszt and P. Nason, QCD at LEP, ibid, Vol. **1**, p. 373.

[18] J.H. Kühn and P.M. Zerwas, Heavy flavours at LEP, ibid, Vol. **1**, p. 267.

[19] G. Altarelli, preprint CERN–TH. 5760/90 (1990).

[20] M. Jacob, Talk given at the 25th Int. Conf. on High Energy Physics, Singapore, 1990.

[21] Yu.L. Dokshitzer, V.A. Khoze and S.I. Troyan in Ref. [9].

[22] V.A. Khoze, Proc. 1989 Int. Symposium on Lepton and Photon Interactions at High Energies, Stanford, 1985 (World Scientific, Singapore, 1990), p. 387.

[23] A. Bassetto, M. Ciafaloni and G. Marchesini, Phys. Lett. **B83** (1979) 207.

[24] W. Furmanski, R. Petronzio and S. Pokorski, Nucl. Phys. **B155** (1979) 253.

[25] K. Konishi, A. Ukawa and G. Veneziano, Nucl. Phys. **B157** (1979) 45.

[26] B.I. Ermolayev and V.S. Fadin, JETP Lett. **33** (1981) 285;
V.S. Fadin, Yad. Fiz. **37** (1983) 408.

[27] A.H. Mueller, Phys. Lett. **B104** (1981) 161.

[28] A.H. Mueller, Nucl. Phys. **B228** (1984) 351;
A.H. Mueller, Nucl. Phys. **B213** (1983) 85 and erratum quoted in Nucl. Phys. **B241** (1984) 141.

[29] B. Andersson, G. Gustafson, and T. Sjöstrand, Phys. Lett. **94B** (1980) 211.

[30] Ya.I. Azimov et al., Phys. Lett. **165B** (1985) 147; Yad. Fiz **43** (1986) 149.

[31] Yu.L. Dokshitzer, V.S. Fadin and V.A. Khoze, Phys. Lett. **115B** (1982) 242; Z. Phys. **C15** (1982) 325 and **C18**(1983) 37.

[32] A. Bassetto et al., Nucl. Phys. **B207** (1982) 189.

[33] R.K. Ellis, G. Marchesini and B.R. Webber, Nucl. Phys. **B286** (1987) 643.

[34] A.E. Chudakov, Izv. Akad. Nauk SSSR Ser. Fiz. **19** (1955) 650.

[35] G. Marchesini and B.R. Webber, Nucl. Phys. **B238** (1984) 1;
B.R. Webber, Nucl. Phys. **B238** (1984) 492.

[36] G. Marchesini and B.R. Webber, Nucl. Phys. **B330** (1990) 261.

[37] B. Andersson, B. Gustafson, G. Ingelman, and T. Sjöstrand, Phys. Rep. **97** (1983) 31;
B. Andersson, Lund preprint LU TP 88–2 (1988);
T. Sjöstrand, Int. J. Mod. Phys. **A3** (1988) 751.

[38] H. Aihara et al., TPC Collaboration, Phys. Rev. Lett. **57** (1986) 945;
P.D. Sheldon et al., Mark II Collaboration, Phys. Rev. Lett. **57** (1986) 1398.

[39] W. Hofmann, Proc. Int. Symposium on Lepton and Photon Interactions at High Energies, Hamburg, 1987, eds. W. Bartel and R. Rükl (North Holland, Amsterdam, 1988), p. 671;
M.J. Shochet, Proc. 24th Int. Conf. on High Energy Physics, Hamburg, 1988, eds. R. Kotthaus and J. Kühn, (Springer-Verlag, Berlin, 1989), p. 18;
P. Mättig, Phys. Rep. **C177** (1989) 141;
K. Sugano, Int. J. Mod. Phys. **A3** (1988) 2249;
W. Hofmann, Ann. Rev. Nucl. Part. Sci **38** (1988) 279.

[40] M.Z. Akrawy et al., OPAL Collaboration, preprint CERN–EP/90–94 (1990).

[41] W. Braunschweig et al., TASSO Collaboration, DESY 90–013 (1990).

[42] D. Decamp et al., ALEPH Collaboration, Phys. Lett. **234B** (1990) 205.

[43] Ya.I. Azimov, Yu.L. Dokshitzer and V.A. Khoze, Proc. 17th Winter School of the LNPI (1982), vol. 1, p. 162.

[44] Yu.L. Dokshitzer, V.A. Khoze, A.M. Mueller and S.I. Troyan, in Ref. [9].

[45] W. Bartel et al., JADE Collaboration, Phys. Lett. **134B** (1984) **275** and **157** (1985) 340;
H. Aihara et al., TCP Collaboration, Z. Phys. **C28** (1985) 31.

[46] DELPHI Collaboration, paper contributed to the 25th Int. Conf. on High Energy Physics, Singapore, 1990.

[47] V.A. Khoze and L. Lönnblad, Phys. Lett. **241B** (1990) 123.

[48] G. Gustafson, Phys. Lett. **175B** (1986) 453;
G. Gustafson and U. Pettersson, Nucl. Phys. **B306** (1988) 746.

[49] L. Lönnblad, ARIADNE–3, A Monte Carlo for QCD cascades, Lund preprint LU TP 89–10 (1989).

[50] Yu.L. Dokshitzer, V.A. Khoze and S.I. Troyan (to be published).

THE PHYSICS PROGRAM AND ACCELERATOR PROPERTIES

OF A B-MESON FACTORY

R.A. Eichler

Institute for Intermediate Energy Physics
ETH Zürich
CH-5232 Villigen PSI, Switzerland

INTRODUCTION

The physics of b-quarks is investigated at several different types of machines. Hadron and e-p colliders probe the production cross section of heavy quarks (charm, beauty and top) and give us insight into the gluon and quark structure functions of nucleons and test QCD calculations. The weak couplings of the Z^o to c-, and b-quarks and τ-leptons are precisely measured at LEP. In e^+e^--colliders operating at the threshold of $B\bar{B}$-meson production (around 10 GeV center of mass) the properties of the weak decays of B-mesons and τ-leptons themselves are best studied. Table 1 gives a comparison of the production rates at various machines. Hadron colliders give the highest rates but the trigger times the reconstruction efficiency at e^+e^--colliders is with present detector and trigger technology several orders of magnitudes superior to hadron colliders. This holds in particular for B-decays with a high charged multiplicity or with neutral particles and for the reconstruction of $B\bar{B}$ meson pairs in the same event.

Table 1. B-meson production rates at existing and planned machines. The production cross section is $\sigma_{b\bar{b}}$, $N_{b\bar{b}}$ the number of $b\bar{b}$ pairs produced in one year (10^7s) and f the fraction of events containing b-quarks from all produced events. For the hadron machines the useful rates for triggering are about 10^{-4} times smaller or in the case of LHC/SSC lower luminosities will be used for B-physics.

		E_{CM} [GeV]	$\mathcal{L}_{peak}$ [cm^{-2}s^{-1}]	$\sigma_{b\bar{b}}$	$N_{b\bar{b}}$ [y^{-1}]	f
B-factory	e^+e^-	10.6	$1\cdot10^{34}$	1.1 nb	10^8	0.23
DORIS	e^+e^-	10.6	$4\cdot10^{31}$	0.9 nb	$2\cdot10^5$	0.19
CESR	e^+e^-	10.6	$9\cdot10^{31}$	1.1 nb	10^6	0.23
LEP	e^+e^-	93	$1.5\cdot10^{31}$	5 nb	$5\cdot10^5$	0.14
HERA	ep	314	$1\cdot10^{31}$	4 nb	$4\cdot10^5$	10^{-3}
TEVATRON	$p\bar{p}$	2000	$1\cdot10^{30}$	30μb	$3\cdot10^8$	$3\cdot10^{-4}$
upgraded		2000	$5\cdot10^{31}$	30μb	$1.5\cdot10^{10}$	$3\cdot10^{-4}$
LHC	pp	15000	$1\cdot10^{34}$	300μb	$3\cdot10^{13}$	$2\cdot10^{-3}$
SSC	pp	40000	$1\cdot10^{33}$	500μb	$5\cdot10^{12}$	$5\cdot10^{-3}$

Z° Physics - Cargèse 1990, Edited by M. Lévy et al.
Plenum Press, New York, 1991

The origin of CP-violation is one of the most intriguing puzzles in elementary particle physics. Within the Standard Model we have now a more or less clear prediction of the size of the effect in a system different from K-decays. The most promising candidate is the decay of neutral B-mesons. It will be shown in these lectures that 10^8 B-mesons have to be produced – and this is important – in a sufficiently clean environment so that the signal is not buried in background. From this last requirement together with the presently available experimental techniques, force us to use e^+e^--colliders. The luminosity needed (several times 10^{33} cm^{-2}s^{-1}) is up to hundred times the performances of existing storage rings (CESR, DORIS, VEPP4M). In recent month, several studies have shown, that such a luminosity is indeed in reach[1-6].

CP-violation is not the only topic which can be studied at such a collider[7]. The scientific program covers a broad spectrum of topics in heavy flavour physics. Since the top-quark is heavier than the W-boson, toponium and T-mesons will have a large width and b-quarks are the heaviest quarks, where quark bound states can be investigated. Decisive contributions are expected to

- the determination of parameters of the Standard Model with high precision (masses, mixing angles ...)

- better understanding the dynamics of the strong interaction at energies of a few GeV, and

- looking for the boundaries of the Standard Model by observing rare decays of B- and D-mesons and τ-leptons.

These lectures start with a short introduction into the physics and technical problems of colliders. At the end, the student should be able to understand the parameter lists of different colliders proposed worldwide.

We then describe the experimental techniques to reconstruct B- and D-mesons and formulate the detector requirements. In the last part some pilot reactions are described such as determination of the Cabibbo-Kobayashi-Moskawa matrix elements and CP-violation in the decay $B^o \rightarrow J/\psi K_S$.

THE PHYSICS OF STORAGE RINGS

The Equation of Motion

The equation of motion of a charged particle in a magnetic guide field can be derived in several ways[8]. An elegant derivation uses the method of Hamiltonian mechanics[9,10].

The Hamiltonian H for a charged particle of mass m, charge e, and momentum p in a magnetic field with vector potential $\vec{A}(x, y, z)$ and electric potential $\varphi(x, y, z)$ is:

$$H = e\varphi + c\sqrt{(\vec{p} - e\vec{A})^2 + m^2c^2} \quad . \tag{1}$$

It is common practise to use local variables, i.e. variables which express the deviations from the design (ideal) orbit (see Fig. 1). The design orbit is a horizontal closed curve with pieces of a circle in bending magnets and straight lines elsewhere. Instead of the time variable t we will use the coordinate s, the path length measured along the particle trajectory. Clearly, $s = v \cdot t$ with v the velocity. We will also drop the electric potential φ which is assumed to be constant. The transformation to the local variables x (horizontal), z (vertical) and s yields

$$H = c\sqrt{m^2c^2 + \left[(p_s - eA_s)^2 \, \frac{\rho}{x+\rho}\right]^2 + (p_x - eA_x)^2 + (p_z - eA_z)^2} \tag{2}$$

with $\rho = \rho(s)$ the local radius of curvature and the conjugate variables (t,H), (s,p_s), (x,p_x) and (z,p_z). With s instead of t as independent variable, we solve (2) for p_s and call the Hamiltonian

$$H = eA_s \pm \left(1 + \frac{x}{\rho}\right)\left\{\left(\frac{p_o}{c}\right)^2 - m^2c^2 - (p_x - eA_x)^2 - (p_z - eA_z)^2\right\}^{1/2} . \quad (3)$$

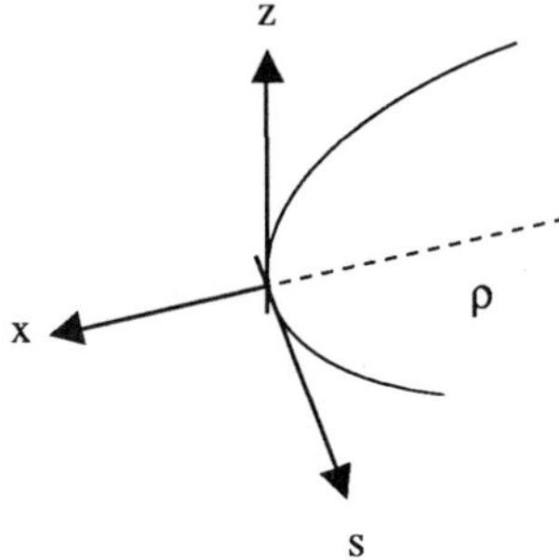

Figure 1. Local coordinates x, z, s defined as deviations from the ideal orbit.

The magnetic vector potential for purely transverse magnetic guide fields $\vec{A}(x,z,s) = (0,0,A_s(x,z,s))$ can be expressed as a sum over $2n$-poles

$$A_s = \sum_{n=1}^{\infty} A_n(x + iz)^n \quad , \quad (4)$$

with $\vec{B} = (B_x, B_z, B_s) = rot\vec{A} = (\frac{\partial A_s}{\partial z}, -\frac{\partial A_s}{\partial x}, 0)$. The real part of (4) corresponds to multipoles symmetric with respect to the x- and z-axes and the imaginary terms to skew multipoles.

We derive for $n = 1$ (dipole)

$$B_x = 0, \quad B_z = -A_1 \quad ,$$

and for $n = 2$ (quadrupole)

$$B_x = -2 \cdot A_2 \cdot z, \quad B_z = -2 \cdot A_2 \cdot x \quad ,$$

and for $n = 3$ (sextupole)

$$B_x = -6 \cdot A_3 \cdot x \cdot z, \quad B_z = -3 \cdot A_3 \cdot (x^2 - z^2) \quad .$$

The magnetic guide field for the stored or accelerated particle consists usually of a series of individual magnets with a multipole field (dipole, quadrupole, sextupole). If we neglect the fringe fields at the end of the magnets (magnets are long compared to the aperture), and keep only terms linear in x and z, i.e. only terms up to $n = 2$ (quadrupole) in eq. (4), the hamiltonian differential equations become simple. In fact, from

$$\frac{\partial p_x}{\partial s} = -\frac{\partial H}{\partial x}, \quad \frac{\partial x}{\partial s} = \frac{\partial H}{\partial p_x}, \quad \frac{\partial p_z}{\partial s} = -\frac{\partial H}{\partial z}, \quad \text{and} \quad \frac{\partial z}{\partial s} = \frac{\partial H}{\partial p_z}$$

we arrive at

$$p_x' = \frac{\partial p_x}{\partial s} = -e\frac{\partial A_s}{\partial x} - \frac{p_o}{\rho}, \quad x' = \frac{\partial x}{\partial s} = (1 + \frac{x}{\rho})\frac{p_x}{p_o}, \quad p_z' = e\frac{\partial A_s}{\partial z} \quad . \quad (5)$$

In a cyclotron $\rho = \frac{p_o}{eB_z}$. A Taylor series expansion for the magnetic field for small x, z is

$$B_z = B_z^o + \frac{\partial B_z}{\partial x} \cdot x \quad \text{and} \quad B_x = \frac{\partial B_x}{\partial z} \cdot z = \frac{\partial B_z}{\partial x} \cdot z \quad ,$$

and using (5)

$$x'' = \left(-\frac{1}{\rho^2} - \frac{e}{p_o}\frac{\partial B_z}{\partial x}\right) \cdot x$$

$$z'' = \frac{e}{p_o}\frac{\partial B_x}{\partial z} \cdot z = \frac{e}{p_o}\frac{\partial B_z}{\partial x} \cdot z \quad . \tag{6}$$

The two equations (6) are completely decoupled and the motion in $x(s)$ and $z(s)$ are independent. Note that this is only true for multipoles $n \leq 2$ and only if the real part in (4) is used, i.e. no skew quadrupoles.

It is useful to define the so-called field index

$$f = -\frac{\partial B_z}{\partial x} \cdot \frac{\rho}{B_z^0} = " - \frac{\Delta B/B}{\Delta \rho/\rho} " \quad . \tag{7}$$

The equations (6) can be rewritten as

$$x'' = -(1-f)\frac{1}{\rho^2} \cdot x$$

$$z'' = -f \cdot \frac{1}{\rho^2} \cdot z \quad . \tag{8}$$

In a cyclotron f and ρ are constants and solutions of (8) are

$$z(s) = a \cdot \cos\left(\frac{s}{\beta} + s_o\right) \qquad \text{for} \quad \frac{1}{\beta^2} = \frac{f}{\rho^2} > 0$$

$$= a \cdot \cosh\left(\frac{s}{\sqrt{-\beta^2}} + s_o\right) \qquad \text{for} \quad \frac{1}{\beta^2} = \frac{f}{\rho^2} < 0 \tag{9}$$

$$= a \cdot s + s_o \qquad \text{for} \quad f = 0 \quad ,$$

and for $x(s)$ accordingly. We have therefore a bounded solution (focussing) only for special values of f. Vertical focussing is achieved for $f > 0$, horizontal focussing for $1 - f > 0$ and focussing in both planes for $0 \leq f \leq 1$.

In a quadrupole magnet $|f|$ is usually bigger than one and we see from (8) that such a quadrupole can either focus horizontally and defocus vertically or vice versa, depending on the sign of A_2 in (4). Around 1950 it was realized [11], that a quadrupole doublet with alternating horizontal focussing and horizontal defocussing and the opposite for the vertical plane can produce a net focussing effect. This is now used in all synchrotrons and storage rings.

The equations (8) can still be used in a synchrotron if we make $\rho(s)$ and $\frac{\partial B_z}{\partial x}(s)$ both functions of s. They are constant within a given magnet but change from magnet to magnet. We then write (8) as

$$x'' = k_x(s) \cdot x$$

$$z'' = k_z(s) \cdot x \tag{10}$$

with $k_x(s)$ the single valued functions of s

$$k_x(s) = -\frac{1}{\rho(s)^2} - \frac{e}{p_o}\frac{\partial B_z}{\partial x}(s) = \frac{(1-f(s))}{\rho(s)^2}$$

$$k_z(s) = \frac{e}{p_o}\frac{\partial B_z}{\partial x}(s) = \frac{f(s)}{\rho(s)^2} \quad . \tag{11}$$

Solutions to Hill's equation

The differential equation (10) is called Hill's Equation. According to Floquet's theorem[12] the solutions of (10) can be written (see also (9))

$$x(s) \; = \; a \; \cdot \; \xi(s) \; \cdot \; cos(\varphi(s) - \theta) \quad . \tag{12}$$

The amplitude $\xi(s)$ and the phase $\varphi(s)$ are functions of s to be determined from (10). With the Ansatz[11]

$$\varphi(s) \; = \; \int_o^s \frac{ds'}{\beta(s')}, \;\; \xi(s) \; = \; \sqrt{\beta(s)} \;\; \text{and therefore} \;\; \varphi'(s) \; = \; \frac{1}{\beta(s)} \quad , \tag{13}$$

we obtain the differential equation for the beta function:

$$\frac{d^2}{ds^2} \sqrt{\beta(s)} \; = \; k(s) \sqrt{\beta(s)} \; + \; \frac{1}{\sqrt{\beta(s)^3}} \quad . \tag{14}$$

A stable particle orbit is obtained, if $\beta(s)$ is limited everywhere. Due to the nonlinear term in (14) the beta function $\beta(s)$ is completely determined including the scale. We can interpret (12) and (13) as an oscillation around the design orbit with a s-dependent amplitude $a \cdot \xi(s) = a\sqrt{\beta(s)}$ and a s-dependent wavelength $\lambda = 2\pi\beta(s)$.

The emittance

From eq. (12) we derive

$$x'(s) \; = \; -\frac{a}{\sqrt{\beta(s)}} \cdot sin\left(\int_o^s \frac{ds'}{\beta(s')} - \theta \right) + \frac{\beta'(s)}{2\beta(s)} \cdot x(s) \quad . \tag{15}$$

If we isolate the cos- and sin-term in (12) and (15), square and add the equations we obtain

$$a^2 \; = \; \left\{ \frac{1}{\beta} - \frac{\beta'^2}{4\beta} \right\} x^2 \; - \; \beta' \, x \cdot x' \; + \; \beta \, x'^2 = \gamma x^2 + 2\alpha x x' + \beta x'^2, \tag{16}$$

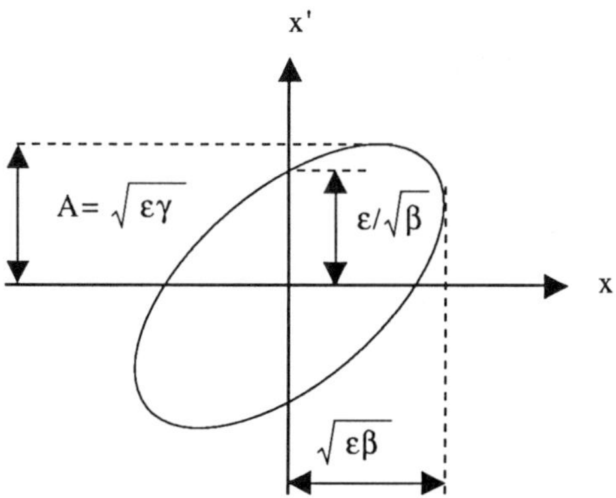

Figure 2. The phase ellipse.

which is the equation of an ellipse in the phase space (x, x') with area πa^2. The quantity $\varepsilon = a^2$ is an invariant of motion (Liouville's theorem of constant phase space volume) and is called the emittance. The ellipse (16) is called the phase ellipse and is depicted in Fig. 2. It describes the horizontal phase space at a particular azimuth s of the storage ring or synchrotron.

Each azimuth has a different ellipse determined by the betafunction. Linear machine elements like free space, dipoles or quadrupoles rotate the ellipse or change the aspect ratio from one azimuth s to the next. Higher multipoles distort the ellipse.

It is sometimes useful to introduce new variables

$$w^2 = x^2/\beta \quad , \quad w'^2 = (\alpha x + \beta x')^2/\beta \tag{17}$$

and in these normalized coordinates the phase ellipse (16) becomes a circle

$$w^2 + w'^2 = \varepsilon. \tag{18}$$

The betatron tune

The betatron tune $Q = \frac{1}{2\pi} \oint \frac{ds'}{\beta(s')}$ describes the phase advance in the betatron oscillation. In a linear machine and in the normalized coordinates w and w' it can be visualized as the number of turns on the phase circle a particle moves in a single revolution around the ring. The tune Q is only a function of β and is therefore completely determined by the magnetic guide field.

As seen from eq. (12), an integer value of Q means, that a particle repeatedly passes a given point in the storage ring with the same phase and any small magnet imperfection will excite a parasitic resonance and the betatron-oscillations will grow till the particle orbit exceeds the machine aperture. Therefore integer Q-values have to be avoided.

As can be seen from eq. 11 and 14, the β-function and therefore Q is energy dependant. Since the beam has an energy spread of a few times 10^{-4} (shown in eq. 22 below), this introduces a tune spread ΔQ and may shift the tune Q in the proximity of resonances.

Energy Oscillations

In the derivation of (8) and (10) we assumed a constant energy of all stored particles. We now work out the equation of motion for a particle with an energy deviation ϵ from the nominal value $E_o \approx cp_o$. For small $\epsilon \ll E_o$ the equations (8) become

$$x'' = k_x(s) \cdot x + \frac{1}{\rho} \frac{\epsilon}{E_o} \quad . \tag{19}$$

We make now the Ansatz

$$x = x_\beta + x_\epsilon \text{ and } x_\epsilon = \eta(s) \frac{\epsilon}{E_o} \quad . \tag{20}$$

Equation (19) is fulfilled if x_β and x_ϵ are solutions to

$$x''_\beta = k_x(s) \cdot x_\beta \text{ and } \eta''(s) = k_x(s) \cdot \eta(s) + \frac{1}{\rho} \quad .$$

In a storage ring $\eta(s)$ is a single valued function and is called the dispersion of the storage ring. It describes the shift of the design orbit for an energy deviation ϵ and is completely given by the parameters $k_x(s)$ and $\rho(s)$, i.e. by the magnetic guide field.

Beam Size

We have seen that the beam envelope as a function of the azimuth s is given by $\sigma_{x_\beta} = \sqrt{\varepsilon\beta(s)}$ (see Fig. 2), where ε is the emittance at injection time. The emittance will decrease with time in electron storage rings due to synchrotron radiation. The particles radiate photons in the direction of motion which is not always tangent to the design orbit due to betatron-oscillations. But in the RF-cavities only longitudinal momentum (along the design orbit) is fed back and therefore the transverse momentum decreases. The emittances ε_x and ε_z will shrink to an asymptotic value proportional to the average emitted synchrotron radiation energy $< u >$, namely $\varepsilon \sim \frac{<u>}{E_o}$. It can be shown[8], that

$$\sigma_{x_\beta} = \sqrt{\beta(s)\varepsilon_x} \sim \sqrt{\beta(s)} \sqrt{\frac{\gamma^2}{<\rho>}} \quad , \tag{21}$$

where $< \rho >$ is the average radius of curvature in the storage ring and γ the Lorentz factor $\gamma = E/m_e c^2$.

σ_{x_β} is not the only contribution to the beam size. The finite energy range of the stored particles introduces via the dispersion function $\eta(s)$ another term. We call it $\sigma_{x_\epsilon} = \eta(s)\frac{\sigma_\epsilon}{E_o}$ where σ_ϵ is the r.m.s. width of the particle energy. We give here without derivation the value for σ_ϵ[8]

$$\left(\frac{\sigma_\epsilon}{E}\right)^2 = \frac{55 \cdot \hbar c}{32\sqrt{3}m_e c^2} \frac{1}{J_\varepsilon} \frac{\gamma^2}{<\rho>} \quad , \tag{22}$$

where J_ε, the energy damping partition number, is a pure number and a function of the magnet lattice only. By comparing (20), (21) and (22) one finds

$$\sigma_x^2 = \sigma_{x_\beta}^2 + \sigma_{x_\epsilon}^2 \simeq const. \frac{\gamma^2}{<\rho>} \quad . \tag{23}$$

Emittance Coupling

In a horizontal flat storage ring the main part of the synchrotron radiation is emitted in the horizontal plane. We have seen, that the asymptotic value of the emittance is given by the quantum fluctuations of the photon emission and is proportional to the radiated energy. Therefore in an ideal ring the natural horizontal emittance ε_x^{nat} will dominate over the vertical one by orders of magnitude. Due to magnet imperfections (field components along the orbit, skew quadrupoles etc.) a coupling between horizontal and vertical betatron oscillations will occur, and the vertical emittance ε_z will be a fraction κ of the horizontal one, $\varepsilon_z = \kappa\varepsilon_x$. Assuming $\varepsilon_z^{nat}=0$ and energy conservation gives the relation

$$\varepsilon_z + \varepsilon_x = \varepsilon_x^{nat} \quad , \tag{24}$$

and we can write consequently

$$\varepsilon_z = \frac{\kappa}{1+\kappa} \varepsilon_x^{nat}; \quad \varepsilon_x = \frac{1}{1+\kappa} \varepsilon_x^{nat} \quad . \tag{25}$$

Typical values for κ are $1 - 5\%$ and 0.1% has been achieved.

Beam-Beam Interaction

The particles in the bunches moving in opposite direction will meet at several places. The bunches penetrate each other and occasionally two particles collide. During the penetration, large electromagnetic forces act on the particles and if they become too large, the stored beam will be lost in only a few revolutions.

Linear tune shift

We calculate first the force on one particle in one bunch due to the field of one particle

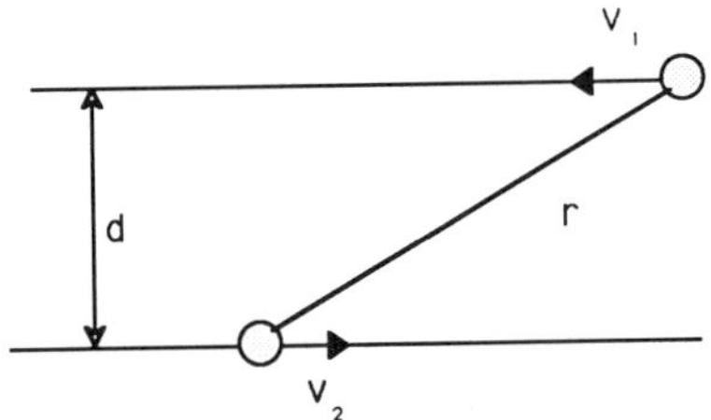

Figure 3. Force between two charged particles with charge e_1 and e_2 and with velocities $\vec{v}_1$ and $\vec{v}_2$.

in the other bunch and sum then the force over all particles in the second bunch. The electric field of a point particle with charge e and velocity $\beta = \frac{v}{c}$ is (see Fig. 3)

$$\vec{E} = \frac{e\vec{r}}{r^3} \frac{1-\beta^2}{(1-\beta^2 sin^2\theta)^{3/2}} \quad .$$

The force on particle one due to the field of particle two is[13]

$$\vec{F} = e_1\vec{E}_2 + \frac{e_1}{c}\left[\vec{v}_2 \times \vec{B}_2\right] = e_1\vec{E}_2 + \frac{e_1}{c}\left[\vec{v}_2 \times \frac{1}{c}\left(\vec{v}_2 \times \vec{E}_2\right)\right] \quad ,$$

and split into parallel and perpendicular forces:

$$F_\parallel = \frac{e_1e_2}{r^2}\frac{(1-\beta^2)cos\theta}{(1-\beta^2 sin^2\theta)^{3/2}}$$

$$F_\perp = \frac{e_1e_2}{r^2}\frac{(1+\beta^2)(1-\beta^2)sin\theta}{(1-\beta^2 sin^2\theta)^{3/2}} \quad . \tag{26}$$

In the limit $\beta \to 1$ the force $F_\parallel \to 0$ and $F_\perp$ is zero except in an angular range of $\Delta\theta = \sqrt{1-\beta^2}$ around $\theta = \frac{\pi}{2}$. The force is transverse during a time $\Delta t = \frac{1}{c}d\Delta\theta$ (d is the nearest distance of the two particles, see Fig. 3) and the particle feels a transverse kick of

$$\Delta p = F_\perp \cdot \Delta t = F_\perp\left(\theta = \frac{\pi}{2}\right)\frac{d\Delta\theta}{c} = -\frac{2e^2}{cd} \quad . \tag{27}$$

An integration over the particle distribution $n_2(x,z)$ in the second bunch yields for particle one at position $[x_o, z_o]$:

$$\Delta\vec{p} = [\Delta p_x, \Delta p_z] = \int \frac{2e^2}{c}\frac{[(x-x_o),(z-z_o)]}{(r-r_o)^2} n_2(x,z)dxdz \quad . \tag{28}$$

With a gaussian beam profile $n_2(x,z) = \frac{N_2}{4\pi\sigma_x\sigma_z}exp\left(-\left(\frac{x^2}{2\sigma_x^2} + \frac{z^2}{2\sigma_z^2}\right)\right)$ the integral (28) gives a $[\Delta p_x, \Delta p_z]$ shown schematically in Fig. 4. For small x_o and z_o respectively the momentum kicks are linear in x_o and z_o

$$\Delta p_x = -\frac{2e^2 N_2}{c\sigma_x(\sigma_x + \sigma_z)} \cdot x_o, \quad \Delta p_y = -\frac{2e^2 N_2}{c\sigma_z(\sigma_z + \sigma_x)} \cdot z_o \quad . \tag{29}$$

In this linear approximation the change in the particle direction is

$$\Delta x' = \frac{\Delta p_x}{p} = \frac{\Delta p_x}{mc\gamma} = -\frac{2r_e N_2}{\gamma\sigma_x(\sigma_z + \sigma_x)} \cdot x_o \text{ with } r_e = \frac{e^2}{m_ec^2} \quad . \tag{30}$$

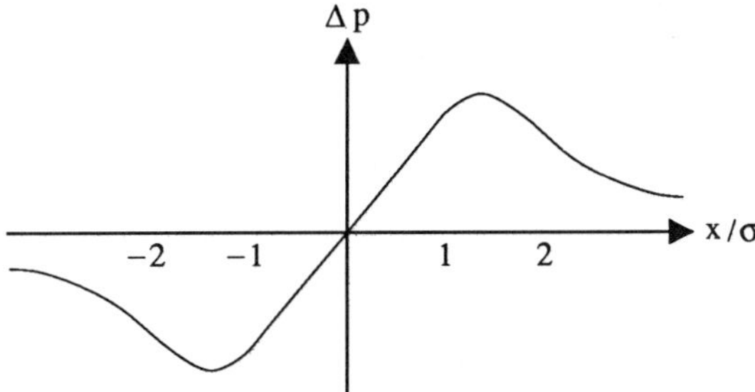

Figure 4. Momentum kick Δp as a function of the transverse separation x/σ of the test particle from the center of the opposite bunch.

The movement of a particle on the phase ellipse eq. (16) in one complete revolution can be described by a 2×2 transport matrix. The kick in eq. (30) due to one beam-beam interaction modifies the transport matrix

$$\begin{pmatrix} x \\ x' \end{pmatrix} = \begin{pmatrix} cos2\pi Q & \beta_x sin2\pi Q \\ -\frac{1}{\beta_x} sin2\pi Q & cos2\pi Q \end{pmatrix} \begin{pmatrix} 1 & 0 \\ \frac{-2r_e N}{\gamma \sigma_x(\sigma_x + \sigma_z)} & 1 \end{pmatrix} \begin{pmatrix} x_o \\ x_o' \end{pmatrix} \quad . \tag{31}$$

The product matrix has eigenvalues

$$\lambda_{1,2} = cos(2\pi(Q + \Delta Q)) \pm i \, sin(2\pi(Q + \Delta Q))$$

where ΔQ can be written in the linear approximation

$$\Delta Q_x^{(1)} = \frac{1}{2\pi} \frac{r_e N_2 \beta_x^{(1)}}{\gamma_1 \sigma_x(\sigma_x + \sigma_z)} = \frac{2r_e \beta_x^{(1)}}{\gamma_1(1 + \frac{\sigma_x}{\sigma_z})} \frac{N_2}{4\pi\sigma_x\sigma_z} \quad . \tag{32}$$

The quantity $N/(4\pi\sigma_x\sigma_z)$ is the particle density D in the bunch. Similar

$$\Delta Q_z^{(1)} = \frac{2r_e \beta_z^{(1)}}{\gamma_1 \left(1 + \frac{\sigma_z}{\sigma_x}\right)} \frac{N_2}{4\pi\sigma_x\sigma_z} \quad . \tag{33}$$

Limits for ΔQ

It is an empirical fact, that in electron-positron storage rings the tune shift ΔQ is limited to about $\Delta Q \leq 0.02 - 0.05$. We give here three ways to visualize these values, assuming $\gamma = 10^4$ and $\beta_x = 3$ cm.

- The maximum ΔQ can be looked at as a maximal particle density D_c in the beam. $\Delta Q < 0.05$ means the critical density is $D_c = 3 \cdot 10^{12} \; mm^{-2}$.

- The tune shift ΔQ from a quadrupole magnet of strength $k[m^{-2}]$ and length ΔS is $\Delta Q = \frac{\beta_x}{4\pi} k \cdot \Delta S$. Assume a bunch length of $\sigma_s = \Delta S = 3$ cm, $\Delta Q = 0.05$ one arrives at $k_{bunch} = \frac{\Delta Q \cdot 4\pi}{\beta_x \cdot \Delta S} = 698 \; m^{-2}$ which is more than two orders of magnitude larger than the strongest quadrupole magnet in any machine.

- The beam divergence at the interaction point is $x' = \sqrt{\frac{\varepsilon}{\beta_x}}$. The beam-beam effect introduces a kink $\Delta x' = \Delta Q \cdot x_o \cdot 4\pi/\beta_x$. If we identify x_o with the beam width $x_o = \sqrt{\varepsilon\beta_x}$ and assume, that the kink $\Delta x'$ is not larger than the beam divergence x', we arrive at $\Delta Q < \frac{1}{4\pi} = 0.08^{14}$.

Another important factor for the limited ΔQ-values are the nonlinearities. ΔQ is not a pure shift of the betatron tune as is the case for the linear regime in Fig. 4, and eq. (32), but rather a spread of shifts and therefore dangerous for resonances. (Resonances occur not only at integer Q values[10]).

Disruption

It was recently rediscovered[15] that the beam-beam force introduces not only a kick $\Delta x'$ (30) expressed by a tune shift $\frac{\Delta x'}{x} = \frac{4\pi\Delta Q}{\beta^*}$, but also a transverse shift $\Delta x = \sigma_s \cdot \Delta x'$ or

$$\mathcal{D} \equiv \frac{\Delta x}{x} = \frac{4\pi\Delta Q}{\beta^*} \cdot \sigma_s \tag{34}$$

where σ_s is the bunch length. The quantity $\mathcal{D}$ was first used in the study of linear colliders and is called disruption. In the spirit of the heuristic arguments given above for the tune shift a similar limit applies for the disruption parameter. The average transverse displacement of a particle due to the beam-beam interaction over the length of the opposing bunch should be at most comparable to the transverse rms beam size. An examination of experimental results drawn from various machines have shown limiting values for $\mathcal{D}$ in the range 0.25-0.3. This limit was first used in the CERN-PSI feasibility study[1] and implies a short bunch length.

Bunch length

The change of energy δU of a particle with RF-phase $\phi = \phi_o + \Delta\phi$ and energy $E = E_o + \epsilon$ is the difference of the accelerating voltage eV and the radiated energy per revolution $U_R(E)$

$$\delta U = eV(\phi) - U_R(E). \tag{35}$$

We can write for small $\Delta\phi$ and ϵ :

$$V(\phi) = V(\phi_o) + \frac{dV}{d\phi}\Delta\phi$$

$$U_R(E) = U_R(E_o) + \frac{dU_R}{dE}\epsilon.$$

Clearly $eV(\phi_o) = U_R(E_o)$. We can divide (35) by the time T_o of one revolution

$$\frac{d\epsilon}{dt} = \frac{\delta U}{T_o} = \frac{e\frac{dV}{d\phi}\Delta\phi - \frac{dU_R}{dE}\epsilon}{T_o}. \tag{36}$$

The change Δs in orbit length L caused by the energy deviation ϵ is defined by the momentum compaction factor α

$$\frac{\Delta s}{L} = \alpha\frac{\epsilon}{E_o}. \tag{37}$$

For $\gamma^2 >> 1/\alpha$ (above the transition energy) the phase change is $\Delta\phi = \omega_{RF} \cdot \Delta s/c = \omega_{RF} \cdot \alpha\frac{\epsilon}{E_o}T_o$ and it follows

$$\frac{d\Delta\phi}{dt} = \alpha\omega_{RF}\frac{\epsilon}{E_o}. \tag{38}$$

Combining (36) and (38) yields :

$$\frac{d^2\Delta\phi}{dt^2} + \frac{\frac{dU_R}{dE}}{T_o} \cdot \frac{d\Delta\phi}{dt} - \frac{\alpha\omega_{RF}e\frac{dV}{d\phi}}{E_o T_o} \cdot \Delta\phi = 0. \tag{39}$$

Note that above the transition energy a stable orbit needs $\frac{dV}{d\phi} < 0$ and (39) is a damped synchrotron oscillation with frequency $\Omega^2 = -\frac{\alpha\omega_{RF}e\frac{dV}{d\phi}}{E_o T_o}$ and damping constant $2\alpha_\epsilon = \frac{dU_R}{dE}/T_o$.

The bunch length σ_s is the amplitude of this oscillation. Since $\alpha_\epsilon << \Omega$ the solutions of (38) and (39) are :

$$\Delta\phi(t) = (\omega_{RF} \cdot \sigma_s/c) \cdot e^{-(\alpha_\epsilon + i\Omega)t}$$

$$\epsilon(t) = -i\frac{E_o\Omega}{\alpha}\frac{\omega_{RF} \cdot \sigma_s}{c} \cdot e^{-(\alpha_\epsilon + i\Omega)t} = \sigma_\epsilon \cdot e^{-(\alpha_\epsilon - i(\pi/2 + \Omega))t}.$$

The amplitudes don't shrink to zero but approach a finite value given by quantum fluctuations in the photon emission. The asymptotic value for σ_ϵ is given in (22) and we can write

$$\sigma_s^2 = \frac{\alpha^2 c^2}{E^2\Omega^2} \cdot (\frac{\sigma_\epsilon}{E})^2 \sim \frac{\alpha}{\omega_{RF}e\frac{dV}{d\phi}} \cdot E^3. \tag{40}$$

A short bunch length needs therefore a small momentum compaction factor α and/or a large $\omega_{RF}\frac{dV}{d\phi} = \frac{dV}{dt}$.

Luminosity

Later on it will be shown that a collider with asymmetric energies is needed. The luminosity of a storage ring with B bunches, N_i particles per bunch in each ring, a revolution frequency f and beam dimensions σ_x, σ_z has a luminosity of

$$\mathcal{L} = f \cdot B \cdot \frac{N_1 \cdot N_2}{4\pi\sigma_x\sigma_z} = f \cdot B \cdot \frac{N_1}{4\pi\sigma_x\sigma_z} \cdot \frac{N_2}{4\pi\sigma_x \cdot \sigma_z} \cdot (4\pi\sigma_x\sigma_z) \ . \tag{41}$$

We know from the previous paragraph, that the density $D = N/(4\pi\sigma_x\sigma_z)$ is limited and therefore a large luminosity is achieved with a large beam cross section and many bunches. This is opposite to linear colliders, where the bunches are thrown away after each collision and much larger critical densities can be tolerated. In the latter case the luminosity is inversely proportional to the beam size.

In (41) we have implicitly assumed, that the two rings have equal circumference and that the beam dimensions are equal at the crossing point. If we further assume, that the tune shift parameters $\Delta Q_x = \Delta Q_z$ are equal and therefore from (32) and (33) $\beta_z/\beta_x = \varepsilon_z/\varepsilon_x$, we can rewrite (41) by substituting $\frac{N_2}{4\pi\sigma_x\sigma_z} \approx \Delta Q^{(1)}$ and $N_1 = \frac{I_1}{ef}$, where I_1 is the bunch current in ring 1

$$\mathcal{L} = \frac{\Delta Q^{(i)}}{2r_e}\gamma_i \cdot I_i \cdot B \left(\frac{1}{\beta_x^{(i)}} + \frac{1}{\beta_z^{(i)}} \right) \ , \ i = 1, 2 \ . \tag{42}$$

Using (21), equation (32) and (33) can be rewritten as

$$I_2 = \frac{e4\pi}{2r_e}\gamma_1 \cdot \Delta Q^{(1)} \cdot f \cdot \varepsilon_x^{(1)} \tag{43}$$

and therefore (42) becomes

$$\mathcal{L} = \frac{\pi}{r_e^2}\gamma_1 \cdot \gamma_2 \cdot \Delta Q^{(1)} \cdot \Delta Q^{(2)} \left(\frac{1}{\beta_x^{(2)}} + \frac{1}{\beta_z^{(2)}} \right) \varepsilon_x^{(1)} \cdot f \cdot B \ . \tag{44}$$

Under the assumption of equal beam dimensions and $\Delta Q_x^{(i)} = \Delta Q_z^{(i)}$ it is easily shown, that (44) has the same value, if the beam indices 1 and 2 are interchanged.

Formula (44) is only valid, if enough RF power is available to compensate the energy loss due to synchrotron radiation. The radiated power $P_{rad}[\text{kW/m}] = 14 \ E^4[\text{GeV}] \ I[\text{A}]/\rho[\text{m}]$ depends on the fourth power of the electron or positron energy of the stored beam.

The losses in the RF accelerating cavities are $P_{cav} = V^2/R_s$ with V the cavity voltage and R_s the shunt impedance. Therefore the cavity loss increases even with the 8th power of the beam energy , if the voltage is given by the synchrotron radiation losses. Two other effects are important. In storage rings with a short bunch length σ_s, the RF-voltage V_{RF}

Table 2. Machine parameters of storage rings.

Accelerator	particles collided	max $\sqrt{s}$ [GeV]	energy resol. [$*10^{-4}$] per beam	circumference [km]	min. bunch spacing [μs]	beam radius [μm] Hor.	Vert.
Tevatron (Fermilab)	$p\bar{p}$	2 × 1000	1.2	6.28	6.98	95./70.	70./50.
HERA (DESY)	$e^{\pm}p$	30 + 820	9.1 + 4.0	6.34	0.096	280./265.	37./84.
LEP (CERN)	e^+e^-	2 × 60	8.2	26.66	22.	312.	12.5
SLC (SLAC)	e^+e^-	2 × 50	2.0	1.45 + 1.47	5560.	1.7	1.7
Tristan (KEK)	e^+e^-	2 × 30	16.4	3.02	5.03	480.	12.
CESR (Cornell)	e^+e^-	2 × 10	5.7	0.768	0.371	500.	11.
DORIS (DESY)	e^+e^-	2 × 6	14.0	0.30	1.0	570.	30.
BFI[1] (CERN)	e^+e^-	3.5 + 8	4.1 + 8.5	0.963	0.01	172	5.5

is mainly determined by the longitudinal focusing needed $V_{RF} \sim \frac{E^3}{\omega_{RF} \cdot \sigma_s^2}$ (see equation (40)).
Short bunches excite through the high frequency fourier components also higher order modes
in the accelerating cavities with a power of $P_{HOM} = k(\sigma_s)\frac{I^2}{f}$ where $k(\sigma_s)$ is roughly inversely
proportional to σ_s. This power is taken from the beam and has to be given back. The power
of the synchrotron radiation, of the focussing and the cavity loss give for a fixed installed
RF power a rather sharp drop in luminosity above a certain beam energy. Table 3 shows the
power requirement for the proposed B-meson factory in the ISR tunnel at CERN. The short
bunch length in the high luminosity version requires superconducting cavities.

Instabilities

Beam instabilities limit the bunch current I and the total current $I \cdot B$ in expression
(42).

Table 3. Some machine parameters and power requirement of the proposed B-meson factory
in the ISR tunnel at CERN (from ref. 1).

Luminosity	[cm^{-2}s^{-1}]	10^{33}		10^{34}	
beam energy	[GeV]	3.5	8	3.5	8
number of bunches		80	80	320	320
bunch length	[cm]	2	2	0.5	0.5
number of cavities		4	20	8	64
synchrotron radiation power	[MW]	0.39	3.1	0.56	6.24
dissipated power per cavity	[kW]	34	60	s.c.	s.c.
HOM losses in vacuum chamber	[kW]	140	27	590	110
HOM losses per cavity	[kW]	8.2	1.6	34	6.5
beam current	[A]	1.28	0.56	2.56	1.12

Single bunch instabilities result from the interaction of the electromagnetic field of the head of a particular bunch with the vacuum chamber. The field is scattered and acts back on the tail of the same bunch. A very low impedance of the vacuum chamber including bellows and other transition pieces help to reduce the effect.

Multibunch instabilities are mostly driven by parasitic resonances in the RF cavities which are excited by the beam and act back on the following bunches. The cure is to install passive dampers on the cavities and an active feedback system with additional cavities to damp these resonances.

Beam separation and interaction region design

To minimize the beam-beam interaction, any parasitic beam crossings have to be avoided. To separate the beams after the collision into two rings, in principle three methods can be employed: electrostatic separation, magnetic separation with unequal beam energies and crossing at an angle.

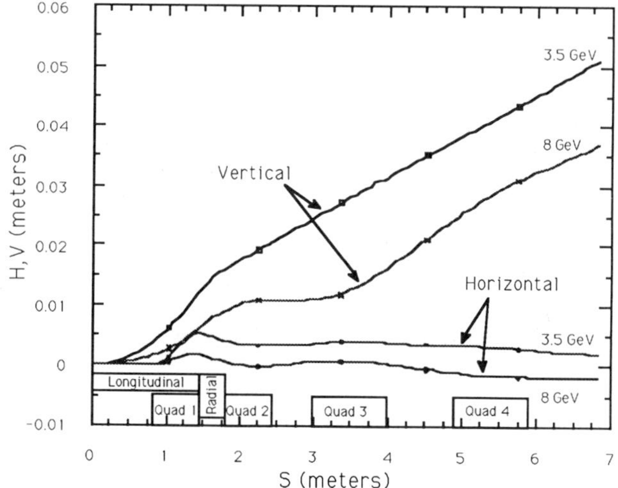

Figure 5. Beam separation around the interaction point with a solenoid field of 1.5 T field tilted by 5 degrees. The longitudinal and radial field component (edge effect) of the solenoid is taken into account.

The DORIS I storage ring at DESY operated with a crossing angle of 12 mrad which caused strong unwanted transverse space charge forces (synchro-betatron resonances) and limited the luminosity considerably[16]. A new trick has been proposed by Oide and Yokoya[17], the so called crab crossing scheme. The two trajectories cross at an angle but the longitudinal axes of the two bunches are rotated with respect to the direction of flight such, that they collide head-on in the center of mass (crab crossing). The angular rotation of the bunch must be produced by RF cavities excited on a deflecting mode and again the problem is to damp the higher order modes (and the fundamental accelerating mode). Electrostatic separators must be fairly long (around 8 m with an electric field of order $5 \cdot 10^6 \text{V/m}$) and the minimum bunch spacing is therefore limited to 16 m. Magnetic bending in the vertical direction with unequal beam energies seems at the moment the simplest solution.

The design of the interaction region is the greatest challenge. Synchrotron radiation from the bends affect the innermost parts of the detectors and have to be effectively shielded. A worked out solution can be found in ref. 1. The solenoid field of the detector is tilted by about 5 degrees with respect to the beam axis producing a dipole component over the length of the coil. A calculation[18] of the optics including the edge effect of the solenoid and focusing quadrupoles is shown in Figure 5. The tilted solenoid field separates the beams over a distance of 4 m. They pass then off axis through the first quadrupole magnet and experience an additional kick. A vertical separation of 1 cm after 2.5 m can be achieved.

The B-meson factory projects around the world

Several projects of B-meson factories are proposed in the world. Table 4 lists the most recent ones. They all have asymmetric energy collisions although the first 5 entries in Table 4 with equal size circumference of the two rings can in principle be converted to a symmetric machine.

Table 4. The proposed B-meson factories in the world.

Location/Country	special feature	reuse of tunnel	beam energies [GeV]	ref.
Novosibirsk/USSR	dual rings		4.0 + 7	2
Cornell/USA	dual rings	CESR	3.5 + 8	3
CERN/Switzerland	dual rings	ISR	3.5 + 8	1
KEK/Japan	dual rings	new	3.5 + 8	4
SLAC/USA	dual rings	PEP	3.1 + 9	5
DESY/Germany	large + small ring	PETRA	2.3 + 12	6
CERN/Switzerland	s.c. linac vs. ring		2.0 + 14	19
CEBAF/USA	s.c. linac vs. ring		2.0 + 14	20

D- AND B-MESON RECONSTRUCTION

Introduction

In this section we explain the reconstruction of D- and B-mesons. A complete reconstruction of the final state of B-mesons is needed to discriminate between B^o, $B^\pm$ or B_s-mesons. This is important in a comparison of lifetimes of B^o and $B^\pm$ and to separate B_s from B^o in measurements of mixing. In some physics questions it is sufficient to tag only the b-quark charge, i.e. to distinguish B from $\overline{B}$.

Since B-mesons have a low branching fraction into any exclusive decay channel, the crucial point is to achieve a large enough signal over the combinatorial background. To reach this goal, a combination of the following four techniques have to be used :

- excellent particle identification π/K, e/π, π^o/γ etc.

- unambiguous assignment of tracks to secondary vertices from B- and D-decays

- excellent invariant mass resolution

- complete reconstruction of one B to reduce the combinatorial background of the other $\overline{B}$ if operating at the $\Upsilon(4S)$-resonance.

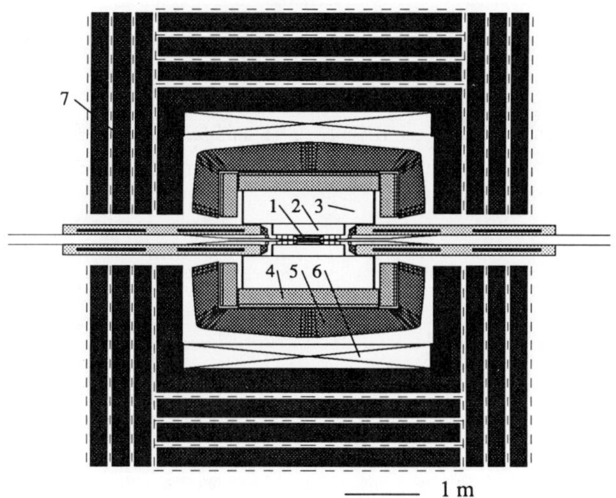

Figure 6. A universal Detector. 1=vertex detector, 2=precision tracking chamber, 3=main drift chamber, 4=ring image cherenkov counter (RICH), 5=CsI(Tl) calorimeter, 6=superconducting coil, 7=instrumented iron for muon detection.

Each of the four methods will be discussed separately. Results of simulation studies are given for the detector shown in Figure 6.

Particle Identification

Particle identification is important in the reduction of the above mentioned combinatorial background and is absolutely necessary in the detection of rare two body decays like $B \rightarrow K\pi,\ \pi\pi,\ p\bar{p},\ K\rho$.

Conventual techniques like time-of-flight or dE/dx-measurements are limited to low momenta ($p \leq 1\ GeV/c$) unless large flight paths or many dE/dx samplings over a long track length are used. An example of a detector with excellent dE/dx measurements also in the relativistic rise region is OPAL[21] with a cylindrical driftchamber of 4 m diameter. A good electromagnetic calorimeter with e.g. CsI-crystals around such a large chamber would be very expensive. Furthermore the dE/dx energy loss is equal for pions and kaons at $\sim 1\ GeV/c$ and for protons and kaons at $\sim 2\ GeV/c$.

A new promising detector is the Ring Imaging Čerenkov Counter (RICH). For a high luminosity storage ring with a short bunch spacing the fast RICH is best suited[22]. The principle of operation is shown in Fig. 7. The Čerenkov radiator is a NaF cylinder of 1 cm thickness. The photons lying on the Čerenkov cone ~ 15 cm apart from the radiator are detected in a MWPC with pad readout. As a photoionising gas TMAE (Tetrakis Dimethylamine Ethylene) or TEA (Tri-Ethyl-Amine) has to be added to the chamber gas. TEA has a vapour pressure of 52 Torr at 20°C and is the preferred choice. TMAE would have to operate at 100°C and adds complication to the design.

The Čerenkov images are pieces of a hyperbola due to total reflection of some of the light in the radiator. With such a detector one can identify pions above $p = 134$ MeV/c, kaons above $p = 472$ MeV/c and protons above $p = 898$ MeV/c.

Figure 8 shows the separation number s_{ij} (number of standard deviations) for particle i separated from particle j

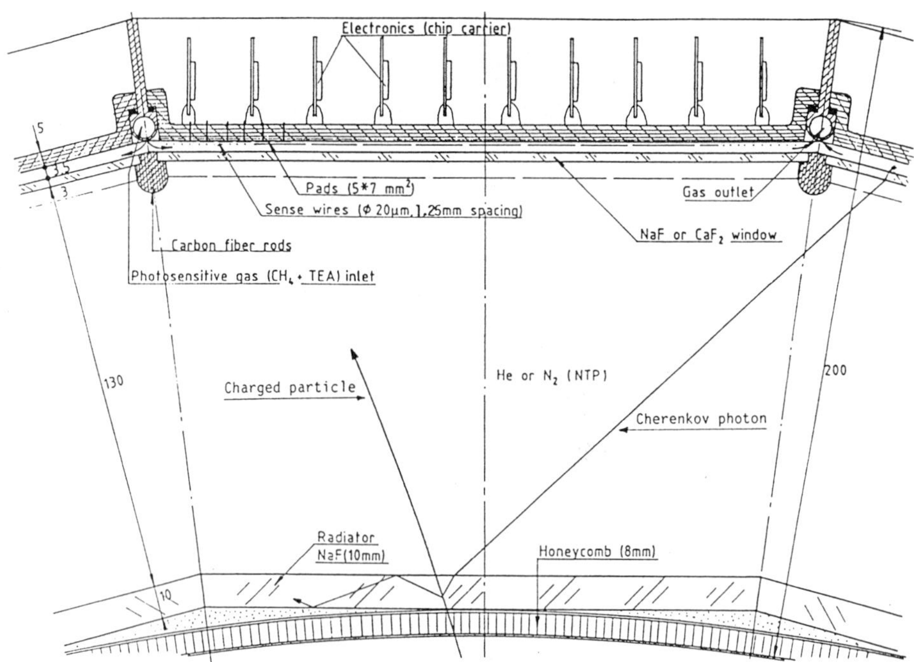

Figure 7. Principle of a fast RICH counter.

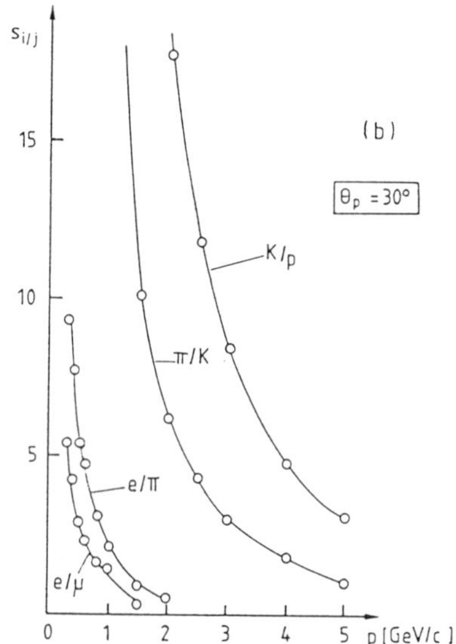

Figure 8. Separation of particles.

$$ s_{ij} = \frac{|\bar{\theta}_c^{(i)} - \bar{\theta}_c^{(j)}|}{\sigma_{\theta_c}^{(i)}} \sqrt{N_{p.e.}^{(i)}} $$

where $\bar{\theta}_c^{(i)}$ is the mean Čerenkov angle, $\sigma_{\theta_c}^{(i)}$ the angular error per photon and $N_{p.e.}^{(i)}$ the number of detected photoelectrons. The π/μ separation, not indicated in Fig. 8, is possible up to about 400 MeV/c. Except for e/π separation, such a RICH detector is excellent for the momentum range up to 2.5 GeV/c which is sufficient for a detector operating near the $\Upsilon(4S)$-resonance.

For the separation of electrons and pions and for the reconstruction of π^o's a fine segmented electromagnetic calorimeter with good energy resolution is necessary. The CLEO II detector and most proposals for detectors at a future B-meson factory have a CsI-calorimeter inside the magnetic coil. The good energy resolution of such a CsI-detector of about $1\%/\sqrt{E}$ can only be maintained, if a RICH detector in front of the calorimeter has a thickness of less than 20% in radiation length. This is not a trivial requirement. The largest amount of material in the RICH counter shown in Figure 7 are the radiator and the photon window in front of the MWPC, both made from NaF. An idea to improve on this, is to omit the window and fill the whole gap between radiator and cathode pads with a gas insensitive to photons. The Čerenkov photons will be converted on the cathode made from a thin layer of CsI with adsorbed TMAE on it. Such a cathode has a quantum efficiency of 45% at $\lambda = 180$ nm and 70% at $\lambda = 150$ nm[23].

The improvement in the signal over combinatorial background with a RICH-detector is clearly seen in Fig. 9, where an attempt is made to reconstruct D-mesons in the decay chain

$$\Upsilon(4S) \;\rightarrow\; B\bar{B}$$
$$\hookrightarrow D + n\pi$$
$$\hookrightarrow K\pi\pi\pi \quad .$$

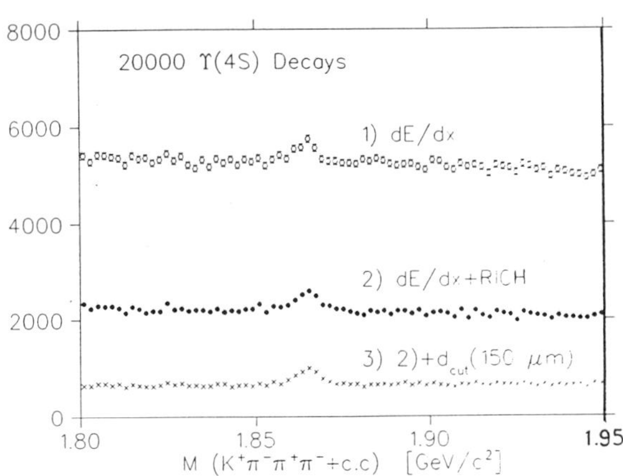

Figure 9. $K^{\pm}\pi^{\pm}\pi^{\pm}\pi^{\pm}$ invariant mass distribution from $\Upsilon(4S)$ decays for three different selection criteria.

The pion/kaon separation helps substantially (curve 2 in Fig. 9). Even more reduction in background (curve 3) is achieved by requiring a maximum distance in space between the kaon and associated pions from the same D-decay with the help of a vertex detector.

Vertex Detection

In events containing a B- and $\overline{B}$-meson several secondary vertices exist, namely the two B decay vertices and the ones from the decay chain $B \rightarrow D$. Table 5 gives the decay length $\lambda = \gamma\beta c\tau$ for several particles at the $\Upsilon(4S)$ and Z^o-resonance. It should be noted, that the B's are produced exclusively at the $\Upsilon(4S)$ whereas at the Z^o many other particles accompany the B-mesons from the fragmentation process. The additional particles originate from the primary vertex and the spectrum peaks at low momenta. Through multiple scattering in beam pipe and vertex detector or in case of π^o, it is often difficult to assign these particles

Table 5. Decay lengths $\lambda = \beta\gamma c\tau$ at the $\Upsilon(4S)$ at rest, at $\Upsilon(4S)$ with a boost of $\beta\gamma = 0.4$, and Z^o resonance. The maximum and minimum values in beam-direction are given.

Particle	$\Upsilon(4S)$ at rest	$\Upsilon(4S)$ boosted	Z^o
B^o	22 μm	130 $\mu m \to$ 180 μm	2.9 mm
B^-	24 μm	130 $\mu m \to$ 180 μm	2.9 mm
D^o	160 μm	-90 $\mu m \to$ 270 μm	1.0 mm
D^-	410 μm	-220 $\mu m \to$ 680 μm	2.2 mm
τ	250 μm	160 $\mu m \to$ 390 μm	2.6 mm

uniquely to a given vertex. A wrongly assigned charged pion with low momentum can then confuse a B^o with a B^+ or vice versa in the analysis.

Invariant Mass Resolution

The signal to background ratio is directly proportional to the invariant mass resolution. A very small resolution can be obtained in the two body decay $\Upsilon(4S) \to B\bar{B}$ through the energy constraint of the B-meson $E_B = \frac{1}{2} m(\Upsilon(4S))$. This is illustrated in the decay $B^+ \to J/\psi K^+ \to \mu^+\mu^- K^{+24}$. The Delphi group at LEP estimates the invariant mass resolution for a detector with a momentum resolution of $\sigma_p/\mathrm{p} = 10^{-3}$ p/GeV for charged particles $\sigma_1(B) = 60$ MeV. The constraint of the two muons to the J/ψ-mass improves this value to $\sigma_2(B) = 30$ MeV. The energy constraint can only be used at the $\Upsilon(4S)$-resonance where the $B\bar{B}$ mesons are produced exclusively and therefore the exact energy of both B-mesons is known within the resolution of the storage ring. This is also true for a boosted $\Upsilon(4S)$ resonance. The method reduces the mass resolution to $\sigma_3(B) = 2.6$ MeV for an assumed energy width of the storage ring of 4 MeV. For zero width of the storage ring energy, $\sigma_3(B) = 1.8$ MeV.

Tagging of B-Mesons

The complete reconstruction of one B (or $\overline{B}$) meson in a particular event reduces the number of particle candidates for the second $\overline{B}$ (or B) meson and lowers in this way the combinatorial background. On the $\Upsilon(4S)$-resonance the particles are then uniquely assigned to the second $\overline{B}$ (or B) meson. As tagging channel one uses one or several decay modes with a large branching fraction. Examples are

$$e^+e^- \; \to \; \Upsilon(4S) \; \to \; B^o \quad \overline{B^o}$$
$$\hookrightarrow \quad D^-\pi^+\pi^+\pi^-\pi^o \quad \text{(to be measured)}$$
$$\hookrightarrow \quad D^{*+}\pi^- \;\; \text{or} \;\; D^*\ell\nu \quad \text{(tag)}$$

$$e^+e^- \; \to \; \Upsilon(4S) \; \to \; B^+ \quad B^-$$
$$\hookrightarrow \quad \tau^-\nu \quad \text{(to be measured)}$$
$$\hookrightarrow \quad \overline{D^o}\pi^+ \;\; \text{or} \;\; D^*\ell\nu \quad \text{(tag)}$$

As suggested in the previous example, the tagging method allows

i) the reconstruction of high multiplicity or rare decay channels,

ii) the measurement of absolute branching fractions

iii) and at least the determination of the quark charge (b or $\bar{b}$) of the tagged meson.

It should be noted, that absolute branching fractions can only be measured at the $\Upsilon(4S)$ resonance through tagging, since in the continuum or at the Z^o-resonance the quark flavour of the light quarks in the B and $\overline{B}$-mesons are not a priori known.

An interesting quantity is the tagging efficiency of a detector, i.e. the efficiency of the complete reconstruction of a B-meson. It is not well defined, since the efficiency depends very much on the signal to background ratio of the tagged decay mode. Depending on the physics question a very background free tag is needed with reduced efficiency or a larger background can be tolerated with a correspondingly higher tagging efficiency. Table 6 shows as an example some reconstruction efficiencies obtained by the ARGUS-Collaboration.

Table 6. Reconstruction Efficiencies for B-mesons from ARGUS. The B, D^*- and D-meson decay channels used are given. The acceptance times reconstruction efficiency is denoted by η_{Det} and the product of η_{Det} and the branching ratios is called tagging efficiency η_{tag}. With a good π^o-detector also the $D^{*-} \to \bar{D}^o \pi^o + \bar{D}^o \gamma$ (45%) could be measured.

Channel	$BR(B \to ...)$	$BR(D^{*-} \to ...)$	η_{Det}	η_{tag}
$B^o \to D^{*-}\pi^+$	0.27%	15.8%	26.6%	0.011%
$D^{*-}\pi^+\pi^o$	1.5%	15.8%	7.8%	0.018%
$D^{*-}\pi^+\pi^+\pi^-$	3.3%	15.8%	11.7%	0.060%
$D^{*-}\ell^+\nu$	14.0%	15.8%	15.3%	0.33%
sum tagging efficiency				0.42%

Channel	D^{*-} branching fraction	D branching fraction
$D^{*-} \to \bar{D}^o \pi^-$	55%	
$\hookrightarrow K^+\pi^-$		3.7%
$K^+\pi^-\pi^o$		11.9%
$K^+\pi^-\pi^+\pi^-$		7.8%
$K_S^o\pi^+\pi^-$		5.3%
sum	55%	28.7%
future $\bar{D}^o$		
$\hookrightarrow K^+\pi^-\pi^o\pi^o$		15.0%
$K^+\pi^-\pi^-\pi^+\pi^o$		4.0%
$K^o\pi^o$		2.7%
K^+K^-		1.2%
$K^oK^+K^-$		1.2%
$\pi^+\pi^-\pi^o$		1.2%
sum		25.3%

The largest η_{tag} is achieved with the semileptonic decay $B \to D^*\ell\nu$. The energy constraint at the $\Upsilon(4S)$ resonance allows the calculation of the missing mass

$$m_\nu^2 = (p_B - p_{D^*} - p_\ell)^2 \ . \tag{45}$$

Since only the absolute value, but not the direction of p_B in (45) is known a priori, the resolution $\sigma(m_\nu)$ is not quite small enough to discriminate between $B \to D^*\ell\nu$ and $B \to D\ell\nu$. This tagging channel can nevertheless predict the b-quark charge from the charge of the lepton.

As seen from Table 6, only decay channels of B-mesons via the D^* are used so far. The mass difference between the D^* and D-meson is 145.3 MeV, close to the pion mass, and the momentum measurement of the pion in the decay $D^* \to D\pi$ allows a precise determination of the difference of the invariant mass $m_{D^*} = m(D\pi)$ and m_D, the latter from $D \to K\pi$ or $D \to K\pi\pi$. The pion momentum in the decay $D^* \to D\pi$ is very low and the reconstruction efficiency depends very crucially on the lowest momentum seen in the detector.

The tagging efficiency presented in Table 6 of about 0.4% can be improved in the future by an order of magnitude with an improved detector with better particle identification, vertex detector and most important, excellent electromagnetic calorimeter for γ and π^o reconstruction. Almost a factor two is gained by using the decay modes $D^{*-} \to \bar{D}^o \pi^o + \bar{D}^o \gamma$. Also other decay modes of the $\bar{D}^o$ shown in the lower part of Table 6 would add another 50%. Particle identification and vertex detectors would reduce some of the combinatorial background and would therefore improve the detector efficiency η_{Det} in Table 6.

SOME SELECTED PHYSICS TOPICS

The Cabibbo-Kobayashi-Maskawa Matrix

The Higgs-fermion coupling

In the Standard Model we believe, that the masses of the elementary fermions, the quarks and leptons, are produced via the Higgs-mechanism. In its simplest form the Higgs-particle forms a weak isospin doublet $\phi(x)$ which can be represented after symmetry breaking by[25]

$$\phi(x) \;=\; \begin{pmatrix} 0 \\ \eta + h(x) \end{pmatrix} \text{ and } \phi_c = -i\tau_2\phi^* = \begin{pmatrix} \eta + h(x) \\ 0 \end{pmatrix} \;, \tag{46}$$

where $h(x)$ is the field amplitude measured from the vacuum expectation value $\eta =< 0/\phi/0 >$. The coupling of the Higgs-field with the elementary fermion fields, namely the lefthanded quark isodoublets $q_L^{(i)} = \begin{pmatrix} u_L^{(i)} \\ d_L^{(i)} \end{pmatrix}$, the right handed quark singlets $u_R^{(i)}$, $d_R^{(i)}$ and similar for the leptons has to provide the mass term in the Lagrangian. The superscript indicates the three generations of quarks and leptons.

The mass term has to be a Lorentz scalar, an isoscalar (T=0) and hypercharge $Y = 0$ where the latter is related to the electric charge Q and the third component of the weak isospin T^3

$$Q \;=\; T^3 + \frac{Y}{2} \;. \tag{47}$$

Two such mass terms can be formed:

$$\mathcal{L}_1 \;=\; \sum_{i,k} g_{ik} \left[\left(\bar{q}_L^{(i)}\phi \right) \cdot d_R^{(k)} + \bar{d}_R^{(k)} \left(\phi^+ q_L^{(i)} \right) \right] =$$

$$= \sum_{i,k} g_{ik}\eta \left[\bar{d}_L^{(i)} d_R^{(k)} + \bar{d}_R^{(k)} d_L^{(i)} \right] = \sum_{i,k} m_{ik}^{(d)} \left[\bar{d}_L^{(i)} d_R^{(k)} + \bar{d}_R^{(k)} d_L^{(i)} \right] \tag{48}$$

$$\mathcal{L}_2 \;=\; \sum_{i,k} h_{ik} \left[\left(\bar{q}_L^{(i)}\phi_c \right) u_R^{(k)} + \bar{u}_R^{(k)} \left(\phi_c^+ q_L^{(i)} \right) \right] =$$

$$= \sum_{i,k} m_{ik}^{(u)} \left[\bar{u}_L^{(i)} u_R^{(k)} + \bar{u}_R^{(k)} u_L^{(i)} \right] \;. \tag{49}$$

The g_{ik} and h_{ik} are at first arbitrarily coupling constants. The fields $q_L^{(i)}$, $d_R^{(i)}$, $u_R^{(i)}$ are the fields which couple in the weak interaction to the Higgs- and gauge bosons $W^\pm, Z$. They are not the fields, which appear in the strong interaction, since $m_{ij}^{(u)}$ and $m_{ij}^{(d)}$ are non-diagonal matrices. Mesons are bound states of $\bar{Q}^{(i)}Q^{(i)}$ quark-pairs with definite masses. We can write the relation between the $q^{(i)}$ and $Q^{(i)}$ fields by defining two 3×3 unitary matrices S, R in flavour space

$$U^{(i)} \;=\; \sum_k S_{ik} u^{(k)} \;, \quad D^{(i)} = \sum_k R_{ik} d^{(k)} \tag{50}$$

where S and R are such, that

$$\left(S^* m^{(d)} S \right)_{ik} \;=\; m_i^{(d)} \delta_{ik} \quad \text{and} \quad \left(R^* m^{(u)} R \right)_{ik} \;=\; m_i^{(u)} \delta_{ik} \;. \tag{51}$$

468

The model has therefore either the 12 coupling constants g_{ik} and h_{ik} (note $g_{ik} = g_{ik}$ and $h_{ik} = h_{ki}$) or equivalently the derived parameters $m_i^{(u)} = m_u, m_c, m_t$, $m_i^{(d)} = m_d, m_s, m_b$ and 3 Euler angles in each of the two unitary matrices S and R respectively. Similar relations hold for the leptons.

Not all 12 parameters are observables. Quarks and leptons can only be observed through their interactions. The Hamiltonian of the three neutral current couplings, namely the gluon G_μ^k, the Z_μ^o and the photon A_μ currents to the fermion fields

$$H^{(g)} \;=\; \sum_i \sum_k G_\mu^k \bar{q}^{(i)} \gamma^\mu \lambda_k q^{(i)} \quad , \lambda_k = SU(3)_{colour} \text{ matrix} \tag{52}$$

$$H^{(Z^o)} \;=\; \sum_i Z_\mu \bar{q}^{(i)} \gamma^\mu \left(c_V - c_A \cdot \gamma^5 \right) q^{(i)} \tag{53}$$

$$H^{(\gamma)} \;=\; \sum_i A_\mu \bar{q}^{(i)} \gamma^\mu q^{(i)} \quad , \tag{54}$$

are all invariant under the transformation S and R, since both, equations (50) and (52-54), connect only quarks with the same electrical charge. The charged current coupling on the other hand transforms ($u^{(i)} = u, c, t; d^{(i)} = d, s, b$)

$$
\begin{aligned}
H^{(W)} \;&=\; \sum_i W_\mu^* \bar{u}^{(i)} \gamma^\mu \frac{1-\gamma^5}{2} d^{(i)} + \sum_i W_\mu \bar{d}^{(i)} \gamma^\mu \frac{1-\gamma^5}{2} u^{(i)} \;= \\
&=\; \sum_i W_\mu^* \left(\bar{U} R^{*-1} \right) \gamma^\mu \frac{1-\gamma^5}{2} \left(S^{-1} D \right) + \sum_i W_\mu \left(\bar{D} S^{*-1} \right) \gamma^\mu \frac{1-\gamma^5}{2} \left(R^{-1} U \right) \;= \\
&=\; \sum_i W_\mu^* \bar{U} \gamma^\mu \frac{1-\gamma^5}{2} \left(R^{*-1} S^{-1} \right) D + \sum_i W_\mu \bar{D} \left(S^{*-1} R^{-1} \right) \gamma^\mu \frac{1-\gamma^5}{2} U \quad .
\end{aligned}
\tag{55}
$$

We conclude, that only the product $V = R^{*-1} \cdot S^{-1}$ is observable. By convention we leave the u-quarks unmixed and write $D^{(i)} = \sum_k V_{ik} d^{(k)}$. (A side remark: for vanishing neutrino masses the corresponding matrix S is arbitrary and can be chosen that $R^{*-1} S^{-1} = 1$). The unitary matrix V is usually parametrized by 3 Euler angles θ, γ, β and one complex phase δ. With the abbreviations $s_\theta = sin\theta, c_\theta = cos\theta$ etc. the matrix can be written as a product of three rotations

$$
V = \begin{pmatrix} 1 & 0 & 0 \\ 0 & c_\gamma & s_\gamma \\ 0 & -s_\gamma & c_\gamma \end{pmatrix} \begin{pmatrix} c_\beta & 0 & s_\beta e^{-i\delta} \\ 0 & 1 & 0 \\ -s_\beta e^{i\delta} & 0 & c_\beta \end{pmatrix} \begin{pmatrix} c_\theta & s_\theta & 0 \\ -s_\theta & c_\theta & 0 \\ 0 & 0 & 1 \end{pmatrix} = \tag{56}
$$

$$
= \begin{pmatrix} c_\theta c_\beta & s_\theta c_\beta & s_\beta e^{-i\delta} \\ -s_\theta c_\gamma - c_\theta s_\beta s_\gamma e^{i\delta} & c_\theta c_\gamma - s_\theta s_\beta s_\gamma e^{i\delta} & c_\beta s_\gamma \\ s_\theta s_\gamma - c_\theta s_\beta c_\gamma e^{i\delta} & -c_\theta s_\gamma - s_\theta s_\beta c_\gamma e^{i\delta} & c_\beta c_\gamma \end{pmatrix} \quad .
$$

The complex phase $e^{i\delta}$ could be attached to any of the three rotations without altering experimental observations and is unobservable if one of the three angles is zero. To change from one representation to another, we multiply each of the six quark states by an arbitrary phase $\phi_u^{(i)}$ for the up-typ quarks and $\phi_d^{(i)}$ for the down typ quarks. The matrix V_{jk} is then replaced by $V_{jk} \cdot exp[i(\phi_u^{(j)} - \phi_d^{(k)})]$.

Properties of the CKM-matrix

Present values of θ, β and γ are[26]

$$\theta = (12.77 \pm 0.17)^\circ \qquad \beta = (0.23 \pm 0.17)^\circ \qquad \gamma = (2.52 \pm 0.80)^\circ \qquad \text{(errors are 90\% C.L.)}$$

Since $cos\beta > 0.999976$ and $cos\gamma = 0.999033$ we can approximate them by 1. Motivated by these experimental values we define new variables A, λ, ρ and η :

$$sin\theta = 0.221 = \lambda, \quad sin\gamma = 0.044 = 0.90 \cdot \lambda^2 = A\lambda^2 \quad, \tag{57}$$

$$sin\beta e^{-i\delta} \lesseqgtr 0.007 e^{-i\delta} = \lambda^3 A \cdot 0.72 \cdot e^{-i\delta} = \lambda^3 A(\rho - i\eta)$$

and write according to Wolfenstein[27]

$$V \cong \begin{pmatrix} 1 & \lambda & A\lambda^3(\rho - i\eta) \\ -\lambda & 1 & A\lambda^2 \\ A\lambda^3(1 - \rho - i\eta) & -A\lambda^2 & 1 \end{pmatrix} \tag{58}$$

The complex phase δ in (56) is responsible for CP-violation in the Standard Model. Unitarity applied to the first and third column of (56) yields

$$V_{ud}V_{ub}^* + V_{cd}V_{cb}^* + V_{td}V_{tb}^* = 0 \quad . \tag{59}$$

Equation (59) is best written in the Wolfenstein representation (58)

$$A\lambda^3(\rho + i\eta) - A\lambda^3 + A\lambda^3(1 - \rho - i\eta) = 0 \tag{60}$$

and dividing by $A\lambda^3$ yields

$$(\rho + i\eta) - 1 + (1 - \rho - i\eta) = \frac{V_{ub}^*}{A\lambda^3} - 1 + \frac{V_{td}}{A\lambda^3} = 0 \quad, \tag{61}$$

which is visualized in the so-called unitarity triangle (Fig. 10). One concludes, that already a precise measurement of $|V_{ub}|$, $|V_{cb}|$ and $|V_{td}|$ determines the complex phase δ completely.

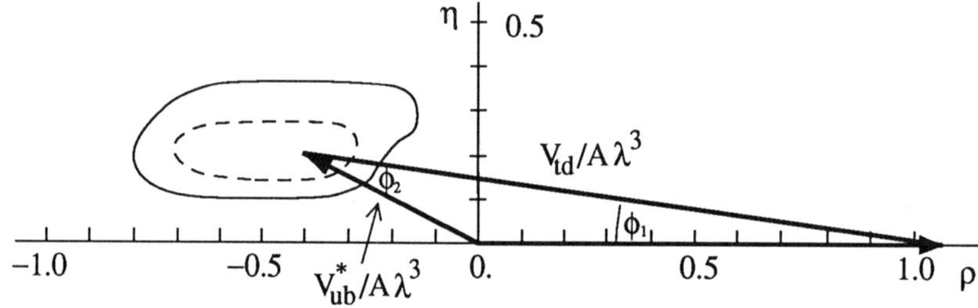

Figure 10. Unitarity triangle in the complex plane (ρ, η). The 68% and 95% confidence limits are shown.

Determination of V_{ud}

The transition $u \rightarrow d$ occurs in pion β-decay $\pi^+ \rightarrow \pi^o e^+ \nu_e$ or in nuclear β-decay where one u-quark decays via $u \rightarrow de^+\nu_e$. The decay is very similar to muon decay. The difference of the amplitude of both process is i) the element V_{ub} at the u-d vertex, ii) the phase space due to different masses, iii) the wave function of the u- resp. d-quark in a nucleus or pion, and iv) radiative corrections.

The trick to overcome the problem iii) is to select $0^+ \rightarrow 0^+$ transitions, where the quark-current matrix element is a pure vector current. The conserved vector-current hypothesis (CVC) assures, that the weak interaction matrix element is not renormalized by the strong interaction and the weak coupling constant measured in muon decay is unchanged. In (62-64) we give the decay width of muon, pion and nuclear beta decay

$$\Gamma\left(\mu^+ \to e^+ \nu_e \bar{\nu}_\mu\right) = \frac{(m_\mu c^2)^5 G_F^2}{192\pi^3(\hbar c)^6}\left\{\left(1 - \frac{8m_e^2}{m_\mu^2}\right)\left(1 + \frac{\alpha}{2\pi}\left(\frac{25}{4} - \pi^2\right)\right)(1 + \Delta_\mu)\right\} \quad (62)$$

$$\Gamma\left(\pi^+ \to \pi^0 e^+ \nu_e\right) = \frac{[(m_{\pi^+} - m_{\pi^0})c^2]^5 G_F^2 |V_{ud}|^2}{30\pi^3(\hbar c)^6}\left\{\left(1 - \frac{(m_{\pi^+} - m_{\pi^0})}{2m_\pi}\right)^3 (1 + \delta_\pi + \Delta_\pi)\right\}$$
$$(63)$$

$$\Gamma\left(0^+ \to 0^+ e\nu_e\right) = \frac{[(M_i - M_f)c^2]^5 G_F^2 |V_{ud}|^2}{30\pi^3(\hbar c)^6}\left\{(1 - \delta_c)(1 + \delta_R)(1 + \Delta_\beta)\right\} \cdot h \quad (64)$$

The expressions in the curly brackets are recoil and radiative corrections, which have to be carefully evaluated[28]. They are of order percent and dominate the error in V_{ud}[28]. The factor h in (64) is the phase space factor including influences of the nuclear charge distribution on the Dirac wave function of the positron and the recoil of the nucleus.

Other elements V_{ik}

Similarly to the previous case semileptonic decays of hadrons have been extensively studied and are well suited for a determination of other V_{ik} elements. Table 7 lists some processes that have been used or could be used in future high luminosity colliders.

For the semileptonic inclusive decays the spectator model describes the decay of a heavy quark Q (b, c, or s-quark) to a lighter one q and the emission of a virtual W^--boson

Table 7. Semileptonic exclusive, inclusive and hadronic decays for determination of the CKM matrix elements. The missing elements V_{td}, V_{ts}, V_{tb} are determined from $B^0 - \bar{B}^0$, $B_s - \bar{B}_s$ mixing and unitarity.

Transition	V_{ud}	V_{us}	V_{ub}
Semileptonic exclusive	$\pi^+ \to \pi^0 e^+ \nu_e$ nuclear $0^+ \to 0^+$	$K \to \pi e \nu_e$ $\Sigma^- \to n e^- \bar{\nu}_e$ $\Xi^- \to \Lambda e^- \bar{\nu}_e$ $\Xi^- \to \Sigma^0 e^- \bar{\nu}_e$ $\Lambda^0 \to p e^- \nu$	$B^+ \to \pi^0 e^+ \nu_e$ $B^0 \to \pi^- e^+ \nu_e$ $B \to \rho \ell \nu$ $B \to \omega \ell \nu$
Semileptonic inclusive			$B \to \mu \nu X_u$
Hadronic			$B^0 \to \pi^+ \pi^-$ $B^+ \to \pi^+ \pi^0$ $B^+ \to \pi^+ \pi^- \pi^+$ $B^0 \to \pi^+ \pi^- \pi^0$ $B \to 4\pi, 5\pi, 6\pi$

Transition	V_{cd}	V_{cs}	V_{cb}
Semileptonic exclusive	$D \to \pi e \nu_e$	$D^0 \to K^- e^+ \nu_e$ $D^+ \to \bar{K}^0 e^+ \nu_e$	$B^0 \to D^- e^+ \nu_e$ $B^+ \to D^0 e^+ \nu_e$ $B \to D\pi e^+ \nu_e$ $B^0 \to D^{*-} e^+ \nu_e$ $B \to D^* \pi e^+ \nu_e$
Semileptonic inclusive	$\nu_\mu d \to c\mu$	$D \to e \nu_e X_s$	$B \to \mu \nu X_c$

which couples to $(\ell\bar\nu_\ell)$

$$\Gamma(Q \to q\ell\bar\nu_\ell) = \frac{(m_Q c^2)^5 G_F^2 |V_{Qq}|^2}{192\pi^3} F(m_q/m_Q)(1+\Delta) \quad . \tag{65}$$

Here is $F(x) = 1 - 8x^2 + 8x^6 - x^8 - 24x^4 \ln x$ and Δ radiative corrections. Formula (65) describes the total semileptonic width and predicts therefore the inclusive decay rate in the limit $m_Q >> \Lambda_{QCD}$. The problem is the strong dependence on partially known quark masses.

The exclusive semileptonic meson decays in Table 7 are either $0^- \to 0^-$ or $0^- \to 1^-$ transitions. The relevant matrix element between two pseudoscalar mesons with momenta p_1 and p_2 contains only the hadronic vector current. The most general lorentz invariant form has two form factors:

$$< M(Q\bar q)|j_\mu^{hadr}|m(q\bar q) > = \frac{1}{\sqrt{2}}\left[(p_1 + p_2)_\mu f_+(q^2) + (p_1 - p_2)_\mu f_-(q^2)\right] . \tag{66}$$

If (66) is multiplied with the leptonic current $< 0 \mid j_\mu^{lept} \mid \ell\nu >= \bar\ell\gamma_\mu(1 - \gamma_5)\nu_\ell$, the second term in (66) becomes $f_-(q^2)m_\ell\bar\ell(1 - \gamma^5)\nu_\ell$ and can be neglected for electrons because of the small electron mass. The exclusive width is then

$$\Gamma(M(Q\bar q) \to m(q\bar q)e\nu_e) = \int dq^2 \frac{G_F^2 |V_{Qq}|^2}{192\pi^3}|f_+(q^2)|^2 \text{ (phase space factor)} \quad . \tag{67}$$

The decay into vector mesons has three form factors, two axialvector and one vector form factor

$$< M(Q\bar q)|j_\mu^{hadr}|m(q\bar q) > = \varepsilon^{*\nu} g_{\nu\mu} f_1^A(q^2) + P_{1\mu} P_{2\nu}\varepsilon^{*\nu} f_2^A(q^2) + i\varepsilon_{\mu\nu\sigma\rho}p_1^\rho p_2^\sigma \varepsilon^{*\nu} f^V(q^2) \quad . \tag{68}$$

Many different parametrizations of form factors have been used[29-34] and a specific investigation of the form factor dependence for $\mid V_{ub}/V_{cb} \mid$ can be found in[33]. These transition form factors are sensitive to the internal quark and gluon dynamics in mesons. The value $f(q^2 = 0)$ is the overlap of initial and final internal meson wave function and varies[34] between 0.25-0.75 for the various decays in Table 7. Uncertainties are larger in the decay of heavier mesons (B, D-mesons), since the overlap (i.e. $f(q^2 = 0)$) is smaller for the heavier quarks compared to the maximal value $f(0) = 1$. Also isospin and $SU(3)_{\text{flavour}}$-symmetry can not be used and the larger q^2-range covered in the decay contributes to the uncertainty.

For the elements V_{cd} and V_{cs} the charm production with ν_μ-beams on isoscalar targets has been successfully applied. The number of valence u- and d-quarks is N(u)=N(d). We define the fraction α and β of $\bar u, \bar d$ and s-sea quarks respectively : $N(\bar u)= N(\bar d)=\alpha N(d)$ and $N(s)=N(\bar s)=\beta N(d)$. Charm particles decay semileptonically

$$BR(c \to s\mu^+\nu) = BR(\bar c \to \bar s\mu^-\nu) = \Gamma(c \to \mu^+\nu)/\Gamma(c \to \text{all})$$

and neutrino production of charm is therefore identified by two leptons in the final state. We can form the four cross sections

$$\begin{aligned}
\sigma(\nu_\mu N \to \mu^+\mu^- X) &= BR \cdot (|V_{cd}|^2 + \beta \cdot |V_{cs}|^2) \\
\sigma(\bar\nu_\mu N \to \mu^-\mu^+ X) &= BR \cdot (\alpha \cdot |V_{cd}|^2 + \beta \cdot |V_{cs}|^2) \\
\sigma(\nu_\mu N \to \mu^- X) &= |V_{ud}|^2 + |V_{cd}|^2 + \tfrac{1}{3}\alpha \cdot |V_{ud}|^2 + \beta \cdot |V_{cs}|^2 \\
\sigma(\bar\nu_\mu N \to \mu^+ X) &= \alpha \cdot |V_{ud}|^2 + \alpha \cdot |V_{cd}|^2 + \tfrac{1}{3} \cdot |V_{ud}|^2 + \beta \cdot |V_{cs}|^2
\end{aligned}$$

where the third line can be read from Fig. 11. The cross section ratio

$$\frac{\sigma(\nu_\mu N \to \mu^-\mu^+ X) - \sigma(\bar\nu_\mu N \to \mu^-\mu^+ X)}{\sigma(\nu_\mu N \to \mu^- X) - \sigma(\bar\nu_\mu N \to \mu^+ X)} = \frac{BR \cdot |V_{cd}|^2}{2/3 + |V_{cd}|^2} \tag{69}$$

gives now the element $|V_{cd}|$.

472

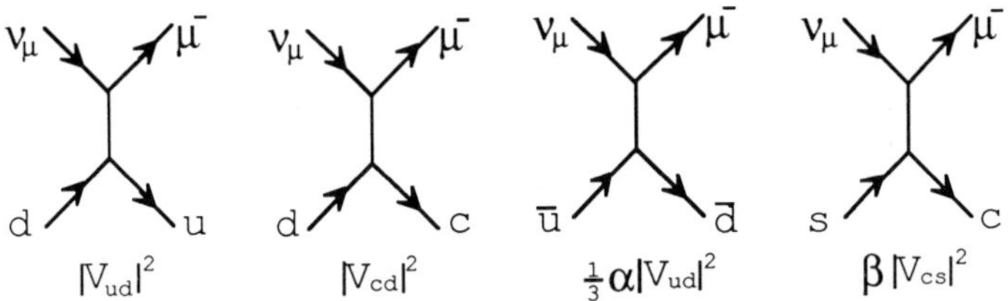

Figure 11. The diagrams contributing to neutrino scattering $\nu_\mu N \to \mu^- X$.

Recent determination of $|V_{ub}/V_{cb}|$

The observation of a nonvanishing CKM parameter V_{ub} is a necessary requirement for CP-violation arising from the CKM matrix phase. Recently, both the Argus[35] and the Cleo collaboration[36] have found indications for a nonzero value of V_{ub}.

The momentum spectrum of the charged lepton in $B \to \ell \nu_\ell X_u$ extends to 2.7 GeV/c whereas in the $B \to \ell \nu_\ell X_c$ decay the maximum lepton momentum is only 2.3 GeV/c due to the larger c-quark mass. The ratio of the semileptonic branching ratios $B_{SL}(\Delta)$ in the two momentum intervals $\Delta_1 = [2.0, 2.3]$ GeV/c and $\Delta_2 = [2.3, 2.6]$ GeV/c in the Argus experiment and $\Delta_1 = [2.2, 2.4]$ GeV/c and $\Delta_2 = [2.4, 2.6]$ GeV/c in the Cleo experiment

$$R\left(\frac{\Delta_2}{\Delta_1}\right) = \frac{B_{SL}[\Delta_2]}{B_{SL}[\Delta_1]} = 4.7 \pm 1.2\% \ (\text{Argus}) \text{ and } = 0.18 \pm 0.05\% \ (\text{Cleo}) \tag{70}$$

is therefore a measure of $|V_{ub}/V_{cb}|^2$.

A number of experimental and theoretical uncertainties have to be solved. On the theoretical side the lepton spectra of a number of exclusive decays has to be calculated

$$R\left(\frac{\Delta_2}{\Delta_1}\right) = \frac{\Gamma(B^+ \to \pi^o \ell^+ \nu) + \Gamma(B^o \to \pi^- \ell^+ \nu) + \Gamma(B^+ \to \rho^o \ell^+ \nu) + \Gamma(B^o \to \rho^- \ell^+ \nu)}{\Gamma(B^+ \to D^o \ell^+ \nu) + \Gamma(B^o \to D^- \ell^+ \nu) + \Gamma(B^+ \to D^{*o} \ell^+ \nu) + \Gamma(B^o \to D^{*-} \ell^+ \nu)}$$

$$= f \left|\frac{V_{ub}}{V_{cb}}\right|^2 . \tag{71}$$

It is believed, that the sum of these exclusive channels represent the vast majority of the semileptonic decays and that e.g. $B \to \omega \ell \nu$ or $B \to D^{**} \ell \nu$ or nonresonant multipion final states can be neglected (see also Table 8). Form factors are crucial for the evaluation of the factor f in (71). For a detailed discussion see ref. 33.

On the experimental side a number of corrections has to be applied to the observed number of leptons (see Table 9).

- subtraction of leptons from the continuum $e^+ e^- \to q\bar{q} \to 2$ jets. These events are recognized by an event shape cut (2 jet-events vs. more spherical events from $\Upsilon(4S) \to B\bar{B}$)

- subtraction of leptons from $B \to \ell \nu X_c$ where the lepton momentum falls into the momentum interval Δ_2 due to the finite resolution of the tracking chamber

- subtraction of leptons from $(B \to J/\psi X, \ J/\psi \to \ell^+ \ell^-)$

- subtraction of hadrons which have been misidentified as leptons.

Table 8. Measured branching ratios and upper limit for semileptonic B-decays. The lepton ℓ is always e or μ and the BR is the sum of the two lepton modes.

BR	Experimental value or limit	Ref.
$B \to \ell\nu X_u$		
$B^o \to \pi^- \ell^+ \nu$	<0.09 %	37
$B^+ \to \pi^o \ell^+ \nu$	<0.22 %	41
$B^+ \to \rho^o \ell^+ \nu$	<0.11 %	37
$B \to \ell\nu X_c$		
$B^+ \to D^o \ell^+ \nu$	$1.6 \pm 0.6^{+0.9}_{-0.6}$ %	38
$B^o \to D^- \ell^+ \nu$	1.8 ± 0.6 %	26
$B^+ \to D^{*o} \ell^+ \nu$	8.0 ± 3.4 %	41
$B^o \to D^{*-} \ell^+ \nu$	9.8 ± 1.5 %	26
Sum of listed exclusive states	21.2 ± 4.1 %	
$B \to \ell\nu(X_u + X_c)$	23.1 ± 1.1 %	26

Table 9. Observed single lepton and dilepton events in the momentum interval $2.3 < p_t < 2.6$ GeV/c and estimated backgrounds (from ref. 35).

	Single Leptons		Dileptons	
	e	μ	e	μ
$\Upsilon(4S)$	31	29	11	10
Backgrounds: Continuum (scaled)	6.8	14.5	1.0	1.0
Lepton momentum resolution in $b \to c$	4.0	4.4	1.1	1.2
$J/\psi \to \ell^+\ell^-$	0.6	0.4	0.2	0.1
Lepton misidentification	0.9	1.7	0.7	1.4
Total background	12.3 ± 2.8	21.0 ± 4.0	3.0 ± 1.0	3.7 ± 1.1
Signal	18.7 ± 6.2	8.0 ± 6.7	8.0 ± 3.5	6.3 ± 3.4

Table 10 summarizes the results. The experimental error is everywhere small compared to the model dependence of 100%.

In current experiments, the large combinatorial background limits the observation of $B \to \rho\ell\nu$ and $\to \omega\ell\nu$. In future colliders with asymmetric beam energies the background can be considerably reduced with the help of vertex information (assign tracks to a vertex and therefore to a unique B-meson) or using the tagging method indicated. In the absence of theoretical uncertainties, 2.10^7 $\Upsilon(4S)$ and correspondingly $6.6 \cdot 10^7$ continuum events would allow to determine $| V_{ub} |$ with an error of 10% for $| V_{ub} |$ as small as 0.001[1]. Theoretical uncertainties are now at the $\pm 40\%$ level (Table 10). Studies on angular and q^2 distributions with hundreds of $B \to \rho, \omega\ell\nu$ events will reduce some of the theoretical uncertainties, but the normalisation error $(f(q^2 = 0))$ will remain. A measurement of all the decay modes listed in Table 7 together with theoretical models will hopefully lead to a consistent description and provide in this way an unambiguous determination of V_{ub}.

Table 10. $|V_{ub}|/|V_{cb}|$ ratio for different models (the errors are non-Gaussian).

Model	endpoint lepton spectrum (Argus)	endpoint lepton spectrum (Cleo)	$B \to \rho\ell\nu, \pi\ell\nu$ (Argus)
ACM[40]	0.10±0.01	0.09±0.01	
WBS[34]	0.12±0.02	0.11±0.01	<0.16
KS [30]	0.09±0.01	0.09±0.02	<0.15
ISGW[31]	0.18±0.02	0.15±0.02	<0.31

CP-Violation

The $\Upsilon(4S)$ resonance decays in about 50% of all cases into a $B^o\bar{B}^o$ meson pair. The weak interaction generates transitions from $B^o \to \bar{B}^o$ and vice versa and eigenstates to the weak interaction with exponential decay and momentum $\vec{p}$ are

$$B_1(t,\vec{p}) = \frac{1}{\sqrt{1+|\alpha|^2}} e^{-i\lambda_1 t} \left[B^o(\vec{p}) + \alpha \bar{B}^o(\vec{p}) \right]$$

and

$$B_2(t,\vec{p}) = \frac{1}{\sqrt{1+|\alpha|^2}} e^{-i\lambda_2 t} \left[B^o(\vec{p}) - \alpha \bar{B}^o(\vec{p}) \right] , \tag{72}$$

where λ_1, λ_2 and α are defined through the effective weak interaction hamiltonian $\hat{H}_W = \hat{M} - i\hat{\Gamma}/2$:

$$< B^o|\hat{M} - i\hat{\Gamma}/2|\bar{B}^o > = M_{12} - i\Gamma_{12}/2$$

$$< B^o|\hat{M} - i\hat{\Gamma}/2|B^o > = < \bar{B}^o|\hat{M} - i\hat{\Gamma}/2|\bar{B}^o > = M - i\Gamma/2$$

$$\alpha = \sqrt{\frac{M_{12}^* - i\Gamma_{12}^*/2}{M_{12} - i\Gamma_{12}/2}} \quad , \lambda_{1,2} = M - i\Gamma/2 \pm \sqrt{(M_{12} - i\Gamma_{12}/2)(M_{12}^* - i\Gamma_{12}^*/2)} \quad . \tag{73}$$

We can write the eigenvalues of $\hat{H}_W$ as $\lambda_{1,2} = m_{1,2} - \frac{i}{2}\gamma_{1,2}$ where

$$Re\lambda_{1,2} = m_{1,2} = M \pm Re\left\{ \sqrt{(M_{12} - i\Gamma_{12}/2)(M_{12}^* - i\Gamma_{12}^*/2)} \right\}$$

$$Im\lambda_{1,2} = \gamma_{1,2} = \Gamma \mp 2Im\left\{ \sqrt{(M_{12} - i\Gamma_{12}/2)(M_{12}^* - i\Gamma_{12}^*/2)} \right\} \quad . \tag{74}$$

This formalism is identical to the kaon system, where CP-violation stems from the fact, that $|\alpha| \neq 1$ and therefore B_1 and B_2 would be no longer eigenstates to CP. In the B-system $|\alpha| - 1 \approx (\gamma_1 - \gamma_2)/(m_1 - m_2)$ arg $(M_{12}/\Gamma_{12}) \approx 0$ since $\Delta\gamma = \gamma_1 - \gamma_2$ is small. The asymmetry $\Upsilon(4S) \to B^o\bar{B}^o \to B^oB^o \neq \bar{B}^o\bar{B}^o$ is of order 10^{-3} and would need an enormous amount of statistics to observe the effect.

A possibly larger effect, CP-violation in the decay amplitude comes about, if the decay B to final state f proceeds via two weak diagrams with amplitudes A_1 and A_2 and strong interaction phase shift δ_1 and δ_2 in the final state

$$\mathcal{A}(B \to f) = A_1 \cdot e^{i\delta_1} + A_2 \cdot e^{i\delta_2}$$

$$\bar{\mathcal{A}}(\bar{B} \to \bar{f}) = A_1^* \cdot e^{i\delta_1} + A_2^* \cdot e^{i\delta_2} \tag{75}$$

Examples are (see Figure 12) :

$$\Upsilon(4S) \to (B^+ \to K^+\rho^o \ , \ B^- \to X) \text{ or } (B^- \to K^-\rho^o \ , \ B^+ \to X)$$

and the difference in the partial decay width is

$$\frac{N(K^+\rho^o) - N(K^-\rho^o)}{N(K^+\rho^o) + N(K^-\rho^o)} = -4 Im(A_1 \cdot A_2^*) \cdot sin(\delta_1 - \delta_2). \qquad (76)$$

CP is violated, if A_1 or A_2 are complex with a different phase and if the strong phase shifts δ_i are different. The size of the asymmetry (76) is uncertain due to the strong phase difference $\delta_1 - \delta_2$, but could be as large as a few percent[42]. Other modes are $B^+ \to \bar{D}\pi^+$ or $B^o \to D^-\pi^+$.

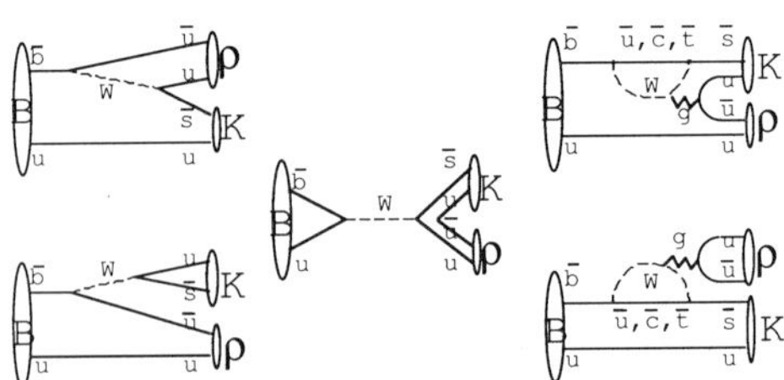

Figure 12. Diagrams for computing the $B^+ \to K\rho$ transition amplitudes. The first three diagrams are double Cabibbo suppresst ($\approx V_{ub} \cdot V_{us}^*$). The main contributions to the asymmetry in (76) are from an interference in the last two diagrams between t- and u,c-quarks in the loop.

A third version of CP-violation and in fact the most promising one is an interference between a single decay amplitude and oscillation. No complication from strong interaction arises.

Consider the decay $\Upsilon(4S) \to B^o \bar{B}^o$ with the two particle wave function Ψ with negative parity and C-parity

$$\Psi(t, \vec{p}, -\vec{p}) = \frac{1}{\sqrt{2}}\left\{ B^o(\vec{p})\,\bar{B}^o(-\vec{p}) - B^o(-\vec{p})\,\bar{B}^o(\vec{p}) \right\}(t) =$$

$$= \frac{1}{\sqrt{2}}\left\{ B_1(t,\vec{p})\,B_2(t,-\vec{p}) - B_2(t,\vec{p})\,B_1(t,-\vec{p}) \right\} = \qquad (77)$$

$$= \frac{1}{\sqrt{2}} e^{-i(\lambda_1+\lambda_2)t}\left\{ B^o(\vec{p})\,\bar{B}^o(-\vec{p}) - B^o(\vec{p})\,\bar{B}^o(-\vec{p}) \right\} \quad .$$

At $t = t_1$ one of the B-mesons decays, say

$$B^o(\vec{p}) \to \ell^+\nu_\ell X \text{ with amplitude } \mathcal{A}(B^o \to \ell^+\nu_\ell X). \qquad (78)$$

The overlap $|< \Psi(t_1) \mid \ell^+\nu_\ell X >|^2 = e^{-2\Gamma t_1} \mid \mathcal{A}(B^o \to \ell^+\nu_\ell X) \mid^2$ is the probability, that B^o survived up to $t = t_1$ and decayed into $\ell^+\nu_\ell X$ at this time. At $t > t_1$, the two particle wave function collapses and we start with a pure $\bar{B}^o(-\vec{p})$ state. The time development is

$$\tilde{\Psi}(t) = \frac{1}{\sqrt{2}}\left\{ B_1(-\vec{p})\,e^{-i\lambda_1(t-t_1)} - B_2(-\vec{p})\,e^{-i\lambda_2(t-t_1)} \right\} \qquad (79)$$

which is at $t = t_1$ indeed a pure $\bar{B}^o(-\vec{p})$-state but develops at later times also a $B^o(-\vec{p})$ component.

We assume now, that at $t = t_2$ the second B-meson decays with an amplitude $a = \mathcal{A}(B^o \to J/\psi K_S)$ and $\bar{a} = \mathcal{A}(\bar{B}^o \to J/\psi K_S)$. The transition amplitude is

$$< \tilde{\Psi}(t_2)|J/\psi K_S > = \frac{1}{2}\left\{e^{-i\lambda_1(t_2-t_1)}(a + \alpha\bar{a}) - e^{-i\lambda_2(t_2-t_1)}(a - \alpha\bar{a})\right\} \qquad (80)$$

and the resulting probability for a semileptonic decay at t_1 and a decay $B \to J/\psi K_S$ at t_2 becomes

$$| < \Psi(t_1)|\ell^+\nu_\ell X > |^2| < \tilde{\Psi}(t_2)/J/\psi K_S > |^2 =$$
$$= \frac{1}{2}e^{-\Gamma(t_2+t_1)}|\mathcal{A}(B^o \to \ell^+\nu_\ell X)|^2\left\{|a|^2 + |\alpha\bar{a}|^2 + \cos\left[(m_2 - m_1)(t_2 - t_1)\right]\left(|\alpha\bar{a}|^2 - |a|^2\right)\right.$$
$$\left. + \sin\left[(m_2 - m_1)(t_2 - t_1)\right](a\bar{a}^*\alpha^* - \bar{a}\alpha a^*)\right\} \qquad . \qquad (81)$$

Since to a very good approximation $|a| = |\bar{a}|$ and $|\alpha| = 1$ (81) becomes equal to

$$e^{-\Gamma(t_2+t_1)}|\mathcal{A}(B^o \to \ell^+\nu_\ell X)|^2|\mathcal{A}(B^o \to J/\psi K_S)|^2\left\{1 + \sin\left[\Delta m(t_2 - t_1)\right]\sin\Delta\phi\right\} \qquad (82)$$

with $\Delta\phi = arg(a) - arg(\alpha\bar{a}) = arg(a/\bar{a}) - arg(\alpha)$.

If on the other hand the first decaying particle is not a B^o as in equation (78) assumed, but a $\bar{B}^o \to \ell^-\nu_\ell X$, the same derivation yields a minus sign in front of $\sin\left[\Delta m(t_2 - t_1)\right]$ in (82). We can now form an asymmetry

$$A(t_1, t_2) = \frac{N(\ell^+\nu X \text{ at } t_1, J/\psi K_S \text{ at } t_2) - N(\ell^-\bar{\nu}X \text{ at } t_1, J/\psi K_S \text{ at } t_2)}{N(\ell^+\nu X \text{ at } t_1, J/\psi K_S \text{ at } t_2) + N(\ell^-\bar{\nu}X \text{ at } t_1, J/\psi K_S \text{ at } t_2)} = \qquad (83)$$
$$= \sin\Delta\phi \cdot \sin(\Delta m)(t_2 - t_1)$$

In the Standard Model both $arg(a/\bar{a})$ and $arg(\alpha)$ are completely determined by CKM matrix elements. This can be seen as follows:

- First $arg(\alpha) \approx -arg(M_{12})$ using the fact, that M_{12} and Γ_{12} have nearly the same phase. The matrix element M_{12} is computed from the box diagram (Fig. 13) and due to $m_t >> m_c, m_u$ and $|V_{bc}| << |V_{td}|$ a good approximation is[43] $M_{12} \approx m_t^2(V_{td} \cdot V_{tb}^*)^2$ and therefore $arg(M_{12})$ is proportional to $2arg(V_{td}V_{tb}^*) \approx 2arg(V_{td})$ since V_{tb}^* is real (see (58)). To the decay width Γ_{12} only the kinematical accessible intermediate states with u and c-quarks contribute and $\Gamma_{12} \approx m_b^2(V_{ud} \cdot V_{ub}^* + V_{cb} \cdot V_{cd}^*)^2$. In this approximation and from (59) follows immediately $arg(M_{12}) = arg(\Gamma_{12})$.

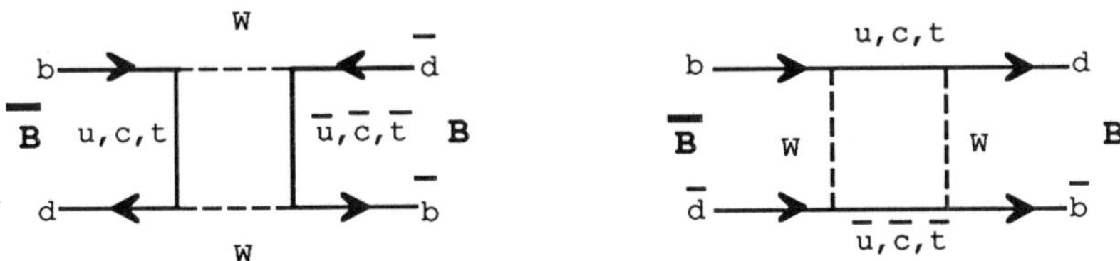

Figure 13. Box diagrams for computing the $B^o \to \bar{B}^o$ transition amplitude M_{12}. It is dominated by the exchange of the heaviest quark, the t-quark.

- Second for the evaluation of $arg(a/\bar{a})$ one computes the diagram of Fig. 14 and

$$arg(a/\bar{a}) = arg\left(\frac{V_{cb}^*V_{cs}}{V_{cb}V_{cs}^*}\right) \approx 0$$

since V_{cs} and V_{cb} are real (see (58)).

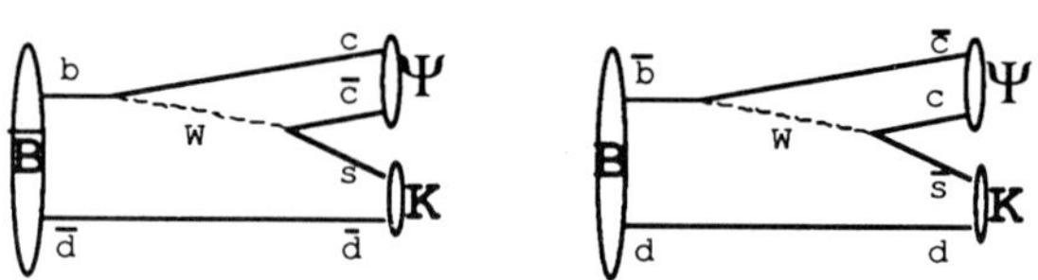

Figure 14. Diagrams for the decay $B \to J/\psi K_s$.

Therefore CP-violation in $B \to J/\psi K_S$ (a $b \to cX$ transition) is proportional $sin\Delta\phi$ with $\Delta\phi = 2arg(V_{td}) = 2\phi_1$ where ϕ_1 is shown in the unitarity triangle Fig. 10. The same analysis yields for the asymmetry in $B \to \pi\pi$ (a $b \to uX$ transition) $\Delta\phi = 2arg(V_{ub}^*) + 2arg(V_{td}) = 2(\pi - \phi_2)$.

Experimental considerations

In the derivation of (83) it was assumed that $t_2 > t_1$, but for $t_2 < t_1$ one obtains the opposite sign and therefore the asymmetry integrated over all times t_1 and t_2 vanishes,

$$\int_o^\infty dt_1 \int_o^\infty dt_2 \ A(t_1, t_2) \ = \ 0. \tag{84}$$

It is therefore important to resolve the two decay times and form a nonvanishing observable

$$A_{obs} = \int_{t_2 > t_1} dt_1 dt_2 \ A(t_1, t_2) \ - \ \int_{t_1 > t_2} dt_1 dt_2 \ A(t_1, t_2). \tag{85}$$

For B-mesons from a $\Upsilon(4S)$- resonance produced at rest the decay length ℓ_i is of order 30 μm (see Table 10) and even the best vertex detector can only measure $t_1 + t_2 = \frac{1}{\beta\gamma c} (\ell_1 + \ell_2)$. Moving $\Upsilon(4S)$ produced by colliding unequal energy beams [44] enable however to observe CP violation of this type[45]. We consider the production of the $\Upsilon(4S)$ resonance in a boosted frame with e^+e^- collisions of energy E_1 and E_2 respectively and with the constraint $E_1 \cdot E_2 = \frac{1}{4} \cdot m^2(\Upsilon(4S)) \approx 28$. This is depicted in Fig. 15. The experiment measures the difference in decay length $\ell_1 - \ell_2$ with $\ell_i = \beta\gamma c t_i$. For a boost variable $\beta\gamma = \frac{E_1}{m(\Upsilon(4S))} - \frac{E_2}{m(\Upsilon(4S))} \gtrsim 0.3$ the quantity $\Delta z = z(J/\psi) - z(tag)$ is proportional to $\Delta\ell$.

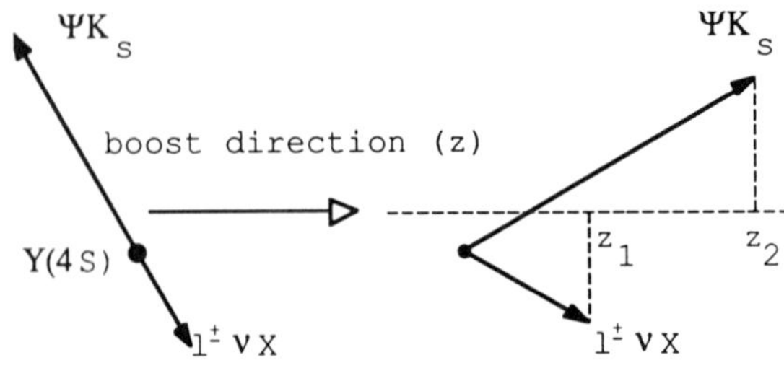

Figure 15. The decay of $\Upsilon(4S) \to B^\circ \bar{B}^\circ$ in a boosted frame.

478

We split now the total number N of observed events into four classes according to their time ordering

$$
\begin{aligned}
n_1 &= \int_{t_2>t_1} dt_1 dt_2 \; N(\ell^+\nu_\ell X \quad at \quad t_1, J/\psi K_S \quad at \quad t_2) \\
n_4 &= \int_{t_2>t_1} dt_1 dt_2 \; N(J/\psi K_S \quad at \quad t_1, \ell^-\bar\nu_\ell X \quad at \quad t_2) \\
n_3 &= \int_{t_2>t_1} dt_1 dt_2 \; N(\ell^-\bar\nu_\ell X \quad at \quad t_1, J/\psi K_S \quad at \quad t_2) \\
n_2 &= \int_{t_2>t_1} dt_1 dt_2 \; N(J/\psi K_S \quad at \quad t_1, \ell^+\nu_\ell X \quad at \quad t_2).
\end{aligned}
\tag{86}
$$

and form a new asymmetry

$$
A_{eff} = \frac{1}{2} \cdot \left(\frac{n_1-n_3}{n_1+n_3} - \frac{n_2-n_4}{n_2+n_4} \right) = \frac{(n_1+n_4)-(n_2+n_3)}{(n_1+n_2)+(n_3+n_4)}.
\tag{87}
$$

The integrals in (86) give

$$
n_{1,2} = \int_o^\infty dt_2 \int_o^{t_2} dt_1 \; e^{-\Gamma(t_1+t_2)} \; [1 \pm sin\Delta\phi \cdot sin\Delta m(t_2-t_1)] = \frac{1}{\Gamma^2}(1 \pm \frac{x \cdot sin\Delta\phi}{1+x^2})
\tag{88}
$$

where the notation $x = \Delta m/\Gamma$ was used. The time integration reduces therefore the expected asymmetry (87)

$$
A_{eff} = \frac{x}{1+x^2} \cdot sin\Delta\phi = 0.47 \cdot sin\Delta\phi \; using \; x = 0.7.
\tag{89}
$$

So far an infinite good experimental time resolution for the separation of $t_2 > t_1$ has been assumed. The experimental accuracy in the determination of the quantity of interest, the value of $sin\Delta\phi$ in (83) clearly depends besides on the total number of observed events $N = n_1 + n_2 + n_3 + n_4$ also on a function $I(\sigma_{\Delta t})$ which expresses the experimental resolution $\sigma_{\Delta t}$ of the time difference $\Delta t = t_1 - t_2$. The maximum information on the CP violation parameter $sin\Delta\phi$ is obtained from a maximum-likelihood fit to the observed CP asymmetry as a function of $t_1 - t_2$. Nakada[46] has shown, that with an energy asymmetry of $\beta\gamma$=0.4 the gain in accuracy with the fit-method slightly overcompensates the loss from the finite vertex resolution of $\sigma_{\Delta t}$=0.48 ps = 0.37 $\cdot\tau_B$ [1].

The semileptonic decay provides a clean flavour tag for the B-meson. In order to avoid confusion from a lepton of the semileptonic D-meson decay simulating a wrong b-flavour tag, only leptons with large momenta, $p > 1.3$ GeV, are used. The tagging efficiency with this method is ~10%. A drastic increase in the tagging efficiency can be obtained using also charged kaons[47]. Wrong tags from $B \to D\bar{D}$ are reduced by the requirement of only one kaon per event.

For an observation of $sin\Delta\phi$ with a significance of S standard deviation $S = \frac{sin\Delta\phi}{\sigma(sin\Delta\phi)}$, we have to produce the following number of boosted $\Upsilon(4S)$ mesons:

$$
N(\Upsilon(4S)) = \frac{S^2}{2 \cdot A_{eff}^2 \cdot B_1 \cdot B_2 \cdot B_3 \cdot B_4 \cdot \eta_{rec} \cdot \eta_{tag} \cdot (1-2\omega)^2} \; ,
\tag{90}
$$

where $B_1 = B(\Upsilon 4S \to B^o\overline{B^o})$, $B_2 = [B(B^o \to J/\psi K_s^o)+B(\overline{B^o} \to J/\psi K_s^o)]/2$, $B_3 = B(J/\psi \to l^+l^-)$, and $B_4 = B(K_s^o \to \pi^+\pi^-)$, η_{rec} is the reconstruction efficiency for $B^0 \to J/\psi K_s^-$, η_{tag} is the reconstruction efficiency for the tagging channels $B(\bar{B}) \to \ell^\pm\nu X \, or \, K^\pm X$, ω is the fraction of wrong flavour tags, and A_{eff} the effective asymmetry defined in (89).

The present standard model expectation for $sin\Delta\phi$ is 0.30±0.09. The estimated 95% confidence range is 0.12 to 0.48. With B_1=0.50, B_2=4·10^{-4}, B_3=2·0.07, B_4=0.67, η_{rec}=0.62, η_{tag}=0.32, and ω=0.07 we obtain $N(\Upsilon 4S)$=6·10^7 for observing CP violation in $B^o \to J/\psi K_s^o$ decays with a significance of 3 standard deviations if $sin\Delta\phi$=0.3. The efficiencies depend on the detector properties and the boost variable $\beta\gamma$. For larger boost, the vertex separation improves but the geometrical acceptance drops and the two effects cancel in first order at $\beta\gamma =$

0.4 (see Table 11); the values chosen here are from the pilot reaction study presented in ref. 1. With an $\Upsilon(4S)$ production cross section of 1.1 nb, the required integrated luminosity corresponds to $6\cdot10^4$/pb. The uncertainty range of $sin\Delta\phi$ between 0.12 and 0.48 as mentioned above translates into integrated luminosities between $2\cdot10^4$ and $4\cdot10^5$/pb, the optimistic value corresponding to a one year's program of a collider with $\mathcal{L}=2\cdot10^{33}\mathrm{cm}^{-2}\mathrm{s}^{-1}$. A factor of 2 or larger in significance can be gained, if several channels like $B \to J/\psi K_s, J/\psi K^*(K^* \to K_s\pi^o), D\bar{D}$ for the determination of $sin(2\phi_1)$ and $B \to \pi\pi, KK, \rho K$ for the determination of $sin(2\phi_2)$ are summed in the experiment.

Table 11. Effect of the beam energy asymmetry on the necessary number of events for the observation of CP violation in $B \to J/\psi K_s^o$. The parameters η_{rec} and η_{tag} are the efficiencies to reconstruct the $J/\psi K_s^o$ and the tagging B, $\sigma_{\Delta\tau}$ is the resolution of the decay time difference. The necessary number of events is normalized to the 8 GeV case.

E_{high} [GeV]	6	7	8	9	10
$\eta_{rec} \cdot \eta_{tag}$	0.23	0.21	0.20	0.18	0.16
$\sigma_{\Delta\tau}$ [ps]	1.47	0.63	0.48	0.36	0.29
N_{events}/N_{events} (8 GeV)	1.86	1.02	1.00	1.04	1.14

Acknowledgement

I acknowledge stimulating discussions with Y. Baconnier, S. Milton, and L. Rivkin on accelerator aspects and T. Nakada for his contribution to the chapter of CP violation.

REFERENCES

1. "Feasibility Study for a B-Meson Factory in the CERN-ISR Tunnel",
      T. Nakada editor, CERN-90-02, PSI-PR-90-08, 1990.
2. "Conceptual Design of a Ring Beauty Factory"
      A.N. Dubrovin, A.N. Skrinsky, G.M. Tumaikin, A.A. Zholents
3. M. Tigner, II. European Part. Acc. Conf., Nice, June 1990 and CLNS 90/999
      and Prospects for a CESR B-Factory Upgrade, CLNS 89/962
4. "Task Force Report on a Asymmetric B-Factory at KEK" , March 1990
5. "Feasibility Study for an Asymmetric B-Factory based on PEP",
      PUB-5244, SLAC-352, Calt-68-1589
6. "The Use of PETRA as a B-Factory", H. Neseman,, W. Schmidt-Parzefall, F. Willeke,
      European Part. Acc. Conf., 1988, Rome, S. Tazzari editor
7. R. Eichler, T. Nakada, K. Schubert, S. Weseler, and K. Wille, SIN PR-86-13 and
      "Proposal for an Electron Positron Collider for Heavy Flavour Particle Physic and
      Synchrotron Radiation", PSI-PR-88-09 (July 1988)
8. M. Sands,"The Physics of Electron Storage Rings. An Introduction" in Proc. of Int.
      School of Physics, Course XLVI, ed. by B. Touschek (1971).
9. J.S. Bell, in "CERN Accelerator School and Advanced Accelerator", CERN 87-03,
      Vol. I, 1987.
10. E.J.N. Wilson, in "CERN Accelerator School and Advanced Accelerator Physics",
      CERN 87-03, Vol. I, 1987.
11. E.D. Courant and H.S. Snyder, Ann. Phys. 3:1 (1958).
12. A.M.G. Floquet, Ann. Ecole Norm. Sup. (2), 12:47 (1883).
13. L.D. Landau and E.M. Lifshitz, The Classical Theory of Fields, Vol. 2.
14. G. Brown, SLAC-Pub-4366 (1987).

15. The study in ref. 1 is the first time that the disruption criterium is explicitly
used in the design of a circular machine. Experimental evidence can be found in
D. Rice, Proc. of the 1989 IEEE Part.Acc.Conf. (1989) 444 and
S. Milton,PSI-PR-90-05, 1990.

16. A. Piwinski, Proc. of the 1977 Part. Acc. Conf., Chicago (1977)

17. K. Oide and K. Yokoya, SLAC-PUB-4832 (1989)

18. S. Milton, private communication

19. P. Grosse-Wiesmann et al., "Linac-Ring-Collider B-Factory", CERN PS/90-50(AR)
and CERN PPE/90-113

20. J.J. Bisognano et al., CEBAF Tech.Note, Nov. 1988

21. H. Breuker et al., OPAL technical Proposal, NIM A260:329(1987).

22. R. Arnold et al., CERN-EP/87-186 ; R. Arnold et al., CRN/HE 88-01.

23. T. Ypsilantis, private communication and J. Seguinot et al., CERN-EP/90-88.
B. Hoeneisen, D. Anderson and S. Kwan, Fermilab-Pub-90/182

24. T. Ruf, SIN Preprint SIN BFP-1 (1986).

25. L. Miani, Proceedings of the 1976 CERN School of Physics, Wépion (Belgium),
June 6-19, 1976, CERN 76-20.

26. Particle Data Group, Phys. Lett. B239:1 (1990).

27. L. Wolfenstein, Phys. Rev. Lett. 51:1945 (1984).

28. A. Sirlin, Phys. Rev. D35:3423 (1987).

29. J. Cline, W.F. Palmer, and G. Kramer, Phys. Rev. 40:793 (1989).

30. J.G. Kömer, and G.A. Schuler, Z. Phys. C38:511 (1988).

31. N. Isgur, D. Scora, B. Grinstein, and M.B. Wise, Phys. Rev. D39:799(1989).

32. K. Hagiwara, A.D. Martin, and M.F. Wade, Nucl. Phys. B327:569 (1989).

33. G. Kramer, and W.F. Palmer, DESY 90-011 (1990).

34. M. Wirbel, B. Stech, and M. Bauer, Z. Phys. C29:637 (1985).

35. H. Albrecht et al., Phys. Lett. B234:409 (1990).

36. R. Fulton et al., Phys. Rev. Lett. 64:16 (1990).

37. H. Albrecht et al., DESY 89-163.

38. S. Stone in Proc. of the XVI Intern. Symp. on Multiparticle Dynamics,Arles,
France; 1988 CLNS88/855.

39. H. Albrecht et al., Phys. Lett. B229:175 (1989).

40. G. Altarelli, N. Cabibbo, G. Corbo, L. Maiani, and G. Martinelli,
Nucl. Phys. B208:365 (1982).

41. Crystal Ball Collaboration, D. Antreasyan et al., DESY 90-038, SLAC-PUB-5250

42. J.-M. Gerard and G. Hou, Phys.Rev.Lett. 62:855 (1989)

43. J.S. Hagelin, Nucl. Phys. B193:123 (1981)

44. P. Oddone, Proc. of UCLA workshop on linear collider B-factory
conceptual design, LA, 1987.

45. I. Dunietz and T. Nakada, Z. Phys. C36:503 (1987). See also
R. Aleksan et al., Phys. Rev. D39:1283 (1989)

46 T. Nakada, AIP Conf. Proc. 196, "Heavy Quark Physics", Ithaca,1989, p. 385

47 R. Aleksan in ref. 45

Unstable Particles

André Martin

CERN

Introduction

The subject of unstable particles is a kind of recurrent subject which comes back every time someone is forced to think about it. In fact one of the most important papers on the subject is that of Maurice Lévy (the Director of the "Institut d'Etudes Scientifique de Cargèse"), published in 1959 [1]. This time the pretext to reexamine the problem is the observation of hundreds of thousands of an unstable particle, the Z_0.

Almost all particles are unstable except the electron, the positron, the neutrinos and also the proton and the antiproton, to the best of our knowledge, in spite of the fact that theoreticians would like them to be unstable. Up to recently the "Particle Data" group made a distinction between "stable particles", decaying only by weak interactions, and "unstable particles" decaying via strong interactions. As a result, the Z_0 was "stable", in spite of its width of 2.42 GeV while the upsilon, with a width of 30 keV was "unstable". I am told that in this year's edition this distinction has been wisely abandoned.

In fact, experiment indicates that one should forget the stable-unstable distinction. It is remarkable that Gell-Mann, at the 1962 Geneva Conference [2] predicted the existence of the "stable" Ω^- particle (which has a very long lifetime of $0.82 \ 10^{-10}$ seconds) from the knowledge of the Δ, Σ^*, Ξ^*, which have widths of the order of 100 MeV. In the time interval between the prediction and the discovery, many distinguished theoreticians [3], playing with the Riemann sheet structure associated to the decay channels of these particles argued that the equal spacing rule could not possibly be valid, but Samios and his collaborators showed that it was nevertheless valid [4].

In spite of that, axiomatic field theory makes a big distinction between stable particles, which have an associated asymptotic field, and unstable particles which do not. However, we have to live with unstable particles. We even have to go further and deal with completely confined particles, the quarks and the gluons. In these lectures, I would like to discuss two problems.

1) The decay of unstable particles which is known to be non exponential, but, nevertheless approximately exponential if one excludes very short times and very long times after the production of the particle. This is often forgotten, and maybe, some people in the audience are not aware of that. Relatively recently these considerations were resurrected to try to explain the apparent absence of proton decay. We shall see that unfortunately the orders of magnitude are such that this "explanation" fails.

2) The association of unstable particles to some pole on some Riemann sheet, and the question to know where is the pole. Naturally this question is essential for Z_0 physics. Essentially everybody believes that a pole is associated to the Z_0, and one could say that the excellent generalized Breit-Wigner fits of the Z_0 confirm this idea. However, as we shall see, it is not perfectly established in theory. Around 1970-71, L. Fonda and his collaborators wrote a series of papers [5] showing that one could simulate the amplitude produced by the pole by an analytic function without pole, at least in the physical region. I shall show, in a very simple case at least, that of pion-pion

Z° Physics - Cargèse 1990, Edited by M. Lévy *et al.*
Plenum Press, New York, 1991

resonances, that this procedure may not work if one takes into account all unitarity and analyticity constraints, and that, with high accuracy measurements, it is possible to decide whether there is a pole or not.

1 The Decay of Unstable Particles

Here, we shall follow partly the presentation of Galindo and Pascual in their Quantum Mechanics course [6], which I recommend to you (there is an English translation by L. Alvarez-Gaumé which will soon appear).

Let us call $|B>$ the unstable state. We assume that B decays into C+D. The Hamiltonian is given by

$$H = H_0 + V \tag{1}$$

B is stable under H_0 and V is responsible for the decay. The probability that B remains B after a time t is

$$P(t) = |A(t)|^2 = \left| < B e^{-iHt} B > \right|^2 \tag{2}$$

H is hermitian, and, for t→0 we have

$$e^{-iHt} = 1 - iHt$$

and hence

$$P(t) \simeq 1 + 0(t^2), \tag{3}$$

instead of $1 - \lambda t$.

Hence we see that for small times we have unavoidably a violation of the exponential decay law.

To find the general structure of the decay law, we insert in A(t) a complete set of states of H:

$$< B \left| e^{-iHt} \right| B > = \sum_n e^{-iE_n t} \left| < B|n > \right|^2$$

In fact these states are scattering states (we assume explicitly that the B+C system has no bound states) and it is more correct to write

$$A(t) = \int_0^\infty d\lambda e^{-i\lambda t} ||E_\lambda|B >||^2, \tag{4}$$

where E_λ represents the projection over the state with eigenvalue. $\lambda = 0$ corresponds to the C+D threshold.

Now, we can define

$$R(t) = \frac{1}{2} \left[A^*(t^*) + A(t) \right], \tag{5}$$

i.e. for real t, the real part of A(t). So, from (4), we have

$$R(t) = \int_0^\infty w(\lambda) \cos(\lambda t) \, d\lambda \tag{6}$$

with

$$w(\lambda) = ||E_\lambda|B >||^2 \geq 0,$$

and then, conversely, by inverse Fourier cosine

$$w(\lambda) = \frac{1}{2\pi} \int_0^\infty R(t) \cos \lambda t \, dt \tag{7}$$

Now, if $|R(t)| \leq C \exp -\gamma t$ (7) shows that $w(\lambda)$ is analytically the strip $|Im\lambda| < \gamma$, because the integral still converges for complex values of λ. This is impossible since $w(\lambda) = 0$ for $\lambda < 0$.

This is more or less the essence of the Paley-Wiener theorem. The essential point is that the spectrum of B+C is <u>lower bounded</u>.

Let me show you, first by hand, that one can get as close as one wishes to exponential decay, for instance $\exp\left(-|t|^{1-\epsilon}\right)$. Take $w(\lambda) = \exp\left(-\frac{1}{\lambda^N} - \lambda\right)$, then

$$A(t) = \int_0^\infty \exp - \left(i\lambda t + \frac{1}{\lambda^N} + \lambda \right) d\lambda \tag{8}$$

the line of integration can be deformed, and instead of integrating over arg λ=0 one can integrate over arg $\lambda = -\phi$

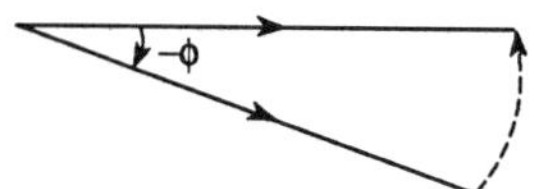

As long as $N\phi < \pi/2$ the contribution from the arc at infinity is zero, and the integral becomes

$$A(t) = \int_0^\infty e^{i\phi}\, dr\, \exp[-rt\,\sin\phi - \frac{1}{r^N}\cos N\phi - r\,\cos\phi + i\,(\text{real terms})] \tag{9}$$

Now the quantity

$$X = rt\,\sin\phi + \frac{1}{r^N}\,\cos N\phi$$

is maximum at

$$r = \left(\frac{N\,\cos N\phi}{t\,\sin\phi}\right)^{\frac{1}{N+1}},$$

and one find that the optimum ϕ is

$$\frac{M}{2(N+1)}$$

then replacing X by its maximum one can carry the integration and get

$$A(t) < \frac{\exp\left(-\frac{N+1}{N}t^{\frac{N}{N+1}}\sin\frac{M}{2(N+1)}\right)}{\cos\frac{M}{2(N+1)}} \tag{10}$$

which is the announced result. Notice, however, that we cannot get the exponential decay by letting N go to infinity because of the factor $\sin(M/2(N+1))$ in the exponential in (10). (10) represents a strict bound, but some minor changes, using the steeped descent method can lead to an estimate very close to the r.h.s. of (10).

In fact, one can approach still slightly closer the exponential decay, for the exact necessary and sufficient condition on $A(t)$ due to the fact that $w(\lambda)$ vanishes for $\lambda < 0$ is [7]

$$\int_{-\infty}^\infty \frac{|\ell n\,|A(t)||\,dt}{1+t^2} < \infty \tag{11}$$

This allows for instance

$$A(t) \sim \exp - \frac{t}{(\log t)^{1+\epsilon}}\ ,\ \text{etc...} \tag{12}$$

However, in practice, reasonable models exhibit much more marked deviations from the exponential law. The weight function $w(\lambda)$ cannot be chosen at will. It is imposed from outside and it is constrained to have a normal threshold behaviour [1] [6].

For

$$w(\lambda) \sim \lambda^\nu f(o),\ f(o) \neq o, \tag{13}$$

one gets

$$\int w(\lambda)\,e^{-i\lambda t}dt \simeq \frac{\Gamma(\nu+1)f(o)}{(it)^{\nu+1}} \tag{14}$$

i.e. a negative power behaviour.

Now we describe the Pascual Galindo model, in which one has a separable interaction between an unstable state $|B>$ and the continuum of scattering states $|\delta(x-x_0)>$ with energy x_0. The main interest of the model is that it allows complete explicit calculations showing deviations from exponential decay as well as approximate exponential decay for intermediate times.

A state is defined as $\{\alpha, f(x)\}$ [13], where α represents the amplitude of the unstable particle and $f(x)$ that of the continuum. We can introduce a norm

$$|\alpha|^2 + \int_0^\infty dx |f(x)|^2 \tag{15}$$

the action of the free hamiltonian on a state is given by

$$H_o\{\alpha, f(x)\} \rightarrow \{\alpha, xf(x)\} \tag{16}$$

Which means that the energy of $|B>$ is taken to be unity. The action of the separable interaction is given by

$$v\{\alpha, f(n)\} \rightarrow \{<v|f>, \alpha v(x)\} \tag{17}$$

Notice that this interaction is <u>Hermitian</u>.

Let us assume that

$$\int \frac{dy|v(y)|^2}{y} < 1 \tag{18}$$

Then the system has <u>no bound state</u> (in fact this is necessary and sufficient). The decay amplitude can be calculated to be

$$P(t) = |A(t)|^2 \tag{19}$$

$$A(t) = \int_0^\infty dy e^{-iyt} \frac{|v(y)|^2}{\left[y - 1 + p\int d\tau \frac{|v(\tau)|^2}{t-y}\right]^2 + \pi^2|v(y)|^4} < 1$$

$$= \int_0^\infty dy e^{-iyt} F(y) \tag{20}$$

We see that $F(y)$ looks like a Breit-Wigner form, if $v(y)$ is small, and sufficiently slowly varying near $y=1$, so that the principal value integral can be neglected or replaced by a constant.

If we make these drastic simplifications we get

$$A(t) \simeq \int_0^\infty dy\, e^{-iyt} \frac{\frac{\Gamma}{2\pi}}{(y - yo)^2 + \left(\frac{\Gamma}{2}\right)^2} \tag{21}$$

with the identification

$$\pi^2|v(1)|^4 = \frac{\Gamma^2}{4}$$
$$yo = 1 - P\int \frac{d\tau|v(\tau)|^2}{\tau-1} \tag{22}$$

Now if we make the further step to replace(19) by an integral running from $-\infty$ to $+\infty$, we get

$$A(t) = \int_{-\infty}^{+\infty} dy \frac{e^{-iyt}\frac{\Gamma}{2\pi}}{(y - yo)^2 + \left(\frac{\Gamma}{2}\right)^2} \tag{23}$$

The integral has 2 poles in the complex plane at $y = yo \pm i\frac{\Gamma}{2}$, and to calculate (23) it is sufficient to replace the integration contour from $-\infty$ for $+\infty$ by a horizontal line in the lower half plane plus a contour around the pole in the lower half plane.

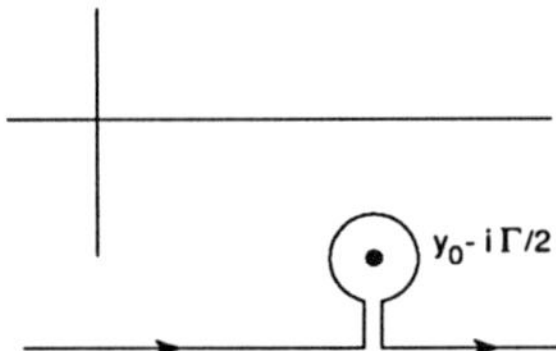

It is easy to see that the integral reduces to the contribution of this pole, which gives:

$$A(t) \simeq -2\pi i \frac{\Gamma/2\pi\, e^{-iyo\, t - \frac{\Gamma t}{2}}}{-i\Gamma}$$
$$\simeq e^{-iyo\, t - \frac{\Gamma t}{2}}$$

and hence

$$B(t) = |A(t)|^2 = \exp - \Gamma t \tag{24}$$

So that we get, with these approximations, the exponential decay law.

However this is certainly not always valid because $v(y)$=constant would lead to the fact that the integral $\int \frac{|v(y)|^2 dy}{y}$ would diverge, and futhermore the integral from $-\infty$ to 0 should not be included. Then it is not anymore possible to deform the contour.

An explicit model has been worked out by Galindo and Pascual. It is given by

$$|v(y)|^2 = \frac{\sqrt{2}}{\pi}g^2 \frac{\sqrt{y}}{1 + y^2} \tag{25}$$

then

$$y - 1 + P \int \alpha\tau \frac{|v(\tau)|^2}{\tau - y} = (y - 1)\left[1 - \frac{g^2}{1 + y^2}\right] \tag{26}$$

then Γ previously defined is $\sqrt{2}g^2$.

The expression

$$y - 1 + \int_0^\infty \frac{d\tau |v(\tau)|^2}{\tau - y},$$

has a pole on the second sheet of the y cut plane if $g^2 < 1$.

We show on Figure 1, Figure 2 and Figure 3 the results of explicit calculations for several values of g^2.

Figure 1 and Figure 2 correspond to $g^2 = 0.01$. The lifetime of the particle is T=70 in dimensionless units, and we see on Figure 1 that the breakdown of the exponential law occurs for $t \geq 30$ T and on Figure 2 that we have also a violation of the exponential law for short times $t < \frac{T}{20}$, with a characteristic quadratic behaviour near t=0.

Figure 3 shows that there are situations where there is no time interval where the exponential decay is even approximately valid. This occurs for $g^2 \geq 0.3$, corresponding to a "lifetime" of T=2.

It is amusing that in the case $g^2 = 0.01$, already for t=3000, i.e. t=40 T, the asymptotic formula (14) is valid, giving

$$P(t) \simeq \frac{2}{\pi^2}(g^2)^2 \frac{(\Gamma(3/2))^2}{t^3} \sim 10^{-15}. \tag{27}$$

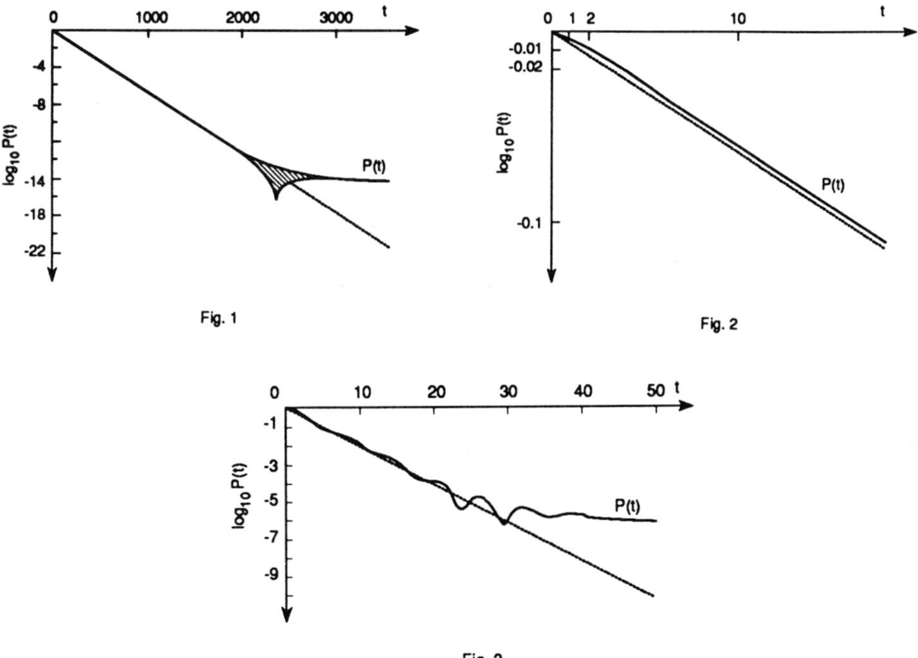

To see when the exponential holds and when it does not hold, let us take a very crude model:

$$A(t) = \int_0^\infty \left(\frac{E}{m}\right)^k \frac{2\pi\Gamma e^{-iEt}}{(E - m)^2 + \frac{\Gamma^2}{4}} dE, \tag{28}$$

k integer.

From the contour argument, already used previously, we have

$$e^{-\Gamma t/2} = \int_{-\infty}^{+\infty} \left(\frac{E}{m}\right)^k \frac{2\pi\Gamma e^{-iEt}}{(E - m)^2 + \frac{\Gamma^2}{4}} dE, \tag{29}$$

Now we try to estimate, the magnitude of what we have added to make (28) calculable, namely

$$\Delta = \int_{-\infty}^{0} \frac{e^{-iEt}\left(\frac{E}{m}\right)^{k} dE\, 2\pi\Gamma}{(m-E)^{2}+\left(\frac{\Gamma}{2}\right)^{2}},$$ (30)

we can rotate the line of integration along the negative imaginary axis and get

$$|\Delta| < \int_{0}^{\infty} \frac{e^{-yt}\left(\frac{y}{m}\right)^{k} 2\pi\Gamma}{m^{2}+\Gamma^{2}} dy$$

$$\leq 2\pi k! \left(\frac{\Gamma}{m}\right)^{k+2}\left(\frac{1}{\Gamma t}\right)^{k+1}$$ (31)

So, we will get the exponential regime if

$$(\Gamma t)^{2k+2} e^{-\Gamma t} \gg 4\pi^{2}(k!)^{2}\left(\frac{\Gamma}{m}\right)^{2k+4}$$ (32)

k, is of course model dependent, we can take

-k=0 then, (30) is satisfied if $t_1 < t < t_2$

$$\begin{cases} \Gamma t_1 &= 2\pi\left(\frac{\Gamma}{m}\right)^2 \\ \Gamma t_2 &= \ln\frac{1}{4\pi^2} + 4\ln\left(\frac{m}{\Gamma}\right) \\ &+ 4\ln 2 + 2\ln\ln\left(\frac{m}{\Gamma}\right) \end{cases}$$

- k=1 $t_1 < t < t_2$

$$\Gamma t_1 = \sqrt{2}\pi\left(\frac{\Gamma}{m}\right)^{3/2}$$

$$\Gamma t_2 = \ln\frac{1}{4\pi^2} + 6\ln\left(\frac{m}{\Gamma}\right)$$

$$+ 4\ln\ln\left(\frac{m}{\Gamma}\right) + 4\ln 6$$ (33)

Let us give now some examples in "real" life. Consider the decay of the lambda particle, here m is of the order of 1 GeV and t is 10^{-10} sec equivalent to 1cm, equivalent to 10^{-14} GeV^{-1}.

With k=1, we get

$$\begin{cases} \Gamma t_1 &= 2\times 10^{-21} \\ \Gamma t_2 &= 200 \end{cases}$$

So that deviations from exponential decay occur at either infinitely too early times or at later times where essentially no particles are left so that the observation is impossible.

Another example which created some excitement some years ago is that of proton decay. Estimates of grand unified theories indicated a lifetime which could be as low as 10^{30} years [8]. On the other hand, no decay was observed and the apparent lifetime was larger than 10^{30} years [9]. On the other hand, the age of the universe is less than 10^{10} years, so that protons were created less than 10^{10} years ago. It would be tempting to say that we are still in the quadratic regime, $P(t) \sim 1 - ct^2$ so that the apparent lifetime is much bigger than the real lifetime.

However, for any lifetime larger than 10^{30} years, we have, with m = GeV $\Gamma/m < 10^{-51}$, and with k=1

$$t_1/T < 10^{-75}$$

i.e. $t_1 < 10^{-27}$ sec, so that we are well inside the first three minutes!

Another case about which one might worry is that of K_0 decay if we take K_{short} decay, with a lifetime of 0.9 10^{-10} seconds if we take m=0.3 GeV we expect deviations to exponential decay around 100 to 200 lifetimes, i.e. 10^{-8} seconds and this is precisely the order of magnitude of the K_{long} decay lifetime. But it must be realized that after 100 lifetimes there will be nothing left of the K_0 short and one should not be impressed by the coincidence of the time after which K_0 short ceases to decay exponentially with the K_0 long lifetime.

To conclued this section, I would like to say that all these considerations can be extended to the "relativistic" Breit-Wigner form which is used for instance to describe the Z_0

$$A(t) = \int_{0}^{\infty} s\frac{\Gamma e^{-itE} dE}{(s-M)^{2}+(M\Gamma)^{2}}$$ (34)

To make the exponential decay law appear, we have to add an integration along the negative imaginary axis:

$$\tilde{A}(t) = \int_{-i\infty}^{0} + \int_{0}^{\infty} s\frac{\Gamma e^{-iEt}dE}{(s-M)^2 + (M\Gamma)^2}, \tag{35}$$

So that a single pole is present in the quadrant. $\tilde{A}(t)$ is a pure exponential and the corrective term is bounded by

$$\frac{\Gamma}{M^2}\int y^2\, e^{-yt}dy.$$

2   Is an Unstable Particle a Pole

In the previous model, we have seen that if $v(y)$ is analytic, for instance if

$$(v(y))^2 = \frac{\sqrt{2}g^2}{M}\frac{\sqrt{y}}{1+y^2} \tag{36}$$

a pole on a second sheet is associated to the unstable particle, given by

$$0 = \lambda - 1 + \frac{g^2}{1+\lambda^2}\left(1 - \lambda - \sqrt{-2\lambda}\right) \tag{37}$$

However, if v is not analytic we cannot make a continuation to the second sheet, and yet, we have an unstable particle.

One might object that this model is rather artificial and that, in real life, unitarity and analyticity force you to associate a pole on the second sheet to an unstable particle.

Long ago it has been "shown" by Zimmerman [10] that one can continue the elastic scattering amplitude $F(s, \cos\phi)$ across the elastic cut. This looks very simple. One projects the amplitude on a certain angular momentum to get the partial wave amplitude:

$$f_\ell(s) = \frac{\sqrt{s}}{2k}e^{i\delta_\ell}\sin\delta_\ell \tag{38}$$

then

$$S_\ell(s) = 1 + \frac{2k}{\sqrt{s}}f_\ell(s) = e^{2i\delta_\ell}(s), \tag{39}$$

and if $S_l(s)$ is analytic in some neighbourhood of the real axis, which is guaranteed by field theory [11] [12] we can define $S_l(s)$ for complex s, $\mathrm{ms}>0$ and we can continue the unitarity relation on the real axis

$$S_\ell(s)\overline{S}_\ell(s) = 1, \tag{40}$$

Where the bar denotes complex conjugation, by

$$S_\ell(s)\overline{S}_\ell(\bar{s}) = 1, \tag{41}$$

(41) defines a continuation of $S_l(s)$ in a domain symmetric with respect to the real axis of the initial analyticity domain of $S_l(s)$. In this way, a zero of S on the first sheet, with $\mathrm{Im}s>0$, gives a pole on the second sheet, with $\mathrm{Im}s<0$

o Zero $S_1(s)$

● Pole $S_1(s) = (\bar{S}_1(\bar{s}))^{-1}$

Then, once this is done for partial waves, one can, if one wants to, resum the partial wave series and obtain a continuation of the full amplitude.

All this seems fine. However some years ago I discovered a difficulty which is considered to be very serious by the experts as shown for instance by the talk of A.S. Wightman at the Vienna Conference in 1968 [13]. I shall try to describe this because, though this result is officially published in a festschrift in honor of N. N. Bogoliubov [14] it has never been distributed to libraries in the West, except to contributors. It is a pity, for this book contains many interesting articles like the one of V. Glaser on positivity.

The point is that without violating any sacred principle you can arrange to have $S_l(s)$ to have an accumulation of zeros along the real axis, the "physical region". You have to do this in a careful way, so that the infinite product over zeros converges, but this can be done, for instance by taking:

$$S_\ell(s) = \prod_{N=1}^{\infty} \prod_{k=0}^{N} \left(\frac{s - s_0 - \frac{k}{N} - \frac{i}{N^3}}{s - s_0 - \frac{k}{N} + \frac{i}{N^3}} \right) \tag{42}$$

on the segment $0<\mathrm{Re}(s\text{-}s_0)<1$ Im s$=1/N^3$, we have N+1 zeros. It is easy to see that, for complex s, the product converges. It is also possible to show that $|S_\ell(s + i\varepsilon)| \to 1$ for $\varepsilon \to 0$ for almost every s. However the real axis is clearly a natural boundary for the function.

Though, as I said, field theory experts consider this to be a catastrophe, one can add an extra postulate, which does not follow from the axioms of axiomatic field theory, which is that $f_l(s)$ or $S_l(s)$ is <u>continuous</u> on the real axis. Then the catastrophe is avoided and the continuation to the second sheet can be carried through. Most physicists will find this acceptable.

Naturally generalisation to N channels can be made without problem. This is essential for the Z_0 problem, because in lowest order in QCD and if hadronization is disregarded, the Z_0 case can be regarded as an N channel problem, the N channels being the lepton-antilepton and quark-antiquark channels. As soon as gluon production is allowed, this is no longer true, and this is not true if one looks at the real final states, made of multitudes of pions, protons etc.

A tough problem is the true 3 body problem. Then, after postulating the continuity of 2 body amplitudes to avoid the catastrophe previously mentioned, J. Bros [15] has been able to prove that this continuation to "a" second sheet through the physical region where 3 body final states are energetically allowed is possible. However, on this second sheet, other singularities than poles can be met, like cuts for instance.

However, in the best of all cases, the scattering amplitude can be continued through the elastic cut or the inelastic cut. This does not say that necessarily a pole on some hidden sheet is associated to an unstable particle. If you look back at the particle hunts of the 60's with bubble chambers, you see that people were content to call "particle" a bump seen in some projection of a Dalitz plot.

Without going to very high mathematics you can see by yourself that you can relatively well reproduce a Breit-Wigner shape by a sum of gaussians. A Breit-Wigner shape

$$|A| = \left| \frac{\Gamma/2}{E - E_0 + i\Gamma/2} \right|$$

correspond to a pole on the second sheet while a gaussian shape can be analytically continued but does not exhibit any pole. Let me give first a very simple example: take a half width $\Gamma/2 = 1$, and let $x = E - E_0$. Compare

$$f = \frac{1}{1 + x^2} \; (\text{Breit} - \text{Wigner}) \; \text{and} \tag{43}$$

$$g = 0.8 \exp - 0.5x^2 + 0.2 \exp - 3x^2 \; (\text{sum of gaussians}) \tag{44}$$

we get

x	0	0.2	0.4	0.6	0.8	1.0	1.2	1.4
$\frac{g}{f}$	1.00	1.00	1.00	1.00	1.00	0.99	0.96	0.89

In fact this is first a preview of a theorem which says that any function defined on the real axis can be approximated with an arbitrary accuracy by an entire function.

These considerations have been made in 1970-71 by L. Fonda and his collaborators in a series of articles [5]. They also point out that one can make a phase shift <u>rise</u> through $\pi/2$ without having an associated pole on the second sheet even through a <u>rising</u> phase shift crossing $\pi/2$ was believed by phase shift analyzers to be the characteristic signal of a resonance.

They also showed that these "particles" exhibit the approximate exponential decay for times which are neither very small, nor very large.

Somehow, little attention was paid by most physicists to the work of these "iconoclasts" and everybody continued to associated unstable particles and poles. However, I was personally very much disturbed and tried to understand what was going on. If you think a little you realize that the price you pay when you replace a Breit-Wigner form by a sum of gaussians is that your amplitude grows very fast in some directions, when you go away from the real axis (physical region) on the physical sheet (i.e. ImE>0 if you start from E+iε, $\varepsilon > 0$). This is unpleasant but there is one simple case where one has bounded the magnitude of the scattering amplitude in the complex plane, inside its domain of analyticity, it is the case of pion-pion scattering.

Let me try to summarize what we know of the pion-pion scattering amplitude.

1) It has a large analyticity domain [16]. If we call F(s, t, u) the amplitude, with one redundant variable, since $s+t+u=4m_\pi^2$, m_π pion mass, s square of the c.m. energy of the s channel. t square of the 4 momentum transfer in the s channel, etc., the amplitude (case of $\pi_0\pi_0 \to \pi_0\pi_0$) is completely symmetric, with 3 physical regions defined by $s> 4m_\pi^2$, $t < 0$, $u < 0$ and permutations, and is analytic in a domain containing the union of the domain s

$$\left\{|t| < 4m_\pi^2 \text{ and } s - 4m_\pi^2 \notin R^+, \text{ and } u - 4m_\pi^2 \notin R^+\right\} \tag{45}$$

and permutations.

2) <u>Inside</u> this domain of analyticity it has been possible, by using a rather tricky combination of analyticity and unitarity to find absolute bounds on the amplitude. Initially the bounds that I obtained for an amplitude normalized in such a way that F(s=4, t=0, u=0) represents the scattering length at threshold, were very poor [17] as indicated on Fig.4, where we show upper and lower bounds at the symmetry point, i.e., at $s=t=u=4m_\pi^2/3$, but with time and efforts, they were considerably improved [18]

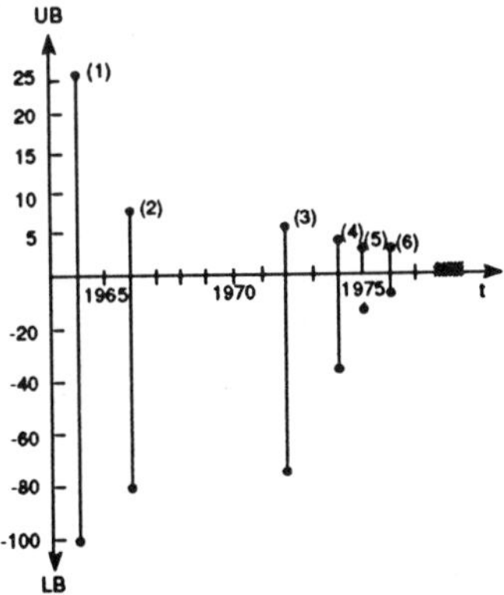

Fig. 4

Perhaps more physically significant is the lower bound on the $\pi_0\pi_0 \to \pi_0\pi_0$ scattering length which was obtained first by Bonnier an Vinh Mau [19] later improved [20] [21] to reach the best possible value in the paper of Caprini and Dità [22] of - 1.65 Compton wave length. This is about a factor 10 times what one observes, but, considering that essentially no dynamical ingredient has been included, except the absence of bound states of the pion-pion system, it is rather remarkable.

Now the initial strategy of Bonnier and Vinh Mau was to use bounds on the full amplitude and to calculate a bound on the S wave by the formula

$$f_0(s) = \frac{2}{s-4} \int_{t=\frac{4-s}{2}}^{0} F(s,t,u)\, dt \tag{46}$$

and since bounds on F exist for any $|t| < 4m_\pi^2$ if one avoids the s and u cuts one sees that one gets a bound on $f_0(s)$ for

$$|s - 4| < 8m_\pi^2,$$

so that the domain in which we have an absolute bound on $f_0(s)$ contains a piece of the physical region

$$4m_\pi^2 < s < 12m_\pi^2,$$

491

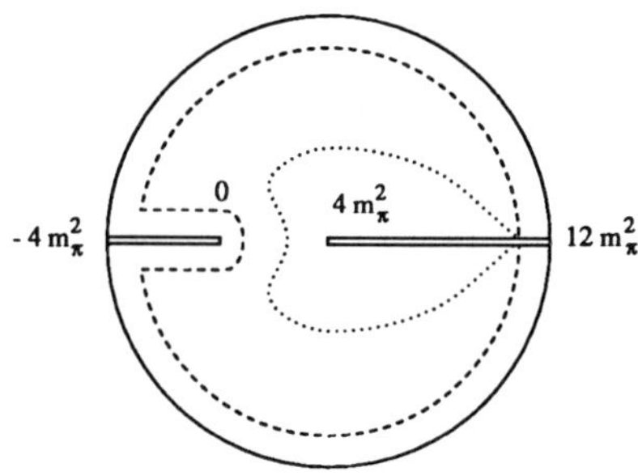

This bound can be translated into a bound on $S = 1 + 2k/\sqrt{s}\, f_l(s)$, with the remark that $|S|$ is bounded by 1 on the segment $4m_\pi^2 < s < 8m_\pi^2$. Naturally the bound becomes infinite on the border of the initial domain but it is <u>finite</u> and <u>calculable</u> on, for instance, the semi continuous contour, with a global upper bound B.

Now we want to use this information to know if the S wave amplitude has, or has not a pole on the second sheet and we want to show that if it is known with <u>sufficient</u> accuracy in the physical region this question is decidable, at least in principle.

Now suppose we have two candidates S_1 and S_2 for the amplitude, (here we take a given isospin state so that we have $|S| = 1$ on the physical region), both known with an uncertainty ε on the physical region $4m_\pi^2 < s < 12m_\pi^2 - \eta$, but that S_1 has a <u>pole</u> on the second sheet, i.e., a <u>zero</u> on the first sheet, while the second has not. Starting from the information that $|S_1 - S_2| < 2\varepsilon$ above and below the physical cut and that $|S_1 - S_2| < B$ along the semi continuous contour, it is possible, using the theory of subharmonic functions, to find "equipotentials" along which $|S_1 - S_2| < |\varepsilon|^\alpha |2B|^{1-\alpha}$, $0 < \alpha < 1$ (see technical note). Such an equipotential E_α is qualitatively drawn by a dotted line on the figure.

Suppose that along this equipotential one has $|S_1| > |\varepsilon|^\alpha |2B|^\alpha$, which will happen if ε is sufficiently small. Then the variation of the phase of S_1 and S_2 along the closed contour made of E_α and the cut $4m_\pi^2 < s < 12m_\pi^2 - \eta$ is the same and therefore according to Rouche's theorem S_1 and S_2 have the <u>same</u> number of zeros. In particular, if S_1 has one zero in this region for Im $s > 0$, S_2 has this zero too and it is possible to know <u>in principle</u> if there is or there is not a pole on the second sheet. In practice, because B is very large, ε must be very small, but not infinitesimally small. Naturally one would like to make this argument work for higher energies and other processes, but it is conforting that at least in a very limited case one can find an answer to this question.

In the case of the Z_0, one might try to construct an analogue of this argument by using the Jost-Lehmann-Dyson representation and the positivity of the $e^+ e^-$ cross section. However, as I said before, one has to remain in the approximate framework where final states are $l^+ l^-$ and $q\bar{q}$ pairs so as to deal with a problem with a finite number of channels.

Technical Note

Let f be an analytic function limited by a domain with a border made of two arcs, Γ_1 and Γ_2, assume that $|f| < B_1$ on $\Gamma_1 |f| < B_2$ on Γ_2. Then construct a condenser with voltage log B_1 on Γ_1 and log B_2 on Γ_2. Then the voltage at any point inside is the logarithm of an upper bound on f at this particular point in the complex plane.

In mathematical terms log f is less than the largest subharmonic function taking the value log B_1 on Γ_1 and log B_2 on Γ_2.

References

[1] M. Lévy Nuovo Cimento **14** (1959) 612 and references therein.

[2] M. Gell-Mann in Proceedings of the 1962 International Conference on High Energy Physics at CERN;
J. Prentki editor (CERN 1962) 805.

[3] R. J. Oakes and C. N. Yang, Phys. Rev. Lett. **11** (1963) 174;
M. Froissart and M. Jacob, Journal de Physics **25** (1964) 313.

[4] V. E. Barnes et al., Phys. Rev. Lett. 12 (1964) 204.

[5] G. Calucci, L. Fonda, G. C. Ghirardi, Phys. Rev. 166 (1968) 1719;
L. Fonda, preprint CTP, Trieste IC/70/62 (1970).

[6] A. Galindo and P. Pascual, Mecanica Cuantica, Eudema editors (1989) 251.

[7] R. Boas, Entire functions, New York, Academic Press. (1954).

[8] H. Georgi and S. L. Glashow, Phys. Rev. Lett. **32** (1974) 438.

[9] See for instance R. Becker-Szendy et al. (IMB-3 group), Phys. Rev. **D42** (1990) 2974.

[10] W. Zimmermann, Nuovo Cimento **21** (1961) 249.

[11] J. Bros, H. Epstein and V. Glaser, Nuovo Cimento **31** (1964) 1265.

[12] A. Martin, Nuovo Cimento **44** (1966) 1219.

[13] A. S. Wightman, in proceedings of the 14th International Conference on High Energy Physics, Vienna, 1968 (CERN, 1968) 431.

[14] A. Martin, in Problems of Theoretical Physics, essays dedicated to N. N. Bogoliubov for his 60th Birthday, publishing house "Nauka" Moscou 1969, p 113.

[15] J. Bros in New developments in Mathematical Physics;
H. Mitter and L. Pittner editors, Springer-Verlag, Vienna New York (1981) and Acta Physica Austriaca supplement XXIII.

[16] A. Martin, Nuovo Cimento **42** (1966) 930, and Ref. [12].

[17] A. Martin, in High Energy Physics and Elementary Particles, ICTP Trieste (I.A.E.A, Vienna 1965) 155;
L. Lukaszuk and A. Martin, Nuovo Cimento **A47** (1967) 265.

[18] J. B. Healy, Phys. Rev. **D8** (1973) 1904;
G. Auberson, L. Epele, G. Mahoux and F.R.A. Simão, Nucl. Phys. **B94** (1975) 311;
B. Bonnier, C. Lopez and G. Mennessier, Phys. Lett. **60B** (1975) 63;
C. Lopez and G. Mennessier, Nucl. Phys. **B118** (1977) 426.

[19] B. Bonnier and R. Vinh Mau, Phys. Rev. **165** (1968) 1923.

[20] B. Bonnier, Nuclear Physics **B95** (1975) 98.

[21] C. Lopez and G. Mennessier, Phys. Lett. **B58** (1975) 437.

[22] I. Caprini and P. Dita, J. Phys. **A13** (1980) 1265 and preprint, Institute of Physics and Nuclear Engineering, Bucharest (I have been unable to find the published version).

THE PARTICLE-COSMOLOGY CONNECTION:
NEUTRINO COUNTING, DARK MATTER AND
LARGE-SCALE STRUCTURE

David N. Schramm

The University of Chicago
5640 S. Ellis Avenue, Chicago, IL 60637
and
NASA/Fermilab Astrophysics Center
Fermi National Accelerator Laboratory
Box 500, Batavia, IL 60510-0500

ABSTRACT

In this series of lectures, three current active areas at the boundary of particle physics and cosmology will be examined. The three are: (1) nucleosynthesis and neutrino counting; (2) the dark matter problems; and (3) the formation of galaxies and large-scale structure. Comments will also be made on the possible implications of the recent solar neutrino experimental results for cosmology.

INTRODUCTION

These lectures will examine three areas of recent activity at the boundary of particle physics and cosmology. The three areas are: (1) nucleosynthesis and neutrino counting; (2) the dark matter problems; and (3) the formation of galaxies and large-scale structure. Comments will also be made on the possible implications of the recent solar neutrino experimental results for cosmology. However, before going into these specific topics, let us first note the strength of the basic big bang framework.

Both particle physics and cosmology have their standard models–which are getting confirmed to remarkable accuracy by the present generation of experiments. The particle model is $SU_3 \times SU_2 \times U_1$ and the cosmological model is the hot big bang.

While Hubble's work in the 1920's established an expanding universe, the establishment of modern physical cosmology and the hot big bang model hinges on two key quantitative observational tests: (1) the microwave background, and (2) big bang nucleosynthesis (BBN) and the light element abundances. This paper will focus on the second of these since that is more directly connected to high energy physics. However, it is worth noting that just as the new COBE[1] results have given renewed confidence in the $3K$ background argument, the LEP collider (along with the SLC) has given us renewed confidence in the BBN arguments. We will return to this point momentarily. Note also that the microwave background probes events at temperatures $\sim 10^4 K$ and

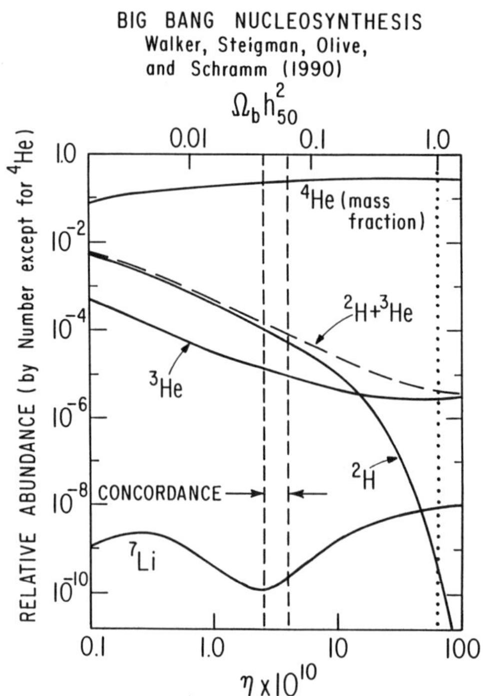

Figure 1. BBN abundances versus the baryon to photon ratio, η, or equivalently the fraction of the critical density, Ω_b.

times of $\sim 10^5$ years, whereas the light element abundances probe the Universe at temperatures $\sim 10^{10} K$ and times of ~ 1 sec. Thus, it is the nucleosynthesis results that played the most significant role in leading to the particle-cosmology merger that has taken place this past decade.

Since the popular press sometimes presents misleading headlines implying *doubts* about the big bang, it is important to note here that the real concerns referred to in these articles are really in regard to observations related to models of galaxy and structure formation. The basic hot big bang model itself is in fantastic shape with high accuracy confirmations from COBE and, as we will discuss, nucleosynthesis. However, there is admittedly no fully developed model for galaxy and structure formation that fits all of the observations. (But, of course, there is also no fully developed first principles model for star formation either.) That we might not really know exactly how to make galaxies and large-scale structure in no way casts doubt on the hot, dense early universe which we call the big bang. (We also have trouble predicting earthquakes and tornadoes, but that hasn't meant that we question celestial mechanics or a round Earth.) We will return to the problems of galaxy and structure formation towards the end of these lectures.

Before going into the specific argument as to the relationship of BBN to neutrino counting, let us review the history of BBN. This will draw heavily on other recent conference proceedings.[2]

HISTORY OF BIG BANG NUCLEOSYNTHESIS

It should be noted that there is a symbiotic connection between BBN and the $3K$ background dating back to Gamow and his associates, Alpher and Herman. The initial BBN calculations of Gamow and his associates[3] assumed pure neutrons as an initial condition and thus were not particularly accurate, but their inaccuracies had little effect on the group's predictions for a background radiation.

Once Hayashi recognized in 1950 the role of neutron-proton equilibration, the framework for BBN calculations themselves has not varied significantly. The work of Alpher, Follin and Herman[4] and Taylor and Hoyle[5], preceeding the discovery of the $3K$ background, and Peebles[6] and Wagoner, Fowler and Hoyle,[7] immediately following the discovery, and the more recent work of our group of collaborators[8,9,10,11,12] all do essentially the same basic calculation, the results of which are shown in Figure 1. As far as the calculation itself goes, solving the reaction network is relatively simple by the standards of explosive nucleosynthesis calculations in supernovae (c.f. the 1965 calculation of Truran *et al.*),[13] with the changes over the last 25 years being mainly in terms of more recent nuclear reaction rates as input, not as any great calculational insight (although the current Kawano/Walker code[11,12] is somewhat streamlined relative to the earlier Wagoner code[7]). With the possible exception of ^7Li yields, the reaction rate changes over the past 25 years have not had any major affect.[9,11,12,13] The one key improved input is a better neutron lifetime determination, a point to which we will also return shortly.

With the exception of the effects of elementary particle assumptions to which we will also return, the real excitement for BBN over the last 25 years has not really been in redoing the basic calculation. Instead, the true action is focused on understanding the evolution of the light element abundances and using that information to make powerful conclusions. In particular, in the 1960's, the main focus was on ^4He which is very insensitive to the baryon density. The agreement between BBN predictions and observations helped support the basic big bang model but gave no significant information at that time with regard to density. In fact, in the mid-1960's, the other light isotopes (which are, in principle, capable of giving density information) were generally assumed to have been made during the T-Tauri phase of stellar evolution,[15] and so, were not then taken to have cosmological significance. It was during the 1970's that BBN fully developed as a tool for probing the universe. This possibility was in part stimulated by Ryter $et\ al.$[16] who showed that the T-Tauri mechanism for light element synthesis failed. Furthermore, ^2H abundance determinations[17,18] improved significantly with solar wind measurements and the interstellar work from the Copernicus satellite. Reeves, Audouze, Fowler and Schramm[19] argued for cosmological ^2H and were able to place a constraint on the baryon density excluding a universe closed with baryons. Subsequently, the ^2H arguments were cemented when Epstein, Lattimer and Schramm[20] proved that no realistic astrophysical process other than the big bang could produce significant ^2H. It was also interesting that the baryon density implied by BBN was in good agreement with the density implied by the dark galactic halos.[21]

By the late 1970's, a complimentary argument to ^2H had also developed using ^3He. In particular, it was argued[22] that, unlike ^2H, ^3He was made in stars; thus, its abundance would increase with time. Since ^3He like ^2H monotonically decreased with cosmological baryon density, this argument could be used to place a lower limit on the baryon density[23] using ^3He measurements from solar wind[17] or interstellar determinations.[24] Since the bulk of the ^2H was converted in stars to ^3He, the constraint was shown to be quite restrictive.[9] Support for this point[25] also comes from the observation of ^3He in horizontal branch stars which, as processed stars still having ^3He on their surface, indicates the survivability of ^3He.

It was interesting that the lower boundary from ^3He and the upper boundary from ^2H yielded the requirement that ^7Li be near its minimum of $^7Li/H \sim 10^{-10}$, which was verified by the Pop II Li measurements of Spite and Spite,[26] hence, yielding the situation emphasized by Yang $et\ al.$[9] that the light element abundances are consistent over nine orders of magnitude with BBN, but only if the cosmological baryon density is constrained to be around 6% of the critical value. It is worth noting that ^7Li alone gives both an upper and a lower limit to Ω_b. However, while its derived upper limit is more than competitive with the ^2H limit, the ^7Li lower limit is not nearly as restictive as the $^2H +^3 He$ limit. Claims that big bang nucleosynthesis can yield Ω_b lower than 0.01 must necessarily neglect the $^3He +^2 H$ limit.

The other development of the 70's for BBN was the explicit calculation of Steigman, Schramm and Gunn,[27] showing that the number of neutrino generations, N_ν, had to be small to avoid overproduction of ^4He. This will subsequently be referred to as the SSG limit. (Earlier work had noted a dependency of the ^4He abundance on assumptions about the fraction of the cosmological stress-energy in exotic particles,[28,5] but had not actually made an explicit calculation probing the quantity of interest to particle physicists, N_ν.) To put this in perspective, one should remember that the mid-1970's also saw the discovery of charm, bottom and tau, so that it almost seemed as if each new detector produced new particle discoveries, and yet, cosmology was arguing against this "conventional" wisdom. Over the years the SSG limit on N_ν improved with ^4He abundance measurements, neutron lifetime measurements and with limits on the lower bound to the baryon density; hovering at $N_\nu \lesssim 4$ for most of the 1980's and dropping to slightly lower than $4^{29,30,10}$ just before LEP and SLC turned on.

BIG BANG NUCLEOSYNTHESIS: Ω_b and N_ν

The power of Big Bang Nucleosynthesis comes from the fact that essentially all of the physics input is well determined in the terrestrial laboratory. The appropriate temperature regimes, 0.1 to $1MeV$, are well explored in nuclear physics labs. Thus, what nuclei do under such conditions is not a matter of guesswork, but is precisely known. In fact, it is known for these temperatures far better than it is for the centers of stars like our sun. The center of the sun is only a little over $1keV$, thus, below the energy where nuclear reaction rates yield significant results in laboratory experiments, and only the long times and higher densities available in stars enable anything to take place.

To calculate what happens in the big bang, all one has to do is follow what a gas of baryons with density ρ_b does as the universe expands and cools. As far as nuclear reactions are concerned, the only relevant region is from a little above $1MeV$ ($\sim 10^{10}K$) down to a little below $100keV$ ($\sim 10^9K$). At higher temperatures, no complex nuclei other than free single neutrons and protons can exist, and the ratio of neutrons to protons, n/p, is just determined by

$$n/p = e^{-Q/T},$$

where

$$Q = (m_n - m_p)c^2 \sim 1.3MeV.$$

Equilibrium applies because the weak interaction rates are much faster than the expansion of the universe at temperatures much above $10^{10}K$. At temperatures much below 10^9K, the electrostatic repulsion of nuclei prevents nuclear reactions from proceeding as fast as the cosmological expansion separates the particles.

Because of the equilibrium existing for temperatures much above $10^{10}K$, we don't have to worry about what went on in the universe at higher temperatures. Thus, we can start our calculation at $10MeV$ and not worry about speculative physics like the theory of everything (T.O.E.), or grand unifying theories (GUTs), as long as a gas of neutrons and protons exists in thermal equilibriuim by the time the universe has cooled to $\sim 10MeV$.

After the weak interaction drops out of equilibrium, a little above $10^{10}K$, the ratio of neutrons to protons changes more slowly due to free neutrons decaying to protons, and similar transformations of neutrons to protons via interactions with the ambient leptons. By the time the universe reaches 10^9K ($0.1MeV$), the ratio is slightly below $1/7$. For temperatures above 10^9K, no significant abundance of complex nuclei can exist due to the continued existence of gammas with greater than MeV energies. Note that the high photon to baryon ratio in the universe ($\sim 10^{10}$) enables significant population of the MeV high energy Boltzman tail until $T \lesssim 0.1 \ MeV$.

Once the temperature drops to about 10^9K, nuclei can exist in statistical equilibrium through reactions such as $n + p \leftrightarrow {}^2H + \gamma$ and $^2H + p \leftrightarrow {}^3He + \gamma$ and $^2H + n \leftrightarrow {}^3H + \gamma$, which in turn react to yield ^4He. Since ^4He is the most tightly bound nucleus in the region, the flow of reactions converts almost all the neutrons that exist at 10^9K into ^4He. The flow essentially stops there because there are no stable nuclei at either mass-5 or mass-8. Since the baryon density at big bang nucleosynthesis is relatively low (much less than $1g/cm^3$) and the time-scale short ($t \lesssim 10^2sec$), only reactions involving two-particle collisions occur. It can be seen that combining the most abundant nuclei, protons and ^4He via two body interactions always leads to unstable mass-5. Even when one combines ^4He with rarer nuclei like ^3H or ^3He, we still get only to mass-7, which, when hit by a proton, the most abundant nucleus around, yields mass-8. (A loophole around the mass-8 gap can be found if $n/p > 1$, so that excess neutrons exist, but for the standard case $n/p < 1$). Eventually, ^3H

radioactively decays to ^3He, and any mass-7 made radioactively decays to ^7Li. Thus, big bang nucleosynthesis makes ^4He with traces of ^2H, ^3He, and ^7Li. (Also, all the protons left over that did not capture neutrons remain as hydrogen.) For standard homogeneous BBN, all other chemical elements are made later in stars and in related processes. (Stars jump the mass-5 and -8 instability by having gravity compress the matter to sufficient densities and have much longer times available so that three-body collisions can occur.) With the possible exception of ^7Li,[9,10,11,12,14] the results are rather insensitive to the detailed nuclear reaction rates. This insensitivity was discussed in reference 9 and most recently using a Monte Carlo study by Krauss and Romanelli.[14] An n/p ratio of $\sim 1/7$ yields a ^4He primordial mass fraction,

$$Y_p = \frac{2n/p}{n/p + 1} \approx \frac{1}{4}.$$

The only parameter we can easily vary in such calculations is the density that corresponds to a given temperature. From the thermodynamics of an expanding universe we know that $\rho_b \propto T^3$; thus, we can relate the baryon density at $10^{10}K$ to the baryon density today, when the temperature is about $3K$. The problem is that we don't know today's ρ_b, so the calculation is carried out for a range in ρ_b. Another aspect of the density is that the cosmological expansion rate depends on the total mass-energy density associated with a given temperature. For cosmological temperatures much above 10^4K, the energy density of radiation exceeds the mass-energy density of the baryon gas. Thus, during big bang nucleosynthesis, we need the radiation density as well as the baryon density. The baryon density determines the density of the nuclei and thus their interaction rates, and the radiation density controls the expansion rate of the universe at those times. The density of radiation is just proportional to the number of types of radiation. Thus, the density of radiation is not a free parameter if we know how many types of relativistic particles exist when big bang nucleosynthesis occurred.

Assuming that the allowed relativistic particles at $1MeV$ are photons, e, μ, and τ neutrinos (and their antiparticles) and electrons (and positrons), Figure 1 shows the BBN yields for a range in present ρ_b, going from less than that observed in galaxies to greater than that allowed by the observed large-scale dynamics of the universe. The ^4He yield is almost independent of the baryon density, with a very slight rise in the density due to the ability of nuclei to hold together at slightly higher temperatures and at higher densities, thus enabling nucleosynthesis to start slightly earlier, when the baryon to photon ratio is higher. No matter what assumptions one makes about the baryon density, it is clear that ^4He is predicted by big bang nucleosynthesis to be around 1/4 of the mass of the universe.

THE SSG LIMIT - COSMOLOGICAL NEUTRINO COUNTING

Let us now look at the connection to N_ν. Remember that the yield of ^4He is very sensitive to the n/p ratio. The more types of relativistic particles, the greater the energy density at a given temperature, and thus, a faster cosmological expansion. A faster expansion yields the weak-interaction rates being exceeded by the cosmological expansion rate at an earlier, higher temperature; thus, the weak interaction drops out of equilibrium sooner, yielding a higher n/p ratio. It also yields less time between dropping out of equilibrium and nucleosynthesis at 10^9K, which gives less time for neutrons to change into protons, thus also increasing the n/p ratio. A higher n/p ratio yields more ^4He. As we will see in the next section, quark-hadron induced variations[31] in the standard model also yield higher ^4He for higher values of Ω_b. Thus, such variants still support the constraint on the number of relativistic species.[32]

In the standard calculation we allowed for photons, electrons, and the three known neutrino species (and their antiparticles). However, following SSG and doing the calculation (see Figure 2) for additional species of neutrinos, we can see when ^4He yields exceed observational limits while still yielding a density consistent with the ρ_b bounds from ^2H, ^3He, and now ^7Li. (The new ^7Li value gives approximately the same constraint on ρ_b as the others, thus strengthening the conclusion.) The bound on ^4He comes from observations of helium in many different objects in the universe. However, since ^4He is not only produced in the big bang but in stars as well, it is important to estimate what part of the helium in some astronomical object is primordial—from the big bang—and what part is due to stellar production after the big bang. The pioneering work of the Peimberts[33] showing that ^4He varies with oxygen has now been supplemented by examination of how ^4He varies with nitrogen and carbon. The observations have also been systematically re-examined by Pagel[34]. The conclusions of Pagel[34], Steigman et $al.$[35] and Walker et $al.$[11] all agree that the ^4He mass fraction, Y_p, extrapolated to zero heavy elements, whether using $N, O,$ or C, is $Y_p \sim 0.23$ with an upper bound allowing for possible systematics of 0.24.

The other major uncertainty in the ^4He production used to be the neutron lifetime. However, the new world average of $\tau_n = 890 \pm 4s (\tau_{1/2} = 10.3\ min)$ is dominated by the dramatic results of Mampe et $al.$[36] using a neutron bottle. This new result is quite consistent with a new counting measurement of Byrne et $al.$[37] and within the errors of the previous world average of $896 \pm 10s$ and is also consistent with the precise C_A/C_V measurements from PERKEO[38] and others. Thus, the old ranges of $10.4 \pm 0.2\ min$, used for the half-life in calculations,[39,9] seem to have converged towards the lower side. The convergence means that, instead of the previous broad bands for each neutrino flavour, we obtain relatively narrow bands (see Figure 2). Note that $N_\nu = 4$ is excluded. In fact, the SSG limit is now $N_\nu < 3.4$.[10,11]

The recent verification of this cosmological standard model prediction by LEP, $N_\nu = 2.98 \pm 0.06$, from the average of ALEPH, DELPHI, L3 and OPAL[40] collaborations as well as the SLC[40] results, thus experimentally confirms our confidence in BBN. (However, we should also remember that LEP and cosmology are sensitive to different things.[41] Cosmology counts all relativistic degrees of freedom for $m_x \lesssim 10 MeV$, with LEP and SLC counting particles coupling to the Z^0 with $m_x \lesssim 45 GeV$.

While ν_e and ν_μ are obviously counted equally in both situations, a curious loophole exists for ν_τ since the current experimental limit $m_{\nu_\tau} < 35 MeV$ could allow it not to contribute as a full neutrino in the cosmology argument.[42] Proposed experiments which push the m_{ν_τ} limit down to less than a few MeV should eliminate this loophole. It might also be noted that if we assume m_{ν_τ} is light so that cosmologically $N_\nu = 3$, we can turn the argument around and use LEP to predict the primordial helium abundance ($\sim 24\%$), or even use limits on ^4He to give an upper limit on Ω_b (also $\lesssim 0.10$). Thus, LEP strengthens the argument that we need non-baryonic dark matter if $\Omega = 1$. In fact, note also that with $N_\nu = 3$, if Y_p is ever proven to be less than ~ 0.235, standard BBN is in difficulty. Similar difficulties occur if Li/H is ever found below $\sim 10^{-10}$. In other words, BBN is a falsifiable theory.

ALTERNATIVE PROPOSALS

As noted above, BBN yields all agree with observations using only one freely

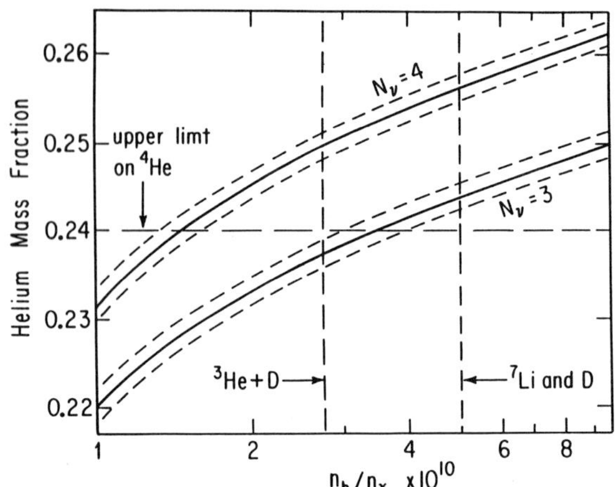

Figure 2. The SSG argument with recent parameter constraints showing the BBN helium mass fraction versus η for $N_\nu = 3$ and 4. Note that 4 is excluded.

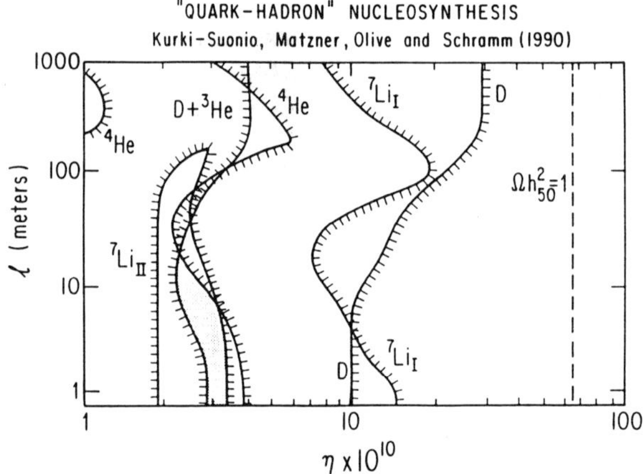

Figure 3. This shows the constraints on η of the various observed abundances in a first-order quark-hadron phase transition with nucleation sites separated by a distance l with density contrast $R \lesssim 10^3$. The Pop II lithium abundance used here is from the compilation of data given by Walker *et al.*[11] and is slightly more restrictive on η than that used in Figure 2 or used in the original Kurki-Suonio *et al.*[32] calculation from which this figure is derived. It should be noted that work by the Tokyo[46] group and by the Livermore group[47] confirms the conclusions on restricting Ω_b to values similar to the standard result even when $R \rightarrow \infty$.

adjustable parameter, ρ_b. Thus, BBN can make strong statements regarding ρ_b if the observed light element abundances cannot be fit with any alternative theory. Before exploring the implications for ρ_b, let us examine alternative proposals which have arisen to try to escape the power of the homogeneous BBN conclusions.

The two alternatives that have recently received interest are:

(1) Decaying particles;[43] and

(2) Quark-hadron transition inspired inhomogeneities.[31]

The first of these notes that if a species of massive particle ($M \gtrsim few\ GeV$) were to decay after traditional BBN, it could redo nucleosynthesis. While previous decaying particle proposals had been made, the new idea[43] emphasizes the importance of the resulting hadron cascade which, they argue, will dominate the yields. While interesting results are obtained, problems with detailed abundance determinations do result. In particular, this class of models seems to predict inevitably that $^6Li/^7Li >> 1$, whereas observations show $^7Li/^6Li \gtrsim 10$. While at first this might seem fatal, it is almost avoidable by noting that ^6Li is much more fragile than ^7Li; thus, it is easy to deplete ^6Li and obtain the observed ratios. However, Brown and Schramm[44] have pointed out that for high surface temperature Pop II stars, the convective zones do not go deep enough to destroy any primordial ^6Li. Pilachowski et al.[45] have now looked at those specific stars and indeed find no ^6Li, again seeing $^7Li/^6Li > 10$. Therefore, unless the Brown and Schramm convection argument can be surmounted, ^6Li seems to contrain this model seriously. Steigman, Audouze and others have noted additional problems with this model for ^3He and ^2H ratios.

Let us now look at the quark-hadron inspired inhomogeneity models.[31] While inhomogeneity models had been looked at previously (c.f. reference 9) and were found to make little difference, the quark-hadron inspired models had the added ingredient of variations in n/p ratios.

The initial claim by Applegate et al., followed by a similar argument from Alcock et al., that $\Omega_b \sim 1$ might be possible, created tremendous interest. Their argument was that if the quark-hadron transition was a first-order phase transition (as some preliminary lattice gauge calculations implied), then it was possible that large inhomogeneities could develop at $T \gtrsim 100 MeV$. The preferential diffusion of neutrons versus protons out of the high density regions could lead to big bang nucleosynthesis occurring under conditions with both density inhomogeneities and variable neutron/proton ratios. In the first round of calculations, it was claimed that such conditions might allow $\Omega_b \sim 1$, while fitting the observed primordial abundances of $^4He, ^2H, ^3He$ with an overproduction of ^7Li. Since ^7Li is the most recent of the cosmological abundance constraints and has a different observed abundace in Pop I stars versus the traditionally more primitive Pop II stars,[26] some argued that perhaps some special depletion process might be going on to reduce the excess ^7Li. Reeves and Audouze each argued against such processes and tried to turn the argument around and use lithium abundances to constrain the quark-hadron transition.

At first it appeared that if the lithium constraint could be surmounted, then the constraints of standard big bang nucleosynthesis might disintegrate. (Although Audouze, Reeves and Schramm emphasized that the number of parameters needed to fit the light elements was somewhat larger for these non-standard models, nonetheless, a non-trivial loophole appeared to be forming.) To further stimulate the flow through the loophole, Mullaney and Fowler showed that, in addition to looking at the diffusion of neutrons out of high density regions, one must also look at the subsequent effect of excess neutrons diffusing back into the high density regions as the nucleosynthesis goes to completion in the low density regions. (The initial calculations treated the two regions separately.) Mullaney and Fowler argued that for certain phase transition parameter values (e.g. nucleation site separations $\sim 10m$ at the time of the transition), this back diffusion could destroy much of the excess lithium. Recent work by Banerjee and Chitre (private communication) suggests that more accurate treatment of the diffusion calculation could reduce the interesting separation distance by several orders of magnitude.

However, Kurki-Suonio, Matzner, Olive and Schramm,[32] the Tokyo group,[46] and the Livermore group[47] have recently argued that in their detailed diffusion models, the back diffusion not only effects ^7Li, but also the other light nuclei as well. They find that for $\Omega_b \sim 1$, ^4He is also overproduced (although it does go to a minimum for similar parameter values as does the lithium). One can understand why these models might tend to overproduce ^4He and ^7Li by remembering that in standard homogeneous big bang nucleosynthesis, high baryon densities lead to excesses in these nuclei. As back diffusion evens out the effects of the initial fluctuation, the averaged result should approach the homogeneous value. Furthermore, it can be argued that any narrow range of parameters, such as those which yield relatively low lithium and helium, are unrealistic since in most realisitic phase transitions there are distributions of parameter values (distribution of nucleation sites, separations, density fluctuations, etc.). Therefore, narrow minima are washed out which would bring the ^7Li and ^4He values back up to their excessive levels for all parameter values with $\Omega \sim 1$. Furthermore, Adams and Freese[48] have argued that the boundary between the two phases may be fractal-like rather than smooth. The large surface area of a fractal-like boundary would allow more interaction between the regions and minimize exotic effects.

Figure 3 shows the results of Kurki-Suonio et al.[32] for varying spacing l with the constraints from the different light element abundances. Notice that the Li and even the ^4He constraint do not allow $\Omega_b \sim 1$. (The ^4He abundance constraint used in Kurki-Suonio et al. was a generous $Y_p \lesssim 0.25$; for the preferred $Y_p \lesssim 0.24$, the ^4He bound is about as tight as the Pop II Li constraint.) Note also that with the Pop II ^7Li constraint, the results for Ω_b are quite similar to the standard model with a slight excess in Ω_b possible if l is tuned to ~ 10.

Furthermore, initially it looked like quark-hadron inspired models might enable leakage[49] beyond mass-7, thus enabling ^9Be, ^{14}N, or maybe even r-process elements to become probes as whether or not the universe had such a transition (even if $\Omega_b \sim 1$). However, Tarasawa and Sato[46] have shown that when full multizone calculations of the type used by Kurki-Suonio et al. are utilized, then no significant leakage occurs.

One possible signature that remains for a first order quark-hadron transition is a slightly larger allowed range for Y_p that is concordant with $N_\nu = 3$ and with the other light element abundances. In particular, if ^4He were ever shown to be definitively $\lesssim 0.23$, it might be evidence for such a quark-hadron induced behavior since the standard homogenous case cannot accomodate such values. Of course, excessively low values for Y_p would still be unallowable.

One can conclude from the failure of the attempts to circumvent the standard BBN results that the results are amazingly robust. Even when many new free parameters are added, as in the quark-hadron case, the bottom line, when one requires concordance with the light element abundances, is essentially the same as the standard result. In other words, $\Omega_b \sim 0.06$ (although with fine-tuning the upper bound might be relaxed a bit to ~ 0.2 rather than 0.1).

LIMITS ON Ω_b AND DARK MATTER REQUIREMENTS

The narrow range in baryon density for which concordance occurs is very interesting. Let us convert it into units of the critical cosmological density for the allowed range of Hubble expansion rates. For the big bang nucleosynthesis constraints,[9,10,11,12,29,30] the dimensionless baryon density, Ω_b, that fraction of the critical density that is in baryons, is less than 0.11 and greater than 0.02 for $0.04 \lesssim h_0 \lesssim 0.7$, where h_0 is the Hubble constant in units of $100 km/sec/Mpc$. The lower bound on h_0 comes from direct observational limits and the upper bound from age of the universe constraints.[49] Note that the constraint on Ω_b means that the universe *cannot be closed with baryonic matter.* If the universe is truly at its critical density, then nonbaryonic matter is required. This argument has led to one of the major areas of research at the particle-cosmology interface, namely, the search for non-baryonic dark matter.

Another important conclusion regarding the allowed range in baryon density is that it is in very good agreement with the density implied from the dynamics of galaxies, *including their dark halos.* An early version of this argument, using only deuterium, was described over fifteen years ago.[21] As time has gone on, the argument has strengthened, and the fact remains that galaxy dynamics and nucleosynthesis agree at about 6% of the critical density. Thus, if the universe is indeed at its critical density, as many of us believe, it requires most matter not to be associated with galaxies and their halos, as well as to be nonbaryonic.

Let us put the nucleosynthetic arguments in context. The arguments requiring some sort of dark matter fall into two separate and quite distinct areas. First are the arguments using Newtonian mechanics applied to various astronomical systems that show that there is more matter present than the amount that is shining. These arguments are summarized in the first part of Table 1. It should be noted that these arguments reliably demonstrate that galactic halos seem to have a mass ~ 10 times the visible mass.

Note, however, that big bang nucleosynthesis requires that the bulk of the baryons in the universe be dark since $\Omega_{vis} << \Omega_b$. Thus, the dark halos could in principle be baryonic.[21] Recently, arguments on very large scales[51] (bigger than clusters of galaxies) hint that Ω on those scales is indeed greater than Ω_b, thus forcing us to need non-baryonic matter. However, until these arguments are confirmed, we must look to the inflation paradigm.

This is the argument that the only long-lived natural value for Ω is unity, and that inflation[52] or something like it provided the early universe with the mechanism to achieve that value and thereby solve the flatness and smoothness problems. Thus, our need for exotica is dependent on inflation and big bang nucleosythesis and not on the existence of dark galatic halos. This point is frequently forgotten, not only by some members of the popular press but occasionally by active workers in the field.

Some baryonic dark matter must exist since from the $^2H +^3 He$ argument we know that the lower bound from big bang nucleosynthesis is greater than the upper limits on the amount of visible matter in the universe. However, we do not know what form this baryonic dark matter is in. It could be either in condensed objects in the halo, such as brown dwarfs and jupiters (objects with $\lesssim 0.08 M_\odot$ so they are not bright shining stars), or in black holes (which at the time of nucleosynthesis would have been baryons). Or, if the baryonic dark matter is not in the halo, it could be in hot intergalactic gas, hot enough not to show absorption lines in the Gunn-Peterson test, but not so hot as to be seen in the x-rays. Evidence for some hot gas

Table 1

"OBSERVED" DENSITIES

$$\left[\Omega \equiv \rho/\rho_c \text{ where } \rho_c = 2\cdot 10^{-29} \text{h}_0^2 \text{g/cm}^3 \text{and } \text{h}_\text{o} \equiv \frac{\text{Ho}}{100 \text{ km/sec/mpc}}\right]$$

Newtonian Mechanics
    (cf. Faber and Gallagher[70])

Visible $\Omega \sim 0.007$
                                                                       (factor of 2 accuracy)

Binaries
Small groups
Extended flat relation curves $\Omega \sim 0.07$
                                                                       (factor of 2 accuracy)

Clusters
Gravitational lenses $\Omega \sim 0.1$ to 0.3

Big Bang Nucleosynthesis (with $t_u \gtrsim 10^{10}$yrs.)
    (c.f. Walker *et al.*[11] and ref. therein) $\Omega_b = 0.065 \pm 0.045$

Preliminary Large Scale Studies

    IRAS red shift study and peculiar velocities $\Omega \gtrsim 0.3$
        (Ref. [51])

    Density redshift counts $\Omega \sim 1 \pm 0.6$
        (Loh and Spillar[71])

Inflation Paradigm
    (Gu$^+$h[52]) $\Omega = 1$

Table 2

MATTER

Baryonic $\Omega_b \sim 0.06$

      VISIBLE $\Omega_{vis} \lesssim 0.01$

      DARK

            Halo
                Jupiters
                Brown Dwarfs
                Stellar Black Holes

            Intergalactic
                Hot gas at $T \sim 10^5 K$
                Stillborn Galaxies

Non Baryonic $\Omega_{nb} \sim 0.94$

      HOT
            $m_{\nu_\tau} \sim 25 eV$

      COLD
            Wimps/Inos $\sim 100 Gev$
            Axions $\sim 10^{-5} eV$
            Planetary Mass Black Holes

is found in clusters of galaxies. However, the amount of gas in clusters would not be enough to make up the entire missing baryonic matter. Another possible hiding place for the dark baryons would be failed galaxies, large clumps of baryons that condense gravitationally but did not produce stars. Such clumps are predicted in galaxy formation scenarios that include large amounts of biasing where only some fraction of the clumps shine.

Hegyi and Olive[53] have argued that dark baryonic halos are unlikely. However, they do allow for the loopholes mentioned above of low mass objects or of massive black holes. It is worth noting that, as Schramm[2] points out, these loopholes are not that unlikely. Furthermore, recent observational evidence,[54], seems to show disk formation is relatively late, occurring at red shifts $z \lesssim 1$. Thus, the first several billion years of a galaxy's life may have been spent prior to the formation of the disk. In fact, if the first large objects to form are less than galactic mass, as many scenarios imply, then mergers are necessary for eventual galaxy size objects. Mergers stimulate star formation while putting early objects into halos rather than disks. Mathews and Schramm[55] have recently developed a galactic evolution model which does just that and gives a reasonable scenario for chemical evolution. (This scenario also provides a natural explanation for the number-versus-redshift relation of low luminosity galaxies found by Cowie.[56] Thus, while making halos out of exotic material may be more exciting, it is certainly not impossible for the halos to be in the form of dark baryons. One application of William of Ockham's famous razor would be to have us not invoke exotic matter until we are forced to do so.

Non-baryonic matter can be divided following Bond and Szalay[57] into two major categories for cosmological purposes: hot dark matter (HDM) and cold dark matter (CDM). Hot dark matter is matter that is relativistic until just before the epoch of galaxy formation, the best example being low mass neutrinos with $m_\nu \sim 25 eV$. (Remember $\Omega_\nu \sim \frac{m_\nu(eV)}{100 h_0^2}$).

Cold dark matter is matter that is moving slowly at the epoch of galaxy formation. Because it is moving slowly, it can clump on very small scales, whereas HDM tends to have more difficulty in being confined on small scales. Examples of CDM could be massive neutrino-like particles with masses, M_x, greater than several GeV or the lightest super-symmetric particle which is presumed to be stable and might also have masses of several GeV. Following Michael Turner, all such weakly interacting massive particles are called "WIMPS." Axions, while very light, would also be moving very slowly[58] and, thus, would clump on small scales. Or, one could also go to non-elementary particle candidates, such as planetary mass blackholes or quark nuggets of strange quark matter, possibly produced at the quark-hadron transition.[59] Another possibility would be any sort of massive topological remnant left over from some early phase transition. Table 2 summarizes the matter options. Note that CDM would clump in halos, thus requiring the dark baryonic matter to be out between galaxies, whereas HDM would allow baryonic halos.

When thinking about dark matter candidates, one should remember the basic work of Zeldovich,[60] resurrected by Lee and Weinberg[61] and others,[62] which showed for a weakly interacting particle that one can obtain closure densities, either if the particle is very light, $\sim 25 eV$, or if the particle is very massive, $\sim 3 GeV$. This occurs because, if the particle is much lighter than the decoupling temperature, then its number density is the number density of photons (to within spin factors and small corrections), and so the mass density is in direct proportion to the particle mass, since the number density is fixed. However, if the mass of the particle is much greater than the decoupling temperature, then annihilations will deplete the particle number since, as the temperature of the expanding universe drops below the rest mass of the particle, Boltzmann suppression prohibits production while the number is depleted via annihilations until the annihilation reacton freezes out. For normal weakly interacting particles, decoupling occurs at a temperature of $\sim 1 MeV$, so higher mass particles are depleted. It should also be noted that the curve of density versus particle mass turns over again (see Figure 4) once the mass of the WIMP exceeds the mass of the coupling boson[63,64,65] so that the annihilation cross section varies as $\frac{1}{M_x^2}$, independent of the mass of the coupling boson. In this latter case, $\Omega = 1$ can be obtained for $M_x \sim 1 TeV \sim (3K \times M_{Planck})^{1/2}$, where $3K$ and M_{Planck} are the only energy scales left in the calculation (see Figure 4). A loophole to this argument occurs if there is a matter-antimatter assymmetry as in the case of baryons. However, such particles would have to be Dirac particles and we will see that they are still severely constrained.

A few years ago the preferred candidate particle was probably a few GeV mass WIMP. However, LEP's lack of discovery of any new particle coupling to the Z° with $M_x \lesssim 45 GeV$, coupled with underground experiments[66] clearly eliminates that candidate.[67,68] Constraints for particles not fully coupled to the Z^0 were discussed by Ellis, Nanopoulos, Roskowski and Schramm[68] and are updated and presented in Figures 5a and 5b. (The inclusion of the Kamiokande II results as well as the newer LEP limits yields an important update over the results of Ellis *et al.*[68] since it closes the loophole for Dirac particles near $12 GeV$.) Note also that the generic constraints of Figure 5 also apply to other hypothetical particles since CDF and UA2 do not see any squarks, sleptons, W' or Z' up to masses significantly greater than M_{Z^0}. Thus, whatever the coupling boson is, it must be greater than M_{Z^0} which means the effective value for $\sin^2 \phi_Z$ is < 1.

Furthermore, as Krauss[67] has emphasized, scaler particles such as sneutrinos interact like Dirac neutrinos so that the Kamiokande II and ionization experimental limits[66] also apply. Since asymmetric candidates are all Dirac particles, the restricted part of Figure 5b constrains asymmetric candidates where $\Omega = 1$ is no longer required to follow the locus shown. Thus, it seems that whether the particle is matter-antimatter symmetric or not, it is required to have an interaction weaker than weak and/or have a mass greater than about $20 GeV$. Future dark matter searches should thus focus on more massive and more weakly interacting particles.

Also, as Dimopoulos[63] has emphasized, the next appealing crossing of $\Omega = 1$ (see Figure 4) is $\gtrsim 1 TeV$ (but, in any case, $\lesssim 340 TeV$ from the unitarity bound[65]), which can be probed by SSC and LHC as well as by underground detectors. After the correct experimental constraints are taken into account, the favoured CDM particle candidate is now either a $10^{-5} eV$ axion or a gaugino with a mass of many tens of GeV. Of course an HDM ν_τ with $m_{\nu_\tau} \sim 20 \pm 10 eV$ is still a fine candidate as long as galaxy formation proceeds by some mechanism other than adiabatic gaussian matter fluctuations.[69,72] This latter candidate becomes particularly attractive if recent hints from the gallium experiment[73] require the solution to the solar neutrino problem to have neutrino mixing with $\nu_e - \nu_\mu$ mass scales of 0.01 to 0.001 eV, making multiple eV mass scales for ν_τ quite plausible from see-saw type models where $m_{\nu_\tau} \sim m_{\nu_\mu}(\frac{M_{f_3}}{M_{f_2}})^2$ and M_{f_i} is an associated fermion mass for the ith generation. For example, if one uses the heavy quark masses, $(\frac{M_t}{M_c})^2 \sim 10^4$, so that ν_τ becomes ideal HDM. Such possibilities also may help late-time phase transition models for producing structure.[72]

STRUCTURE FORMATION

Perhaps the most outstanding problem in physical cosmology today is that of

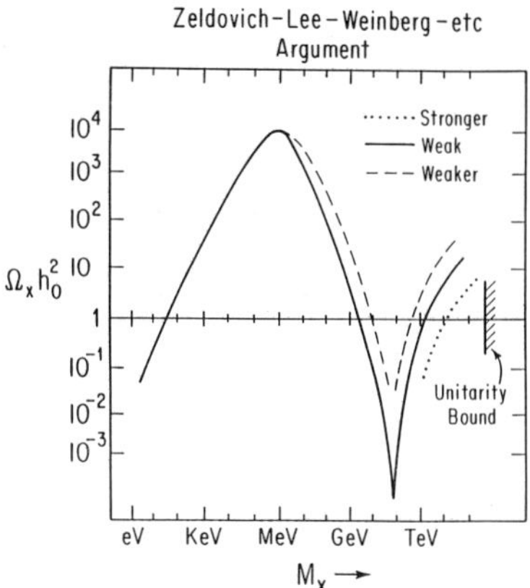

Figure 4. $\Omega_x h_0^2$ versus M_x for weakly interacting particles showing three crossings of $\Omega h_0^2 = 1$. Note also how the curve shifts at high M_x for interactions weaker or stronger than normal weak interaction (where normal weak is that of neutrino coupling through Z^0). Extreme strong couplings reach a unitarity limit at $M_x \sim 340 TeV$.

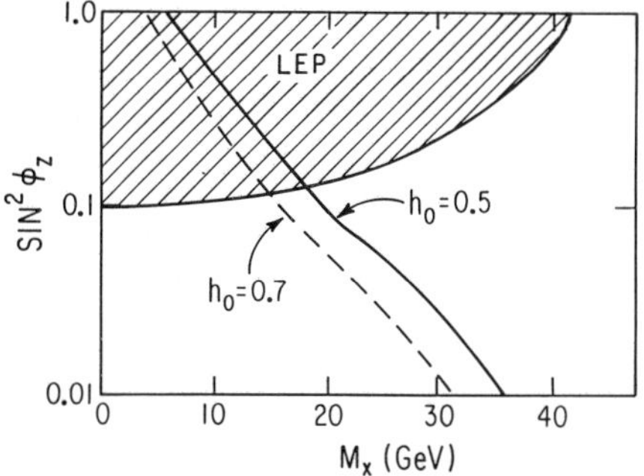

Figure 5a. Constraints on WIMPS of mass M_x versus $\sin^2 \phi_z$, the relative coupling to the Z^0. The constraints are shown assuming Majorana particles (p-wave interactions). The diagonal lines show the combinations of M_x and $\sin^2 \phi_z$ that yield $\Omega = 1$. The cross-hatched region is what is ruled out by the current LEP results. Note that $\Omega = 1$ with $h_0 = 0.5$ is possible only if $M_x \gtrsim 20 GeV$ and $\sin^2 \phi_z < 0.1$. This figure is revised from that of Ref.[68] using latest LEP results.

the formation of structure. Let us review the basic framework of structure formation in the universe. In particular, let us note that structure formation requires that density fluctuations grow. In order for this to occur, $\rho_{m(atter)}$ must be greater than $\rho_{r(adiation)}$. If we define T_{eq} as the temperature where $\rho_m = \rho_r$, then for an $\Omega = 1$ universe with h_0 equal to 0.5, equality is approximately 10^4 times the present temperature T_o. The horizon mass at T_{eq} is $\sim 5 \times 10^{16} (\frac{0.5}{h_0})^4 M$ which gives a present comoving scale of $\sim 60 (\frac{0.5}{h_0})^2 Mpc$. The recombination epoch T_{rec} for an $\Omega = 1$ universe occurs slightly after matter domination. At $T_{rec} \sim 1100 T_o$ baryon fluctuations begin to grow after recombination and the horizon mass at recombination is about $10^{18} (\frac{0.5}{h_0}) M_\odot$ with a comoving scale of $200 (\frac{0.5}{h_0}) Mpc$. We also know that the fluctuations in the microwave background temperature at the time of recombination are less than a few parts in 10^5. [74] Thus, in traditional models with primordial fluctuations existing prior to matter domination, growth begins at matter domination with the limits from $\frac{\delta T}{T}$ forcing $\frac{\delta \rho}{\rho}$ to be less than the order of 10^{-4} since

$$\frac{\delta \rho_m}{\rho} \lesssim 3 \frac{\delta T}{T} \lesssim 10^{-4}.$$

Since small fluctuation $\delta \rho$ grows linearly with $1 + z$, this would mean that fluctuations could reach the order of unity only at the present epoch. Non-linear growth, and thus true structure formation, does not begin until $\frac{\delta \rho}{\rho}$ has reached unity (see Figure 6). Thus, in the standard model, the existence of objects at $z > 1$ (see for example Gunn, Schneider, and Schmidt[75]) requires that there be fluctuations far larger than the average in order that these objects currently exist. As Efstathiou and Rees[76] point out, the gaussian fluctuation model for primordial fluctuations would not allow a large number of quasar-like objects to form at $z \gtrsim 5$.

All models for structure formation require at least two basic ingredients for that structure:

(1) the matter,
(2) the seeds.

In traditional models, the seeds are random fluctuations in the density field generated at the end of the GUT phase transition, presumably accompanying inflation.[77,78]

As mentioned in the previous section, the matter in any model of galaxy formation with $\Omega = 1$ consists of normal baryonic matter with Ω the order of 0.06 and some non-baryonic matter, either hot or cold, with Ω the order of 0.94.[79]

The seeds which clump the matter to form objects may be divided into two broad categories (see Table 3) which can further be subdivided. The two broad categories would be (1) random gaussian seeds, presumably induced by quantum fluctuations at the end of a phase transition, and (2) topological defects produced in a vacuum phase transition. For the random gaussian seeds, the traditional assumption has been that the phase transition is the one associated with inflation[77,78]. However, it has been shown that similar kinds of fluctuations can also be generated in late-time phase transitions.[15,16] Similarly, for the topological defects, they could be formed either at the end of a GUT phase transition ($\sim 10^{15} GeV$) or in some late-time transition.[73,82,83] In some sense this current division of random versus topological replaces the old division of adiabatic versus isothermal (or isocurvature). In fact, the current "random gaussian" are indeed "adiabatic" and the topological are isothermal and isocurvature. However, the latter have the new added feature of also being non-gaussian.

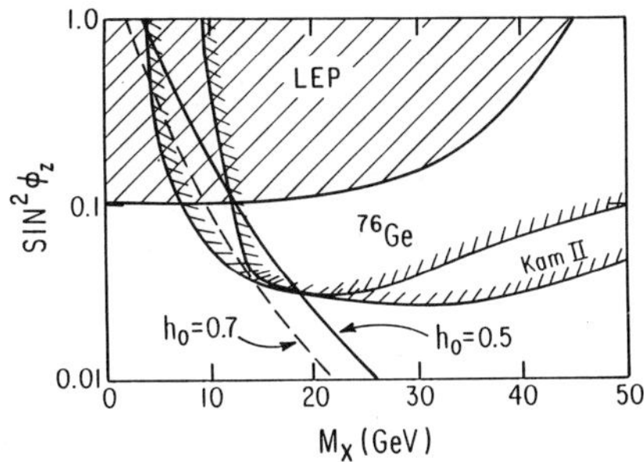

Figure 5b. This is the same as 5a but for Dirac particles (s-wave interactions). The ^{76}Ge region is that ruled out by the Caldwell *et al.* double-β decay style experiments. This figure is revised from that of Ref.[68] using latest LEP results and using new Kamiokande limits which closed a possible loophole near $M_x \sim 10 GeV$. The current results require $M_x \gtrsim 20 GeV$ and $\sin^2 \phi_z \lesssim 0.03$ for matter-antimatter symmetric particles and also exclude the entire cross hatched region for asymmetric particle candidates.

Table 3

SEEDS

I RANDOM GAUSSIAN, Quantum Fluctuations

  A. End of Inflation

  B. LTPT

II TOPOLOGICAL DEFECTS

  A. GUT

  B. LTPT

Let's note that all models for galaxy formation require new fundamental physics beyond

$$SU_3 \times SU_2 \times U_1.$$

In particular, all non-baryonic dark matter, whether hot or cold, requires new physics, and similarly, all seeds, whether GUT scale or late-time and whether random gaussian or topological, require vacuum phase transitions. No model exists that does not invoke new physics. In fact, the existence of structure in the universe is one of the most important clues to the existence of physics beyond the standard model.

We should also note that not all combinations of seeds and matter are possible. For example, if one uses random gaussian seeds, then the non-baryonic matter must be cold, whereas if one uses topological seeds, the non-baryonic matter can be either hot or cold. One should also note that baryonic halos would require hot dark matter and hence topological seeds. Thus, searches for the dark baryons will also help constrain the non-baryonic candidates.

All seed models require some form of vacuum phase transition. Thus, let us explore what possible phase transitions might occur (see Table 4). It should be noted in looking at Table 4 that of the three general classifications of cosmological phase transitions—the early, intermediate and late—the only ones that we absolutely know must have occurred are in the intermediate category when there is a horizon problem, namely that the horizon at the time of that transition is too small to generate galactic sized structure, and yet, the transition is not accompanied by significant inflation. The traditional early transitions have been used in the past because, while their horizon is small, inflation can amplify the effects to large scales. The other option is that of a late-time transition, where the universe waits until the horizon is sufficiently large that the physics of the phase transition directly yields the structures without having to use inflation to avoid the horizon problem.

POTENTIAL OBSERVATIONS TO BE EXPLAINED

In the last couple of years there have been a number of observations affecting galaxy formation and large-scale structure that have been a potential problem for traditional models which invoked early random gaussian fluctuations. However, because each of these observations is new and has not stood the test of time, in this discussion we refer to these as potential observations. In particular, many of the advocates of gaussian fluctuations and cold dark matter have tried to argue that these observations are statistical flukes that have yet to be established. Obviously, if these potential observations continue to hold up and are verified and are shown to be ubiquitous rather than statistical rareties, then the traditional models are in serious trouble. Table 5 summarizes these potential observations. Perhaps the most potentially damning would be observations of microwave anisotropies $\frac{\Delta T}{T}$ at levels significantly below 10^{-5}. However, at the present time, observations of small scale anisotropy are at the level of a couple times 10^{-5}. Observations on angular scales of degrees or more are also approaching a few 10^{-5}. As this paper is being written, the measurements have not yet reached the point of ruling out the model of random fluctuations. However, as noted by Smoot[74], within the not too distant future, COBE may be able to achieve limits as low as 3×10^{-6} on scales of a few degrees and larger, and antarctic studies may also push to similar levels on somewhat smaller scales, as might the baloon studies of Meyer *et al.* at MIT.

The next observation that can be a potential problem for traditional models is the existence of structures with scales greater than the order of $100 Mpc$. In particular,

Table 4

VACUUM PHASE TRANSITIONS

EARLY (Small horizon but inflation)

$$\sim 10^{19} GeV \text{ - T.O.E.}$$

$$\sim 10^{16} GeV \text{ - GUT}$$

INTERMEDIATE (Known to occur but horizon problem)

$$\sim 10^{2} GeV \text{ - Electroweak}$$

$$\sim 1 GeV \text{ - QCD}$$

LATE (Horizon large)

Table 5

POTENTIAL OBSERVATIONS

1. $\frac{\delta T}{T} \lesssim 10^{-5}$

2. Structures $\gtrsim 100 Mpc$

3. Large coherent velocity flows

4. Objects existing at $z \gtrsim 5$

5. Large cluster - cluster correlations

the great wall observed by Geller and Huchra[84] shows that there is at least one such wall in the universe. The observations of Broadhurst *et al.*[85,86] show evidence for a multiplicity of such great walls with the characteristic spacing comparable to the size of the Geller-Huchra wall itself. While much debate has been made about whether or not the multiple walls of Broadhurst *et al.* are periodic or quasi-periodic, it does seem clear from their observations, as well as the work reported by Szalay,[87] that there is significant structure in the universe on scales of $\sim 100 Mpc$. This is thoroughly supported by the large coherent velocity flows where the Seven Samurai[88] and others have found evidence for the existence of an object they call the "Great Attractor" towards which the Virgo cluster and the Hydro-Centaurus cluster all seem to be flowing with a velocity $\sim 600 km/sec$. This again seems to indicate evidence of structures on the scales of at least $60 Mpc$.

Perhaps most constraining of the traditional astronomical measurements is the existence of objects at very large redshifts. In particular, Gunn, Schneider, and Schmidt[75] have found a quasar with a redshift of 4.73. As Efstathiou and Rees[76] have noted, if such objects are ubiquitous, this would be fatal for primordial gaussian fluctuation models. Similarly, if one ever finds a quasar-type object at much larger redshifts, that would also be fatal.

Another potentially fatal observation for gaussian fluctuation models comes from the work of Bahcall and Soneira,[89] and Klypin and Khlopov[90] where they find that clusters of galaxies seem to be more strongly correlated with each other than galaxies are correlated with each other. While Primack and Dekel[91] have warned of the dangers of projection effects on such observations, it seems difficult to understand how projection effects would give the fractal-like behavior[92]. Furthermore, the southern hemisphere work of Huchra[93] also seems to support high cluster correlations. Most recently Vandenburg and West[94] have also found similar correlations for the CD galaxies observed at cluster centers. These CD's should not have the projection effect problems because redshifts are known. Even Primack and Dekel now acknowledge that there seems to be some excess in cluster correlations. If such large correlations turn out to be real, they too cannot be easily explained in the gaussian model, and, as Szalay and Schramm[93] note, they seem to be best fit by some sort of fractal-like pattern, as one might get from topological defects induced by a phase transition.

LATE-TIME TRANSITIONS

By late-time transition we will mean any non-linear growth occurring shortly after recombination. As mentioned above, such non-linear growth can be related either to a gaussian pattern or to a topological pattern such as walls, strings or textures. It is also possible that some normal random gaussian pattern from the very early universe could be triggered to undergo non-linear growth by some sort of phase transition or related phenomenon occurring after recombination. An example of this latter case would be the neutrino flypaper model of Fuller and Schramm.[95]

In general we will see that these late-time transitions can give the smallest possible $\frac{\Delta T}{T}$ for a given size structure. They can produce non-gaussian structural patterns, fractal-like with large velocity flows. It might be noted that the co-moving horizon at the time of the transition is not too different than the scale associated with the largest structures observed. No model of primordial fluctuations naturally imbeds this horizon scale onto the structural pattern. If some non-linear growth is associated with the patterns, the horizon scale can be imposed on the structure.

Another very dramatic advantage of late-time transitions, illustrated in Figure

6, is that it can produce structure with $\frac{\delta\rho}{\rho} \gtrsim 1$ at $z \gtrsim 10$. Thus, one could have significant structure and a significant number of objects at high redshift, which is a problem in any normal model with the seeds forming prior to recombination.

Let us now explore the possible physics that might give rise to a late-time transition, that is, a transition with a critical temperature between $0.001eV$ and $1eV$. It might be noted that in some sense it is a "hierarchy" rather than a "fine-tuning" problem to obtain a transition in this temperature range. We are trying to find a small mass scale somewhat analogous to how one would like to find the mass scale of the electron, or, for that matter, the Z^0 boson, when the natural mass scales to the problem are closer to $10^{19}GeV$, as in superstring models, or to 0. The hierarchy problem of trying to find the intermediate scale of the electroweak interaction of somewhere between the quark-lepton scale and the GUT or Planck scale has traditionally been approached with either a supersymmetric solution or a dynamical solution ("technicolor"). This supersymmetric solution, in some sense, is analogous to the model proposed in the appendix of Hill, Schramm and Fry,[72] denoted as HSF, which is an adaptation of the Hill-Ross[96] mechanism. A dynamical solution which has been proposed by Dimopoulos[97] involves a shadow $SU3$. The scale of a physics that might be associated with an HSF mechanism was relating to the MSW mixing solution to the solar neutrino problem.

The MSW[98,99] mixing solution to the solar neutrino problem is achieved if the neutrino mass difference squared, δm^2, is of the order of 10^{-4} to $10^{-7}eV^2$, or, in other words, neutrino masses of the order of a fraction of an electron volt. If we assume, following HSF, that the neutrino masses are generated by a pseudo-Nambu-Goldstone boson mechanism with mass

$$m_\phi \sim \frac{m_\nu^2}{f}$$

and with a transition occurring at $T_{crit} \sim m_\nu$, and if we further assume that the coupling f is related to the GUT scale, since we want to imbed this in some sort of unified theory, then the Compton wavelength $\lambda_\phi \sim 1Mpc$, in other words, a galactic scale. The density of the ϕ field at the time of the transition is the order of the cosmological density, in other words,

$$\frac{\rho_\phi}{\rho} \sim 1.$$

(Note that this is natural for phase transitions, whereas the requirement for primordial transitions to have small fluctuations, as inflation requires, is a fine tuning requirement.) Furthermore, the average spacing of the nucleation sights, L, can be estimated from Coleman's theory on spontaneous nucleation to yield spacings today that are interesting:

$$\frac{R_H}{L} \sim Log(\frac{M_p}{T_{crit}})$$

$$L_co \equiv L(1 + z_{crit}) = \frac{6000}{Log\frac{(M_p)}{T_{crit})}}(\frac{0.5}{H_0})(1 + z_c)^{-1/2}$$

where

$$z_{crit} = z_c = (T_{crit}/T_0),$$

R_H is the horizon radius at z_c' and $M_p \sim 10^{19}GeV$. This yields for $T_{crit} \sim 10^{-2}eV\,\mathrm{to}\,10^{-3}eV/L_{co(moving)} \sim 40\,\mathrm{to}\,140Mpc$.

As we mentioned previously, recent impetus for new physics at this energy scale has come from the SAGE experiment which detects neutrinos from the PP chain in the sun. The previous solar neutrino experiments, the chlorine and the Kamiokande experiments, are mainly sensitive to the rare ^8B branch of the solar energy generating reactions. It is well established that the ^8B experiments have seen fluxes at levels somewhat below theoretical predictions.[100] However, there has always been the worry that the ^8B channel may be supressed due to astrophysical effects since its yield is very temperature sensitive. However, the PP chain that produces the neutrinos to be detected by SAGE must work if the sun is burning by fussion. Thus, the report[101] of no significant counts above background after five months of running the gallium experiment when they expected nineteen counts for the standard model implies that something is happening to the neutrinos on their way between emission and arrival at earth. Of course, the present results are very preliminary. Questions with regard to estimates of background, counting efficiencies, systematics, statistics, etc., remain, but the tantalizing hint that the ν_e's mixed into some other species of neutrino on their way out of the sun is certainly exciting. The final state of this experiment will not be known for several years. In 1991 we will begin to have results from a similar gallium experiment operated by the GALLEX collaboration in the Grand Sasso Tunnel in Italy. Their chemistry is somewhat cleaner and we will have an independent check. Furthermore, both of these gallium experiments will be callibrated using ^{51}Cr sources of MeV neutrinos. Thus, one will have a true check of their counting efficiencies, etc., and both of these experiments will run for a long-enough time that the statistics will reach significant levels. If the neutrinos really are mixing on their way out of the sun, then the MSW solution is probably valid and we are in the realm discussed above.

It might also be noted that a simple application of the Gell-Mann–Ramond–Slansky see-saw model[102] for neutrino masses yields some interesting implications. If we assume that there is a mass hierarchy in the neutrinos with the electron neutrino having negligible mass, the μ the intermediate mass and the τ the heaviest, and we assume that the mixing of the ν_e in the sun goes to its nearest neighbor family, the ν_μ, then the ν_μ is carrying most of the mass of the MSW δ_m^2. The see-saw mechanism argues that

$$m_{\nu_i} \sim \frac{m_{f_i}^2}{M}$$

for a given family, or, in other words,

$$m_{\nu_\tau} \sim m_{\nu_\mu} \left(\frac{m_{f_3}}{m_{f_2}}\right)^2.$$

If we use lepton masses for the fermion masses, this yields a ν_τ mass in the neighborhood of a few eV. However, if we use heavy quark masses, then, since the top quark mass is $\gtrsim 100$ times that of the charm quark, this yields ν_τ masses in the neighborhood of 10 to 100 eV, making it perfect hot dark matter. It might also be noted that the see-saw mass scale, M, in this picture, ends up being the order of 10^9 to $10^{12}GeV$, which happens to be the only window allowed for the DFS-axion[103] scale. It might further be noted that if the non-baryonic dark matter is indeed the τ neutrino, then one is required to dismiss primordial gaussian fluctuations.

Note that even if the MSW mixing is $\nu_e - \nu_\tau$, the LTPT possibility is still there, but then all neutrinos would be light and could not serve as HDM. It is interesting that in this latter case the see-saw M is the GUT scale.

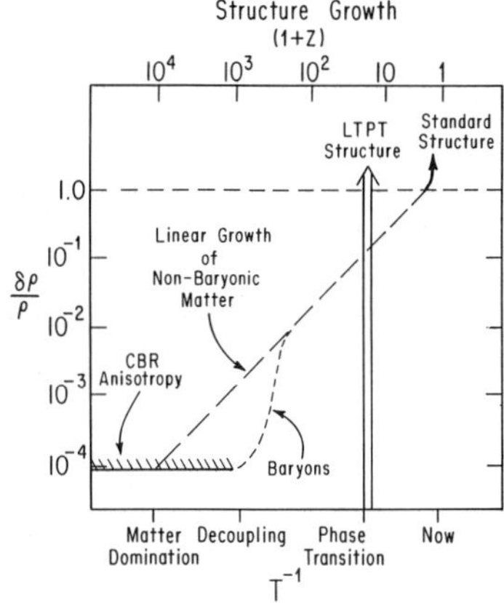

Figure 6. The growth of density fluctuations with the expansion of the universe. Note that LTPT can yield non-linear growth and thus structure at epochs much earlier than standard primordial models.

WALL NETWORK

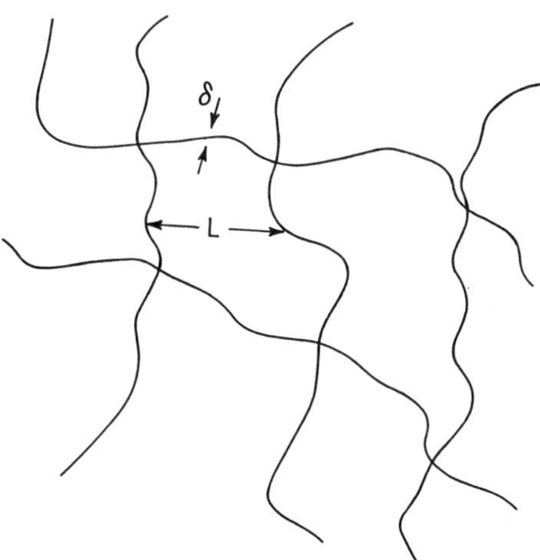

Figure 7a. A generic wall network defining the wall thickness δ and the characteristic spacing of structure L.

SEED NETWORK
(Bag, balls-of-wall, Textures, etc)

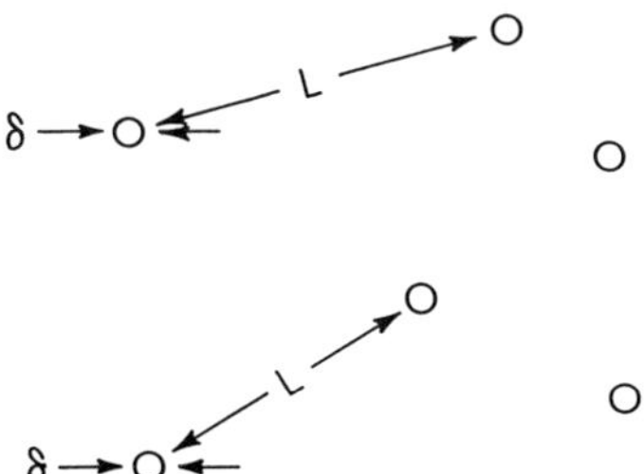

Figure 7b. A generic network for seed generation with seed size, δ and seed separation L.

STRUCTURE FROM LTPT

LTPT can produce vacuum fluctuations of the random gaussian character just as could be generated at the end of inflation.[76] However, as emphasized in references 80 and 81, these structures will have a quantum scale that is the order of a galaxy size, and the bosons associated with the fluctuations might even serve as the dark matter of the universe.

The other alternative for LTPT is to produce topological structures. Just as early universe phase transitions can produce strings and/or textures, LTPT can also produce such objects. Furthermore, LTPT can produce walls which are a problem for primordial phase transitions. However, there is a problem for some walls, depending on the nature of the interaction potential. LTPT that have a $\lambda\phi^4$ potential will end up with one wall dominating as was demonstrated in references 104, 105 and 106. However, this problem of one wall dominating can be surmounted in a variety of ways which have varying degrees of attractiveness, depending on the eyes of the beholder. For example, in the HSF phase transition, the walls are sine-Gordon rather than $\lambda\phi^4$. As Widrow has shown,[107] the sine-Gordon walls can yield "bags" of wall or "balls" of wall which survive several expansion times. These bags or balls can then serve as seeds in galaxy formation, and thus, it is their amplitude that becomes a deciding factor for $\frac{\Delta T}{T}$ limits as opposed to the energy scale of the infinite walls which can be made quite small. This latter point was emphasized by Hill, Schramm and Widrow[82]. Another way of avoiding single wall dominance is the decaying wall model of Kawano[108] where the walls serve as seeds and then decay away. It is also possible to escape one-wall domination with a large number of minima in the potential. Perhaps the most dramatic way of escaping one-wall domination, thus keeping a network of walls, as shown in Figure 7A, is if the walls have friction with the ambient medium, whether it be neutrinos or the remaining baryonic and/or non-baryonic matter in the universe[109]. Alessandro Massarotti has shown that friction can in many reasonable cases slow the walls down sufficiently that they do not evolve to the one-ball domination situation. In this case, one retains a complex network with L for the wall being much less than the horizon size.

It might be noted that long walls gravitationally repel rather than attract[110,111], whereas balls of wall are attractive seeds. Thus, a combined network of balls and slowed-down long walls can yield a complex structure which may be even of a fractal character in agreement with the claims of Schramm and Szalay[92] from cluster correlations.

In addition to walls, LTPT can also produce textures[112] or non-topological solitons[113]. In these latter cases, or with the bags of wall dominating, one will have networks more closely ressembling Figure 7B and Figure 7A. It should be noted that the parameters L and δ and the nature of the structures generated are dependent on the model for the LTPT. It should also be noted that questions of the detailed physics of imbedding the LTPT into some larger GUT or TOE are dependent on the unification model. HSF have shown that a reasonable toy model can be constructed which can give a phase transition. These phase transitions in many ways are quite analogous to the axion-producing phase transition which has a coupling at a scale near to the order of $10^{11} GeV$, far above the QCD phase transition scale of the order of GeV. And like the axion, the particle involved in the LTPT of HSF has a pseudo-Nambu-Goldstone boson. However, instead of being related to the strong interaction and quarks, in the LTPT case it is related to the neutrinos and probably to family symmetry.

Generating seeds at an LTPT might be advantageous for producing the multiple walls of Broadhurst *et al.*[85,86]. In particular, Icke and Weygaert[114], and Coles[115] have independently demonstrated that the phenomenological Vornoi tessalations of the intersection of expanding rarefaction shells give a very good fit to large scale structure if the nodes of these tessalations are fit to the Abell clusters. In particular, they note that one gets quasi-periodic walls at $\sim 130 Mpc$ with cluster correlation functions that are quite strong and follow the fractal behavior of Schramm and Szalay. However, the seed distribution required to give this tessalation causes a conflict with the microwave background radiation, if the seeds are generated prior to the decoupling. However, an LTPT could remedy that. Similarly, an LTPT can provide the seeds to enable hot dark matter to work as a galaxy formation model (see, for example, reference 116). It might be noted that the typical bag of wall can easily yield a galaxy or a quasar-forming seed.

We can estimate the mass associated with a wall in the following way:

Let σ = energy density per unit area, that is:

$$\sigma \equiv \rho_w \delta \sim \frac{4 \times 10^{-5} s \rho_o \delta}{h_0^2} (1 + z_c)^4$$

where

$$\delta \equiv \text{thickness}$$

$$\rho_w = \text{density of the wall}$$

$$s = \frac{\rho_w}{\rho_r} \text{ at } z_c$$

$$\rho_0 = 3 \times 10^{11} h_0^2 M_\odot / Mpc^3$$

then

$$M_w \sim \pi \sigma L^2 \sim 3 \times 10^7 (1 + z_c)^4 (\frac{\delta}{Mpc})(\frac{L}{Mpc})^2 M_\odot$$

and for stable walls

$$\Omega_w(z) \sim \frac{3}{4} \frac{\sigma}{\rho_0 L (1 + z)^3}.$$

Note that Ω_w at the present epoch, can be the order of unity. Wall domination can occur at present epoch if

$$z_c \gtrsim 11 (\frac{L}{\delta})^{\frac{1}{4}} (\frac{ho^2}{s})^{\frac{1}{4}} - 1$$

for stable walls. It might be noted that if wall domination occurs at the present epoch, as long as there are multiple walls, rather than just one wall dominating, one has the interesting situation where the expansion of the universe is no longer following the normal matter-dominated relationship, and, in particular, one can achieve ages greater than $\frac{1}{h_o}$. Such a situation may be a solution to the age-Hubble constant problem if h_o is ever shown to be greater than 0.7.

It might also be noted for topological structure generated by LTPT that the structure is relatively independent of whether the non-baryonic dark matter is hot or cold.

MICROWAVE ANISOTROPIES

Since LTPTs provide no fluctuations on the surface of last scattering, all fluctuations from the microwave background must be due to the differential redshift-blueshift non-cancellation due to a changing potential in the transparent medium or due to scattering of the microwave photons off of moving objects. One can estimate the potential change due to the ϕ field itself generated in the phase transition and by the dynamic motion of the structures and the Doppler shift thereby produced. One can also do the classical Rees-Sciama and Sachs-Wolf calculations for the $\frac{\Delta T}{T}$ generated by existing objects[118,119]. We can estimate its effects roughly in the following way: The static effects will dimensionally go as

$$\frac{\delta T}{T} \sim \Omega_w (\frac{L}{R_H})^2 \sim G\sigma L.$$

The time-changing effects can be estimated by multiplying the static effect by $\frac{V}{c}$. While different people remember different formulations of these things, one can show that because of the nature of walls and other topological systems, the effects can be reduced to the form $G\sigma L$ times $\frac{V}{c}$ or $\frac{V^2}{c^2}$. Since any walls or topological seeds we ever see must be moving with $V < c$, the dominant effect will in general go like $G\sigma L$, which can be shown to yield the result:

$$\frac{\delta T}{T} \sim 10^{-8} (\frac{1 + z_c}{10})^4 \ s(\frac{\delta}{Mpc})(\frac{L}{Mpc})$$

$$\sim 10^{-6} \text{ for } L \sim 100 Mpc, \ \delta \sim 1 Mpc, \ z_c \sim 10.$$

Note that this yields $\frac{\delta T}{T} \sim 10^{-6}$ even for an L of $100 Mpc$. The distribution, however, of these fluctuations depends very much on the detailed topological nature of the structures produced. In particular, Turner, Watkins and Widrow[119] have shown that balls of wall tend to produce spikes very similar in nature to the spikes that textures produce. A general formalism showing the wide range of structures in non-gaussian microwave background fluctuations has been developed by Goetz and Noetzold[120].

In general, one can see that if structure of size L is generated by a late-time phase transition, and L is the maximum size of structure produced in that transition, then the late-time transition does give the minimum $\frac{\delta T}{T}$ for that structure. Of course the question is what is the characteristic size L of structure generated in a transition. For $\lambda\phi^4$ structures, L goes to the horizon size, in which case $\frac{\delta T}{T}$ gets larger than current observational limits. However, as mentioned above, many other possibilities can be generated in LTPT, with L at present being a somewhat freely adjustable parameter, depending on the model, the amount of friction, the decay of the walls, etc.

CONCLUSION

In these lectures we have seen that the basic big bang model is in excellent shape with the recent collider results helping to confirm it in the same way they've helped with $SU_3 \times SU_2 \times U_1$. We've also seen that the prediction that the bulk of the matter in the universe is in the form of some exotic non-baryonic species obviously remains to be confirmed. Furthermore, we've examined the current problems of generating structure in the universe, mentioning the traditional primordial scenarios as well as the new exotic idea of a late-time phase transition.

All in all, we've seen that cosmology is tremendously active with new data coming from both astronomical and particle physics techniques. We've also seen that some of the best current indications for particle physics beyond $SU_3 \times SU_2 \times U_1$ are coming from astrophysical arguments.

ACKNOWLEDGMENTS

I would like to thank my recent collaborators, Chris Hill, John Ellis, Savas Dimopoulos, Gary Steigman, Keith Olive, Michael Turner, Rocky Kolb and Terry Walker, for many useful discussions. I would also like to acknowledge useful communications with Denys Wilkinson, Walter Mampe, Jim Byrne and Stuart Friedman on the neutron lifetime and with Bernard Pagel and Jay Gallagher on helium abundances. This work was supported in part by NSF grant AST 88-22595 and by NASA grant NAGW 1321 at the University of Chicago and by the DoE and NASA grant NAGW 1340 at the NASA/Fermilab Astrophysics Center.

REFERENCES

1. Mather, J. *et al.*, 1990, COBE preprint, Goddard Space Flight Center *Astrophys. J.*, in press.
2. Schramm, D., 1990, in: *Proc. of the 1990 Rencontre de Physique at La Thuile, March 1990*, M. Greco, ed., in press.
   Schramm, D., 1990, in: *Proc. 1990 Nobel Symposium, Gräftvållen, Sweden*, in press.
   Schramm, D., 1990, in: *Proc. of the Moriond Astrophysics Meeting on Particle Astrophysics at Les Arcs, Savoie, France, March 1990*, T. Piran, ed., in press.
3. Alpher, R.A.,Bethe, H., and Gamow, G., 1948, *Phys. Rev.* 73:803.
4. Alpher, R.A., Follin, J.W., and Herman, R.C., 1953, *Phys. Rev.* 92:1347.
5. Taylor, R. and Hoyle, F., 1964, *Nature* 203:1108.
6. Peebles, P.J.E., 1966, *Phys. Rev. Lett.* 16:410.
7. Wagoner, P., Fowler, W.A., and Hoyle, F., 1967 *Astrophys. J.* 148:3.
8. Schramm, D.N. and Wagoner, R.V., 1977, *Ann. Rev. of Nuc. Sci.* 27:37;
   Olive, K., Schramm, D.N., Steigman, G., Turner, M., and Yang, J., 1981, *Astrophys. J.* 246:557;
   Boesgaard, A. and Steigman, G., 1985, *Ann. Rev. of Astron. and Astrophys.* 23:319.
9. Yang, J., Turner, M., Steigman, G, Schramm, D.N., and Olive, K., 1984, *Astrophys. J.* 281:493.
10. Olive, K., Schramm, D.N., Steigman, G., and Walker, T., 1990, *Phys. Lett B.* 236:454.
11. Walker, T., Steigman, G., Schramm, D.N., Kang H.-S., and Olive, K., 1990, *Astrophys. J.*, submitted.
12. Kawano, L., Schramm, D.N., and Steigman, G., 1988, *Astrophys. J.* 327:750.
13. Truran, J.W., Cameron, A.G., and Gilbert, A., 1966, *Canadian Journal of Physics* 44:563.
14. Krauss, L. and Romanelli, P., 1990, Yale University preprint and *Ap.J.*, in press.

15. Fowler, W., Greenstein, J., and Hoyle, F., 1962, *Geophys. J.R.A.S.* 6:6.

16. Ryter, C., Reeves, H., Gradstajn E., and Audouze, J., 1970, *Astron. and Astrophys.* 8:389.

17. Geiss, J. and Reeves, H., 1971, *Astron. and Astrophys.* 18:126;
Black, B., 1971, *Nature* 234:148.

18. Rogerson, J. and York, D., 1973, *Astrophys. J.* 186:L95.

19. Reeves, H., Audouze, J., Fowler, W.A., and Schramm, D.N., 1973, *Astrophys. J.* 179:909.

20. Epstein, R., Lattimer J., and Schramm, D.N., 1976, *Nature* 263:198.

21. Gott, III, J.R., Gunn, J., Schramm D.N., and Tinsley, B.M., 1974, *Astrophys. J.* 194:543.

22. Rood, R.T., Steigman, G., and Tinsley, B.M., 1976, *Astrophys. J.* 207:L57.

23. Yang, J., Schramm, D.N., Steigman, G., and Rood, R.T., 1979, *Astrophys. J.* 227:697.

24. Wilson, T., Rood, R.T., and Bania, T., 1983, in: *Proc. of the ESO Workshop on Primordial Healing*, P. Shaver and D. Knuth, eds., European Southern Observatory, Garching.

25. Hartoog, M.R., 1979, *Ap.J.* 231:161;
Ostriker, J. and Schramm, D.N., 1991, FNAL/Princeton preprint.

26. Spite, J. and Spite, F., 1982, *Astron. and Astrophys.* 115:357;
Rebolo, R., Molnaro, P., and Beckman, J., 1988, *Astron. and Astrophys.* 192:192.
Hobbs, L. and Pilachowski, C., 1988, *Astrophys. J.* 326:L23.

27. Steigman, G., Schramm, D.N., and Gunn, J. *Phys. Lett.* 66B:202.

28. Schvartzman, V.F., 1969, *JETP Letters* 9:184;
Peebles, P.J.E., 1971, *Physical Cosmology*, Princeton University Press.

29. Schramm, D.N. and Kawano, L., 1989, *Nuc. Inst. and Methods* A284:84.

30. Pagel, B., 1990, *Proc. of 1989 Rencontre de Moriond*.

31. Scherrer, R., Applegate, J., and Hogan, C., 1987, *Phys. Rev. D* 35:1151;
Alcock, C., Fuller, G., and Mathews, G., 1987, *Astrophys. J.* 320:439;
Fowler, W.A. and Malaney, R., 1988, *Astrophys. J.* 333:14.

32. Kurki-Suonio, H., Matzner, R., Olive, K., and Schramm, D.N., 1990, *Astrophys. J.* 353:406;

33. Peimbert, M. and Torres-Peimbert, S., 1974, *Astrophys. J.* 193:327.

34. Pagel, B., 1990, in *ESO/CERN Proc.*

35. Steigman, G., Schramm, D.N., and Gallagher, J., 1989, *Comments on Astrophys.* 14:97.

36. Mampe, W., Ageron, P., Bates, C., Pendlebury, J.M., and Steyerl, A., 1989, *Phys. Rev. Lett.* 63:593.

37. Byrne, J., 1990, *et al. Phys. Lett. B*, submitted.

38. Abele, H., Arnould, M., Borel, H.A., Dohner, J. Dubbers, D., Freedman, S., Last, J., and Reichert, I., 1989, in: *Proc. Grenoble Workshop on Slow Neutrons*.

39. Steigman, G., Olive, K., Schramm, D.N., and Turner, M., 1986, *Phys. Lett. B* 176:33.

40. ALEPH, DELPHI, L3, OPAL and MARK II collaboration papers, 1990, in: *Proc. of the Int. High Energy Conference, Singapore*.

41. Schramm, D.N. and Steigman, G., 1984, *Phys. Lett. B* 141:337.

42. Kolb, E. and Scherrer, R., 1982, *Phys. Rev. D* 25:1481;
Kolb, E., Turner, M.S., Chakravorty, A., and Schramm, D.N., 1991, Fermilab PUB-91/28-A.

43. Dimopoulos, S., Esmailzadeh, R., Hall, L. and Starkman, G., 1988, *Astrophys. J.* 330:545.
44. Brown, L. and Schramm, D.N., 1988, *Ap.J.* 329:L103.
45. Pilachowski, C., Hobbs, L., and De Young, D., 1989, *Ap.J. Lett.* 345:L39.
46. Reeves, H., Sato, K., and Tarasawa, M., 1989, U. of Tokyo preprint.
    Tarasawa, N. and Sato, K., 1990, U. of Tokyo preprint.
47. Alcock, C., Fuller, G., Mathews, G., and Meyer, B., 1990, Livermore preprint.
48. Adams. F. and Freese, K., 1990, MIT preprint.
49. Kawano, L., Fowler, W., and Malaney, R., 1990, Caltech-Kellogg preprint.
    Applegate, J., 1989, Columbia University preprint.
50. Freese, K. and Schramm, D.N., 1984, *Nucl. Phys.* B233:167.
51. Strauss, M., Davis, M., and Yahil, A., 1989, U.C. Berkeley preprint;
    Kaiser, N., and Stebbins, A., 1990, CITA preprint;
    Bertschinger, E., Dekel, A., and Yahil, A., 1990, in: *Proc. Blois Symposium on the Microwave Background*;
    Rowan-Robinson. M. and Yahil, A. 1989, *Proc. Rencontres de Moriond*.
52. Guth, A., 1981, *Phys. Rev. D* 23:347; see also Olive, K., 1990, *Physics Reports* 190:309-403 or Lindei, A., 1990, *Particle Physics and Inflationary Cosmology*, Harwood, New York.
53. Hegyi, D. and Olive, K., 1986, *Astrophys. J.* 303:56.
54. Gunn, J., 1988, Talk at ITP Santa Barbara;
    York, D., 1988, Talk at University of Chicago.
55. Mathews, G. and Schramm, D.N., 1990, *Astrophys. J.*, submitted.
56. Cowie, L., 1990, in: *Proc. IUPAP Conference on Primordial Nucleosynthesis and Evolution of Early Universe, Tokyo, September 1990*, in press.
57. Bond, R. and Szalay, A., 1982, *Proc. Texas Relativistic Astrophysical Symposium, Austin, Texas*.
58. Turner, M., Wilczek, F., and Zee, A., 1983, *Phys. Lett. B* 125:35; and 125:519.
59. Crawford, M. and Schramm, D.N., 1982, *Nature* 298:538.
    Witten, E., 1984, *Phys. Rev. D* 30:272; see also Alcock, C. and Olinto, A., 1988, *Ann. Rev. Nuc. Part. Phys.* 38:161.
60. Zeldovich, Ya., 1965, *Adv. Astron. and Astrophys.* 3:241.
61. Lee, B. and Weinberg, S., 1977, *Phys. Rev. Lett.* 39:165.
62. Chiu, H.-Y., 1966, *Phys. Rev. Lett.* 17:712.
    Hut, C.P., 1977, *Phys. Lett. B* 69:85;
    Sato, K. and Koyayashi, H., 1977, *Prog. Theor. Phys.* 58:1775.
63. Dimopoulos, S., Esmailzadeh, R., Hall, L., and Tetradis, N., 1990, *Nucl. Phys. B*, submitted.
64. Brahm, D. and Hall, L., 1990, *Phys. Rev. D* 41:1067.

65. Griest, K., Kamionkowski, M., and Turner, M., 1990 FNAL preprint.
    Olive, K. and Srednicki, M., 1989, *Phys. Lett. B* 230:78.
66. Caldwell, D. *et al.*, 1990, in: *Proc. La Thuile*, M. Greco, ed.; see also Kamiokande II Collaboration, 1990, in: *Proc. IUPAP Sumposium on Primodial Nucleosynthesis and Evolution of Early Universe, Tokyo, September 1990*, in press.
67. Krauss, L., 1990, *Phys. Rev. Lett.*, in press;
    Griest, K. and Silk, J., 1990, U.C. Berkeley preprint.
68. Ellis, J., Nanopoulos, D., Roskowski, L., and Schramm, D.N., 1990, *Phys. Lett. B* 245:251-257.

69. Schramm, D.N., 1989, in: *Proc. 1988 Berkeley Workshop on Particle Astrophysics*, E. Norman, ed., (World Scientific, Singapore.

70. Faber, S.M. and Gallagher, J.S., 1979, *Ann. Rev. Astron. and Astrophys.* 17:135.

71. Loh, E. and Spillar, E., 1988, *Astrophys. J.* 329:24.

72. Hill, C., Schramm, D.N., and Fry, J., 1989, *Comments Nucl. Part. Phys.* 19:25.
Schramm, D., 1990, in: *Proc. IUPAP Symposium on Primordial Nucleosynthesis and Evolution of Early Universe, Tokyo, September 1990*, in press.

73. SAGE collaboration, 1990, in: *Proc. Int. High Energy Physics Meeting, Singapore, August 1990*, in press.

74. Smoot, G., 1990, "The Microwave background," in: *Proceedings of the Tokyo meeting on Primordial Nucleosynthesis, September 1990*.

75. Schneider, D., Schmidt, M., and Gunn, J., 1989, "A Quasar at $z = 4.73$," *Astron. J.* 98:1951-1958.

76. Efstathiou, G. and Rees, M.J., 1988, "High red-shift quasars in the Cold Dark Matter cosmogony," *Mon. Not. Royal Astro. Soc.*, 230:5-11.

77. Olive, K., 1990, "Inflation," *Physics Reports*, 190:309-403.

78. Kolb, E. and Turner, M., 1989, *The Early Universe*, Addison Wesley, San Francisco.

79 Schramm, D.N.,1990, "Big bang nucleosynthesis: the standard model and alternatives," in: *Proc. of the 1990 Nobel Symposium, Graäftåvalen, Sweden*.

80. Wasserman, I., 1986, "Late phase transitions and spontaneous generation of cosmological density perturbations," *Phys. Rev. Lett.*, 57:2234.

81. Press, W., Ryden, B., and Spergel, D., 1990, "Single mechanism for generating large-scale structure and providing dark missing matter," *Phys. Rev. Lett.*, 64:1084.

82. Hill, C.T., Schramm, D.N., Widrow, L.M., 1990, "Late-time phase transitions and large scale structure," in: *Proc. of the XXVth Rencontres de Moriond at Les Arcs, Savoie, France, March 1990*, in press, and *Science*, submitted.

83. Frieman, J., Hill, C. and Watkins, R., 1990, "Late-time phase transitions," Fermilab Preprint, in press.

84. Geller, M. and Huchra J., 1989, "Mapping the universe," *Science*, 246:897.

85. Broadhurst, T., Ellis, R., Koo, D., and Szalay, A., 1990, "Large-scale distribution of galaxies at the Galactic poles," *Nature*, 343:726.

86. Koo, D.C. and Kron, R.G., 1988, "A deep redshift survey of field galaxies," *Towards Understanding Galaxies at Large Redshifts*, F.G. Kron and A. Renzini, eds., Kluwer Academic Publishers, Dordrecht, p.209.

87. Szalay, A., 1990, "Correlations of galaxies over cosmic scales," in: *Proc. IUPAP Conf. Primordial Nucleosynthesis and Evolution of Early Universe, Tokyo, September 1990*.

88. Faber, S. and Dressler, A., 1988, "The Great attractor," in: *Proc. Texas Symposium on Rel. Astrophysics, Dallas*.

89. Bahcall, N. and Soneira, R., 1983, "The Spatial correlation function of rich clusters of galaxies," *Astrophys. J.*, 270:20-38.

90. Klypin, A. and Kopylov, A., 1983, "Cluster correlations," *Soviet Astr. Lett.*, 9:41-46.

91. Primack, J. and Dekel, A., 1990, U. of California at Santa Cruz preprint.

92. Szalay, A. and Schramm, D.N., 1985, "Are galaxies more strongly correlated than clusters?" *Nature*, 314:718-719.

93. Huchra, J., 1990, Harvard University preprint.

94. Vandenburgh, S. and West, J., 1990, Dominion Observatory preprint.

95. Fuller, G. and Schramm, D.N., 1990, "Neutrino flypaper and the formation of structure in the universe," University of California at San Diego preprint.

96. Hill, C. and Ross, G., 1988, "Pseudo-Goldstone bosons and new macroscopic forces," *Phys. Lett. B*, 203:125.

97. Dimopoulos, S., 1990, private communication.

98. Mikheyev, S.P. and Smirnov, A.Yu.., 1986, "Resonant amplification of ν oscillations in matter and solar-neutrino spectroscopy," *Nuovo Cimento*, 9C:17.

99. Wolfenstein, L., 1978, "Neutrino oscillations in matter," *Phys. Rev. D*, 17:2369.

100. Bahcall, J., 1989, *Neutrino Astrophysics*, Cambridge University Press, Cambridge, England and New York.

101. SAGE colloboration, 1990, Int. High Energy Meeting, Singapore, 1990.

102. Gell-Mann, M, Ramond, P. and Slansky, R.,, 1979, "Complex spinars in unified theories," in: *Supergravity*, P. Van Nieuwenhuizen and D. Freedman, eds., North Holland Pub. Co., Dordrecht.

103. Dine, M., Fischler, W., and Srednicki, M., 1981, "A Simple solution to the strong CP problem with a harmless axion," *Phys. Lett. B*, 104:199.

104. Press, W., Ryden, B., and Spergel, D., 1989, "Dynamical evolution of domain walls in an expanding universe," *Astrophys. J*, 397:590.

105. Hodges, H., 1988, "Domain walls with bound Bose condensates," *Phys. Rev. D*, 37:3052-5.

106. Stebbins, A.J. and Turner, M.S., 1989, "Is the great attractor really a great wall?" *Astrophys. J.*, 339:L13.

107. Widrow, L., 1990, "Wall interactions," CfA/Harvard preprint.

108. Kawano, L., 1990, "Evolution of domain walls in the early Universe," *Phys. Rev. D*, 41:1013.

109. Massarotti, A., 1990, "Evolution of light domain walls interacting with dark matter," Fermilab Pub-90/77-A, and *Phys. Rev. D*, in press.

110. Ipser, J. and Sikivie, P., 1984, "Gravitationally repulsive domain walls," *Phys. Rev. D*, 30:712.

111. Götz, G., 1990, "Scalar field configurations with planar and cylindrical symmetry: Thick domain walls and strings," Fermilab PUB-90/35-A.

112. Turok, N., 1989, "Textures," *Phys. Rev. Lett.*, 63:2625.

113. Lee, T.D., 1987, "Soliton stars and the critical masses of black holes," *Phys. Rev. D*, 35:3637.

114. Icke, V. and van de Weygaert, R., 1990, "Vornoi Cosmology," *Quarterly Journal of the Royal Astronomical Society*, in press.

115. Coles, P., 1990, "Understanding recent observations of large-scale structure," *Mon. Notices Royal Astronomical Society*, submitted.

116. Bertschinger, E., Scherrer, R., Vilenson, J., and VanDalen, A., 1990, "Seeded hot dark matter," Ohio State preprint.

117. Rees, M. and Sciama, D., 1968, "Cosmological background radiation," *Nature*, 217:511.

118. Sachs, R.K. and Wolf, A.M., 1967, "Perturbations of a cosmological model and angular variations of the microwave background," *Astrophys. J.*, 147:74.

119. Turner, M., Watkins, R., and Widrow, L., 1990, "Microwave distortions from collapsing domain-wall bubbles," *Astrophys. J. Lett.*, submitted.

120. Götz, G. and Nötzold, D., 1989, "On thick domain walls in General Relativity," Fermilab PUB-89/235-A and, 1990) Physical Review D, submitted; and Götz, G. and Nötzold, D., 1989, "An exact solution for a thick domain wall in general relativity" Fermilab PUB-89/236-A, and 1990, *Phys. Rev. D*, submitted.

INDEX

ALEPH, 5, 13–17, 32–38, 40–41, 44, 49,
  52–53, 57, 59, 64–65, 72, 78, 80–81,
  85, 91, 94–95, 115–116, 280
Anomalies, 254–259
Aplanarity, 54–55, 212
ARGUS, 37
Asymmetry
  forward-backward, *see* Charge asymme-
    try
  left-right, 120–125, 192–199

$B^0\overline{B^0}$ mixing, 37, 44, 114, 207, 285–288,
  476–480
B-factory, 449–480
Bhabha scattering, 15–18, 24, 361–364
Big bang, 495, 520
B-meson reconstruction, 462–463
Bose-Einstein correlations, 221–225
Bremsstrahlung, 327–337, 421
b' quark mass, 75

Cabibbo-Kobayashi-Maskawa matrix, 254,
  285–288, 305–306, 450, 468–480
Charge asymmetry, 4, 11, 119, 121, 201–
  203, 311
  leptons, 11–13, 16, 203–204, 354–361
  quarks, 28–29, 32, 36–37, 40–41, 44,
    205–209
CLEO, 37
Cluster algorithms, 399–400
Color coherence, 419–445
Compositeness, 99–103
CP-violation, 44, 285, 288, 296, 450,
  475–480

Dark matter, 505–509, 519

DELPHI, 5, 14–15, 38–39, 44, 52–53, 57–
  59, 61–65, 72-73, 78-79, 88, 92, 96–
  98, 101, 115, 127, 280, 385, 402,
  405–406
D-meson reconstruction, 31, 38, 462–468

Energy-energy correlation, 48, 57–59, 213–
  222, 373, 398
Event generators, 369–373, 382–384
  ALIBABA, 17
  ARIADNE, 49, 52–55, 59–66, 382
  BABAMC, 17, 364
  HERWIG, 49, 52–55, 59–66, 382, 394–
    396, 401, 405
  JETSET, 49, 52–55, 58, 60–66, 383, 392,
    400–401, 405, 411
  LUND, 13, 210–212, 219, 225, 386–387
  ZFITTER, 17
  ZSHAPE, 322–323
Evolution equations, 380

Fragmentation, *see* Hadronization

Galaxies, 495, 510–514, 519
Gauge theory, 238–246

Hadronization, 47, 49–50, 56, 376, 385–396
Helicity amplitudes, 309–312, 328–329
Higgs mass, 19, 22–23, 28, 69, 74, 76, 78–
  79, 81, 115, 123, 237, 281–284, 289–
  294

Intermittency, 65, 404–406

Jets, 367–413, 419–445
  gluon jet, 44, 216–221

Jets (continued)
    intrajet coherence, 433–439
    jet fragmentation, 392–394
    jet type separation, 216–221, 407–409
    quark jet, 44, 216–221
    2-jet events, 39, 372–373
    3-jet events, 213–216, 373–377, 384
    4-jet events, 62–64, 373–377, 402
    5-jet events, 379

LEP, 5, 449, 460
    beam momentum, 7–11
    future program, 108–114
    luminosity, 5–7, 11, 111
    polarization, 115–137
L3, 5, 13, 15, 32–37, 44, 57, 62, 64, 72, 78–79, 86, 92, 101, 115, 280, 385

MARK II, 52–53, 141, 151–153, 163, 166, 175, 181–182, 207
Multihadron production, 420–422
Multiplicity distributions, 64–65, 403

Neutrinos
    mass, 176–179, 507–508, 515–516
    number of, 23, 102–103, 167–168, 495–504
    solar, 296, 495, 515
Nucleosynthesis, 495–505

Oblateness, 48, 212
OPAL, 5, 14–15, 23, 32–37, 42, 44, 52–64, 72–75, 78–79, 87–88, 92, 101, 103, 115, 280, 385

Particle searches, 69–114
    Higgs, 76–83, 175, 179–183
    neutral leptons, 83, 100
    SUSY particles, 89–99
    top quark, 72–76
    W'-boson, 235
    Z'-boson, 235

QCD, 209–216, 370–385
    coupling constant, 47–49, 55–59, 66, 259–267, 368, 375–378
    tests, 47–66 384–385, 412–413
Quark
    couplings, 27–28, 119, 155, 309
    tagging, 29–32, 44, 114

Radiative corrections, 106, 194, 270–274, 277–279, 301–364
    higher order, 337–342
    real, 327–337
    virtual, 326–327
Renormalization, 268–270, 312–316

SLC, 141-151, 449, 460
    luminosity, 163–165
    polarization, 184–199
Sphericity, 39, 212, 398–399
Spherocity, 51
Spontaneous symmetry breaking, 242–245
Standard model, 2, 23–28, 112, 115, 123, 129, 153–155, 175–176, 237–296, 307–309
    minimal supersymmetric, 69, 91
Storage rings, 450–462
Structure functions, 348–358
Sudakov form factor, 381
Supersymmetry, 89, 107

TASSO, 52, 61, 65, 405–406
TRISTAN, 201–225, 385
Thrust, 40, 48, 51, 54–55, 212, 374, 398–399
t-quark mass, 19, 22–23, 28, 69, 74–75, 116, 237, 280–289, 292–294, 301–305

Unitarity, 250–253
Unstable particles, 483–492

W-boson
    mass, 168–174, 233
    production, 229–232
    width, 168–174, 233

Z-boson
    decays
        in leptons, 13–14, 115, 163
        in $q\bar{q}$, 13–14, 163
        line shape, 1–4, 16–19, 153–157, 166, 274–284, 343–347
        mass, 3, 5, 10, 17–24, 115, 154, 163, 167, 280
        width, 3, 5, 17, 20–24, 155, 163, 167, 280, 317–320
            hadronic, 20–24, 28, 32, 35–39, 43, 156–157, 280, 320, 378
            invisible, 22–24, 157, 320
            leptonic, 19, 21, 24, 156, 280, 320